Relation: A relation is a set of ordered pairs. (A collection of ordered pairs of numbers)

Function Given a set of ordered pairs:

A function is a relation in which no two distinct ordered pairs have the same first component; or a function is a set of ordered pairs in which for any two different ordered pairs, the first elements are different.

or

Given an equation of the function:

If x and y are two variables, then we say that y is a function of x if there is a rule which gives just one corresponding value of y for a given value of x.

One to-one function Given a set of ordered pairs:

A **one-to-one** (1-1) function is a set of ordered pairs in which for any two ordered pairs, the first elements are different from each other and the second elements are also different from each other;

or

Given an equation of the function

A function $f(x)$ is one-to-one if whenever $f(x_1) = f(x_2), x_1 = x_2$. If we let $y_1 = f(x_1)$, and $y_2 = f(x_2)$, then $y_1 = y_2$ **must** imply that $x_1 = x_2$, otherwise, $f(x)$ is not one-to-one.

Factoring: 1. $a^2 - b^2 = (a + b)(a - b)$;   **2.** $a^3 + b^3 = (a + b)(a^2 - ab + b^2)$.
    3. $a^3 - b^3 = (a - b)(a^2 + ab + b^2)$.

Complex numbers: **1.** $i = \sqrt{-1}$; $i^2 = -1$;   **2.** If $x^2 = -4$, then $x = \pm 2i$.

Logarithms: $\log_b x = y$ if and only if $b^y = x$; $\log_2 16 = 4$, because $2^4 = 16$.

Absolute value: $|x| = \begin{cases} x \text{ if } x \geq 0 \\ -x \text{ if } x < 0 \end{cases}$

Absolute value equations: If $|x - 2| = 6$, then $x - 2 = 6$ or $-(x - 2) = 6$; and from which $x = 8$ or $x = -4$.

Absolute Value Inequalities 1. $|ax + b| < c$. is equivalent to $-c < ax + b < c$; (or
$$-c < ax + b \text{ and } ax + b < c)$$

    2: $|ax + b| > c$ is equivalent to either $ax + b > c)$ or $-(ax + b) > c$

Equation of the **circle** with center at (h, k) and radius r is given by $(x - h)^2 + (y - k)^2 = r^2$

Distance Formula: $d = \sqrt{(x_2 - x_1)^2 + (y_2 - y_1)^2}$

Midpoint of the line connecting the points $P_1(x_1, y_1)$ and $P_2(x_2, y_2)$ has the coordinates given by the following formulas: The x-coordinate, x_m, of the midpoint is given by $x_m = \dfrac{x_1 + x_2}{2}$

The y-coordinate, y_m of the mid-point is given by $y_m = \dfrac{y_1 + y_2}{2}$

De Moivre's Theorem: $z^n = [r(\cos\theta + i\sin\theta)]^n = r^n(\cos n\theta + i\sin n\theta)$

Remainder Theorem: The remainder theorem states that if a polynomial $P(x)$ is divided by $x - r$, (where r is a constant) then the remainder $R = P(r)$.

Factor Theorem: Given a polynomial $P(x)$, if $P(r) = 0$, then $x - r$ is a factor of $P(x)$.

True: If you can remember a needed information, you make decisions faster, you work faster, and you are more productive.

Trigonometric Identities

Reciprocal Identities

1. $\sec\theta = \dfrac{1}{\cos\theta}$

2. $\csc\theta = \dfrac{1}{\sin\theta}$

3. $\cot\theta = \dfrac{1}{\tan\theta}$

Ratio Identities

1. $\tan\theta = \dfrac{\sin\theta}{\cos\theta}$

2. $\cot\theta = \dfrac{\cos\theta}{\sin\theta}$

Pythagorean Identities

1. $\sin^2\theta + \cos^2\theta = 1$

2. $1 + \tan^2\theta = \sec^2\theta$

3. $1 + \cot^2\theta = \csc^2\theta$

Sum and Difference Identities (Addition and Subtraction Formulas)

1. $\cos(A - B) = \cos A\cos B + \sin A\sin B$;   4. $\sin(A - B) = \sin A\cos B - \sin B\cos A$

2. $\cos(A + B) = \cos A\cos B - \sin A\sin B$;   5. $\tan(A + B) = \dfrac{\tan A + \tan B}{1 - \tan A\tan B}$

3. $\sin(A + B) = \sin A\cos B + \sin B\cos A$,   6. $\tan(A - B) = \dfrac{\tan A - \tan B}{1 + \tan A\tan B}$

Double Angle Identities

1. $\sin 2\theta = 2\sin\theta\cos\theta$;   2. $\cos 2\theta = \cos^2\theta - \sin^2\theta$;   3. $\cos 2\theta = 2\cos^2\theta - 1$;

4. $\cos 2\theta = 1 - 2\sin^2\theta$;   5. $\tan 2\theta = \dfrac{2\tan\theta}{1 - \tan^2\theta}$

Half Angle Identities

1. $\cos\dfrac{1}{2}\theta = \pm\sqrt{\dfrac{1 + \cos\theta}{2}}$;   2. $\sin\dfrac{1}{2}\theta = \pm\sqrt{\dfrac{1 - \cos\theta}{2}}$;

3. $\tan\dfrac{1}{2}\theta = \dfrac{1 - \cos\theta}{\sin\theta} = \dfrac{\sin\theta}{1 + \cos\theta}$

The $\pm$ sign indicates the sign to use depending on the quadrant location of $\dfrac{\theta}{2}$

Product or Product as a Sum Formulas

1. $\cos(A - B) + \cos(A + B) = 2\cos A\cos B$

2. $\cos(A - B) - \cos(A + B) = 2\sin A\sin B$ }Derived from sum and difference identities}

3. $\sin(A + B) + \sin(A - B) = 2\sin A\cos B$

4. $\sin(A + B) - \sin(A - B) = 2\cos A\sin B$ }Derived from sum and difference identities}

Sum Identities

1. $\cos A + \cos B = 2\cos\dfrac{A + B}{2}\cos\dfrac{A - B}{2}$

2. $\cos A - \cos B = -2\sin\dfrac{A + B}{2}\sin\dfrac{A - B}{2}$

3. $\sin A + \sin B = 2\sin\dfrac{A + B}{2}\cos\dfrac{A - B}{2}$

4. $\sin A - \sin B = 2\sin\dfrac{A - B}{2}\cos\dfrac{A + B}{2}$

Cofunction Relationships

$\sin A = \cos(90° - A)$
$\cos A = \sin(90° - A)$ }cofunctions.

$\tan A = \cot(90° - A)$
$\cot A = \tan(90° - A)$ }cofuntions.

$\sec A = \csc(90° - A)$
$\csc A = \sec(90° - A)$ }cofuntions.

Other identities (Useful when finding functional values of negative angles)

For any θ,

1. $\cos(-\theta) = \cos\theta$ 4. $\sec(-\theta) = \sec\theta$

2. $\sin(-\theta) = -\sin\theta$ 5. $\csc(-\theta) = -\csc\theta$

3. $\tan(-\theta) = -\tan\theta$ 6. $\cot(-\theta) = -\cot\theta$

FREMPONG'S STEP-BY-STEP SERIES IN MATHEMATICS

College Algebra and Trigonometry (Pre-Calculus)

Second Edition

Includes Sample Problems with

Step-by-Step Solutions
plus
Practice Problems with Answers

A.A. FREMPONG

College Algebra & Trigonometry (Pre-Calculus)

ISBN 978-1-946485-51-9

Printed in the United States of America

Faculty/Student Suggestions for Future Editions

Send Suggestions to the Publisher

Copy and complete the following. This survey will help improve future editions of this book.

1. Which Topics would you like to be added to future Editions? You may include sample problems (and solutions).

....
....
....
....
....
....

2. Which Topics in this book do you think need more or better coverage?.

....
....
....
....
....
....
....
....
....
....
....
....

3. Which coverage in this book impressed you most?

....
....
....
....
....
....
....
....
....
....
....
....
....
....
....

. How useful was this book in preparing for and taking the Final Exam?

Check one: A (Excellent) ; B (Good) ; C (Average) ;   D (Fair)

In Memory of My Parents

Mom:
She was a devoted mother, sharing, kind, kinder to strangers and generous to a fault. She never cursed, she never hated; she never cheated, and she never envied. She never lied, and she never got angry. Once, she nursed an almost dying stranger renting a room in her house back to good health to the extent that the relatives of this renter later travelled one hundred miles just to thank mom. She was always peaceloving and forever forgiving. An angel once lived on this earth to serve others.

Dad:
A great dad, kind, generous and forgiving. He emphasized and was an example of both formal education and self-education. A veterinarian, a bacteriologist, an Associate of the Institute of Medical Laboratory Technology (UK), a Fellow of the Royal Society of Health (UK); an incorruptible civil servant; his book on ticks has always inspired me to write whenever the need arises.

NOTE TO THE STUDENT

This book was written with you in mind at all times. You may use this book as the course textbook or as a review book since the book gets to the point quickly on all relevant topics and yet covers these topics in detail.

Begin to master the definitions and the solutions of the sample problems thoroughly. (You have mastered a sample problem if you can solve the sample problem and similar problems without any reference to this book or any other source. For some problems, two or more methods are presented. Read the various methods and decide which methods you would like to remember; but always be aware of the existence of the other methods, in case the need arises. After having mastered the sample problems, try the exercise problems. The answers to these problems are presented immediately after the problems. You may cover the answers with paper before you attempt these problems, if the answers are too obvious. You may refer back and forth to the solved problems when you do not remember how to proceed.

You may also attempt some of the sample problems first, if you have been exposed to the topics previously, before reading the solution methods, and in this approach, the sample problems become more practice problems for you.

As a reminder, in any book, do not dwell on the few inadvertent errors you may find, but rather concentrate on what is useful to you.

For this book to be useful both as the course textbook, as well as review for exams, it is **important** to **Understand, Remember, Apply**, and $\boxed{\textbf{Remember}}$ the material covered.

Wishing you Good Luck on all the exams
A.A.Frempong

Books in the series by the author: Integrated Arithmetic; Elementary Algebra; Intermediate Algebra, Elementary Mathematics; Intermediate Mathematics; Elementary & Intermediate Mathematics (combined); College Algebra; College Trigonometry; **College Algebra & Trigonometry** and Calculus 1 & 2.

PREFACE TO THE SECOND EDITION

This second edition retains the spirit of the first edition and continues to emphasize the importance of this level of mathematics for various branches of mathematics.
Some topics have been rewritten. However, the pagination of the previous edition has been retained.

The topics rewritten include the transformation of functions and relations (p.225), end behavior of polynomial functions. and applications of quadratic functions to maxima and minima problems (p.257). Other topics include applications of synthetic division, factor theorem and the converse of factor theorem (p.440-442).

In the transformation of functions, a general transformation equation $y - k = af[b(x - h)]$ has been presented and the parameters k, a, b and h have been discussed, It is suggested that the student constructs tables for the variables to obtain ordered pairs which are then plotted and connected. The student should study the general example on page 226. Having mastered this example, the student will find the subsequent examples much easier. Students would be amazed that having sketched the graph of the parabola
$y + 4 = 3\left([-2x + 2]^2\right)$, p.226, from the graph of $y = x^2$, they could similarly sketch trigonometric graphs such as $y - 1 = 3\sin(4x + \pi)$ (p.610) from the graph of $y = \sin x$; and the graph of $y = \text{Arccos } 2x + \frac{\pi}{2}$ (p.635-6) from the graph of $y = \text{Arccos } x$.

A.A.Frempong
New York, July 2011

PREFACE TO THE FIRST EDITION

The idea of producing this book first came to me some time when I was working with students on one-to-one basis. My experience at that time was that I could not find that one book which was easily read and understood on all the required topics for College Algebra and Trigonometry. On occasion, I could recommend a book only for one or two topics, and then I had to search in the library for other recommendable books (to explain the other topics) but without much success. In recent times, more and more students need this level of mathematics in order to graduate from college or to study more mathematics. In particular, four-year nursing students, business majors, liberal arts majors, engineering and science majors need this level of mathematics.

Another observation of mine was that there was a group of students who inspite of all the hard work could not understand their textbooks well. These students had done well at the elementary and intermediate levels of algebra and trigonometry. Often, after having helped a student understand a material, the student would remark" Why doesn't the book state it so?".

This book is an attempt to help such a student. This book could be used for self-study, especially by the working student or the continuing student with so much demand on his or her time. It could also be used as a reference book since each topic has been covered very well. This is not a book for the lazy student. It is a book for the serious student who inspite of all the hard work finds his or her textbooks rather difficult to understand or follow.

 This book could be used as a textbook for a **pre-calculus** course. To the instructor, this book should be a relief so that more time could be spent on the applications of the definitions and principles.
 For those students who will take calculus in the future, if you master this level of mathematics, the techniques and applications of the first two courses of calculus would be found to be much easier than you thought. In fact, most of the difficulties that students studying calculus have are all problems of **precalculus and** not of calculus.

This book was written with the student in mind at all times, and at times, some of the explanations may seem redundant, however, this is intentional. Analogies from everyday life are presented whenever they help to explain a principle. At times, the book may be found to be rather informal, but this also is intentional, because the main objective is to communicate.

The following concepts have been treated in more detail than most current textbooks do at this level:

Decreasing functions, i**ncreasing** functions, **continuous** functions, discontinuous functions, positive functions, negative functions, **translation** of axes, contraction and expansion of curves, symmetry, **asymptotes,** critical points, and **maximum** and minimum points. The author recommends that as soon as possible, the above mentioned concepts be mastered. The early mastery of these concepts will help the student read and understand subsequent material much more quickly than otherwise. The student would also be able to see the beauty in studying mathematics as a system and would be able to appreciate the unifying concepts among the various functions and relations.

On the solutions of equations, the concepts of "lost roots" and extraneous roots have been discussed
 On **parabolas,** an early distinction is made between the parabola as a function, as a relation, and as a conic. The different methods (and their relative merits) of sketching the graphs of parabolas are discussed. A new concept of relating simple continuous functions and their reciprocals has been presented, perhaps for the first time (by the author).

In t**rigonometry,** the concept of the period of a function is explained as the simultaneous occurrence of two events.
 On i**nequalities,** different methods of solving both linear and quadratic inequalities are presented. Also presented are different methods of solving systems of linear inequalities.

A step-by-step approach is used throughout the book.

This book can also be used for short term programs such as mini-sessions, immersion programs, workshops, as well as in distance learning programs.

My sincere appreciation goes to those who wrote before me, and without whom this book would have been inconceivable.

A. A. Frempong
New York, April, 2006

CONTENTS

page

CHAPTER 1

Lesson 1: Definitions, **Terminology, and Types of Numbers** 1

CHAPTER 2   4

Lesson 2: **Exponents and Radicals (Review)**

Mnemonic Examples; Definitions and Simplification of Radicals; Addition, Multiplication and Division of Radicals; Rationalization of Denominators; Fractional Exponents Reduction of Indices

CHAPTER 3   20

Equations (Review)

Lesson 3: Types of Equations 20
Lesson 4: **Principles for Solving Equations; Extraneous and Lost Roots**

CHAPTER 4   26

Quadratic Equations; Higher Polynomial Equations; and Systems Containing Nonlinear Equations

Lesson 5: **Solving Quadratic Equations (Review};** Discriminant; 26
       Roots r_1, r_2 and the Constants a, b, and c of the Quadratic Equation;
       Deriving Quadratic Equations; Quadratic Formula to aid Factoring 27
Lesson 6: **Solving Equations Quadratic in Form** 38
Lesson 7: **Higher Polynomial Equations; Systems of Nonlinear Equations** 40

CHAPTER 5   44

Lesson 8 **Solving Radical Equations (Review)** 44

CHAPTER 6   48
FUNCTIONS

Lesson 9: **Sets, Relations, Functions,** Comparison of Relations and Functions 48
Lesson 10A: **Functional Notation; Defined Functions; Excluded Values, Domain and Range** 52
Lesson 10B: **Algebra of Functions** 68, 652
Lesson 11: **One-to-One Functions, Composite Functions** 69
Lesson 12: **Inverse Functions and Inverse Relations** 73

CHAPTER 7   83
Straight Lines

Lesson 13: **Linear Equations; Graphing; Points on Lines** 83
Lesson 14: **Intercepts; Slopes** 92
Lesson 15: **Equations of Straight Lines** 96
Lesson 16 :**Perpendicular and Parallel Lines** 106
Lesson 17: **Midpoint of a Line; Distance Formula** 110

CHAPTER 8   118
Variation (Review)

Lesson 18: **Direct, Inverse,** Joint and Combined Variation 118

CHAPTER 9   123

Inequalities (Review)

Lesson 19: **Set Notation and Operations; Interval Notation** 123
Lesson 20: **Solving Linear Inequalities Containing a Single Variable; Compound Inequalities** 129
Lesson 20 Extra: **Review of Inequality Operations** 134

CHAPTER 10   135

Lesson 21: **Absolute Value Equations** 135
Lesson 22: **Absolute Value Inequalities** 143

CHAPTER 11   146

Lesson 23: **Graphing Linear Inequalities in two Variables** 146
Lesson 24: **Solving Systems of Linear Inequalities** 164
Lesson 25: **Linear Programming ; Deriving Systems of Inequalities; Applications of Linear Programming** 166

CHAPTER 12   175

Lesson 26: **Solving Quadratic Inequalities 175**
Lesson 27: **Solving Factored Inequalities; Inequalities Involving** 184 **Rational Expressions and Absolute Values** 184

CHAPTER 13   192

Lesson 28: **Domain Specification; Piecewise** (Multipart) **Functions** 192
Lesson 29: **Miscellaneous Graphs Involving Absolute Values** 196
Lesson 30 :**Special Functions: Greatest Integer Function; Step Function;** 207 **Unit Step Function; Sign Function**

CHAPTER 14   210

Lesson 31: **Positive, Negative, Increasing, and Decreasing Functions** 210
Lesson 32: **Concavities of Curves; Critical Points** 214
Lesson 33: **Reflection of Points, Lines and Curves** 218
Lesson 34: **Introduction to Transformation of Functions and Relations** **225**
    Translation of Points, Relations, Functions and Axes **228**
Lesson 35: **Vertical Stretching and Shrinking of Graphs** 232
    Horizontal Stretching and Shrinking 234
Lesson 36: **Symmetry; Even and Odd Functions** 236

CHAPTER 15   241

POLYNOMIAL FUNCTIONS

Lesson 37: **The x- and y-intercepts of a line or a curve** 241
    Graphs of Polynomial Functions
Lesson 38A: **Parabola: The Graph of** $y = f(x) = ax^2 + bx + c$); 244
    Graph of the Inverse Relation of $y = ax^2 + bx + c$

Lesson 38B: Applications of Quadratic Functions; Maxima /Minima Problems **257**
Lesson 39: **Graphs of other Polynomial Functions** 259
    End Behavior of Polynomial Functions 264
Lesson 40: **Graphs of Factorable Polynomial Functions** 265

CHAPTER 16   270

RATIONAL FUNCTIONS

Lesson 41: **Continuous and Discontinuous Functions** 270
Lesson 42: **Asymptotes** 273
Lesson 43: **Graphs of Rational Functions** 283

CHAPTER 17   298

Logarithms

Lesson 44: **Definitions, Basic Properties and Applications** 298
Lesson 45: **Evaluation of Logarithms; Common and Natural Logarithms** 301

CHAPTER 18   305

Lesson 46 **Exponential Equations** 305
Lesson 47: **Logarithmic Equations** 307

CHAPTER 19   311

Lesson 48 **Graphs of Exponential Functions &**
    Applications of Exponential Functions 311
Lesson 49: **Graphs of Logarithmic Functions &**
    Applications of Logarithmic Functions 317

CHAPTER 20   322

Lesson 50: **General System of Linear Equations** 322

Lesson 51: **Matrix and Matrix Methods** 327

Lesson 52: **Determinants and Applications** 331

CHAPTER 21   343
Complex Numbers

Lesson 53: **Basic Definitions; Basic Operations with Complex Numbers** 343

Lesson 54: **Equality of Complex Numbers; Roots of Complex Numbers;** 351
Equations Involving Complex Numbers

Lesson 55: **Graphical Representation and** Addition of Complex Numbers 355

Lesson 56: **Polar (Trigonometric) Form of Complex Numbers** 359

Lesson 57: **Powers of Complex Numbers; De Moivre's Theorem; Roots of**
Complex Numbers 368

CHAPTER 22   375

Lesson 58: **Graphing Polar Coordinates** 375

Lesson 59: **Graphing Polar Equations** 381

CHAPTER 23A   390
Conic Sections

Lesson 60: **Circles** 391

Lesson 61: **Parabolas** 397

Ellipses (See Chapter 23B)

Hyperbolas (See Chapter 23B)

CHAPTER 23B   412
Conic Sections

Lesson 62: Ellipse 412

Lesson 63: Hyperbola 423

Circle and Parabola (see Chapter 23A)

CHAPTER 24   438
Lesson 64: **Synthetic Division, Remainder and Factor Theorems** 438

Lesson 65: **Descartes' Rules;** Rational Roots of Polynomial Equations 443

CHAPTER 25   453
Partial Fractions
Lesson 66: **Decomposition of Fractions into Partial Fractions** 453

CHAPTER 26   458
Sequences and Series
Lesson 67: **Specification of a Sequence; Finite and Infinite Sequence** 458
     Finding a Formula the general term of a Sequence 460

Lesson 68: **The Summation Notation, Σ, Properties and Shifting of the
Summation Index** 464

Lesson 69: **Arithmetic Sequence; Definition of Series; Arithmetic Series** 468

Lesson 70: **Geometric Sequence, Geometric Series and Applications** 473

CHAPTER 27   479
Lesson 71: **Factorial Notation, $n!$; Operations Involving $n!$** 479

Lesson 72: **The Binomial Theorem and Applications** 481

CHAPTER 28   486
Combinations and Permutations
Lesson 73: Fundamental Principle of Counting; Introduction to
Permutations and Combinations; Tree Diagrams 486

Lesson 74: **Permutations; Linear Permutations; Cyclical Permutations** 489

Lesson 75: **Combinations** 491

CHAPTER 29   492
Lesson 76: **Mathematical Induction** 492

Trigonometry

CHAPTER 30   498

Introduction to Trigonometry, Right Angle Trigonometry

Lesson 77: **Basic Review for Geometry** 499

Lesson 78: **Right Triangle Trigonometry** 504

Lesson 79: **Angles of Elevation and Depression; Bearing; Linear Interpolation** 520

CHAPTER 31   526

Trigonometrical Functional Value of any Angle

Lesson 80: **Definitions; Functional Value Given the Measure of the Angle** 526

Lesson 81: **Given a Point on the Terminal Side of the Angle** 532

Lesson 82: **Trigonometric Functional Values of Quadrantal Angles** 536

CHAPTER 32   542

Finding Other Trigonometric Functional Values

Lesson 83: **Given a Functional Value and the Quadrant of the Angle** 542

Lesson 84: **Given a Functional Value and no Specification of the** 544 **Quadrant of the Angle**

CHAPTER 33   546

Trigonometry of Oblique Triangles

Lesson 85: **The Law of Cosine; The Law of Sines** 546
Lesson 86: **Applications of the Laws of Cosine and Sines** 546

CHAPTER 34   557

Application of Trigonometry to Vectors   557

Lesson 87: **Basic Definitions; Representation of Vectors**   557

Lesson 88: **Addition (Sum, Resultant, or Composition) of Vectors** 557

CHAPTER 35   570

Trigonometry of Real Numbers

Lesson 89: **Definitions; Radian Measure; Arc Length; Reference Number** 570
Lesson 90: **Trig Functional Values of Angles and of Real Numbers** 579

CHAPTER 36   581
Graphs of Trigonometric Functions

Lesson 91: **Introduction; Labeling the Coordinate Axes;Illustration of the** 581
Periodicity of Trigonometric Functions

Lesson 92: **Sketching the Graph of** $y = \sin x$ 586

Lesson 93: **Sketching the Graph of** $y = \cos x$ 591

Lesson 94: **Graphs of Discontinuous Trigonometric Functions:** 594
$y = \tan x$; $y = \csc x$; $y = \sec x$; $y = \cot x$

Lesson 95: **Direct Procedure for Sketching the Graphs of** $y - k = a \sin(bx - h)$,
and $y - k = a \cos(bx - h)$ 602

CHAPTER 37   623
From Trigonometric Functions to
Inverse Trigonometric Functions

Lesson 96: **Introduction; Definitions; Operations Involving Inverse** 623
Trigonometric Functions

Lesson 97: **How to Sketch the Graph of** $y = \text{Arcsin } x$ 630

Lesson 98: **How to Sketch the Graph of** $y = \text{Arccos } x$ 634

Lesson 99: **How to Sketch the Graph of** $y = \text{Arctan } x$ 637

CHAPTER 38   639

Lesson 100: **Trigonometric Identities** 639

Lesson 101: **Proving Trigonometric Identities**; 643
Applications of Sum and Difference Identities

CHAPTER 39   648

Lesson 102: Solutions of Trigonometric Equations 648

Appendix A   655
Changing Decimals to Rational numbers

Appendix B   659
Definitions of Inverse Trigonometric Functions

Appendix C   661
**Area and Perimeter of a Circle; Arc Length and Sector of a Circle;
Area and Perimeter of Composite Figures**

Appendix D   667
About Measurements
Standard Unit, Error, Rounding-off Numbers, Significant Digits, Scientific
Notation

Index   673

CHAPTER 1

Lesson 1

Definitions, Terminology, and Types of Numbers

The basic elements we deal with in our study of mathematics are numbers. It is important that we obtain a good understanding of the types of numbers we deal with in mathematics. It is also important that we are able to distinguish between the different kinds of numbers and their associated terminology. A very good grasp of the terminology will help us read and understand subsequent material much more quickly than otherwise.

As shown in the number flow chart below, all the numbers we deal with in mathematics can be divided into two main sets, namely the set of **real numbers** and the set of **non-real numbers**.
(We must note that the terms "real", "non-real", and similar terms are only names we use in mathematics to distinguish between numbers and we must note that the real numbers are **not literally** more real than the non-real numbers. These terms are only names for convenient distinction between some numbers.) The real numbers are also divided into two main sets, namely the **rational numbers** and the **irrational numbers.**

The rational numbers are further subdivided into **integers** (strictly integers) and **fractions** (strictly fractions). The integers consist of the **negative integers, zero,** and the **positive integers** (natural numbers); and the positive integers consists of the **prime numbers** and the **composite numbers**.

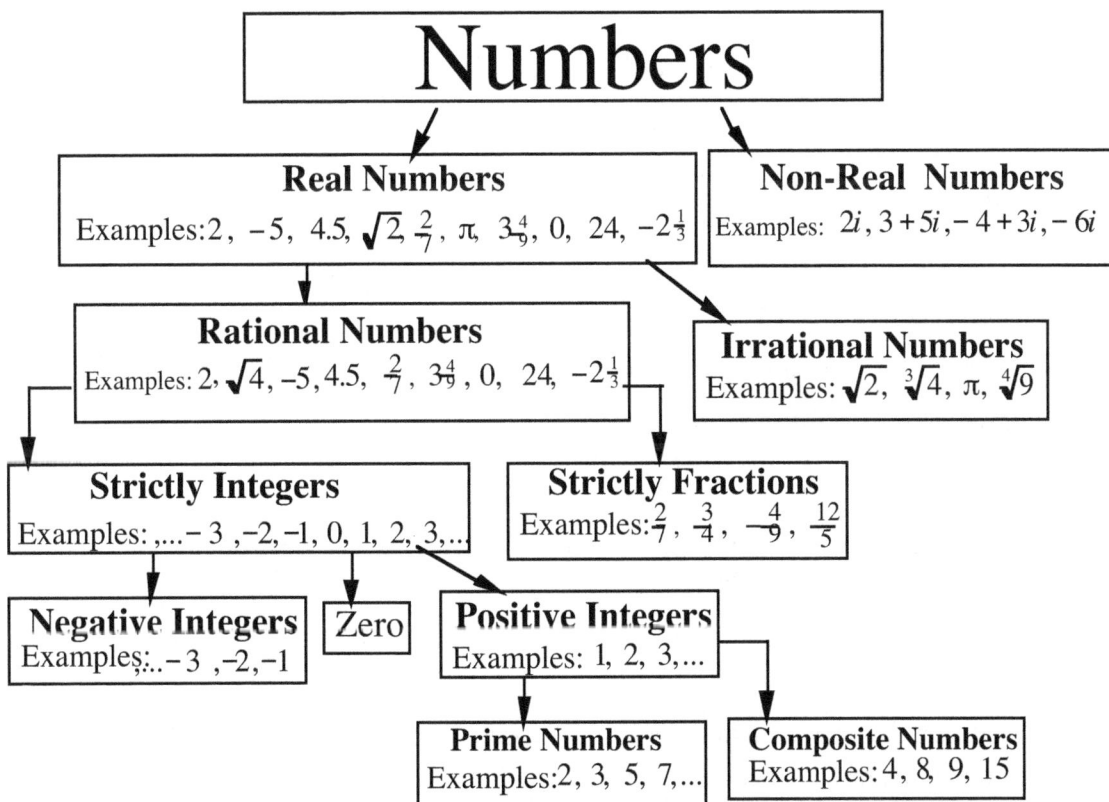

Numbers

Real Numbers
Examples: $2,\ -5,\ 4.5,\ \sqrt{2},\ \frac{2}{7},\ \pi,\ 3\frac{4}{9},\ 0,\ 24,\ -2\frac{1}{3}$

Non-Real Numbers
Examples: $2i,\ 3+5i,\ -4+3i,\ -6i$

Rational Numbers
Examples: $2,\ \sqrt{4},\ -5, 4.5,\ \frac{2}{7},\ 3\frac{4}{9},\ 0,\ 24,\ -2\frac{1}{3}$

Irrational Numbers
Examples: $\sqrt{2},\ \sqrt[3]{4},\ \pi,\ \sqrt[4]{9}$

Strictly Integers
Examples: $,...-3,-2,-1,0,1,2,3,...$

Strictly Fractions
Examples: $\frac{2}{7},\ \frac{3}{4},\ -\frac{4}{9},\ \frac{12}{5}$

Negative Integers
Examples: $...-3,-2,-1$

Zero

Positive Integers
Examples: $1, 2, 3,...$

Prime Numbers
Examples: $2, 3, 5, 7,...$

Composite Numbers
Examples: $4, 8, 9, 15$

Number Flow Chart

Lesson 1: Definitions, Terminology, and Types of Numbers

We shall now define some terms mentioned in the number flow chart, above.

We define a **set of numbers** as a well-defined collection of numbers.

The set of the **natural numbers** consists of the numbers $1, 2, 3, 4, 5, 6, 7, 8, 9, 10, 11, 12, 13,...$, If we know a natural number, to obtain the next natural number we add 1. The smallest natural number is 1, but we do not know the largest natural number, since given any large natural number, we can always obtain the next natural number by adding 1. The natural numbers are also known as the **counting numbers** or the **positive integers**.

The set of **whole numbers** consists of the numbers $0, 1, 2, 3, 4, 5, 6, 7, 8, 9, 10, 11, 12, 13,...$, If we know a whole number, to obtain the next whole number we add 1. The smallest whole number is 0, but we do not know the largest whole number, since given any large whole number, we can always obtain the next whole number by adding 1.

If we take the opposites of the set of natural numbers also called positive integers, we obtain a set of numbers called **negative integers** (such as -1, -2, and -12). If we combine the whole numbers with the negative integers we obtain a set of numbers called the **integers**. The set of integers therefore consists of the set of numbers $...,-7, -6, -5. -4, -3, -2, -1, 0, 1, 2, 3, 4, 5, 6, 7,...$

(The three dots preceding the -7 on the left indicates that the numbers continue to decrease to the left and the three dots after the 7 on the right indicates that the numbers continue to increase to the right)

Ratio: The ratio of a is to b is the fraction $\frac{a}{b}$. **Example:** The ratio of 3 is to 4 is the fraction $\frac{3}{4}$.

Rational number: A rational number (a fraction) is a number which **can** be written as the ratio of two integers. The word **rational** pertains to the word **ratio.**

Examples are (a) $\frac{2}{3}$;   (b) $\frac{1}{5}$;   (c) 4   (since $4 = \frac{4}{1}$)

(d) 0   (since $0 = \frac{0}{7} = \frac{0}{3}$ ... or $0 = \frac{0}{b}$, where b is an integer and b $\neq$ 0)

(e) $\sqrt{4}$ (because $\sqrt{4} = 2 = \frac{2}{1}$)

A rational number can also be written either as a terminating decimal or as a repeating decimal.

Examples of terminating decimals: $\frac{1}{4} = .25$; $\frac{13}{2} = 6.5$; $\frac{37}{8} = 4.625$.

Examples of repeating decimals: $\frac{1}{3} = .333...$ or $.\overline{3}$ and $\frac{1}{6} = .1666...$ or $.1\overline{6}$; $\frac{2}{3} = .66...$ or $.\overline{6}$.

Note the bar (vinculum) placed over the repeating digit or block of digits.
We may also regard a terminating decimal as non-terminating if we attach zeros to the right of the decimal. Examples are $.25 = .250000....$, $.5 = .5000...$

We define an **irrational number** as a number which **cannot** be written as the ratio of two integers. However, we can approximate irrational numbers as closely as we wish by rational numbers or decimals.

Examples of irrational numbers are $\sqrt{2}, \sqrt[3]{4}$, and π (pi). We can for example, approximate $\sqrt{2}$ by 1.414, and π by $\frac{22}{7}$ or 3.142.

When written in decimal form, an irrational number is non-repeating and non-terminating.

If we combine the natural numbers, the fractions, decimals, irrational numbers, the negatives of these numbers, and zero, we obtain the set of **real numbers**. Simply, the real numbers consists of the rational numbers and the irrational numbers.

Lesson 1: Definitions, Terminology, and Types of Numbers

We can represent real numbers by points on a horizontal line called the **real number line** (Fig.1)   3

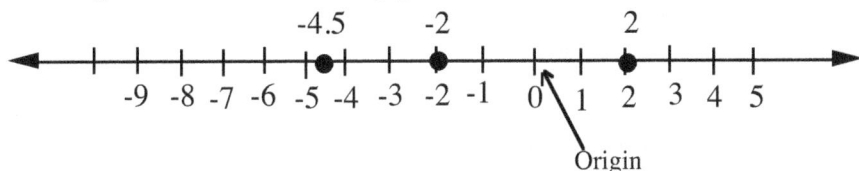

Figure 1

The **real number line** is a horizontal straight line with equally spaced intervals as in Figure 1 above. We label a point called the origin, 0 (zero). Points to the right of the origin are labeled positive and points to the left of the origin are labeled negative. The numbers increase as one moves from the left to the right on the real number line. Roughly speaking, a real number is a number that can be represented by a point on the real number line. The real numbers consists of the integers, fractions, mixed numbers, decimals, and radicals. In Figure 1, if the real numbers, -4.5, -2, and 2 are of interest, we can represent them by the dots shown. Every point on this line is associated with a real number; and every real number is associated with a point on this line. We can also say that the set of real numbers consists of the signed numbers and zero.

Summary for Some Number Terminology

Positive integers, or natural numbers or counting numbers $= \{1, 2, 3, 4, ...\}$

Negative integers $= \{..., -4, -3, -2, -1\}$

Non-negative integers or **whole numbers** $= \{0, 1, 2, 3, 4, ...\}$

Non-positive integers $= \{..., -4, -3, -2, -1, 0\}$

Non-negative real numbers consist of 0 and the positive real numbers.
Examples are $0, 4, 7.5, 3\frac{1}{2}, \frac{1}{4}, 6,$ and $\sqrt{2}$

Non-positive real numbers consist of 0 and the negative real numbers.
Examples are $-\sqrt{11}, -4, -3\frac{1}{2}, -2, -1, -\frac{1}{4}, -.126,$ and 0.

We must understand thoroughly the above terms because they will be used over and over in the future. Anytime we meet any of these terms, we should try to form a quick mental picture of representative examples.

Signed Numbers

A signed number is a number with either a plus sign " +" or a minus sign "-" preceding it.
 If there is no sign preceding a number, we will assume that the number has a plus sign.

Absolute Value

The absolute value of a signed number may be defined as the number without its sign. The absolute value of zero is zero.

Note: The absolute value of a signed number is also its distance from zero on the number line.

Absolute value defined more formally:

The absolute value of a real number x is x if x is a positive number or zero, but it is $-x$ if x is negative number .   (i.e. the negative of a negative number).

CHAPTER 2

Lesson 2
Review
Exponents and Radicals

Mnemonic Examples; Definitions and Simplification of Radicals; Addition, Multiplication and Division of Radicals; Rationalization of Denominators; Fractional Exponents Reduction of Indices

Mnemonic Examples

1. $x^2 x^3 = x^5$

2. $\dfrac{x^7}{x^4} = x^{7-4} = x^3$

3. $(x^4)^2 = x^8$

4. $(xy)^5 = x^5 y^5$

5. $\left(\dfrac{x}{y}\right)^6 = \dfrac{x^6}{y^6}$

6. $x^{-2} = \dfrac{1}{x^2}$

7. $x^0 = 1$

8. $\dfrac{x^4}{x^7} = x^{4-7} = x^{-3} = \dfrac{1}{x^3}$

9. $9^{1/2} = \sqrt{9}$

10. $\sqrt[3]{xy} = \sqrt[3]{x}\,\sqrt[3]{y}$

11. $8^{1/3} = \sqrt[3]{8} = 2$

12. $8^{2/3} = (\sqrt[3]{8})^2 = \sqrt[3]{8^2} = 4$

13. $(\sqrt[4]{8})^4 = (\sqrt[4]{8^4}) = 8$

14. $\sqrt[3]{\dfrac{x}{y}} = \dfrac{\sqrt[3]{x}}{\sqrt[3]{y}}$

Extra: Show that $x^{-2} = \dfrac{1}{x^2}$.   **Note:** $x^{-2} = \dfrac{x^{-2}}{1} \bullet \dfrac{x^2}{x^2} = \dfrac{x^{-2+2}}{x^2} = \dfrac{x^0}{x^2} = \dfrac{1}{x^2}$.

Example 1 Evaluate 7^{-2}

  Step 1:   Express with positive exponents.

$$7^{-2} = \dfrac{1}{7^2}$$

  Step 2:   Simplify. $\dfrac{1}{7^2} = \dfrac{1}{49}$

Example 2 Simplify, leaving answer with $\dfrac{x^{-3}y^2z^6}{x^2y^{-3}z}$ only positive exponents

Method 1 (Strictly, using the rules of exponents)

$$\frac{x^{-3}y^2z^6}{x^2y^{-3}z} = x^{-3-2}y^{2+3}z^{6-1}$$

$$= x^{-5}y^5z^5$$

$$= \frac{y^5z^5}{x^5}$$

Method 2 Taking powers across the division bar, and changing the signs of exponents, followed by cancellation

$$\frac{x^{-3}y^2z^6}{x^2y^{-3}z} = \frac{y^2y^3z^6 z^5}{x^2x^3 z}$$

(change x^{-3} to x^3 and write it in the denominator.

Change y^{-3} to y^3 and write it in the numerator)
If the exponent is already positive, leave the power where it was originally. The changes are for the powers with negative exponents).

$$= \frac{y^5z^5}{x^5}$$

A note about cancellation in Method 2 above: In order to apply cancellation (as done in arithmetic), the exponents of the powers involved in the cancellation must have the same sign; preferably, the exponents must be positive (or made positive by following the instructions outlined above) .

Example 3 Simplify $(4x)^0 - 4x^0$

Solution

$(4x)^0 - 4x^0 = 4^0x^0 - 4x^0$ **Note:** $x^0 = 1$, $(4)^0 = 1$
$\qquad\qquad\quad = (1)(1) - 4(1)$
$\qquad\qquad\quad = 1 - 4$
$\qquad\qquad\quad = -3$

Example 4 Multiply $\left(\dfrac{-8a^2b}{5}\right)\left(\dfrac{9c}{3abc}\right)\left(\dfrac{-15}{27c}\right)$

Solution $\dfrac{-8(9)(-15)a^2bc}{5(3)(27)abc^2}$

$= + \dfrac{8(9)(15)a^2bc}{5(3)(27)abc^2}$

$= \dfrac{8a}{3c}$ (After canceling the common factors in the numerator and the denominator)

Definitions and Simplification of Radicals

Rational number : A rational number (a fraction) is a real number which **can** be written as the ratio of two integers. The word **rational** pertains to the word **ratio.**

Examples are (a) $\frac{2}{3}$; (b) $\frac{1}{5}$; (c) 4 (since $4 = \frac{4}{1}$)

(d) 0 (since $0 = \frac{0}{7} = \frac{0}{3}$ ... or $0 = \frac{0}{b}$, where b is an integer and b ≠ 0)

(e) $\sqrt{4}$ (because $\sqrt{4} = 2 = \frac{2}{1}$)

Irrational number: An irrational number is a real number which **cannot** be written as the ratio of   6
two integers. However, we can approximate irrational numbers as closely as
we wish by rational numbers or decimals.

Examples of irrational numbers are $\sqrt{2}, \sqrt[3]{4}$, and π (pi). We can for example, approximate
$\sqrt{2}$ by 1.414, and π by $\frac{22}{7}$ or 3.142. Except otherwise instructed, in simplifying radical expressions, we
prefer to leave the irrational numbers in **radical** forms.

Radicals and Roots

Consider two real numbers r and A. We denote the **principal _n_th root** of A (n being a positive integer)

by $\sqrt[n]{A}$. Also $\sqrt[n]{A}$ is called a **radical**. We define the principal nth root of A as follows:

$\sqrt[n]{A} = r$ if $r^n = A$ (i.e., the nth root of A = r if r^n = A) with the following qualifications:

1. Any root of zero is zero (i.e. $\sqrt[n]{0} = 0$)
2. If A is a positive number, then the principal nth root of A is the positive nth root of A.
3. If A is a negative number, then the principal nth root of A is the negative nth root of A.

Note from above that a **radical** is an expression used for indicating the root of a number.

We call n the index of the radical (n indicating the type of root being considered); A is called the radicand

(the number of which a root is being taken). The radicand is thus the expression under the symbol " $\sqrt[n]{\ }$ " .

Thus, the radical consists of the index n with root symbol " $\sqrt[n]{\ }$ " and the radicand.

If $n = 2$, we obtain $\sqrt[2]{A}$, the square root of A, We usually omit the "2" and write $\sqrt{A}$.

Therefore, $\sqrt[2]{A}$ $= \sqrt{A}$. However, if explicit indication helps you to understand a problem, write the 2.

When $n = 3$, we obtain $\sqrt[3]{A}$, the cube root A, but in this case, we have to write the "3". Similarly, for $n = 4$

and higher indices, we must always write the index. Some higher orders of the root of A are $\sqrt[4]{A}$, $\sqrt[5]{A}$,

and $\sqrt[8]{A}$. Some specific radicals are $\sqrt{2}$, $\sqrt[3]{4}$, $\sqrt{10}$, and $\sqrt{5}$.

Example 1 Name the parts of the radical $\sqrt{32}$

Solution   The index is understood to be 2, since,$\sqrt{32}$ $= \sqrt[2]{32}$
The radicand is 32.

Example 2   Name the parts of the radical $\sqrt[3]{16}$. **Solution** The index is 3, and the radicand is 16

Square Roots

Example $\sqrt{9}$ or $\sqrt[2]{9}$

This radical is read as the square root of 9. The square root of 9 is 3 (because $3^2 = 9$).

The following definition is useful in finding the **square root** of a number:

The **principal square root** of a number (nonzero number) is one of the two equal positive factors of that
number.

Thus, if we can "break up" a number into two equal positive factors, then, one of the positive factors is the square root.(
Note that we exclude negative roots).

Square root of zero: $\sqrt{0} = 0$, because $0^2 = 0$

Examples   (a) $\sqrt{9} = \sqrt{(3)(3)} = 3$;   (b) $\sqrt{64} = \sqrt{(8)(8)} = 8$ (one of (8)(8) is 8)

Cube Roots

Example $\sqrt[3]{8}$
The above radical consists of the index 3, the radical sign and the radicand 8.
This radical is read " the cube root of 8". The cube root of 8 is 2. (because $2^3 = 8$)

The following definition is also useful in finding the cube root of a number:

The **cube root** of a number is **one of the three equal** factors of that number. Here, we do not specify positive root, since the cube root may be positive or negative, depending on whether the given number is positive or negative.

Examples:(a) $\sqrt[3]{8} = \sqrt[3]{(2)(2)(2)} = 2$ (Check: $2^3 = 8$); (b) $\sqrt[3]{-8} = \sqrt[3]{(-2)(-2)(-2)} = -2$ (Check: $(-2)^3 = -8$)

Note: For **even roots** such as the square root of a, $(\sqrt{a})$, the fourth root of a, $(\sqrt[4]{a})$, the sixth root of a,

$(\sqrt[6]{a})$, $\sqrt[n]{a}$ implies the positive root.

For **odd roots** such as the cube root $(\sqrt[3]{a})$, the fifth root $(\sqrt[5]{a})$, or the seventh root $(\sqrt[7]{a})$, the root may be positive or negative according to whether the given number is positive or negative.

More Examples

(a) $\sqrt[4]{16} = 2$ or $\sqrt[4]{16} = \sqrt[4]{(2)(2)(2)(2)} = 2$
because $2^4 = 16$

(b) $\sqrt[5]{32} = 2$ or $\sqrt[5]{32} = \sqrt[5]{(2)(2)(2)(2)(2)} = 2$
because $2^5 = 32$

(c) $\sqrt[5]{-32} = -2$ or $\sqrt[5]{-32} = \sqrt[5]{(-2)(-2)(-2)(-2)(-2)} = -2$
because $(-2)^5 = -32$

Simplifying Radicals

Two approaches are considered: One approach depends on guessing correctly a perfect power which is a factor of the radicand. The other approach is the application of prime factorization, and the grouping of the factors into n equal factors: for example, for square root, we would like to have two equal **positive** factors. For cube root, we would like to have three equal factors.

Example Simplify the following: **(a)** $\sqrt{32}$; **(b)** $\sqrt{\frac{49}{64}}$; **(c)** $\sqrt[3]{80}$.

Solutions

(a) **Method 1** $\sqrt{32} = \sqrt{(16)(2)}$
$= \sqrt{16}\sqrt{2}$
$= 4\sqrt{2}$ ($\sqrt{16}$ is **rational** but $\sqrt{2}$ is **irrational**)

Method 2 $\sqrt{32} = \sqrt{(2)(2)(2)(2)(2)}$
$= \sqrt{(4)(4)(2)}$ (**Rule :** $\sqrt[n]{ab} = \sqrt{a}\sqrt{b}$)
$= 4\sqrt{2}$

(b) $\sqrt{\dfrac{49}{64}} = \dfrac{\sqrt{49}}{\sqrt{64}} = \dfrac{7}{8}$ $\left(\text{Rule: } \sqrt[n]{\dfrac{a}{b}} = \dfrac{\sqrt[n]{a}}{\sqrt[n]{b}}\right)$

(c) $\sqrt[3]{80} = \sqrt[3]{(2)(2)(2)(2)(5)}$ **Scrapwork:**

$= \sqrt[3]{(2)(2)(2)}\,\sqrt[3]{10}$ $80 = 2 \times 2 \times 2 \times 2 \times 5$

$= \sqrt[3]{8}\,\sqrt[3]{10}$

$= 2\sqrt[3]{10}$ (We leave the product of the "extra" 2 and the 5 under the radical sign as 10)

Simplify the following: **1.** $\sqrt{x^3 y^2}$; **2.** $\sqrt{27}$; **3.** $\sqrt{18x^3 y^4}$; **4.** $\sqrt{64x^7 y^{14} z^3}$; **5.** $\sqrt{\dfrac{8}{9}}$

Solution: 1. $\sqrt{x^3 y^2} = \sqrt{x^2 y^2 x} = xy\sqrt{x}$

2. $\sqrt{27} = \sqrt{(9)(3)} = \sqrt{9}\sqrt{3} = 3\sqrt{3}$

3. $\sqrt{18x^3 y^4} = \sqrt{(9)(2)x^2 xy^4} = \sqrt{9x^2 y^4 \cdot 2x} = 3xy^2\sqrt{2x}$

4. $\sqrt{64x^7 y^{14} z^3} = \sqrt{64x^6 y^{14} z^2 \cdot xz} = 8x^3 y^7 z\sqrt{xz}$;

5. $\sqrt{\dfrac{8}{9}} = \dfrac{\sqrt{8}}{\sqrt{9}} = \dfrac{2\sqrt{2}}{3}$

Addition of Radicals (Like Radicals or Similar Radicals)

Like radicals have the same index **and** the same radicand.
Examples: **1.** $4\sqrt{3}$ and $\sqrt{3}$ are like radicals
(index = 2, the square root index; radicand = 3).

2. $5\sqrt[3]{4}$ and $2\sqrt[3]{4}$ are like radicals
(index = 3, radicand = 4).

Unlike radicals: Unlike radicals have different indices **or** radicands or both.
Examples **1.** $\sqrt{3}$ and $\sqrt{2}$ are unlike radicals. (They have different radicands.)

2. $\sqrt[3]{2}$ and $\sqrt[4]{2}$ are unlike radicals (They have different indices.)

Like radicals can be combined into a single radical (i.e., added or subtracted in much the same way as like terms of polynomials are added or subtracted).

Unlike radicals cannot be combined into a single radical (i.e., cannot be added). However, in some cases, by simplifying the given radical(s), like radicals may be obtained, and which then may be added.

To **add like radicals**, add the non-radical parts (coefficients) and keep the radical part.
Example 1 Add: $3\sqrt{11} + 5\sqrt{11}$ (same index = 2; same radicand =11)

Solution $3\sqrt{11} + 5\sqrt{11}$
$= (3 + 5)\sqrt{11}$
$= 8\sqrt{11}$

Note that $3\sqrt{11}$ means 3 times $\sqrt{11}$; but $\sqrt[3]{11}$ means the cube root of 11.

Example 2   Simplify : $3\sqrt{19} + 8\sqrt{10} + 6\sqrt{19} - 2\sqrt{10}$

Solution $3\sqrt{19} + 8\sqrt{10} + 6\sqrt{19} - 2\sqrt{10}$

$= 3\sqrt{19} + 6\sqrt{19} + 8\sqrt{10} - 2\sqrt{10}$ <-------- You may skip this step.

$= (3 + 6)\sqrt{19} + (8 - 2)\sqrt{10}$

$= 9\sqrt{19} + 6\sqrt{10}$

Example 3 Add : $\sqrt{50} + \sqrt{72}$

Solution Step 1: Simplify the radicals first to see if there are any like radicals

$\sqrt{50} = \sqrt{(25)(2)} =$ or $\sqrt{50} = \sqrt{(5)(5)(2)}$
$\sqrt{50} = 5\sqrt{2}$ $= 5\sqrt{2}$

$\sqrt{72} = \sqrt{(36)(2)}$ or $=\sqrt{(3)(3)(2)(2)(2)}$
$\sqrt{72} = 6\sqrt{2}$ $= (3)(2)\sqrt{2} = 6\sqrt{2}$

Step 2: Now, we have like radicals and therefore, we can add.

$\sqrt{50} + \sqrt{72}$
$= 5\sqrt{2} + 6\sqrt{2}$
$= (5 + 6)\sqrt{2}$
$= 11\sqrt{2}$

In the future, Step 1 could be considered as scrapwork and show only Step 2.

Example 4   Simplify: $2\sqrt{75} - 4\sqrt{27}$

Solution

$= 2\sqrt{75} - 4\sqrt{27}$

$= 2\sqrt{(25)(3)} - 4\sqrt{(9)(3)}$

$= 2(5)\sqrt{3} - 4(3)\sqrt{3}$

$=10\sqrt{3} - 12\sqrt{3}$

$= (10 - 12)\sqrt{3}$

$= -2\sqrt{3}$

Example 5 Simplify: $4x\sqrt{25y} - 6x\sqrt{16y}$.

Solution

$4x\sqrt{25y} - 6x\sqrt{16y}$

$= 4x\,(5)\sqrt{y} - 6x(4)\sqrt{y}$ Scrapwork: $\sqrt{25} = 5; \sqrt{16} = 4$

$= 20x\sqrt{y} - 24x\sqrt{y}$

$= (20x - 24x)\sqrt{y}$ (Adding the coefficients of $\sqrt{y}$.)

$= -4x\sqrt{y}$

Multiplication of Radicals

Compared to the conditions in the addition of radicals, the only condition here is that the radicals to be multiplied must have the **same index**. Thus, the radicands may be different or the same, but the index must be the same. Note however that radicals with different indices can always be changed to radicals with a common index by using fractional exponents (see p. 15 Examples 6 & 7, bottom of page) and then multiplying the resulting radicals (The common index will be the LCM of the different indices.)

Example 1   Multiply $\sqrt{3}$ and $\sqrt{5}$

Solution

$$(\sqrt{3})(\sqrt{5}$$
$$=\sqrt{(3)(5)}$$
$$=\sqrt{15}$$

Example 2 Multiply $7\sqrt{3}$ and $9\sqrt{5}$

Solution

Procedure: First multiply the non-radical parts (coefficients) and then multiply the radical parts and simplify if possible.

$$(7\sqrt{3})(9\sqrt{5})$$
$$= (7)(9)\sqrt{(3)(5)}$$
$$= 63\sqrt{15}$$

Example 3   Multiply $4\sqrt{3}$ and $5\sqrt{3}$

Solution

$$(4\sqrt{3})(5\sqrt{3}) = (4)(5)\sqrt{(3)(3)}$$
$$= 20\sqrt{9}$$
$$= 20(3)$$
$$= 60$$

Example 4 Find the product of $2\sqrt{3}$ and $5\sqrt{2} - 4\sqrt{3}$

Solution $2\sqrt{3} \ (5\sqrt{2} - 4\sqrt{3})$<----------The parentheses are important.

$$= (2\sqrt{3})(5\sqrt{2}) \ - (2\sqrt{3})(4\sqrt{3}) \ \text{<--------Application of the distributive rule.}$$
$$= (2)(5)\sqrt{(3)(2)} \ - (2)(4)\sqrt{(3)(3)}$$
$$= 10\sqrt{6} \ - 8\sqrt{9}$$
$$= 10\sqrt{6} \ - 8(3)$$
$$= 10\sqrt{6} - 24$$
$$= -24 + 10\sqrt{6}$$

Note: In Example 4, you may skip writing lines 2 and 3.

Example 5 Simplify: $\sqrt{8}\sqrt{24}$

Method 1

$$\sqrt{8}\sqrt{24} = \sqrt{(8)(24)}$$
$$= \sqrt{(2)(2)(2)(3)(2)(2)(2)}$$
$$= \sqrt{(2)(2)(2)\ (2)(2)(2)\ (3)}$$
$$= \sqrt{(8)(8)3} = 8\sqrt{3}$$

Method 2 $\sqrt{8}\sqrt{24}$

Step 1:   Simplify each radical first.

$\sqrt{8}\sqrt{24}$ **Scrapwork:**
$= (2\sqrt{2})(2\sqrt{6})$ **1.** $\sqrt{8} = 2\sqrt{2}$;   **2.** $\sqrt{24} = 2\sqrt{6}$

Step 2: Multiply the radicals.

$$= (2)(2)(\sqrt{2})(\sqrt{6})$$
$$= 4\sqrt{12}$$
$$= 4\sqrt{4}\sqrt{3}$$
$$= 4(2)\sqrt{3}$$
$$= 8\sqrt{3}$$

Example 6 $(\sqrt{5})(\sqrt{5}) = \sqrt{25} = 5$

Example 7 $(\sqrt{6})(\sqrt{6}) = \sqrt{36} = 6$

Division of a Radical by a Rational Number

Example   Simplify: $\dfrac{6 - \sqrt{12}}{2}$

$$= \frac{6}{2} - \frac{\sqrt{12}}{2}$$ Scrapwork: $\sqrt{12} = \sqrt{(2)(2)(3)} = = 2\sqrt{3}$

$$= 3 - \frac{{}^1\cancel{2}\sqrt{3}}{\cancel{2}_1}$$

$$= 3 - \sqrt{3}$$

Rationalization of Denominators

To rationalize a denominator, we change a given "fraction" to an "equivalent fraction" so that there are no radicals in the denominator. The equivalent fraction may have radicals in the numerator (but not in the denominator).

Example 1 Rationalize the denominator: $\dfrac{3}{\sqrt{6}}$

Solution
We will multiply both the denominator and the numerator by a radical such that the radicand in the denominator becomes a perfect square.
We will multiply both the denominator and the numerator by $\sqrt{6}$

$$\frac{3}{\sqrt{6}} = \frac{3}{\sqrt{6}}\frac{\sqrt{6}}{\sqrt{6}}$$

$$= \frac{3\sqrt{6}}{\sqrt{36}}$$

$$= \frac{3\sqrt{6}}{6}$$

$$= \frac{{}^1\cancel{3}\sqrt{6}}{\cancel{6}_2}$$

$$= \frac{\sqrt{6}}{2}$$

Lesson 2: Review: Exponents and Radicals

A motivation for rationalizing denominators:

Note above that in practical applications, $\dfrac{\sqrt{6}}{2}$ is more convenient to use than $\dfrac{3}{\sqrt{6}}$; for example, it is easier to divide the decimal approximation of $\sqrt{6}$ by 2 than to divide 3 by the decimal approximation of $\sqrt{6}$.

Example 2   Rationalize the denominator: $\sqrt{\dfrac{7}{3}}$ (one-term denominator)

Solution   We shall multiply both the denominator and the numerator by $\sqrt{3}$.

$$\sqrt{\dfrac{7}{3}} = \sqrt{\dfrac{7\cdot 3}{3\cdot 3}} \quad \text{or} \quad \dfrac{\sqrt{7}}{\sqrt{3}}\cdot\dfrac{\sqrt{3}}{\sqrt{3}}$$

$$= \dfrac{\sqrt{21}}{\sqrt{9}} \quad \text{or} \quad \dfrac{\sqrt{21}}{\sqrt{9}}$$

$$= \dfrac{\sqrt{21}}{3} \quad \text{or} \quad \dfrac{\sqrt{21}}{3}$$

Example 3   Rationalize the denominator: $\sqrt[3]{\dfrac{5}{4}}$

$$\sqrt[3]{\dfrac{5}{4}} = \sqrt[3]{\dfrac{5}{2\cdot 2}}$$

$$= \sqrt[3]{\dfrac{5\cdot 2}{2\cdot 2\cdot 2}} \longleftarrow \text{-------One more 2 will make the denominator a perfect cube.}$$

$$= \dfrac{\sqrt[3]{10}}{\sqrt[3]{8}}$$

$$= \dfrac{\sqrt[3]{10}}{2} \;.$$

Example 4 Rationalize the denominator: $\sqrt[3]{\dfrac{8}{9}}$

Method 1 $\sqrt[3]{\dfrac{8}{9}} = \sqrt[3]{\dfrac{(2)(2)(2)}{(3)(3)}}$

$$= \sqrt[3]{\dfrac{(2)(2)(2)(3)}{(3)(3)(3)}} \longleftarrow \text{(one more 3}$$
$$\text{will make the denominator a perfect cube)}$$

$$= \dfrac{2\sqrt[3]{3}}{3}$$

Method 2 $\sqrt[3]{\dfrac{8}{9}} = \dfrac{\sqrt[3]{8}}{\sqrt[3]{9}}$

$$= \dfrac{2}{\sqrt[3]{9}} \qquad (\sqrt[3]{8}=2)$$

$$= \dfrac{2\sqrt[3]{3}}{\sqrt[3]{9}\,\sqrt[3]{3}}$$

$$= \dfrac{2\sqrt[3]{3}}{\sqrt[3]{27}}$$

$$= \dfrac{2\sqrt[3]{3}}{3}$$

Method 3   In the following approach, simplifying the numerator becomes more involved than Methods 1 & 2:   13

$$\sqrt[3]{\frac{8}{9}} = \frac{\sqrt[3]{8}}{\sqrt[3]{9}} \cdot \frac{\sqrt[3]{9}(\sqrt[3]{9})}{\sqrt[3]{9}(\sqrt[3]{9})} = \frac{\sqrt[3]{8 \cdot 9 \cdot 9}}{\sqrt[3]{9 \cdot 9 \cdot 9}} = \frac{\sqrt[3]{8 \cdot 9 \cdot 9}}{9}$$

$$= \frac{\sqrt[3]{2 \cdot 2 \cdot 2 \cdot 3 \cdot 3 \cdot 3 \cdot 3}}{9} = \frac{2 \cdot 3\sqrt[3]{3}}{9} = \frac{2\sqrt[3]{3}}{3}$$

Example 5 Rationalize the denominator: $\dfrac{\sqrt{5}}{\sqrt[3]{2}}$

Solution $\dfrac{\sqrt{5}}{\sqrt[3]{2}} = \dfrac{\sqrt{5}}{\sqrt[3]{2}} \cdot \dfrac{\sqrt[3]{2}(\sqrt[3]{2})}{\sqrt[3]{2}(\sqrt[3]{2})} = \dfrac{\sqrt{5} \cdot \sqrt[3]{4}}{2}$ or $\dfrac{\sqrt[6]{2000}}{2}$ (See also page 15 Examples 6 & 7, bottom of page)

Definition : The **conjugate** of a given binomial is another binomial that differs from the given binomial only in the sign of one of the terms. The conjugate of $a + b$ is $a - b$; and the conjugate of $a - b$ is $a + b$.

Example 6 Rationalize the denominator $\dfrac{5}{\sqrt{3} + \sqrt{2}}$ (two-term denominator).

Procedure: Multiply both the denominator and the denominator by the conjugate of $\sqrt{3} + \sqrt{2}$.

The conjugate of $\sqrt{3} + \sqrt{2}$ is $\sqrt{3} - \sqrt{2}$. (You may also multiply by $-\sqrt{3} + \sqrt{2}$; try it later on.)

Step 1: $\dfrac{5}{\sqrt{3} + \sqrt{2}} = \dfrac{5 \ (\sqrt{3} - \sqrt{2})}{(\sqrt{3} + \sqrt{2})(\sqrt{3} - \sqrt{2})}$

Step 2: $= \dfrac{5(\sqrt{3} - \sqrt{2})}{(\sqrt{3})(\sqrt{3}) - (\sqrt{3})(\sqrt{2}) + (\sqrt{2})(\sqrt{3}) - (\sqrt{2})(\sqrt{2})}$

$= \dfrac{5(\sqrt{3} - \sqrt{2})}{\sqrt{9} + 0 - \sqrt{4}}.$ Note: $(-\sqrt{3})(\sqrt{2}) + (\sqrt{2})(\sqrt{3}) = 0$

$= \dfrac{5(\sqrt{3} - \sqrt{2})}{3 - 2}$

$= \dfrac{5(\sqrt{3} - \sqrt{2})}{1}$

$= 5(\sqrt{3} - \sqrt{2})$ or $5\sqrt{3} - 5\sqrt{2}$

Note above: As was suggested in the procedure, you could also have multiplied by $-\sqrt{3} + \sqrt{2}$; and obtained the same result. Therefore, it is not critical which of the terms differ in sign.

Example 7   Rationalize the denominator : $\dfrac{1}{\sqrt{8} - 2}$

Procedure: Multiply both the denominator and the numerator by the conjugate of $\sqrt{8}$ - 2.

The conjugate of $\sqrt{8}$ - 2 is $\mathbf{\sqrt{8} + 2}$ (You may also multiply by $-\sqrt{8}$ - 2; try it later on.)

$$\frac{1}{\sqrt{8} - 2} = \frac{1}{(\sqrt{8} - 2)} \frac{(\mathbf{\sqrt{8} + 2})}{(\mathbf{\sqrt{8} + 2})}$$

$$= \frac{\sqrt{8} + 2}{(\sqrt{8})(\sqrt{8}) + 2\sqrt{8} - 2\sqrt{8} - 4} \qquad \textleftarrow\text{----After some practice, you may skip writing this step.}$$

$$= \frac{\sqrt{8} + 2}{\sqrt{64} + 0 - 4} \qquad\qquad (2\sqrt{8} - 2\sqrt{8} = 0)$$

$$= \frac{\sqrt{8} + 2}{8 - 4}$$

$$= \frac{2\sqrt{2} + 2}{4}$$

$$= \frac{2(\sqrt{2} + 1)}{4}$$

$$= \frac{\sqrt{2} + 1}{2} \quad \text{or} \quad \frac{1}{2} + \frac{\sqrt{2}}{2}$$

Note above that $\dfrac{1}{\sqrt{8} - 2} = \dfrac{1}{-2 + \sqrt{8}}$

Extra

The following are worth knowing

1. $x\sqrt{x + 1} = \sqrt{x^2(x + 1)} = \sqrt{x^3 + x^2}$

2. $\dfrac{\sqrt{x + 1}}{x} = \dfrac{\sqrt{x + 1}}{\sqrt{x^2}} = \sqrt{\dfrac{x + 1}{x^2}}$

* Perhaps, it would be better to say "a" conjugate instead of "the" conjugate since it does not matter which terms differ in sign, so far as the rationalization is concerned.

Fractional Exponents

Algebraic operations involving radicals can sometimes be made easy by changing the radicals involved to their equivalent exponential forms, and then applying the laws of exponents.

Interconversion between radicals and exponential expressions

Examples

1. $25^{1/2} = \sqrt[2]{25} = \sqrt{25} = 5$

2. $9^{1/2} = \sqrt{9} = 3$

3. $8^{1/3} = \sqrt[3]{8} = 2$

4. $8^{2/3} = (8^{1/3})^2 = (\sqrt[3]{8})^2 = 2^2 = 4$

5. $16^{3/4} = (16^{1/4})^3 = (\sqrt[4]{16})^3 = 2^3 = 8$

Rules: **1.** $\sqrt[n]{a} = a^{1/n}$

     2. $\sqrt[n]{a^m} = a^{m/n}$

     3. $\sqrt[n]{a^m} = (\sqrt[n]{a})^m$

Example 1   Simplify the following:

1. $a^{1/3} \cdot a^{2/3}$;   **2.** $b^{2/3} \cdot b^{1/2}$;   **3.** $4^{-1/2}$;   **4.** $2^{2/5} \cdot 2^{1/4}$ **5.** $\sqrt[5]{x^2} \cdot \sqrt[5]{x^3}$

6. $\sqrt{x} \cdot \sqrt[3]{x^2}$;   **7.** $\sqrt[5]{3^2} \cdot \sqrt[4]{3}$

Solution

1. $a^{1/3} \cdot a^{2/3} = a^{1/3 + 2/3} = a^1 = a$

2. $b^{2/3} \cdot b^{1/2} = b^{2/3 + 1/2} = b^{7/6}$ $(2/3 + 1/2 = 7/6)$

3. $4^{-1/2} = \dfrac{1}{4^{1/2}} = \dfrac{1}{\sqrt{4}} = \dfrac{1}{2}$

4. $2^{2/5} \cdot 2^{1/4} = 2^{2/5 + 1/4} = 2^{13/20}$

5. $\sqrt[5]{x^2} \cdot \sqrt[5]{x^3} = x^{2/5} x^{3/5} = x^{5/5} = x^1 = x$ or $\sqrt[5]{x^2} \cdot \sqrt[5]{x^3} = \sqrt[5]{x^2 \cdot x^3} = \sqrt[5]{x^5} = x$

6. $\sqrt[5]{3^2} \cdot \sqrt[4]{3} = 3^{2/5} \cdot 3^{1/4} = 3^{13/20} = \sqrt[20]{3^{13}}$ <---(Multiplying radicals with different indices)

7. $\sqrt{x} \cdot \sqrt[3]{x^2} = x^{1/2} \cdot x^{2/3} = x^{7/6} = \sqrt[6]{x^7} = x\sqrt[6]{x}$ <---(Multiplying radicals with different indices)

Example 2 Evaluate. $1000^{-1/3}$ 16

Step 1: Express with positive exponents.

$$1000^{-1/3} = \frac{1}{1000^{1/3}}$$

Step 2: Change the exponential form to radical form.

$$\text{then, } \frac{1}{1000^{1/3}} = \frac{1}{\sqrt[3]{1000}} = \frac{1}{10}$$

Example 3 Evaluate. $81^{-1/2}$

$$81^{-1/2} = \frac{1}{81^{1/2}}$$
$$= \frac{1}{\sqrt{81}}$$
$$= \frac{1}{9}$$

Example 4 Evaluate $\left(-\frac{8}{27}\right)^{-5/3}$

We will cover two methods.
Method 1

$$\left(-\frac{8}{27}\right)^{-5/3} = \left(\frac{-8}{27}\right)^{-5/3}$$ <--- (We can give the minus sign to either the "8" or the "27")
$$= \frac{1}{\left(\frac{-8}{27}\right)^{5/3}}$$
$$= \frac{1}{\left(\sqrt[3]{\frac{-8}{27}}\right)^5}$$
$$= \frac{1}{\frac{(\sqrt[3]{-8})^5}{(\sqrt[3]{27})^5}}$$
$$= \frac{1}{\frac{-32}{243}} \quad \text{or} \quad 1 \div \frac{-32}{243}$$
$$= -\frac{243}{32}$$

Scrapwork: **1.** $\sqrt[3]{-8} = -2$

2. $(-2)^5 = -32$

3. $\sqrt[3]{27} = 3$

4. $3^5 = 243$

Method 2   (Much faster than Method 1)

$$\left(-\frac{8}{27}\right)^{-5/3} = \left(\frac{-8}{27}\right)^{-5/3}$$

$$= \left(\frac{27}{-8}\right)^{5/3} \quad \text{(Interchange the numerator and the denominator and change the sign of the exponent)}$$

$$= \frac{(\sqrt[3]{27})^5}{(\sqrt[3]{-8})^5} = \frac{(3)^5}{(-2)^5} = \frac{243}{-32}$$

$$= -\frac{243}{32}$$

Reduction of Indices

Example 1   Reduce the index: $\sqrt[4]{36}$

Step 1:   Change the radical to exponential form (power form).

Then, $\sqrt[4]{36} = \sqrt[4]{6^2} = 6^{2/4}$

Step 2: Reduce the exponent $\frac{2}{4}$ to lowest terms.

then $\frac{2}{4} = \frac{1}{2}$

Step 3: $6^{2/4} = 6^{1/2}$

Step 4:   Change back to the radical form.

then $6^{1/2} = \sqrt{6}$ **Note:** $6^{1/2} = \sqrt[2]{6} = \sqrt{6}$

Perfect Powers

Examples: **1.** 9 is a perfect square.
2. 8 is a perfect cube .
3. $16y^2$ is a perfect square .
4. $\frac{4}{9}$ is a perfect square .

What is meant by to simplify a radical?

1. It may mean remove all perfect powers from the radicand (page 7); or

2. It may mean rationalize the denominator (page 11), or

3. It may mean reduce the index of the radical (page 17), or

4. It may mean perform a combination of the above.

Lesson 2 Exercises

A Simplify: **1.** $16^{1/2}$ **2.** $\sqrt[3]{x^6 y^{12}}$ **3.** $64^{1/3}$;   **4.** $27^{2/3}$;   **5.** $(\sqrt[4]{16})^3$ **6.** $\sqrt[3]{\dfrac{x^6}{y^{15}}}$

Answers: 1. 4 ; **2.** $x^2 y^4$;   **3.** 4 ;   **4.** 9;   **5.** 8;   **6.** $\dfrac{x^2}{y^5}$

B Simplify t **1.** 3^{-3}; **2.** $100^{-1/2}$; **3.** $\left(\dfrac{4}{9}\right)^{-3/2}$;   **4.** $\dfrac{x^{-4} y^3 z^2}{x^4 y^{-2} z}$;   **5.** $(6x)^0 - 6x^0$

   6. $\left(\dfrac{-4a^3 b}{3b}\right)\left(\dfrac{9c}{2bc}\right)\left(\dfrac{-12}{18c}\right)$ **7.** $81^{-1/4}$

Answers: 1. $\dfrac{1}{27}$;   **2.** $\dfrac{1}{10}$;   **3.** $\dfrac{27}{8}$;   **4.** $\dfrac{y^5 z}{x^8}$;   **5.** -5;   **6.** $\dfrac{4a^3}{bc}$;   **7.** $\dfrac{1}{3}$

C Simplify: **1.** $\sqrt{18}$;   **2.** $\sqrt{50}$;   **3.** $\sqrt{24}$;   **4.** $\sqrt{72}$;   **5.** $\sqrt{288}$; **6.** $\sqrt{48}$;   **7.** $\sqrt{\dfrac{81}{100}}$

Answers: 1. $3\sqrt{2}$; **2.** $5\sqrt{2}$; **3.** $2\sqrt{6}$; **4.** $6\sqrt{2}$;   **5.** $12\sqrt{2}$;   **6.** $4\sqrt{3}$; **7.** $\dfrac{9}{10}$

D Simplify **1.** $\sqrt{16x^6}$;   **2.** $\sqrt{9x^8 y^2}$; **3.** $\sqrt{27x^9 y^4}$ **4.** $\sqrt{18x^3 y^5}$; **5.** $\sqrt{32x^8 y}$;   **6.** $\sqrt{8x^4 y^2 z^3}$

Answers: 1. $4x^3$;   **2.** $3x^4 y$;   **3.** $3x^4 y^2 \sqrt{3x}$;   **4.** $3xy^2\sqrt{2xy}$;   **5.** $4x^4\sqrt{2y}$; **6.** $2x^2 yz\sqrt{2z}$

E. Add or subtract: **1.** $2\sqrt{3} + 4\sqrt{3}$;   **2.** $3\sqrt{5} + \sqrt{98} + \sqrt{20} + 8\sqrt{2}$;   **3.** $4\sqrt{48} - \sqrt{12} + \sqrt{75}$

   4. $2x\sqrt{9y} - 5x\sqrt{4y}$

Answers: 1. $6\sqrt{3}$;   **2.** $5\sqrt{5} + 15\sqrt{2}$;   **3.** $19\sqrt{3}$;   **4.** $-4x\sqrt{y}$

F Multiply and simplify: **1.** $2\sqrt{5}$ and $3\sqrt{2}$;   **2.** $3\sqrt{5}$ and $4\sqrt{10}$;   **3.** $4\sqrt{2}$ and $5\sqrt{2}$

   4. $4\sqrt{3}$ and $4\sqrt{3}$;   **5.** $2x\sqrt{3y}$ and $\sqrt{6y}$

Answers: 1. $6\sqrt{10}$; **2.** $60\sqrt{2}$; **3.** 40 ; **4.** 48 ; **5.** $6xy\sqrt{2}$

G **1.** $\dfrac{8 + \sqrt{48}}{4}$ **2.** $\dfrac{12 - \sqrt{18}}{3}$

Answers: 1. $2 + \sqrt{3}$;   **2.** $4 - \sqrt{2}$

H Rationalize the denominators:

1. $\dfrac{1}{\sqrt{3}}$;   **2.** $\dfrac{2}{\sqrt{5}}$;   **3.** $\dfrac{5}{\sqrt{2}}$;   **4.** $\dfrac{6}{\sqrt{5} - \sqrt{3}}$;   **5.** $\dfrac{\sqrt{3}}{4 + \sqrt{3}}$;

6. $\sqrt[3]{\dfrac{4}{9}}$;   **7.** $\sqrt[3]{\dfrac{8}{9}}$.   **8.** Is $\dfrac{\sqrt{3}}{\sqrt{2}} = \dfrac{3}{\sqrt{6}}$?

Answers: 1. $\dfrac{\sqrt{3}}{3}$; **2.** $\dfrac{2\sqrt{5}}{5}$; **3.** $\dfrac{5\sqrt{2}}{2}$; **4.** $3\sqrt{5} + 3\sqrt{3}$; **5.** $\dfrac{-3 + 4\sqrt{3}}{13}$; **6.** $\dfrac{\sqrt[3]{12}}{3}$; **7.** $\dfrac{2\sqrt[3]{3}}{3}$; **8.** Yes

Lesson 2: Review: Exponents and Radicals

I Simplify the following: **1.** $x^{3/4} \cdot x^{1/4}$ **2.** $x^{2/3} \cdot x^{3/4}$ **3.** $3^{1/2} \cdot 3^{2/3}$

Answers**: 1.** x ; **2.** $x^{17/12}$; **3.** $3^{7/6}$

J Evaluate the following: **1.** $100^{-1/2}$; **2.** $25^{-1/2}$; **3.** $\left(\frac{16}{81}\right)^{-1/4}$

Answers: **1.** $\frac{1}{10}$; **2.** $\frac{1}{5}$; **3.** $\frac{3}{2}$

K Reduce the index: **1.** $\sqrt[4]{x^2}$; **2.** $\sqrt[6]{x^8}$; **3.** $\sqrt[4]{25}$; **4.** $\sqrt[4]{4}$

Answers: 1. $\sqrt{x}$; **2.** $\sqrt[3]{x^4}$; **3.** $\sqrt{5}$; **4.** $\sqrt{2}$.

CHAPTER 3

Review
Equations

Lesson 3: Types of Equations

Lesson 4: **Principles for Solving Equations; Extraneous and Lost Roots**

Lesson 3
Types of Equations

Definition: An equation is a statement of the equality between two mathematical expressions.

Examples: 1. $2x + 3 = x - 10$;   **2**. $x^2 + 6x + 9 = 0$;

   3. $\sqrt{x-2} = 5$;   **4**. $\log x + \log(x-3) = 1$;

   5. $\sin x = \frac{1}{2}$;   **6**. $\sin^2 x + \cos x = -1$.

The equality symbol "=' breaks up an equation into two sides or members, namely the left-hand side and the right-hand side of the equation. To solve an equation involving a single variable, say x, means we are to find values of x which satisfy the equation. **A value of x is said to satisfy an equation if this value when substituted in the equation makes both the left-hand side and the right-hand side of the equation equal to each other.**
A value of x which satisfies a given equation is said to be a solution or a root of the given equation.

To obtain a value for the variable, x, we will get x by itself alone on one side of the equation .
We agree that a value of x has been obtained if we have x by itself alone on one side of the equation and all the other quantities on the other side of the equation do **not** involve x.
 To get x by itself (i.e., isolate x), we will use inverse operations. Addition and subtraction are inverse operations. Multiplication and division are inverse. Root finding and power finding are inverse operations. Finding the logarithm of a number and finding the anti-logarithm of a number are inverse operations. Inverse operations "undo" (reverse the action of) each other. For example, to undo multiplication
by a number, we will use division by the same the number. We will keep the above discussion in mind when we solve equations.

 We must note that a value of x obtained should be tested (for good practice) in the given equation to determine if it satisfies the equation. If it does, it is a solution. If it does not, it is not a solution. Not every equation has a solution. An equation may have only real solutions; it may have only non-real solutions (complex roots, see page 344). or it may have both real and, non-real solutions. We must be careful to distinguish between an equation having non-real solutions and an equation having no solutions at all. Thus, if we are only interested in finding only the real solutions and we obtain non-real solutions, then, we will say that the equation has no solution. However, if we are interested in finding either a real solution or a non-real solution, then we will say that we have a solution if we obtain either a real solution, or a non-real solution .

Sometimes, we also say that if an equation has solution, then it is consistent. If it has no solution, it is inconsistent. There are many types of equations that a student should be able to handle with confidence by the time a student has successfully completed a pre-calculus course. By the time a student gets through a pre-calculus course, the student should be able to solve the various types of equations listed below.

Most students have covered types **(A)**, **(B)**, **(C)**, **(E)**, **(F)** and **J**; and perhaps even **G** and **H** in previous courses.

A. Simple or linear equation in a single variable

Examples:
1. $5x + 1 = 16$
2. $\frac{2x}{7} = 10$
3. $\frac{x-2}{2} + \frac{2}{3} = \frac{x}{6}$

B. A system of simultaneous linear equations containing two variables

Examples:

1. $\begin{cases} 3x + 8y = 0 \\ 6x + 7y = 5 \end{cases}$

2. $\begin{cases} 5x - 3y = -2 \\ 2x + 5y = 8 \end{cases}$

C. Second degree equation in one variable (quadratic equations)

Examples:
1. $x^2 + 6x + 9 = 0$
2. $2x^2 + 4x = -3$
3. $x^2 - 6x = 0$

D. A system of quadratic equations in two variables

Example: $\begin{cases} x^2 + y^2 = 9 \\ 16x^2 + y^2 = 16 \end{cases}$

E. A system of a quadratic equation and a linear equation

$\begin{cases} x^2 + y^2 = 25 \\ x + y = 1 \end{cases}$

F. Rational equations

Examples 1. $\frac{9}{x + 4} = 1$

2. $\frac{2}{x - 3} = \frac{5}{x - 4}$

G. Radical equations (Irrational equations)

Examples:

1. $\sqrt{x-2} = 5$ or $(x-2)^{\frac{1}{2}} = 5$

2. $\sqrt{3x-1} = 2 + \sqrt{4x+7}$

H. Exponential and Logarithmic equations (See Chapter 18)

Examples

1. $3^{2x} = 81$

2. $\log x + \log(x-3) = 1$

I. Trigonometric equations (See Chapter 39)

Examples:

1. $\sin x = \dfrac{1}{2}$

2. $\sin^2 x + \cos x = -1$

J. Literal equations

To solve a literal equation for a specified variable, say A, we will obtain A by itself alone on one side of the equation so that the other side does not contain A.

Example   Solve for A: $B = AC + CE$

We will obtain A by itself on one side of the equation.

$$B = AC + CE$$
$$B - CE = AC + CE - CE \qquad \text{(subtracting } CE \text{ from both sides)}$$
$$B - CE = AC$$
$$\frac{B - CE}{C} = \frac{AC}{C} \qquad \text{(dividing both sides by } C\text{)}$$
$$\frac{B - CE}{C} = A$$
$$\therefore A = \frac{B - CE}{C} \text{ or } \frac{B}{C} - \frac{CE}{C} = \frac{B}{C} - E$$

Lesson 4   23

Principles for Solving Equations; Extraneous and Lost Roots

Principles for Solving Equations

The procedure for obtaining the solutions to the above equations depends on two main principles, namely inverse operations and operational rules of equality on both sides of an equation.
Inverse operations are operations that undo (reverse the action of) each other.

Examples

1. Addition and subtraction are inverse operations.

Example 1 Solve for x:
$$x + 4 = 0$$
$$\underline{-4 \quad -4} \qquad \text{(Subtracting 4 from (or adding - 4 to) both sides of the equation)}$$
$$x = 2$$

Example 2 Solve for x:
$$x - 5 = 12$$
$$\underline{+5 \quad +5} \qquad \text{(Adding 5 to both sides of the equation)}$$
$$x = 17$$

2. Multiplication and division are inverse operations.

Example 1 Solve for x:: $5x = 60$

Solution To solve for x, we will use division by 5 to undo the 5 that multiplies the x.

$$\text{From } 5x = 60, \quad x = \frac{60}{5} = 12$$

Example 2 Solve for x: $\frac{x}{6} = 9$

Solution

$$\frac{(6)x}{6} = 9(6) \qquad \text{(Multiplying to undo the 6 which is dividing)}$$

$$x = 54$$

3. Raising a number to a power and finding a root of a number are inverses of each other.

Example 1 Solve x $x^2 = 9$.
To solve for x, we use "root extraction" since the x is being raised to a power.

$$\text{Then } \sqrt{x^2} = \pm\sqrt{9}$$
$$x = \pm\sqrt{3}$$

Example 2 Solve for x: $\sqrt{x} = 6$.
We raise both sides of the equation to a power to undo the root.

$$\left(\sqrt{x}\right)^2 = (6)^2$$
$$x = 36$$

4. Logarithms and exponentials are inverses of each another.

Example 1: If $\log_{10} 100 = 2$, then anti-log$2 = 100$ or $10^2 = 100$.

In general, if $\log_{10} x = y$, then $10^y = x$, or anti-log $y = x$.

Operational rules of equality on both sides of an equation

The basic rules of equality are for addition, subtraction, multiplication, and division, root finding, power finding, taking logarithms and taking antilogarithms. We may perform the same operation on both sides of an equation. However, if we introduce a variable or an expression containing a variable, then, we may obtain extraneous roots, and in this case, we must check the solution.

Extraneous Roots and Lost Roots

In performing the basic operations of multiplication, division, power finding (for example, squaring) and root extraction on both sides of an equation, we must be careful not to lose roots or introduce "additional roots",

Extraneous Roots

If both sides of a given equation are multiplied by the same variable or an expression containing the variable in the equation, then the resulting equation may have more solutions than the original equation. We must check any roots obtained in the original equation.

Example   Solve for x: $\sqrt{2x - 5} = x - 4$

Solution

Step 1: Square both sides of the equation.
$$(\sqrt{2x - 5})^2 = (x - 4)^2$$
$$2x - 5 = x^2 - 8x + 16$$
$$0 = x^2 - 10x + 21$$

Step 2: Solving by factoring, since the factors are easily recognizable.
$$x^2 - 10x + 21 = 0$$
$$(x - 3)(x - 7) = 0, \text{ and from which}$$
$$x = 3 \text{ or } x = 7 \text{ (Setting each factor equal to zero and solving for } x)$$

Step 3: Checking for $x = 3$ in $\sqrt{2x - 5} = x - 4$
$$\sqrt{2(3) - 5} \overset{?}{=} 3 - 4$$
$$\sqrt{6 - 5} \overset{?}{=} -1$$
$$\sqrt{1} \overset{?}{=} -1$$
$$1 = -1 \text{ False   (LHS \textbf{not} equal to RHS)}$$

Therefore, 3 is not a solution. It is an **extraneous** root.

Checking for $x = 7$:
$$\sqrt{2(7) - 5} \overset{?}{=} 7 - 4$$
$$\sqrt{14 - 5} \overset{?}{=} 3$$
$$\sqrt{9} \overset{?}{=} 3$$
$$3 = 3 \text{   True   (RHS = LHS)}$$

Therefore, 7 is a solution but 3 is an **extraneous** solution.

Lost Roots

Just as introducing an expression containing the variable into both sides of an equation may result in extraneous roots, dividing out (canceling) common factors containing the variable on both sides of an equation may result in the loss of some of the roots or solutions of the original equation.

Example A common **wrong** approach in solving the quadratic equation

$3x^2 - 4x = 0$ is as follows:

$$3x^2 - 4x = 0$$

$$3x^2 = 4x$$

$$3x = 4$$ (Canceling, that is dividing out an x on both sides of the equation ; such a cancellation excludes $x = 0$ as a solution, which is one of the solutions)

$$x = \frac{4}{3}$$ (Of course, you still obtain one of the solutions but lose the other solution, 0)

Note that the solutions to $3x^2 - 4x = 0$ are 0 and $\frac{4}{3}$.

(obtained by factoring: and solving for x.

CHAPTER 4

Quadratic Equations; Higher Polynomial Equations; and Systems Containing Nonlinear Equations

Lesson 5: **Solving Quadratic Equations (Review};** Discriminant;
Roots r_1, r_2 and the Constants a, b, and c of the Quadratic Equation;
Deriving Quadratic Equations; Quadratic Formula to aid Factoring

Lesson 6: **Solving Equations Quadratic in Form**

Lesson 7: **Higher Polynomial Equations; Systems of Nonlinear Equations**

Standard form of the quadratic equation: $ax^2 + bx + c = 0$, where a, b, and c are constants and $a \neq 0$.

Examples 1. $x^2 - 3x - 28 = 0$<---------in standard form

2. $x^2 - 2x = 0$ <--------- in standard form

3. $x^2 = 5x + 14$<---------**not** in standard form

4. $5x^2 = 8x$ <---------**not** in standard form

Lesson 5

Solving Quadratic Equations (Review}

We will consider four methods:

Method 1: **By factoring** (For easily recognizable factors and for cases in which the constant term is missing, i.e., $c = 0$)

Method 2: **By the square root method** (For cases in which the x-term is missing i.e., $b = 0$)

Method 3: **By the quadratic formula** (This always works)

Method 4: **By completing the square** (If asked to use this method, otherwise choose from Methods 1, 2, and 3 above).

Note that Method 4 also always works. **Note also** that the quadratic formula can be derived from the quadratic equation **by completing the square** and solving for x.

Solving Quadratic Equations by Factoring

Principle of zero products: If $ab = 0$ then either $a = 0$, or $b = 0$ (or both $= 0$).

Example 1 Solve by factoring
$$x^2 - 3x - 28 = 0$$

Step 1: Factor the quadratic trinomial.
$$(x + 4)(x - 7) = 0$$

Step 2: Set each factor equal to zero and solve each equation for x.

$$x + 4 = 0 \qquad \text{or} \qquad x - 7 = 0$$
$$\underline{-4 \ -4} \qquad \qquad \underline{+7 \ +7}$$
$$x = -4 \qquad \qquad x = 7$$

$$\therefore \ x = -4, \text{ or } x = 7$$

The solutions are -4 and 7.

Example 2 Solve by factoring.

$$3x^2 - 4x = 0$$

Solution

Step 1: Factor. $3x^2 - 4x = 0$

$$x(3x - 4) = 0 \qquad \text{(performing common monomial factoring)}.$$

Step 2: Set each factor equal to zero and solve for x.

$$x = 0 \quad \text{or} \quad 3x - 4 = 0$$

$$\underline{+ 4 \quad +4}$$

$$\frac{3}{3}x = \frac{4}{3}$$

$$x = \frac{4}{3}$$

$$\therefore x = 0, \text{ or } x = \frac{4}{3} \qquad\qquad \textbf{Note} \text{ that 0 is also a solution}.$$

The solutions are 0 and $\frac{4}{3}$

Note: A common **wrong** approach in the above problem which has been previously covered:

$$3x^2 - 4x = 0$$

$$3x^2 = 4x$$

$3x = 4$ (Canceling, that is dividing out an x on both sides of the equation ; such a cancellation excludes
$x = 0$ as a solution.)

$x = \frac{4}{3}$ (Of course, you still obtain one of the solutions but lose the other solution)

Example 3 Solve for x by factoring.

$$\frac{x^2}{2} + \frac{10x}{3} + 2 = 0$$

Step 1: Undo the denominators by multiplying the equation by the LCM of 2 and 3 which is 6.

$$(6)\frac{x^2}{2} + (6)\frac{10x}{3} + 2(6) = 6(0)$$

$$\overset{3}{\cancel{(6)}}\frac{x^2}{\cancel{2}} + \frac{2\cancel{(6)}10x}{\cancel{3}} + 2(6) = 0$$
$$\underset{1}{} \qquad \underset{1}{}$$

$$3x^2 + 20x + 12 = 0 \dots\dots\dots\dots\dots\dots\dots\dots(A)$$

Step 2:   Factor by the substitution method or otherwise

$$3(3x^2) + 20x(3) + 12(3) = 0$$

$$9x^2 + 20(3x) + 36 = 0$$

$$(3x)^2 + 20(3x) + 36 = 0 \dots\dots\dots\dots\dots(B)$$

Step 3:   Let $3x = s$ in equation (B), and factor.

$$s^2 + 20s + 36 = 0$$

$$(s + 2)(s + 18) = 0\dots\dots\dots\dots\dots\dots(C)$$

Step 4:   Replace s by $3x$ in equation (C)

$$(3x + 2)(3x + 18) = 0 \text{ ....(D)}$$

Step 5:   Set each factor equal to zero and solve for x.

$$3x + 2 = 0 \qquad\qquad 3x + 18 = 0$$
$$\underline{-2 \quad -2} \qquad\qquad \underline{-18 \quad -18}$$
$$\frac{3}{3}x = \frac{-2}{3} \qquad\qquad \frac{3}{3}x = \frac{-18}{3}$$
$$x = -\frac{2}{3} \qquad\qquad x = -6$$

The solutions are $-\frac{2}{3}$ and -6.   or solution set $= \{-\frac{2}{3}, -6\}$

Note that in the above problem , since it is an equation, it was not necessary to factor completely the left-hand side of equation (D) of Step 4.

Factorability of a quadratic trinomial

A quadratic trinomial is factorable if $b^2 - 4ac$ is a perfect square.

Solving by the Square Root Method
(For cases in which $b = 0$, i.e., the x-term is missing)

Principle: If $x^2 = k$, then $x = \pm\sqrt{k}$ (i.e., $x = +\sqrt{k}$ or $x = -\sqrt{k}$)

Example 1 Solve by the square root method.
$$x^2 - 9 = 0$$

Solution $x^2 - 9 = 0$
$$\underline{+9 \quad +9}$$
$$x^2 = 9$$
$$x = \pm\sqrt{9}$$
$$x = \pm 3 \quad \text{(i.e. } x = 3 \text{ or } x = -3)$$
The solutions are -3 and 3.

Example 2 Solve by the square root method.
$$9x^2 - 36 = 0$$

Solution $9x^2 - 36 = 0$
$$\underline{+36 \quad +36}$$
$$9x^2 = 36$$
$$\frac{9}{9}x^2 = \frac{36}{9}$$
$$x^2 = 4$$
$$x = \pm\sqrt{4}$$
$$x = \pm 2$$
The solutions are -2 and 2.

Solving by Completing the Square

Example 1   Solve by completing the square.

$$x^2 - 12x + 8 = 0$$

Step 1: Eliminate the constant term (the "8")
from the left -hand side.   (Note that the 8 ends up on the right-hand
side of the equation as -8).

$$
\begin{array}{r}
x^2 - 12x + 8 = 0 \\
\underline{- 8 \quad - 8} \\
x^2 - 12x = - 8
\end{array}
$$

Step 2:   Add the square of half the coefficient of the x-term to both sides of the equation.

(i.e., add the square of $\frac{b}{2}$ to both sides of the equation)

$$x^2 - 12x + \left(\frac{-12}{2}\right)^2 = - 8 + (-6)^2 \quad (b = -12, \ \frac{b}{2} = -6; \text{ and the square of } \frac{b}{2} = (-6)^2$$

$$x^2 - 12x + (-6)^2 = - 8 + (-6)^2$$

Step 3:   Complete the square on the left-hand side of the equation

$$
\begin{array}{ll}
(x - 6)^2 & = - 8 + 36 \\
(x - 6)^2 & = 28
\end{array}
$$

Step 4:   " Take the square root " of both sides of the equation.

$$x - 6 = \pm \sqrt{28}$$

Step 5: Solve for x and simplify right-hand side.

$$
\begin{array}{r}
x - 6 = \pm \sqrt{28} \\
\underline{+6 \quad +6} \\
x = 6 \pm \sqrt{28}
\end{array}
$$

$$\mathbf{x = 6 \pm 2\sqrt{7}}$$

Scrapwork:

$$\sqrt{28} = \sqrt{4}\sqrt{7} = 2\sqrt{7}$$

Example 2   Solve by completing the square.   30

$$3x^2 - 9x - 2 = 0$$

Step 1: $$3x^2 - 9x = 2$$

Step 2:   Divide the equation by the coefficient of the x^2-term (we want this coefficient to be 1, for this method)

$$\frac{3x^2}{3} - \frac{9x}{3} = \frac{2}{3}$$

$$x^2 - 3x = \frac{2}{3}$$

Step 3:   Add the square of the coefficient of the x-term to both sides of the equation, and complete the square.

$$x^2 - 3x + \left(\frac{-3}{2}\right)^2 = \frac{2}{3} + \left(\frac{-3}{2}\right)^2$$ **Note:** $\frac{b}{2} = \frac{-3}{2}$; $\left(\frac{b}{2}\right)^2 = \left(\frac{-3}{2}\right)^2$

$$\left(x - \frac{3}{2}\right)^2 = \frac{2}{3} + \frac{9}{4}$$ Multiply this out

Scrapwork:

$$\left(x - \frac{3}{2}\right)^2 = \frac{35}{12}$$ $\frac{2}{3} + \frac{9}{4} = \frac{35}{12}$

$$x - \frac{3}{2} = \pm\sqrt{\frac{35}{12}}$$

Step 4: Solve for x and simplify right-hand side.

$$x - \frac{3}{2} = \pm\sqrt{\frac{35}{12}}$$
$$+\frac{3}{2} \qquad +\frac{3}{2}$$
_____

$$x = +\frac{3}{2} \pm \sqrt{\frac{35}{12}}$$

$$x = \frac{3}{2} \pm \frac{\sqrt{105}}{6}$$ Scrapwork: $\sqrt{\frac{35}{12}} = \sqrt{\frac{35 \cdot 3}{12 \cdot 3}}$

Solution is $\left\{\frac{3}{2} \pm \frac{\sqrt{105}}{6}\right\}$ or $\left\{\frac{3}{2} + \frac{\sqrt{105}}{6} , \frac{3}{2} - \frac{\sqrt{105}}{6}\right\}$

Solving by the Quadratic Formula

Example 1   Solve for x by the quadratic formula

$$3x^2 - 6x - 2 = 0$$

Step 1: $a = 3, \ b = -6, \ c = -2$

Step 2: $x = \dfrac{-b \pm \sqrt{b^2 - 4ac}}{2a}$

Step 3: $x = \dfrac{-(-6) \pm \sqrt{(-6)^2 - 4(3)(-2)}}{2(3)}$ (Substituting for $a, b,$ and c)

$$= \dfrac{+6 \pm \sqrt{36 + 24}}{6}$$

$$= \dfrac{6 \pm \sqrt{60}}{6}$$

$$= \dfrac{6}{6} \pm \dfrac{\sqrt{60}}{6}$$

$$= 1 \pm \dfrac{2\sqrt{15}}{6}$$

$$x = 1 \pm \dfrac{\sqrt{15}}{3}$$

Solution set is $\left\{ 1 - \dfrac{\sqrt{15}}{3} \ , \ 1 + \dfrac{\sqrt{15}}{3} \right\}$

Example 2   Solve by the quadratic formula:

$$x^2 - 16 = 8 - 2x$$

Step 1:   Place the equation in standard form (i.e. rewrite the equation so that the only term on the right-hand side is zero)

$$x^2 - 16 \quad = 8 - 2x$$
$$ +2x \qquad +2x$$

$$x^2 + 2x - 16 = 8$$
$$ -8 \quad -8$$

$$x^2 + 2x - 24 = 0 \text{ <-----------(Standard Form)}$$

Step 2: $a = 1$, $b = 2$, $c = -24$

Step 3: $x = \dfrac{-b \pm \sqrt{b^2 - 4\ ac}}{2a}$ <----------The quadratic formula

Step 4 : $x = \dfrac{-2 \pm \sqrt{(2)^2 - 4(1)(-24)}}{2(1)}$ (Substituting for a, b, and c)

$$x = \dfrac{-2 \pm \sqrt{4 + 96}}{2}$$

$$x = \dfrac{-2 \pm \sqrt{100}}{2}$$

$$x = \dfrac{-2 \pm 10}{2}$$

$$x = \dfrac{-2 + 10}{2} \text{ , or } x = \dfrac{-2 - 10}{2}$$

$$x = \dfrac{8}{2}, \text{ or } x = \dfrac{-12}{2}$$

$$x = 4 \text{ or } x = -6$$

The solutions are 4 and -6.

Solving by any method

Example   Solve by any method: $x^2 = 6x$

$$x^2 = 6x$$

Step 1: Write the equation in standard form.

$$x^2 - 6x = 0$$

Step 2: We solve by factoring (since $c = 0$, the quadratic is easily factorable)

$$x(x - 6) = 0$$

Step 3: Set each factor equal to 0 and solve for x:

$$x = 0 \text{ , or } \quad x - 6 = 0$$
$$\underline{+6 \quad +6}$$
$$x = + 6$$

$$x = 0, \text{ or } x = 6$$

The solutions are 0 and 6; or the solution set is {0,6}.

Discriminant of the quadratic equation: Nature of the roots and graphs

The expression $b^2 - 4ac$ is called the **discriminant** of the quadratic equation. The value of the discriminant determines the nature of the roots (or solutions) of the quadratic equation:

1. If $b^2 - 4ac > 0$, the equation $ax^2 + bx + c = 0$ has two real and unequal roots. Also if $b^2 - 4ac$ is a perfect square, the roots are rational. Graphically, the curve crosses the x-axis at two different points.

2. If $b^2 - 4ac = 0$, the quadratic equation has real equal roots (double root or repeated root). We therefore have only one real solution. Graphically, the curve touches the x-axis but does not cross it.

3. If $b^2 - 4ac < 0$, the equation has two non-real (complex) roots. Graphically, the curve does not touch or cross the x-axis. The curve is either entirely above the x-axis or entirely below the x-axis.

Note also that the quadratic equation is factorable if $b^2 - 4ac$ is a perfect square (see page 28)

Relationship between the roots of the quadratic equation, $ax^2 + bx + c = 0$, and the constants a, b, and c of the equation

If the quadratic equation $ax^2 + bx + c = 0$ has roots r_1, r_2 then

$$(x - r_1)(x - r_2) = 0 \implies x^2 - r_1 x - r_2 x + r_1 r_2 = 0 \implies x^2 - (r_1 + r_2)x + r_1 r_2 = 0$$

By comparing $x^2 + \frac{b}{a}x + \frac{c}{a} = 0$ with   ($x^2 + \frac{b}{a}x + \frac{c}{a} = 0$ is obtained from $ax^2 + bx + c = 0$)

$$x^2 - (r_1 + r_2)x + r_1 r_2 = 0 \text{ , we obtain}$$

1. $r_1 + r_2 = -\dfrac{b}{a}$ (Sum of the roots)

2. $r_1 r_2 = \dfrac{c}{a}$ (Product of the roots)

Note also that $r_1 = \dfrac{-b + \sqrt{b^2 - 4ac}}{2a}$ and $r_2 = \dfrac{-b - \sqrt{b^2 - 4ac}}{2a}$

We may use the above property relationships to derive a quadratic equation, knowing the roots.

Deriving a quadratic equation from its roots

Example Find a quadratic equation whose roots are 3 and -2.

Solution   The general form $ax^2 + bx + c$ can be factored as $a(x - r_1)(x - r_2)$, where r_1, r_2 are the roots of $ax^2 + bx + c = 0$

If $r_1 = 3$, $r_2 = -2$ in the factored equation, we obtain
$$a(x - 3)(x - (2)) = 0$$
$$a(x - 3)(x + 2) = 0 \qquad (1)$$

If $a = 1$, then equation (1) becomes

$$1(x - 3)(x + 2) = 0$$
$$x^2 - x - 6 = 0$$
Similarly, if $a = 2$, we obtain $2x^2 - 2x - 12 = 0$

Generally, we will determine the value of a uniquely from additional information, say, from the graph of the equation.

Using the quadratic formula to aid factoring

If r_1 and r_2 are the roots of $ax^2 + bx + c = 0$, then we can factor it as
$$ax^2 + bx + c = a(x - r_1)(x - r_2).$$

Example Factor by first finding the roots: $18x^2 - 63x + 40 = 0$

Solution By the quadratic formula,
$$x = \frac{+63 \pm \sqrt{(63)^2 - 4(18)(40)}}{36}$$
$$= \frac{63 \pm 33}{36}$$
$$x = \frac{8}{3} \text{ or } x = \frac{5}{6}$$
Recalling that $ax^2 + bx + c = a(x - r_1)(x - r_2)$
$$18x^2 - 63x + 40 = 18(x - \tfrac{8}{3})(x - \tfrac{5}{6})$$
$$= 18\left(\frac{3x - 8}{3}\right)\left(\frac{6x - 5}{6}\right)$$
$$= 18 \cdot \frac{(3x - 8)(6x - 5)}{18}$$
$$18x^2 - 63x + 40 = \mathbf{(3x - 8)(6x - 5)}.$$

Lesson 5 Exercises   35

A Solve by factoring: **1.** $x^2 - 11x + 18 = 0$;   **2.** $x^2 - 5x - 36 = 0$; **3.** $x^2 - 18x = 0$;   **4.** $3x^2 = 24x$

  Solutions **1.** $\{2, 9\}$;   **2.** $\{-4, 9\}$;   **3.** $\{0,18\}$;   **4.** $\{0,8\}$

B **1.** $x^2 - 4x - 21 = 0$ **2.** $x^2 + 4x - 45 = 0$ **3.** $x^2 - 25 = 0$

  4. $9x^2 - 6x = 0$ **5.** $14t = 7t^2$ **6.** $x^2 = 11x - 18$

  7. $2x^2 + 2x - 144 = 0$ **8.** $9x^2 - 36 = 0$ **9.** $ax^2 = -bx$

  10. $x^2 - mx + nx - mn = 0$

Answers: 1. $\{-3, 7\}$; **2.** $\{-9, 5\}$; **3.** $\{-5, 5\}$; **4.** $\{0, \frac{2}{3}\}$; **5.** $\{0, 2\}$; **6.** $\{2, 9\}$; **7.** $\{-9, 8\}$; **8.** $\{-2, 2\}$;

  9. $\{0, -\frac{b}{a}\}$;   **10.** $\{m, -n\}$

C   Solve by the square root method: **1.** $x^2 - 16 = 0$; ;   **2.** $4x^2 - 20 = 0$;   **3.** $x^2 - 49 = 0$;

  Read page 344 before attempting Problem 4:   **4.** $x^2 + 49 = 0$

Answers: **1.** $\{-4, 4\}$;   **2.** $\{-\sqrt{5}, \sqrt{5}\}$; **3.** $\{-7, 7\}$; 4. $\{-7i, 7i\}$

D Solve the following for x:

  1. $x^2 - 9 = 0$;   **2.** $-8 + x^2 = 0$; b **3.** $ax^2 - c = 0$ **4.** $x^2 + 9 = 0$. **5.** $s = ax^2$ **6.** $x^2 - \frac{1}{4} = 0$

Answers: 1. $\{-3, 3\}$;   **2.** $\{-2\sqrt{2}, 2\sqrt{2}\}$;   **3.** $\{-\sqrt{\frac{c}{a}}, \sqrt{\frac{c}{a}}\}$;   **4.** $\{-3i, 3i\}$;

  5 $\{-\sqrt{\frac{s}{a}}, \sqrt{\frac{s}{a}}\}$;   **6.** $\{-\frac{1}{2}, \frac{1}{2}\}$

E   Solve by completing the square: **1.** $x^2 + 12x + 10 = 0$;   **2.** $x^2 - 4x + 6 = 0$;

  3. $x^2 - 3x - 8 = 0$;   **4.** $x^2 - 12x - 35 = 0$

Answers: **1.** $\{-6 + \sqrt{26}, -6 - \sqrt{26}\}$; **2.** $\{2 + i\sqrt{2}, 2 - i\sqrt{2}\}$; **3.** $\{\frac{3}{2} + \frac{\sqrt{41}}{2}, \frac{3}{2} - \frac{\sqrt{41}}{2}\}$;

  4. $\{6 + \sqrt{71}, 6 - \sqrt{71}\}$

F   Solve by completing the square:

1. $x^2 - 11x + 18 = 0$;   **2.** $2x^2 + 5x - 12 = 0$;   **3.** $x^2 - 4x - 1 = 0$;   **4.** $x^2 - 4x + 8 = 0$.

5. $4x^2 + 48x + 40 = 0$;   **6.** $2x^2 - 7x - 12 = 0$;   **7.** $\frac{1}{2}at^2 + bt + k = 0$; **8.** $ax^2 + bx + c = 0$

Solutions: **1.** $\{2, 9\}$; **2.** $\{-4, \frac{3}{2}\}$; **3.** $\{2 + \sqrt{5}, 2 - \sqrt{5}\}$; **4.** $\{2 + 2i, 2 - 2i\}$; **5.** $\{-6 + \sqrt{26}, -6 - \sqrt{26}\}$;

  6. $\{\frac{7}{4} + \frac{\sqrt{145}}{4}, \frac{7}{4} - \frac{\sqrt{145}}{4}\}$; **7.** $\{-\frac{b}{a} \pm \frac{\sqrt{b^2 - 2ak}}{a}\}$; **8.** $x = -\frac{b}{2a} \pm \frac{\sqrt{b^2 - 4ac}}{2a}$

Lesson 5: Solving Quadratic Equations (Review)

G Solve by the quadratic formula:

1. $x^2 - 11x + 18 = 0$;   **2.** $2x^2 + 5x - 12 = 0$;   **3.** $x^2 - 4x - 1 = 0$; **4.** $x^2 - 4x + 8 = 0$.

5. $x^2 + 12x + 10 = 0$;   **6.** $x^2 + 6 - 4x = 0$ **7.** $x^2 = 3x + 8$

Solutions: **1.** $\{2, 9\}$;   **2.** $\{-4, \frac{3}{2}\}$;   **3.** $\{2 + \sqrt{5}, 2 - \sqrt{5}\}$;   **4.** $\{2 + 2i, 2 - 2i\}$; **5.** $\{-6 \pm \sqrt{26}\}$

6. $\{2 \pm i\sqrt{2}\}$; **7.** $\{\frac{3 \pm \sqrt{41}}{2}\}$

H   Solve by factoring: **1.** $x^2 - 11x + 18 = 0$;   **2.** $x^2 - 5x - 36 = 0$;

 3. $x^2 - 18x = 0$;   **4.** $3x^2 = 24x$

Answers: **1.** $\{2, 9\}$;   **2.** $\{-4, 9\}$;   **3.** $\{0,18\}$;   **4.** $\{0,8\}$

I   **1.** $x^2 - 4x - 21 = 0$ **2.** $x^2 + 4x - 45 = 0$ **3.** $x^2 - 25 = 0$

4. $9x^2 - 6x = 0$ **5.** $14t = 7t^2$ **6.** $x^2 = 11x - 18$

7. $2x^2 + 2x - 144 = 0$ **8.** $9x^2 - 36 = 0$ **9.** $ax^2 = -bx$

10. $x^2 - mx + nx - mn = 0$

Answers: **1.** $\{-3, 7\}$; **2.** $\{-9, 5\}$; **3.** $\{-5, 5\}$; **4.** $\{0, \frac{2}{3}\}$; **5.** $\{0, 2\}$; **6.** $\{2, 9\}$; **7.** $\{-9, 8\}$; **8.** $\{-2, 2\}$;

9. $\{0, -\frac{b}{a}\}$;   **10.** $\{m, -n\}$

J Solve by the square root method:

1. $x^2 - 16 = 0$; **2.** $4x^2 - 20 = 0$;   **3.** $x^2 - 49 = 0$;

Read page 344-345 before attempting Problem 4:   **4.** $x^2 + 49 = 0$

Answers: **1.** $\{-4, 4\}$;   **2.** $\{-\sqrt{5}, \sqrt{5}\}$; **3.** $\{-7, 7\}$; **4.** $\{-7i, 7i\}$

K Solve for x: **1.** $x^2 - 9 = 0$;   **2.** $-8 + x^2 = 0$;   **3.** $ax^2 - c = 0$

4. $x^2 + 9 = 0$;   **5.** $s = ax^2$;   **6.** $x^2 - \frac{1}{4} = 0$

Answers: **1.** $\{-3, 3\}$;   **2.** $\{-2\sqrt{2}, 2\sqrt{2}\}$;   **3.** $\{-\sqrt{\frac{c}{a}}, \sqrt{\frac{c}{a}}\}$;   **4.** $\{-3i, 3i\}$;

5 $\{-\sqrt{\frac{s}{a}}, \sqrt{\frac{s}{a}}\}$;   **6.** $\{-\frac{1}{2}, \frac{1}{2}\}$

L Solve by completing the square: 1. $x^2 + 12x + 10 = 0$;;2. $x^2 - 4x + 6 = 0$;   **3.** $x^2 - 3x - 8 = 0$

4. $x^2 - 12x - 35 = 0$

Answers: **1.** $\{-6 + \sqrt{26}, -6 - \sqrt{26}\}$; **2.** $\{2 + i\sqrt{2}, 2 - i\sqrt{2}\}$; **3.** $\{\frac{3}{2} + \frac{\sqrt{41}}{2}, \frac{3}{2} - \frac{\sqrt{41}}{2}\}$;

4. $\{6 + \sqrt{71}, 6 - \sqrt{71}\}$

M Solve by completing the square:

1. $x^2 - 11x + 18 = 0$;   **2.** $2x^2 + 5x - 12 = 0$;   **3.** $x^2 - 4x - 1 = 0$;   **4.** $x^2 - 4x + 8 = 0$.

5. $4x^2 + 48x + 40 = 0$;   **6.** $2x^2 - 7x - 12 = 0$;   **7.** $\frac{1}{2}at^2 + bt + k = 0$;   **8.** $ax^2 + bx + c = 0$

Solutions: **1.** $\{2, 9\}$; **2.** $\{-4, \frac{3}{2}\}$; **3.** $\{2 + \sqrt{5}, 2 - \sqrt{5}\}$; **4.** $\{2 + 2i, 2 - 2i\}$; **5.** $\{-6 + \sqrt{26}, -6 - \sqrt{26}\}$;

6. $\{\frac{7}{4} + \frac{\sqrt{145}}{4}, \frac{7}{4} - \frac{\sqrt{145}}{4}\}$;   **7.** $\{-\frac{b}{a} \pm \frac{\sqrt{b^2 - 2ak}}{a}\}$;   **8.** $x = -\frac{b}{2a} \pm \frac{\sqrt{b^2 - 4ac}}{2a}$

N Solve by the quadratic formula:

1. $x^2 - 11x + 18 = 0$;   **2.** $2x^2 + 5x - 12 = 0$;   **3.** $x^2 - 4x - 1 = 0$; **4.** $x^2 - 4x + 8 = 0$.

5. $x^2 + 12x + 10 = 0$;   **6.** $x^2 + 6 - 4x = 0$ **7.** $x^2 = 3x + 8$

Solutions: **1.** $\{2, 9\}$;   **2.** $\{-4, \frac{3}{2}\}$;   **3.** $\{2 + \sqrt{5}, 2 - \sqrt{5}\}$;   **4.** $\{2 + 2i, 2 - 2i\}$; **5.** $\{-6 \pm \sqrt{26}\}$

6. $\{2 \pm i\sqrt{2}\}$;   **7.** $\{\frac{3 \pm \sqrt{41}}{2}\}$

O Solve by any applicable method:

1. $x^2 - 6x + 4$;   **2.** $6x^2 + 11x - 10 = 0$;   **3.** $2x^2 - 3x + 6 = 0$;   **4.** $2x(x - 4) = 6$

Solutions: **1.** $\{3 + \sqrt{5}, 3 - \sqrt{5}\}$;   **2.** $\{\frac{2}{3}, -\frac{5}{2}\}$;   **3.** $\{\frac{3 + i\sqrt{39}}{4}, \frac{3 - i\sqrt{39}}{4}\}$ **4.** $\{2 + \sqrt{7}, 2 - \sqrt{7}\}$

P First, try to factor by the usual methods, and then redo the problems by using the quadratic formula as an aid:

   1. $x^2 - 4x - 21$;   **2.** $18x^2 + 3x - 10$;   **3.** $10x^2 - 31x + 24$;   **4.** $15x^2 - 122x + 240$.

   5. Comment on the relative merits of the two methods used in each of the Problems 1-4.

Answers: **1.** $(x - 7)(x + 3)$;   **2.** $(6x + 5)(3x - 2)$;   **3.** $(2x - 3)(5x - 8)$;   **4.** $(5x - 24)(3x - 10)$.

Q Derive a quadratic equation for each of the given roots:

   1. $r_1 = 2$; $r_2 = 3$;   **2.** $r_1 = \frac{3}{4}, r_2 = \frac{3}{4}$;   **3.** $r_1 = -\frac{2}{5}, r_2 = \frac{2}{3}$.

   4. Find a quadratic equation with roots -5 and 4.

Answers: **1.** $x^2 - 5x + 6 = 0$;   **2.** $16x^2 - 24x + 9 = 0$;   **3.** $15x^2 - 4x - 4 = 0$; **4.** $x^2 + x - 20 = 0$.

R Determine if the given equation is factorable:

   1. $x^2 - 6x - 2 - 0$;   **2.** $x^2 - 11x + 18 = 0$;   **3.** $5x^2 - 6x + 4 = 0$.

Answers : **1.** Not factorable;   **2.** Factorable;   **4.** Not factorable

Lesson 6

Solving Equations Quadratic in Form (quadratic-like equations)

If by substituting a new variable in an equation, the equation can be reduced to a quadratic form, then we say that the equation is quadratic in nature.

Example 1 Solve for x: $x^4 - 7x^2 + 12 = 0$ (1)

Solution

Step 1: Let $x^2 = u$. Then equation (1) becomes

$$u^2 - 7u + 12 = 0 \qquad (2)$$

Equation (2) is quadratic in u, and since it is easily factorable, we solve by factoring.

Step 2: $(u - 3)(u - 4) = 0$

$u - 3 = 0$ or $u - 4 = 0$

$u = 3$ or $u = 4$ (3)

Step 3: Replace u by x^2 in (3).

Then $x^2 = 3$ or $x^2 = 4$

$x = \pm\sqrt{3}$ or $x = \pm 2$

The solution set is $\{\sqrt{3}, \ -\sqrt{3}, \ 2, \ -2\}$

Note above that we could have factored equation (1) without substitution as $(x^2 - 3)(x^2 - 4) = 0$ and then solve. This approach would be faster than that by substitution.

Example 2 Solve for x: $x^4 + 7x^2 + 6 = 0$ (1)

Solution

Step 1: Let $x^2 = u$. Then equation (1) becomes

$$u^2 + 7u + 6 = 0 \qquad (2)$$

Equation (2) is quadratic in u, and since it is easily factorable, we solve by factoring.

Step 2: $(u + 1)(u + 6) = 0$

$u + 1 = 0$ or $u + 6 = 0$
$u = -1$, or $u = -6$ (3)

Step 3: Replace u by x^2 in (3).

Then $x^2 = -1$ or $x^2 = -6$.

$x = \pm i$ or $x = \pm i\sqrt{6}$

The solution set is $\{i, \ -i, \ i\sqrt{6}, \ -i\sqrt{6}\}$.

Note above that we could have factored equation (1) without substitution as $(x^2 + 1)(x^2 + 6) = 0$ and then solve. This approach would be faster than that by substitution.

In the above problem (Example 2), we obtain only the complex solutions.

Example 3 Solve for x: $x - 5\sqrt{x} + 6 = 0$

Solution We use a substitution method, but see also Chapter 5 for another approach.

Step 1: Let $u = \sqrt{x}$. Then $u^2 = x$,   (since $\left(\sqrt{x}\right)^2 = x$) and equation (1) becomes

$$u^2 - 5u + 6 = 0 \qquad (2)$$

Step 2: We solve by factoring (We could also solve by the quadratic formula):
Then $(u - 3)(u - 2) = 0$ and from which
$$u = 3 \text{ or } u = 2 \qquad (3)$$

Step 3: Replace u by $\sqrt{x}$ in (3).
Then $\sqrt{x} = 3$ or $\sqrt{x} = 2$ (4)

Step 4: Square both sides of (4)
$$\left(\sqrt{x}\right)^2 = (3)^2 \text{ or } \left(\sqrt{x}\right)^2 = (2)^2$$
$$x = 9 \qquad \text{or} \qquad x = 4$$
After testing these values in the original equation, (why), the solutions are 4 and 9.

Lesson 6 Exercises

Solve for x: **1.** $x^4 - 7x^2 + 12 = 0$;   **2.** $x^4 + 7x^2 + 6 = 0$ **3.** $x - 5\sqrt{x} + 6 = 0$
      4. $3x^4 - 2x^2 - 5 = 0$;   **5.** $x - \sqrt{x} - 20 = 0$

Answers 1. $\{\sqrt{3}, -\sqrt{3}, 2, -2\}$; **2.** $\{i, -i, i\sqrt{6}, -i\sqrt{6}\}$; **3.** $\{4, 9\}$; **4.** $\left\{\pm\sqrt{\frac{5}{3}}, \pm i\right\}$;

     5. 25   Note: $x = 16$ is extraneous.

Lesson 7

Higher Polynomial Equations; Systems Containing Nonlinear Equations

Solving easily factorable higher polynomials

Example 1 Solve for x: $x^3 - 4x = 0$
Solution

Step 1: Factor the left-hand side.

$$x(x^2 - 4) = 0$$

$$x(x + 2)(x - 2) = 0$$

Step 2: Equate each factor to zero and solve for x.

$x = 0$ or $(x + 2) = 0$ or $(x - 2) = 0$

$x = 0$ or $x = -2$ or $x = +2$

The solutions are 0, -2 and 2.

Example 2 Solve for x: $(x - 1)(x + 1)(x^2 + 9) = 0$

Solution

Equate each factor to zero and solve for x:

For $(x - 1) = 0$, $x = 1$

For $(x + 1) = 0$, $x = -1$

For $(x^2 + 9) = 0$, $x = \pm 3i$ (imaginary roots)

There are two real solutions: -1, and 1; and two non-real solutions (complex solutions) $3i$, and $-3i$.

Solving systems containing nonlinear equations in x and y

Example 1 Solve the system consisting of one linear equation and one quadratic equation, simultaneously (This is a quadratic-linear system)

$$\begin{cases} x + y = 1 & (1) \\ x^2 + y^2 = 25 & (2) \end{cases}$$

Solution We use the substitution method.

Step 1: Solve the linear equation for one of the variables,(usually, the one that is easily solved for), say y.
Solving for y from the linear equation (1), we obtain
$$y = (1 - x) \qquad (3)$$
Step 2: Substitute for y in the other equation, equation (2):

Then $x^2 + (1 - x)^2 = 25$ (4)

$x^2 + 1 - 2x + x^2 = 25$

$2x^2 - 2x - 24 = 0$ (5)

Step 3: Solve equation (5) by factoring or by formula.
Solving, $x = -3, x = 4$.

Step 4: For each value of x, calculate the corresponding y-value.
  Substituting $x = -3$ in (1), the linear equation,
    $-3 + y = 1$, and $y = 4$ (solving for y)
Similarly, substituting $x = 4$ in equation (1),
    $4 + y = 1$, and $y = -3$ (solving for y)

Conclusion: The solutions are when $x = -3, y = 4$ and when $x = 4, y = -3$.
  The solutions can be written as $(-3, 4)$, and $(4, -3)$.

Graphically, these solutions are the points of the intersections of **the graphs of**
$x^2 + y^2 = 25$ (a circle) and $x + y = 1$ (a straight line).

Example 2 Solve the system algebraically
$$\begin{cases} x^2 + 4y^2 = 13 & (1) \\ xy = 3 & (2) \end{cases}$$

Solution We use a substitution method.

Step 1: Solve the linear equation for one of the variables,(usually, the one that is easily solved for), say x.
  Solving for x from equation (2), we obtain
$$x = \frac{3}{y} \qquad (3)$$

Step 2: Substitute for $x = \frac{3}{y}$ in the other equation, equation (1):
$$\left(\frac{3}{y}\right)^2 + 4y^2 = 13$$
$$\frac{9}{y^2} + 4y^2 = 13$$
$$9 + 4y^4 = 13y^2$$
$$4y^4 - 13y^2 + 9 = 0 \qquad (3)$$

Step 3: Let $y^2 = u$. Then equation (3) is quadratic in u, and we obtain:
$4u^2 - 13u + 9 = 0$
$$u = \frac{13 \pm \sqrt{169 - 4(4)(9)}}{8}$$
(applying the quadratic formula)

Step 4: Replace u by y^2.
$$y^2 = \frac{13 \pm 5}{8} \text{ or } y^2 = \frac{13 - 5}{8};$$
$$y = \pm\frac{3}{2}, \text{ or } y = \pm 1$$

Or Step 3, Method 2, by factoring
$$4y^4 - 13y^2 + 9 = 0 \qquad (3)$$
$$(4y^2 - 9)(y^2 - 1) = 0$$
$4y^2 - 9 = 0$ or $y^2 - 1 = 0$
$4y^2 = 9$ or $y^2 = 1$
$$y^2 = \frac{9}{4} \qquad y = \pm 1$$
$$y = \pm\sqrt{\frac{9}{4}}$$
$$y = \pm\frac{3}{2}$$

Step 5: Substitute the y-values in equation (2) and calculate the corresponding x-values.
  When $y = +\frac{3}{2}$, $x = 2$; When $y = -\frac{3}{2}$, $x = -2$
  When $y = +1$, $x = +3$; When $y = -1$, $x = -3$

The solutions are the ordered pairs $\left(2, \frac{3}{2}\right)$, $\left(-2, -\frac{3}{2}\right)$, $(3,1)$, $(-3,-1)$.

Example 3 Solve the system algebraically.   42

$$\begin{cases} x^2 + y^2 = 16 & \quad (1) \\ -2x^2 + y^2 = 10 & \quad (2) \end{cases}$$

Solution The above system is linear in x^2 (not in x) and linear in y^2, because by replacing x^2 by u, and y^2 by v, the equations will become linear. However, we can solve the above system either by substitution or by addition or subtraction. (Addition method because there are like terms, unlike the previous example)

By substitution method.

Step 1: Solve for y^2 in equation (2).

   Then $y^2 = 10 + 2x^2$.

Step 2: Substitute $(10 + 2x^2)$ for y^2 in equation (1):

   Then $x^2 + (10 + 2x^2) = 16$.

   Thus $3x^2 = 6$.

   $x^2 = 2$

   $x = \pm\sqrt{2}$

Step 3: Substitute $x = \pm\sqrt{2}$ in (1).

   When $x = +\sqrt{2}$, $\left(\sqrt{2}\right)^2 + y^2 = 16$

   $2 + y^2 = 16$

   $y = \pm\sqrt{14}$

   When $x = -\sqrt{2}$, $\left(-\sqrt{2}\right)^2 + y^2 = 16$

   $2 + y^2 = 16$

   $y = \pm\sqrt{14}$

The solutions are the ordered pairs $\left(\sqrt{2}, \sqrt{14}\right)$, $\left(\sqrt{2}, -\sqrt{14}\right)$, $\left(-\sqrt{2}, \sqrt{14}\right)$, and $\left(-\sqrt{2}, -\sqrt{14}\right)$

By addition or subtraction method

$$\begin{cases} x^2 + y^2 = 16 & \quad (1) \\ -2x^2 + y^2 = 10 & \quad (2) \end{cases}$$

Solution: Subtract equation (2) from (1). (To subtract, change the sign and add) Then we obtain

   $3x^2 = 6$

   $x^2 = 2$

   $x = \pm\sqrt{2}$

Again, we obtain the same solution for x. By following Step 3 of the substitution method, we will obtain the other solutions.

Lesson 7 Exercises

A Solve the following for x:

1. $x^4 - 5x^2 + 4 = 0$;   2. $x - 10\sqrt{x} + 9 = 0$;   3. $x^4 - 10x^2 + 16 = 0$;

4. $x^{\frac{1}{2}} - 3x^{\frac{1}{4}} = 4$;   5. $\dfrac{1}{x^2} - \dfrac{1}{x} = 12$

Solution sets:

1. $\{-1, -2, 1, 2\}$;   2. $\{1, 81\}$;   3. $\{\pm 2, \pm 2\sqrt{2}\}$;   4. $\{256\}$, $x = 1$ is extraneous;   5. $\left\{-\dfrac{1}{3}, \dfrac{1}{4}\right\}$

B   Solve the following systems of equations algebraically

1. $\begin{cases} x^2 + y^2 = 25 \\ x - y = 4 \end{cases}$ 2. $\begin{cases} 4y = x^2 \\ x + 2y = 8 \end{cases}$ 3. $\begin{cases} x^2 + y^2 = 25 \\ x^2 - y = 5 \end{cases}$

Answers: **1.** When $x = 2 + \dfrac{\sqrt{34}}{2}, y = -2 + \dfrac{\sqrt{34}}{2}$; when $x = 2 - \dfrac{\sqrt{34}}{2}, y = -2 - \dfrac{\sqrt{34}}{2}$.

2. when $x = -1 + \sqrt{17}$, $y = \dfrac{9 - \sqrt{17}}{2}$; when $x = -1 - \sqrt{17}$, $y = \dfrac{9 + \sqrt{17}}{2}$

3. When $x = \pm 3$, $y = 4$; when $x = 0, y = -5$.

CHAPTER 5

Lesson 8
Review
Solving Radical Equations

We define **a radical equation** as an equation in which one of the variables occurs under one or more radical signs.

Examples

1. $\sqrt{x} = 6$

2. $\sqrt{x - 3} = 4$

3. $\sqrt{2x + 4} = \sqrt{x + 2}$

Principle of powers: If $a = b$

Then, $a^n = b^n$. However, this principle does not always produce equivalent equations, and therefore, if we use this principle to solve an equation, we must check the solutions in the original equation.

Procedure   The technique here involves the elimination of the radical signs by raising the radical expressions to integral (integer) powers: both sides of an equation are to be raised to the same integral power. We will then solve the resulting non-radical equation for the variable. If there are more than one radical, we may sometimes have to "undo" the radical signs in a number of steps. It will also be necessary that we test any solutions in the original equation for **extraneous solutions**. Before raising each side of an equation to an integral power, the radical whose radical sign we want to eliminate must be on one side of the equation by itself.

Example 1   Solve for x: $\sqrt{x} = 6$.

Solution

Step 1:   Square both sides of the equation to undo the square root symbol (i.e., raise both sides of the equation to the second power).
Then, $(\sqrt{x})^2 = (6)^2$
$x = 36$.
Step 2: Check the solution in the original equation.
$\sqrt{36} \overset{?}{=} 6$
$6 = 6$ True (left-hand side equals right-hand side, i.e., LHS = RHS).
$\therefore$ the solution is 36.

The above problem was so simple that we could have asked: If the square root of a number is 6, what is the number? Of course, the number is 36.

Example 2 Solve for x: $\sqrt{x-3} = 4$ 4 5

Solution

Step 1: Square both sides of the equation to undo the square root symbol.

Then, $(\sqrt{x-3})^2 = (4)^2$

$$x - 3 = 16$$

Step 2: Solve for x.

$$x - 3 = 16$$
$$\underline{\quad +3 \quad +3}$$
$$x = 19$$

Step 3: Check the solution 19 in the original equation.

$$\sqrt{19-3} \overset{?}{=} 4$$
$$\sqrt{16} \overset{?}{=} 4$$
$$4 = 4 \text{ True} \quad (\text{LHS} = \text{RHS})$$

$\therefore$ the solution is 19.

Note above that squaring the square root of a number eliminates the square root symbol.

Example 3   Solve for x:

$$\sqrt{2x+4} = \sqrt{x+2}$$

Solution

Step 1: Square both sides of the equation.

$$(\sqrt{2x+4})^2 = (\sqrt{x+2})^2$$

$$2x + 4 = x + 2$$

Step 2: Solve for x.

$$2x + 4 = x + 2$$
$$\underline{-x \qquad -x}$$
$$x + 4 = 2$$
$$\underline{\quad -4 \quad -4}$$
$$x = -2$$

Step 3: Check for $x = -2$ in the original equation

$$\sqrt{2(-2)+4} \overset{?}{=} \sqrt{(-2)+2}$$
$$\sqrt{-4+4} \overset{?}{=} \sqrt{0}$$
$$\sqrt{0} \overset{?}{=} \sqrt{0}$$
$$0 = 0 \qquad \text{True} \quad (\text{RHS} = \text{LHS})$$

$\therefore$ the solution is -2.

Example 4   Solve for x: $\sqrt{2x - 5} = x - 4$

Solution

Step 1:   Square both sides of the equation.
$$(\sqrt{2x - 5})^2 = (x - 4)^2$$
$$2x - 5 = x^2 - 8x + 16$$
$$0 = x^2 - 10x + 21$$
or $x^2 - 10x + 21 = 0$

Step 2:   Solve the quadratic equation by any method.
We will solve by factoring, since the factors are easily recognizable.
$$x^2 - 10x + 21 = 0$$
$$(x - 3)(x - 7) = 0$$
$$x - 3 = 0 \text{ or } x - 7 = 0$$
$$x = 3 \text{ or } x = 7 \quad \text{(Setting each factor equal to zero and solving for } x)$$

Step 3: Checking for $x = 3$:
$$\sqrt{2(3) - 5} \overset{?}{=} 3 - 4$$
$$\sqrt{6 - 5} \overset{?}{=} -1$$
$$\sqrt{1} \overset{?}{=} -1$$
$$1 = -1 \text{ False} \quad \text{(LHS \textbf{not} equal to RHS)}$$

Therefore 3 is not a solution. It is an extraneous root.

Checking for $x = 7$:
$$\sqrt{2(7) - 5} \overset{?}{=} 7 - 4$$
$$\sqrt{14 - 5} \overset{?}{=} 3$$
$$\sqrt{9} \overset{?}{=} 3$$
$$3 = 3 \quad \text{True} \quad \text{(RHS = LHS)}$$

Therefore, 7 is a solution.
Since 3 is an extraneous root , the only solution is 7.

Example 5 Solve for x: $\sqrt{2x - 2} + 2 = \sqrt{4x + 3}$ (1)

Solution

Step 1: Squaring both sides of equation (1)

$$(\sqrt{2x - 2} + 2)^2 = (\sqrt{4x + 3})^2$$
$$2x - 2 + 4\sqrt{2x - 2} + 4 = 4x + 3$$
$$4\sqrt{2x - 2} = 2x + 1 \qquad (2)$$

Step 2: Squaring equation (2),

$$16(2x - 2) = 4x^2 + 4x + 1$$
$$32x - 32 = 4x^2 + 4x + 1$$
$$0 = 4x^2 - 28x + 33$$

Step 3: Solving the quadratic equation by any method, we obtain $x = \frac{3}{2}$ and $x = \frac{11}{2}$.
Now, we check the solutions in the original equation.

Letting $x = \frac{3}{2}$ in equation (1), we obtain

$$\sqrt{2\left(\frac{3}{2}\right) - 2} + 2 \overset{?}{=} \sqrt{4\left(\frac{3}{2}\right) + 3}$$

$$3 = 3 \qquad \text{True}$$

Similarly, letting $x = \frac{11}{2}$ in equation (1), we obtain

$$\sqrt{2\left(\frac{11}{2}\right) - 2} + 2 \overset{?}{=} \sqrt{4\left(\frac{11}{2}\right) + 3}$$

$$5 = 5 \qquad \text{True}$$

Therefore, the solutions are $\frac{3}{2}$ and $\frac{11}{2}$, or the solution set is $\left\{\frac{3}{2}, \frac{11}{2}\right\}$.

We should note above that even though we squared the equations twice, we did **not** obtain any extraneous roots. Therefore, we do not always obtain extraneous roots when we square equations.

Example 5 Extra: If given $\sqrt{2x - 2} - \sqrt{4x + 3} = -2$, in the first step, we will get one of the radicals by itself alone on one side of the equation as in Example 5. Thus we could rewrite $\sqrt{2x - 2} - \sqrt{4x + 3} = -2$ as $\sqrt{2x - 2} + 2 = \sqrt{4x + 3}$ or as $\sqrt{2x - 2} = \sqrt{4x + 3} - 2$ before we square both sides of the equation.

Example 6 Solve for x: $x - 5\sqrt{x} + 6 = 0$ (We solved this problem in Example 3 of Lesson 6)

Solution $\quad x - 5\sqrt{x} + 6 = 0$

$$-5\sqrt{x} = -x - 6$$
$$5\sqrt{x} = x + 6$$
$$25x = x^2 + 12x + 36 \quad \text{(squaring both sides of the equation)}$$
$$0 = x^2 - 13x + 36$$
$$(x - 9) = 0 \ \text{ or } \ (x - 4) = 0$$
$$x = 9 \ \text{ or } \ x = 4$$

After testing these values in the original equation, (why), the solutions are 4 and 9.

Lesson 8 Exercises

A   Solve for x: **1.** $\sqrt{x} = 7$;   **2.** $\sqrt{x - 4} = 5$ **3.** $\sqrt{3x + 5} = \sqrt{x + 9}$

Answers: 1. $x = 49$;   **2.** $x = 29$;   **3.** $x = 2$;

B   Solve and check: **1** $\sqrt{x - 5} + 6 = 2$; **2..** $2\sqrt{x} = 6$;   **3.** $\sqrt{x^2 - 8x} = 3$;

4.. $3\sqrt{x + 2} = 4\sqrt{x - 5}$; **5.** $\sqrt{x + 5} - \sqrt{x} = 3$;;   **6.** $\sqrt[3]{x + 1} = 2$;   **7.** $\sqrt{3y - 11} - 1 = \sqrt{2y + 9}$;
8. $x - 10\sqrt{x} = -9$

Answers: **1.** No solution (21 is extraneous) ;   **2.** { 9 } ;   **3.** { -1, 9 } ;   **4.** { 14 } ;
5. No solution. ($\frac{4}{9}$ is extraneous) ;   **6.** {7}; **7.** $25 + 2\sqrt{55}$; **8.** {1, 81}

CHAPTER 6
FUNCTIONS

Lesson 9: Sets, Relations, Functions, Comparison of Relations and Functions

Lesson 10A: Functional Notation; Defined Functions; Excluded Values, Domain and Range

Lesson 10B: Algebra of Functions

Lesson 11: One-to-One Functions, Composite Functions

Lesson 12: Inverse Functions and Inverse Relations

Lesson 9
Sets, Relations, Functions, Comparison of Relations and Functions

Ordered Pair

An **ordered pair** of numbers is an arrangement of two numbers in a specified order. In an x-y rectangular coordinate system of axes, the first element (or component) is the x-value and the second element is the y-value.

Example 1   (a) $(1, 2)$ <--- $(x = 1, y = 2)$
           (b) $(2, 3)$ <---- $(x = 2, y = 3)$
           (c) $(5, -1)$ <---- $(x = 5, y = -1)$

Note that each ordered pair represents a point in an x-y coordinate system of axes.

Set of numbers

A **set of numbers** is a well-defined collection of numbers. The numbers are called the elements or members of the set.

Example 2: If we denote the set of the numbers 2, 5 and 6 by A, then we may write $A = \{2, 5, 6\}$

Example 3: The set B of the ordered pairs $(1, 2), (2, 3),$ and $(5,-1)$ is given by
       $B = \{(1, 2), (2, 3), (5,-1)\}$.

Relation

A **relation** is a set of ordered pairs. (A collection of ordered pairs of numbers)

Example 4: The set $E = \{(2, 3), (2, 5), (4, 6)\}$ is a relation, <---There are three ordered pairs.

Example 5: The set $C = \{(6, 2), (7,4), (11, 5)\}$ is a relation.

Definition of a **Function**

A function may be defined in a number of ways, namely,

(a) in terms of ordered pairs; (b) in terms of a rule involving two variables;
(c) in terms of a rule for inputs and outputs; (d) in terms of correspondence of two sets; (e) as a graph

Definition 1: In terms of ordered pairs

A **function** is a relation in which no two distinct ordered pairs have the same first component; or a function is a set of ordered pairs in which for any two different ordered pairs, the first elements are different. The set in Example 5, above, is a function but the set in Example 4 is not a function, because the first two ordered pairs have the same first element, namely 2.

The set of all the first elements of the ordered pairs is called the **domain** of the function; and the set of all the second elements is called the **range** of the function.

Example 6: In the function $C = \{(6, 2), (7, 4), (11, 5)\}$. The domain, $D = \{6, 7, 11\}$ (first elements)
                              The range, $R = \{2, 4, 5\}$. (second elements)

Other definitions of a function

Definition 2: If x and y are two variables. then we say that y is a function of x if there is a rule which gives just one corresponding value of y for **each** value of x. The variable x is called the independent variable, and a variable y is called the dependent variable. The rule may be specified in the form of a set, in the form of a graph, in the form of a table, or in the form of an equation or formula.

The **domain** of a function is the set of numbers that can be assigned to x (the independent variable). The **range** of a function is the set of all the corresponding numbers y (the dependent variable) associated by the function (rule) with the numbers, x, in the domain .

We symbolize that f is a function of x by $f(x)$, where x is called the independent variable, y is called the dependent variable. **Note:** $f(x)$s is read as f of x.

The following are examples of how the rules for functions may be specified:

(*a*) In the form of an equation or a formula: $y = 2x$.
(*b*) In the form of a set: $\{(2, 3), (1, 4), (7, 5)\}$.
(*c*) In the form of a table for x and y: See Table 1.
(*d*) In the form of a graph.   See Figure

Table 1:: $y = 2x$

$x =$	0	1	2	3	4
$y =$	0	2	4	6	8

Figure: Graph of $y = 2x$

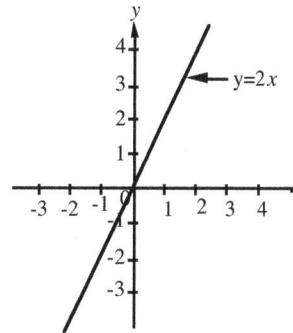

Definition 3 In terms of the correspondence of two sets (Fig. 1)

A function is a correspondence between a first set, say set A and a second set, say Set B such that **each** element of set A corresponds to exactly one element of set B. The set of all the elements of set A is a called the **domain** of the function, and the set of all the corresponding elements of set B is called the **range** of the function.

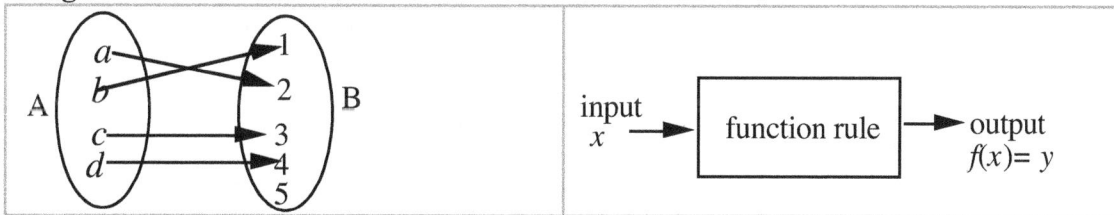

Fig 1   **Fig 2**

Definition 4   In terms of a rule for inputs and outputs (Fig. 2)

A function is a rule which assigns to each input number exactly one output number. The set of all input numbers that the rule is applicable to is called the **domain** of the function; and the set of all the corresponding output numbers is called the **range** of the function.

A variable representing an input number is called the independent variable, and a variable representing an output number is called the dependent variable.

Given a graph   (Vertical line test)

A given graph is that of a function if every possible vertical line drawn to intersect the graph intersects (cuts) the graph exactly once (i.e., at one point only).

Comparison of a Function and a Relation   50

Similarities: Each is a set of ordered pairs.
Differences: In a relation, two or more ordered pairs may have the same first component; but in a function, no two distinct ordered pairs may have the same first component.
Example: The set $D = \{(1, 6), (3, 4), (3, 5), (4, 6)\}$ is only a relation and **not** a function. because the second and third ordered pairs have the same first component, which is 3.

Example: The set $E = \{(1, 2), (2, 3), (4, 5), (7, 5)\}$ is a function (even though the second components of the third and fourth ordered pairs are the same).

A function is a relation, but a relation is not necessarily a function.

Determining if a given graph is a relation or a function

We will use the so-called **vertical line test**.

Procedure

Step 1: Draw as many vertical lines as possible (This can be done visually.) to intersect the graph.

Step 2: If any of the possible lines intersects (cuts) the graph at more than one point, then the given graph is not a function but a relation. However, if each of the possible vertical lines intersects the graph only once (at one point only), then the graph represents a function. Figure.. is a graph which is a relation but not a function.

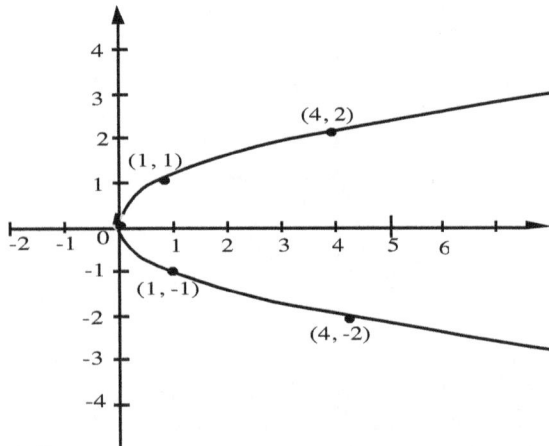

Figure: Graph of $y = \pm\sqrt{x}$ or $x = y^2$
This graph is a relation but not a function

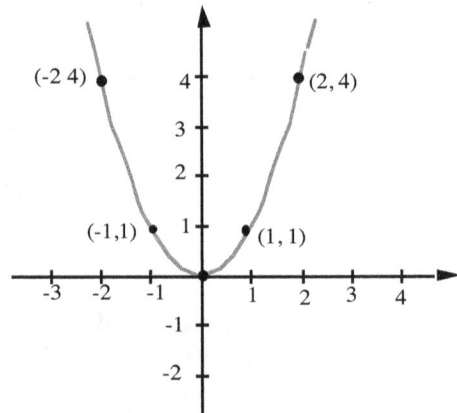

Figure: Graph of $y = x^2$.
This graph is that of a function.

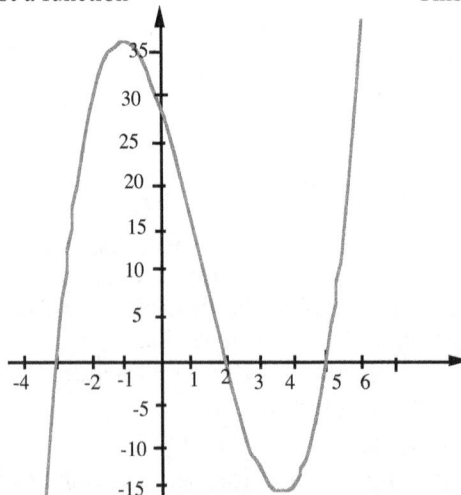

Figure: Graph of $y = (x - 5)(x - 2)(x + 3)$. This graph is that of a function.

Lesson 9 Exercises

Determine which of the following are graphs of functions.

Figure (a)

Figure (b)

Figure (c)

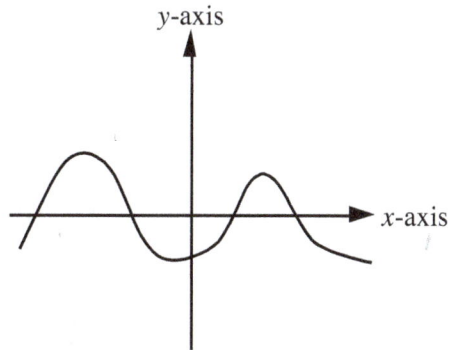

Figure (d)

Answers: (a) A function; (b) Not a function; (c) Not a function; (d) A function

Lesson 10A

Functional Notation; Defined Functions; Excluded Values; Domain and Range

Functional Notation

Let a function $f(x)$ be specified by the rule $f(x) = x^2 + 3$.   (1)

In equation (1), $f(x)$ is read "f of x" or f is a function of x.

To evaluate a function for a particular value of x, we substitute that value of x in the rule that defines $f(x)$. Note that $f(x)$ does not mean f times x but that $f(x)$ is written as a symbol.

Example 1   Given that $f(x) = x^2 + 3$, find (a) f(-1).; (b) $f(x_0 + h) - f(x_0)$

Solution: (a) $f(x) = x^2 + 3$

$$f(-1) = (-1)^2 + 3 \qquad\qquad \text{(replacing } x \text{ in the given equation by -1)}$$
$$= 1 + 3$$
$$f(-1) = 4$$

(b) (Replace x in the given equation by $(x_0 + h)$ and x_0, accordingly in $f(x) = x^2 + 3$

$$f(x_0 + h) \ - \ f(x_0) = [(x_0 + h)^2 + 3] - (x_0^2 + 3)$$
$$= [x_0^2 + 2hx_0 + h^2 + 3] - x_0^2 - 3$$
$$= x_0^2 + 2hx_0 + h^2 + 3 - x_0^2 - 3$$
$$= 2hx_0 + h^2$$

Example 2   Find f(2), given that $f(x) = x + 7$

Solution

$$f(x) = x + 7$$
$$f(2) = 2 + 7$$
$$= 9$$

Example 3   If $f(x) = 2 - \dfrac{1}{x - 4}$, find (a) f(-3); (b) $f(x_0 + h)$; (b) $f(-x)$.

Solution   (a) $f(x) = 2 - \dfrac{1}{x - 4}$

$$f(\text{-}3) = 2 - \frac{1}{(\text{-}3)\,\text{-}4} \qquad\qquad \text{(replacing } x \text{ in the given equation by -3)}$$
$$= 2 - \frac{1}{\text{-}7}$$
$$= 2 + \frac{1}{7}$$
$$= 2\tfrac{1}{7}$$

(c)   (Replace x in the given equation by $x_0 + h$

$$f(x) = 2 - \frac{1}{x - 4} \quad <\,-\,-\,-\,- \text{given equation}$$

$$f(x_0 + h) = 2 - \frac{1}{(x_0 + h) - 4} \quad <\,-\,-\,-\text{replacing } x \text{ by } x_0 + h$$

$$= 2 - \frac{1}{x_0 + h - 4}$$

$$= \frac{2(x_0 + h - 4) - 1}{x_0 + h - 4}$$

$$= \frac{2x_0 + 2h - 8 - 1}{x_0 + h - 4}$$

$$= \frac{2x_0 + 2h - 9}{x_0 + h - 4}$$

(c)   (Replace x in the given equation by $-x$)

$$f(x) = 2 - \frac{1}{x - 4} \qquad <\,-\,-\,-\,- \text{given equation}$$

$$f(-x) = 2 - \frac{1}{(-x) - 4} \qquad <\,-\,-\,-\,-\,-(\textit{replacing } x \textit{ by } - x)$$

$$= 2 - \frac{1}{-x - 4}$$

$$= \frac{2(-x - 4) - 1}{-x - 4}$$

$$= \frac{-2x - 8 - 1}{-x - 4}$$

$$= \frac{-2x - 9}{-x - 4}$$

$$= \frac{-(2x + 9)}{-(x + 4)} \qquad (\textit{factoring out} - 1)$$

$$= \frac{2x + 9}{x + 4}$$

Note above that in (b) the final result contains x. This is so, because we replaced x by $-x$. In the case of (a), we replaced x by the integer -3 , and the final result was purely a numerical value. Furthermore, in Example 3, $f(-a) = \frac{2a + 9}{a + 4}$

Defined Real-Valued Function

Meaning of a defined function of x

A real-valued function $f(x)$ is said to be defined for a variable x if the following conditions are satisfied:

      1. The x and $f(x)$ must be real (i.e., x and $f(x)$ should not be the square root or an even root of a negative number). Thus, a value such as $\sqrt{-4}$ or $2i$ is not allowed.

        For example, in $f(x) = \sqrt{x-4}$, we have to make sure that $x \geq 4$, since otherwise, we obtain imaginary numbers.

      2. The value of x when substituted in the functional equation should yield specific real numbers. (i.e., the value of x when substituted in the functional equation should **not** make the denominator become zero.)

Condition (2) implies that the function should not become undefined when the value of x is substituted in the functional equation. When the function involves a denominator, we have to make sure that the denominator is not allowed to be zero. A particular example of this function occurs when the given function is the ratio of two polynomial functions. (We call such functions rational functions.) In this book, it is agreed that a function is real-valued unless otherwise specified.

Examples of rational functions are: (a) $f(x) = \dfrac{x^2 + 4}{x - 1}$; (b) $f(x) = \dfrac{1}{(x - 3)(x + 4)}$

Excluded Values, Domain and Range

The **excluded values** are (usually) the values which when substituted in the functional equation make the function either undefined or imaginary.

The **domain** (say , D) of a function $f(x)$ consists of those real values of the independent variable say, x, for which $f(x)$ is real and defined.

The **range** (say, R) of the function consists of the corresponding values which $f(x)$ assumes for x in the domain of the function .

Specification of Domain and Range

The domain and the range of a given function may be specified in several ways, namely in the form of a set, in the form of a table, in the form of an equation, in the form of an inequality, or in the form of a graph.

(In some of the examples that follow, we will use a graphing calculator or a computer grapher to generate graphs which will be useful in determining the range of some of the functions. Determining the range of some of the functions analytically may require advanced methods (which we do not cover in this book) , and therefore, we use graphing as an aid.)

Example 1: In the form of a set

Find the domain and range of the function specified by $\{(1, 2), (3, 2), (4, 4), (5, 5)\}$.

Solution: The domain consists of the first components of the ordered pairs.
        The domain, $D = \{1, 3, 4, 5\}$.
        The range consists of the second components of the ordered pairs.
        The range, $R = \{2, 4, 5\}$.

Example 2: In the form of a table of values

Find the domain and range of the function specified by the table of values below.

x	y
1	2
3	2
4	4
5	5

Solution: The domain consists of (the x-values) $1, 3, 4$, and 5.

           The range consists of the (y-values) $2, 4$ and 5.

Note that this form is the set form written in a different format.

Example 3:   In the form of an equation:

The domain consists of those real values of x for which the function is defined. The corresponding values of $f(x)$ form the range of the function.

Example 4: In the form of a graph

Find the domain and range of the function specified by the graph below.

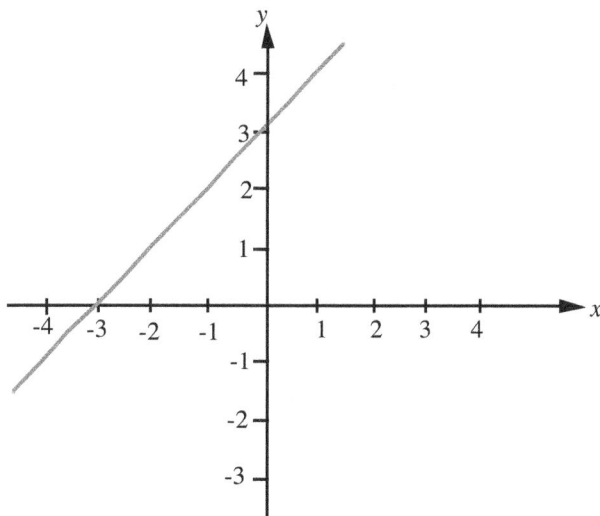

Figure: The graph of $y = x + 3$

Solution: From the graph, the function is defined for all real values of x.

           The domain consists of all real x-values. The range consists of all real y-values.

Implicit and Explicit Specification of the Domain of a Function

The domain of a function may be specified either implicitly or explicitly.

Consider the function $f(x) = x^2$

As it stands, the domain is implicitly specified. The above function is that of a polynomial and as such the domain consists of all real numbers. Here, we assume the largest possible domain.

Now, consider $f(x) = x^2$ $\qquad 0 \le x \le 5$

In this case, the inequality written to the right of the function specifies explicitly and restricts the domain. The domain of this function is such that x is between 0 and 5, including 0 and 5.

If the inequality to the right had not been indicated, the domain would have consisted of all real x-values.

The restrictions on the domains are very important and useful in sketching the graphs of functions.

Other functions with explicitly specified domains are

(a) $\quad y = \sin x$ $\qquad 0 \le x \le 2\pi$

(b) $\quad y = \cos x$ $\qquad -2\pi \le x \le 2\pi$

Determining the Excluded Values, Domain and Range of a Function

Case 1: Polynomial functions

Example 1 (a) For what values of x is the following function not defined? (b) what is the domain?

(c) what is the range? $f(x) = x^2 - 3x + 1$

Solution: All polynomial functions are defined for all real values of the independent variable.

(a) Since the given function is a polynomial function, it is defined for all real values of x. We may note that the right-hand side of the equation does not involve denominators or square roots (or even roots) of the independent variable. There are **no** excluded values.

(b) The domain consists of all real values of x. Set-builder notation: Domain = $\{x \mid x \text{ is a real number}\}$
Using interval notation: Domain $= (-\infty, +\infty)$

(c) Generally, determining the range of a polynomial function may require advanced methods. However, since the given function is a quadratic function, we may apply the range inequality formula, $y \geq \dfrac{4ac - b^2}{4a}$

From the function, $a = 1, b = -3$, and $c = 1$. Substituting these values,

$y \geq \dfrac{4(1)(1) - (-3)^2}{4(1)} = -\dfrac{5}{4}$ (i.e., $y \geq -\dfrac{5}{4}$)

Set-builder: $\{y \mid y \geq -\dfrac{5}{4}\}$

The range consists of all real numbers greater than or equal to $-\dfrac{5}{4}$.

(Note that any horizontal line drawn below the point $(\dfrac{3}{2}, -\dfrac{5}{4})$ will **not** intersect the curve, but any horizontal line through or above this point will intersect the curve)

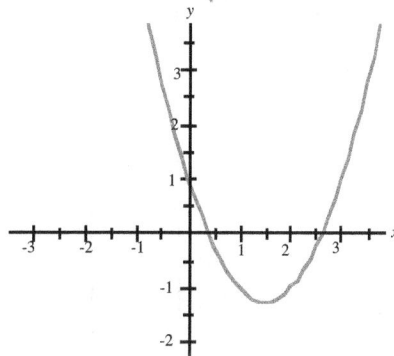

Figure: Graph of $f(x) = x^2 - 3x + 1$

Case 2: Rational functions

Note: A rational function is a function which is the ratio of two polynomial functions.

Example 2 (a) For what values of x is the following function not defined? (b) what is the domain? (c) what is the range?

$$f(x) = \frac{3x - 2}{x - 1}$$

Solution   Step 1: Setting the denominator to zero,
$$x - 1 = 0$$
Step 2: Solving for x, $x = 1$.

(a) The function is not defined when $x = 1$. (The excluded value of x is 1.)

(b) The domain is all real values of x, except 1. same as $\{x \mid x$ is a real number and $x \neq 1\}$

(c) The range (from graph) if found by being guided by the horizontal asymptote, $y = 3$. (see p.273)
The range is given by the set $\{y \mid y < 3$ or $y > 3\}$ or simply $\{y \mid y \neq 3\}$
(Note that a horizontal line drawn through $(0, 3)$ will **not** intersect the curve)

Checking for $x = 1$, $f(1) = \frac{3(1) - 2}{1 - 1} = \frac{3 - 2}{0} = \frac{1}{0}$ which is undefined .(The right-hand side is division by zero).

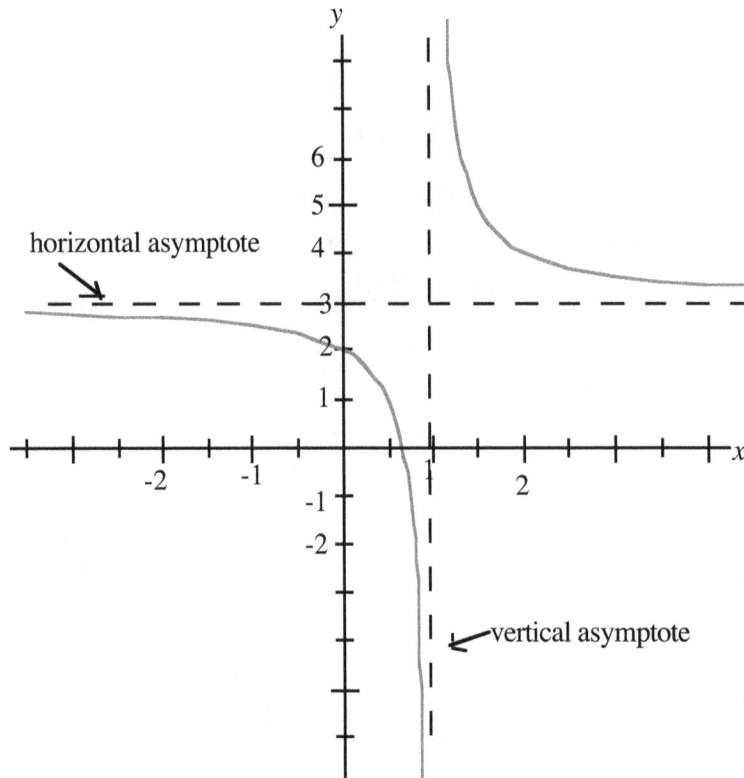

Figure: Graph of $f(x) = \dfrac{3x - 2}{x - 1}$

Example 3   (a) For what values of x is the given function not defined? (b) what is the domain?
(c) what is the range?

$$f(x) = \frac{2(x-1)}{(x-2)(x+4)}$$

Solution   Step 1: Setting the denominator to zero,
$$(x-2)(x+4) = 0$$

Step 2: Solving for x, $x = 2$, or $x = -4$.

(a) The function is not defined when $x = 2$ and -4. (The excluded values of x are 2 and -4.)

(b) The domain is all real values of x, except 2 and -4.

Set-builder notation: $\{x \mid x$ is a real number and $x \neq -4$, $x \neq 2\}$
(c) The range (from graph) is all real y.
Set-builder notation: $\{y \mid y$ is a real number$\}$
(Note that any horizontal line drawn through the y-axis will intersect the curve)

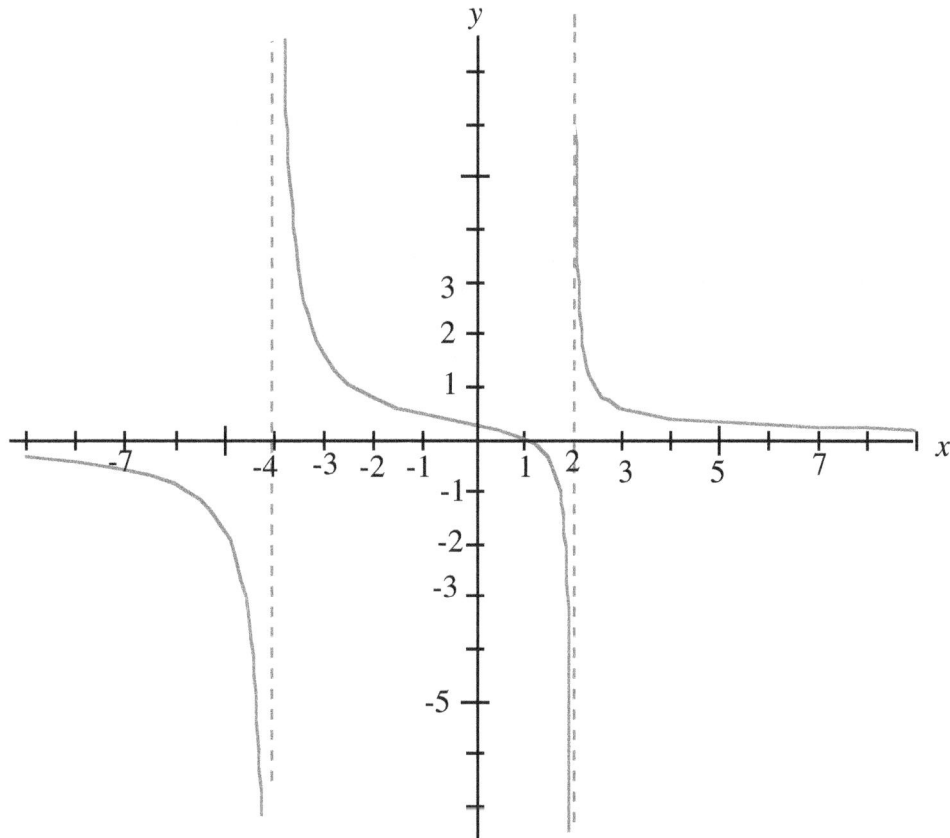

Figure: Graph of $f(x) = \dfrac{2(x-1)}{(x-2)(x+4)}$

Example 4 (a) For what values of x is the given function not defined? (b) what is the domain? 60

(c) what is the range?

$$f(x) = \frac{1}{x}$$

Solution Setting the denominator to zero,

$x = 0$

(a) The function is not defined when $x = 0$. (The excluded value of x is 0.)

(b) The domain is all real values of x, except 0.

Set-builder notation: $\{x \mid x$ is a real number and $x \neq 0 \}$

(c) The range (from graph) is given by the set $\{y \mid y < 0$ or $y > 0\}$ or simply $\{y \mid y \neq 0 \}$

(the horizontal asymptote is $y = 0$)

(Note that any horizontal line drawn through the y-axis, except through the point $(0, 0)$, will intersect the curve.

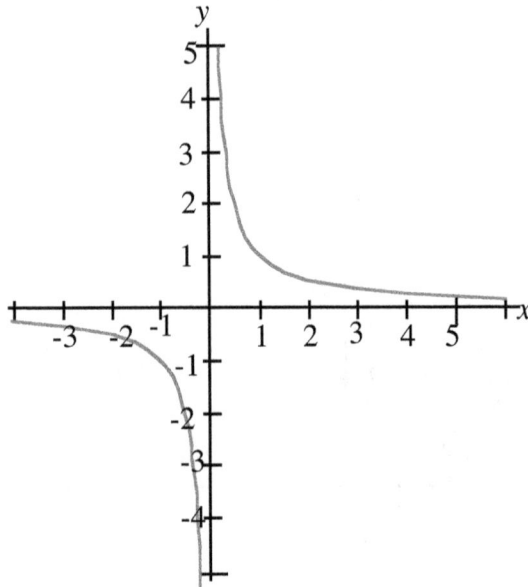

Figure: Graph of $f(x) = \frac{1}{x}$

Example 5 (a) For what values of x is the given function not defined? (b) what is the domain?
(c) what is the range?

$$f(x) = \frac{x^3 + x^2 + 2}{x^2 - 16}$$

Solution: Setting the denominator to zero and solving,

$$x^2 - 16 = 0$$

$$(x + 4)(x - 4) = 0$$

$$x = -4 \text{ or } x = 4$$

The function is not defined when $x = -4$ or 4. (The excluded values of x are -4 and 4.)

(a) The domain is all real x except -4 and 4.

{ $x \mid x$ is a real number and $x \neq -4$, $x \neq 4$ }

(b) The range (from graph) is all real values of y. same as { $y \mid y$ is a real number}
(Note that any horizontal line drawn through the y-axis will intersect the curve)

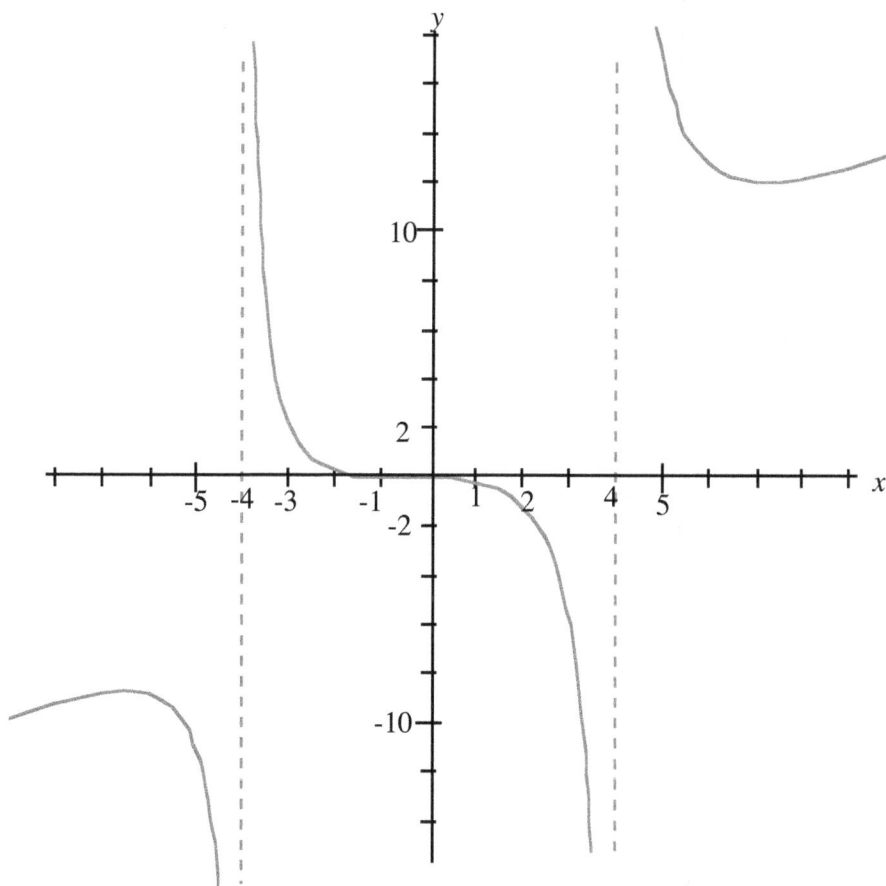

Figure: Graph of $\dfrac{x^3 + x^2 + 2}{x^2 - 16}$

Example 6 (a) For what values of x is the given function not defined? (b) what is the domain?   6 2

       (c) what is the range?

$$f(x) = \frac{8}{x^2 - 4}$$

Solution   Step 1: Setting the denominator to zero,

$$x^2 - 4 = 0$$
$$(x + 2)(x - 2) = 0.$$

       Step 2: Solving for x, $x = 2$, or $x = -2$.

(a) The function is not defined when $x = 2$ and -2. (The excluded values of x are 2 and -2.)

(b) The domain is all real values of x except 2 and -2.

    Same as $\{x \mid x$ is a real number and $x \neq -2$, $x \neq 2\}$

(c) The range (from graph) is given by the set $\{y \mid y \leq -2$ or $y > 0\}$

  (Note that any horizontal line drawn through or below $y = -2$ or above $y = 0$ will intersect the curve)

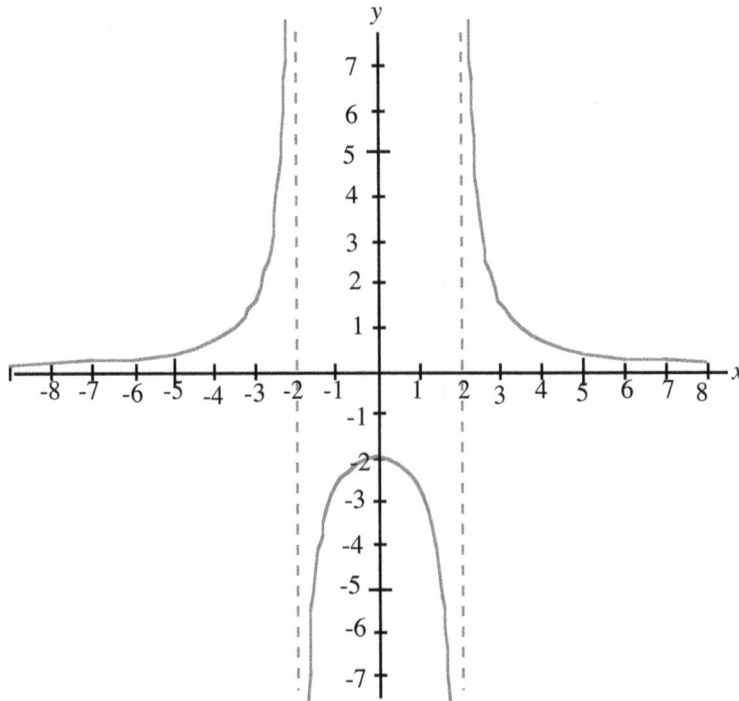

Figure: Graph of $f(x) = \dfrac{8}{x^2 - 4}$

Example 7 (a) For what values of x is the given function not defined? (b) what is the domain?
(c) what is the range?

$$f(x) = \frac{8}{x^2 + 4}$$

Solution Step 1: Setting the denominator to zero, and solving, we obtain non-real values. Since we are dealing with real-valued functions, we conclude that

Step 2: (a) There are **no** excluded values.

$x^2 + 4$ is positive for all real values of x and never zero, since the square of any nonzero real number is always positive.

(b) The function is defined for all real values of x.

Domain: $= \{x \mid x$ is a real number$\}$

(c) The range (from graph) is given by the set $\{y \mid 0 < y \le 2\}$

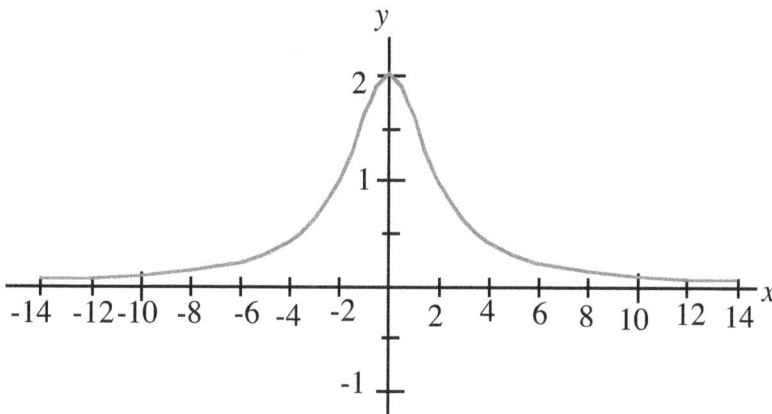

Figure: Graph of $f(x) = \dfrac{8}{x^2 + 4}$

Case 3: Functions containing even roots (e.g., square root) **of polynomials**

Example 8 (a) For what values of x is the function $f(x) = \sqrt{x - 5}$ not real?

(b) What is the domain? What is the range?

Solution (a) For real roots, the radicand, $(x - 5)$ must be positive or zero.

Symbolically, $x - 5 \ge 0$
$x \ge 5$ (solving for x).

The function is not real when $x < 5$ (because the square root would be that of a negative number. The square root of a negative number is imaginary. Any x-value less than 5 is excluded.

Checking for some specific values of x:

1. When $x = 5$, $f(5) = \sqrt{x - 5}$ $= \sqrt{5 - 5} = 0$, which is real.   (i.e., for $x = 5$).

2. When $x = 6$, $f(6) = \sqrt{6 - 5} = 1$, which is real.   (i.e., for $x > 5$).

3. When $x = 4$, $f(4) = \sqrt{4 - 5}$ $= \sqrt{-1}$, which is imaginary (non-real (i.e., for $x < 5$).

(b) The domain consists of all real values of x such that $x \ge 5$.

(c) The range for $y = \sqrt{x - 5}$ is from $y = 0$ to $y = \infty$.

(The range is obtained by considering $x \ge 5$ ($x = 5$ and $x > 5$) in $y = \sqrt{x - 5}$, or from graph (next page Fig.2)

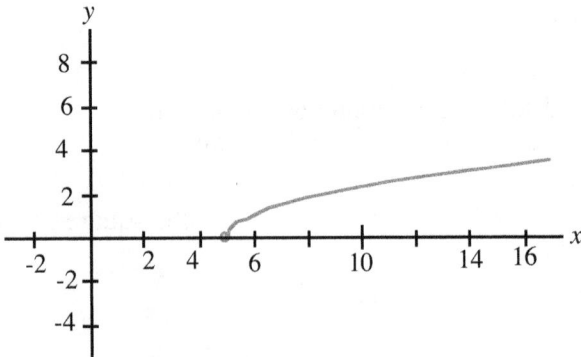

Figure 2: Graph of $f(x) = \sqrt{x-5}$

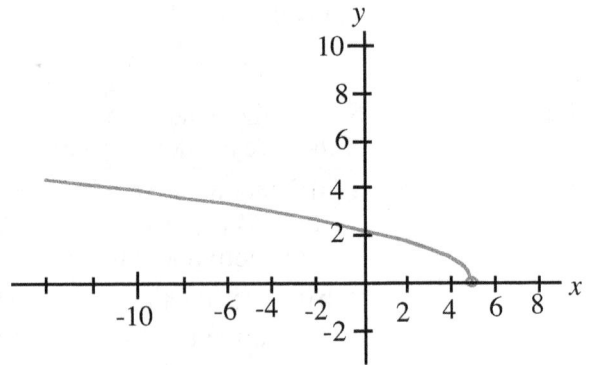

Figure 3: Graph of $f(x) = \sqrt{5-x}$

Example 9 (a) For what values of x is $f(x) = \sqrt{5-x}$) not real?

(b) What is the domain? What is the range?

Solution

(a) $5 - x \geq 0$ (for real values of $f(x)$, the radicand must be positive or zero).

   $x \leq 5$ (solving for x)

The function is not real if $x > 5$. (Any x-value greater than 5 is excluded).

(b) The domain is all real x such that $x \leq 5$.

(c) The range is from $y = 0$ to $y = \infty$ (obtained by considering $x \leq 5$ ($x = 5$ and $x < 5$) in $y = \sqrt{5-x}$., or (from graph)

Example 10 Find (a) the domain and (b) the range of $f(x) = \dfrac{\sqrt{x+6}}{x-1}$

Solution   Here, two conditions must be satisfied.

Condition 1: $x + 6 \geq 0$ and from which $x \geq -6$ (For real root, the radicand must be positive or zero)

Condition 2: $x - 1 \neq 0$ and from which $x \neq 1$.

(a) The domain consists of all real numbers greater than or equal to -6, except 1, that is $x \geq -6, x \neq 1$.

(b) The range (from graph) is given by the set $\{y \mid y$ is a real number$\}$. Note that when $x = -6$, $y = 0$

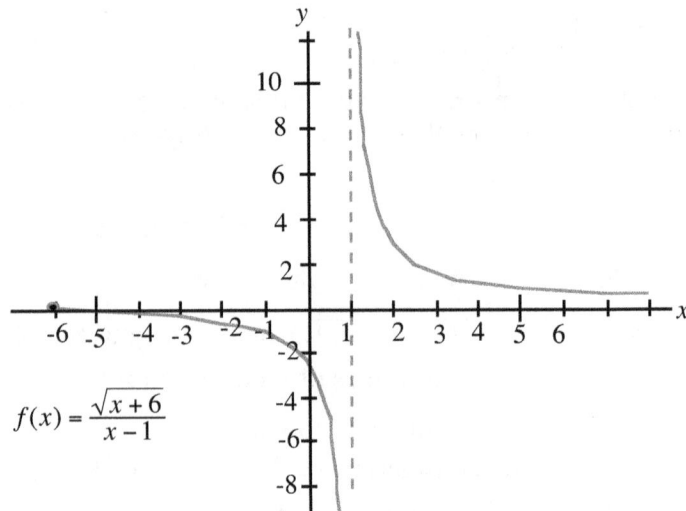

Fig.4: $f(x) = \dfrac{\sqrt{x+6}}{x-1}$

Example 11 Determine (a) the excluded values of x ; (b) the domain ; (c) the range of

$$f(x) = \sqrt{\frac{x+2}{x-5}}$$

Solution

For real and defined values of $\sqrt{\frac{x+2}{x-5}}$, $\frac{x+2}{x-5} \geq 0$ ($x + 2 \geq 0$ and $x - 5 > 0$ or $x + 2 \leq 0$ and $x - 5 < 0$)

We will use a sign diagrams to solve this problem. (See also chapter 12)

	Factor	Signs of the intervals		
Row 1	$x + 2$	$-$	$+$	$+$
Row 2	$x - 5$	$-$	$-$	$+$
Row 3	$\frac{x+2}{x-5}$	$+$	$-$	$+$

In Row 3, Columns 1 and 3 (with the "+" signs), we read that $\frac{x+2}{x-5} \geq 0$ if $x \leq -2$ or $x > 5$

(a) For these columns,, the excluded values are such that $-2 < x \leq 5$.
(b) The domain is such that $x \leq -2$ or $x > 5$
(c) The range (from graph) is given by $\{y \mid 0 \leq y < 1$ or $y > 1\}$.

Figure: Graph of $f(x) = \sqrt{\frac{x+2}{x-5}}$

Lesson 10A Exercises   66

1. If $(x) = x^2 - 5x + 2$, find (a) $f(-2)$; (b) $f(-1)$, (c) $f(-x)$, (d) $f(a+h)$, (e) $\dfrac{f(a+h) - f(a)}{h}$;

A: 2. If $(x) = 4 - \dfrac{1}{x-2}$, find (a) $f(-3)$; (b) $f(-x)$; (c) $f(a)$.

3. If $(x) = x^3 - x^2 - x - 1$, find $f(-1)$.

Answers: **1.** (a) 16; (b) 8; (c) $x^2 + 5x + 2$; (d) $a^2 - 5a + 2ah - 5h + h^2 + 2$; (e) $2a + h - 5$

2. (a) $4\frac{1}{5}$; (b) $\dfrac{4x+9}{x+2}$; (c) $\dfrac{4a-9}{a-2}$; **(3)** -2

B 1. Determine the domain of $f(x) = \dfrac{\sqrt{x+2}}{\sqrt{x-5}}$.

Hint: .The following conditions must be satisfied simultaneously:
a. $x + 2 \geq 0$ or $x \geq -2$ (so that the square root is not imaginary)
b. $x - 5 \geq 0$ or $x \geq 5$ (so that the square root is not imaginary)
c $x - 5 \neq 0$ or $x \neq 5$ (Since otherwise the function is undefined)

1. Domain: $x > 5$. (see the graph in Example 11 above for $x > 5$)

Note above: $\dfrac{\sqrt{x+2}}{\sqrt{x-5}} \neq \sqrt{\dfrac{x+2}{x-5}}$ (For real numbers a and b, $\dfrac{\sqrt{a}}{\sqrt{b}} = \sqrt{\dfrac{a}{b}}$ if $a \geq 0$ and $b > 0$)

C 1. What is meant by the domain of a function?
2. What conditions do you look out for in determining the domain of a function?
3. What is meant by the range of a function?

Find the domain and range in each of the following:
4. $A = \{(2,4),(3,9),(4,16),(5,25)\}$; **5.** $B = (2,1), (5,1), (6,1), (7,3)$
6. $y = x + 2$; **7.** $y = x^2 + 3x + 7$; **8.** $y = -x - 4$; **9.** $y = x^3 + x^2 - 5$

Answers: **4.** Domain : $\{2,3,4,5\}$;, range: $\{4,9,16,25\}$

5. Domain : $\{2,5,6,7\}$;, range: $\{1, 3\}$
6. Domain : All real values of x: range: all real values of y.

7. Domain : All real x; range: all real y such that $y \geq 4.75$.
8. Domain : All real values of x: range: all real values of y.
9. Domain : All real values of x.

D Determine the excluded values (if any) for the following:
1. $y = \dfrac{2}{x-3}$; **2.** $f(x) = \dfrac{2x-4}{x^2-1}$; **3.** $f(x) = \dfrac{3x}{x^2+1}$; **4.** $f(x) = \dfrac{1}{x}$; **5.** $f(x) = \dfrac{(x-1)}{(x-1)(x+2)}$

In Problems 6 & 7 below, for what values of x is the function not real?

6. $y = \sqrt{x-3}$; **7.** $y = \sqrt{4-x}$ **8.** Find the domain of $f(x) = \sqrt{\dfrac{x+2}{x-5}}$.

9. Find the domain of $f(x) = x - \dfrac{1}{x}$

Extra: Why is it important to note the excluded values in dealing with functions and solving equations?

Answers: **1.** 3; **2.** $\{-1, 1\}$; **3.** None; **4.** 0; **5.** $\{-2, 1\}$; **6.** $x<3$; **7.** $x>4$; **8.** $x \leq -2$ or $x > 5$
9. All real x except 0.

Lesson 10B
Algebra of Functions

In the past, we covered the addition, subtraction, multiplication and division of algebraic expressions. Below, we similarly cover the same basic operations on functions.

Let f and g be functions of x *such that* x is in the domains of both f and g. Then

1. Sum: $(f + g)(x) = f(x) + g(x)$ (Domain: All real numbers common to the domains of f and g)

2. Difference: $(f - g)(x) = f(x) - g(x)$ Domain: All real numbers common to the domains f and g)

3. Product: $(f \bullet g)(x)$ or $(fg)(x) = f(x) \bullet g(x)$ or $f(x)g(x)$ (Domain: All real numbers common to the domains f and g

4. Quotient $\left(\frac{f}{g}\right)(x) = \frac{f(x)}{g(x)}$ (Domain: All real numbers common to the domains f and g and $g(x) \neq 0$.

Examples

If $f(x) = x^2 + 2$ *and* $g(x) = x - 2$; find (a) $(f + g)(x)$; (b) $(f - g)(x)$; (c) $(f \bullet g)(x)$; (d) $\left(\frac{f}{g}\right)(x)$;

(e) $(f + g)(3)$; (f) $\left(\frac{f}{g}\right)(3)$; (g) $\left(\frac{f}{g}\right)(2)$.

Solution

(a) $(f + g)(x) = f(x) + g(x)$
$$= (x^2 + 2) + (x - 2)$$
$$= x^2 + 2 + x - 2$$
$$= x^2 + x$$

(b) $(f - g)(x) = f(x) - g(x)$
$$= (x^2 + 2) - (x - 2)$$
$$= x^2 + 2 - x + 2$$
$$= x^2 - x + 4$$

(c) $(f \bullet g)(x) = f(x) \bullet g(x)$
$$= (x^2 + 2)(x - 2)$$
$$= x^3 - 2x^2 + 2x - 4$$

(d) $\left(\frac{f}{g}\right)(x) = \frac{f(x)}{g(x)}$
$$= \frac{x^2 + 2}{x - 2} \text{ or } x + 2 + \frac{6}{x - 2} \text{ (long division)}$$

(e) From (a), $(f + g)(x) = f(x) + g(x)$
$$= x^2 + x$$
$$(f + g)(3) = (3)^2 + 3 \ - 12$$

Method 2: $(f + g)(3) = f(3) + g(3)$
$$f(3) = (3)^2 + 2 = 11$$
$$g(3) = 3 - 2 = 1$$
$$(f + g)(3) = 11 + 1 = 12$$

(f) From (d), $\left(\frac{f}{g}\right)(x) = \frac{f(x)}{g(x)}$
$$= \frac{x^2 + 2}{x - 2}$$
$$\left(\frac{f}{g}\right)(3) = \frac{(3)^2 + 2}{(3) - 2}$$
$$= \frac{9 + 2}{3 - 2}$$
$$= 11.$$

Method 2: $f(3) = (3)^2 + 2 = 11$
$$g(3) = 3 - 2 = 1$$
$$\left(\frac{f}{g}\right)(3) = \frac{f(3)}{g(3)} = \frac{11}{1} = 11$$

(g) From (d), $\left(\frac{f}{g}\right)(x) = \frac{x^2 + 2}{x - 2}$
$$\left(\frac{f}{g}\right)(2) = \frac{(2)^2 + 2}{(2) - 2}$$
$$= \frac{6}{0} \text{ is undefined.}$$

Therefore, $\left(\frac{f}{g}\right)(2)$ is undefined. Note: $g(2) = 0$.

On the next page, we cover the domains for algebraic of functions.

Domains for Algebra of Functions 68

In the first step, we determine the common domain (intersection of domains) of the functions. In the second step, we check for the domain of the resulting function. (sum. difference, product or quotient). We add any excluded values to those from Step 1 and if there is an overlap in the domains, we determine the intersection of the domains from Step 1 and Step 2. Pay attention to radical functions such as $f(x) = \sqrt{x - 2}$, determine the domain and find the intersection for the overall domain.

Example

Given $f(x) = \frac{3}{x}$ and $g(x) = \frac{3x-4}{x+2}$, find the domains of the following:

(a) $(f + g)(x)$; (b) $(f - g)(x)$; (c) $(f \bullet g)(x)$; (d) $\left(\frac{f}{g}\right)(x)$;

Solution:

Domain of $f : \{x \mid x$ is a real number and $x \neq 0\}$

Domain of $g : \{x \mid x$ is a real number and $x \neq -2\}$.

For **(a), (b), and (c):**

domain of $f + g =$ domain of $f - g =$ domain of $f \bullet g = \{x \mid x$ is a real number and $x \neq 0, x \neq -2\}$

(d) $\left(\frac{f}{g}\right)(x) = \frac{f(x)}{g(x)}$

$= \frac{3}{x} \div \frac{3x - 4}{x + 2}$ $(x \neq 0, x \neq -2)$

$= \frac{3}{x} \bullet \frac{x + 2}{3x - 4}$

$(x \neq \frac{4}{3}$. See $= \frac{3(x + 2)}{x(3x - 4)}$ below)

Note: $\frac{f(x)}{g(x)} = \dfrac{\frac{3}{x}}{\frac{3x - 4}{x + 2}}$

Setting $x(3x - 4) = 0$

$x = 0$ or $3x - 4 = 0$ and from which $x = \frac{4}{3}$

The excluded values are $-2, 0$ and $\frac{4}{3}$.

Domain of $\left(\frac{f}{g}\right)(x)$: $\{x \mid x$ is a real number and $x \neq 0, x \neq -2$ and $x \neq \frac{4}{3}\}$

Extra

Example: The intersection of the domains of $f(x) = \sqrt{x - 2}$ and $g(x) = \frac{x - 4}{x + 3}$ is

$\{x \mid x$ is a real number and $x \geq 2\}$ or $[2, \infty)$ (from $\{x \mid x \geq 2\} \cap \{x \mid x \neq -3\}$)

It seems the radical function is "King".

Lesson 10B Exercises

If $f(x) = x^2 + 2$ *and* $g(x) = x - 2$; find   (a) $(f + g)(x)$;   (b) $(f - g)(x)$; (c) $(f \bullet g)(x)$; (d) $\left(\frac{f}{g}\right)(x)$;

(e) $(f + g)(3)$; (f) $\left(\frac{f}{g}\right)(3)$; (g) $\left(\frac{f}{g}\right)(2)$; (h) $(f - g)(4)$.

Ans: (a) $x^2 + x$; (b) $x^2 - x + 4$; (c) $= x^3 - 2x^2 + 2x - 4$; (d) $\frac{x^2 + 2}{x - 2}$ or $x + 2 + \frac{6}{x - 2}$; (e) 12; (f) 11

(g) undefined; (h) 16.

Lesson 11
One-to-One Functions, Composite Functions

One-to-One Functions

(In Lesson 12, we will learn that if a function is one-to-one, then it has an inverse function.)
We consider three main cases according to how the function is specified.

Case 1: **Given a set of ordered pairs** (or a table of x- and y-values)

A one-to-one (1-1) function is a set of ordered pairs in which for any two ordered pairs, the first elements are different from each other and the second elements are also different from each other.

Example The set, $A = \{(3,2),(4,7),(1,5),(2,3)\}$ is a one-to-one function. However.

The set, $B = \{(3,2),(4,7),(1,5),(5,2)\}$ is **not** a one-to-one function because the first and the last ordered pairs have the same second elements, namely, 2.

Case 2: Given the graph of the function

By the so called **horizontal line test**, the graph of a function is that of a one-to-one function if every possible horizontal line drawn to intersect the graph cuts (intersects) the graph only once (at one point only). **Figures** 2 and 3, p. 75, are graphs of one-to-one functions but **Figure 1** is not one-to-one.

Case 3: Given the equation of the function

A function $f(x)$ is one-to-one if whenever $f(x_1) = f(x_2), x_1 = x_2$. If we let $y_1 = f(x_1)$, and $y_2 = f(x_2)$, then $y_1 = y_2$ **must** imply that $x_1 = x_2$, otherwise, $f(x)$ is not one-to-one.

Example 2: Determine if $f(x) = 2x + 3$ is one-to-one.

Solution

Step 1: $f(x_1) = 2x_1 + 3$
$\qquad f(x_2) = 2x_2 + 3$
Equate RHS of $f(x_1)$ to RHS of $f(x_2)$
(That is, let $f(x_1) = f(x_2)$):
$\qquad 2x_1 + 3 = 2x_2 + 3 \qquad (1)$

Step 2: Solve for x_1. (You may also solve for x_2)
$$2x_1 + 3 = 2x_2 + 3$$
$$2x_1 = 2x_2$$
$$x_1 = x_2 \qquad (2)$$

Since from above, whenever $f(x_1) = f(x_2)$ (from equation (1)) $x_1 = x_2$ (from equation (2))
$f(x) = 2x + 3$ is one-to-one. (You may check by sketching its graph and using the horizontal line test)

Example 3 Determine if $f(x) = \sqrt{25 - x^2}$ is one-to-one.

Solution

Step 1: $f(x_1) = \sqrt{25 - x_1^2}$
$\qquad f(x_2) = \sqrt{25 - x_2^2}$
Equate RHS of $f(x_1)$ to RHS of $f(x_2)$
(That is, let $f(x_1) = f(x_2)$.)
$$\sqrt{25 - x_1^2} = \sqrt{25 - x_2^2} \qquad (1)$$

Step 2: Solve for x_1.
$$\sqrt{25 - x_1^2} = \sqrt{25 - x_2^2}$$
$$25 - x_1^2 = 25 - x_2^2$$
$$-x_1^2 = -x_2^2$$
$$x_1^2 = x_2^2$$
$$x_1 = \pm\sqrt{x_2^2}$$
$$x_1 = +x_2 \text{ or } -x_2 \qquad (2)$$

Since from above, whenever $f(x_1) = f(x_2), x_1 = -x_2$, (That is for the same y-value, x is not unique: $x_1 = +x_2$, and **also** $x_1 = -x_2$ (from equation (2))

$f(x) = \sqrt{25 - x^2}$ is **not** one-to-one. (You may check by sketching its graph and using the horizontal line test)

Example 4: Determine if $f(x) = |x - 2|$ is one-to-one.

Solution

Step 1: $f(x_1) = |x_1 - 2|$

$\qquad f(x_2) = |x_2 - 2|$

Equate RHS's of $f(x_1)$ to $f(x_2)$

(That is, let $f(x_1) = f(x_2)$):

$\qquad |x_1 - 2| = |x_2 - 2| \qquad$ (1)

Step 2: Solve for x_1:

If both $x_1 - 2$ and $x_2 - 2$ are positive,

$\qquad x_1 - 2 = x_2 - 2$ and from which $x_1 = x_2$

(same result as if both $x_1 - 2$ and $x_2 - 2$ are negative)

However, if $x_1 - 2$ is positive and $x_2 - 2$ is negative,

$\qquad x_1 - 2 = -(x_2 - 2)$

$\qquad x_1 - 2 = -x_2 + 2$

$\qquad x_1 = -x_2 + 4$. That is, $x_1 \neq x_2$

Since from above, whenever $f(x_1) = f(x_2), x_1 \neq x_2$

$f(x) = |x - 2|$ is **not** one-to-one. (You may check by sketching its graph and using the horizontal line test)

Another method for determining if a function is one-to-one

The author proposes the following definition for a one-to-one function.

Definition : A function is one-to-one if its inverse relation is a function.

This definition provides another method for determining if a given function is one-to-one.

Example 5

Determine if $f(x) = 2x + 6$ is one-to-one

Solution

Given: $f(x) = 2x + 6$

Required: To determine if $f(x) = 2x + 6$ is one-to-one.

Plan: If it can be shown that $f(x) = 2x + 6$ has an inverse relation which is function, then

$\qquad f(x) = 2x + 6$ is one-to-one.

Determination:

Step 1: Let $f(x) = y$ to obtain $y = 2x + 6$.

Step 2: Interchange x and y to obtain the inverse relation $x = 2y + 6$

Step 3: Solve for y.

$\qquad 2y = x - 6$

$\qquad y = \frac{1}{2}x - 3$

Since clearly, for a given value of x there is exactly one corresponding value of y.

the inverse relation, $y = \frac{1}{2}x - 3$, is a function

and therefore, $f(x) = 2x + 6$ is one-to-one.

(You may also check that $y = \frac{1}{2}x - 3$ is a function by sketching its graph and applying the vertical line test)

Example 6

Determine if $f(x) = x^2$ is one-to-one.

Solution

Given: $f(x) = x^2$

Required: To determine if $f(x) = x^2$ is one-to-one.

Plan: If it can be shown that $f(x) = x^2$ has an inverse relation which is function, then

$\qquad f(x) = x^2$ is one-to-one.

Determination

Step 1: Let $f(x) = y$ to obtain $y = x^2$

Step 2: Interchange x and y to obtain the inverse relation $x = y^2$

Step 3: Solve for y.

$\qquad$ If $y^2 = x$

$\qquad y = \pm\sqrt{x}$ (that is, $y = +\sqrt{x}$ or $y = -\sqrt{x}$

Clearly, for a given value of x there are two different corresponding y-values. Therefore, y is **not** a function of x, and the inverse relation $y^2 = x$ is **not** a function and therefore , the given function $f(x) = x^2$ is **not** one-to-one.

(You may also check that $y^2 = x$ or $y = \pm\sqrt{x}$ is **not** a function by sketching its graph and applying the vertical line test)

Composite Functions

Some authors refer to composite functions as "product functions". This alternative terminology may be misleading because a reader might be inclined to multiply the given functions. Perhaps, a better alternative terminology for a composite function is " a function within another function".

Use of Composition of Functions:
The principle of composition of functions can be used to determine algebraically if two given functions are inverses of each other.

Definition: If two functions f and g are such that the range of g is in the domain of f, then the composite function of f with g, symbolized $f \circ g$, is specified by $(f \circ g)(x) = f[g(x)]$ which is read "f of g of x". That is, the output of g becomes the input for f.

Similarly, if the range of f is in the domain of g, then the composite function of g with f, symbolized $g \circ f$ is specified by $g \circ f = g[f(x)]$ which is read "g of f of x." That is, the output of f becomes the input for g.

We must **note** that generally, $f \circ g \neq g \circ f$ (i.e., generally, $f \circ g$ is not equal to $g \circ f$)

Example 1

If $f(x) = x^2$, and $g(x) = x + 1$, find (a) $f[g(x)]$; (b) $g[f(x)]$.

Solution

Recall that if, for example,

$$f(x) = x^2$$

then $f(3) = (3)^2$.

Similarly, (a) $f[g(x)] = f[x + 1]$

$$= [x + 1]^2$$

$$f[g(x)] = x^2 + 2x + 1$$

Thus, (in the above problem) wherever there is x in the equation for $f(x) = x^2$, write (substitute) $x + 1$.

(b) $g[f(x)]$

$$= g[x^2]$$

$$= (x^2) + 1 \quad \text{(substitute } x^2 \text{ for } x \text{ in the equation for } g(x) = x + 1)$$

$$= x^2 + 1$$

Thus, wherever there is x in the equation for $g(x) = x + 1$, we substitute x^2.

Example 2: If $f(x) = 2(x + 10)$, and $g(x) = x - 2$, find (a) $f[g(x)]$; (b) $g[f(x)]$.

Solution

(a) $f[g(x)] = 2[(x - 2) + 10]$

$$= 2[x - 2 + 10]$$

$$= 2[x + 8]$$

$$f[g(x)] = 2x + 16$$

(Substituting $x - 2$ in the equation for $f(x)$.

(b) $g[f(x)] = 2(x + 10) - 2$

$$= 2x + 20 - 2$$

$$g[f(x)] = 2x + 18$$

(Substituting $2(x + 10)$ in the equation for $g(x)$.

Domains of Composite Functions 72

The approach here is similar to that for domains of algebra of functions. However, here, in the first step, we check only for the domain of the "inside" function. In the second step, we check for the domain of the resulting composite function. We add any excluded values to those from Step 1 and if there is an overlap in the domains we use the intersection of the domains from Step 1 and Step 2. Pay attention to radical functions such as $f(x) = \sqrt{x-2}$,.

Given $f(x) = \frac{3}{x}$ and $g(x) = \frac{3x-4}{x+2}$, find the domains of the following:

find (a) $f[g(x)]$; (b) $g[f(x)]$.

(a) Step 1: We check the domain of the "inside function", $g(x) = \frac{3x-4}{x+2}$

Domain of g : $\{x \mid x$ is a real number and $x \neq -2\}$.

Step 2: Form the composite function and check its domain.

$f[g(x)] = \dfrac{3}{\frac{3x-4}{x+2}}$ $(f(x) = \frac{3}{x})$

$= \dfrac{3}{1} \cdot \dfrac{x+2}{3x-4}$

$f[g(x)] = \dfrac{3(x+2)}{3x-4}$

Set $3x - 4 = 0$ to obtain $x = \frac{4}{3}$. Therefore $x \neq \frac{4}{3}$.

Combining the domains from Step 1 and Step 2, the excluded values are -2, and $\frac{4}{3}$.; and

the domain of $f[g(x)] = \dfrac{3(x+2)}{3x-4}$ is

$\{x \mid x$ is a real number and $x \neq -2$ and $x \neq \frac{4}{3}\}$

(a) Step 1: We check the domain of the "inside function", $f(x) = \frac{3}{x}$

Domain of f : $\{x \mid x \neq 0\}$.

Step 2: Form the composite function and check its domain.

$g[f(x)] = \dfrac{3(\frac{3}{x})-4}{\frac{3}{x}+2}$ $(g(x) = \frac{3x-4}{x+2})$

$g[f(x)] = \dfrac{-4x+9}{2x+3}$

Set $2x + 3 = 0$ to obtain $x = -\frac{3}{2}$.

Therefore $x \neq -\frac{3}{2}$.

Combining the domains from Step 1 and Step 2, the excluded values are 0, and $-\frac{3}{2}$.; and

the domain of $g[f(x)] = \dfrac{-4x+9}{2x+3}$ is

$\{x \mid x$ is a real number and $x \neq -0$ and $x \neq -\frac{3}{2}\}$

Application of Composition of Functions

If two functions f_1 and f_2 are inverses of each other, then, the following two conditions must be satisfied simultaneously. 1. $f_1[f_2(x)] = x$ and 2. $f_2[f_1(x)] = x$. For examples, see page 80.

Lesson 11 Exercises

A Show and determine which of the following functions are one-to-one

1. The set $\{(4,5), (3,4), (1,2)\}$; 2. The set $\{(2,2), (4,3), (5,7)\}$; 3. $f(x) = 3x - 2$;

4. $f(x) = \sqrt{x+3}$; 5. $f(x) = \sqrt{16 - x^2}$; 6. $f(x) = |x|$; 7. $f(x) = |x - 2|$; 8. $f(x) = x^3 + 2$.

Answers: 1. Yes; 2. Yes; 3. Yes; 4. Yes; 5. No; 6. No; 7. No.; 8. Yes

B 1. Given that $f(x) = x + 2$, $g(x) = x - 1$, find (a) $f[g(x)]$; (b) $g[f(x)]$

2. Given that $f(x) = 3(x + 2)$, $g(x) = \frac{1}{x}$ find (a) $f[g(x)]$; (b) $g[f(x)]$

3. Given that $f(x) = (x+1)^2 + 3$, $g(x) = -x$, find (a) $f[g(x)]$; (b) $g[f(x)]$

Answers: 1. (a) $x + 1$; (b) $x + 1$; 2. (a) $\frac{3}{x} + 6$; (b) $\frac{1}{3x+6}$; 3. (a) $x^2 - 2x + 4$; (b) $-x^2 - 2x - 4$

Lesson 12   73

Inverse Functions and Inverse Relations
(Exchange is no Robbery)

Recalling the definitions of a relation and a function (page 48), a relation is a set of ordered pairs and function is a relation in which no two distinct ordered pairs have the same first elements (components).

Inverse Relation

Given a relation which is specified by the set of ordered pairs (x, y), the inverse relation of this given relation is the set of ordered pairs (y, x). This inverse relation is obtained by interchanging the first and second elements of each ordered pair.

Example 1: Find the inverse relation of the function specified by the set A:
$$A = \{(1, 3), (3, 2), (5, 7) \}$$

Solution The inverse relation is obtained by interchanging the first and second elements of each ordered pair
The inverse relation of A is the set
$$\{(3, 1), (2, 3), (7, 5)\}$$

From the definition of the inverse, we can conclude that when we form the inverse of a relation, the domain and the range of the original relation are interchanged. Thus, the domain of the original relation becomes the range of the inverse, and the range of the original relation becomes the domain of the inverse. The inverse may be a relation or a function. (See page 48)

Inverse function

Let a function be specified by the set of ordered pairs $\{(x, y)\}$. Then the inverse relation of this function is the set of ordered pairs $\{(y, x)\}$. If this inverse relation is also a function, then for the inverse relation ,we symbolize $f^{-1}(x)$, which is read "f inverse of x" and we say that $f(x)$ and $f^{-1}(x)$ are inverse functions of each other.

Example 2 Find the inverse function of the function specified by the set $B,$:
$$B = \{(a, b), (c, d), (e, f)\}$$
Solution: The inverse function of B is the set, denoted by B^{-1}, is given by
$$B^{-1} = \{(b, a), (d, c), (f, e)\}$$

 Sometimes, authors ask the question :" Does this function have an inverse?" Such a question may sometimes be misunderstood, since the inverse is found by interchanging the roles of x and y, which is always possible. What is implied in such a question is whether or not the inverse relation obtained is **also** a function. Perhaps, an unambiguous form of the question should be "Does this function have an **inverse function**?" If the inverse relation of a given function is also a function, then we also say that the given function is invertible

Note that it is possible that a given relation which is not a function may have an inverse relation which is a function.

Let us elucidate how the terms, " relation, function and inverse", are connected by using the terms "inverse relation" and "inverse function" in the following statements:

Every relation has an inverse relation. This inverse relation may or may not be a function.
Every function has an inverse relation. Some functions have inverse relations which are (inverse) functions.
A function is also a relation , but a relation is not necessarily a function.

Determining if a function has an inverse function

Necessary and sufficient condition for a function to have an inverse function: A function has an inverse function if and only if it is a one-to-one function.

We will consider the different forms in which the rules for specifying a function may be given and determine if the function has an inverse function. We will consider the set form, the tabular form, the graphical form, and the equation form.

Case 1: Given the set form of the function

A function has an inverse function if for any two different ordered pairs, the second elements are different. If a function has an inverse function it is said to be invertible.

The necessary and sufficient condition is a consequence of the definitions of a function (page 48) and of an inverse function. We may note here that the condition for invertibility refers to the differences in the second components while the condition for being a function refers to the differences in the first components.

Example 3 Given the function specified by the set

$A = \{(1,2),\ (2,3), (4,5), (6,7)\}$, determine if the inverse relation of this set is a function.

Solution Since the second components are all different from one another, the inverse relation is a function.

In fact, the inverse of the set A is given by $A^{-1} = \{(2,1),(3,2),(5,4),(7,6)\}$,
which is clearly a function, since the first components are all different from one another.

An example of a function whose inverse relation is **not** a function is the set $\{(3,4),(5,4),(6,5)\}$, because the first and the second ordered pairs have the same second component , 4.

Case 2: Given the tabular form of the function

If any two y-values are the same, then the inverse relation is not a function (but only a relation) otherwise, it is a function. This case corresponds to the set form of a function.

Example 4 **Table 3** represents a function whose inverse relation is a function, while **Table 4** represents a function whose inverse relation is **not** a function.

Tables 3

x	y
1	2
2	3
3	4
4	5
5	6
6	7

Table 4

x	y
1	2
2	**3**
3	**3**
4	5
5	7
6	8

} same two y-values

Case 3: Given the graphical form of the function

The horizontal line test

If any **horizontal line** drawn to intersect the graph intersects the graph at only one point, then the inverse relation of the function (graph) is a function. However, if a horizontal line meets the graph in more than one point, then the inverse of the graph is not a function but only a relation. See Figs 1, 2 and 3 below/ .

Example 3 Figure **1** is **not** the graph of a one-to-one function, Figures **2** and **3** are the graphs of one-to-one functions.

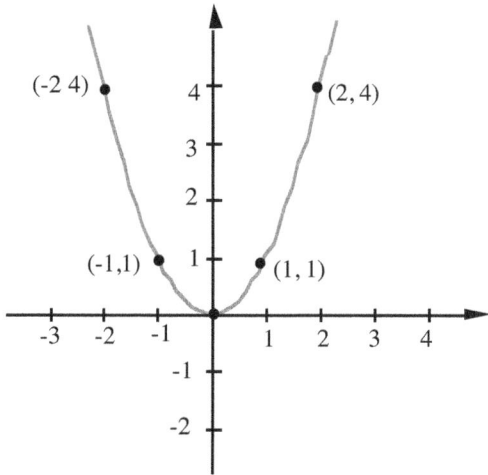

Figure 1: The graph of $y = x^2$,
The inverse relation of this graph is **not** a function

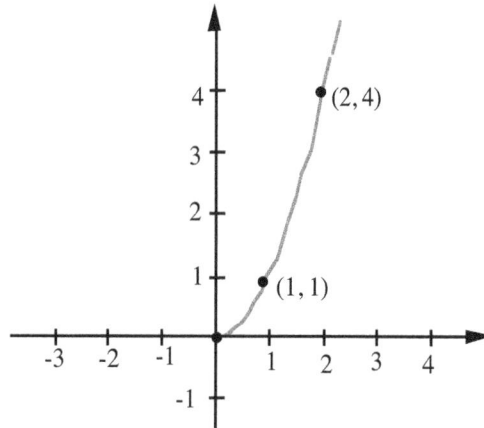

Figure 2: The graph of $y = x^2$, $x \geq 0$.
The inverse relation of this graph is a function
(This is the graph of a one-to-one function)

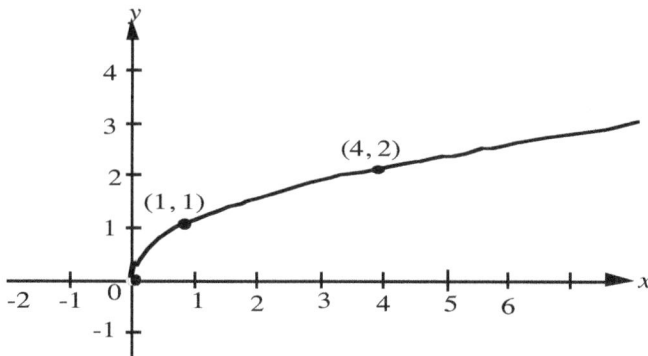

Figure 3: The inverse relation of this graph is a function. (This is the graph of a one-to-one function)

Case 4: Given the equation of the function

By definition, y is a function of x if there is a rule which gives only one corresponding value of y for a given value of x.

Example Determine if the inverse relation of $y = x^2$ is a function.
Method 1: By interchanging the roles of x and y in the given equation, we obtain the inverse relation
$$x = y^2 \text{ or } y^2 = x. \text{ Solving } y^2 = x \text{ for } y, \text{ we obtain, } y = +\sqrt{x} \text{ or } y = -\sqrt{x}$$
Since for the same x-value (say, 4) we have two different y-values (+2 and -2),. the inverse relation of $y = x^2$ is not a function. However, we could make the inverse relation become a function by restricting the domain of this function (see Figures 1 and 2, above)
For some functions, we will graph the function and use the graphical method.

Method 2: Let $y = f(x)$

Step 1: $f(x_1) = x_1^2$

$f(x_2) = x_2^2$

Equate RHS of $f(x_1)$ to RHS of $f(x_2)$ (That is, let $f(x_1) = f(x_2)$):

$x_1^2 = x_2^2$ (1)

Step 2: Solve for x_1.   (You may also solve for x_2)

$x_1^2 = x_2^2$

$x_1 = \pm\sqrt{x_2^2}$

$x_1 = +x_2 \text{ or } -x_2$ (2)

Since from above, $f(x_1) = f(x_2)$, does **not** imply $x_1 = x_2$ (from equation (2)), $f(x)$ is **not** one-to-one and therefore does **not** have an inverse function.

Finding the inverse of a function specified by an equation

Example 3. Find the inverse function of $f(x) = 3x + 2$.   (1)

Solution

Step 1: Let $f(x) = y$. Then $y = 3x + 2$.

Interchange x and y in the given equation.

Then, we obtain $x = 3y + 2$.   (2)

By tradition, we want to keep the x-axis horizontal and express y as a function of x.

Step 2: Solve equation (2) for y.

Then from $x = 3y + 2$.

$\dfrac{x - 2}{3} = y$

or $y = \dfrac{x - 2}{3}$

The inverse $f^{-1}(x) = \dfrac{x - 2}{3}$

Alternatively,

Step 1: Solve $y = 3x + 2$. for x.

Then $x = \dfrac{y - 2}{3}$

Step 2: Interchange x and y (by definition of the inverse)

$y = \dfrac{x - 2}{3}$

The inverse $f^{-1}(x) = \dfrac{x - 2}{3}$

We can observe from above that Steps 1 and 2 are interchangeable.

Since the inverse $y = \dfrac{x - 2}{3}$ is also a function, we can say that $y = 3x + 2$. and $y = \dfrac{x - 2}{3}$ are inverse functions (of each other). We may also add that $f(x) = 3x + 2$. is invertible.

Example 4 Find the inverse function of $f(x) = x^2 + 6$.

Solution

Step 1: Let $f(x) = y$

Step 2: Solve for x .

$$y = x^2 + 6.$$
$$x = \pm\sqrt{y - 6} \qquad\qquad (2)$$

Step 3: Interchange x and y in equation (2).

Then $y = \pm\sqrt{x - 6}$ (same as $y = +\sqrt{x - 6}$ or $y = -\sqrt{x - 6}$)   (3)

Clearly, equation (3) does not represent a function, since for a given value of x, there are two y-values. We can test this by graphing and using the so called **vertical line test** (see page 50). We can say that the given function has an inverse relation specified by $y = \pm\sqrt{x - 6}$. However, the given function does not have an inverse function.

In this example, if we were asked to find $f^{-1}(x)$, we would say that there is no $f^{-1}(x)$ (no inverse function) for the given function.

Example 5 Find the inverse function of $f(x) = x^2 + 6$ $x \geq 0$

Solution

Step 1: Let $f(x) = y$

Then $y = x^2 + 6$ $x \geq 0$

Step 2: Solve for x .

Then $x = \pm\sqrt{y - 6}$ (That is, $x = +\sqrt{y - 6}$ or $x = -\sqrt{y - 6}$)   (1)

Since we are only interested in the domain $x \geq 0$, we reject the negative part of equation (1).

Then, we obtain $x = +\sqrt{y - 6}$ (2)

Step 3: Interchange (to obtain the inverse) x and y in equation (2)

Then $y = +\sqrt{x - 6}$ (3)

Equation (3) is the inverse of $y = x^2 + 6$ $x \geq 0$

Therefore, $f^{-1}(x) = \sqrt{x - 6}$ $x \geq 6$

The graph of this inverse (Figure **2**) indicates that the inverse is a function.

Similarly, if the given function were $f(x) = x^2 + 6$ $x \leq 0$,

the inverse relation would be $f(x) = -\sqrt{x - 6}$ and this also is a function.

ALTERNATIVELY

Step 1: Let $f(x) = y$.   Then $y = x^2 + 6$ $x \geq 0$

Step 2: Interchange x and y and solve for y.

$$x = y^2 + 6 \qquad\qquad y \geq 0$$
$$y = \pm\sqrt{x - 6} \qquad\qquad y \geq 0$$
$$y = \sqrt{x - 6} \quad\text{since } y \geq 0$$

Therefore, $f^{-1}(x) = \sqrt{x - 6}$ $x \geq 6$

We may note from Examples 2 and 3 that, sometimes, by redefining (or restricting) the domain of a given function, a function which does not have an inverse function can be made to have an inverse function. See Figures 1, 2 and 3.

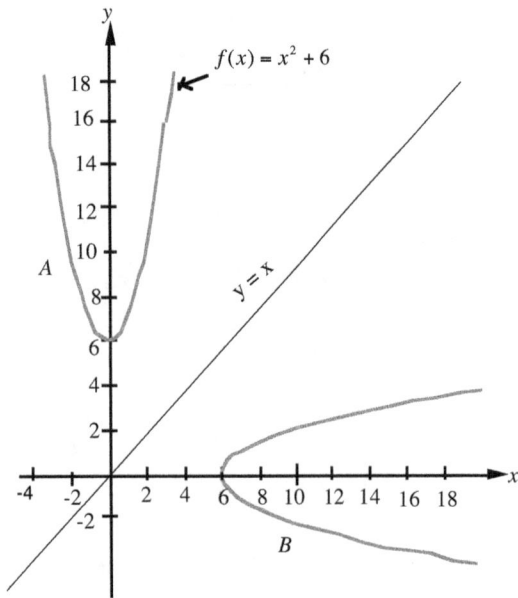

Figure 1: B, the inverse relation of A, is **not** a function .

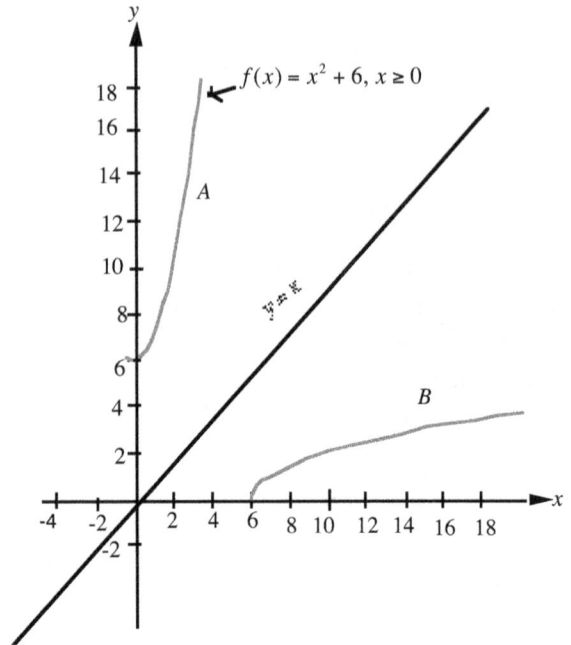

Figure 2: B, the inverse relation of A, is a function (by the vertical line test.)

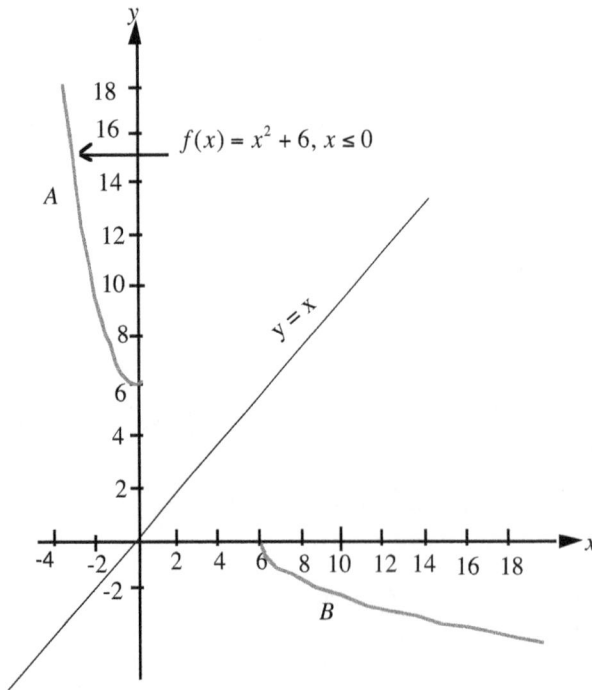

Figure 3: B, the inverse relation of A, is a function (by the vertical line test)

Given the graph of a function, how to sketch the inverse of the graph 79

Example 5 Given the graph of $y = 3x + 2$, sketch the graph of the inverse of this function.

Geometrically, a function (or relation) and its inverse are symmetric with respect to the line $y = x$. To obtain the inverse graph, we will reflect (see page 223) the given graph about the line $y = x$. The given function is a straight line and so, we will reflect two points about the line $y = x$, and then connect these points by a straight line.

Step 1: Interchange the x- and y-coordinates of any two points on the given line, say, $(1, 5)$ becomes $(5, 1)$; and $(-2, -4)$ becomes$(-4, -2)$.

Step 2: Plot the points $(5, 1)$ and $(-4,- 2)$ on the same coordinate system of axes (as the given graph)

Step 3: Connect the points from Step 2 by a straight line to obtain the inverse graph (Fig.)

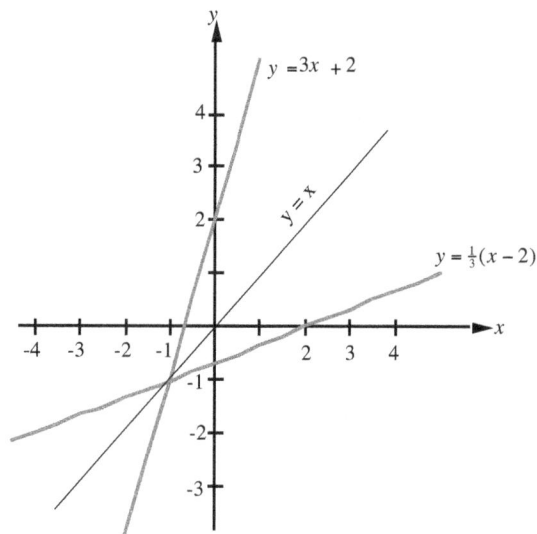

Figure 4: The graphs of $y = 3x + 2$ and $y = \frac{1}{3}(x - 2)$

Note above that, we could also find the equation of the graph, find its inverse equation and then use this equation to sketch the inverse graph.

Determining if two functions are inverses of each other

80

Case 1: Given the equations of the functions

Example Are the functions $y = 4x + 2$ and $y = \dfrac{x-2}{4}$ inverses of each other?

Solution

Method 1 We will use the principle of composition of functions which states that:

Two functions f_1 and f_2 are inverses of each other if

1. $f_1[f_2(x)] = x$ and **2.** $f_2[f_1(x)] = x$ (Note that inverse functions reverse the action of each other.)

Let $f_1 = 4x + 2$ and $f_2 = \dfrac{x-2}{4}$

$$f_1[f_2(x)] = 4[\dfrac{x-2}{4}] + 2$$

$$= 4[\dfrac{x-2}{4}] + 2$$

$$= x - 2 + 2$$

$$= x$$

$$f_2[f_1(x)] = [\dfrac{(4x+2)-2}{4}]$$

$$= \dfrac{4x+2-2}{4}$$

$$= \dfrac{4x}{4}$$

$$= x$$

Since $f_1[f_2(x)] = f_2[f_1(x)] = x$,

$y = 4x + 2$ and $y = \dfrac{x-2}{4}$ are inverses of each other.

Method 2 Find the inverse of one of the functions and compare this inverse with the other function.

Step 1: We find the inverse of $y = 4x + 2$ (by interchanging x and y and solving for y)

$$x = 4y + 2$$

$$y = \dfrac{x-2}{4} \quad \text{(solving for } y\text{)}$$

Step 2: Clearly, this inverse is identical with the other function.

Therefore, $y = 4x + 2$ and $y = \dfrac{x-2}{4}$ are inverses of each other.

Case 2: Given the graphs of the functions

Procedure: Fold the page along the line $y = x$ and if the two graphs coincide, then the functions are inverses of each other. See Figures 2 and 3 (page 78, 79,)

Lesson 12 Exercises

A Find the inverse relation in each of the following and state if the inverse relation is a function:

1. $A = \{(2,3), (4,5), (6,7)\}$ **2.** $B = \{(0,1), (2,5), (4,10)\}$ **3.** $C = \{(b,a), (c,d), (a,b)\}$

4. $\{(1,3),(2,3),(4,5)\}$

5. Given a table for a function, determine if the inverse relation represents an inverse function.

x	2	4	8
y	5	6	10	1

Determine if the inverse relation in each of the following graphs represents an inverse function:

6.

7.

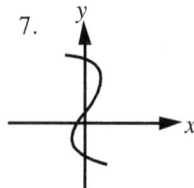

Answers: 1. $\{(3,2),(5,4),(7,6)\}$. It is a function; **2.** $\{(1,0),(5,2),(10,4)\}$. It is a function.

3. $\{(a,b),(d,c),(b,a)\}$. It is a function; **4.** $\{(3, 1),(3, 2),(5, 4)\}$. It is not a function.

5. The inverse relation is a function; **6.** The inverse relation is **not** a function;

7. The inverse relation is a function; even though the given relation is not a function.

B Find the inverse relation of the following and indicate which of the inverse relations are functions.

1. $y = -4x + 1;$ **2.** $y = x^2 - 3$ **3.** $y - 5 = x^2;$ **4.** $y = x^3;$ **5.** $x - 2y = 6$

Determine which of the following pairs are inverses of each other.

6. $y = \dfrac{1}{x + 2} - 3$ and $y = \dfrac{1}{x + 1} - 4;$ **7.** $y = 5x + 2$ and $y = \dfrac{x - 2}{5};$ **8.** $y = x^3$ and $y = x^{\frac{1}{3}}$

Answers: **1.** $y = -\dfrac{x}{4} + \dfrac{1}{4}$ (a function); **2.** $y = \pm\sqrt{x + 3}$ (**not** a function); **3.** $y = \pm\sqrt{x - 5}$ (**not** a function);

4. $y = x^{\frac{1}{3}}$ (a function); **5.** $y = 2x + 6$ (a function); **6.** No; **7.** Yes; **8.** Yes.

C. In each of the following graphs, sketch its inverse by reflecting the graph in the line $y = x$:

1.

2.

Answers:--

1.

2.

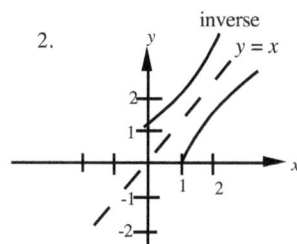

Introductory Theme for Chapter 7

(Next Chapter)
Straight Line

Theme: Two points

1. Why two points? Two points, because given or knowing two points, a straight line can be drawn by connecting the two points, using a straight edge and pencil.

2. Why two points? Two points, because given (knowing) two points; the slope, m, of the line segment connecting the two points $P_1(x_1, y_1)$ and $P_2(x_2, y_2)$ can be found by applying $m = \dfrac{y_2 - y_1}{x_2 - x_1}$

3a. Why two points? Two points, because if we know the **slope, m,** and the **y-intercept, b,** of the line, we can obtain two points and draw the graph of the line

3b **Note:** y–intercept, b implies the point $(0, b)$, By choosing a point (x, y) on a line, we have two points, and the slope (as well as an equation) of the line connecting the two pints $(0, b)$ and (x, y)

is given by $m = \dfrac{y - b}{x - 0}$ <--**slope = slope**

$mx = y - b$ or $\boxed{y = mx + b}$ <------**slope-intercept form** of the equation of a line

4. Why two points? Two points, because given or knowing two points
an equation of the line segment connecting the two points $P_1(x_1, y_1)$ and $P_2(x_2, y_2)$ can be found by

applying $\boxed{y - y_1 = \left(\dfrac{y_2 - y_1}{x_2 - x_1}\right)(x - x_1)}$ (from $\dfrac{y - y_1}{x - x_1} = \dfrac{y_2 - y_1}{x_2 - x_1}$ <-- **is slope = slope**

or $\boxed{y - y_1 = m(x - x_1)}$ <------**point-slope form,** where $m = \dfrac{y_2 - y_1}{x_2 - x_1}$

5. Why two points? Two points, because given or knowing the two-intercept points $(a, 0)$, $(0, b)$
an equation of the line segment connecting the two points $P_1(a, 0)$ and $P_2(0, b)$ can be found by

applying $y - y_1 = \left(\dfrac{y_2 - y_1}{x_2 - x_1}\right)(x - x_1)$ to obtain $y - 0 = \dfrac{b - 0}{0 - a}(x - a)$; from $\dfrac{y - 0}{x - a} = \dfrac{b - 0}{0 - a}$)

or $y = \dfrac{b}{-a}(x - a)$ or $y = \dfrac{b}{-a}x + (\dfrac{b}{-a})(-a)$ or $\boxed{y = -\dfrac{b}{a}x + b}$ also $\dfrac{x}{a} + \dfrac{y}{b} = 1$ **(Two intercept form 6.**

Why two points? Two points, because given the graph (picture) of a line , we are given
infinitely many points from which we can read the coordinates of any two points on the line
and write an equation of a line by applying **3, 4** or **5** above. A picture is worth a thousand words

7. Why two points? Two points, because given or knowing two points; the **midpoint** of the line
segment connecting the points $P_1(x_1, y_1)$ and $P_2(x_2, y_2)$ is given by x-coordinate, $x_m = \dfrac{x_1 + x_2}{2}$,

and the y-coordinate, $y_m = \dfrac{y_1 + y_2}{2}$

8. Why two points? Two points, because given or knowing) two points; $P_1(x_1, y_1)$, $P_2(x_2, y_2)$,
the distance, d, between the two points on a line **in a plane** is given by

$$d = \sqrt{(x_2 - x_1)^2 + (y_2 - y_1)^2}$$

9. Why two points? Two points, because given or knowing) two points, $P_1(x_1, y_1)$, $P_2(x_2, y_2)$ on each
of two lines, the slopes m_1, m_2 as well as parallelism or perpendicularity can be determined.

The above theme summarizes Chapter 7

CHAPTER 7
Straight Lines

Lesson 13: **Linear Equations; Graphing; Points on Lines**
Lesson 14: **Intercepts; Slopes**
Lesson 15: **Equations of Straight Lines**
Lesson 16 :**Perpendicular and Parallel Lines;**
Lesson 17: **Midpoint of a Line; Distance Formula**

Lesson 13
Linear Equations; Graphing; Points on Lines

Linear Equations

A linear equation is an equation of form

$$y = ax + b$$ where a and b are constants.

We will call this equation the standard form of the equation of a straight line. This form is also the slope-intercept form of the equation of a straight line. We must note that the above equation is a linear equation in two variables and can be written more generally as

$$ax + by = c$$, where a, b, and c are constants.

Examples of linear equations are:

1. $ax - by = 20$
2. $6x - 2y = 15$
3. $y = 3x + 9$

Special cases:

6. $y = x$
5. $y = 2$
6. $y = 0$
7. $x = 0$
8. $x = 5$

The domain of the linear equation $y = ax + b$ is all real numbers. The range of $y = ax + b$ consists of all real numbers.

Graphing in the Rectangular Coordinate System of Axes

We begin by means of an example. Let us consider the equation $y = 3x + 2$ (1)

If we let $x = 1$ in equation (1) then $y = 3(1) + 2$
$$= 5$$

Thus, when $x = 1$, $y = 5$

If we let $x = 2$ in equation (1) m then $y = 3(2) + 2$
$$= 8$$

Similarly, when $x = 2$, $y = 8$.

From the above example, we can say that for each value of x we choose, there is a corresponding value of y. By agreement, we can write the $x-$ and $y-$values values as an ordered pair (x, y), where x is the first component and y is the second component.

Then for the solutions to $y = 3x + 2$, when $x = 1$, $y = 5$, we can represent the solution as the ordered pair $(1, 5)$.

For $x = 2$, $y = 8$, we can write the ordered pair $(2, 8)$

Similarly, when $x = 4$, $y = 14$, giving us the ordered pair $(4, 14)$.

The equation $y = 3x + 2$ has infinitely many solutions. The solution set consists of all ordered pairs. If an ordered pair is a solution of a given equation, then these numbers when substituted in the equation for the unknowns, should make both sides of the equation equal. We should note that we could have chosen values for y and then calculate the corresponding values of x. In fact, in some instances, it may be more convenient to choose y and calculate x

Rectangular Coordinate System of Axes (Cartesian Coordinate System)

Consider a horizontal number line (scale) labeled as in Figure. On this line, all numbers to the right of the origin are positive, and all numbers to the left of the origin are negative.

Figure 1: Horizontal real number line

Consider also a vertical line **(Figure 2)**. On this line, all numbers above the origin are positive, and all numbers below the origin are negative.

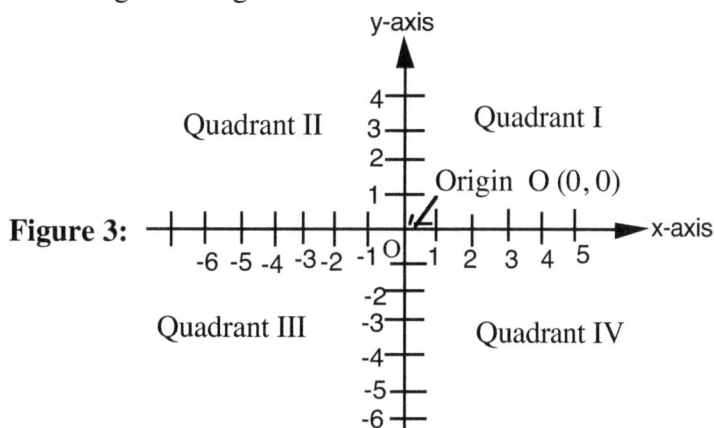

Figure 2:

Vertical line

Figure 3:

If we combine the horizontal and vertical lines at right angles so that the zero points (the origins) coincide, then we obtain what is called the rectangular coordinate system of axes. **(Figure 3).** We call the intersection of these lines, the origin and we label it O(0, 0). We call the horizontal number scale the x–axis, and we call the vertical number scale the y–axis. The two axes divide the plane (area) into four quadrants. These quadrants are numbered counterclockwise as quadrants I, II, III, and IV (i.e., first quadrant, second quadrant, third quadrant and fourth quadrant).

Graphing Ordered Pairs in a Rectangular Coordinate Plane

Geometrically, the first component (element) of each ordered pair is called the x-coordinate or the abscissa of the point. The second component (element) of the ordered pair is called the y-coordinate or ordinate of the point. The x-coordinate is the directed (positive or negative) distance from the y–axis and the y-coordinate is the directed distance from the x–axis. In quadrant I, both the x- and y-coordinates are positive. In quadrant II, the x–coordinate is negative but the y–coordinate is positive, In quadrant III, both coordinates are negative, In quadrant IV, the x–coordinate is positive, but the y–coordinate is negative.

Geometrically, each ordered pair represents a **point** in an x–y coordinate system of axes.

Plotting Points

In plotting points, we will be guided as follows:

1. We always count from the origin $O(0, 0)$.

2. If the x–coordinate is positive, we count horizontally to the right from the origin, but if the x–coordinate is negative, we count horizontally to the left from the origin.

3. If the y–coordinate is positive, we count vertically upwards from the origin, but if the y–coordinate is negative, we count vertically downwards from the origin. The counting is done visually, and we do not make any marks on the graph paper as we count.

Example .Graph the following ordered pairs (points) in a rectangular coordinate system of axes.   8 6
$A(1,5)$. $B(-2,4)$, $P_1(-3,-4)$. $P_2(1,-2)$.

Solution

For $A(1,5)$: From the origin, count horizontally one unit to the right ($x=1$) and stop.
From this point, count 5 units ($y=5$) vertically upwards and stop. Place a dot
here and label this point as $A(1,5)$.

For $B(-2,4)$: From the origin, count horizontally 2 units to the left ($x=-2$) and stop,
From this point, count 4 units ($y=4$) vertically upwards and stop. Place a
dot here and label this point as $B(-2,4)$.

For $P_1(-3,-4)$: From the origin, count horizontally 3 units to the left ($x=-3$) and
stop. From this point, count 4 units vertically downwards ($y=-4$) and
stop. Place a dot here and label this point as $P_1(-3,-4)$.

Similarly, for $P_2(1,-2)$, count horizontally 1 unit to the right ($x=1$) and 2 units vertically downwards ($y=-2$) and stop. Place a dot here and label this point as $P_1(-3,-4)$.

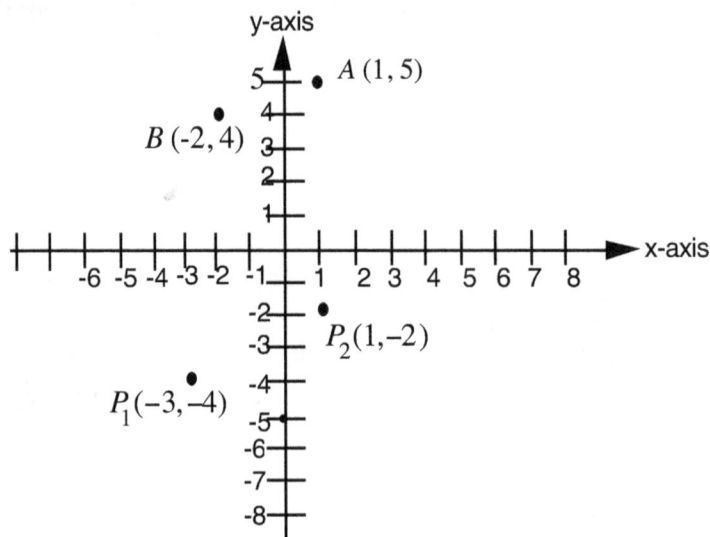

Figure 4

Drawing the Graph of a Straight Line

Example   Draw the graph of the line whose equation is given by : $y = 3x - 5$

We will cover three methods, namely a general method; the x- and y-intercepts method; and the slope-intercept method.

Method 1: General method

Step 1: Choose three convenient x-values and calculate the corresponding y-values to obtain ordered pairs.

(Actually, two ordered pairs will be sufficient; the third pair is used as a check: all **three** points must be in line)

choosing $x = 1$, $y = 3(1) - 5$ ordered pairs

$$\left. \begin{array}{l} y = 3 - 5 \\ y = -2 \end{array} \right\} \text{-------} > (1, -2)$$

choosing $x = 0$, $y = 3(0) - 5$ $\left. \begin{array}{l} \\ y = -5 \end{array} \right\} \text{-------} > (0, -5)$

choosing $x = -2$, $y = 3(-2) - 5$

$$\left. \begin{array}{l} y = -6 - 5 \\ y = -11 \end{array} \right\} \text{-------} > (-2, -11)$$

Step 2: Plot the points $(1,-2), (0,-5)$ and $(-2,-11)$ on a rectangular
coordinate system of axes (on graph paper) and connect the points by a straight line.

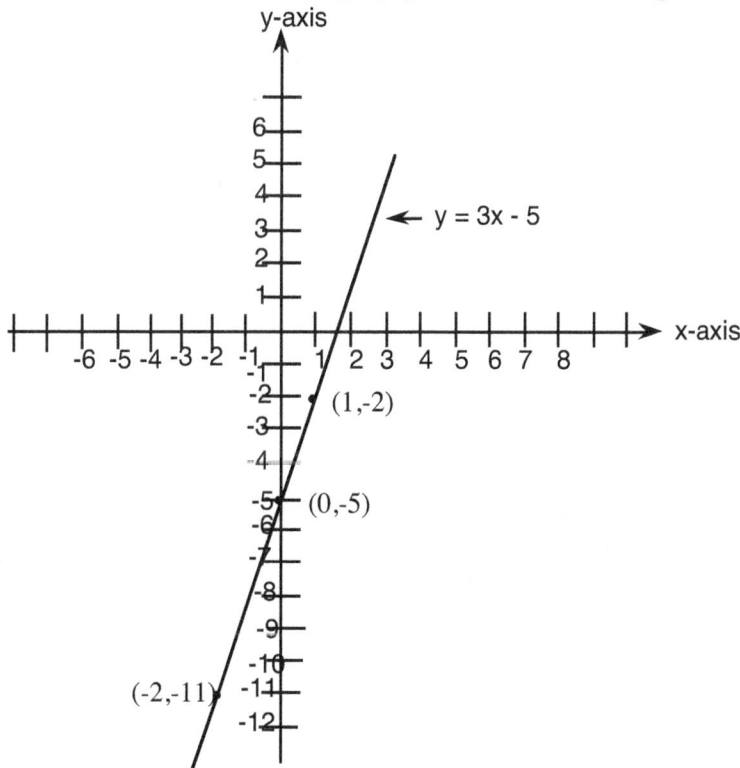

Note that from the given equation, the line has a positive slope and therefore the direction (see also page 98)
of the line (it **leans to the right**) in the figure is appropriate.

Note also that the above general method can be extended to draw the graphs of nonlinear equations such as $y = x^2$.

Method 2: The *x*- and *y*-intercepts method

We need two points to plot and connect by a straight line. The coordinates of the *x*-intercept yield one point and the coordinates of the *y*-intercept yield a second point..

Step 1: Find the *x*-intercept by letting $y = 0$ in the equation, $y = 3x - 5$, and solving for *x*.

Then $0 = 3x - 5$ and from which $x = \frac{5}{3}$. This step yields the point $(\frac{5}{3}, 0)$.

Step 2: Find the *y*-intercept by letting $x = 0$ in the equation and solving for *y*.

$$y = 3(0) - 5$$
$$y = -5$$

This yields the point $(0, -5)$

Step 3: Plot these two points from Steps 1 and 2 and connect them by a straight line.

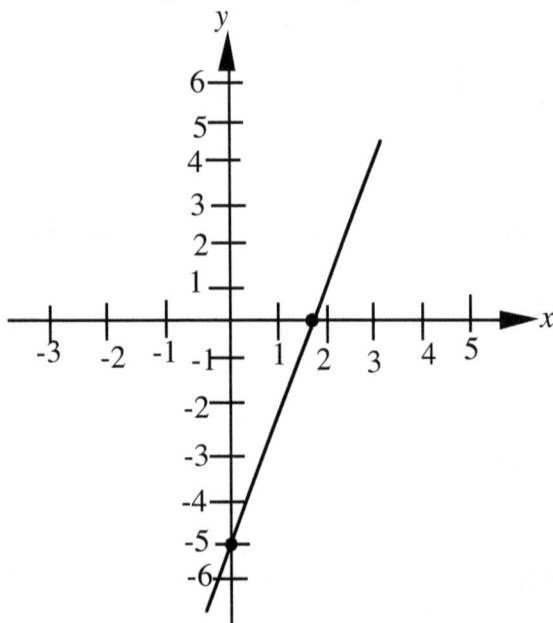

Figure: Graph of $y = 3x - 5$

Note that in Method 2, if the *y*-intercept is 0 (i.e., the equation is of the form $y = mx$), then we need to obtain another point by choosing another *x*-value (other than 0) , calculate the corresponding *y*-value to obtain an ordered pair which we plot to locate a second point.

An example of such a case is given by the line $y = 3x$. Note also that if the *y*-intercept is 0, the line passes through the origin.

Method 3: Slope-Intercept Method   89

Here also, we need two points to plot and connect by a straight line.
The y-intercept locates one point, and we use the slope to locate a second point.

Step 1: Solve the equation for y, and read the y-intercept.
  Since the equation has already been solved for y, we read the intercept, -5.
  This step locates the point $(0, -5)$.

Step 2: Graph the point $(0, -5)$.

Step 3: Since the slope $= +3 = \frac{3}{1}$, the vertical change is $+3$ and the horizontal change is $+1$.

  From the point $(0, -5)$, the y-intercept, count 3 units vertically upwards and 1 unit horizontally to the right, stop and place a dot here; this is a second point. Connect the two points by a straight line.

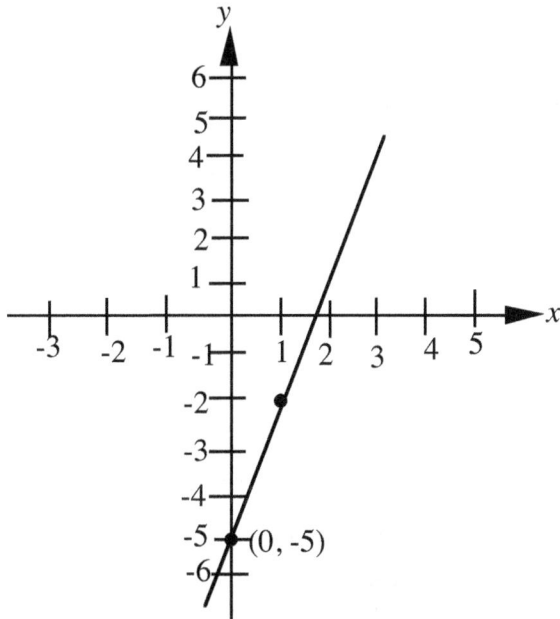

Figure: Graph of $y = 3x - 5$

Note: In Step 3, we adopt the following convention:

1. Write the slope as a fraction if it is not a fraction. and if the slope is negative, give the minus sign to the numerator.

   For example, a slope of -3 is expressed as $\frac{-3}{1}$; $-\frac{4}{5}$ is expressed as $\frac{-4}{5}$; $-\frac{1}{2}$ is expressed as $\frac{-1}{2}$

2. In counting from the y-intercept, if the numerator (change in y) is positive, we count vertically upwards, but if it is negative, we count vertically downwards.
   For the denominator (change in x) we always count to the right since we have agreed to give the minus sign to the numerator.

   Examples: For a slope of $\frac{-2}{3}$, starting from the y-intercept, we count 2 units vertically downwards (negative numerator) and 3 units horizontally to the right. For a slope of $\frac{-1}{2}$, starting from the y-intercept, we count 1 unit vertically downwards (negative numerator) and 2 units horizontally to the right. For a slope of $\frac{4}{5}$, starting from the y-intercept, we count 4 units upwards (positive numerator) and 5 units to the right.

Points on a Line   90

Determining if a point is on a line whose equation is given

Example Which of the points $(5,2)$ and $(3,11)$ is on the line whose equation is given by
$$y = 3x + 2.$$
We will substitute each ordered pair in turn in the given equation. The ordered pair whose substitution makes the left-hand side of the equation **equal** to the right-hand side **is on** the line.

Step 1:   To check for the point $(5,2)$, Substitute the ordered pair $(5, 2)$ in

$\quad\quad y = 3x + 2$ $\quad\quad\quad\quad (x = 5, y = 2)$

The question mark "?" above the equality symbol

$\quad\quad$ then, $2 \overset{?}{=} 3(5) + 2$ $\quad\quad$ is there to show that at this step, we do not yet know
if the left-hand side and the right-hand side of the

$\quad\quad\quad\quad 2 \overset{?}{=} 15 + 2$ $\quad\quad$ equation are equal. In checking solutions, it is a good
practice to use the question mark.

$\quad\quad\quad\quad 2 \overset{?}{=} 17$ No   (or $2 = 17$ is False)

The left-hand side is **not** equal to the right-hand side of the equation and therefore the point $(5,2)$ is **not** on the given line.
We also say (algebraically) that the ordered pair $(5,2)$ is **not** a solution of the
the equation $y = 3x + 2$

Step 2:   Checking for the point $(3,11)$

$\quad\quad\quad\quad$ Substitute $(3, 11)$ in $y = 3x + 2$

$\quad\quad\quad\quad\quad$ then, $11 \overset{?}{=} 3(3) + 2$

$\quad\quad\quad\quad\quad\quad\quad 11 \overset{?}{=} 9 + 2$

$\quad\quad\quad\quad\quad\quad 11 = 11$ True   (At this step we no longer use the
question mark since it is obvious
that the left-hand side equals the
right-hand side of the equation)

 Since the left-hand side **is equal to** the right-hand side of the equation, the point $(3,11)$ **is on** the line $y = 3x + 2$. Algebraically, we also say that the ordered pair $(3,11)$ **is a solution** of the equation $y = 3x + 2$. In fact, the last example (question) could have been posed as : Which of the following ordered pairs is a solution of the equation $y = 3x + 2$? (a) $(5,2)$, (b) $(3,11)$

Solution: Proceed exactly as steps 1 and 2 of example above.

In calculations, the following have the same interchangeable implications:

1. A line passes through a given point.

2. A line contains a given point.

3. A given point is on a line.

4. The coordinates of a given point satisfy an equation of a line.

5. The coordinate pair of a given point is a solution of an equation of a line.

For example, the problem "find an equation of the line **passing through** the points.
 $(2, 3)$ and $(6, 8)$" is **equivalent to** "find an equation of the line **containing** the points $(2, 3)$ and $(6, 8)$".

Lesson 13 Exercises

A Draw the graphs of the following lines whose equations are given:

    1. $y = -3x + 5$ **2.** $6y = 18 - 3x$ **3.** $2y - 6x = -3$

Answers: Graphs of the lines **1.** $y = -3x + 5$;   **2.** $6y = 18 - 3x$;   **3.** $2y - 6x = -3$

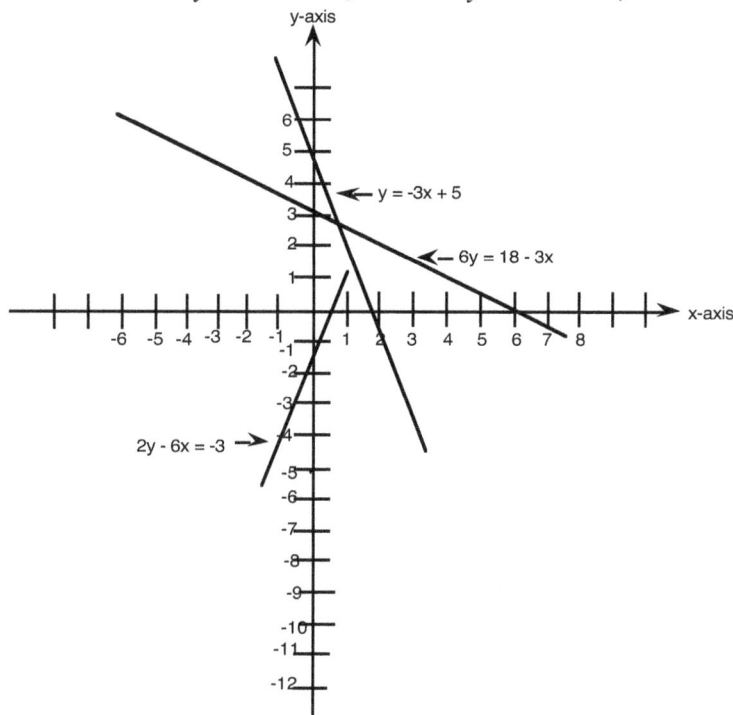

B

Determine which of the points (2,4), (-2,5), (-2,-16) are on the line whose equation is given by
$$y = 5x - 6$$

Answer: The points (2,4), and (-2,-16) are on the given line

Lesson 14
Intercepts and Slopes of Lines

The x- and y-intercepts of a line

The **x-intercept** of a line is the x-coordinate of the point where the line crosses or meets the x-axis. At this point, $y = 0$. (Geometrically, we should also say that the x-intercept is the point where the line intersects the x-axis, and in which case, we must specify two coordinates, with the y-coordinate always being zero.) See Figure 1 below.

The **y-intercept** of a line is the y-coordinate of the point where the line crosses or meets the y-axis. At this point, $x = 0$. (Geometrically, we should also say that the y-intercept is the point where the line intersects the y-axis, and in which case, we must specify two coordinates, with the x-coordinate always being zero.) See Figure 1 below.

Example
In Figure 1 below, the x-intercept is -3; and the y-intercept is 2.
The x- and y--intercepts are at (-3,0) and (0,2) respectively.

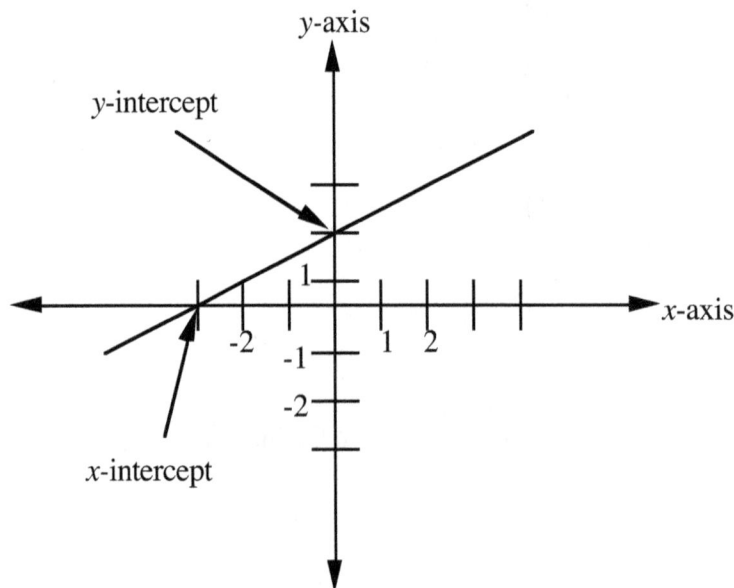

Figure 1

Example   Find the x- and y-intercepts for the graph of $3x + 2y = 12$

For the x-intercept: Let $y = 0$ in $3x + 2y = 12$.
Then $3x + 2(0) = 12$; $3x = 12$; and from which $x = 4$.
The x-intercept is 4.

For the y-intercept: Let $x = 0$ in $3x + 2y = 12$.
Then $3(0) + 2y = 12$; $2y = 12$; and from which $y = 6$.
The y-intercept is 6.

Finding the Slope of a Straight Line

Given the coordinates of two points on the line

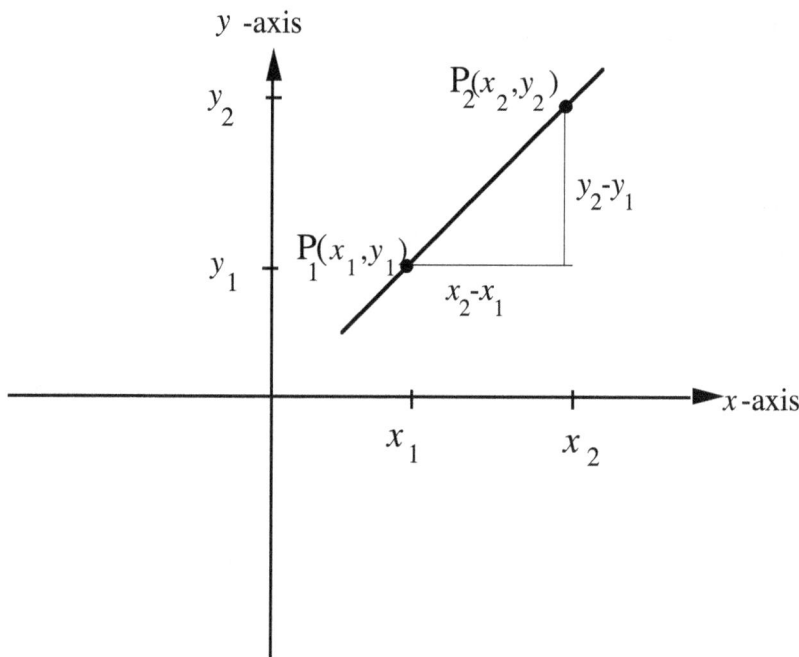

The slope, m, of the line segment connecting the points $P_1(x_1, y_1)$ and $P_2(x_2, y_2)$ is given by

$$m = \frac{y_2 - y_1}{x_2 - x_1} \qquad (= \text{the ratio: } \frac{\text{change on } y}{\text{change in } x} = \frac{\text{vertical change}}{\text{horizontal change}})$$

Example Find the slope of the line passing through the points (2,3) and (6,8).

Solution : Identify the first point as $P_1(2,3)$ and the second point as $P_2(6,8)$

Then $x_1 = 2, y_1 = 3;$ and $x_2 = 6, y_2 = 8$

Applying the slope formula, $m = \dfrac{y_2 - y_1}{x_2 - x_1}$

$$m = \frac{8 - 3}{6 - 2}$$

$$m = \frac{5}{4}$$

The slope is $\dfrac{5}{4}$.

Special Cases of the Slopes of Lines
The special cases are for horizontal and vertical lines

Slope of a Horizontal Line

The **slope** of a **horizontal** line is **zero**, since the vertical change is zero. For example, the slope, m, of the horizontal line in Figure **1**, below, by the slope formula is $m = \frac{3-3}{5-2} = \frac{0}{3} = 0$.

$\left(\text{Note that } m = \frac{\text{vertical change}}{\text{horizontal change}} = \frac{\text{change in } y}{\text{change in } x}\right)$

Slope of a Vertical Line

The **slope** of a **vertical** line is **undefined** since the horizontal change is zero. For example, the slope, m, of the vertical line in Figure **2**, below, by the slope formula is $m = \frac{2-(-3)}{4-4} = \frac{5}{0}$ is undefined.

Figure 1

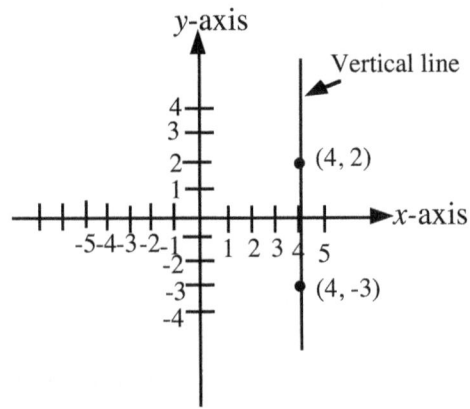

Figure 2

Finding the slope and the *y*-intercept of a line, given the equation of the line

Example 1   Find the slope of the line whose equation is given by $y = -5x + 6$

The slope-intercept form of the equation of a straight line is given by $y = mx + b$, where m is the slope and b is the *y*-intercept.

Compare $y = -5x + 6$ to   **Note:** (The slope is the coefficient of the *x*- term
$\quad\quad y = mx + b$,   if the equation is in slope-intercept form.)

then, the slope, $m = -5$,
and the *y*-intercept $= 6$

Example 2   Find the slope and the y-intercept of the line whose equation is given by
$$5y + 4x = 20$$

Step 1: Solve the equation for y (that is, write the equation in slope-intercept form).
$$5y + 4x = 20$$
$$\underline{\quad -4x \quad\quad\quad -4x}$$
$$5y = -4x + 20$$
$$\frac{5}{5}y = \frac{-4}{5}x + \frac{20}{5}$$
$$y = -\frac{4}{5}x + 4$$

Step 2: By comparison with $y = mx + b$, the slope is $-\frac{4}{5}$ (the coefficient of the *x*-term).
and the *y*-intercept is +4 or 4.

Lesson 14 Exercises 9 5

A Find the slope of each line passing through the given points:

1. (2, 3) and (6, 4)   **2.** (3, -2) and (2, -4)   **3.** (5, 2) and (1, -1)

4. (2, 2) and (-3, 3)   **5.** (-4, 2) and (5, 2)   **6.** (-3, -5) and (-3, -6)

7. Find c so that the slope of the line passing through the points $(c, 2c)$ and $(-6, 8)$ is 1.

Answers: **1.** $\frac{1}{4}$;   **2.** 2;   **3.** $\frac{3}{4}$;   **4.** $-\frac{1}{5}$;   **5.** 0;   **6.** undefined;   **7.** $c = 14$

B **1.** Find the slope of the line passing through the points (3,2) and (5,9).

2. The points (2,2) and (-5,6) are on a line whose slope we want to determine. What is the slope of this line?

3. Find the slope of the line containing the points (1,6) and (2,8).

4. Find the slope of the line passing through the points (1,-1), (3,3) and (-2,-7).

Answers: **1.** slope $= \frac{7}{2}$ **2.** slope $= -\frac{4}{7}$ **3.** slope $= 2$;   **4.** slope $= 2$

C Find the slope and the y-intercept of the following:

1. $y = -6x + 4$;   **2.** $-7 + 5x = y$;   **3.** $y = -\frac{x}{2} + 8$

Answers: **1.** slope $= -6$, y-intercept $= 4$; **2.** slope $= 5$, y-intercept $= -7$; **3.** slope $= -\frac{1}{2}$, y-intercept $= 8$

D Find the slope and y-intercept: **1.** $3x + 8y = 32$ **2.** $2y - 6x = 21$

3. $y = -3x + 4$ **4.** $y = \frac{1}{2}x - 3$ **5.** $3y = 6x + 12$

6. $y = \frac{x}{3} + 5$ **7.** $2y + 5x = 8$ **8.** $4x + 3y = 12$

1. slope $= -\frac{3}{8}$, y-intercept $= 4$;   **2.** slope $= 3$, y-intercept $= \frac{21}{2}$;   **3.** slope $= -3$; y-intercept $= 4$;

4. slope $= \frac{1}{2}$, y-intercept $= -3$;   **5.** slope $= 2$, y-intercept $= 4$;   **6.** slope $= \frac{1}{3}$, y-intercept $= 5$;

7. slope $= -\frac{5}{2}$, y-intercept $= 4$;   **8.** slope $= -\frac{4}{3}$, y-intercept $= 4$;

Lesson 15

Finding Equations of Straight Lines

There are a number of approaches that we can use in finding equations of straight lines. Each approach depends upon what we are given.

Generally, we can easily write down an equation of a line if we know any of the following pairs of properties or characteristics (information about) of the line.

The formulas presented below are based on finding two different expressions for the slope of a line and equating these expressions to each other.

1. If we know the **slope, m**, and the **y-intercept, b,** of the line,
    we can apply the slope-intercept form, $y = mx + b$.

2. If we know the slope, m, and the coordinates (x_1, y_1) of one point on the line,
    we can apply the point-slope form, $(y - y_1) = m(x - x_1)$.

3. If we know the coordinates (x_1, y_1) and (x_2, y_2) of two points on the line,

    we can apply $(y - y_1) = \dfrac{y_2 - y_1}{x_2 - x_1}(x - x_1)$.

4. If we are given the graph (picture), we can determine any of the above pairs of
    properties (information) from the graph, and then apply the formulas in the above cases;
    however, if we want to memorize one more formula, we can apply the equation

    $y = -\dfrac{b}{a}x + b$, where a is the x-intercept and b is the y-intercept. (If $b = 0$ or $a = 0$ use the other methods.)

    (This equation is another form of the two-intercept form: $\dfrac{x}{a} + \dfrac{y}{b} = 1$)

We will now cover the above **four** cases in detail with examples.

Case 1: Equation of a line given the slope and the y-intercept of the line

Example 1   Find an equation of the line with slope 3 and y-intercept of 4.

Solution   We will apply the slope-intercept form of the equation
            of a straight line, $y = mx + b$.
            Substituting the slope, $m = 3$, and the y-intercept, b $= 4$ in this equation,
                    $y = 3x + 4$

Example 2   Find an equation of the line with slope - 2 and y-intercept of - 5.

Solution   Substituting $m = -2, b = -5$ in $y = mx + b$, we obtain
            the equation $y = -2x - 5$

Case 2: **Equation of a line given the slope and the coordinates of a point on the line**

Example 1   Find an equation of the line passing through the point   (3,-2) and having a slope 4.
Solution. We will cover two methods.

Method 1

Step 1: Find the y-intercept, b, by substituting $x_1 = 3$, $y_1 = -2$, $m = 4$ in

$$y = mx + b$$
$$\text{then, } -2 = 4(3) + b$$
$$-2 = 12 + b$$
$$-14 = b$$

Step 2: Now, since we know that $m = 4$, $b = -14$, we can apply $y = mx + b$ (as in Case 1, above)
and then, $y = 4x - 14$

Method 2

We will use the point-slope form of the equation of a straight line which is
given by

$$y - y_1 = m(x - x_1) \text{ <------point-slope form.} \tag{1}$$

Substituting the coordinates, $x_1 = 3$, $y_1 = -2$ and slope, $m = 4$ in
equation (1), we obtain

$$y - (-2) = 4(x - 3)$$

$$y + 2 = 4(x - 3) \tag{2}$$

Equation (2) is the point-slope form of the required equation.

By solving equation (2) for y, we obtain

$$y = 4x - 14 \quad \text{<----------- slope-intercept form} \tag{3}$$

Equation (3) is the slope-intercept form of the required equation.
In this form, we can, by inspection, determine the slope and the y-intercept.

The author recommends the slope-intercept form (for this course), unless otherwise specified.

Case 3: Equation of a line given the coordinates of two points on the line

Example   Find an equation of the line passing through the points (2, 1) and (-3, -4).

Step 1: Find the slope, m, with $x_1 = 2$, $y_1 = 1$, $x_2 = -3$ $y_2 = -4$

$$m = \frac{y_2 - y_1}{x_2 - x_1}$$

Scrapwork

$$m = \frac{-4 - 1}{-3 - 2}$$

$$\frac{-4 - 1}{-3 - 2} = \frac{-5}{-5} = 1$$

$$m = 1$$

Now, $m = 1$, $x_1 = 2$, $y_1 = 1$, $x_2 = -3$, $y_2 = -4$, and we can apply the procedure in Case 2.

Step 2: Applying Method 2 of Case 2 and
Substituting $m = 1$, $x_1 = 2$, $y_1 = 1$ in $y - y_1 = m(x - x_1)$, we obtain
$$y - 1 = 1(x - 2)$$
$$y = x - 1$$

Also, If we substitute $m = 1, x_2 = -3, y_2 = -4$ in $y - y_2 = m(x - x_2)$, we obtain

$$y - (-4) = 1(x - (-3))$$

$$y + 4 = x + 3$$
$$y = x - 1$$

Again, we obtain the same equation. An equation of the line is $y = x - 1$.
We conclude also that any of the two given points can be used in finding an equation of the straight line.
From the above solution, we can state a general formula for Case 3 as

$$(y - y_1) = \frac{y_2 - y_1}{x_2 - x_1}(x - x_1)$$

Case 4: Equation of a line given the graph (picture) of the line

If we are given the graph (picture), we can determine any of the above pairs of
properties (information) from the graph (see p.96, case 4).

It is also useful to be able to tell immediately from the graph if the line has a positive slope, a negative
slope, a zero slope, or an undefined slope.

The signs of the slopes of lines

The lines in Figure 1 have positive slopes.　　The lines in Figure 2 have negative slopes.

Figure 1

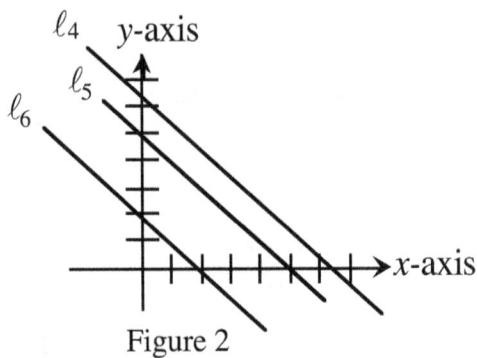

Figure 2

Lines ℓ_1, ℓ_2, and ℓ_3 have positive slopes.
(Simply, these lines **lean** to the **right** in the
page; or these lines rise as one moves ones
head from the left to the right in the page.)

Lines ℓ_4, ℓ_5 and ℓ_6 have negative slopes.
(Simply, these lines **lean** to the **left** in the page;
or these lines fall as one moves ones head from
the left to the right in the page.)

Note: The slope of a **horizontal line** is zero. The slope of a **vertical line** is undefined.
(For the equations of horizontal and vertical lines, see page 100-102)

Example 1   Find an equation of the line whose graph is given below.

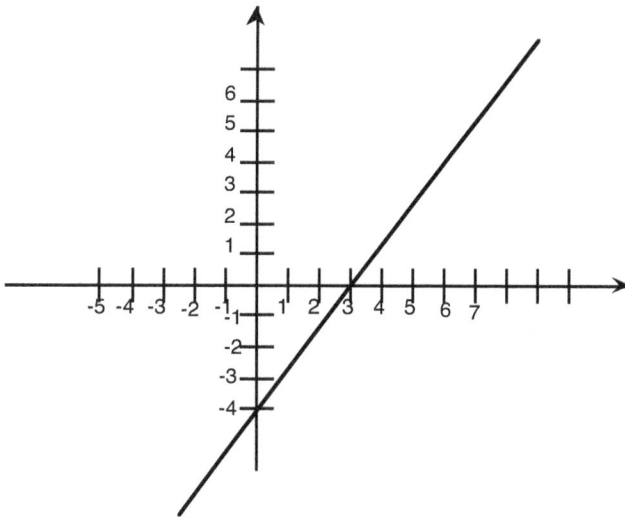

Solution: Apply $y = -\dfrac{b}{a}x + b,$ where a is the x-intercept and b is the y-intercept. (see also p.92)

Note that $m = -\dfrac{b}{a}$, where m is the slope. Note also that a and b may be positive or negative.

Step 1: From the graph, we read the values of a and b.
  $a = 3, b = -4$

Step 2: Substitute $a = 3, b = -4$ in $\;\; y = -\dfrac{b}{a}x + b,$

  Then $y = -\dfrac{-4}{3}x + (-4)$ (Make sure you take into account the **minus sign** that comes with the formula.)

  $y = +\dfrac{4}{3}x - 4$ (Two minus signs make the x-term positive)

  $y = \dfrac{4}{3}x - 4.$

Example 2   Find an equation of the line whose graph is given below.

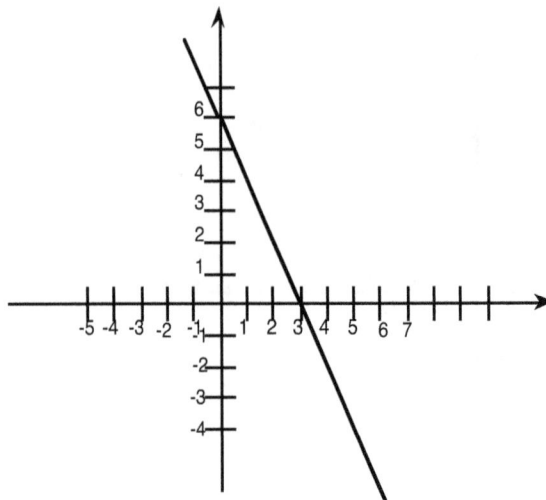

Solution: Apply $y = -\dfrac{b}{a}x + b$ where a is the x-intercept and b is the y-intercept.

Step 1: From the graph, $a = 3, b = 6$

Step 2: Substitute $a = 3, b = 6$ in $y = -\dfrac{b}{a}x + b$

Then $y = -\dfrac{6}{3}x + 6$

$y = -2x + 6$ < --------slope-intercept form of the
equation of a straight line.

Special Cases of the Equations of Straight Lines and their Graphs

The equation $y = mx + b$, (with m defined and, $m \neq 0$) geometrically represents oblique lines
(i.e., lines which are neither vertical nor horizontal). The special cases of the equation of a straight line
are for horizontal and vertical lines. In these cases, either the x- or the y-term is missing.

Equation and Graph of a Horizontal Line

The slope, $m = 0$, since the vertical change is zero.

Substituting $m = 0$, in $y = mx + b$,

$y = 0x + b$

$y = b$ (1)

Equation (1) means that as x varies, y remains unchanged.

Note that an equation of a horizontal line is of the form $y = b$,. where b is the y –intercept .

Note also that an equation of a horizontal line represents a function (a constant function). since the graph
passes the vertical line test.

Examples: Sketch the graphs of the following lines:
  1. $y = 3$,   **2.** $y = -4$.   **3.** $y = 0$.

Solution: The line $y = 3$ is the horizontal line passing through $(0,3)$. See Figure 1, below.

The line $y = -4$ is the horizontal line passing through $(0,-4)$

The line $y = 0$ is the horizontal line along the x-axis (i.e., the line $y = 0$ is the x-axis)

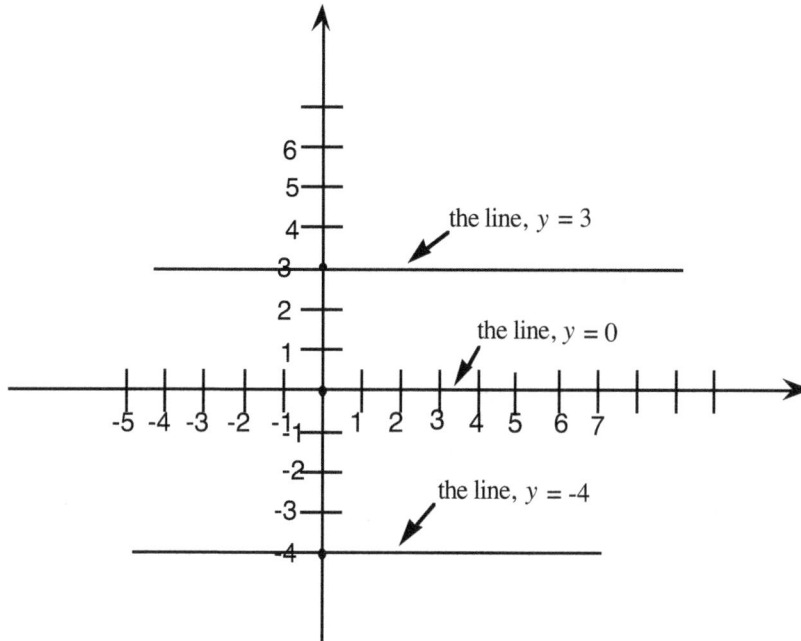

the line, $y = 3$

the line, $y = 0$

the line, $y = -4$

Figure 1

Equation and Graph of a Vertical Line

The slope of a vertical line is undefined since the horizontal change is zero.
The equation of a vertical line is of the form $x = a$, where a is the x–intercept .This form of the equation means that as y varies, x remains unchanged. Below, we sketch the graphs of some vertical lines.
Note also that an equation of a vertical line does **not** represent a function (see p. 50)

Examples: Sketch the lines with the following equations:

$$1. \, x = 2, \qquad 2. \, x = 0, \qquad 3. \, x = -4$$

Solution: See Figure 2, below.

1. The line $x = 2$ is the vertical line passing the point $(2, 0)$.

2. The line $x = 0$ is the vertical line along the y-axis (i.e. , the line $x = 0$ is the y-axis).

3. The line $x = -4$ is the vertical line passing through the point $(-4, 0)$.

4. The line $x = -2$ is the vertical line passing through the point $(-2, 0)$.

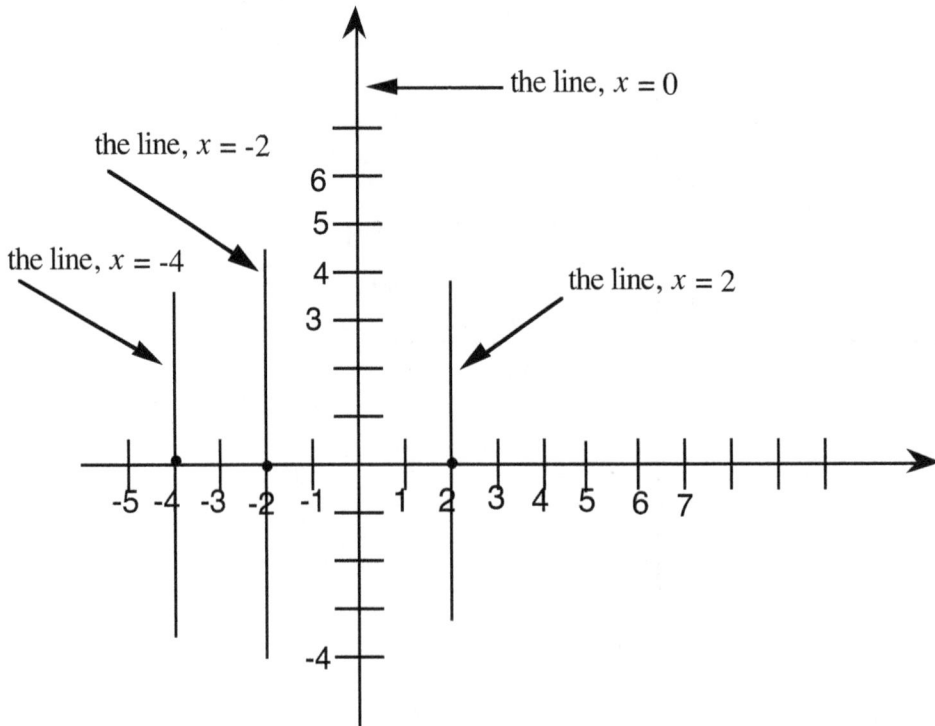

Figure 2

Lesson 15 Exercises

A .**1.** Find an equation of the line with slope 7 and *y*-intercept -2.

2. Find an equation of the line with slope $\frac{2}{3}$ and *y*-intercept 5.

3. A line has a *y*-intercept -3 and a slope of 6. Find an equation for this line.

Answers: 1. $y = 7x - 2$;   **2.** $y = \frac{2}{3}x + 5$;   **3**. $y = 6x - 3$

B **1.** Find an equation of the line with slope -4 and passing through the point (3,-2).

2. A line passes through the point (-1,-7) and has a slope of 5. Find an equation for this line.

Answers: **1.** $y = -4x + 10$;   **2.** $y = 5x - 2$

C **1.** Find an equation of the line passing through the points (2,2) and (-5, 6).

2. If the points (2, -5) and (-3, 1) are on a certain line, find an equation for this line.

3. Find an equation of the line with slope -3 and y-intercept 8.

4. Find an equation of the line with slope 2 and passing through the point (1, 6)

Answers: **1.** $y = -\frac{4}{7}x + \frac{22}{7}$;   **2.** $y = -\frac{6}{5}x - \frac{13}{5}$;   **3.** $y = -3x + 8$; **4.** $y = 2x + 4$;

D **1.** Find an equation of the horizontal line passing through the point (3,-4).

2. Sketch the graph the line in Problem 1.

3. Does the line in Problem 1 have a *y*-intercept ? If yes, find it.

4. Sketch the graph of the line $y = -5$.

5. Sketch the graph of the line $y = 2$.

Answers: **1.** $y = -4$; **2.** See Fig 1, above and imitate. ; **3.** Yes. y-intercept = - 4 ; **4. & 5.** Imitate Fig. 1 above

E **1.** Find an equation of the vertical line passing through the point (2,-3).

2. Sketch the graph the line in Problem **1**.

3. Does the line in Problem **1** have a *y*-intercept ? If yes, find it.

Answers: **1.** $x = 2$; **2.** See Fig. 2 and imitate ;   **3.** No.

F **1.** Draw the graph of the line $x = 4$.;   **2.** Draw the graph of the line $x = -1$.

Hint: See Fig. 2 and imitate.

G **1.** Find an equation of the line with slope -5 and *y*-intercept 2.

2. Find an equation of the line whose *y*-intercept is -3 and whose slope is $\frac{1}{2}$.

3. Find an equation of the line whose slope is -2 and which passes through the point (4, -5).

4. A line passes through the point (-3, 2) and has slope 4. Find an equation for this line.

5. Find an equation of the line passing through the points (5, 4) and (8, 6).

Answers: **1.** $y = -5x + 2$; **2.** $y = \frac{1}{2}x - 3$;   **3.** $y = -2x + 3$;   **4.** $y = 4x + 14$;   **5.** $y = \frac{2}{3}x + \frac{2}{3}$

H Find the equations (slope-intercept forms) of the lines ℓ_1, ℓ_2, ℓ_3, ℓ_4

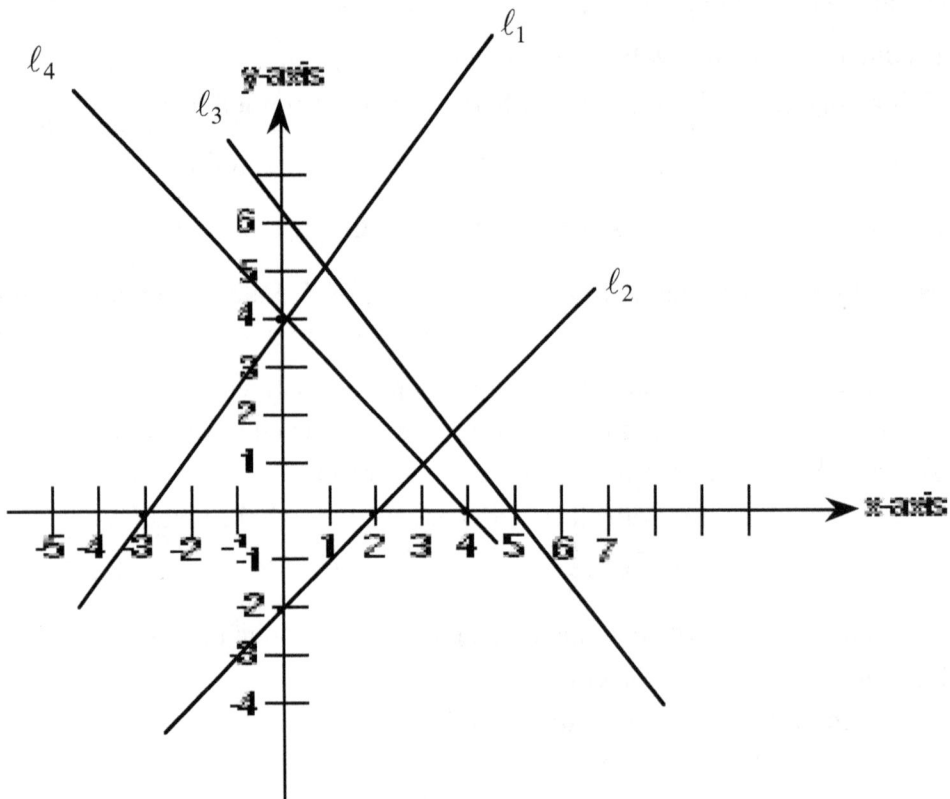

Answers: For ℓ_1: $y = \dfrac{4}{3}x + 4$. For ℓ_2: $y = x - 2$.   For ℓ_3: $y = -\dfrac{5}{4}x + \dfrac{25}{4}$ For ℓ_4: $y = -x + 4$

I   Sketch the graphs of the following straight lines:

1. $y = -3x - 2$;   **2.** $y = 5x + 1$;   **3.** $y = x - 5$;   **4.** $y = -\dfrac{1}{2}x + 3$;   **5.** $4y = x$;

6. $y - 1 = 2(x - 3)$;   **7.** $x - y = 1$;   **8.** $12(x + y) = 8$;   **9.** $y = x - 1$;   **10.** $y = \dfrac{1}{3} - 2$;

11. $y = 3$;   **12.** $y = -2$;   **13.** $x = -5$;   **14.** $x = 3$.

Answers:

$y = -3x - 2$

$y = 5x + 1$

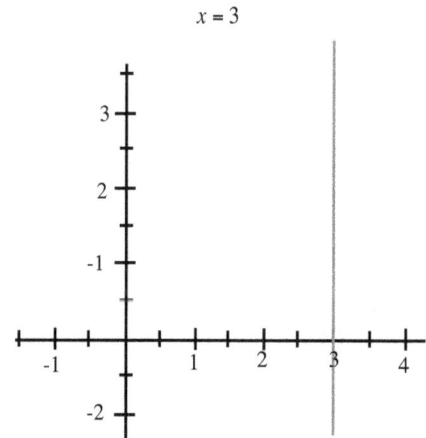

$y = x - 5$

$y = -\frac{1}{2}x + 3$

$y = \frac{1}{4}x$

$y - 1 = 2(x - 3)$

$y = 3$

$y = -2$

$x = 3$

Lesson 16

Parallel lines and Perpendicular Lines : Slopes and Equations

Parallel Lines: Slopes and Equations

If two lines are parallel, then they have the **same slope**. The following are equations of two parallel lines:

  (a) $y = 6x + 5$ <----The slope of this line is 6.

  (b) $y = 6x - 4$ <----The slope of this line is 6.

Example 1 Find an equation of the line passing through the point (2, -5), and parallel to the line $y = -3x + 2$.

Solution

The slope of the line whose equation is given = -3.
The slope of the line whose equation we want to find = -3 (since parallel lines have the same slope).
Applying $y - y_1 = m(x - x_1)$; with $m = -3, x_1 = 2, y_1 = -5$

$$y - (-5) = -3(x - 2)$$
$$y + 5 = -3x + 6$$
$$y = -3x + 1 \quad \text{<------slope-intercept form.}$$

Example 2 Find an equation of the line parallel to the line $2y + 1 = 12x$ and having the same y-intercept as the line $y + 5 = 2x$

Solution. We can readily write an equation of a line if we know its slope and its y-intercept.

Step 1: To find the slope, solve $2y + 1 = 12x$ for y.

Solving, $y = 6x - \dfrac{1}{2}$

The slope of this line is 6 and the slope of the required line is also 6 (since parallel lines have the same slope).

Step 2: Solve $y + 5 = 2x$ for y.

Solving, $y = 2x - 5$.

The y-intercept of this line is - 5 and the y-intercept of the required line is also - 5.
Therefore, the line whose equation we want to find has a slope of 6 and a y-intercept of - 5.

Step 3: Substitute $m = 6$ and $b = -5$ in $y = mx + b$.
Then $y = 6x - 5$, which is the required equation.

Perpendicular Lines

If two lines are perpendicular, then **their slopes** are negative reciprocals of each other; or symbolically, if m_1 and m_2 are their slopes, then $m_1 = -\dfrac{1}{m_2}$ or $m_2 = -\dfrac{1}{m_1}$ or $m_1 \cdot m_2 = -1$

Example   Find an equation of the line passing through the point (-2, 4) and perpendicular to the line $y = 3x - 5$.

Solution

Step 1: Determine the slope of the line whose equation we want to find.

The slope of $y = 3x - 5$ is 3

The slope of the line whose equation we are to find is $-\dfrac{1}{3}$ (Since the two lines are perpendicular).

Step 2: Apply $y - y_1 = m (x - x_1)$

with $m = -\dfrac{1}{3}$ and $x_1 = -2, y_1 = 4$

Then, $y - 4 = -\dfrac{1}{3} (x - (-2))$

$$y - 4 = -\frac{1}{3} (x + 2)$$

$$y - 4 = -\frac{1}{3}x - \frac{2}{3}$$

$$y = -\frac{1}{3}x - \frac{2}{3} + 4$$

$$y = -\frac{1}{3} x + \frac{10}{3} \quad \text{<---------slope-intercept form.}$$

A **note** about finding the slope in the above problem:

The slope of the given line is 3. To find the slope of the other (perpendicular line) line,

Step 1. Invert 3 to obtain $\dfrac{1}{3}$

Step 2. Change the sign. Since $\dfrac{1}{3}$ has a plus sign, after the change, we obtain $-\dfrac{1}{3}$

Similarly, if the slope of one line were -5, the slope of the other perpendicular line would be $+\dfrac{1}{5}$.

If the slope of one line were $-\dfrac{1}{4}$, the slope of the other perpendicular line would be +4.

Summary: Invert and change the sign, or change the sign and invert.

Distinction between the point $x = a$ and the line $x = a$ 108

We must distinguish between, for example, the point $x = 2$ and the line $x = 2$. We use a dot to locate the point $x = 2$ on a number line.

For the graph of the line $x = 2$, we draw a vertical line through the point $(2, 0)$

Figure: The graph of the point $x = 2$.

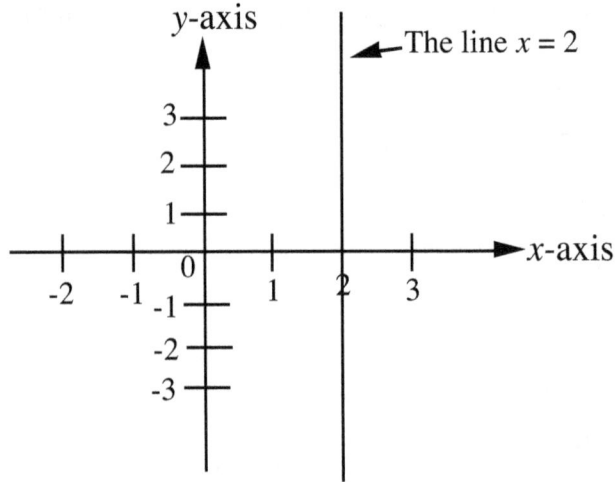

Figure: The graph of the line $x = 2$.

Similarly, the point $y = 2$ and the line (horizontal line) $y = 2$ are shown in Figures .. and respectively.

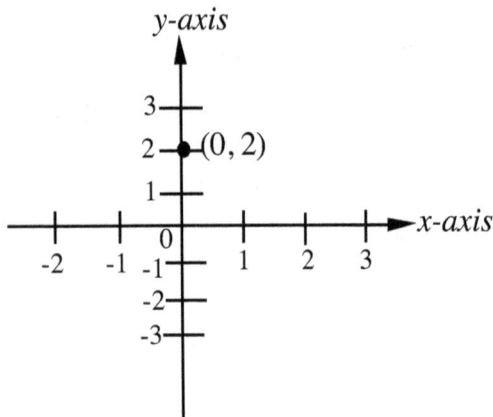

Figure: The graph of the point $y = 2$.

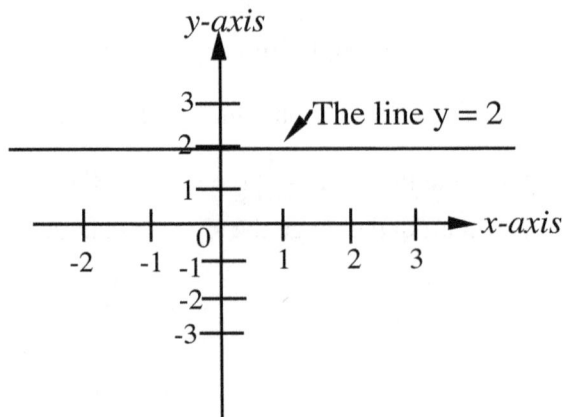

Figure: The graph of the line $y = 2$.

Lesson 16 Exercises   109

A 1. Find an equation of the line through the point $(3,-4)$ and perpendicular to the line $y = 2x + 5$.

2. Find an equation of the line perpendicular to the line $2y + 3x = 12$ and passing through $(4,1)$.

3. Find an equation of the line perpendicular to the line $x - y = 7$ and passing through $(1,-2)$.

4. Find an equation of the line having y-intercept -5 and perpendicular to $3y + 1 = 6x$.

Answers: **1.** $y = -\frac{1}{2}x - \frac{5}{2}$; **2.** $y = \frac{2}{3}x - \frac{5}{3}$; **3.** $y = -x - 1$; **4.** $y = -\frac{1}{2}x - 5$

B 1. Find an equation of the line with the same y–intercept as the line $2y = 3x + 8$ and parallel to the line $y = -4x + 2$.

2. Find an equation of the line passing through the point $(2,3)$ and parallel to the line $y = 4x + 5$.

3. Find an equation of the line parallel to the line $3x + y = 12$ and passing through $(-1,-3)$

4. Find an equation of the line parallel to the line $2y + 6 = 10x$ and passing through $(-3, 4)$.

Answers: **1.** $y = -4x + 4$; **2.** $y = 4x - 5$; **3.** $y = -3x - 6$; **4.** $y = 5x + 19$;

C 1. **By** graphing, distinguish between the point $x = 2$ on the number line and the line $x = 2$.

2. By graphing, distinguish between the point $y = 2$ on the y-axis and the line $y = 2$.

1.

2.

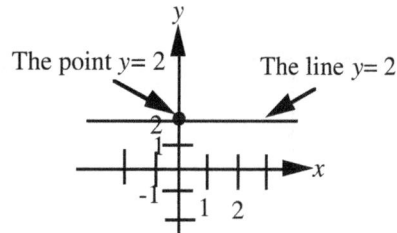

Lesson 17

Midpoint of a Line; Distance between Points; Distance Formula

Midpoint of a Line

The **midpoint** of (Figure below), the line, P_1P_2, connecting the points $P_1(x_1, y_1)$ and $P_2(x_2, y_2)$ has the coordinates given by the following formulas:

The x-coordinate, x_m, of the midpoint is given by $x_m = \dfrac{x_1 + x_2}{2}$

The y-coordinate, y_m of the midpoint is given by $y_m = \dfrac{y_1 + y_2}{2}$

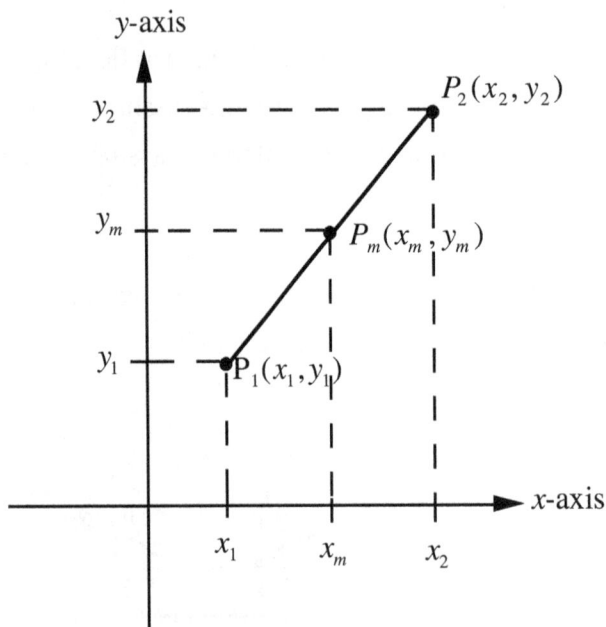

Example Find the coordinates of the mid-point of the line connecting the points $(3, 2)$ and $(-9, 4)$.

Solution

$$\text{The } x\text{-coordinate of the mid-point} = \frac{x_1 + x_2}{2} = \frac{3 + (-9)}{2} \qquad (x_1 = 3, x_2 = -9)$$

$$= \frac{-6}{2}$$

$$= -3$$

$$\text{The } y\text{-coordinate of the mid-point} = \frac{y_1 + y_2}{2} = \frac{2 + 4}{2} \qquad (y_1 = 2, y_2 = 4)$$

$$= \frac{6}{2}$$

$$= 3$$

Therefore $(x_m, y_m) = (-3, 3)$ where x_m and y_m are the coordinates of the mid-point.
The mid-point is at $(-3, 3)$.

Distance Between two Points on a Line

The distance between two points A and B on a line is equal to the number of units (equal intervals) between A and B. If we denote the distance between A and B by $d(A, B)$. Then, we can write

$$d(A, B) = |A - B|$$

that is, the distance between A and B is the absolute value of the difference between A and B.

Figure: Distance on a horizontal line

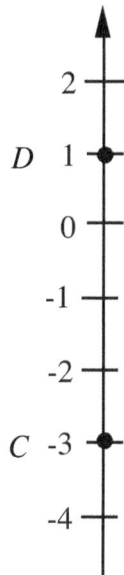

Figure: Distance on a vertical line

Example

$$d(\text{-}3, 1) = |-3 - 1|$$
$$= |-4|$$
$$= 4$$

We conclude also that the distance between two points on the same **horizontal line** or the same **vertical line** can be found algebraically.

However, the distance between two points in a plane cannot be found algebraically, but must be found geometrically as discussed in the next section.

Distance Between two Points on a Line in a Plane: Distance Formula

The distance, d, between the points $P_1(x_1, y_1)$ and $P_2(x_2, y_2)$ on a line **in a plane** (Fig.1) is given by

$d = \sqrt{(x_2 - x_1)^2 + (y_2 - y_1)^2}$ (By applying the Pythagorean theorem to the right triangle, and solving for d)

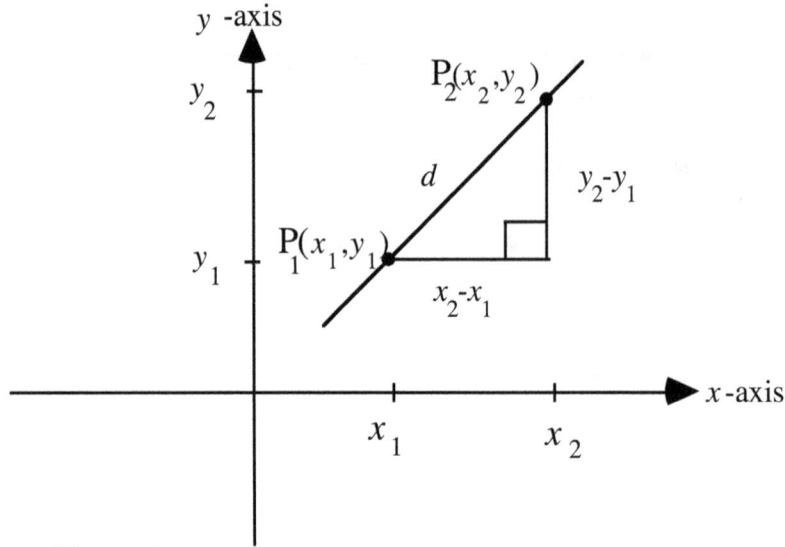

Figure 1

Example Find the distance between the points (-2,3) and (4, -5) .

Solution Apply the distance formula:

$$d = \sqrt{(x_2 - x_1)^2 + (y_2 - y_1)^2} \qquad (1)$$

(where d is the distance between the points $P_1(x_1, y_1)$ and $P_2(x_2, y_2)$).

Substituting $x_1 = -2, y_1 = 3, x_2 = 4, y_2 = -5$ in equation (1) above,

$$d = \sqrt{(4 - (-2))^2 + (-5 - 3)^2}$$
$$d = \sqrt{(4 + 2)^2 + (-8)^2}$$
$$= \sqrt{(6)^2 + (-8)^2}$$
$$= \sqrt{36 + 64}$$
$$= \sqrt{100}$$
$$= 10$$

$\therefore$ the distance between the given points is 10 units.

Application 1: Perpendicular distance from a point to a line, whose equation is given 1 1 3

Find a formula for the distance, d, from the point $A(x_0, y_0)$ to the line whose equation is given by $y = mx + b$.

Solution

Method 1

Given: The point $A(x_0, y_0)$ and the line $y = mx + b$.

Required: To find a formula for the distance, d, between the point $A(x_0, y_0)$ and the line $y = mx + b$.

Construction: Draw the perpendicular line from $A(x_0, y_0)$ to meet the line $y = mx + b$ (line ℓ_2) at E.

Also, draw $\overline{AC} \perp$ to the x-axis.

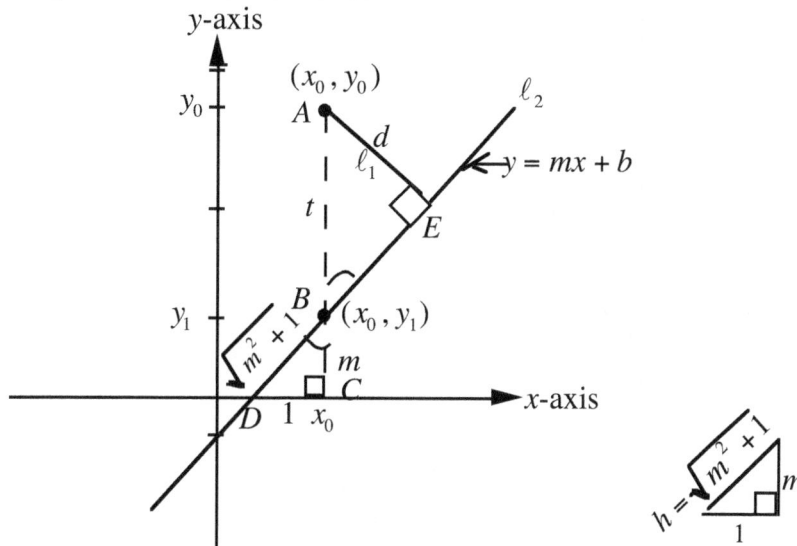

Step 1: Let $h = DB$. If the slope of $y = mx + b$ is $m = \frac{m}{1}$, then $BC = m$, $DC = 1$, and $h = \sqrt{m^2 + 1}$.

$\triangle ABE$ and $\triangle DBC$ are similar: (Two angles of $\triangle ABE$ are congruent to two angles of $\triangle DBC$)

$\therefore \dfrac{AE}{DC} = \dfrac{AB}{DB}$ (Corresponding sides of similar triangles are in proportion)

$\dfrac{d}{1} = \dfrac{t}{\sqrt{m^2 + 1}}$ (1)   ($AE = d$, $AB = t$, $DC = 1$, $DB = \sqrt{m^2 + 1}$)

Step 2:

$t = y_0 - y_1$

$t = y_0 - (mx_0 + b)$ ($y_1 = mx_0 + b$ is obtained by substituting x_0 in $y = mx + b$)

$t = y_0 - mx_0 - b$

Step 3: Substitute for t in equation (1)

$\dfrac{d}{1} = \dfrac{y_0 - mx_0 - b}{\sqrt{1 + m^2}}$

$d = \dfrac{y_0 - mx_0 - b}{\sqrt{1 + m^2}}$

$\therefore d = \dfrac{|y_0 - mx_0 - b|}{\sqrt{m^2 + 1}}$. (We take the absolute value of the numerator in case the numerator is negative)

Method 2

Given: The point $A(x_0, y_0)$ and the line $y = mx + b$.

Required: To find a formula for the distance, d, between the point $A(x_0, y_0)$ and the line $y = mx + b$.

Construction: Draw the perpendicular, l_1, from the point $A(x_0, y_0)$ to meet the line $y = mx + b$, l_2 at E

.Step 1: Let lines l_1, and ℓ_2 meet at (r, s). (Figure below)

Let the distance between the points (x_0, y_0) and $(r, s) = d$.

Then $d^2 = (x_0 - r)^2 + (y_0 - r)^2$ or $d = \sqrt{(x_0 - r)^2 + (y_0 - r)^2}$

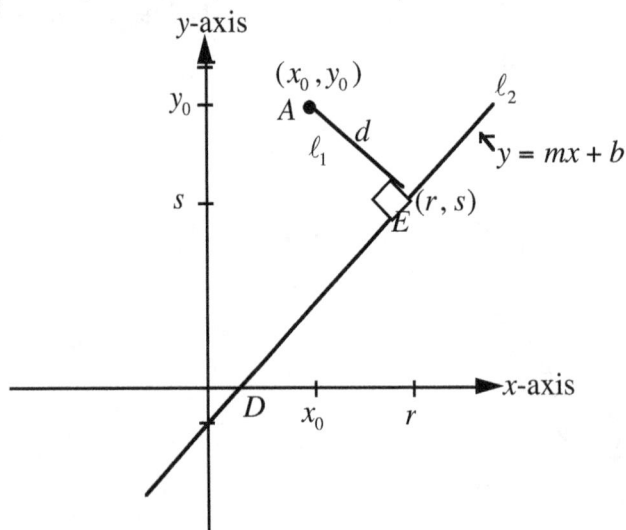

Step 2: We shall now express d in terms of x_0, y_0, m, and b only.

(That is we shall eliminate r, and s from the formula)

The slope of line $l_1 = \dfrac{y_0 - s}{x_0 - r}$. $\qquad$ (1)

The slope of line $l_2 = m$. (from $y = mx + b$)

Therefore slope of $l_1 = -\dfrac{1}{m}$ $\qquad$ (2) $\qquad$ (Since l_1 and l_2 meet at right angles at E, the

slopes are negative reciprocals of each other)

Equating right-hand-sides of (1) and (2) to each other, we obtain

$\dfrac{y_0 - s}{x_0 - r} = -\dfrac{1}{m}$

$my_0 - ms = -x_0 + r$ (cross-multiplying)

$r = my_0 + x_0 - ms$ $\qquad$ (3) $\qquad$ (solving for r)

Since (r, s) is on the line $y = mx + b$

$s = mr + b$ $\qquad$ (4) $\qquad$ (substituting r for x and s for y.)

Step 3: We now solve equations (3) and (4) simultaneously for r and s.

Substitute for s from (4) in (3): $my_0 + x_0 - m(mr + b) = r$

Now, we solve for r: $my_0 + x_0 - m^2 r - mb = r$

$my_0 + x_0 - mb = m^2 r + r$

$my_0 + x_0 - mb = r(m^2 + 1)$

$$r = \frac{my_0 + x_0 - mb}{m^2 + 1} \qquad (5)$$

Substitute for r from (5) in (4). Then we obtain

$$s = m\left[\frac{my_0 + x_0 - mb}{m^2 + 1}\right] + b$$

$$= \frac{m^2 y_0 + mx_0 - m^2 b + m^2 b + b}{m^2 + 1}$$

$$s = \frac{m^2 y_0 + mx_0 + b}{m^2 + 1} \qquad (6)$$

Step 4: We now substitute right-hand sides of equations (5) and (6) for r and s respectively in the formula $d^2 = (x_0 - r)^2 + (y_0 - r)^2$. Then we obtain

$$d^2 = \left\{x_0 - \left[\frac{my_0 + x_0 - mb}{m^2 + 1}\right]\right\}^2 + \left\{y_0 - \left[\frac{m^2 y_0 + mx_0 + b}{m^2 + 1}\right]\right\}^2$$

$$= \frac{\{m^2 x_0 + x_0 - my_0 - x_0 + mb\}^2}{(m^2 + 1)^2} + \frac{\{m^2 y_0 + y_0 - m^2 y_0 - mx_0 - b\}^2}{(m^2 + 1)^2}$$

$$= \frac{\{m^2 x_0 - my_0 + mb\}^2}{(m^2 + 1)^2} + \frac{\{y_0 - mx_0 - b\}^2}{(m^2 + 1)^2}$$

$$= \frac{\{-m(-mx_0 + y_0 - b)\}^2}{(m^2 + 1)^2} + \frac{\{y_0 - mx_0 - b\}^2}{(m^2 + 1)^2}$$

$$= \frac{\{(-m)^2(-mx_0 + y_0 - b)\}^2}{(m^2 + 1)^2} + \frac{\{y_0 - mx_0 - b\}^2}{(m^2 + 1)^2}$$

$$= \frac{m^2(y_0 - mx_0 - b)^2}{(m^2 + 1)^2} + \frac{(y_0 - mx_0 - b)^2}{(m^2 + 1)^2}$$

$$= \frac{m^2(y_0 - mx_0 - b)^2 + (y_0 - mx_0 - b)^2}{(m^2 + 1)^2}$$

$$= \frac{(y_0 - mx_0 - b)^2(m^2 + 1)}{(m^2 + 1)^2}$$

$$= \frac{(y_0 - mx_0 - b)^2(m^2 + 1)}{(m^2 + 1)^2}$$

$$d^2 = \frac{(y_0 - mx_0 - b)^2}{(m^2 + 1)}$$

$$d = \sqrt{\frac{(y_0 - mx_0 - b)^2}{(m^2 + 1)}}$$

$$= \frac{\sqrt{(y_0 - mx_0 - b)^2}}{\sqrt{m^2 + 1}}$$

$$\therefore d = \frac{|y_0 - mx_0 - b|}{\sqrt{m^2 + 1}} \qquad \text{(noting that } \sqrt{a^2} = |a|)$$

and the derivation is complete.

Application 2: Perpendicular Bisector of a Line

Example

Find the slope-intercept form of the equation of line L_2 (Figure below) which passes through the mid-point C of line L_1 given that line L_1 passes through the two points A (1, 4) and B (7, 8)

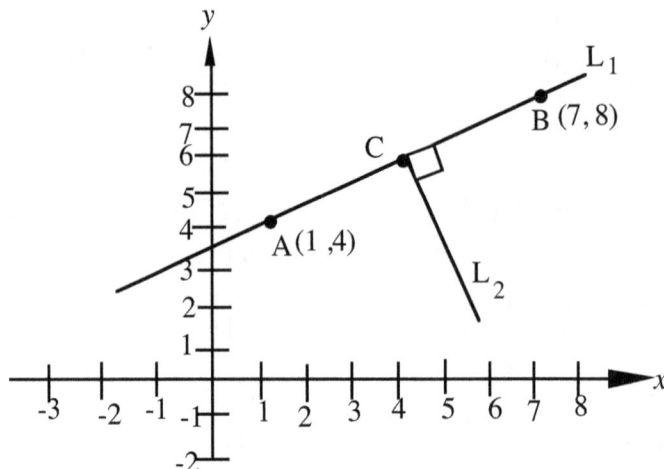

Solution

Step 1 Find the slope of line L_1

$$m = \frac{y_2 - y_1}{x_2 - x_1}$$

$$= \frac{8 - 4}{7 - 1}$$

$$= \frac{4}{6}$$

$$m = \frac{2}{3}$$

Step 2: Find the slope of L_2

Since L_1 is perpendicular to L_2, the slope of L_2 is $-\frac{3}{2}$ ($m_2 = -\frac{1}{m_1}$. (The lines are perpendicular)

Step 3 : Find the coordinates of the midpoint C

$$x_m = \frac{1+7}{2} = 4 \quad y_m = \frac{4+8}{2} = 6$$

The mid-point is at (4, 6)

Step 4: Now, find the slope-intercept form of the equation of the line L_2 with slope $-\frac{3}{2}$ and passing through the point (4, 6).

Applying $y - y_1 = m(x - x_1)$, we obtain

$y - 6 = -\frac{3}{2}(x - 4)$ <-------- point-slope form

$y - 6 = -\frac{3}{2}x + 6$

$y = -\frac{3}{2}x + 6 + 6$

$y = -\frac{3}{2}x + 12$ <-----slope-intercept form.

Lesson 17 Exercises

A   **1.** Find the distance between the points (3,4) and (5, -1) .

2. Find the distance between the points (-4, 2) and (-6, -3)

Answers: **1.** $\sqrt{29}$; **2.** $\sqrt{29}$

B Find the distance between each given pair of points.

1. (2, 3) and (−7, 4);   **2.** (−4, 3) and (6, 2);   **3.** (−3, 5) and (7, 10)

4. (a, b), and (c, d);   **5.** (−3, 5), and (4, 7)

6. (a) Find the directed distance from (4,−5) to (4,−7);

 (b) Find the distance from (4,−5) to (4,−7)

Find the coordinates of the mid–point of each of the following points :

7. (2,−4), and (4, 0);   **8.** (3, 2), and (4, 6).

9. (−4, − 4), and (5, 3);   **10.** (5, 12), and (0, 0).

Answers: **1.** $\sqrt{82}$; **2.** $\sqrt{101}$; **3.** $5\sqrt{5}$; **4.** $\sqrt{a^2 + b^2 + c^2 + d^2 - 2ac - 2bd}$; **5.** $\sqrt{53}$; **6.** (a) −2 ; (b) 2.
7. $(3, -2)$; **8.** $(\frac{7}{2}, 4)$; **9.** $(\frac{1}{2}, -\frac{1}{2})$; **10.** $(\frac{5}{2}, 6)$.

CHAPTER 8

Lesson 18
Variation
Direct, Inverse, Joint and Combined Variation

In mathematics, a change in a physical quantity, say x, may be accompanied by a corresponding change in another physical quantity, say y. Sometimes, there is a simple functional relationship between the two quantities involved. This type of relationship occurs very often in the arithmetic of everyday life. We will call this relationship **variation** (or proportion). Note that problems stated in terms of variation can be stated in terms of proportion and conversely. In this chapter, we will cover only problems stated in terms of variation. Basically, there are two main **types** of **variation**:

1. **Direct variation** or simply, **variation**.
2. **Inverse variation** (or indirect variation).

Direct variation

It is this type that you meet very often. In a direct **variation**, as one quantity increases, the other quantity also increases; or as one quantity decreases, the other quantity also decreases. When one quantity is zero, the other quantity is also zero. Uniform or constant changes in one quantity results in uniform changes in the other quantity.

Example
The cost of oranges we buy varies directly as (or is directly proportional to) the number of oranges we buy (the cost of an orange being constant). As we buy more oranges , we pay more money and as we buy less oranges we pay less money. (i.e., more oranges, more money; and less oranges, less money).

Inverse variation (or indirect variation)
In inverse variation, as one quantity increases, the other quantity decreases; or as one quantity decreases, the other quantity increases.

Example 1
The time taken by a number of people to do a piece of work varies **inversely as** (or is inversely proportional to) the number of people (assuming that each person works at the same rate as everyone else). Thus, as the number of people increases, the time taken to do the work decreases and as the number of people decreases, the time taken increases (i.e., more people, less time; and less people, more time).

Example 2
At constant temperature, the volume of a given mass of a gas is inversely proportional to the pressure on the gas. (This relationship is known as Boyle's law.)

We will cover direct variation, indirect (inverse) variation; joint and combined variation.

Direct Variation

There are a number of ways of explicitly specifying direct variation. If y is a function of x, then the following are equivalent to one another:

(a) y varies directly as x, or simply,
(b) y varies as x, (The word "directly" is omitted.)
(c) y is directly proportional to x, or simply
(d) y is proportional to x.
Symbolically, for each of the above statements we write
   $y = kx$, where k is a positive constant . We call k the constant of variation (or proportion)
The value of k depends on the particular problem under consideration.

Example 1: y varies directly as x. When $y = 48, x = 3$. Find x when $y = 96$

Solution

Step 1: $y = kx$ (1)

Step 2: Substitute $y = 40, x = 2$ in equation (1) and solve for k.

$48 = k(3)$

$48 = 3k$

$16 = k$

Substitute $k = 16$ in equation (1) above to obtain

the formula $y = 16x$ (2)

Step 3: Now, to find x when $y = 96$, replace y by 96 in (2)

Then $96 = 16x$

$6 = x$

Therefore when $y = 96, x = 6$.

Example 2 The distance y an object falls from rest is directly proportional to the square of the time x. When $y = 128, x = 4$. (*a*) Find x when $y = 200'$ (*b*) Find y when $x = 6$.

Solution

Step 1: $y = kx^2$ ....(1)

Step 2: Substitute $y = 128, x = 4$ in equation (1) and solve for k.

Then $128 = k(4)^2$

$128 = 16k$

$k = 8$

Substitute $k = 8$ in equation (1) above to obtain

the formula $y = 8x^2$ .... (2)

Step 3: (a) To find x when $y = 200$. Replace y by 200 in equation (2) and solve for x.

$200 = 8x^2$

$\dfrac{200}{8} = x^2$

$25 = x^2$

$5 = x$

$x = 5$ (Since the time taken must be positive in this problem, we reject -5 as a solution)

(b) To find y when $x = 6$, replace x by 6 in equation (2) and evaluate.

$y = 8(6)^2$

$y = 8(36)$

$y = 288$.

Therefore when $x = 6, y = 288$.

Inverse variation

If y is a function of x, then the following are equivalent to one another:

(a) y varies inversely as x, or simply,
(c) y is inversely or indirectly proportional to x.

Symbolically, for each of the above statements we write

$$y = \frac{k}{x}$$ where k is a positive constant . We call k the constant of variation (or proportion)

The value of k depends on the particular problem under consideration.

Example : The number of people required to sort a quantity of letters varies inversely as the time taken to sort the letters. If 8 people can sort a quantity of letters in 3 hours, how many people would be required to sort the same quantity of letters in 2 hours?

Solution

Let the number of people required to sort the letters in x hours be y

Step 1: Then $y = \dfrac{k}{x}$ (1)

Step 2: When $x = 3, y = 8$. Substitute these values in equation (1) and solve for k.

$$8 = \frac{k}{3}.$$

Solving, $k = 24$.
Substitute for $k = 24$ in equation (1) above to obtain

the formula $y = \dfrac{24}{x}$ ....(2)

Step 3. when $x = 2, y = \dfrac{24}{2}$ (Replacing x by 2 in equation (2))

$$y = 12$$

Therefore, to sort the letters in 2 hours would require 12 people.

Joint and Combined Variation

If a quantity varies **directly** as **two** or more other quantities (i.e. as the product of two or more other quantities), we call such a variation **joint variation**. However, if a quantity varies **directly** as one quantity (or as two or more quantities) and **inversely** as another quantity (or other quantities), we call such a variation **combined variation**.

Example: If z varies directly as x and y, then
$$z = kxy \quad \text{(joint variation)}$$

However, if z varies directly as x and inversely as y, then

$$z = k\frac{x}{y} \text{ or } \frac{kx}{y} \quad \text{(combined variation)}$$

Implicit Specification of Variation Problems

The implicit specification does not contain phrases such as " varies as", " is proportional to". The type of variation in a particular problem has to be deduced from the worded problem and/or from life experience. For example, if 15 oranges cost \$3, what is the cost of 10 oranges?. From life experience, we know that as the **number** of oranges **increases**, the **cost** of oranges **increases**. Therefore the relationship in this problem is that of direct variation (or direct proportion).

Note: For comprehensive coverage of this chapter, see the author's book "Elementary Mathematics" or "Elementary Algebra"

Lesson 18 Exercises

A **1.** y varies as x. When $y = 12, x = 4$. Find y when $x = 10$.

2. The distance S an object falls from rest is directly proportional to the square of the time t. If when $S = 64, t = 2$, find S when $t = 4$.

3. W is proportional to V. When $W = 120, V = 3$, find W when $V = 8$.

4.. S is directly proportional to t.. When $S = 64, t = 2$. Find t when $S = 288$.

5. Express by means of an equation: $(a - b)$ is proportional to $(c - d)$.

6. The distance S an object falls from rest varies directly as the square of the time t. If when $S = 32, t = 1$, find S when $t = 5$.

Answers: **1.** 30; **2.** 256; **3.** 320; **4.** 9; **5.** $a - b = k(c - d)$, where k is a constant; **6.** 800.

B

1. At constant pressure, the volume V of a perfect gas varies as the absolute temperature T. If when the temperature is 2730 Absolute, the volume of one gram of a gas is 1.7 cm^3, find the volume when the absolute temperature is 4830°.

2. According to Hooke's law, the force required to stretch a spring is directly proportional to the elongation of the spring. If a 20 lb force stretches a spring 6 in., what force will be required to stretch it 8 in.?

3. The cost C of gasoline is proportional to the number of gallons N of gasoline. If 10 gallons of gasoline cost $12, what is the cost of 25 gallons of gasoline? How many gallons of gasoline can $18 purchase?

Answers: **1.** 3.0 cm^3; **2.** 27 lb; **3.** $30; 15 gallons

C 1. At constant pressure, the volume V of a perfect gas varies as the absolute temperature T. If when the temperature is 2730° Absolute, the volume of one gram of a gas is 1.7 cm^3, find the volume when the Absolute temperature is 4830°.

2. According to Hooke's law, the force required to stretch a spring is directly proportional to the elongation of the spring. If a 20 lb force stretches a spring 6 in., what force will be required to stretch it 8 in.?

3. The cost C of gasoline is proportional to the number of gallons N of gasoline. If 10 gallons of gasoline cost $12, (a) what is the cost of 25 gallons of gasoline? and (b) how many gallons of gasoline can $18 buy?

Answers: **1.** 3.0 cm^3; **2.** $26\frac{2}{3}$ lb; **3.**(a) $30, (b) 15 gallons

D **1.** y varies inversely as x. If when $y = 6, x = 2$, determine y when $x = 8$.

2.. F is indirectly proportional to the square of D. When $F = 30, D = 3$. Find F when $D = 2$.

3.. W is indirectly proportional to V. When $W = 15, V = 45$ find W when $V = 60$.

Answers: **1.** $\frac{3}{2}$; **2.** $67\frac{1}{2}$; **3.** $11\frac{1}{4}$;

E

1. According to Boyle's law, at constant temperature, the volume of a given mass of a gas varies inversely as the pressure on the gas. When the volume of a gas is 50 cu. in, the pressure is 10 lb per sq.in. What is the pressure (a) when the volume is 80 cu. in.? ; (b) when the volume is 15 cu. in.?; (c) What is the volume when the pressure is 40 lb per sq. in.?

2. If 8 people can complete a piece of work in 40 days, how many people will complete the same piece of work in I0 days? Assume that all the people work at the same rate as one another.

3. At constant temperature, the volume of a given mass of a gas is inversely proportional to the pressure on the gas. When the volume is 150 cu. in,. the pressure is 30 lb per sq. in. Determine the pressure (a) when the volume is 60 cu. in?; (b) when the volume is 300 cu. in.

AAns:**1.** (a) $6\frac{1}{4}$ lb; (b) $33\frac{1}{3}$ lb per in.^{2}; (c) $12\frac{1}{2}$ in.^{3} **2.** 32 people; **3.** (a) 75 lb per in.^{2} .; (b) 15 lb per in.^{2} .

F **1.** Z varies jointly as x and y. If when $x = 2, y = 3, Z =16$, what is Z when $x = 5$ and $y = 4$?

2. V is indirectly proportional to P and directly proportional to T. When $V = 30, T = 25, P = 15$. Find V when $T = 40$, and $P = 3$.

3. F is directly proportional to G and varies inversely as L. When $F = 256, G= 36$, and $L = 30$. What is G, when $F = 128$ and $L = 15$?

4. The kinetic energy E of a moving object varies as the mass M. of the object and the square of the velocity V. If when $E = 50, M = 4, V = 5$, find E when $M = 16$ and $V = 3$.

5. If 12 people working 2 days a week can complete a piece of work in 4 weeks, how many people working 3 days per week can complete the same piece of work in 2 weeks? (Assume that all the people involved work at the same rate as one another)

Answers: **1.** $53\frac{1}{3}$; **2.** 240; **3.** 9; **4.** 72; **5.** 16 people;

CHAPTER 9
Inequalities

Lesson 19: Set Notation and Operations; Interval Notation
Lesson 20: Solving Linear Inequalities Containing a Single Variable;
   Compound Inequalities

Lesson 19

Set Notation and Operations; Intervals and Interval Notation

Sets, set notation

Set A set is a well-defined collection of objects or things. The objects or things are called the elements or members of the set. A set may contain many elements; it may contain only one element; or it may contain no element. We call the set which contains no element the empty set or the null set.
(**Note**: From the above, a set of **numbers** is a well-defined collection of numbers.)

Representation of sets: We will cover the roster method and the set-builder notation.

Roster method: In the roster method, we list the elements of the set and enclose them by braces.
 Example: We can denote the set of numbers 2, 3, 4, as { 2, 3, 4}.
 We separate the elements by commas. Note that the order in which the elements are listed does not matter. We may **name** a set by a capital letter . For example, if we denote the set of numbers 2, 3, 4 by A, then we write $A = \{ 2, 3, 4\}$.
We read this as "A is the set whose elements are 2, 3, 4.

Set-builder notation: In the set builder notation, we state the conditions which the elements of the set must satisfy.
Example: Let E be the set of all odd integers between 4 and 12. Then if we use x to represent an arbitrary element of this set, we write $E = \{x \mid x$ is an odd integer between 4 and 12$\}$.
 This is read " E is the set of numbers (elements)x such that x is an odd integer between 4 and 12". We read the vertical line " |" as "such that".
 (Note that by the roster method, the set E in this example would be represented by $E = \{5, 7, 9, 11\}$

We denote the empty set by { } or Ø.

We will use the **set-builder notation** in stating the solutions of inequalities.

Definition

An **inequality** is a mathematical statement that two expressions are not equal.

An inequality, like an equation, has three parts, namely the left-hand side, the inequality symbol which may be " >, <, ≥, ≤ or ≠", and the right-hand side.

Examples 1. $2x - 5 > 6x + 3$ read " $2x$ minus 5 **is greater than** $6x$ plus 3".

2. $2x + 3 \geq 6x + 7$ read " $2x$ plus 3 **is greater than or equal to** $6x$ plus 7".

3. $4x + 5 < 8x - 2$ read "$4x$ plus **five is less than** $8x$ minus 2".

4. $3x - 1 \leq 5x + 7$ read "$3x$ minus 1 **is less than or equal to** $5x$ plus 7".
5. $x > 6$ read " x is greater than 6"

Sense of an inequality (or direction of an inequality)
 The sense of an inequality refers to whether the inequality symbol is the greater than symbol " >" or the less than symbol "<".

Set Operations Involving Inequalities

Union of two sets

The union of two sets A and B, written $A \cup B$, is the set containing all the elements that belong to either set A or set B (or belong to both sets).

Symbolically, $A \cup B = \{x \mid x \in A, \text{ or } x \in B\}$

Example 1   If $A = \{x \mid x < -3\}$ and $B = \{x \mid x > 4\}$ then
$$A \cup B = \{x \mid x < -3\} \cup \{x \mid x > 4\}$$
$$= \{x \mid x < -3 \text{ or } x > 4\}$$

Graph for $\{x \mid x < -3 \text{ or } x > 4\}$

Intersection of two sets

The intersection of two sets A and B, written $A \cap B$ is the set containing the elements that are common to A and B. Thus the elements of the intersection are the elements that belong to both sets simultaneously. Symbolically, $A \cap B = \{x \mid x \in A \text{ and } x \in B\}$

Example 2 If $A = \{x \mid x > -3\}$ and $B = \{x \mid x < 2\}$ then

$$A \cap B = \{x \mid x > -3\} \cap \{x \mid x < 2\}$$

$$A \cap B = \{x \mid -3 < x < 2\}$$

Example 3   If $A = \{x \mid x < -3\}$ and $B = \{x \mid x > 4\}$ then

$$A \cap B = \{x \mid x < -3\} \cap \{x \mid x > 4\}$$

$$A \cap B = \{\ \}, \text{ the empty set.}$$

(Since a number cannot be less than -3 and greater than 4 at the same time.)

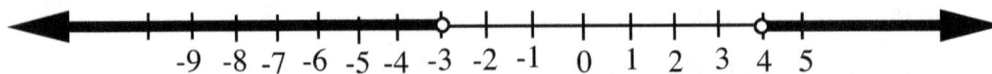

Compare the solutions to Examples 1 and 3, above, and note that even though they have identical graphs, Example 1 has a solution but Example 3 has no solution.

The Real Number Line

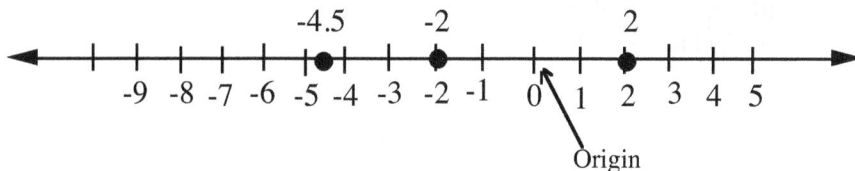

Figure 1

The real number line is a horizontal straight line with equally spaced intervals as in Figure 1 above. We label a point called the origin, 0 (zero). Points to the right of the origin are labeled positive and points to the left of the origin are labeled negative. The numbers increase as one moves from the left to the right on the real number line. Every point on this line is associated with a real number; and every real number is associated with a point on this line.

Interval   An interval is a set of numbers between two numbers. (Geometrically, we draw a line segment between the two points on the line, noting that every point on the line represents a real number)
There are a number of kinds of intervals, namely, finite intervals, infinite intervals, open intervals, closed intervals, half-open and half-closed intervals.

Consider two points represented by a and b (above figure) . On this line, a is less than b ; and this comparison is written $a < b$. The set of all numbers between a and b is called the **open interval** from a to b. The open interval does not include the end points a and b.

 There are a number of ways of symbolically indicating intervals.
One convention uses the familiar inequality symbols " > " and " < ", while another convention uses the grouping symbols " ( ) ", the parentheses, and " [ ] ' the brackets. For example, the open interval between a and b may be symbolized as (a, b) or $a < x < b$, while the closed interval between a and b may be symbolized as $[a, b]$ or $a \le x \le b$.
 (The convention employing parentheses may sometimes be confused with the ordered pair (x, y) representing a point in an x-y coordinate plane. To avoid confusion, it would be preferred that we specify whether a point or an interval is being symbolized; for example, the interval (a, b) or the ordered pair (x, y)).

Finite Intervals

Open interval The **open** interval (Figure) from a to b , written (a, b) or $a < x < b$, (where x is real number between a and b) is the set of all numbers between a and b but excluding the end-points a and b. (" open" means the endpoints are **excluded**)

Closed interval The **closed** interval from a to b, written $[a, b]$ or $a \leq x \leq b$ (where x is

real number between a and b) is the set of all numbers between a and b and including the
end-points a and b. (" closed " means the endpoints are **included**)

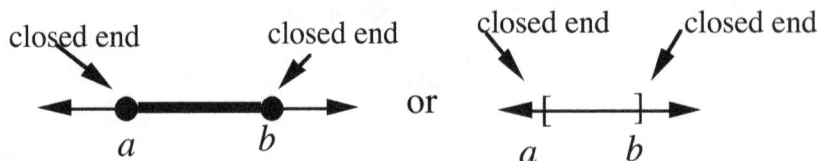

Half-closed and half-open Interval The half-closed and half-open interval from a to b is the set of all
numbers between a and b, including either the endpoint a or the endpoint b but not both.

In Figure the interval half-open on the left (or half-closed on the right) is the set of all numbers between
a to b, including the endpoint b but excluding the endpoint a.. This interval is symbolized $(a, b]$ or
$a < x \leq b$.

Figure 1:

In Figure **1**, the interval half-open on the right (or half-closed on the left) is the set of all numbers
between a to b, including the endpoint a but excluding the endpoint b.
 This interval is symbolized $[a, b)$ or $a \leq x < b$.

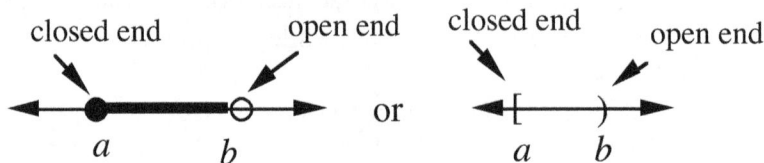

Figure 2

Note: Within any interval, there are infinitely many rational numbers and infinitely many irrational numbers.

Infinite Intervals

The infinite interval, symbolized $(a, +\infty)$ or $a < x < +\infty$ is the set of all numbers x such that $x > a$

The infinite interval, symbolized $[a, +\infty)$ or $a \leq x < +\infty$ is the set of all numbers x such that $x \geq a$

The infinite interval, symbolized $(-\infty, a)$ or $-\infty < x < a$ is the set of all numbers x such that $x < a$

The infinite interval, symbolized $(-\infty, a]$ or $-\infty < x \leq a$ is the set of all numbers x such that $x \leq a$
The infinite interval, symbolized $(-\infty, +\infty)$ is the entire number line, noting that the
symbols $+\infty$ (read: plus infinity) and $-\infty$ (read: minus infinity) are **not** numbers.

Finite Interval Notation

The following equivalent notations will be used in stating the solution sets of linear inequalities.

Algebraic Notation   **Graphical Illustration**

Set notation   Interval notation

$\{x \mid a < x < b\} = (a, b)$

$\{x \mid a \le x \le b\} = [a, b]$

$\{x \mid a < x \le b\} = (a, b]$

$\{x \mid a \le x < b\} = [a, b)$

Infinite Intervals

$\{x \mid x > a\} = (a, +\infty) = a < x < +\infty$

$\{x \mid x \ge a\} = [a, +\infty) = a \le x < +\infty$

$\{x \mid x < a\} = (-\infty, a) = -\infty < x < a$

$\{x \mid x \le a\} = (-\infty, a] = -\infty < x \le a$

$\{x \mid x \text{ is a real number }\} = (-\infty, +\infty)$

Example 1 Draw the graph for $\{x \mid x > 2\} = (2, +\infty)$

Solution

Graph for $\{x \mid x > 2\} = (2, +\infty)$

Example 2 Draw the graph for $\{x \mid x \ge 2\} = [2, +\infty)$

Solution

Graph for $\{x \mid x \ge 2\} = [2, +\infty)$

Example 3 Draw the graph for $\{x \mid x \le -8\} = (-\infty, -8]$

```
◄──┼─┼─●─┼─┼─┼─┼─┼─┼─┼─┼─┼─┼─┼─┼──►
   -9 -8 -7 -6 -5 -4 -3 -2 -1  0  1  2  3  4  5
```

Graph for $\{x \mid x \le -8\} = (-\infty, -8]$

Lesson 19 Exercises

Sketch the graphs of the following intervals

1. $(-3, 2)$; 2. $(-4, -2)$; 3. $(-\infty, 3)$; 4. $(-\infty, 0)$; 5. $[-2, 1)$; 6. $[-3, 2]$;

7. $(4, 6]$; 8. $(2, 3) \cap (-2, 5)$; 9. $(2, 3) \cap (1, 4)$; 10. $(1, 3) \cup (\frac{1}{2}, 3)$;

11. $(-\infty, -2) \cup (2, \infty)$; 12. $(-\infty, -2) \cap (2, \infty)$.

Use interval notation to represent the following inequalities: Example: $1 < x < 2 = (1, 2)$

13. $1 \le x < 3$; 14. $-6 < x \le 4$; 15. $-3 < x < 5$; 16. $0 \le x < +\infty$; 17. $3 < x \le 6$

Answers:

1. (graph) -3 to 2   or   (graph) -3 to 2

2. (graph) -4 to -2   or   (graph) -4 to -2

3. (graph) 3   or   (graph) 3

4. (graph) 0   or   (graph) 0

5. (graph) -2 to 1   or   (graph) -2 to 1

6. (graph) -3 to 2   or   (graph) -3 to 2

7. (graph) 4 to 6   or   (graph) 4 to 6

8. (graph) 2 to 3   or   (graph) 2 to 3

9. (graph) 2 to 3   or   (graph) 2 to 3

10. (graph) 1 to 3   or   (graph) 1 to 3

11. (graph) -2 to 2

12. The empty set. (No solution)

13. $[1, 3)$; 14. $(-6, 4]$; 15. $(-3, 5)$; 16. $[0, +\infty)$; 17. $(3, 6]$

Lesson 20

Solving Linear Inequalities; Solving Compound Inequalities

Solving Linear Inequalities

The solution set of a linear inequality is the set of real numbers each of which when substituted for the variable makes the inequality true.

The techniques for solving linear inequalities are similar to the techniques for solving linear equations, except that when an inequality is divided or multiplied by a negative number, the sense of the inequality must be reversed as follows: The symbol " > " when reversed becomes " < "; the symbol " < " when reversed becomes " > "; the symbol " ≥ " when reversed becomes " ≤ "; and the symbol " ≤ " when reversed becomes " ≥ ".

Example 1 Solve and graph the solution set of the inequality:

$$3x - 4 > 0$$

Step 1: To undo the -4, add + 4 to both sides of the inequality.

$$
\begin{array}{r}
3x - 4 > 0 \\
\underline{+\,4 \;\; +4} \\
3x > 4
\end{array}
$$

Step 2: Divide both sides of the inequality by 3.

$$\frac{3x}{3} > \frac{4}{3}$$

$$x > \frac{4}{3}$$

The solution set is $\{x \mid x > \frac{4}{3}\}$. (The solution set is all real numbers greater than $\frac{4}{3}$.)

Graph for $x > \frac{4}{3}$:

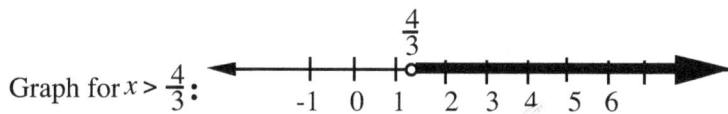

Note: The hollow circle at $\frac{4}{3}$ indicates that $\frac{4}{3}$ is **not** part of the solution set.

Example 2 Solve and graph the solution set of the inequality

$$3x - 4 \geq 0$$

Step 1: To undo the -4, add + 4 to both sides of the inequality.

$$
\begin{array}{r}
3x - 4 \geq 0 \\
\underline{+4 \;\; +4} \\
3x \geq 4
\end{array}
$$

Step 2: Divide both sides of the inequality by 3.

$$\frac{3x}{3} \geq \frac{4}{3}$$

$$x \geq \frac{4}{3}$$

The solution set is $\{x \mid x \geq \frac{4}{3}\}$. (The solution set is all real numbers greater than or equal to $\frac{4}{3}$)

Graph for $x \geq \frac{4}{3}$:

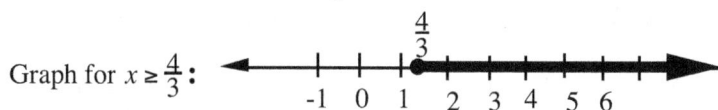

Note: The solid circle at $\frac{4}{3}$ indicates that $\frac{4}{3}$ is part of the solution set.

Example 3   Solve: and graph the solution set.: $5x + 5 > 9x - 6$ 130
Solution

Step 1: We undo the $9x$ by adding $-9x$ to both sides of the inequality.

$$5x + 5 > 9x - 6$$
$$\underline{-9x \qquad - 9x}$$
$$-4x + 5 > -6$$

Step 2: To undo the 5, add -5 to both sides of the inequality.

$$-4x + 5 > -6$$
$$\underline{\quad -5 \quad -5}$$
$$-4x > -11$$

Step 2: Divide both sides of the inequality by -4 and **reverse the sense** of the inequality.

$$\frac{-4x}{-4} < \frac{-11}{-4} \quad \text{(change this "} > \text{' to that "} < \text{")}$$

$$x < \frac{11}{4}$$

The solution set is $\{x \mid x < \frac{11}{4}\}$.   (The solution set is all real numbers less than $\frac{11}{4}$.)

Graph for $x < \dfrac{11}{4}$

Example 4   Solve for x: $2(2x + 1) \geq 5(x + 2)$
Solution

$$2(2x + 1) \geq 5(x + 2)$$
$$4x + 2 \geq 5x + 10$$
$$\underline{-5x \qquad\quad -5x}$$
$$-x + 2 \geq 10$$
$$\underline{\quad -2 \quad -2}$$
$$-x \geq 8$$
$$x \leq -8$$

(Multiplying or dividing both the left-hand side and the right-hand side by -1 and reversing the sense of the inequality or changing the signs of both sides and reversing the sense of the inequality)

The solution set is $\{x \mid x \leq -8\}$.

Graph for $x \leq -8$

Compound Inequalities

A compound inequality is formed by connecting two inequalities with the words "**and**" or " **or**".

Example We will call the following type an "**AND**" problem. (More formally, a **conjunction** problem)

$$3x - 2 < 4 \text{ and } 2x + 10 > 4$$

Because of the connective "and", we want a solution set (a set of numbers, if any) which is common to both solution sets of the two inequalities (i.e., the intersection of the solution sets). Any solution set must satisfy both inequalities simultaneously.

Lesson 20: Solving Linear Inequalities and Compound Inequalities

Example 1   Solve for x: $3x - 2 < 4$ and $2x + 10 > 4$

Procedure: Solve each inequality separately and find the intersection (common set of numbers) of the solutions.

$$3x - 2 < 4 \quad \text{and} \quad 2x + 10 > 4$$
$$\underline{+2\ +2} \qquad\qquad \underline{-10\ -10}$$
$$3x < 6 \qquad\qquad 2x > -6$$
$$x < 2 \quad \text{and} \qquad x > -3$$

The solution is the intersection of $\{x \mid x < 2\}$ and $\{x \mid x > -3\}$ i.e., $\{x \mid x < 2\} \cap \{x \mid x > -3\}$
Graph the solutions to determine if the solutions intersect.
The solution is $\{x \mid -3 < x < 2\}$

Graph for $-3 < x < 2$

Another "**AND**" problem

Example 2 Solve for x: $-5 < 2x + 1 \le 7$

Solution: The above inequality is equivalent to the compound inequality
$-5 < 2x+1$ and $2x+ 1 \le 7$.
We can solve this compound inequality using the method used in Example 1 above, however, we will use a faster method called the "**Condensed Method**" and proceed as follows:

$$-5 < 2x + 1 \le 7$$
$$\underline{-1 \qquad\quad -1 \quad -1} \qquad \text{(Adding -1 to all three sides of the inequality)}$$
$$-6 < 2x \qquad \le 6$$

$$\frac{-6}{2} < \frac{2x}{2} \le \frac{6}{2} \qquad \text{(Dividing each of all the three sides of the inequality by 2)}$$
$$-3 < x \le 3$$

Graph for $-3 < x \le 3$ (Note the hollow circle at -3 and the solid circle at 3)

Example 3   Solve for x: $-6x > 12$ and $x + 1 > 5$

Solution: For the first inequality, divide by -6 and reverse the sense of the inequality.
For the second inequality, add -1 to both sides of the inequality.
$$-6x > 12 \quad \text{and} \quad x + 1 > 5$$
$$x < -2 \quad \text{and} \qquad x > 4$$
Graph the solution set for $x < -2$ and for $x > 4$ and determine if the two graphs intersect. If they intersect, there is a solution, but if they do not intersect, then there is no solution.
$x < -2$ $x > 4$

From the graph, the two solution sets do not intersect. (i.e., there are no numbers which belong to both solutions). Therefore, there is no solution or the solution set is the empty set.

An "OR " Compound Inequality Problem (More formally, a **disjunction** problem) 1 3 2

Example 4 Solve for x: $2x - 3 > 5$ or $3x + 1 < -8$

Because of the connective "or" we want the solution set (if any) to be the union of the solution sets of the two inequalities. The solution is the set of real numbers which satisfies either inequality or both.

$$2x - 3 > 5 \text{ or } 3x + 1 < -8$$
$$\underline{+3 +3} \qquad \underline{-1 \quad -1}$$
$$2x > 8 \qquad 3x < -9$$
$$x > 4 \quad \text{or} \quad x < -3$$

Same as $x < -3$ or $x > 4$

Graph for $x < -3$ or $x > 4$

Solution set is $\{x \mid x > 4\} \cup \{x \mid x < -3\} = \{x \mid x < -3 \text{ or } x > 4\}$.

Comparison of Example 3 and Example 4
Note that even though the two problems have similar graphs, Example 3 has no solution because it is an " **and**" problem but Example 4 has a solution because it is an " **or** " problem".

Applications (Word Problems)

Example During the semester James scored 85, 70, 75 on the first three tests. He wants his class average for the term to be at least 80. There is one more test to take. What is the score he needs on the last test ?

Solution Let x be the last score.
Then, the total score for the four tests will be $85 + 70 + 75 + x$.

$$\text{Average score} = \frac{\text{Total score}}{\text{Number of tests}} = \frac{85 + 70 + 75 + x}{4}$$

Now, we want this average score to be at least 80 (i.e., 80 or more).

$$\frac{85 + 70 + 75 + x}{4} \geq 80 \qquad \qquad \text{Scrapwork}$$

$$\frac{230 + x}{4} \geq 80 \qquad \qquad 85+70+75=230$$

We solve for x $\quad \frac{4(230 + x)}{4} \geq 80(4)$

$$230 + x \geq 320$$
$$\underline{-230 \qquad \quad -230}$$
$$x \geq 90$$

$\therefore$ James needs at least a score of 90 on the last test. (i.e., 90 or more).

Some words to help translate inequality word problems

1. " is at least " translates to " $\geq$ ".	**7.** " is no more than" translates to " $\leq$ ".
2. " is no less than" translates to " $\geq$ ".	**8.** " is at most " translates to " $\leq$ ".
3 " is greater than or equal to " translates to " $\geq$"	**9.** " is less than or equal to " translates to " $\leq$ ".
4 " is more than" translates to " $>$ ".	**10** " is less than " translates to " $<$ ".
5. " is greater than" translates to " $>$ ".	**11.** " is under " translates to " $<$ "
6. " is over " translates to " $>$ "	

Also: x is between a and b translates to " $a < x < b$ ".

   x is greater than a but less than b translates to " $a < x < b$ ".

Note the difference between "**more than**" and "**is more than** " as in the following:

Example: 3 **more than** twice a number **is more than** 5 less than the number

   If the number is x, the translation is $2x + 3 > x - 5$. ("Is more than" translates to " > ")

Lesson 20 Exercises

A Solve and graph the solution set:

   1. $3x - 6 > 0$;   2. $-4x + 1 > 13$;   3. $2x - 1 \le 7x - 11$;   4. $2(x + 4) \ge 3(x - 2)$

   5. $x + 4 > -5$;   6. $x + 4 < 6$;   7. $x + 4 \le 6$;   8. $x + 4 \ge 6$;   9. $-4x > 12$

   10. $-5x \le -30$;   11. $\frac{-x}{4} \le 9$;   12. $\frac{x}{4} - 6 > 14$;   13. $2(x - 4) + 3 \le 17$

Answers; **1.** $\{x \mid x > 2\}$;   **2.** $\{x \mid x < -3\}$;   **3.** $\{x \mid x \ge 2\}$;   **4.** $\{x \mid x \le 14\}$;   **5.** $\{x \mid x > -9\}$;   **6.** $\{x \mid x < 2\}$;
7. $\{x \mid x \le 2\}$; **8.** $\{x \mid x \ge 2\}$; **9.** $\{x \mid x < -3\}$; **10.** $\{x \mid x \ge 6\}$; **11.** $\{x \mid x \ge -36\}$; **12.** $\{x \mid x > 80\}$; **13.** $\{x \mid x < 11\}$;

B Solve for: **1.** $2x - 3 < 1$ and $3x + 5 > 2$;   **2.** $-4 < 3x - 1 < 5$;   **3.** $2x - 3 < 1$ and $3x - 5 > 7$

   4. $-1 \le 2x + 5 < 3$;   **5.** $3x + 7 < 1$ and $2x - 2 < 5x + 7$

Answers: **1.** $\{x \mid -1 < x < 2\}$;   **2.** $\{x \mid -1 < x < 2\}$ **3.** No solution: $x < 2$ and $x > 4$ do not intersect.
   4. $\{x \mid -3 \le x < -1\}$;   **5.** $\{x \mid -3 < x < -2\}$

C Solve for x: **1.** $3x - 2 > 4$ or $3x - 1 > -7$;   **2.** $3x + 4 > 1$ or $2x + 1 < -9$;   **3.** $2x - 3 < 1$ or $3x - 5 > 7$

   4. $2x + 3 < 5$ or $x + 1 > 4$;   **5.** $3x - 2 > 4$ or $x - 2 < 6$;   **6.** $x - 1 < \frac{1}{2}$ or $x + 3 > 5$

Answers: **1.** $\{x \mid x > -2)$;   **2.** $\{x \mid x < -5$ or $x > -1\}$;   **3.** $\{x \mid x < 2$ or $x > 4\}$; **4.** $\{x \mid x < 1$ or $x > 3\}$;

5. All real numbers ; **6.** $\{x \mid x < \frac{3}{2}$ or $x > 2\}$;

D During the semester, Betty scored 75 and 78 on the first two tests. She wants her class average for the term to be at least 80. There is one more test to take. What is the score she needs on the last test ?

   Answer: Betty needs at least a score of 87 on the last test

E Solve for x:

   1. $\frac{x}{2} + 1 < x - 14$;   2. $\frac{x - 3}{2x + 1} < 0$;   3. $\frac{1}{x + 5} > \frac{1}{8}$;   4. $3x - 2 < \frac{10 - x}{-2}$;

   5. $\frac{1}{x + 1} > \frac{2}{x}$;   6. $\frac{x + 3}{2} \le \frac{2 + x}{-2}$

Answers

   1. $x > 30$;   **2.** $-\frac{1}{2} < x < 3$;   **3.** $-5 < x < 3$;   **4.** $x < -\frac{6}{5}$;

   5. $\{x \mid x < -2\} \cup \{x \mid -1 < x < 0\}$;   **6.** $x \le -\frac{5}{2}$

Lesson 20 Extra
Review of Inequality Operations

The operations on inequalities are similar to the operations on equations (equalities), except for: **1.** When both sides of an inequality are multiplied by or divided by a **negative** number, or if both sides (having the same sign) of an inequality are inverted, the sense of the inequality must be reversed.

2. To operate on **two inequalities** simultaneously, the sense (direction or order) of both inequalities must be the same.

The properties below are useful for the epsilon-delta proofs in the next lesson.

Transitivity	Addition	Multiplication
1. If $a > b$ and $b > c$. then $a > c$ Example: If $7 > 5$ and $5 > 2$ then $7 > 2$ (true)	**2.** If $a > b$ and $c > d$. then $a + c > b + d$ Example: If $8 > 4$ and $3 > 2$ then $8 + 3 > 4 + 2$ or $11 > 6$ (True)	**2.** If $a > b$ and $c > 0$. then $ac > bc$ but If $a > b$ and $c < 0$. then $ac < bc$ (sense reversed)

Subtraction	Multiplication	Division
4. If $a > b$ and $c > d$. We **cannot** subtract as we do with **equations.** Step 1: $-c < -d$ (reverse sense) Step 2: $-d > -c$ (rewriting) If $a > b$ and $-d > -c$ then (same sense) $a - d > b - c$ (adding) ------------------------------- **Inversion** of both sides of the same sign. If diffent signs, **no** inversion Invert both sides and reverse the sense of the inequality. If $a > b > 0$ e.g.($10 > 2$) Then $\frac{1}{a} < \frac{1}{b}$ ($\frac{1}{10} < \frac{1}{2}$) **No** inversion: $10 > -2$; $\frac{1}{10} > -\frac{1}{2}$	**5.** If $a > b > 0$ and $c > d > 0$. then $ac > bd$ **Note:** $a, b, c, d,$ are all positive **Example 1:** If $8 > 4$ and $3 > 2$ then $8 \times 3 > 4 \times 2$ or $24 > 8$ (True) **Example 2:** If $|x - 3| < \frac{\varepsilon}{7}$ and $|x + 3| < 7$, then $|x - 3||x + 3| < \frac{\varepsilon}{7}(7)$ $|(x + 3)(x - 3)| < \varepsilon$	**5.** If $a > b > 0$ and $c > d > 0$. We **cannot** divide as we do with **equations.** Step 1: If $\frac{1}{c} < \frac{1}{d}$ Step 2: $\frac{1}{d} > \frac{1}{c}$.(rewriting) If $a > b$ and $\frac{1}{d} > \frac{1}{c}$ $\frac{a}{d} > \frac{b}{c}$ (Multiplying) **Note**: In Step 1, we inverted both sides and reversed the sense of the inequality.

Powers: Given $a > b > 0$
If m and n are positive integers then
1. Then $a^m > b^m$
Like positive powers of unequal quantities a re unequal in the same sense.
Example We can square both sides of an inequality.
If $4 > 3$, then $4^2 > 3^2$ or $16 > 9$
Note above that a aand b re positive.

Roots: If $a > b > 0$, then
$\sqrt[n]{a} > \sqrt[n]{b}$ or $a^{\frac{1}{n}} > b^{\frac{1}{n}}$
Like positive roots of unequal quantities are unequal in the same sense
Example We can take the square root both sides of an inequality
Example 1. If $16 > 9$, then
$\sqrt{16} > \sqrt{9}$ and $4 > 3$ (true).

CHAPTER 10

Lesson 21: **Absolute Value Equations**
Lesson 22: **Absolute Value Inequalities**

Lesson 21
Absolute Value Equations

Geometric definition: The **absolute value** of a real number is the distance between the number and zero on the number line.

Algebraic definition: The **absolute value** of x symbolized, $|x|$, is defined by a two-part rule as follows:

$$|x| = \begin{cases} x \text{ if } x \geq 0 \\ -x \text{ if } x < 0 \end{cases}$$

In words, the absolute value of a real number x is x if x is a positive number or zero; but it is $-x$ if x is a negative number. The above two-part rule in the definition requires us to consider two cases when solving absolute value equations. Understanding the above definition very well will save us from having to memorize the solution methods for absolute value equations.

Absolute value equations: Absolute value equations are equations in which the absolute value terms contain variables.

Example l Solve for x: $|x| = 2$

We will consider two methods.

Method 1 (Using the geometric definition)

$|x| = 2$ means the distance between x and 0 on the number line is 2. Therefore $x = 2$ or -2 Check: $|2| = 2$; $|-2| = 2$.

The solution set is $\{-2, 2\}$

Method 2 If $x \geq 0$, then $|x| = x = 2$
If $x < 0$, then $|x| = -x = 2$, and from which $x = -2$.
Again the solution set is $\{-2, 2\}$

Graph for $\{-2, 2\}$

Example 2 Solve for x: $|x - 3| = 4$

Method 1 (Using the geometric definition)

$|x - 3| = 4$ means, on the number line, the distance between x and 3 is 4.

Step l: Locate the point 3 on the number line. (To obtain 3, set $x - 3$ to zero and solve for x.)

Step 2: From this point, count 4 units to the right and stop, the coordinate of this stopping point, 7, is one of the solutions. Similarly from the point 3, count 4 units to the left and stop, the coordinate of this stopping point is -1, and it is another solution. The solution set is $\{-1, 7\}$

Graph for $\{-1, 7\}$

Method 2 (Using the algebraic definition)

Solution $|x - 3| = 4$ is equivalent to
$$x - 3 = 4 \text{ or } -(x - 3) = 4$$

Solve each equation for x:

$x - 3 = 4$ or $-(x - 3) = 4$ (or $-x + 3 = 4$ and from which $x = -1$)

$x - 3 = 4$ or $(x - 3) = -4$ (Multiplying each side of the second equation by -1)

$x - 3 = 4$ or $x - 3 = -4$

$\underline{+3 + \quad 3}$ $\underline{+3 \quad +3}$

$x = 7$ or $x = -1$

Check # 1: $|7 - 3| = |4| = 4$; Check # 2: $|-1-3| = |-4| = 4$.

The solution set is $\{-1, 7\}$

Example 3a Solve for x: $4 - |3x - 4| = -7$

Solution: $4 - |3x - 4| = -7$

$\quad \underline{-4 \qquad\qquad\qquad -4}$

$\quad\quad -|3x - 4| = -11$

$\quad\quad\quad |3x - 4| = 11$ (Multiplying both sides of the equation by -1)

Now, we apply the algebraic definition of the absolute value:

Then, $3x - 4 = 11$ or $-(3x - 4) = 11$

$\quad\quad \underline{+4 \quad +4}$ $3x - 4 = -11$ (Multiplying both sides of the equation by -1)

$\quad\quad\quad 3x = 15$ $\quad\quad \underline{+4 \qquad +4}$

$\quad\quad\quad\quad x = 5$ $\quad\quad\quad 3x = -7$ and $x = -\dfrac{7}{3}$

Checking for $x = 5$:

$4 - |3(5) - 4| \stackrel{?}{=} -7$

$4 - |15 - 4| \stackrel{?}{=} -7$

$4 - |11| \stackrel{?}{=} -7$

$4 - 11 \stackrel{?}{=} -7$

$-7 = -7$ True (LHS =RHS)

Checking for $x = -\dfrac{7}{3}$: $4 - |3 \cdot (-\dfrac{7}{3}) - 4| \stackrel{?}{=} -7$

$4 - |-7 - 4| \stackrel{?}{=} -7$

$4 - |-11| \stackrel{?}{=} -7$

$4 - 11 \stackrel{?}{=} -7$ (note: $|-11| = 11$)

$-7 = -7$ True

The solution set is $\{-\dfrac{7}{3}, 5\}$.

Example 3b: Solve for x: $|x - 2| = |x + 6|$. (A) (See Appendix A, p.658 for algebraic approach)
(We apply the geometric definition which says that the absolute value of a real number is its distance from zero on the number line)

Applying the geometric definition, either the two expressions of the equation are the same number, or are opposites. For examples, $|3| = |3|, |-3| = |-3|$ (same number) or $|-3| = |3|, |3| = |-3|$ opposites.

Case 1: If the two expressions represent the same number, then $x - 2 = x + 6$	**Case 2:** If the two expressions represent
$x - 2 = x + 6$	(A) opposite numbers,
Solving, $-2 = 6$,	then $x - 2 = -(x + 6)$
which is a contradiction.	$x - 2 = -x - 6$, and from which
There is therefore no solution for **Case 1.**	$x = -2$
	We check for $x = -2$. in (A)
	Then $|-2 - 2| \stackrel{?}{=} |-2 + 6|$;
	$|-4| \stackrel{?}{=} |4|$; $4 \stackrel{?}{=} 4$. Yes.
	The solution is -2.

Note above that sometimes, there may be solutions to both cases.

Extra: Examples Involving two Absolute Value Terms
(Consult the course syllabus. These extras may **not** be included in your syllabus)

Example 4 Solve for x: $|x + 2| + |x - 3| = 11$ (A)   (We apply the algebraic definition)

By definition, $|x + 2| = \begin{cases} x + 2 & \text{if } x + 2 \geq 0, \text{ or } x \geq -2 \\ -(x + 2) & \text{if } x + 2 < 0, \text{ or } x < -2 \end{cases}$

and

$|x - 3| = \begin{cases} x - 3 & \text{if } x - 3 \geq 0, \text{ or } x \geq 3 \\ -(x - 3) & \text{if } x - 3 < 0, \text{ or } x < 3 \end{cases}$

We will rewrite the original equation considering the different interpretations of the absolute value terms. From the original equations, we consider four cases:

Case 1 Both expressions within the absolute value bars are considered to be positive.

$(x + 2) + (x - 3) = 11$ Solving, $x = 6$	Subject to the intersection (if any) of the domains $x \geq -2$ and $x \geq 3$. The intersection of these domains is $x \geq 3$. Any solution to Case 1 must satisfy the condition $x \geq 3$.

Since 6 is in the domain $x \geq 3$, 6 is a possible solution which we will check, later, in the original equation.

Case 2 Both expressions within the absolute value bars are considered to be negative.

$-(x + 2) - (x - 3) = 11$ $-x - 2 - x + 3 = 11$ $x = -5$	Subject to the intersection (if any) of the domains $x < -2$ and $x < 3$. The intersection of these domains is $x < -2$. Any solution to Case 2 must satisfy the condition $x < -2$.

Since -5 is in the domain $x < -2$, -5 is a possible solution which we will check, later, in the original equation.

Case 3 The **first** expression within the absolute value bars is **positive** and the **second** expression within the absolute value bars is **negative.**

$(x + 2) - (x - 3) = 11$ $x + 2 - x + 3 = 11$ $\qquad 5 = 11$ False	Subject to the intersection (if any) of the domains $x \geq -2$ and $x < 3$. The intersection of these domains is $-2 \leq x < 3$. Any solution to Case 2 must satisfy the condition $-2 \leq x < 3$.

Since we obtain an inconsistent equation (false statement) in Case 3, this case has no solution.

Case 4 The **first** expression within the absolute value bars is **negative** and the **second** expression within the absolute value bars is **positive .**

$-(x + 2) + (x - 3) = 11$ Solving, $-x - 2 + x - 3 = 11$ $\qquad -5 = 11$ False	Subject to the intersection (if any) of the domains $x < -2$ and $x \geq 3$. The intersection of these domains is empty. There is no domain for this case, since a number cannot be less than - 2 and greater than or equal to 3 simultaneously.

Similarly, as in Case 3, above, this case also does not have any solution.
The solutions to the original equation are from Cases 1 and 2. Therefore, the solutions are -5 and 6.

It is good practice to check each solution in the original equation.

Checking for -5: $|-5 + 2| + |-5 - 3| \overset{?}{=} 11$

$$|-3| + |-8| \overset{?}{=} 11$$

$$3 + 8 \overset{?}{=} 11$$

$$11 = 11 \text{ Yes.}$$

Checking for 6: $|6 + 2| + |6 - 3| \overset{?}{=} 11$

$$|8| + |3| \overset{?}{=} 11$$

$$8 + 3 \overset{?}{=} 11$$

$$11 = 11 \text{ Yes.}$$

The solution set is $\{-5, 6\}$

Example 5 Solve for x: $|x+2| + |x-3| = 5$ (A)

By definition, $|x+2| = \begin{cases} x+2 & \text{if } x+2 \geq 0, \text{ or } x \geq -2 \\ -(x+2) & \text{if } x+2 < 0, \text{ or } x < -2 \end{cases}$

and

$\qquad |x-3| = \begin{cases} x-3 & \text{if } x-3 \geq 0, \text{ or } x \geq 3 \\ -(x-3) & \text{if } x-3 < 0, \text{ or } x < 3 \end{cases}$

We will rewrite the original equation considering the different interpretations of the absolute value terms. From the original equation, we consider four cases:

Case 1 **Both** expressions within the absolute value bars are considered to be positive..

$\qquad x+2+x-3 = 5$ Subject to the intersection (if any) of the domains $x \geq -2$ and $x \geq 3$.

The intersection of these domains is $x \geq 3$. Any solution to Case 1 must satisfy the condition $x \geq 3$.

We now solve $x+2+x-3 = 5$.

$$2x - 1 = 5$$
$$2x = 6$$
$$x = 3$$

Since 3 is in the domain $x \geq 3$, 3 is a possible solution which we shall check, later, in the original equation.

Case 2 **Both** expressions within the absolute value bars are considered to be negative

$\qquad -(x+2) - (x-3) = 5$ Subject to the intersection (if any) of the domains $x < -2$ and $x < 3$.

The intersection of these domains is $x < -2$. Any solution to Case 2 must satisfy the condition $x < -2$.

Solving,

$$-(x+2) - (x-3) = 5$$
$$-x - 2 - x + 3 = 5$$
$$-2x + 1 = 5$$
$$-2x = 4$$
$$x = -2$$

Since -2 is **not** in the domain $x < -2$, -2 is rejected as a solution to Case 2.

Case 3 The expression within the **first** absolute value bars is **positive** and the expression within the **second** absolute value bars is **negative.**

$\qquad x+2 - (x-3) = 5$ Subject to the intersection (if any) of the domains $x \geq -2$ and $x < 3$.

The intersection of these domains is $-2 \leq x < 3$. Any solution to Case 2 must satisfy the condition $-2 \leq x < 3$.

Solving,

$$x+2 - x + 3 = 5$$
$$5 = 5$$

Since we obtain an identity, all values of x in the domain $-2 \leq x < 3$ is a solution to Case 3. Therefore, the solution to Case 3 is $-2 \leq x < 3$.

(Note that if the domain for Case 3 had been all real x., then the solution would have been all real x. However, the solution to Case 3 is subject to the restricted domain $-2 \leq x < 3$.)

Case 4 The expression within the **first** absolute value bars is **negative** and the expression
within the **second** absolute value bars is **positive.**

$-(x+2)+x-3=5$ Subject to the intersection (if any) of the domains $x<-2$ and $x\geq 3$.
 The intersection of these domains is empty. There is no domain for this case,
 since a number cannot be less than - 2 and greater than or equal to 3
 simultaneously

There is therefore, no solution for Case 4, and we do not solve the equation for this case.

The solutions to the original equation are obtained by combining the solutions to Case 1 ($x=3$) and Case
3 ($-2\leq x<3$).

Therefore, the solution to the original equation is $\{x \mid -2\leq x\leq 3\}$. It is good practice to check each
solution in the original equation.

We will check for the endpoints -2 and 3 and also for 0.

Checking for -2 in Checking for 3 in : Checking for 0 in

$|x+2|+|x-3|=5,|x+2|+|x-3|=5,|x+2|+|x-3|=5,$

$|-2+2|+|-2-3|\overset{?}{=}5|3+2|+|3-3|\overset{?}{=}5|0+2|+|0-3|\overset{?}{=}8$

$|0|+|-5|\overset{?}{=}5|5|+|0|\overset{?}{=}5|2|+|-3|\overset{?}{=}5$

$5=55=55=5$

The solution set is $\{x \mid -2\leq x\leq 3\}$.

Discussion

In solving absolute value equations involving two or more absolute value terms, we must pay careful
attention to the domains (inequalities) which the absolute value terms are subject to. Ignoring the domains
will sometimes lead to incomplete or false solutions.

Let us apply a **short-cut method** to solve the last two examples the nature of whose solutions are
different from each other.

**1.** Solve for x: $|x+2|+|x-3|=11$ <------------Example 4, above
**2.** Solve for x: $|x+2|+|x-3|=5$. <-----------Example 5, above

Solution
1. In $|x+2|+|x-3|=11$, $a=-2$, $b=3$, $c=11$ (From the general equation $|x-a|+|x-b|=c$)

Step 1: Set $x+2=0$, and solve to obtain $x=-2$.
Also set $x-3=0$, and solve to obtain $x=3$.

Step 2: Add the solutions from Step 1:
-2 + 3 = +1.

Step 3: (a) Add 11 ($c=11$) to +1 (from Step 2) to obtain 12 and divide this 12 by 2 to obtain 6, which
is one of the solutions. (This is the same as counting 11 units to the right from +1 on the number
line to obtain 12, and then dividing by 2 to obtain 6)

(b) Subtract 11 ($c=11$) from +1 (from Step 2) to obtain -10 and divide this -10 by 2 to
obtain -5, which is one of the solutions. (This is the same as counting 11 units to the left
from +1 on the number line to obtain -10, and dividing by 2 to obtain -5)

The solution is $\{-5,\ 6\}$. This is the same solution as the approach in Example 4, above.

So far, so good for the short-cut method!

Now, let us also apply the same **short-cut method** to solve Example 5,

$$|x + 2| + |x - 3| = 5.$$

In $|x + 2| + |x - 3| = 5$, $a = -2$, $b = 3$, $c = 5$ (From the general equation $|x - a| + |x - b| = c$)

Step 1: Set $x + 2 = 0$, and solve to obtain $x = -2$.

   Also set $x - 3 = 0$, and solve to obtain $x = 3$.

Step 2: Add the solutions from Step 1:
   -2 + 3 = +1.

Step 3: (a) Add 5 ($c = 5$) to +1 (from Step 2) to obtain 6 and divide this 6 by 2 to obtain 3, which
      is one of the solutions. (This is the same as counting 5 units to the right from +1 on the number
      line to obtain 6, and then dividing by 2 to obtain 3)

   (b) Subtract 5 ($c = 5$) from +1 (from Step 2) to obtain -4 and divide this -4 by 2 to
      obtain -2, which is one of the solutions. (This is the same as counting 5 units to the left
      from +1 on the number line to obtain -4, and then dividing by 2 to obtain -2)

From Step 3, we might conclude that the solution to $|x + 2| + |x - 3| = 5$ is $\{-2, \ 3\}$. However, this is not
the complete solution. The complete solution by the approach in Example 5, above is $\{x \,|\, -2 \le x \le 3\}$, an
interval solution.

We conclude that the short-cut method may sometimes lead to incomplete or false solutions. We should
therefore always use the formal approach in Examples 4 and 5, since this approach avoids incomplete or
false solutions.

Moreover, the author experimented with groups of some absolute value equations and their solutions, and came up with
three groups of absolute value equations. They are presented only for your enrichment and **not** as an approach to solving
absolute value equations.

Group 1. If either $|a| + |b| \ne |c|$, or $|a| + |b| = |c|$ but a and b are preceded by the **same** sign, in the
      general equation, $|x - a| + |x - b| = c$, we generally have solutions (if any) such as $x = k$, $x = l$.

Examples:

   1. $|x + 2| + |x - 3| = 11$; Solution: $x = -5$, $x = 6$ (Note that $|-2| + |3| \ne |11|$).

   2. $|x - 3| + |x + 4| = 9$; Solution: $x = -5$, $x = 4$ (Note that $|3| + |-4| \ne |9|$).

   3. $|x - 3| + |x - 4| = 5$; Solution: $x = 1$, $x = 6$ (Note that $|3| + |4| \ne |5|$).

   4. $|x - 3| + |x - 5| = 8$; Solution: $x = 0$, $x = 8$ (Note that $|3| + |5| = |8|$ but a and b have the same sign.)

   5. $|x + 3| + |x + 5| = 8$; Solution: $x = 0$, $x = -8$ (Note that $|3| + |5| = |8|$ but a and b have the same sign.)

   6. $|x - 2| + |x - 3| = 5$; Solution: $x = 0$, $x = 5$ (Note that $|2| + |3| = |5|$ but a and b have the same sign.)

   7. $|x + 2| + |x + 3| = 5$; Solution: $x = 0$, $x = -5$ (Note that $|2| + |3| = |5|$ but a and b have the same sign.)

Lesson 21: Absolute Value Equations

Group 2:. If $|a| + |b| = |c|$ and a and b are preceded by **opposite** signs, in the general equation, 1 4 1 $|x - a| + |x - b| = c$, we generally have interval solutions such as $k \le x \le l$.

Examples:

1. $|x + 2| + |x - 3| = 5$; Solution: $-2 \le x \le 3$ (Note that $|-2| + |3| = |5|$ but a and b have opposite signs.)

2. $|x - 3| + |x + 4| = 7$; Solution: $-4 \le x \le 3$ (Note that $|3| + |-4| = |7|$ but a and b have opposite signs.)

3. $|x + 3| + |x - 5| = 8$; Solution: $-3 \le x \le 5$ (Note that $|-3| + |5| = |8|$ but a and b have opposite signs.)

Group 3. Given $|x - a| - |x - b| = c$, if $|a| + |b| = |c|$ with a and b being of **opposite** signs, there are generally solutions such as $x \ge k$, or $x \le l$ as in Examples 1, 2 and 3 below; however if a and b are of the same sign, there are generally no solutions, as in Examples 4 and 5, below. Also, if $|a| + |b| \ne |c|$ and a and b have the same sign, there may be interval solutions such as $x \ge k$, or $x \le l$..

Examples:

1. $|x + 2| - |x - 3| = 5$; Solution: $x \ge 3$ (note that $|-2| + |3| = |5|$)

2. $|x + 2| - |x - 3| = -5$; Solution: $x \le -2$ (note that $|-2| + |3| = |5|$)

3. $|x - 3| - |x + 2| = 5$; Solution: $x \le -2$ (note that $|3| + |-2| = |5|$)

4. $|x - 2| - |x - 3| = 5$; There is **no** solution: (note that $|2| + |3| = |5|$)

5. $|x - 3| - |x - 2| = 5$; There is **no** solution: (note that $|3| + |2| = |5|$).

6. $|x - 3| - |x - 5| = 2$; Solution: $x \ge 5$ (note that $|3| + |5| \ne |2|$, and a and b have the same sign)

7. For $|x - 2| - |x - 3| = 1$; solution: $x \ge 3$ (note that $|2| + |3| \ne |1|$).

After studying the above three groups of equations, the student should be convinced that the method in Examples 4 and 5 is the approach to use in solving absolute value equations, since otherwise the student would have to recall the various forms of the equation and the nature of the solutions, and even then, there may be some unknown exceptions to the rule for the group of equations.

Lesson 21 Exercises

A Solve for x: 1. $|x - 5| = 2$; 2. $|2x - 1| = 5$;   3. $|2x + 1| = -4$;   4. $3 - |2x - 1| = -8$

5. $|x + 4| = 3$;   6. $|3x - 2| = 4$;   7. $6 - |x - 3| = -2$;   8. $|x - 10| = -5$

Answers: **1.** $\{3, 7\}$; **2** $\{-2, 3\}$; **3.** $\{-\frac{5}{2}, \frac{3}{2}\}$; **4.** $\{-5, 6\}$; **5.** $\{-1, -7\}$; **6.** $\{-\frac{2}{3}, 2\}$; **7.** $\{-5, 11\}$; **8.** No solution

Lesson 21: Absolute Value Equations

B 1. $|x+2|+|x-3|=11$;   2. $|x-3|+|x+4|=9$;   3. $|x-3|+|x-4|=5$;

4. $|x-3|+|x-5|=8$;   5. $|x+3|+|x+5|=8$;   6. $|x-2|+|x-3|=5$;   7. $|x+2|+|x+3|=5$;

8. $|x+2|+|x-3|=5$;   9. $|x-3|+|x+4|=7$;   10. $|x+3|+|x-5|=8$;   11. $|x+2|-|x-3|=5$;

12. $|x+2|-|x-3|=-5$;   13. $|x-3|-|x+2|=5$;   14. $|x-3|-|x-5|=2$;   15. $|x-2|-|x-3|=5$;

16. $|x-3|-|x-2|=5$;   17. $|x-2|-|x-3|=1$.

Solutions

1. $x=-5,\ x=6$; 2. $x=-5,\ x=4$;   3 $x=1,\ x=6$;   4. $x=0,\ x=8$; 5. $x=0,\ x=-8$;

  6 $x=0,\ x=5$;   7. $x=0,\ x=-5$; 8. $-2\le x\le 3$;   9. $-4\le x\le 3'$ 10. $-3\le x\le 5$;

11. $x\ge 3$; 12. $x\le -2$; 13. $x\le -2$; 14 $x\ge 5$; 15. No solution: 16. No solution; 17. $x\ge 3$.

Lesson 22
143
Absolute Value Inequalities

We will consider two cases

Case 1. $|ax + b| < c$. where c is positive. Example: $|x - 2| < 1$. (There is no solution if c is negative)

Case 2. $|ax + b| > c$, where c may be positive or negative. Example: $|x - 2| > 1$.

Examples

      1. $|x - 2| < 1$ means the distance from x to 2 is less than 1.

      2. $|x - 2| > 1$ means the distance from x to 2 is greater than 1.

Case 1, $|ax + b| < c$. is equivalent to

$$\boxed{-c < ax + b < c}, \text{ the " \textbf{and}" problem. } (\text{or } -c < ax + b \text{ and } ax + b < c)$$

Therefore, we can solve this inequality using the condensed method.
Generally, the solution is a one-interval solution. Case 1 is a conjunction problem.

Case 2, $|ax + b| > c$ *is* equivalent to

$$\text{either } \boxed{ax + b > c} \quad \text{or} \quad \boxed{-(ax + b) > c}$$

Here, we solve an " or" problem. Generally, we have a two-interval solution.
Note that $-(ax + b) > c$ *is* equivalent to $ax + b < -c$. (obtained by multiplying by -1 and reversing the sense)
The form $-(ax + b)$ is easier to recall since it follows immediately from the absolute value definition.
(see page 135). Case 2 is a disjunction problem.

Example 1: Solve for x: $|x - 2| < 3$.

Solution $|x - 2| < 3$ is equivalent to

$$-3 < x - 2 < 3$$
$$\underline{+2 \quad +2 \quad +2}$$
$$-1 < x \quad < 5 \quad\quad\quad \text{(Solving for } x \text{ by adding +2 to all three sides of the inequality)}$$

Graph for $\{-1 < x < 5\}$

Example 2 Solve for x: $|x - 2| > 3$

Solution $|x - 2| > 3$ is equivalent to

$(x - 2) > 3$ or $-(x - 2) > 3$

$x - 2 > 3$ or $\quad x - 2 < -3 \quad$ or $-x + 2 > 3$

$\underline{+2 \ +2} \qquad \underline{+2 \quad +2} \qquad \underline{-2 \ -2}$

$\quad x > 5 \quad$ or $\quad x < -1 \qquad \quad -x > 1$

$\qquad \qquad \qquad \qquad \qquad \qquad \qquad x < -1$

Solution set is $\left\{ x \mid x < -1 \text{ or } x > 5 \right\}$

Graph for $\left\{ x \mid x < -1 \text{ or } x > 5 \right\}$

Example 3 Solve for x: $|x + 3| \geq 4$

Solution $\qquad |x + 3| \geq 4$ is equivalent to

$(x + 3) \geq 4 \quad$ or $\quad -(x + 3) \geq 4$

$x + 3 \geq 4$ or $\quad x + 3 \leq -4$

$\underline{-3 \ -3} \qquad \qquad \underline{-3 \quad -3}$

$\quad x \geq 1 \quad$ or $\qquad x \leq -7$

Solution set is $\left\{ x \mid x \leq -7 \text{ or } x \geq 1 \right\}$

Graph for $\left\{ x \mid x \leq -7 \text{ or } x \geq 1 \right\}$

Extra: Solve for x:: $|3 - 7x| < 17$

We cover two methods

Method 1: (Rewriting after solving)

$|3 - 7x| < 17$

$-17 < 3 - 7x < 17$

$\underline{-3 \quad -3 \qquad \quad -3}$

$-20 < -7x < 14$

$\dfrac{-20}{-7} > \dfrac{-7}{-7}x > \dfrac{14}{-7}$ (Reversing the direction or sense)

$\dfrac{20}{7} > x > -2 \qquad (x < \dfrac{20}{7} \text{ and } x > -2)$

Rewriting the lower limit first, $-2 < x < \dfrac{20}{7}$

$$\left\{ x \mid -2 < x < \dfrac{20}{7} \right\}$$

Method 2: Relatively, a popular method (Rewriting the absolute value term before solving)

Note for example that $|7 - 2| = |2 - 7|$, and generally, $|b - a| = |a - b|$; Replacing $|3 - 7x|$ by $|7x - 3|$, the original inequality becomes

$$|7x - 3| < 17$$

Now, we solve the equivalent inequality,

$|7x - 3| < 17$

$-17 < 7x - 3 < 17$

$\underline{+3 \qquad \quad +3 \quad +3}$

$-14 < 7x < 20$

$\dfrac{-14}{7} < \dfrac{7}{7}x < \dfrac{20}{7}$

$-2 < x < \dfrac{20}{7}$

$$\left\{ x \mid -2 < x < \dfrac{20}{7} \right\}$$

Lesson 22 Exercises

A Solve for x: **1.** $|x - 4| < 2$; **2.** $|x - 4| > 2$; **3.** $|x + 5| < -4$

Answers: 1. $\{x \mid 2 < x < 6\}$; **2.** $\{x \mid x < 2 \text{ or } x > 6\}$; **3.** No solution

B Solve for x:: **1.** $|x - 5| < 2$; **2.** $|x - 5| > 2$; **3.** $|x - 5| \geq 2$; **4.** $|x - 5| = 2$

5. $|2x + 3| \geq 3$

Answers: 1. $\{x \mid 3 < x < 7\}$; **2.** $\{x \mid x < 3 \text{ or } x > 7\}$; **3.** $\{x \mid x \leq 3 \text{ or } x \geq 7\}$; **4.** $\{3, 7\}$;
5. $\{x \mid x \leq -3 \text{ or } x \geq 0\}$

C By applying the appropriate definitions, represent the following using interval notation

1. $|x| \leq a$; **2.** $|x| \geq b$; **3.** $|x| \leq 3$; **4.** $|x| \leq 2.5$; **5.** $|x| > 5$; **6.** $|x| \geq 4$;

Write an absolute value inequality that has the given solution:

7. $[-4, 4]$; **8.** $[-\infty, -4] \cup [-2, +\infty)$; **9.** $(1, 3)$.

Answers:
1. $[-a, a]$; **2.** $(-\infty, -b] \cup [b, +\infty)$; **3.** $[-3, 3]$; **4.** $[-2.5, 2.5)$; **5.** $(-\infty, -5] \cup (5, +\infty)$;
6. $(-\infty, -4] \cup [4, +\infty)$; **7.** $|x| \leq 4$; **8.** $|x + 3| > 1$; **9.** $|x - 2| < 1$

For absolute value inequalities involving rational expressions see page 184.

CHAPTER 11

Lesson 23: **Graphing Linear Inequalities in two Variables**

Lesson 24: **Solving Systems of Linear Inequalities**

Lesson 25: **Linear Programming ; Deriving Systems of Inequalities; Applications of Linear Programming**

Lesson 23

Graphing Linear Inequalities in two Variables

Geometrically, the solution set of a linear inequality in two variables is a section of the *x-y* plane. The plane may be open or closed. If the plane is closed, the symbol used would be " ≥ " or " ≤ "; but if plane is open, the symbol would be " > " or "<". The solution set for the closed half-plane includes the boundary line, but the solution set for the open half-plane does not include the boundary line. We will learn how to graph the solution sets of some inequalities.

General steps in graphing linear inequalities

Step 1: Replace the inequality symbol by the equality symbol " = ".

Step 2: Using convenient points, graph the equation (now a linear equation) from Step 1. To do this, choose convenient *x*-values and calculate corresponding y-values. Connect the points by a straight solid line if the inequality is the symbol " ≥ " or " ≤ " ; but use broken (dashed or dotted) lines if the given inequality is the symbol " > " or " < ".

Note: You may also use any other specialized methods (for graphing straight lines) such as the "slope-intercept method or the *x*- and *y*-intercepts method.

Step 3: We will determine which side of the straight line forms the solution. Choose any convenient point not on the boundary (of the straight line drawn) and substitute its *x*- and *y*- coordinates in the original inequality. If these values satisfy the inequality, then the chosen point is in the solution set of the inequality, and consequently, that side of the line (half-plane) where this point is located is the solution set. However, if the tested point does not satisfy the original inequality, then it is the other side of the line (the half-plane) which is the solution set We must note that the straight line drawn in Step 2 always divides the *x-y* plane (area or surface) into two halves.

Step 4: Shade the half-plane which is the solution set.

Example 1   Graph the solution set for the inequality
$$3x + 2y \geq 4$$

Solution

Step 1: Replace the inequality symbol by the equality symbol and solve for y.
$$3x + 2y = 4$$
$$y = -\tfrac{3}{2}x + 2$$

Step 2: Choose convenient *x*-values and calculate corresponding *y*-values
(To avoid fractional values of y, solve for *y* first, and choose *x*-values divisible by the denominator of the *x*-term)

For $x = 0, y = 2 \Leftrightarrow (0, \ 2)$

For $x = 2, y = -1 \Leftrightarrow (2, -1)$

Step 3: Plot the ordered pairs $(0, 2)$ and $(2, -1)$. See Figure .

Step 4: Connect the two points by a **solid straight line** since the inequality symbol is " $\geq$ "
(That is, the half-plane is closed)

Step 5: Now, determine which half-plane to shade (or is the solution set). Let us choose the origin $(0, 0)$ to test the inequality. Substituting $x = 0, y = 0$ in the original inequality.

$$3(0) + 2(0) \overset{?}{\geq} 4$$

$$0 \overset{?}{\geq} 4 \qquad \textbf{No} \quad \text{(False inequality)}$$

Thus, the origin and the half-plane in which the origin is located **is not** in the solution set. Therefore, the solution set is **the other** half-plane. The solution contains, for example, the point $(3, 4)$.

Step 5: Shade this half-plane (the solution half-plane) as in Figure.

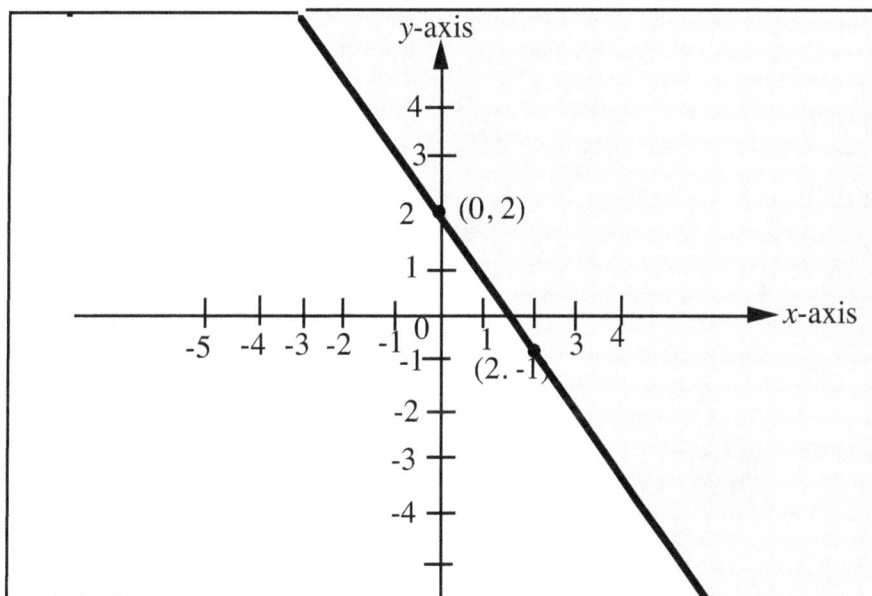

Figure Graph of $3x + 2y \geq 4$ (Note the solid boundary line)

The use of the origin $(0 , 0)$ in determining the solution half-plane

Using the origin, $x = 0, y = 0$ makes the checking relatively easy and fast because the product of zero and a number is zero; and also when zero is added to a number, the number remains unchanged. However, if the boundary line passes through the origin $(0, 0)$, then choose some other convenient point for testing. (Do not use $(0, 0)$ for testing if there are inequalities such as $y \geq ax$, $y \leq ax$, $x \geq 0$ and $y \geq 0$.)

Example 2 Graph the solution set for the inequality: $3x + 2y > 4$ 148

Steps 1, 2, 4, and 5 are the same as those in the last example (Example 1) except that in Step 3, we use **broken** (dashed) **lines in** connecting the two points , since the inequality symbol is " > ". and consequently, the solution half is open. See Figure 44.

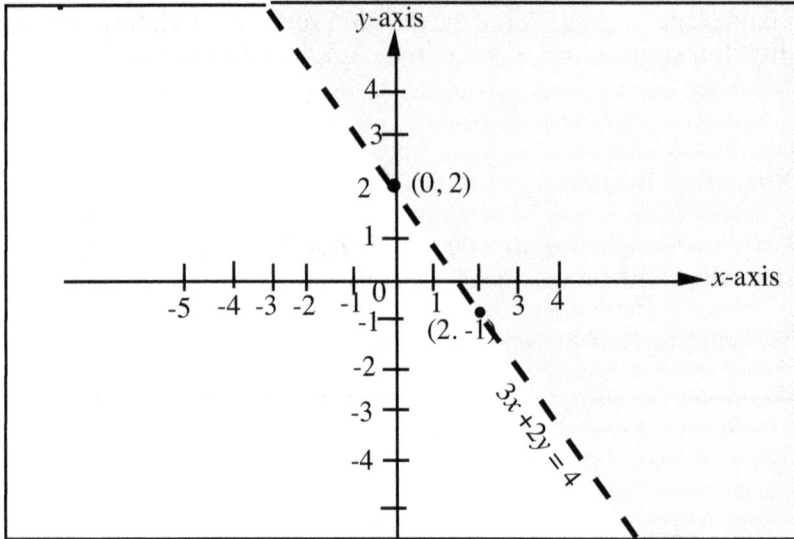

Figure   The graph of $3x + 2y > 4$ (Note the broken boundary line)

Example 3 Graph the solution set for the inequality: $3x + 2y \leq 4$

Steps 1, 2, 3, and 4, are the same as those in Example 1, except that in Step 5. we have

$$3(0) + 2(0) \overset{?}{\leq} 4$$

$$0 \overset{?}{\leq} 4 \qquad \textbf{Yes} \quad \text{(True inequality)}$$

The true statement shows that the origin $(0, 0)$ **is in** the solution set. We therefore shade the solution half-plane containing $(0, 0)$.. See Figure

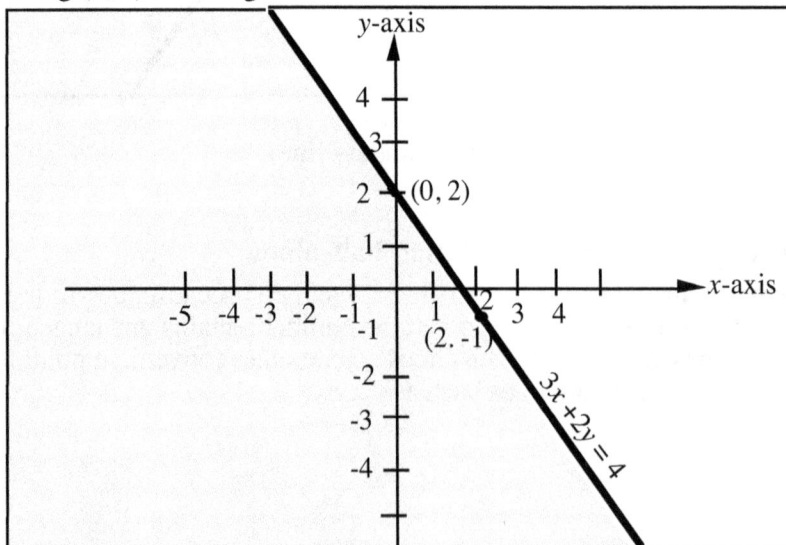

Figure: The graph of $3x + 2y \leq 4$

Example 4 From Example 3, the solution set for the inequality $3x + 2y < 4$ is

graphed in Figure 46. Note the broken line (compared to the solid line in Example 3).

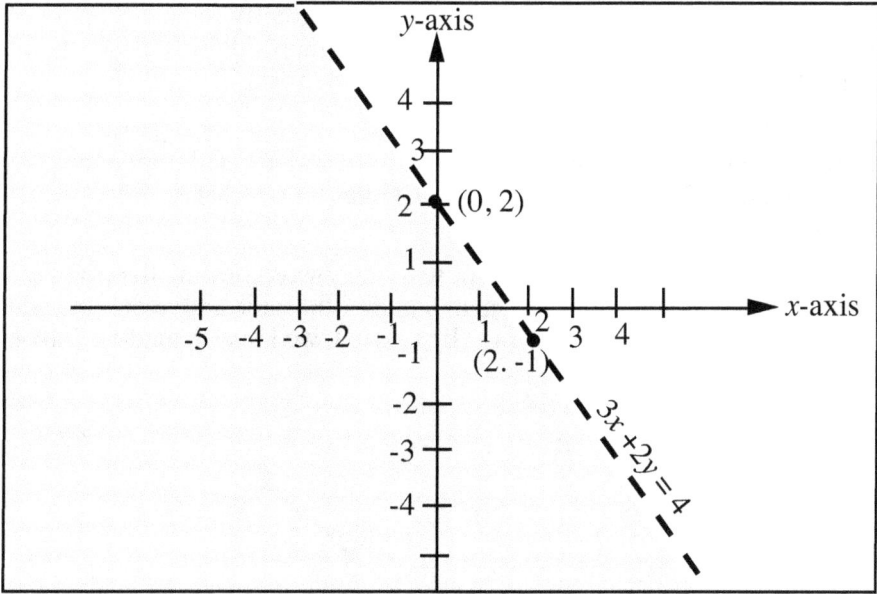

Figure: The graph of $3x + 2y < 4$

Lesson 23: Graphing Linear Inequalities in two Variables

Another method of determining the solution half-plane:

150

"The Seesaw-View method"

This method depends on two main principles. First, we solve the inequality for y. Secondly we should be able to point out which side of the boundary line is " above", and which side is "below" the boundary line. (When the inequality is solved for y, the inequality would be of the form

$$y \geq ax + b, \ y \leq ax + b, \ y < ax + b, \ \text{or} \ y > ax + b)$$

The symbols " > and ≥" will imply " above "; and the symbols " ≤ and < " will imply "below ". To determine the " above" and the " below " of the boundary line (the straight line drawn) we use the following mnemonic device:

Assume the line, ℓ, and the y-axis form a seesaw system, with the y-axis forming the pivot at where the axis meets the line. We will call the side where one sits or stands " the side above the line and we will call the other side " the side below the line ". In Figures (a), (b), (c), (d), we have shaded the half-planes above the lines for clarity.

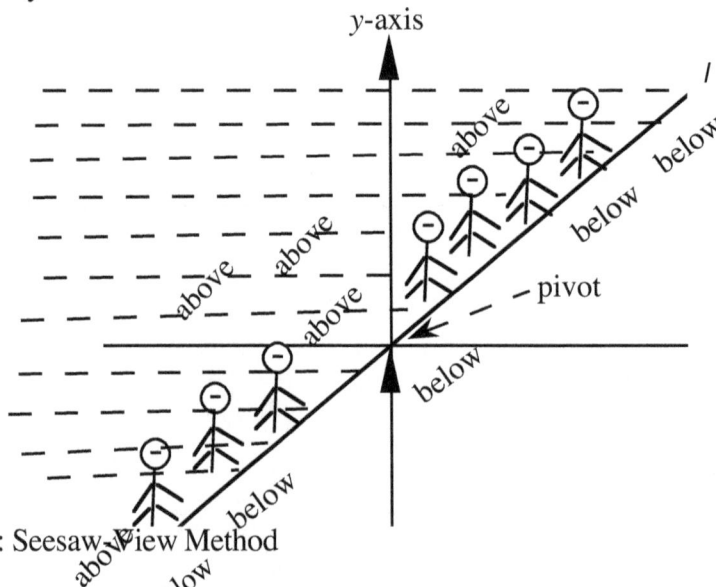

Figure: Seesaw-View Method

Note: The seesaw-view method is an original method devised by the author.

Figure: Seesaw-View Method

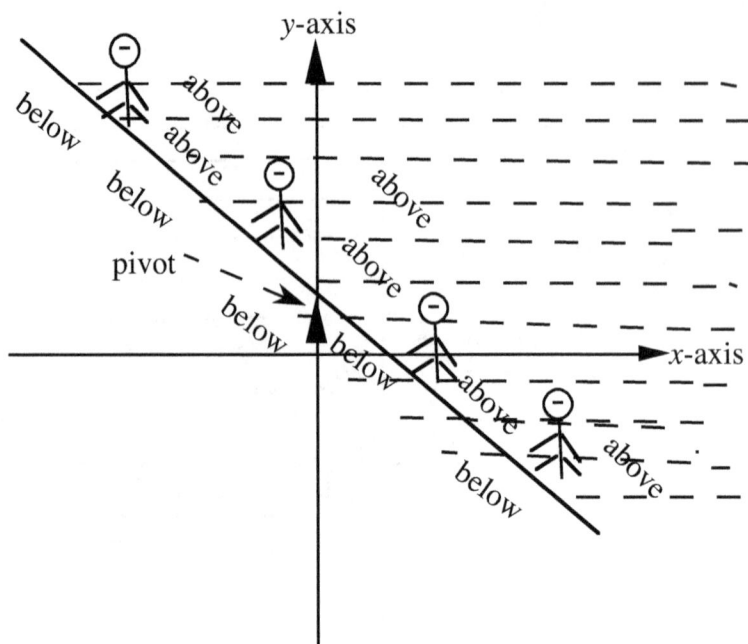

Figure: Seesaw-View Method

We will redo the previous examples using the " seesaw view" method, noting that::
 the symbols " **>** and **≥** " (greater than symbols) will imply " **above the line**" and the
symbols "**<** and **≤** (less than symbols) will imply "**below the line** ". The inequality must be placed in the
form $y ≥$ (or $>, ≤, <$) $ax + b$. (That is, solve the inequality for y first before using this method.)

Example 5 (Example 1 redone)

 Graph the solution set for the inequality

$$3x + 2y ≥ 4$$

Solution (The solution is a closed half-plane including the boundary because of the " ≥ ".)

Step 1: Solve the inequality for y.

$$3x + 2y ≥ 4$$

$$y ≥ -\tfrac{3}{2}x + 2$$

Step 2: Replace the inequality symbol by the equality symbol and find two convenient ordered pairs
to graph the straight line.

$$y = -\tfrac{3}{2}x + 2$$

For $x = 0, y = 2 ⇔ (0,\ 2)$

For $x = 2, y = -1 ⇔ (2, -1)$

Step 2: Plot the ordered pairs (0, 2) and (2, -1). See Fig. 43.

Step 3: Connect the two points by a **solid straight line** since the inequality symbol is " ≥ "
     (That is, the half-plane is closed)

Step 4: Since the sense of the inequality with y solved for is " > " , the **greater than** symbol,
we shade half-plane **above** the line . See Figure. 43
The solution set is a closed half-plane including the boundary.

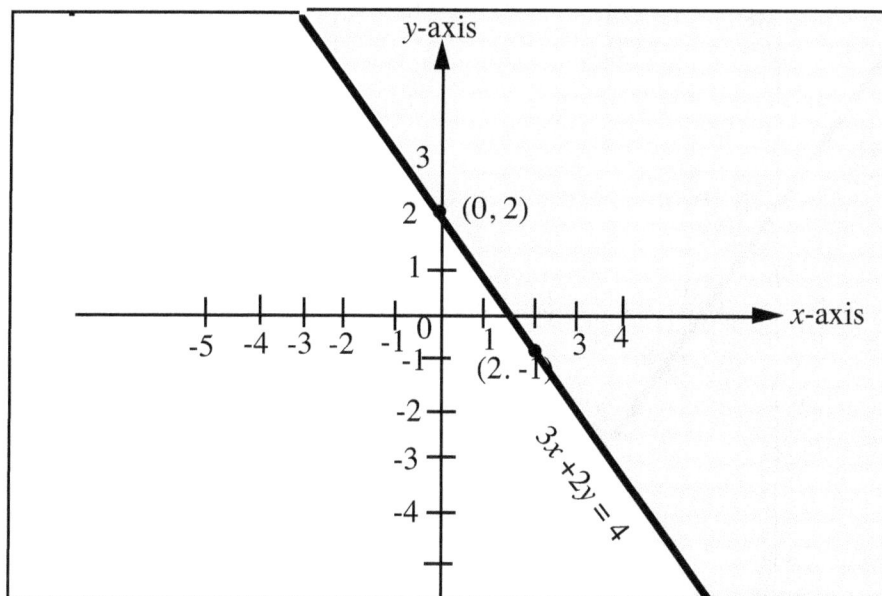

Figure: Graph of $3x + 2y \geq 4$

Example 6 (Example 3 redone)

Graph the solution set for the inequality

$$3x + 2y \leq 4$$

Step 1: Solve the inequality e for y.

$$3x + 2y \leq 4$$
$$y \leq -\tfrac{3}{2}x + 2$$

Step 2: Replace the inequality symbol by the equality symbol and find two convenient ordered pairs to graph the straight line.

$y = -\tfrac{3}{2}x + 2$

For $x = 0, y = 2 \Leftrightarrow (0, 2)$

For $x = 2, y = -1 \Leftrightarrow (2, -1)$

Step 2: Plot the ordered pairs $(0, 2)$ and $(2, -1)$. See Figure. 43.

Step 3: Connect the two points by a **solid straight line** since inequality symbol is " $\geq$ "
(That is, the half-plane is closed)

Step 4: Since the sense of the inequality with y solved for is" $<$ " , the **less than** symbol , we shade half-plane **below** the line $y = -\tfrac{3}{2}x + 2$. See Figure
The solution set is a closed half-plane including the boundary.

We will now use the " seesaw view method " to graph more inequalities without explanation

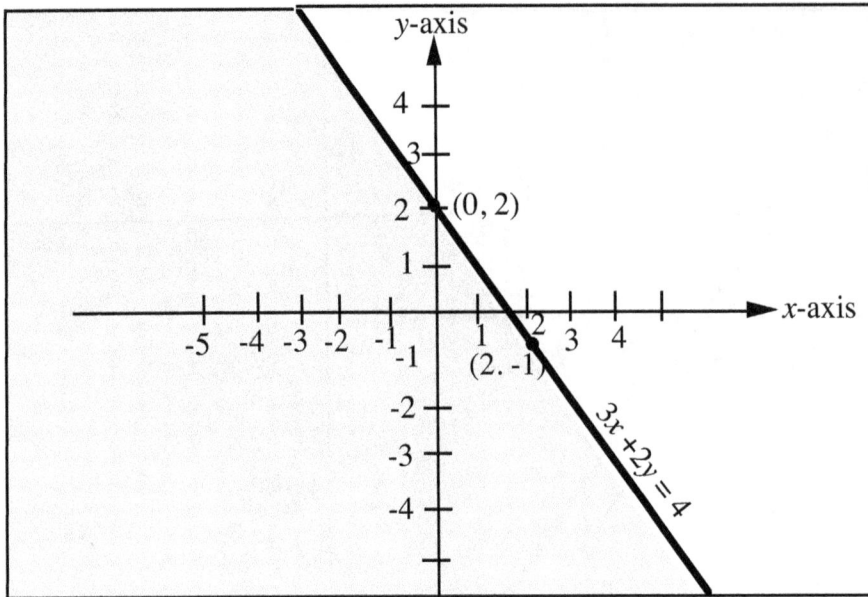

Figure Graph of $3x + 2y \leq 4$

Example 7 Graph the solution set for the inequality $y + 2x \geq 6$.

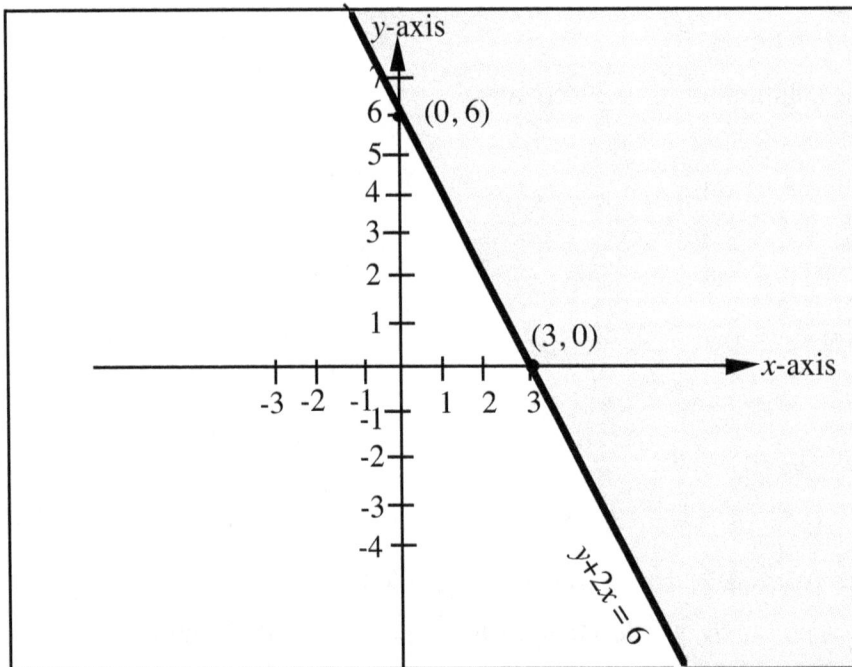

Figure : The graph of $y + 2x \geq 6$.

Example 8: Graph the solution set for the inequality $y + 2x \le 6$

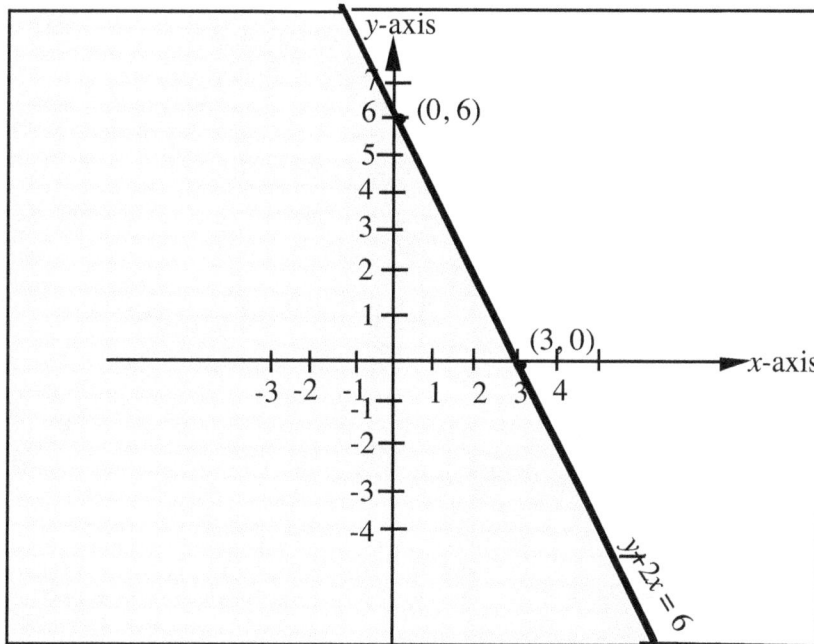

The graph shows the line $y + 2x = 6$ passing through $(0, 6)$ and $(3, 0)$, with the region shaded.

Figure The solution set for the inequality $y + 2x \le 6$

Example 9 Graph the solution set for the inequality

$$4x - 2y \le -5 \qquad \text{(A)}$$

Step 1: Solve the inequality for y.

$$4x - 2y \le -5$$

$$-2y \le -5 - 4x$$

$$y \ge \frac{-5}{-2} - \frac{4x}{-2} \qquad \text{(The sense of the inequality reverses because of division by a negative number)}$$

$$y \ge \frac{5}{2} + 2x$$

$$\boxed{y \ge 2x + \frac{5}{2}} \qquad \text{(B)}$$

Step 2: Replace the inequality symbol by the equality symbol and find two convenient ordered pairs to graph the line.

$y = 2x + \frac{5}{2}$

For $x = 0, y = \frac{5}{2}$, and we obtain the ordered pair $(0, \frac{5}{2})$

For $y = 0, x = -\frac{5}{4}$, and we obtain the ordered pair $(-\frac{5}{4}, 0)$

Step 2: Plot the ordered pairs $(0, \frac{5}{2})$ and $(-\frac{5}{4}, 0)$ See Fig. 49.

Step 3: Connect the two points by a **solid straight line** (boundary included).

Step 4: Since the sense of the inequality with y solved for is the " **greater than** "symbol inequality B) we shade half-plane **above** the line $y = 2x + \frac{5}{2}$ See Figure . For a quick check, test with $(0, 0)$

Note: In Step 4, do not use implicitly inequality (A) but use inequality (B).

Lesson 23: Graphing Linear Inequalities in two Variables

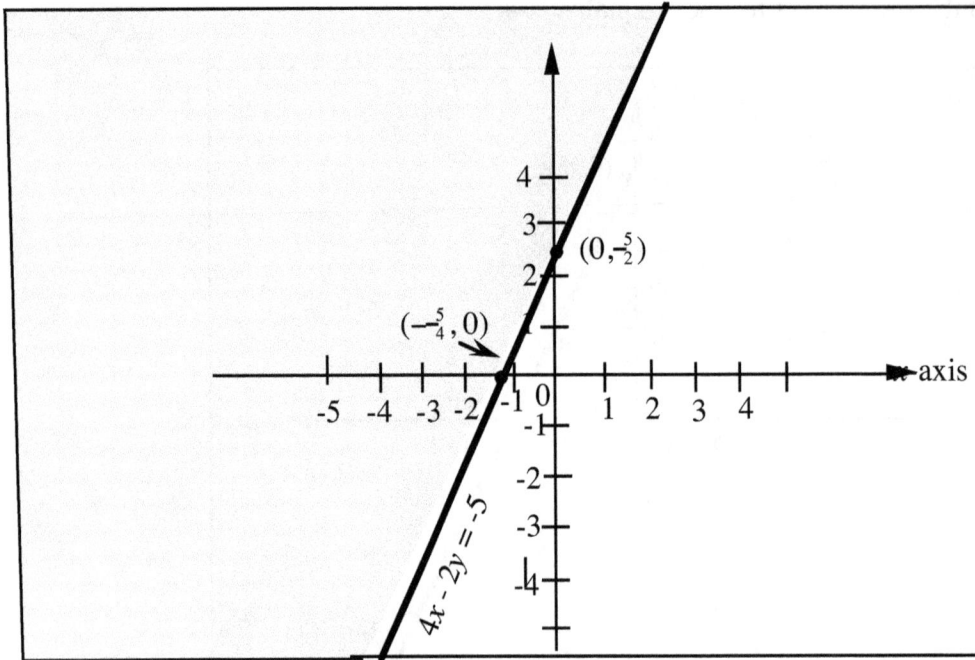

Figure : Graph for $4x - 2y \leq -5$

Special Cases of Linear Inequalities and Distinction between a Point, a Set of Points, a Line, and a Half-plane

We must distinguish between, for example, the set of points $x \geq 4$ on a number line and the half-plane, $x \geq 4$ just as we distinguish between the point $x = 4$ on a number line and the line $x = 4$ in an x-y.

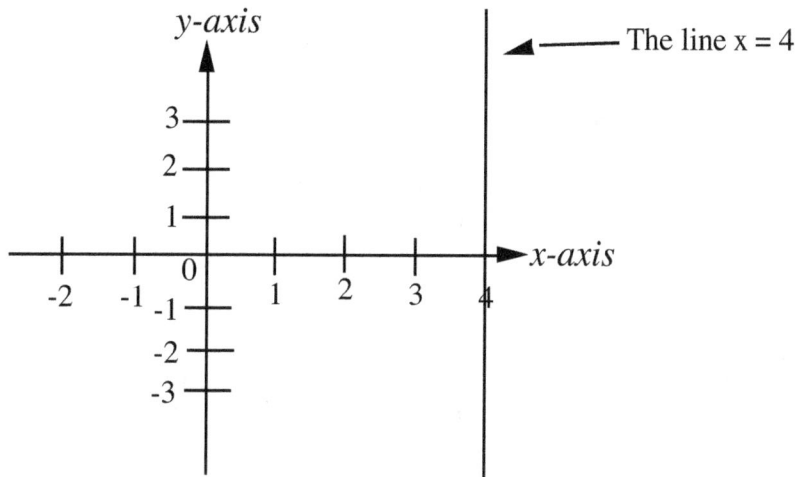

The graph of the point $x = 4$:

The graph of the set of points $x \geq 4$:

Figure : The graph of the line $x = 4$

Example 8 Graph the linear inequality $x \geq 4$.

Step 1: Replace " > " by " = ".
Then $x = 4$

Step 2: Draw the vertical line $x = 4$ using a solid line (Fig. 50)
To determine the half-plane which is the solution set, either choose a point and test or use the sense of the inequality.

Point test: Substitute $x = 0, y = 0$, (the origin) in $x \geq 4$, then $0 \geq 4$, which is false. Therefore, the origin is not in the solution. The solution set is the other side of the line $x = 4$

Sense test: Since the sense of the inequality symbol is " >", the solution half-plane is the half-plane to the right of the line $x = 4$ (noting that $x > 4$). The solution includes the boundary line.

Step 4: Shade the half-plane to the right of the line $x = 4$.

The graph of the half-plane $x \geq 4$:

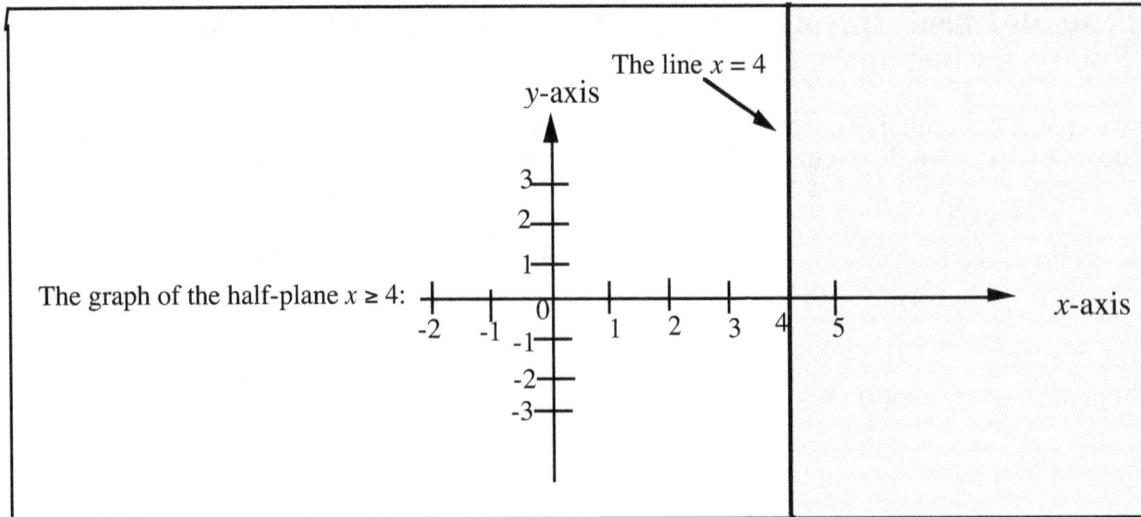

We present, below, similar special cases without the procedures:

(*a*) For the linear plane $x \leq 4$, the solution plane is the half-plane to the left of the line $x = 4$ (noting that $x < 4$), including the boundary. (Figure)

The graph of the half-plane $x \leq 4$:

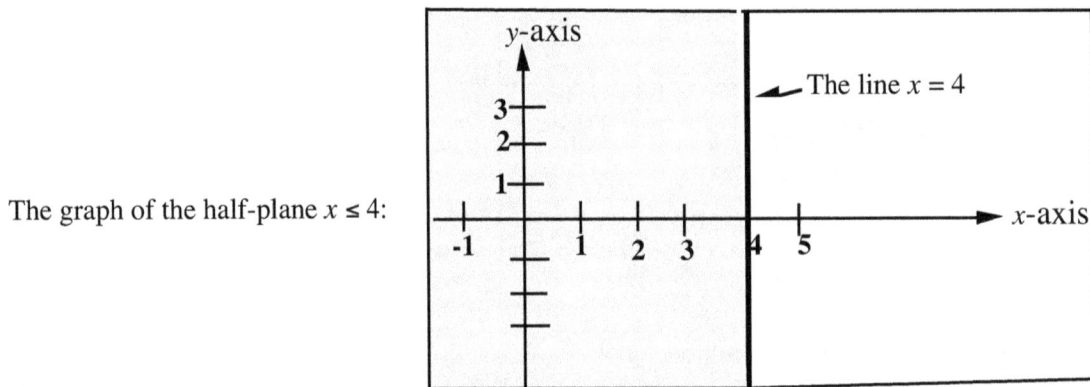

(b) For the linear plane $y > 2$, the solution plane (Figure ..) is the half-plane above (for the **greater** than symbol) the horizontal line $y = 2$, excluding the boundary; and we use a broken (dotted) line to draw the boundary, since the inequality symbol does **not include** the equality symbol.

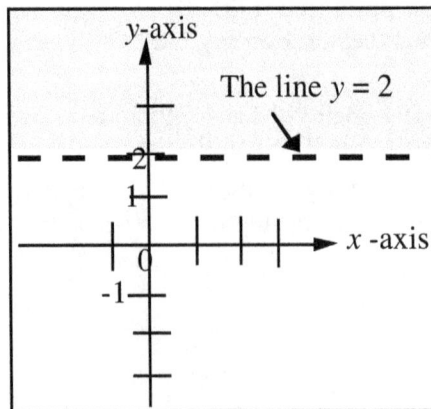

Figure

(c) For the linear plane $y \leq -4$, the solution plane (Fig. 53) is the half-plane below (for the **less** than symbol) the horizontal line $y = -4$, including the boundary and we use a solid line to draw the boundary, since the inequality symbol **includes** the equality symbol.

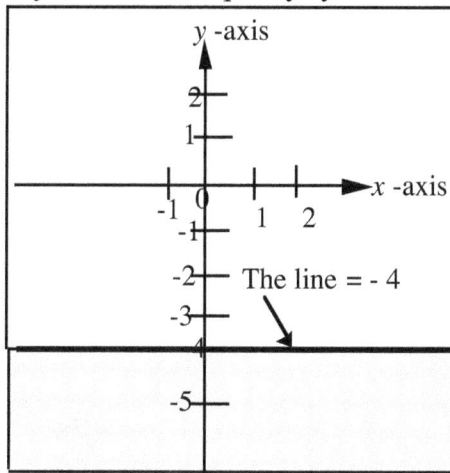

Figure: The graph of the linear plane $y \leq -4$

Lesson 23 Exercises

Questions 1-16:   Graph the solution-plane of each of the following:

1. $y + 4x < 2$;   2. $x + 5y + 5 \geq 0$;   3. $-4y + x \leq 12$;   4. $-x + y \geq 0$;

5. $3x - 2y < 6$; 6. $y \geq |x|$;   7. $y \leq |x|$; 8. $y \leq -|x|$; 9. $y < x$ 10. $y \geq x$ 11. $y \geq 4$;   12. $y \geq -2$;

13. $y \leq 4$;   14. $x \leq -2$;   15. $|y| \geq 2$;   16. $|y| - 2 \leq -|x|$;

Questions 17-20: Draw the graphs for each of the following:

17. The point $x = 3$;   18. The line $x = 3$;

19. The set of points $x \geq 3$;   20. The half–plane $x \geq 3$.

Answers:

1. $y + 4x < 2$

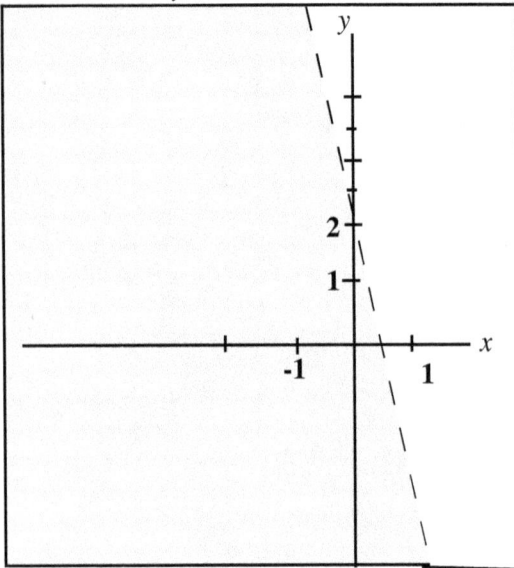

2. $x + 5y + 5 \geq 0$

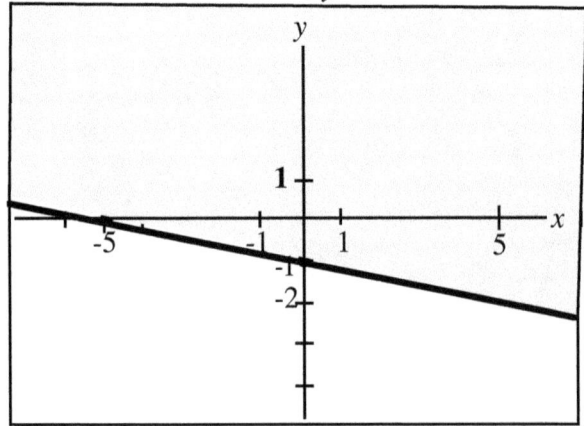

3. $-4y + x \leq 12$

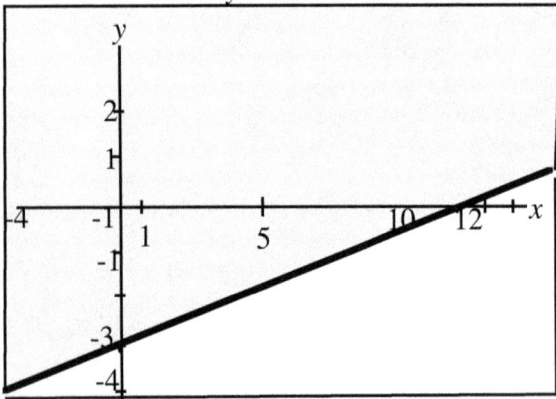

4. $-x + y \geq 0$

5.

$3x - 2y < 6$

6.

$y \geq |x|$

7.

$y \leq |x|$

8.

$y \leq -|x|$

9.

$y < x$

10.

$y \geq x$

11. $y \geq 4$

12. $y \geq -2$

13. $y \leq 4$

14. $x \leq -2$

15. $|y| \geq 2$

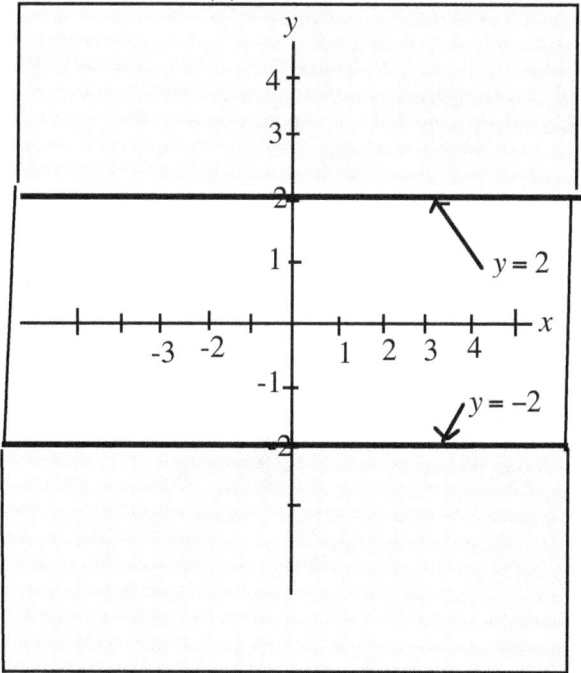

16. $|y| - 2 \leq -|x|$ or $|y| + |x| \leq 2$

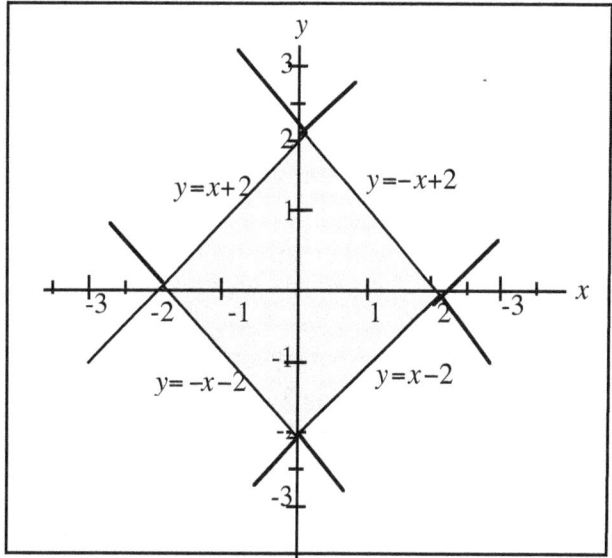

For solutions to Problems 17-20, see page 157-158 and imitate.

Lesson 24

Solving Systems of Linear Inequalities

Many relations in nature are not equalities (equations) but inequalities. The solution set of a system of linear inequalities is the intersection of the half-planes of the inequalities. Thus, the region (area) common to all the inequalities in the system is the solution set. We will learn the solution methods by an example.

Example   Solve the system of inequalities:

$$\begin{cases} 3y + 2x > 6 & \text{(A)} \\ y + 3 \geq 2x & \text{(B)} \end{cases}$$

Step 1: Graph the inequality $3y + 2x > 6$ or $y > -\frac{2}{3}x + 2$

Let $y = -\frac{2}{3}x + 2$

When $x = 0$, $y = 2$, and we obtain the ordered pair $(0, 2)$
When $x = 3$, $y = 9$, and we obtain the ordered pair $(3, 0)$.
Plot these points and connect them by a broken line. Determine the solution half-plane for $y > -\frac{2}{3}x + 2$ either by testing points or using the" seesaw view" method. Shade (- - -) the appropriate half-plane. (Figure 55)

Step 2: To graph the inequality $y + 3 \geq 2x$, solve for y to obtain $y \geq 2x - 3$

Let $y = 2x - 3$

When $x = 0$,. $y = -3$, and we obtain the ordered pair $(0, -3)$
When $x = 1$, $y = -1$ and we obtain the ordered pair $(1, -1)$

Step 3: Plot the ordered pairs (from Step 2) on the same set of axes as the first line (Figure 55) and connect the points by a solid line. Determine solution half-plane for $y + 3 \geq 2x$ either by testing points or by the " seesaw view" method. The solution is the half-plane above the line $y = 2x - 3$. Use a shading ("| | |") different from the shading for the first inequality.

Step 4: The solution to the system of inequalities is the region or area of the intersection of the solution planes of the two inequalities. This region has " + +" shading.. The solution set includes the boundary for $y + 3 \geq 2x$ but excludes the boundary for $3y + 2x > 6$ (Figure 55).

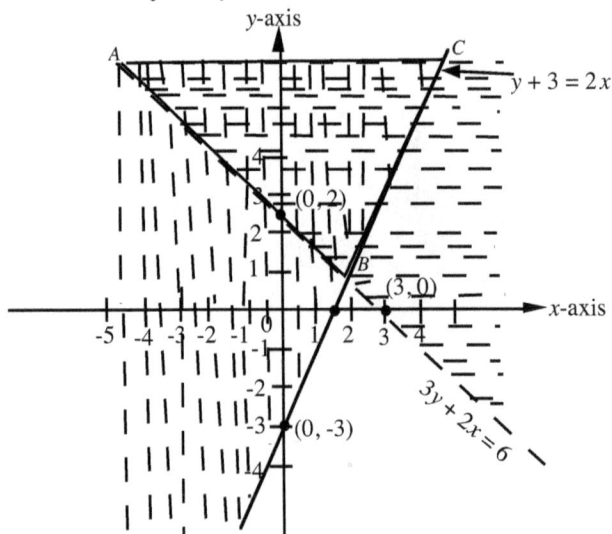

Figure: The solution set is the plane shaded with " ⊞ " (Triangle ABC)

Solve each system of inequalities graphically

1. $\begin{cases} x \le 3 \\ y > 5 \end{cases}$
2. $\begin{cases} x + y > 2 \\ x > 4 \end{cases}$
3. $\begin{cases} y \le x \\ x < 1 \end{cases}$
4. $\begin{cases} 2x - y < 3 \\ x + y < -2 \end{cases}$
5. $\begin{cases} 2x + 5y > -4 \\ 4x + 3y \le 6 \end{cases}$

6. $\begin{cases} -3x + 4y \ge 12 \\ 2x + 3y > -6 \end{cases}$
7. $\begin{cases} 2x + 3y \le 6 \\ x \ge 0 \\ y \ge 0 \end{cases}$
8. $\begin{cases} 3x + y \le 12 \\ x + 2y \le 12 \\ x \ge 0 \\ y \ge 0 \end{cases}$

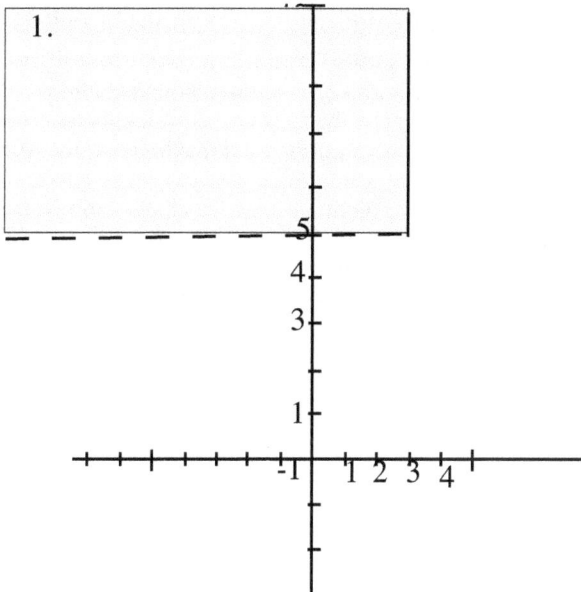

1.

Lesson 25

Linear Programming; Deriving Systems of Inequalities ; Applications of Linear Programming

Linear Programming

There are certain problems which arise in business management (for example, maximization and minimization problems), economics, engineering and logistics.

These problems involve two or more variables with restrictions (constraints) on the values of these variables in the form of linear inequalities.

In linear programming, we try to optimize certain results such as maximizing profits, or minimizing costs which are expressed as linear functions. The finding of the proper values (optimum values) is known as linear programming.

Let us consider a business problem involving two variables x and y. Let us choose x and y so that a certain linear function $P(x,y)$, expressed in the form of an equation will have either a maximum value or a minimum value, with restrictions (constraints) on the values of x and y. These restrictions are expressed in the form of linear inequalities. The function $P(x,y)$ which we want to minimize or maximize is called the **objective function.** Any ordered pair (x, y) satisfying the system of **constraints** is a feasible solution.

If we graph the inequalities of the constraints, and the solution set of this system forms a polygon (a bounded region), then the objective function has both a minimum value and a maximum value. However, if the solution set of the inequalities does not form a polygon (the solution set is unbounded), there may be a minimum value but not a maximum value. If there is no solution set for the system of inequalities, the programming problem has no solution.

The corners (vertices) of the polygon or the intersections of the boundaries of the solution set are feasible solutions. If there is a minimum or maximum value, that value will occur at one of the vertices (corners) of the polygon, or at an intersection point of the boundary. On substituting the x- and y-coordinates of this vertex in the equation for the objective function, we obtain the minimum or maximum value. We can also find the coordinates of the vertices of the polygon by solving the system of equations (arising from the inequalities) two by two (i.e., two equations at a time).

Example Given that $x + 2y \le 8$

$$5x + 3y \le 30$$

$$x \ge 0$$

$$y \ge 0,$$

maximize $P(x, y) = 2x + 5y$

Solution The procedure here is to graph the solution for the system of inequalities, determine the coordinates of the vertices (corners) of the polygon; and substitute in turns, the ordered pairs in $P(x, y) = 2x + 5y$. The largest value of $P(x, y)$ is the maximum value.

Step 1: Graph $x + 2y \le 8$ or $y \le -\frac{1}{2}x + 4$

Let $y = -\frac{1}{2}x + 4$

When $x = 0, y = 4$ and we obtain the ordered pair $(0, 4)$.

When $y = 0, x = 8$ and we obtain the ordered pair $(8, 0)$.

Plot these points and connect them by a solid line (Figure).

Using the seesaw view method or point testing, determine the solution half-plane for $x + 2y \le 8$, and shade the solution half-plane.

Step 2: Graph $5x + 3y \le 30$ or $y \le 10 - \frac{5}{3}x$

Let $y = 10 - \frac{5}{3}x$

When $x = 0, y = 10$ and we obtain the ordered pair $(0, 10)$. When $y = 0, x = 6$, and we obtain the ordered pair $(6, 0)$. Plot these ordered pairs and connect them by a solid line. Shade the solution half-plane (mentally).

Step 4: Graph $x \geq 0$.

Draw the line $x = 0$ (along the y-axis) and shade the half-plane $x \geq 0$

(That is, shade half-plane to the right of the y-axis).

Step 5: Graph $y \geq 0$.

Draw the line $y = 0$ (along the x-axis) and shade the half-plane above the x-axis (Fig.)

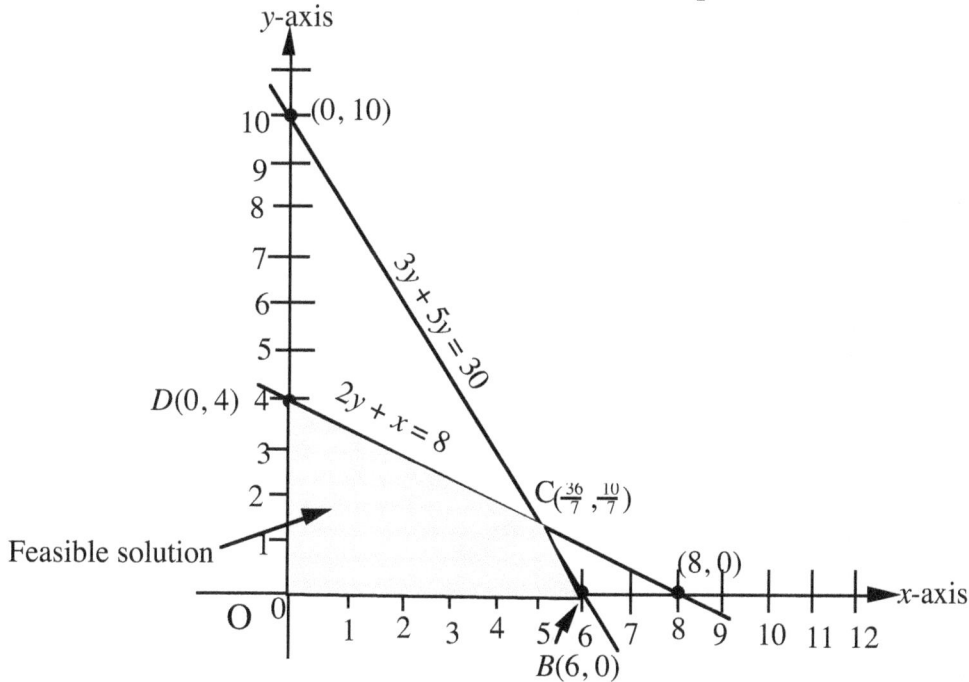

Figure: Graph of the system of inequalities:

$$2y + x \leq 8$$
$$3y + 5x \leq 30$$
$$x \geq 0$$
$$y \geq 0$$

In the figure, **solution plane** or region for the system of inequalities is the area (region) enclosed by polygon OBCD. Only the **intersection** of all the half-planes is shaded. The shadings for the individual half-planes are not shown for drawing convenience.

Step 6: Read the coordinates of the vertices of the polygon from the graph.

The coordinates of the vertices are $O(0,0)$, $B(6, 0)$, $C\left(\left(\frac{36}{7}, \frac{10}{7}\right)\right.$ and $D(0, 4)$.

Step 7: Substitute, in turns, the coordinates of the vertices in the objective function $P(x, y) = 2x + 5y$

For the vertex $(0,0)$: $P(0,0) = 2(0) + 5(0) = \mathbf{0}$

For the vertex $(6,0)$: $P(6,0) = 2(6) + 5(0) = \mathbf{12}$

For the vertex $\left(\frac{36}{7}, \frac{10}{7}\right)$: $P\left(\frac{36}{7}, \frac{10}{7}\right) = 2\left(\frac{36}{7}\right) + 5\left(\frac{10}{7}\right) = \mathbf{17.43}$

For the vertex $(0, 4)$: $P(0, 4) = 2(0) + 5(4) = \mathbf{20}$

Step 8: Find the maximum value.

From Step 7, the maximum value (largest value) of $P(x, y)$ is 20, and it occurs when $x = 0, y = 4$.

Note that generally, graphical solutions are approximate. Sometimes, to obtain the exact coordinates of the optimum solution, we will solve, simultaneously, the equations of the lines forming the vertex of the optimum solution . In the present problem, the vertex $(0, 4)$ has exact coordinates.

Derivation of Systems of Inequalities (Given the vertices of the convex polygon)

Example 4: Derive a system of inequalities for the graph of a convex polygon ABCD which has the following coordinates for its vertices: $A (3, 0)$, $B(6, 0)$, $C(6, 4)$, $D(3, 6)$.

Solution

Step 1: Plot the coordinate pairs of the given points.

Step 2: By means of straight lines, connect A and B, B and C, C and D, D and A (Fig.)

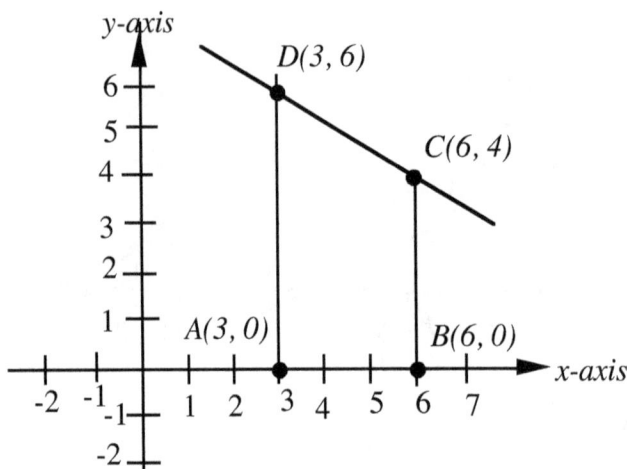

Figure:

Step 3: Write down the equations of the straight lines AB, BC, CD and DA. (See Chapt 7, Lesson 15)

> For AB : $y = 0$
>
> For BC : $x = 6$
>
> For DC : $y - 4 = \frac{6-4}{3-6}(x - 6)$ or $3y + 2x = 24$
>
> For DA : $x = 3$

Step 4: In Step 3, replace the equality symbol, " $=$ " by " $\geq$ " or " $\leq$" depending upon which half- plane intersects the solution plane of the graph.

> For AB : $y \geq 0$ (for the horizontal line $y = 0$ or the x–axis)
>
> For BC : $x \leq 6$ (for the vertical line $x = 6$)
>
> For DC : $3y + 2x \leq 24$ (applying $y - y_1 = \frac{y_2 - y_1}{x_2 - x_1}(x - x_1)$)
>
> For DA : $x \geq 3$ (for the vertical line $x = 3$)

The system of inequalities is

$$\begin{cases} y \geq 0 \\ x \geq 3 \\ x \leq 6 \\ 2x + 3y \leq 24 \end{cases}$$

The solution procedure in Example 4 can be viewed as the reversal of the procedure in Examples (1), (2) & (3), above.

Applications of Linear Programming

Example 1

A math test consists of sections A and B. In section A, each question is worth 6 points, and in section B, each point is worth 10 points.

A student spends 2 minutes to answer each of the questions in section A and 5 minutes to answer each of the questions in section B. A student is allowed to answer no more than a total of 24 questions from both sections. The time allowed for the test is one hour. How many questions should the student attempt from each of the sections A and B in order to get the best score. Assume that the student gets the correct answer to every question attempted.

Solution

Step 1: Determine what is required from the word problem. In this problem, we want to find the number of questions from section A and section B for maximization.

Let the number of questions attempted in section $A = x$
Let the number of questions attempted in section $B = y$
Let the total score $= S(x, y)$ (Where S is a function of x and y)

Step 2: Write an equation for the objective function (the quantity that we want to maximize). In this problem, we want to maximize the total score, $S(x, y)$

$S(x, y)$ = No. of A-questions × worth of each A-question + No. of B-questions × worth of each B-question
$S(x,y) = (x)(6) + (y)(10)$

$S(x,y) = 6x + 10y \leftarrow$ The objective function

Step 3: Write inequalities for the constraints (restrictions)

$x \geq 0$ (This means the number of A-questions answered is non-negative; i.e., zero or a positive number
$y \geq 0$ This means the number of B-questions answered is non-negative; i.e., zero or a positive number

Step 4: There is a constraint on the time.
Total time = No. of A-questions × time spent on each A-question + No. of B-questions × time spent on each B-question.
Total time $= 2x + 5y$

The total time allowed is either less than or equal to 60 minutes (1 hour).
Therefore, $2x + 5y \leq 60$

Step 5: There is a constraint on the total number of questions that may be attempted.
$x + y \leq 24$

Step 6: Summarize the objective function and the constraints.
Maximize $S(x,y) = 6x + 10y \leftarrow$ The objective function   (1)

Subject to $\begin{cases} 2x + 5y \leq 60 & (2) \\ x + y \leq 24 & (3) \\ x \geq 0 & (4) \\ y \geq 0 & (5) \end{cases}$

Step 7: Draw the graph of the system of inequalities (2), (3), (4), (5) to determine the coordinates of the vertices of polygon involved.

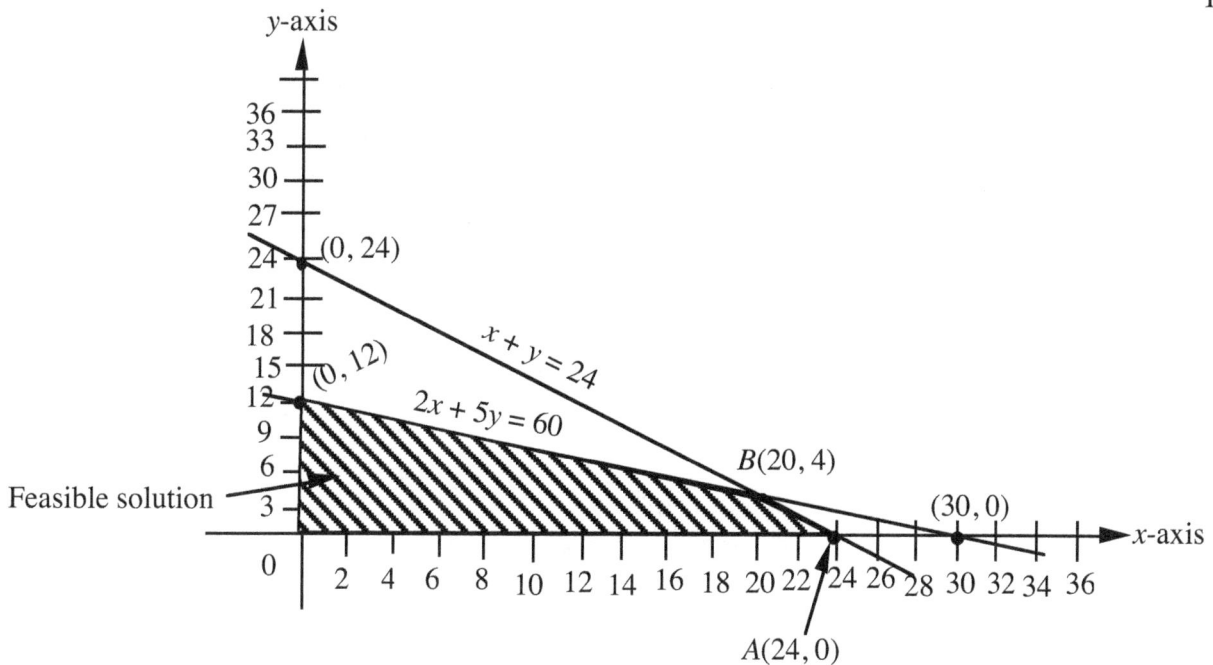

Figure :Graph of the solution set for the system of inequalities:

$$\begin{cases} 2x + 5y \le 60 & (2) \\ x + y \le 24 & (3) \\ x \ge 0 & (4) \\ y \ge 0 & (5) \end{cases}$$

Step 8: From the graph, read the coordinates of the vertices of the solution polygon. The solution polygon is the region of intersection of inequalities (2), (3), (4), (5) (i.e., region common to the solution sets of the inequalities.)

Step 9: Substitute each coordinate pair of the vertices of the polygon, in turns, in the objective function $S(x,y) = 6x + 10y$. See Table 2.

Table 2->

Coordinates of Vertices	$S(x, y) = 6x + 10y$
$O(0,0)$	$S = 0 + 0$
$A(24,0)$	$S = 6(24) + 10(0) = 144$
$B(20,4)$	$S = 6(20) + 10(4) = 160$
$C(0,12)$	$S = 6(0) + 10(12) = 120$

From Table 2, the maximum value of S= 160.
This value is obtained when $x = 20$, and $y = 4$. Hence, the best score the student is 160. To get this score, the student must answer 20 questions from section A and 4 questions from section B.
In the above solution, if we want to know if the values $x = 20$, $y = 4$ are exact, then we could solve the equations from inequalities (2) and (3) algebraically to obtain the exact coordinates of the point of intersection.

Lesson 25: Linear Programming; Systems of Inequalities, and Applications

Example 2 A business problem

A bag manufacturer produces two types of bags. One type is made of leather and the other type is made of vinyl. Each of the leather bags requires 2 hours of cutting and 4 hours of sewing, while each vinyl bag requires 1 hour of cutting and 3 hours of sewing. The cutting machine can be used up to 20 hours per day, and the sewing machine can be used up to 16 hours per day. On each leather bag, the profit is $40, and on each vinyl bag, the profit is $25. Assuming that all the bags manufactured are sold, how many of each type of bags must be produced per day to obtain maximum profit?

Solution

Step 1: Determine what is required. In this problem, we want to know the number of leather bags and the number of vinyl bags to be produced to obtain maximum profit.

Let the number of leather bags $= x$

Let the number of vinyl bags $= y$

Let the total profit $= P(x, y)$ (where P is a function of x and y.)

Step 2: Write an equation for the objective function (the quantity that we want to maximize).

Total profit = No. of leather bags $\times$ Profit on each leather bag) + No. of vinyl bags $\times$ profit on each vinyl bag.

$$P(x,y) = (x)(40) + (y)(25)$$

$$P(x,y) = 40x + 25y \leftarrow \text{The objective function}$$

Step 3: Constraint on the length of time allowed for the cutting machine

$$2(x) + 1(y) \leq 20 \qquad \text{(total cuting time is at most 20 hours)}$$

$$2x + y \leq 20$$

Step 4: Constraint on the length of time allowed for sewing machine.

$$4x + 3y \leq 16 \qquad \text{(total sewing time is at most 16 hours)}$$

Step 6: Summarize Steps 1, 2, 3, and 4.

Maximize $P(x,y) = 40x + 25y \leftarrow$ The objective function

Subjective to:

$$\begin{cases} 2x + y \leq 20 \\ 4x + 3y \leq 16 \\ \quad x \geq 0 \\ \quad y \geq 0 \end{cases} \qquad \text{(A)}$$

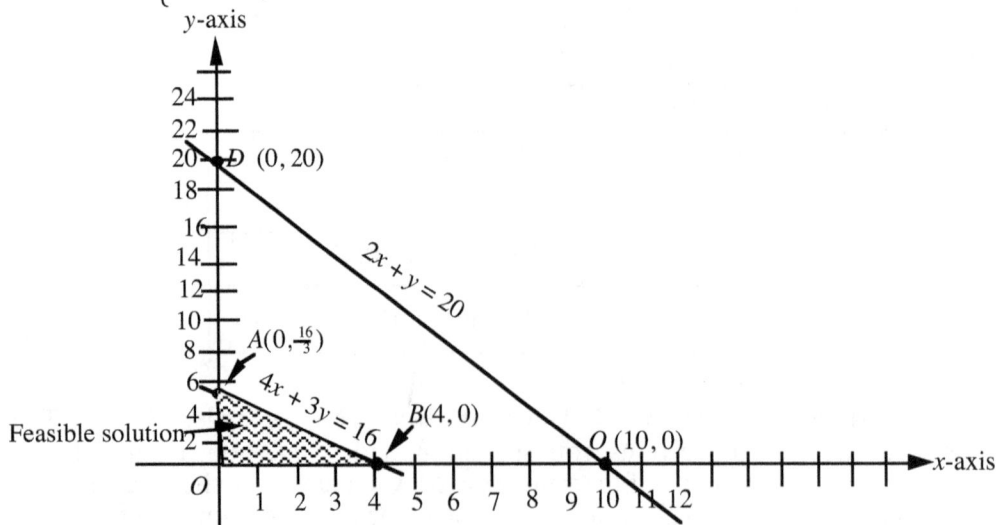

Figure : Graph of the system of inequalities (A)

Step 8: Shade the intersection of all the half-planes of the system of inequalities. The solution polygon is the convex polygon AOB (also a triangle). We should note that the solution polygon is **not** ABCD.

Step 9: Read the coordinates of the vertices (corners) of polygon AOB. Then we obtain

$A(0, \frac{16}{3}),)$ $O(0,0)$, $B(4,0)$.

Step 10: Substitute the coordinates of each coordinate pairs in turns, in the objective function

$P(x,y) = 40x + 25y \leftarrow$ The objective function

$P(0,\frac{16}{3}) = 40(0) + 25(\frac{16}{3}) = 133\frac{1}{3}$

$P(0,0) = 40(0) + 25(0) = 0$

$P(4,0)) = 40(4) + 25(0) = 160$

Step 11: The maximum value of $P(x, y)$ is 160, and it occurs at $x = 4, y = 0$.
Therefore, the manufacturer should produce 4 leather bags and no vinyl bags per day.

We must note in Step 11 that the decision made is based solely on maximum profit considerations and that in practical applications, a manufacturer may decide to produce some vinyl bags, based on other considerations (such as market situation).

Lesson 25 Exercises

1. Maximize $P = 3x + y$ subject to:

$$2x + y \leq 6$$
$$x + 3y \leq 8$$
$$x \geq 0$$
$$y \geq 0$$

2. Maximize $P = 16x - 2y$ subject to:

$$6x + 8y \leq 48$$
$$x + 3y \leq 8$$
$$0 \leq y \leq 4$$
$$0 \leq x \leq 7$$

3. Maximize $P = 5x + 6y$ subject to:

$$x + y \leq 10$$
$$x + 3y \leq 8$$
$$y \leq 3$$
$$x \leq 4$$
$$x \geq 0$$
$$y \geq 0$$

4. Maximize $P(x,y) = 6x + 4y$ subject to:

$$2x + 5y \leq 60$$
$$x + y \leq 24$$
$$x \geq 0$$
$$y \geq 0$$

Answers: 1. Maximum $= 9$ at $(3,0)$; **2.** Maximum $= 112$ at $(7, 0)$;

3. Maximum $= 28$ at $(4, \frac{4}{3})$; **4.** Maximum $= 144$ at $(24, 0)$

B A math test consists of sections A and B. In section A, each question is worth 5 points and in section B, each question is worth 8 points. A student takes 3 minutes to answer each of the questions in section A, and 5 minutes to answer each of the questions in section B.
The total number of questions a student can answer from both sections is 16. The time allowed for the test is 62 minutes. How many questions should the student attempt from each section in order to maximize his score. Assume that every question attempted is correct.

Answer: 9 questions from section A and 7 questions from section B. Max score $=101$

CHAPTER 12

Lesson 26: Solving Quadratic Inequalities
Lesson 27: Solving Factored Inequalities; Inequalities Involving
Rational Expressions and Absolute Values

Lesson 26
Solving Quadratic Inequalities

We will cover five methods:

Method 1: By point-testing of intervals.

Method 2: By rules of signs.

Method 3: By sign diagrams.

Method 4: By completing the square.

Method 5: By graphing.

Method 1: Solving Quadratic Inequalities by point-testing of intervals

Example 1 Solve the quadratic inequality. $x^2 + 2x - 15 \leq 0$ (1)

Solution

Step 1. Change the inequality sign "$\leq$" to the equality sign "$=$".

Then we obtain $x^2 + 2x - 15 = 0$
Solve for x by factoring if easily factorable, otherwise solve by the quadratic formula.
By factoring, $(x - 3)(x + 5) = 0$ and from which $x = 3$, or $x = -5$.

Step 2: Draw a number line and graph the points $x = 3$, $x = -5$ (Figure.)

Figure :

Since there are **two** endpoints on the number line, we consider **three** (**2+1**) intervals for testing.
The intervals are:

     1. All numbers less than -5;
     2. All numbers between -5 and 3;
     3. All numbers greater than 3.

We will include the endpoints, - 5 and 3 in the solution set because of the "$\leq$" symbol.

Of course we can test the endpoints in the inequality, but this is not necessary since in practice, we can determine if the endpoints are included solely by the type of the inequality symbol.

Step 3. Consider each of the above intervals and test for convenient numbers from each interval.

    Testing for all numbers less -5 (or $x < -5$) :
      Choose say $x = -6$ and substitute it in the original inequality (1)

$$(-6)^2 + 2(-6) - 15 \overset{?}{\leq} 0$$

$$9 \overset{?}{\leq} 0 \quad \text{(9 is not less than 0 and the inequality is \textbf{not} true when } x = -6)$$

Therefore, the interval, $x < -5$ is **not** satisfied and the interval "all numbers less than -5 is **not** a solution of the inequality.

Testing for all numbers between -5 and 3 $(-5 < x < 3)$.

Choose say $x = 0$ and substitute it in the original given inequality.

Then $(-0)^2 + 2(0) - 15 \overset{?}{\le} 0$

$-15 \overset{?}{\le} 0$ True statement

Thus $x = 0$ satisfies the given inequality and thus any number between -5 and 3 is a solution, including - 5 and 3.

Testing for all numbers greater than 3 (or $x > 3$):

Choose say $x = 4$ and substitute in $x^2 + 2x - 15 \le 0$

$(4)^2 + 2(4) - 15 \overset{?}{\le} 0$

Then we obtain $9 \overset{?}{\le} 0$ False statement   (9 is **not** less or equal to 0)

Therefore any number greater than 3 does **not** satisfy the inequality, and therefore is not a solution to the given inequality.

Conclusion: The interval satisfying the given inequality is the interval consisting of the set of numbers between -5 and 3 and including -5 and 3.

Symbolically, the solution is $[-5, 3] = \{x | -5 \le x \le 3\}$.

Method 2: Solving Quadratic Inequalities by Rules of Signs

Basis: The product of two factors of the same sign is positive but the product of two factors of different signs is negative.

Example 2 Solve the quadratic inequality. $x^2 + 2x - 15 < 0$.   (1)

Step 1: Factor the inequality (1) to obtain

$(x - 3)(x + 5) < 0$ ($(x - 3)(x + 5) < 0$ means $(x - 3)(x + 5)$ is negative)

Step 2: If $(x - 3)(x + 5) < 0$ then **either** (first factor is positive and second factor is negative: i.e.,$+ - = -$)

   Case 1: $x - 3 > 0$ **and** $x + 5 < 0$, which on solving yields $x > 3$ and $x < -5$.
         There is no solution for this case since a number cannot be greater than 3 and less than - 5 simultaneously.

                    or   (first factor is negative and second factor is positive: i.e.,$- + = -$)

   Case 2: $x - 3 < 0$ **and** $x + 5 > 0$, which on solving yields $x < 3$ and $x > -5$.
         There is solution for this case since a number can be less than 3 and greater than - 5 simultaneously . The solution is $\{x | -5 < x < 3\}$

Note: The factor, $x - 3 > 0$ means $x - 3$ is positive; and $x + 5 < 0$ means $x + 5$ is negative.

Example 3 Solve for x by rules of signs: $x^2 + 2x - 15 \le 0$

Solution: The procedure is the same as that in Example 2 above except that the endpoints are included in the solution.

By including the endpoints in the solution set (because of the " $\le$ " sign), we obtain the solution set $\{x | -5 \le x \le 3\}$ which is the same as that for Example 1 above.

Method 3: Solving Quadratic Inequalities by sign diagrams.

This method has the same basis as Method 2. The product of two factors of the same sign is positive but the product of two factors of different signs is negative. We will illustrate this method by redoing Example 1.

Solve the quadratic inequality $x^2 + 2x - 15 \leq 0$ (A)

Step 1: Find the linear factors either by factoring or by the quadratic formula.
From inequality (A), $(x - 3)(x + 5) \leq 0$ (B)

Step 2: To find where each factor is positive, assume that each binomial factor is positive and solve for x. (Note that saying that $x - 3$ is positive is the same as saying $x - 3$ is greater than 0)
Then for $x - 3 > 0$, we solve to obtain $x > 3$.. Thus, any number greater than 3 makes the linear factor $(x - 3)$ positive and any number less than 3 makes the linear factor $(x - 3)$ negative. We show these results on a sign diagram (Figure.).
Similarly, for $(x + 5) > 0$, we solve to obtain $x > -5$. Thus, any number greater than -5 makes $(x + 5)$ positive, and any number less than -5 makes $(x + 5)$ negative.

Step 3: Draw a number line and mark the points $x = 3, x = -5$.
The two points divide the line into three intervals (Figure) . For each linear factor, from Step 2, write plus signs ("+") over those values of x which make the linear factor positive; and write minus signs ("-") over those values of x which make the linear factor negative.

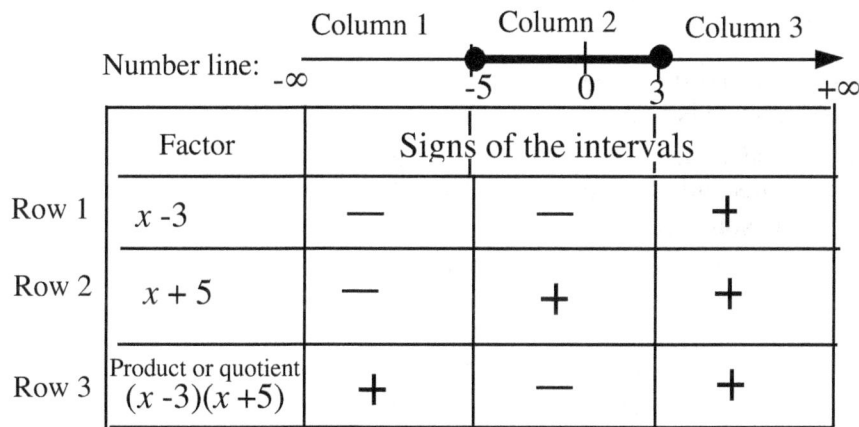

	Factor	Signs of the intervals		
Row 1	$x - 3$	—	—	+
Row 2	$x + 5$	—	+	+
Row 3	Product or quotient $(x - 3)(x + 5)$	+	—	+

For $(x - 3)$, we write plus signs to the right of the point $x = 3$ and minus signs to its left.
For $(x + 5)$, we write plus signs to the right of the point $x = -5$ and minus signs to its left.
The linear factors are in the first two rows and the product of the factors is in the third row.
The signs in the last row are the results of multiplying the signs in the columns above it.
For example, in column 1, the product of the minus signs gives a plus sign in row 3 in that column. For column 2, the product of the minus sign in Row 1 and the plus sign in Row 2 gives a minus sign as a product (column 2, row 3). Finally in column 3, the product of the two plus signs gives a plus sign (column 3, row 3).

Going back to the original inequality,
$x^2 + 2x - 15 \leq 0$ or
$(x - 3)(x + 5) \leq 0$,

we want the inequality to be negative for a solution (because of the " $\leq$ "). We inspect the row for the product (row 3) to determine which intervals (have minus signs) satisfy the given inequality. Inspection of the columns in row 3, shows that column 2 (column 2, row 3) has a minus sign (is negative) and therefore satisfies the given inequality.

Lesson 26: Solving Quadratic Inequalities

Conclusion: The interval satisfying the given inequality is the interval consisting of the set of numbers between -5 and 3 and including -5 and 3.
Symbolically, the solution is $[-5, 3] = \{x | -5 \le x \le 3\}$

Figure: Graph of the solution set to $x^2 + 2x - 15 \le 0$

Example 4 Solve for x: $x^2 + 2x - 15 < 0$

Solution Same solution as in Method 3 (previous example), except that the end points -5 and 3 are not included in the solution set, because of the change from " $\le$ " to "$<$."
Symbolically, the solution is $\{x | -5 < x < 3\}$ or $(-5, 3)$.

Example 5 Use the same sign diagram in Method 3 above to solve the inequality.
$$x^2 + 2x - 15 > 0$$

Solution: The inequality $x^2 + 2x - 15 > 0$ has the same linear factors as the inequality in Method 3. The only difference is in the sense of the inequality symbol.

Since we want the product of the linear factors in $x^2 + 2x - 15 > 0$ to be positive, we look along row 3, and choose the columns with the plus signs. The intervals with the plus signs are in column 1 and column 3. The **solution** is $(-\infty, -5) \cup (3, +\infty)$

If the inequality in Example 5 had been $x^2 + 2x - 15 \ge 0$, the end points -5 and 3 would have been included in the solution set.

We should note that if the factors of the inequality are not easily recognizable, we can use the quadratic formula to find the roots, say a_1, a_2. The linear factors to consider then would be $(x - a_1)(x - a_2)$.

Lesson 26: Solving Quadratic Inequalities

Example 6 Solve the inequality

$$2x^2 - 3x + 3 \le x^2 + 1 \qquad \text{(A)}$$

Step 1: Eliminate all the terms on the right-hand side (That is, move all the terms to the left-hand side)

Then inequality (A) becomes

$$x^2 - 3x + 2 \le 0 \qquad \text{(B)}$$

Step 2: Factor the left-hand side of (B) into linear factors.

Then (B) becomes $(x-1)(x-2) \le 0 \qquad$ (C)

Step 3: To find where each linear factor is positive, assume the linear factor is positive; solve for x and draw a sign diagram as was done in Method 3 of Example 1.

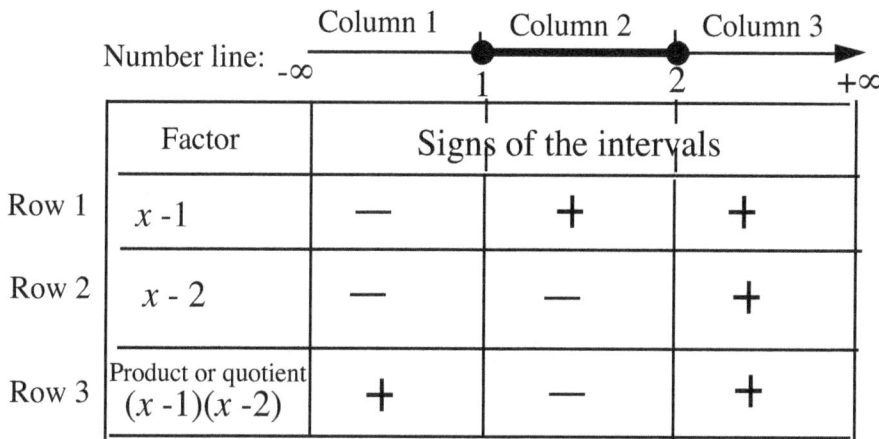

		Column 1	Column 2	Column 3
Number line:	$-\infty$	1	2	$+\infty$

	Factor	Signs of the intervals		
Row 1	$x - 1$	$-$	$+$	$+$
Row 2	$x - 2$	$-$	$-$	$+$
Row 3	Product or quotient $(x-1)(x-2)$	$+$	$-$	$+$

Figure: Sign diagram for $2x^2 - 3x + 3 \le x^2 + 1$
or $(x-1)(x-2) \le 0$

Since for a solution, the given inequality (in simplified form and standard form) (B) must be negative, the intervals having minus signs are in the solution set. From the sign diagram (Figure), this interval is in column 2, row 3. Thus, the solution set is $\{x \mid 1 \le x \le 2\}$. (Note that the endpoints are included.)

Example 7 Use the sign diagram in Example 6 to solve

$$x^2 - 3x + 2 \ge 0$$

Solution: Since the left-hand side of above inequality must be positive or zero, from the sign diagram (Figure..).We look for the column(s) in Row 3 which have plus signs "+".
Column 1 and Column 3 have "+" signs in Row 3.

The solution set $= (-\infty, 1] \cup [2, +\infty)$ which is the same as

$$\{x \mid x \le 1 \text{ or } x \ge 2\} \quad \text{(Figure. 146)}$$

Example 8 Solve the inequality: $(2 - x)(x + 3) > 0$ (A)   180

Solution   From $2 - x > 0$, we obtain $x < 2$ (solving for x)

Thus any number less than 2 makes the factor $(2 - x)$ positive and any number greater than 2 makes it negative. Similarly, for $x + 3 > 0$, we obtain $x > -3$

We now draw a sign diagram for $x < 2$ and for $x > -3$, noting that for $x < 2$, we write minus signs to the right of $x = 2$ (on the sign diagram); and plus signs to the left of the point $x = 2$. Note in particular, the " reversal " of the plus and minus signs (compared to previous examples) for $2 - x > 0$.

However, for $x + 3 > 0$, we write plus signs to the right of the point $x = -3$ and minus signs its left (Figure.): Try some numerical values in the solution set: That is

If $x = 0$, $(2 - 0)((0 + 3) = 2(3) = 6 > 0$, which is true.

The solution is $-3 < x < 2$

Number line:		Column 1	Column 2	Column 3

Number line: $-\infty$ ——o—— ——o—— $+\infty$ with points -3 and 2

	Factor	Sighs of the intervals		
Row 1	$2 - x$	$+$	$+$	$-$
Row 2	$x + 3$	$-$	$+$	$+$
Row 3	Product or quotient $(2 - x)(x + 3)$	$-$	$+$	$-$

Sign diagram for $(2 - x)(x + 3) > 0$

Solution graph for $(2 - x)(x + 3) > 0$

Method 4: Solving Quadratic Inequalities by Completing the Square and using the Absolute Value Definition

The principle involved in this method depends upon the relationship between the absolute value of an algebraic expression and the square of an algebraic expression. The absolute value of a non-zero number is always positive and the square of any nonzero number is also always positive. In addition, we will make use of the definitions of the absolute value inequality. Thus

    1. $|x| < b$ means $-b < x < b$

    2. $|x| > b$ means $x > b$ or $x < -b$

Example 9 Solve the inequality by completing the square

$$x^2 - 3x + 2 \le 0$$

Step 1: Complete the square on the x-term.

$$x^2 - 3x \le -2$$

$$x^2 - 3x + \left(-\frac{3}{2}\right)^2 \le -2 + \left(-\frac{3}{2}\right)^2$$

$$(x - \tfrac{3}{2}) \le -2 + \frac{9}{4}$$

$$(x - \tfrac{3}{2})^2 \le \frac{-8 + 9}{4}$$

$$(x - \tfrac{3}{2})^2 \le \frac{1}{4}$$

$$(x - \tfrac{3}{2})^2 \le \left(\tfrac{1}{2}\right)^2$$

$$\sqrt{(x - \tfrac{3}{2})^2} \le \sqrt{\left(\tfrac{1}{2}\right)^2}$$

$$\left|x - \tfrac{3}{2}\right| \le \tfrac{1}{2} \quad \text{Note: } \sqrt{(x - \tfrac{3}{2})^2} = \left|x - \tfrac{3}{2}\right|$$

Step 2: Apply "$|x| < b$ means $-b < x < b$

$$\left|x - \tfrac{3}{2}\right| \le \tfrac{1}{2} \qquad (\text{Note}: |x - a| \le b \text{ implies}$$

$$-\tfrac{1}{2} \le x - \tfrac{3}{2} \le \tfrac{1}{2} \qquad -b \le x - a \le b)$$

$$+\tfrac{3}{2} \qquad +\tfrac{3}{2} \qquad \tfrac{3}{2}$$

$$1 \le x \le 2$$

<----Note this relationship for the future

Example 10 Solve the inequality by completing the square $\quad x^2 - 3x + 2 \ge 0$

Step 1: Complete the square on the x-term.

$$x^2 - 3x \ge -2$$

$$x^2 - 3x + \left(-\tfrac{3}{2}\right)^2 \ge -2 + \left(-\tfrac{3}{2}\right)^2$$

$$(x - \tfrac{3}{2})^2 \ge -2 + \frac{9}{4}$$

$$(x - \tfrac{3}{2})^2 \ge \frac{1}{4}$$

$$(x - \tfrac{3}{2})^2 \ge \left(\tfrac{1}{2}\right)^2$$

$$\sqrt{(x - \tfrac{3}{2})^2} \ge \sqrt{\left(\tfrac{1}{2}\right)^2}$$

$$\left|x - \tfrac{3}{2}\right| \ge \tfrac{1}{2}$$

Step 2: $|x| > b$ means $x > b$ or $x < -b$

$$\left|x - \tfrac{3}{2}\right| \ge \tfrac{1}{2} \qquad (\text{Note}: |x - a| \ge b \text{ mplies}$$

$$x - \tfrac{3}{2} \ge \tfrac{1}{2} \text{ or } x - \tfrac{3}{2} \le -\tfrac{1}{2} \quad x - a \ge b \text{ or } x - a \le -b)$$

$$+\tfrac{3}{2} \quad +\tfrac{3}{2} \qquad +\tfrac{3}{2} \quad +\tfrac{3}{2}$$

$$x \ge 2 \qquad \text{or} \quad x \le 1$$

The solution set is $\{x \mid x \le 1 \text{ or } x \ge 2\}$

Note that the solution is the same as before in Example 7

Method 5: Solving Quadratic Inequalities Graphically

We can solve quadratic inequalities graphically. The procedure is as follows.

Step 1: Assume the inequality is of the form $f(x) > 0$ or $f(x) \geq 0$ or $f(x) < 0$ or $f(x) \leq 0$
and then graph $y = f(x)$

Step 2: Read the solution set for the inequality as follows:
   (*a*) If the inequality from Step 1 is of the form $f(x) \geq 0$ or $f(x) > 0$
      then the solution set consists of those x-values for which the graph is above the x-axis.
   (*b*) If the inequality from Step 1 is of the form $f(x) \leq 0$, then the solution set consists of those
      x-values for which the graph is below the x-axis.

Example 11 Solve graphically the following quadratic inequalities:
$$(a) \quad x^2 + 2x - 15 \geq 0; \qquad (b) \quad x^2 + 2x - 15 \leq 0$$

Solution

Step 1: Sketch the graph of $y = x^2 + 2x - 15$

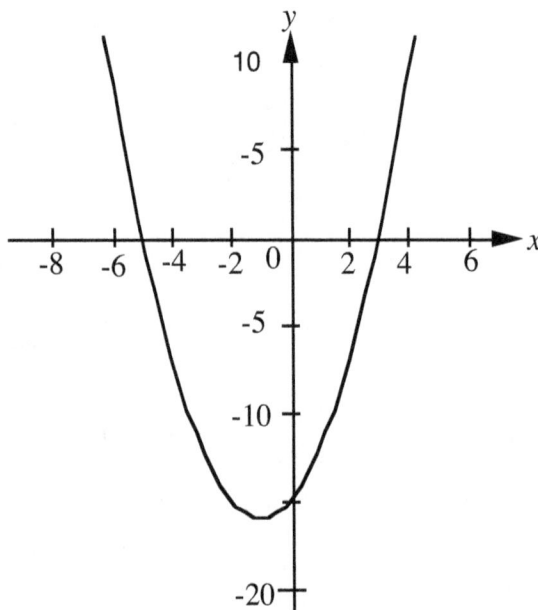

Step 2: Read the solution from the graph.
   Note that since we are dealing with inequalities, the solution is set is (usually) a set of infinite
   x-values (compared to the solution of a corresponding quadratic equation which have at most two roots).
   We use the roots of the corresponding quadratic equation as the endpoints of the solution set.

(a) Since the given inequality contains the " $>$ " symbol, we are interested in the part of the curve
   which is **above** the x-axis (i.e., where the corresponding quadratic function is positive: $f(x) > 0$)
   For the corresponding x-values of the parts of the curve above the x-axis, the solution set is
   $\{x | x \leq -5 \text{ or } x \geq 3\}$ Note that each x-value has a corresponding y-value.

(b) Since the given inequality contains the " $<$ " symbol, we are interested in the part of the curve
   which is below the x-axis (i.e., where the corresponding quadratic function is negative: $f(x) < 0$)
   For the corresponding x-values for the parts of the curve above the x-axis, the solution set is
   $\{x | x - 5 \leq x \leq 3\}$. Note that each x-value has a corresponding y-value.

Lesson 26 Exercises   183

1 Solve the quadratic inequality. $x^2 + 2x - 15 \leq 0$;

2. Solve the quadratic inequality $x^2 + 2x - 15 < 0$

3. $2x^2 - 3x + 3 \leq x^2 + 1$;   **4.** $x^2 - 3x + 2 \geq 0$;   **5.** $(2 - x)(x + 3) > 0$

6. $x^2 - 3x + 2 \leq 0$;   **7.** $x^2 - 3x + 2 \geq 0$

Answers: See Lesson 26

Note the following about the methods for solving quadratic inequalities

Note 1:

When studying, apply each method to do the same problem and then note the relative merits of each method for each type of problem. In the examples we cover in the future, we will apply the method of "sign diagrams". more often, However, try to do each problem using Method 1 (point testing of intervals) as well as other methods.

Note 2 (about using sign diagrams):

If there is a factor such as x^2 or $(x + 5)^2$ we note that each square is non-negative and we write plus signs only for the row for each squared factor. However we can also consider x^2 as two linear factors x and x and create two rows for them. Similarly, we can consider $(x + 5)^2$ as $x + 5$ and $x + 5$ and create two different rows for them. We can also extend this note and consider $(x - 2)^3$ as $(x - 2)(x - 2)(x - 2)$ and write three rows for them

Example Consider the inequality $x(x - 3)(x + 5)^2 \leq 0$ and note the two different ways of illustrating the factor $(x + 5)^2$ as in figures 1 and 2 below.

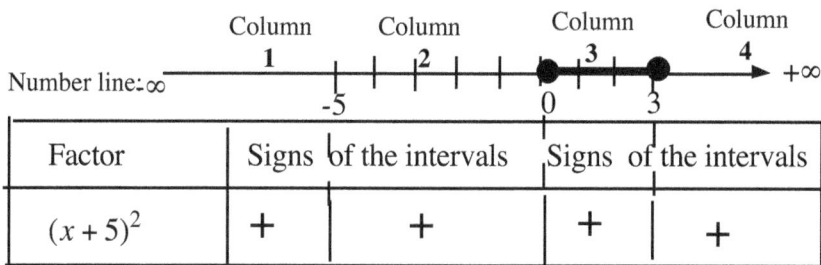

Fig, 1

Factor	Signs of the intervals		Signs of the intervals	
$(x + 5)^2$	+	+	+	+

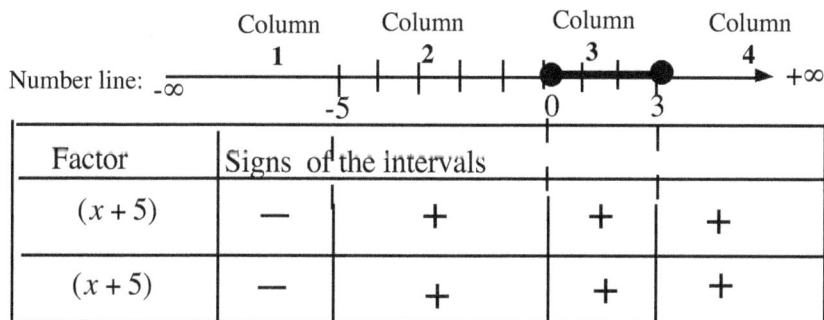

Fig, 2

Factor	Signs of the intervals			
$(x + 5)$	—	+	+	+
$(x + 5)$	—	+	+	+

Lesson 27 184

Solving Factored Inequalities; Inequalities Involving Rational Expressions and Absolute Values

Solving Factored Inequalities

Example 12 Solve the inequality $(x-2)(x-3)(x+1) \geq 0$

Step 1: As before, we determine where each linear factor is positive.

Then $(x-2) > 0$ and we obtain $x > 2$

$(x-3) > 0$ and we obtain $x > 3$

$(x+1) > 0$ and we obtain $x > -1$

Step 2: Draw a number line and plot the points $x = 2$, $x = 3$ and $x = -1$

Since there are three points, the number line would be divided into four (3+1) intervals.

Step 3: Complete a sign diagram for the factors as shown in Figure.

Factor	Intervals and signs		Intervals and signs	
Row 1 $x - 2$	—	—	+	+
Row 2 $x - 3$	—	—	—	+
Row 3 $x + 1$	—	+	+	+
Row 4 Product or quotient $(x - 2)(x - 3)(x+ 1)$	—	+	—	+

Sign diagram for $(x-2)(x-3)(x+1) \geq 0$

The given inequality must be positive and so, from the sign diagram, the intervals with the plus signs will yield the solution set.

The solution set is $\{x | -1 \leq x \leq 2 \text{ or } x \geq 3\}$

Figure Graph of the solution set for $(x-2)(x-3)(x+1) \geq 0$.

Note that we can solve the above factored inequality graphically, by imitating the graphical method for solving quadratic inequalities.

Inequalities Involving Rational Expressions

Example 13 Solve the inequality $\dfrac{x^2(x-1)(x+2)}{(2x+1)(x-3)} \le 0$.

We will a sign diagram to solve the above inequality.

Step 1: Determine where each linear factor is positive .

x^2 is a special case because it is positive for all non-zero real numbers.
For $(x-1) > 0$, $x > 1$.

For $(x+2) > 0$, $x > -2$.

For $(2x+1) > 0$, $x > -\frac{1}{2}$.

For $(x-3) > 0$, $x > 3$.

Step 2: Draw a number line and a sign diagram for the factors (Figure. ..).

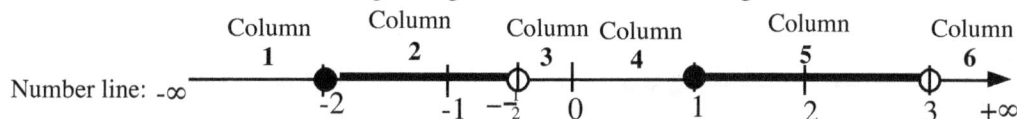

	Factor	Signs of the intervals				Signs of the intervals	
Row 1	x^2	+	+	+	+	+	+
Row 2	$x-1$	−	−	−	−	+	+
Row 3	$x+2$	−	+	+	+	+	+
Row 4	$2x+1$	−	−	+	+	+	+
Row 5	$x-3$	−	−	−	−	−	+
Row 6	$\dfrac{x^2(x-1)(x-2)}{(2x+1)(x-3)}$	+	−	+	+	−	+

Sign diagram for $\dfrac{x^2(x-1)(x-2)}{(2x+1)(x-3)} \le 0$

The solution is $\{x \mid -2 \le x < -\frac{1}{2} \text{ or } 1 \le x < 3\}$

Graph of the solution set for $\dfrac{x^2(x-1)(x-2)}{(2x+1)(x-3)} \le 0$.

Note that $-\frac{1}{2}$ and 3 are excluded values, otherwise the denominator becomes zero.

Example 14 Solve for x..

$$\frac{x^3(x-2)(x+1)}{(x-3)(x-1)} \geq 0$$

Step 1: x^3 is positive when x is positive and negative when x is negative.

For $(x-2) > 0,\ x > 2$.

For $(x+1) > 0,\ x > -1$.

For $(x-3) > 0,\ x > 3$.

For $(x-1) > 0,\ x > 1$.

Step 2: Draw a sign diagram for the above factors (Figure.)

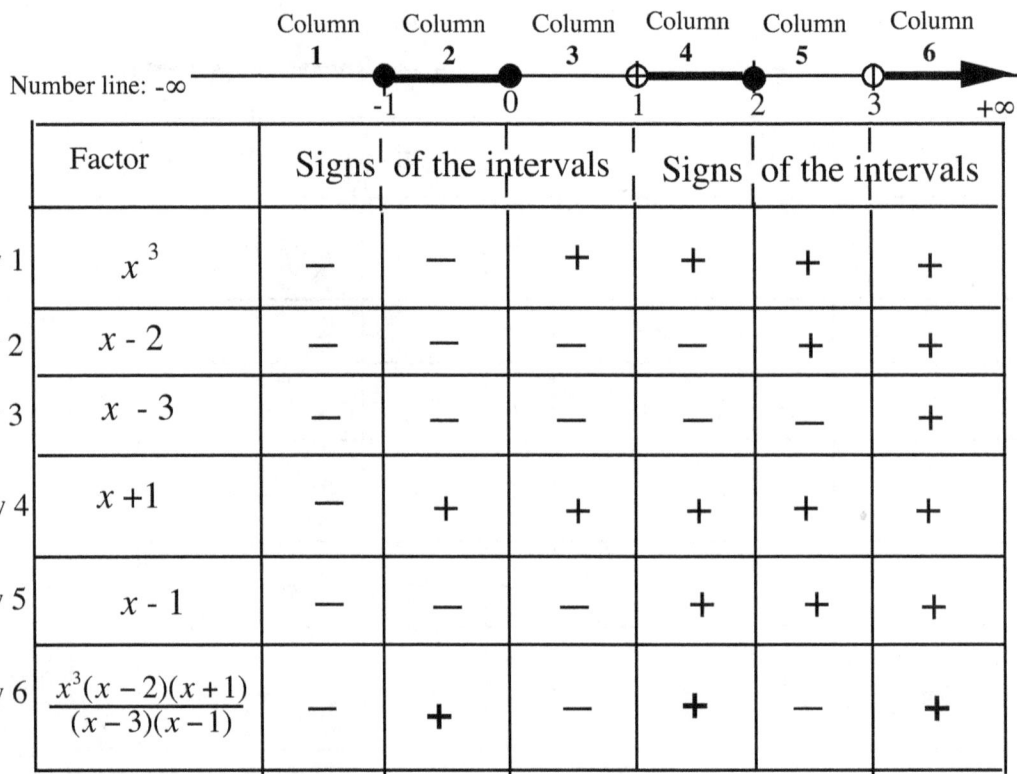

		Column 1	Column 2	Column 3	Column 4	Column 5	Column 6
	Factor	Signs of the intervals			Signs of the intervals		
Row 1	x^3	−	−	+	+	+	+
Row 2	$x-2$	−	−	−	−	+	+
Row 3	$x-3$	−	−	−	−	−	+
Row 4	$x+1$	−	+	+	+	+	+
Row 5	$x-1$	−	−	−	+	+	+
Row 6	$\dfrac{x^3(x-2)(x+1)}{(x-3)(x-1)}$	−	+	−	+	−	+

The solution is $\{x \mid -1 \leq x \leq 0 \text{ or } 1 < x \leq 2 \text{ or } x > 3\}$

Graph of the solution set of $\dfrac{x^3(x-2)(x+1)}{(x-3)(x-1)} \geq 0$

Note that 1 and 3 are excluded. When in doubt about a particular value of x, test it in the inequality.

Example 15 Solve for x

$$\frac{8}{x} > 3 \qquad (A)$$

Solution Method 1: **Using a sign diagram** .

$$\frac{8}{x} - 3 > 0$$

$$\frac{8 - 3x}{x} > 0$$

We now draw a sign diagrams for the factors.

For $(8 - 3x) > 0$, $x < \frac{8}{3}$ (plus signs to the left of $\frac{8}{3}$.

For $x > 0$, we have plus signs to the right of $x = 0$, and minus signs to the left of $x = 0$.

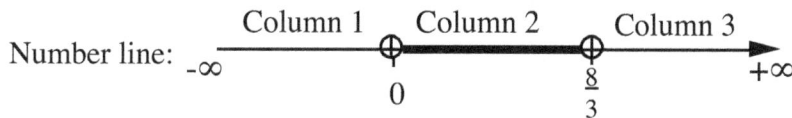

Number line:

	Factor	Column 1	Column 2	Column 3
		Signs of the intervals		
Row 1	$8 - 3x$	$+$	$+$	$-$
Row 2	x	$-$	$+$	$+$
Row 3	Quotient: $\frac{8 - 3x}{x}$	$-$	$+$	$-$

Sign diagram for $\frac{8}{x} > 3$

The solution is $\left\{ x \mid 0 < x < \frac{8}{3} \right\}$.

Method 2 Without sign diagram

For true inequality, the left-hand side of inequality (A) must be positive, since the right-hand side is positive. Therefore $x > 0$. (Note that if x were negative, the left-hand side of the inequality would be negative, and we know that a negative number is never greater than a positive number and therefore x must be positive so that $\frac{8}{x}$ is positive)

$$\frac{8}{x} > 3$$

$$(x)\frac{8}{x} > 3(x) \qquad \text{if } x > 0$$

$$8 > 3x$$

$$\frac{8}{3} > x \qquad \qquad \text{(dividing both sides of the inequality by 3)}$$

or $x < \frac{8}{3}$

Since x must be positive and less than $\frac{8}{3}$, the solution is $0 < x < \frac{8}{3}$.

Example 16 Solve the inequality

$$\frac{x-4}{x-2} \le 5 \qquad\qquad \text{(A)}$$

Step 1: Eliminate (move) the term on the right-hand side and simplify the resulting left-hand side.

$$\frac{x-4}{x-2} - 5 \le 0 \qquad\qquad \text{(B)}$$

$$\frac{x-4-5x+10}{x-2} \le 0$$

$$\frac{-4x+6}{x-2} \le 0$$

$$\frac{6-4x}{x-2} \le 0$$

Step 2: Draw a sign diagram for $(6 - 4x)$ and $(x - 2)$.

For $(6 - 4x) > 0$, $x < \frac{3}{2}$ (plus signs to the left of $\frac{3}{2}$ and minus signs to its right.
For $x - 2 > 0$, $x > 2$ (plus signs to the right of $x = 2$ and minus signs to its left).

Step 3: Since the left-hand side of inequality (B) must be negative or zero, the intervals with the negative signs (Figure.) constitute the solution.

	Factor	Signs of the intervals		
Row 1	$6 - 4x$	$+$	$-$	$-$
Row 2	$x - 2$	$-$	$-$	$+$
Row 3	Quotient: $\frac{6-4x}{x-2}$	$-$	$+$	$-$

Number line: Column 1, Column 2, Column 3 — $-\infty$, $\frac{3}{2}$, 2, $+\infty$

Sign diagram for $\frac{6-4x}{x-2} \le 0$

The solution is $\{x | x \le \frac{3}{2} \text{ or } x > 2\}$

We must note that the end point $x = 2$ is not included in the solution (even though the inequality contains the "=" symbol), since otherwise $\frac{-4x+6}{x-2}$ or $\frac{x-4}{x-2}$ would be undefined.

Inequalities Involving Absolute Values of Rational Expressions

Example 17   Solve for x.

$$\left|\frac{8}{x}\right| > 2$$

We will consider the two cases of the absolute value definition.

Case 1:

$$\frac{8}{x} > 2 \quad (x \neq 0)$$

$$\frac{8 - 2x}{x} > 0$$

Case 2:

$$-\left(\frac{8}{x}\right) > 2 \quad (x \neq 0)$$

$$\frac{8}{x} < -2 \quad (x \neq 0)$$

$$\frac{8 + 2x}{x} < 0$$

We use sign diagrams to solve the above inequalities.

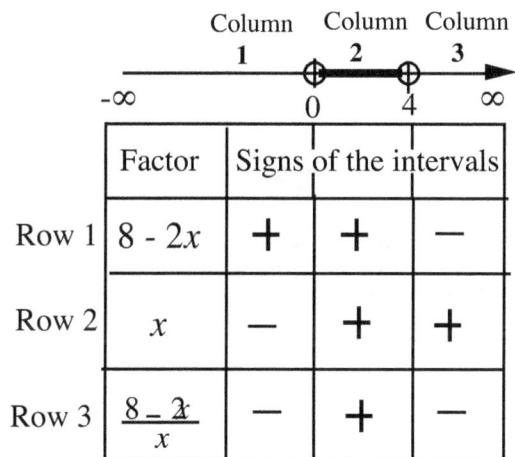

	Factor	Signs of the intervals		
Row 1	8 - 2x	+	+	−
Row 2	x	−	+	+
Row 3	$\frac{8 - 2x}{x}$	−	+	−

Column 1 / Column 2 / Column 3 correspond to intervals $-\infty$, 0, 4, ∞.

	Factor	Signs of the intervals		
Row 1	8 + 2x	−	+	+
Row 2	x	−	−	+
Row 3	$\frac{8 + 2x}{x}$	+	−	+

Column 1 / Column 2 / Column 3 correspond to intervals $-\infty$, -4, 0, ∞.

Solution for Case 1 is $0 < x < 4$

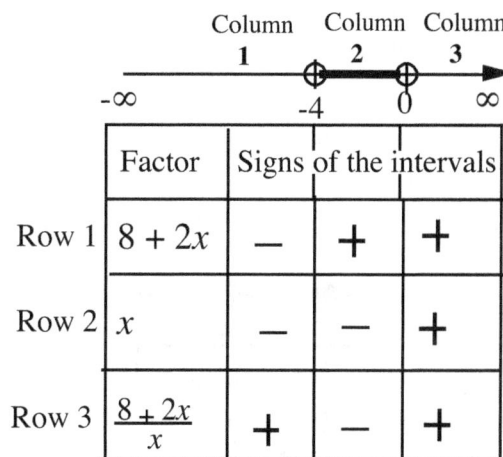

The solution for Case 2 is $-4 < x < 0$

The **complete solution** to the original problem is the union of the solutions for Cases 1 and 2.

The solution is $\{x | -4 < x < 0\} \cup \{0 < x < 4\}$, or simply, $\{x | -4 < x < 4, x \neq 0\}$

$$\xleftarrow{\hspace{1cm}} \overset{-6 \ -5 \ -4 \ -3 \ -2 \ -1 \ 0 \ 1 \ 2 \ 3 \ 4 \ 5 \ 6}{\longrightarrow}$$

Figure : Graph of the solution set $\{x | -4 < x < 0\} \cup \{0 < x < 4\}$

Example 18 Solve the inequality $\left|\dfrac{2x+3}{x-2}\right| \leq 1$

Interpret the given inequality as

$$-1 \leq \frac{2x+3}{x-2} \leq 1 \quad \text{(same as } -1 \leq \frac{2x+3}{x-2} \text{ and } \frac{2x+3}{x-2} \leq 1)$$

(**Note**: $|x| < c \Rightarrow -a < x < c$)

Thus we have two cases to consider. The solution set will be the **intersection** of the solutions in the two cases.

Case 1		Case 2
$\dfrac{2x+3}{x-2} \leq 1$	and	$\dfrac{2x+3}{x-2} \geq -1$
$\dfrac{2x+3}{x-2} - 1 \leq 0$	and	$\dfrac{2x+3}{x-2} + 1 \geq 0$
$\dfrac{2x+3-x+2}{x-2} \leq 0$	and	$\dfrac{2x+3+x-2}{x-2} \geq 0$
$\dfrac{x+5}{x-2} \leq 0$	and	$\dfrac{3x+1}{x-2} \geq 0 \quad (x \neq 2)$

We will draw a sign diagram for $(x + 5)$ and $(x - 2)$. (Figure)

We will draw a sign diagram for $(3x + 1)$ and $(x - 2)$. (Figure b)

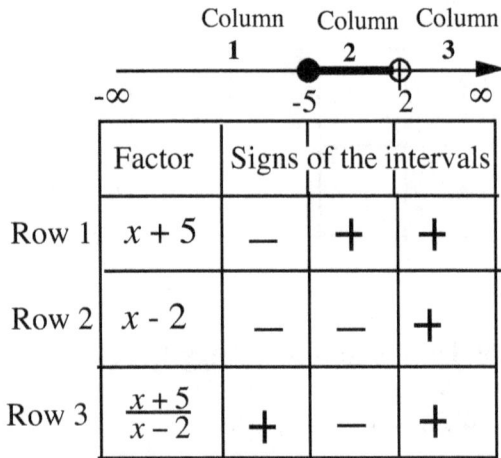

	Factor	Signs of the intervals		
Row 1	$x + 5$	$-$	$+$	$+$
Row 2	$x - 2$	$-$	$-$	$+$
Row 3	$\dfrac{x+5}{x-2}$	$+$	$-$	$+$

Column 1, Column 2, Column 3; $-\infty$, -5, 2, ∞

Solution for case 1 is $-5 \leq x < 2$

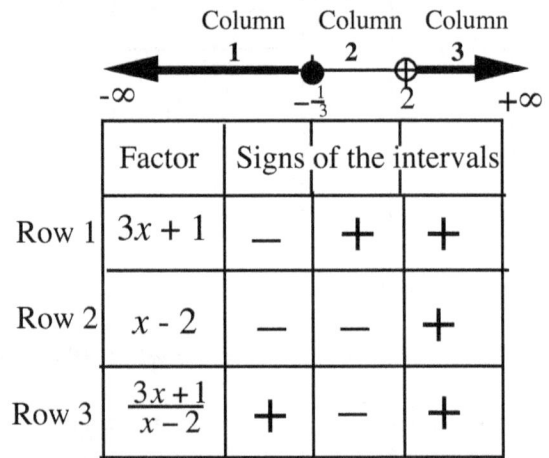

	Factor	Signs of the intervals		
Row 1	$3x + 1$	$-$	$+$	$+$
Row 2	$x - 2$	$-$	$-$	$+$
Row 3	$\dfrac{3x+1}{x-2}$	$+$	$-$	$+$

Column 1, Column 2, Column 3; $-\infty$, $-\frac{1}{3}$, 2, $+\infty$

Solution for Case 2 is $x \leq -\frac{1}{3}$ or $x > 2$

Conclusion: The solution to the original inequality is the **intersection** of the solutions for the above two cases. The solution is $-5 \leq x \leq -\frac{1}{3}$ (the intersection of the solutions to Cases 1 and 2)

-6 -5 -4 -3 -2 -1 0 1 2 3 4 5 6

Figure: Graph of the solution set $\left\{x \mid -5 \leq x \leq -\frac{1}{3}\right\}$

Lesson 27 Exercises

A Solve the following inequalities:

1. $(x-1)(x+2) < 0$;   **2.** $(2x+3)(x-1) > 0$;   **3.** $x^2 - 3x - 10 > 0$;   **4.** $x^2 + 2x + 1 > 0$;

5. $x^2 - 5x + 6 > 0$;   **6.** $x^2 < 3 - 2x$;   **7.** $x + 1 < x^2 - 1$;   **8.** $x^2(x-3) < 0$;

9. $3x^2 - 2x - 1 > 0$;   **10.** $(x-3)(x-4)(x+1) < 0$;   **11.** $(x+1)^2(x-2) > 0$;

Answers:

1. $\{x \mid -2 < x < 1\}$;   **2.** $\{x \mid x < -\frac{3}{2} \text{ or } x > 1\}$;   **3.** $\{x \mid x < -2 \text{ or } x > 5\}$;   **4.** $\{x \mid x \neq -1\}$;

5. $\{x \mid x < 2 \text{ or } x > 3\}$;   **6.** $\{x \mid -3 < x < 1\}$;   **7.** $\{x \mid x < -1 \text{ or } x > 2\}$;   **8.** $\{x \mid x < 0, \text{ or } 0 < x > 3\}$;

9. $\{x \mid x < -\frac{1}{3} \text{ or } x > 1\}$;   **10.** $\{x \mid x < -1, \ 3 < x < 4\}$;   **11.** $\{x \mid x > 2\}$

B   Solve the following inequalities:

12. $\dfrac{x+4}{x-5} \leq 2$;   **13.** $\dfrac{x^2(x-1)(x+2)}{(x+3)(x+1)} \geq 0$;   **14.** $\dfrac{4}{x} > 5$;   **15.** $\dfrac{7}{x} < 3$;

16. $\left|\dfrac{5}{x}\right| > 3$;   **17.** $\left|\dfrac{x+3}{x-1}\right| \leq 2$

Answers:

12. $\{x \mid x < 5 \text{ or } x \geq 14\}$;   **13.** $\{x \mid x < -3, \ -2 \leq x < -1, \ x \geq 1\}$;   **14.** $\{x \mid 0 < x < \frac{4}{5}\}$;

15. $\{x \mid x < 0 \text{ or } x > \frac{7}{3}\}$;   **16.** $\{x \mid -\frac{5}{3} < x < \frac{5}{3}, \ x \neq 0\}$;   **17.** $\{x \mid x \leq -\frac{1}{3} \text{ or } x \geq 5\}$.

C   **1.** $x^2 < (x-2)(2x+3)$;   **2.** $|4x+3| \geq 5$;

Answers   **1.** $\{x \mid x < -2 \text{ or } x > 3\}$;   **2.** $\{x \mid x \leq -2 \text{ or } x \geq \frac{1}{2}\}$;

CHAPTER 13

Lesson 28: **Domain Specification; Piecewise** (Multipart) **Functions**

Lesson 29: **Miscellaneous Graphs Involving Absolute Values**

Lesson 30 :**Special Functions: Greatest Integer Function; Step Function; Unit Step Function; Sign Function**

Lesson 28

Domain Specification; Piecewise (Multipart) Functions

Explicit Specification of the restricted domain

Consider the function given by

$$f(x) = x^2 + 4x - 1 \qquad\qquad 0 \le x \le 5$$

Although the function is defined for all real numbers with the same assignment rule, the domain is restricted by the inequality $0 \le x \le 5$. Without that restriction (being written to the right of the given function), the domain would have implicitly consisted of all real numbers, since the given function is a polynomial function. With the restricted domain indicated explicitly, the domain of the above function consists of only those real numbers between 0 and 5 and including 0 and 5. The restricted domain in this case indicates also the domain which is of interest to us.

Analogy: Suppose you are having lunch and you are offered an apple without any comment (restriction), you may eat the whole apple if you wish. However, if you were offered the apple with the comment (restriction) "do not eat more than half of the apple", you would obviously not eat the whole apple, but at most only half of it.

Implicit and Explicit Specification of Domains

Let us consider the function given by
$$f(x) = x^2$$

As the function is specified, the domain consists of all real numbers, the function being a polynomial. We say that the domain in the above example was **implicitly** specified (i.e. understood).

Now, let us consider the same function with some additional qualification.

$$f(x) = x^2 \qquad x > 0 \qquad \text{(domain of interest or restricted domain)}$$

The additional qualification means that if we were to find solutions to the above equation $f(x) = x^2$, we would choose only values of x greater than zero (i.e. only positive x-values) and calculate corresponding values of $f(x)$. We must note that even though, the function can be defined for negative x-values ($x < 0$), we are not interested in that part of the given function where x is negative.

To sketch the graph of the given function, we would sketch it for only positive x-values (figure).

Similarly, if we were given $f(x) = x^2 \qquad x < 0$, we would choose only negative x-values both in finding solutions to the equation and in graphing (Figure.)

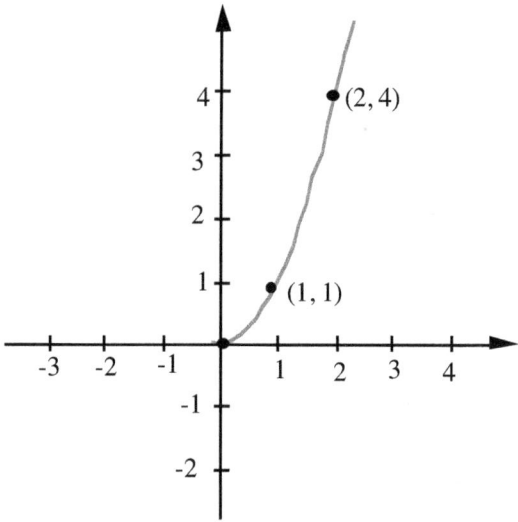

Figure: Graph of $f(x) = x^2$, $x > 0$ (constraint)

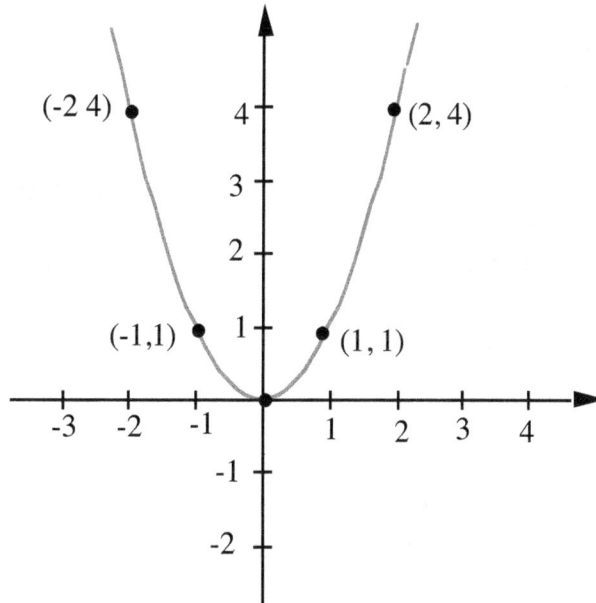

Figure: Graph of $f(x) = x^2$, $x < 0$ (constraint)

Figure: Graph of $f(x) = x^2$ (no constraint)

Specification of Functions by Piecewise Rules 194

A function may be specified by a number of different partial rules, each rule with its own domain (restricted domain). However, we must note that the partial rules altogether specify only one function. Example

$$f(x) = \begin{cases} x^2 + 1 & x > 0 \\ -x + 1 & x < 0 \end{cases}$$

This specification means that when we choose positive x-values ($x > 0$) we must use the upper expression $x^2 + 1$, but if we choose negative x-values ($x < 0$), we must use the lower expression $-x + 1$. It would be **wrong** to choose, say, $x = -2$ and substitute it in the upper expression, $x^2 + 1$. It would also be wrong to choose, say, $x = 3$ and substitute it in the lower expression, $-x + 1$.

Moreover, $f(x) = x^2 + 1$ is a quadratic function, but $f(x) = -x + 1$ is a linear function.

Analogy: A method of undertaking a journey from your house in New York to London airport is by car from your house to New York airport and by plane to London airport.

$$\text{From house in New York to London airport} = \begin{cases} \text{from house to New York Airport} & \text{by car} \\ \text{from New York Airport to London Airport} & \text{by plane} \end{cases}$$

N

ote above that the two means of transportation, namely, by car and by plane together specify the means of getting from your house in New York to London airport.

Let us prepare a **table** for the function.

$$f(x) = \begin{cases} x^2 + 1 & x \geq 0 \\ -x + 1 & x < 0 \end{cases}$$

Table for $x^2 + 1, x \geq 0$

x	0	1	2	3	4	5
$y = x^2 + 1$	1	2	5	10	17	26

Table for $-x + 1, x < 0$

x	-1	-2	-3
$y = -x + 1$	2	3	4

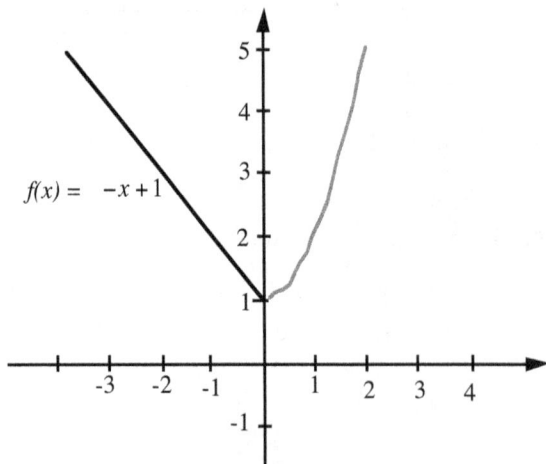

Figure: Graph of $f(x) = \begin{cases} x^2 + 1 & x \geq 0 \\ -x + 1 & x < 0 \end{cases}$

Lesson 28 Exercises

Determine the domain:

1. $f(x) = 4x^2 + 3x - 3$;   **2.** $f(x) = x^2 + 5x + 9$;   **3.** $f(x) = \dfrac{2}{x-3}$;   **4.** $f(x) = \dfrac{1}{x^2}$;

5. $f(x) = \dfrac{x-3}{(x+1)(x+3)(x+5)}$;   **6.** $f(x) = \dfrac{1}{x-5}$;   **7.** $f(x) = \dfrac{x+7}{x+2}$;

8. $f(x) = \sqrt{x-6}$;   **9.** $\sqrt{5-x}$.

Answers: **1.** $\{x \mid x \text{ is a real number}\}$,   **2.** $\{x \mid x \text{ is a real number}\}$
3. $\{x \mid x \text{ is a real number, and } x \neq 3\}$ **4.** $\{x \mid x \text{ is a real number, and } x \neq 0\}$;
5. $\{x \mid x \text{ is a real number, and } x \neq -1, x \neq -3, x \neq -5\}$ **6.** $\{x \mid x \text{ is a real number, and } x \neq 5\}$
7. $\{x \mid x \text{ is a real number, and } x \neq -2\}$;   **8.** $\{x \mid x \geq 6\}$;   **9.** $\{x \mid x \leq 5\}$

Lesson 29

Miscellaneous Graphs Involving Absolute Values

Preliminaries:

Absolute Value Functions and Piecewise Rules

Let us recapitulate the inequality symbols (page 123)

>: $x > 0$ means x is positive (or x is greater than zero).

≥: $x \geq 0$ means x is positive or zero.

<: $x < 0$ means x is negative (i.e. x is less than zero).

≤: $x \leq 0$ means x is negative or zero.

Absolute Value: The absolute value of x symbolized, $|x|$, is defined by a two-part rule as follows:

$$|x| = \begin{cases} x \text{ if } x \geq 0 \\ -x \text{ if } x < 0 \end{cases}$$

In words, the absolute value of a real number x is x if x is a positive number or zero; but it is $-x$ if x is a negative number. The above two-part rule in the definition requires us to consider two cases when solving absolute value equations and inequalities. Understanding the above definition very well will save us from having to memorize the solution methods for absolute value equations.

Examples: **1.** The absolute value of -4 , symbolized $|-4| = 4$.

2. The absolute value of 6, symbolized $|6| = 6$.

3. The absolute value of 0 is 0.

The absolute value of a real number is always positive or zero.

The domain of the absolute value function consists of all real numbers. The range of the absolute value function consists of zero and all positive real numbers.

More examples:

1. The absolute value of $x + 2$ is denoted by $|x + 2|$, and would be defined by a two part rule as:

$$|x + 2| = \begin{cases} x + 2 & x + 2 \geq 0, \text{ or } x \geq -2 \\ -(x + 2) & x + 2 < 0, \text{ or } x < -2 \end{cases}$$

We should note above that the domain is specified by using the binomial $x + 2$.
We must also note how the domain is specified by using the whole expression within the absolute value bars and then solving the resulting inequality for x.

2. The absolute value of the quadratic expression $x^2 + 2x - 15$ is denoted by $\left|x^2 + 2x - 15\right|$, and is defined by a two part rule as follows:

$$\left|x^2 + 2x - 15\right| = \begin{cases} x^2 + 2x - 15 & \text{if } x^2 + 2x - 15 \geq 0. \\ -(x^2 + 2x - 15) & \text{if } x^2 + 2x - 15 < 0. \end{cases}$$

Note above that in this case, the domain is specified by using a quadratic trinomial. The domain is later on broken down to linear factors and the solutions found.

Case 1: Graphs Involving the absolute value of only one Variable

Example 1 Sketch the graph of $y = |x|$

By definition, $|x| = \begin{cases} x & x \geq 0 \\ -x & x < 0 \end{cases}$

Step 1: Sketch the graph of $y = x$ for $x \geq 0$ (i.e., for $x = 0$ and positive values of x)

Step 2: Sketch the graph of $y = -x$ for $x < 0$ (i.e., for negative x-values) on the same axes as in Step 1
Note that the graph is V-shaped. The graph is shown in Figure

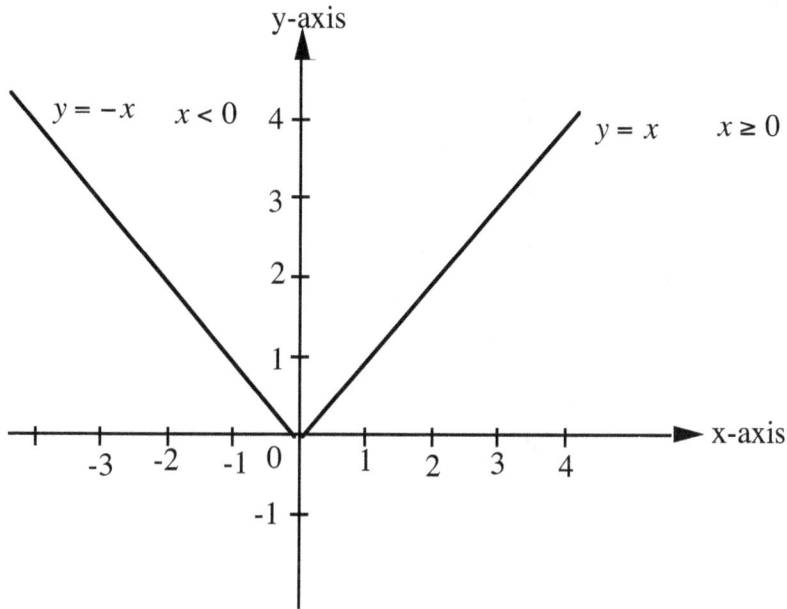

Figure: The graph of $y = |x|$

Example 2 Sketch the graph of $y = |x + 1|$

Method 1

$$y = |x + 1| = \begin{cases} x + 1 & \text{if } x + 1 \geq 0, \text{ or } x \geq -1 \\ -(x + 1) & \text{if } x + 1 < 0, \text{ or } x < -1 \end{cases}$$

Step 1: Sketch the graph of $y = x + 1$ for $x \geq -1$; and

Step 2: Sketch the graph of $y = -(x + 1)$ or $-x - 1$ for $x < -1$,
to obtain the characteristic V-shape.

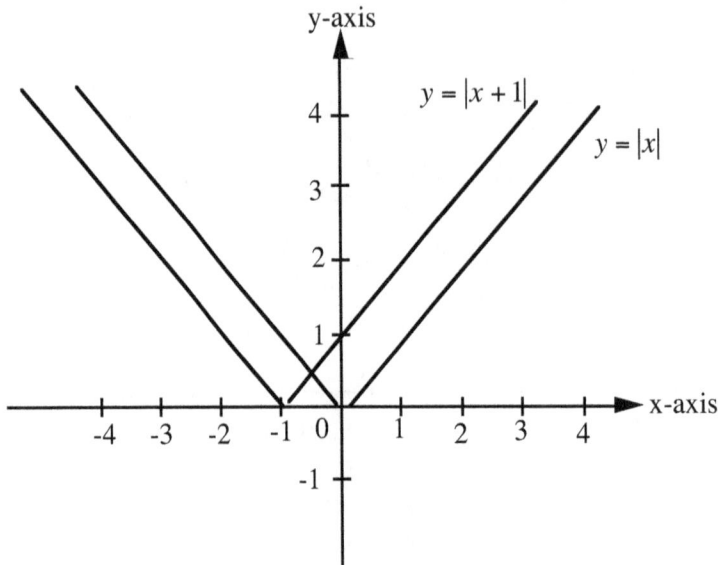

Figure: The graph of $y = |x + 1|$

Method 2

We can obtain the graph of $y = |x + 1|$ from that of $y = |x|$ by the proper translations (see Chapt. 14).

Step 1: Using broken lines sketch the graph of $y = |x|$

Step 2: On the line $y = x$ shift each of any two points one unit to the left. Similarly, on $y = -x$ shift each of any two points one unit to the left and connect each pair of points by a solid line, to obtain the characteristic " V " shape. Figure

Example 3   Sketch the graph of $y = |x - 1|$

We can obtain this graph by shifting the graph of $y = |x|$ one unit to the right, keeping the y-values unchanged. OR alternatively, we could sketch $y = x - 1$ for $x \geq 1$, and then sketch $y = -(x - 1)$ for $x < 1$, and obtain the characteristic V-shape.

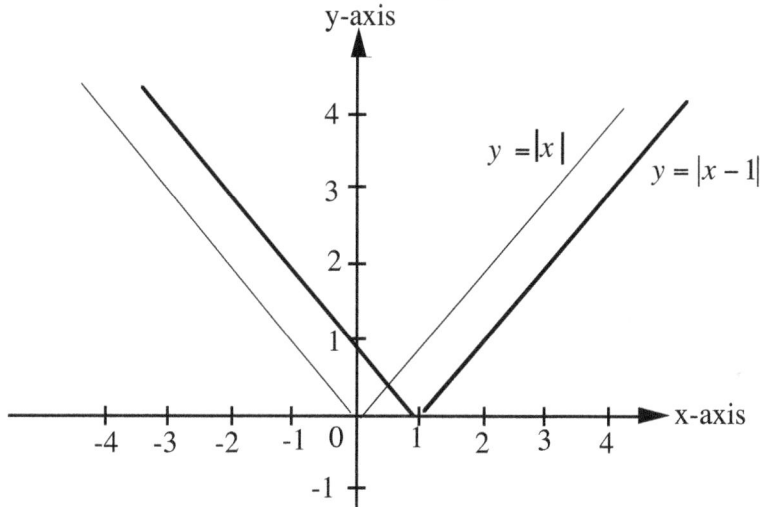

Example 4   Sketch the graph of $y = |x| + 2$.

Method 1   From $y = |x| + 2 = \begin{cases} x + 2 & \text{if } x \geq 0 \\ -x + 2 & \text{if } x < 0 \end{cases}$

Step 1. Sketch $y = x + 2$ for $x \geq 0$, and

Step 2: Sketch $y = -x + 2$ for $x < 0$, to obtain the characteristic V-shape.
(Figure)on the same axes as in Step 1.

Note, above, how the domain was specified using x and **not** $x + 2$ because the absolute value applies to the x and not to $x + 2$.

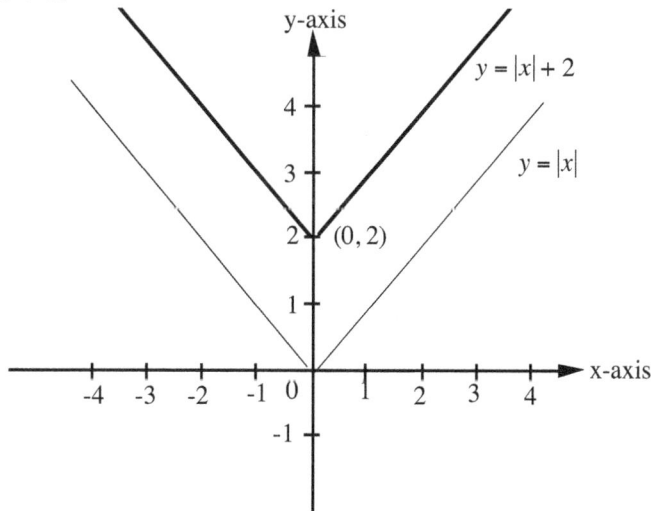

Method 2

In standard form, the equation is $y - 2 = |x|$

We can sketch this graph by shifting the graph of $y = |x|$ 2 units up, keeping the x-values unchanged.

If necessary, we would obtain additional points, say $(1, 3)$ and $(-1, 3)$ by choosing $x = 1$, $x = -1$ and calculating corresponding y-values .

Example 5 Sketch the graph of $y = 3|x|$

From $y = 3|x| = \begin{cases} 3x & \text{if } x \geq 0 \\ -3x & \text{if } x < 0 \end{cases}$

Step 1: Sketch $y = 3x$ for $x \geq 0$; and

Step 2: Sketch $y = -3x$ $x < 0$, to obtain the characteristic V-shape.

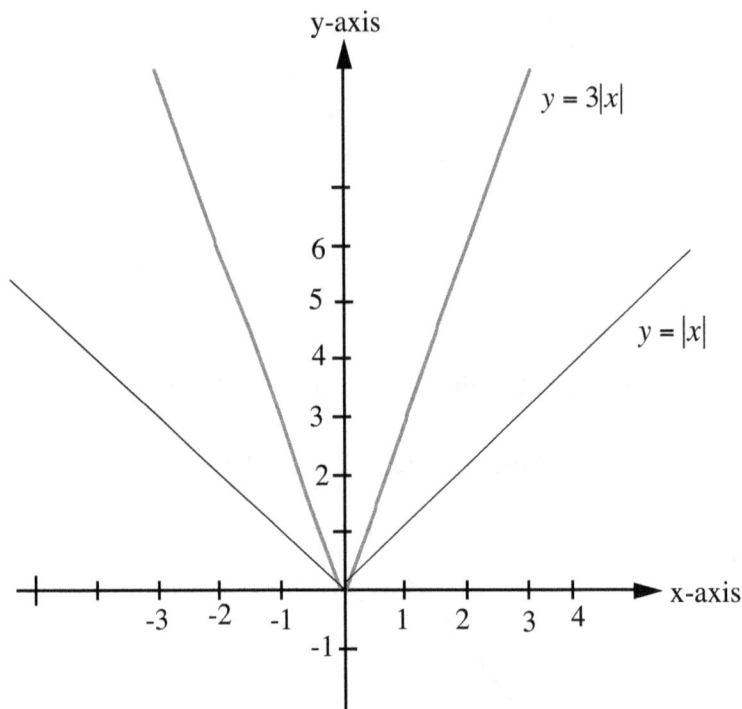

Figure: Graph of $y = 3|x|$

Example 6 Sketch the graph of $y = \frac{1}{3}|x|$

From $y = \frac{1}{3}|x| = \begin{cases} \frac{1}{3}x & \text{if } x \geq 0 \\ -\frac{1}{3}x & \text{if } x < 0 \end{cases}$

Step 1: Sketch $y = \frac{1}{3}x$ for $x \geq 0$; and

Step 2: Sketch $y = -\frac{1}{3}x$ $x < 0$, to obtain the characteristic "V" shape.

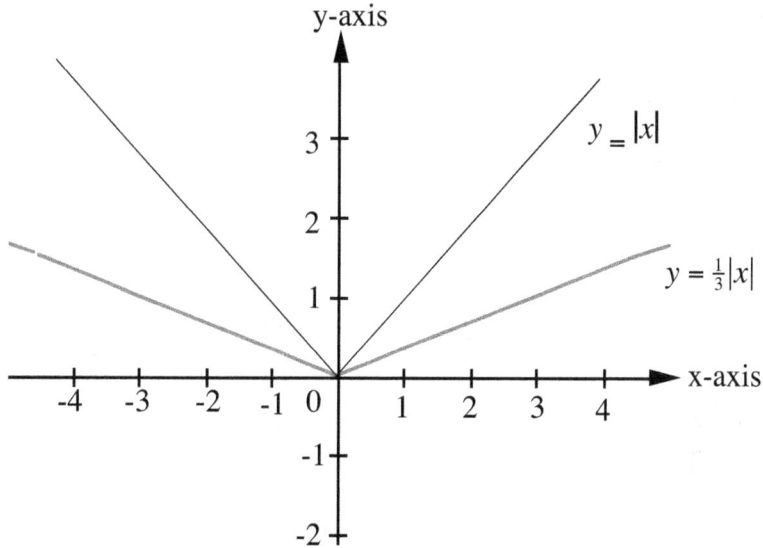

Figure: Graph of $y = \frac{1}{3}|x|$

Case 2: Graphs Involving the Absolute Value of Quadratic Trinomial

Example   Sketch the graph of $f(x) = \left| x^2 + 2x - 15 \right|$ (A)

Solution   Let $y = f(x)$

         Then $y = \left| x^2 + 2x - 15 \right|$

Applying the absolute value definition to the trinomial, we obtain the two parts as follows:

$$y = \begin{cases} x^2 + 2x - 15 & \text{for } x^2 + 2x - 15 \geq 0. \\ -(x^2 + 2x - 15) & \text{for } x^2 + 2x - 15 < 0. \end{cases} \quad \text{(B)}$$

Step 1: Solve the domain inequalities.

$$x^2 + 2x - 15 \geq 0 \qquad \text{(C)}$$

$$(x-3)(x+5) \geq 0$$

Using sign diagrams (Figure.), solution to (C) is

$$x \leq -5 \text{ or } x \geq 3$$

Similarly, and using the same sign diagram (Figure) we solve the other domain inequality.

$$x^2 + 2x - 15 < 0$$

Since this inequality must be negative, we choose those intervals for which $(x - 3)(x + 5)$ has minus signs (Figure). The solution is therefore $-5 < x < 3$

Interval line:

Factor	Signs of the intervals		
Row 1 $(x - 3)$	$-$	$-$	$+$
Row 2 $(x + 5)$	$-$	$+$	$+$
Row 3 $(x - 3)(x + 5)$	$+$	$-$	$+$

Now, rewriting System (B) with the domain inequalities solved for, we obtain

$$y = \begin{cases} x^2 + 2x - 15 & \text{for } x \leq -5 \text{ or } x \geq 3 \quad \text{(D)} \\ -(x^2 + 2x - 15) & \text{for } -5 < x < 3 \quad \text{(E)} \end{cases}$$

Step 2: Sketch the graph of $y = x^2 + 2x - 15$ subject to domain $x \leq -5$ or $x \geq 3$.

    Using dotted lines, first sketch the graph of $y = x^2 + 2x - 15$; and then for the part of the curve for which $-5 < x < 3$, use solid lines to sketch (to indicate the domain of interests).

Step 3: Similarly, sketch $y = -(x^2 + 2x - 15)$ subject to the constraint $-5 < x < 3$,
or just reflect the graph of $y = x^2 + 2x - 15$ (flip over) about the x-axis, noting that the whole
graph is above the x-axis. The dotted parabola is to help visualize the part of $y = x^2 + 2x - 15$
below the x-axis, which is of course **not** part of the required graph. Note also that because we
are dealing with absolute values, $f(x)$ is either zero or a positive number and never negative
and that is why the whole curve is above the y-axis.

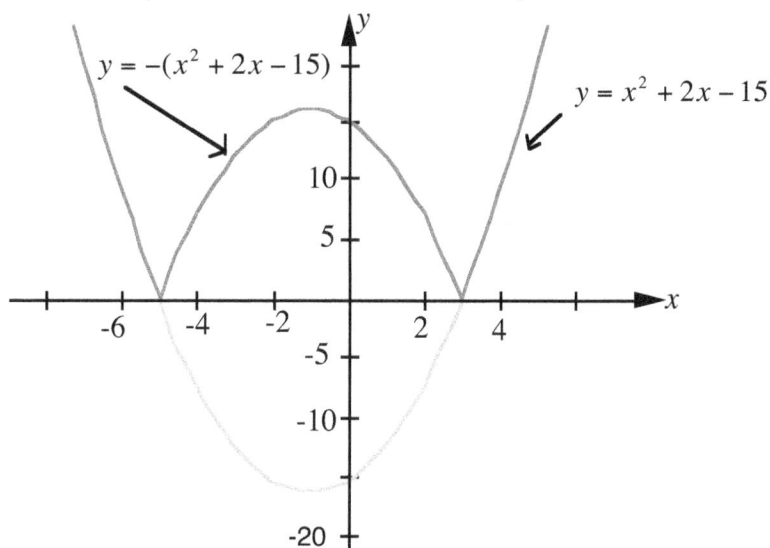

Figure: The graph of $y = |x^2 + 2x - 15|$

Case 3: Graphs Involving the Absolute Values of two Variables

Example 1 Sketch the graph of $|x| + |y| = 2$

Step 1

$$|x| = \begin{cases} x & \text{if } x \geq 0 \\ -x & \text{if } x < 0 \end{cases}$$

$$|y| = \begin{cases} y & \text{if } y \geq 0 \\ -y & \text{if } y < 0 \end{cases}$$

Step 2: Draw a rectangular coordinate system of axes. The plane would be divided into four
quadrants. The origin $x = 0$, $y = 0$ will divide the x-axis into two intervals: $x < 0$, $x > 0$;
and the y-axis into two intervals: $y < 0$, $y > 0$. We obtain the following possible lines
by applying the cases of absolute value definitions for $|x|$ and $|y|$ to $|x| + |y| = 2$.

Case 1:	$x + y = 2$	for $x > 0, y > 0$	(1st quadrant)	(C)
Case 2:	$-x + y = 2$	for $x < 0, y > 0$	(2nd quadrant)	(D)
Case 3:	$-x - y = 2$	for $x < 0, y < 0$	(3rd quadrant)	(E)
Case 4:	$x - y = 2$	for $x > 0, y < 0$	(4th quadrant)	(F)

Step 3. On the same coordinate axes, graph each of the four equations (C), (D), (E) and (F)
subject to the corresponding specified domains (Figure)

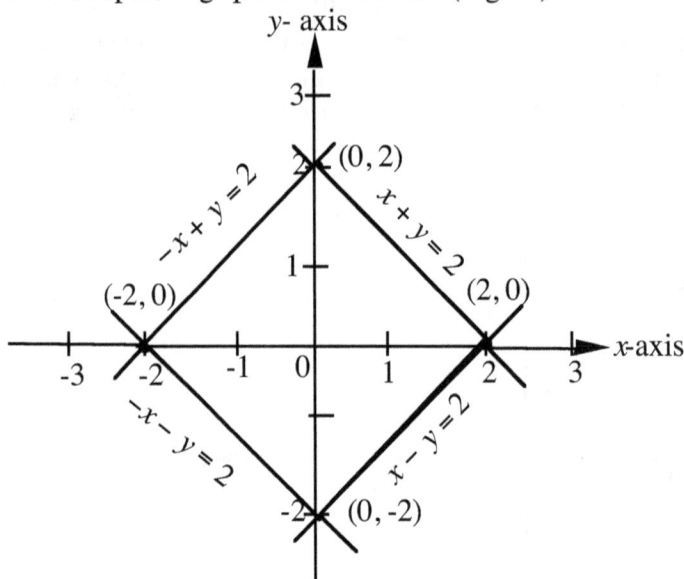

Figure: The graph of $|x| + |y| = 2$

Case 4: Graphs of Inequalities Involving the Absolute Values of two Variables

Example 2. Graph the inequality $|x| + |y| \leq 2$

Solution: We repeat Steps 1, 2 and 3 as in Example 1, above, and **shade** the interior of the figure
as shown in Figure. **or**
We may also imitate Steps 1 and 2, sketch the half-plane for each linear inequality and
find the intersection of the half-planes.

Graph of the inequality $|x| + |y| \leq 2$:

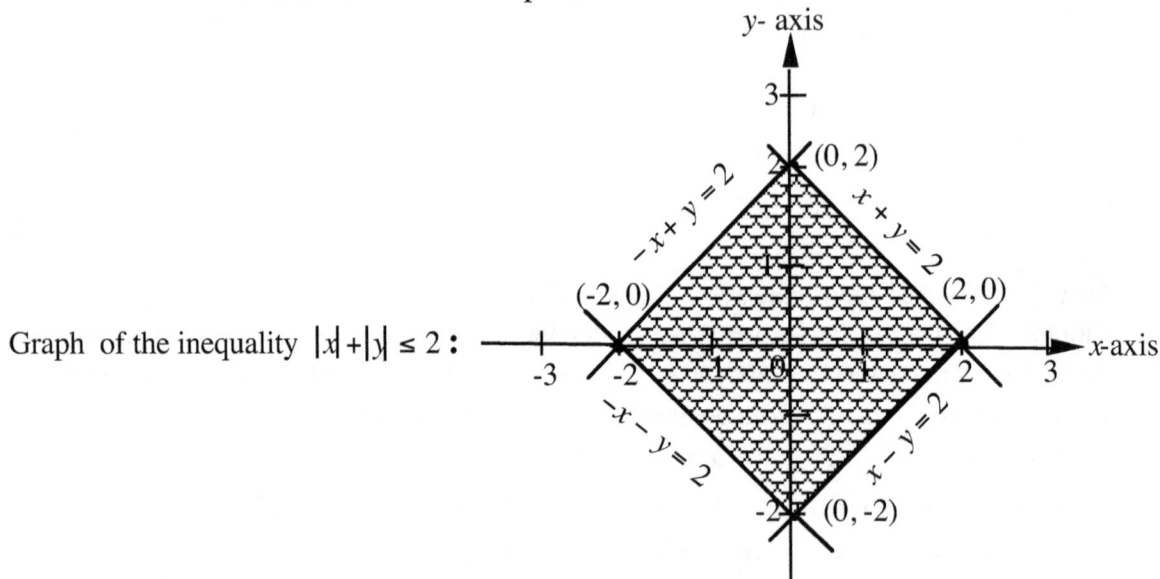

Lesson 29 Exercises

A Sketch the graphs of the following

1. If $f(x) = \begin{cases} -x^2 - 1 & \text{if } x \geq 0 \\ x + 1 & \text{if } x < 0 \end{cases}$ **2.** $f(x) = \begin{cases} x & \text{if } x \geq 2 \\ 2x - 5 & \text{if } x < 2 \end{cases}$,

3. $f(x) = \begin{cases} x - 1 & \text{if } x \geq 2 \\ x + 6 & \text{if } 0 \leq x < 4 \\ -x - 2 & \text{if } x < 0 \end{cases}$ Complete the following: **4.** $|x + 3| = \begin{cases} \ldots\ldots\text{if}\ldots\ldots \\ \ldots\ldots\text{if}\ldots\ldots \end{cases}$

5. $|(x - 1)(x + 2)| = \begin{cases} \ldots\ldots\text{if}\ldots\ldots \\ \ldots\ldots\text{if}\ldots\ldots \end{cases}$ **6.** $|x^2 + 8x + 15| = \begin{cases} \ldots\ldots\text{if}\ldots\ldots \\ \ldots\ldots\text{if}\ldots\ldots \end{cases}$

Answers:

1.

2.

3.

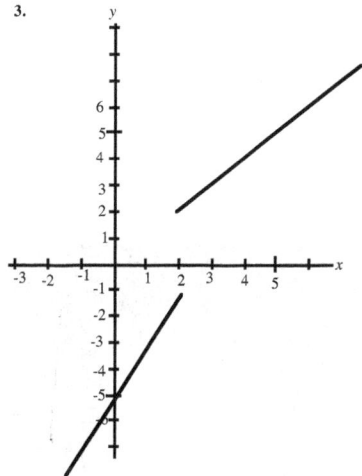

B Sketch the graphs of the following functions: **1.** $f(x) = |x^2 + 8x + 15|$; **2.** $f(x) = |x^2 - x - 12|$

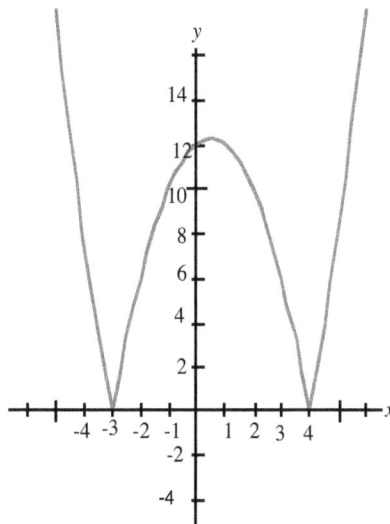

C Sketch the graphs of the following:

1. $|x| + |y| = 4;$ **2.** $|x| + |y| \leq 1;$ **3.** $|x| + |y| \geq 1;$ **4.** $|x| - |y| \geq 0$ for $-2 < x < 2$

$|x| + |y| = 1$

$|x| + |y| < 1$

$|x| + |y| \geq 1$

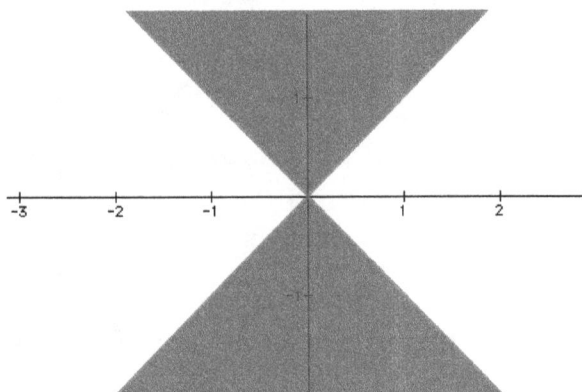

$|x| - |y| \geq 0$ $-2 < x < 2$

Lesson 30   207
Special Functions: Greatest Integer Function; Step Function; Unit Step Function; Sign Function

Greatest Integer Function, $[x]$

The greatest integer function, denoted by $[x]$, is the largest integer that is less than or equal to x.

Symbolically, $[x] = n$ if $n \le x < n+1$, where n is an integer. That is, if the given number is an integer, then the given number is the greatest integer; but if it is not an integer, then on the number line, the nearest integer on its left is the greatest integer.

Examples $[0] = 0$; $[2] = 2$; $[5] = 5$; $[1.5] = 1$; $[-1.5] = -2$; $[2.5] = 2$; $[-4.5] = -5$

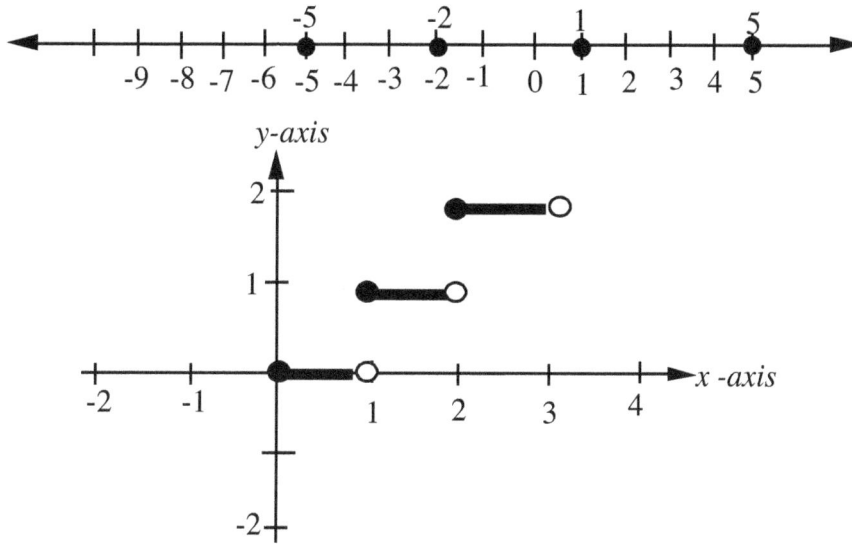

Figure: Graph of $y = [x]$

Step Function

The step function (sometimes referred to as the postage stamp function) is so called because its graph resembles a set of steps.

Example. Let $f(x)$ be the fee charged by the post office to mail a package of weight x oz.

Then, the fee and the weight can be written in functional notation as:

$$f(x) = \begin{cases} \ \ldots & \text{if } \ < x \ \le \ldots \\ \ \ldots & \text{if } \ < x \ \le \ldots \\ \ \ldots & \text{if } \ < x \ \le \ldots \\ \ \ldots & \text{if } \ < x \ \le \ldots \end{cases}$$

Unit Step Function   208

A unit step function of time denoted by $u(t - t_0)$ is defined as follows:

$$u(t - t_0) = \begin{cases} 1 & \text{if} \quad t > t_0 \\ 0 & \text{if} \quad t \le t_0 \end{cases}$$

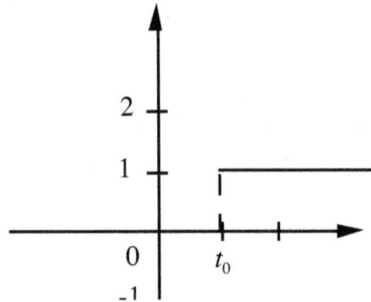

Figure: Graph of unit step function, $u(t - t_{0)} = \begin{cases} 1 & \text{if} \quad t > t_0 \\ 0 & \text{if} \quad t \le t_0 \end{cases}$

Sign Function

The sign function (or signum function) denoted by sgn x (read "sign of x") is defined by a piecewise rule as follows:

$$\text{sgn } x = \begin{cases} +1 & \text{if} \ x > 0 \\ 0 & \text{if} \ x = 0 \\ -1 & \text{if} \ x < 0 \end{cases}$$

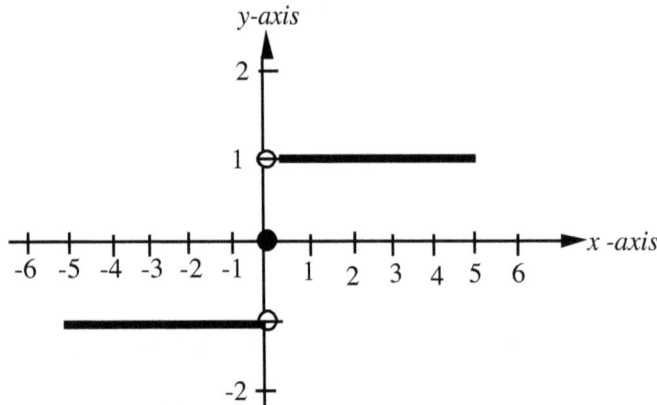

Figure: Graph of sgn x $-5 \le x \le 5$

We must be careful not to confuse "sign of x" with the trigonometric function sin x (sine of x). They are just names for different types of functions.

The **sign function** and the **absolute value function** are related as follows:

For any real number a, a sgn $a = |a|$. Example: -3.5 sgn $- 3.5 = -3.5(-1) = 3.5$ and $|-3.5| = 3.5$

Also, if $y = $ sgn x, then the slope of $y = |x|$ is sgn(x), $x \ne 0$.

Lesson 30 Exercises

A Find the value of each of the following

    1. $[6]$;   2. $[3.5]$;   **3.** $[-3.5]$;   **4.** $[-6.5]$;

                  Answers: 1 . 6;   **2** . 3;   **3** . -4;   **4** . -7

B Evaluate: 1. sgn 6;   2. sgn 3.5;   **3.** sgn -3.5;   **4.** sgn 0;

                  Answers: 1 . 1;   **2** . 1;   **3** . -1;   **4** . 0.

C Graph the following: **1.** $y = (x + 2)\operatorname{sgn} x$ $-4 \le x \le 4$;   **2.** $y = x \operatorname{sgn}(x - 2)$ $-3 \le x \le 3$

Answers:

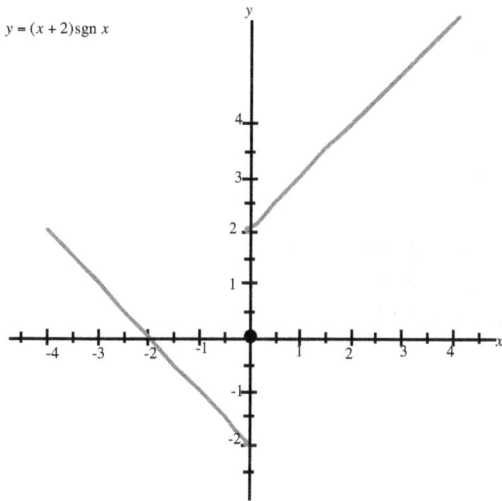

$y = (x + 2)\operatorname{sgn} x$

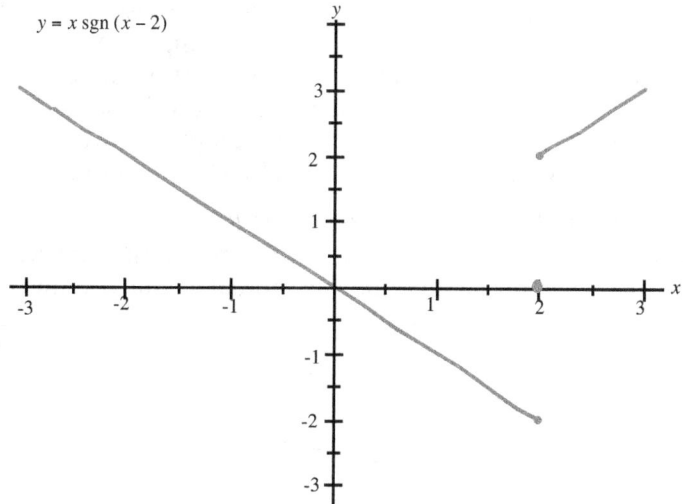

$y = x \operatorname{sgn}(x - 2)$

CHAPTER 14

Lesson 31: **Positive, Negative, Increasing, and Decreasing Functions**
Lesson 32: **Concavities of Curves; Critical Points**
Lesson 33: **Reflection of Points, Lines and Curves**
Lesson 34: **Translation of Points, Relations, Functions and Axes**
Lesson 35: **Contraction and Expansion of Curves**
Lesson 36: **Symmetry; Even and Odd Functions**

Lesson 31

Positive, Negative, Increasing, and Decreasing Functions

Introduction

We will need the above terms in describing the behavior of functions. A conceptual understanding of these terms will facilitate a "unified " or " lumped" approach in covering functions. We will make use of the concept of a positive or a negative function in sketching the graphs of polynomial functions, rational functions, exponential functions, logarithmic functions, and trigonometric functions. For example, in order to draw the correct connection (graph) between, say, any two zeros (roots) of a polynomial function, a knowledge of whether the function is positive or negative between the zeros will be invaluable.

Positive and Negative Functions

When we say that a function is **positive** on a certain interval, say , the interval from $x = a$ to $x = b$, we mean that, algebraically, all the y-coordinate values on this interval from a to b are **positive**, and graphically, the entire curve lies above the x-axis, on this interval (Figure).

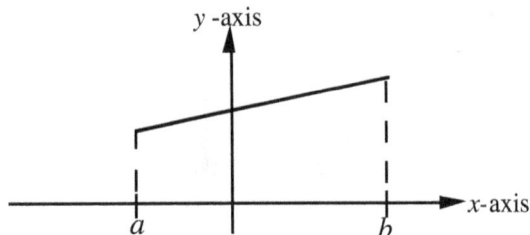

Figure

Similarly, when we say that a function is **negative** on the interval from $x = a$ to $x = b$, we mean that all the y-coordinate values are negative (have minus signs) on this interval from a to b, and that, graphically, the entire curve lies below the x-axis (Figure)

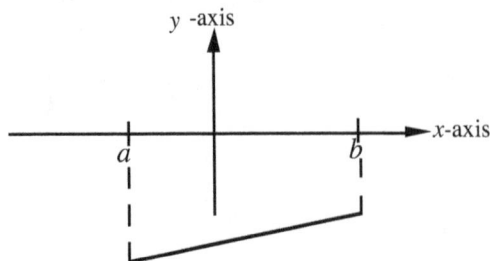

Figure

Decreasing and Increasing Functions

We use the above terms in describing critical points, namely, minimum and maximum points, and points of inflection. We will use these terms in describing the behavior of functions as well as in sketching their graphs. For example, we will use the term " concavity " (e.g. concave up or down) in defining critical points (turning points); in describing the behavior of polynomial, rational, trigonometric, exponential and logarithmic functions as well as conic relations, namely the hyperbola and the ellipse.

Increasing Function (strictly increasing function)

Geometrically, a strictly increasing function is one whose graph (curve or line) rises from left to right, as one reads from left to right (or simply, the curve leans to the right). That is, as we move our eyes horizontally to the right in the coordinate plane, we encounter higher and higher values of y. Thus, any point on the curve is higher than any other point (on the curve) to its left. (Recall from elementary math that a straight line leaning to the right in the page has a positive slope).

More formally, a **function** f is increasing (or strictly increasing) on an interval containing the numbers x_1, x_2 if whenever $x_2 > x_1$, $f(x_2) > f(x_1)$ (See Figure)

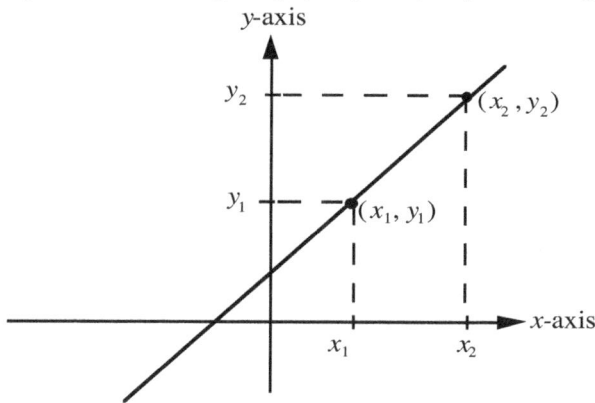

Figure: Graph of a increasing function

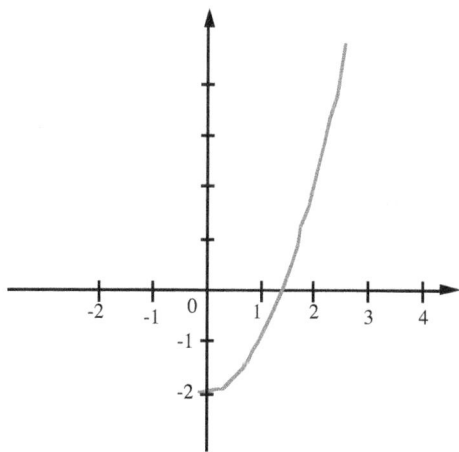

Figure : The graph of an increasing function

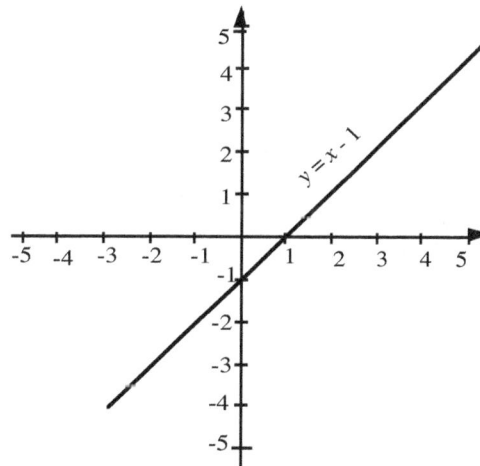

Figure : Graph of $y = x - 1$ (an increasing function)

Example The function given by $f(x) = x - 1$ is an increasing function. (Fig.)   212

> Let us test for $x_1 = 2, x_2 = 5$ (that is $x_2 > x_1$) in $f(x) = x - 1$
> Then $f(x_1) = f(2) = 2 - 1 = 1$; and $f(x_2) = f(5) = 5 - 1 = 4$
>
> Clearly, $5 > 2$ implies $f(5) > f(2)$ (since $f(5) = 4$ and $f(2) = 1$

Note: The above is **no** proof, but is only an illustration

Decreasing Function (strictly decreasing function)

Geometrically, a strictly **decreasing function** is one whose graph (curve or line) falls (drops) from left to right as one reads from the left to the right,
(or simply, the curve leans to the left). That is, as we move our eyes horizontally from the left to the right in the coordinate plane, we encounter lower and lower values of y. Thus, any point on the curve is lower than any other point (on the curve) to its left (Figure). The curve leans to the left in the page.

More formally, a **function** f is decreasing (or strictly decreasing) on an interval containing the numbers x_1, x_2 if whenever $x_2 > x_1$, $f(x_2) < f(x_1)$
(That is, as the x-values increase, the y-values decrease) (See Figure)

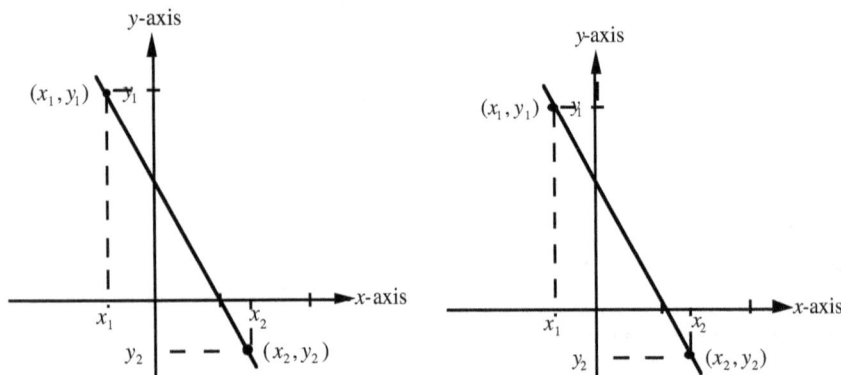

Figure: Graphs of decreasing functions

Example

The function given by $f(x) = -x - 1$ is decreasing on the interval $[-5, -2]$
Let us test for $x_1 = -5, x_2 = -2$ (that is, $x_2 > x_1$) in $f(x) = -x - 1$
Then $f(x_1) = f(-5) = -(-5) - 1 = 5 - 1 = 4$

$\quad\quad f(x_2) = f(-2) = -(-2) - 1 = 2 - 1 = 1.$ (We could also use -4 or -3 for testing)

Clearly, $-2 > -5$ implies $f(-2) < f(-5)$ (since $f(-2) = 1$ and $f(-5) = 4$)

Note: The above is **no** proof, but is only an illustration

Some functions are neither increasing nor decreasing. A function which is neither increasing nor decreasing is called a **constant function.** More formally, a function is **constant** on an interval containing the numbers x_1, x_2 if whenever $x_2 > x_1$, $f(x_2) = f(x_1)$. Example; see below.

Figure: The graph of a constant function:
$$f(x) = 5$$

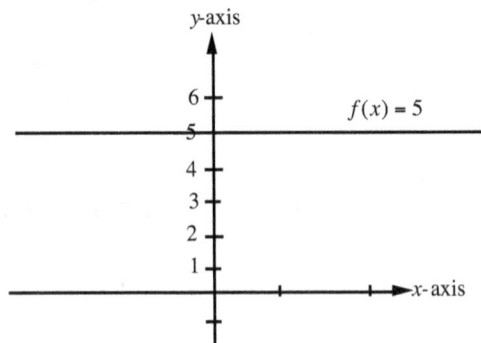

Lesson 31 Exercises

Give both graphical and algebraic answers (with examples) to the following:

1. What is meant by saying that a function is positive on an interval?

2. What is meant by saying that a function is negative on an interval?

3. What is meant by saying that a function is increasing on an interval?

4. What is meant by saying that a function is decreasing on an interval.

5. What is a constant function?

Answers: See text.

About endpoints of increasing intervals and decreasing intervals from graphs

In writing the intervals of increasing or decreasing functions from graphs, we will include the endpoints whenever the endpoints are defined. For example, with reference to the diagram below,
$f(x)$ is increasing on the intervals $[-2, -1)$ and $[1, 2]$,, and decreasing on the intervals $(-1, 1]$, and $[2, 3]$.

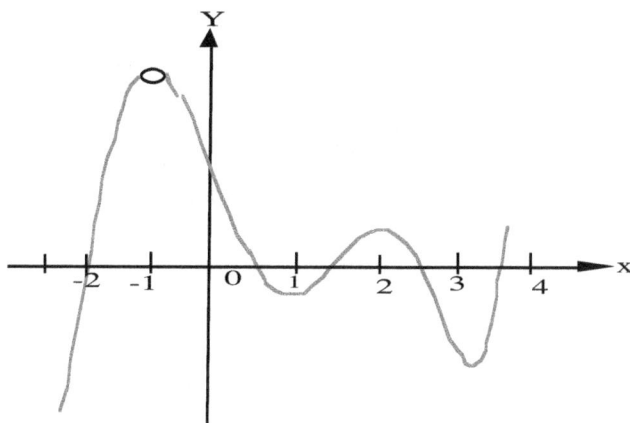

Other Terminology
Nonincreasing function (monotone decreasing function):

A function f is nonincreasing on an interval containing the numbers x_1, x_2
  if whenever $x_2 > x_1$. $f(x_2) \leq f(x_1)$

(Note that nonincreasing implies "does not increase" and therefore, it is either constant or decreasing on that interval) Note: Using nonincreasing is consistent with terms such as a "nonpositive" integer which is either zero or a negative integer.

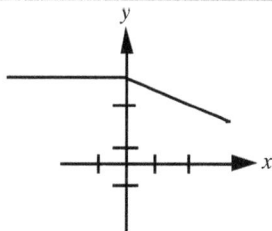

Graph of a nonincreasing function

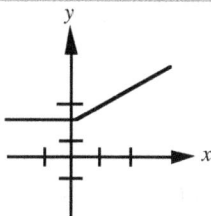

Graph of a nondecreasing function

Nondecreasing function (monotone increasing function):

A function f is nondecreasing on an interval containing the numbers x_1, x_2

if whenever $x_2 > x_1$, $f(x_2) \geq f(x_1)$ (Note that nondecreasing implies "does not decrease" and therefore, it is either constant or increasing on that interval)

Lesson 32

Concavities of Curves; Critical Points

Concavities of Curves

The following meanings given to concavities will be useful in describing the turning behavior of curves at certain points or locations. (In calculus, we will learn more formal definitions of concavities). In particular, we will illustrate the "sense of concavity " (direction of the concavity).

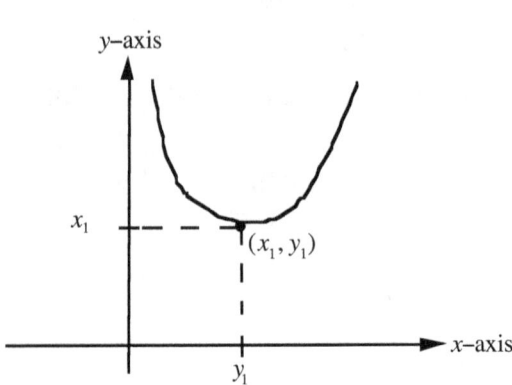

Figure 1: Curve is concave up
(opens upwards) at (x_1, y_1)

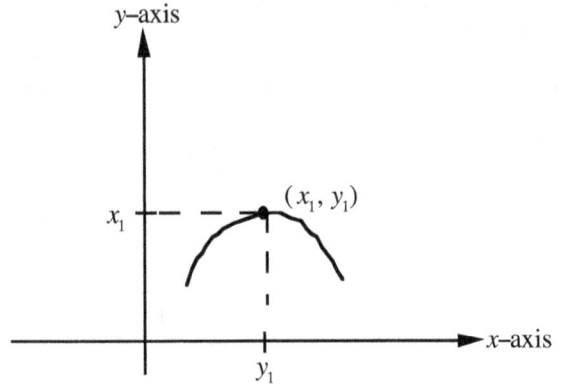

Figure 2: Curve is concave down
(opens downwards) at (x_1, y_1)

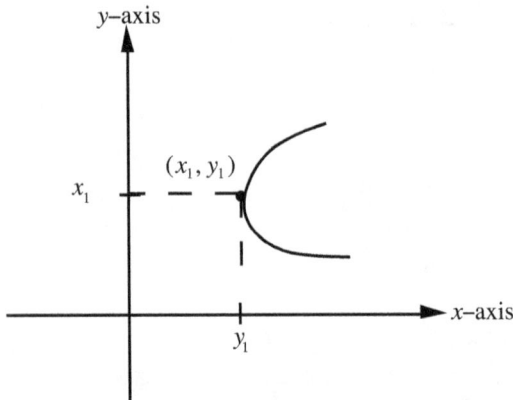

Figure 3: Curve is concave to the right
(opens to the right) at (x_1, y_1))

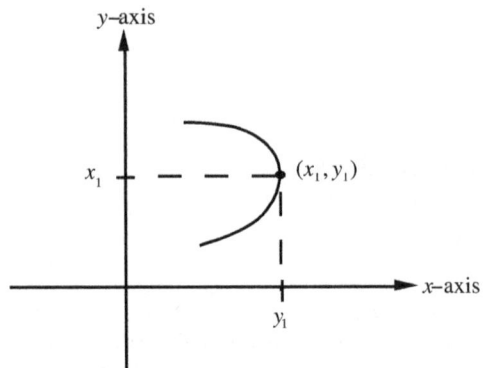

Figure 4: Curve is concave to the left
(opens to the left) at (x_1, y_1))

Note above that Figures 1 and 2 are graphs of functions; but Figures 3 and 4 are **not** functions by the vertical line rule.

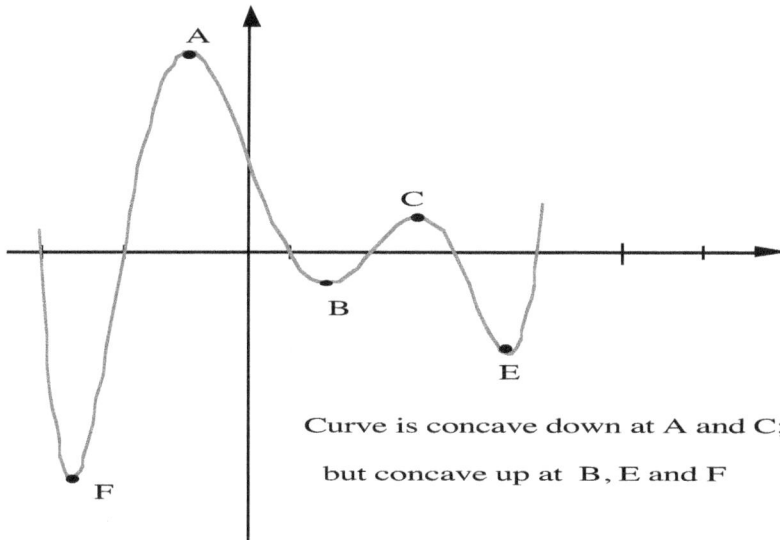

Curve is concave down at A and C;

but concave up at B, E and F

Extra:

The author also proposes the following new system:

1. Concave north (concave up)   **2.** Concave south (concave down)

3. Concave east (concave to the right).   **4.** Concave west (concave to the left)

5. Concave North-East (N-E);   **6.** Concave North-West (N-W);

7. Concave South-West (S-W)   **8.** Concave South-East (S-E)

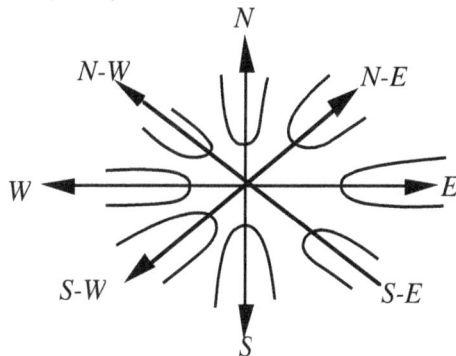

Figure: Concavity directions based on the North-South Line

Note above that only concave north and concave south are graphs of functions.

Critical Points 216

Turning Points, Minimum Points, Maximum Points, Inflection Points

Maximum Point: A given point (turning point) on a curve is a maximum point of the curve if the given point is higher than any other point on the curve in the immediate vicinity (on both sides of the point). The y-coordinate of this point is called a maximum value of the curve. At a maximum point, the curve is concave downwards. Sometimes, a maximum point of a curve may not be the highest point on the curve. When this happens, this maximum point is higher than only points sufficiently near it and there are other points on the curve which are higher than this point. In this case, we call this maximum point a relative maximum. However, if this maximum point is the highest point on the curve (irrespective of the domain of the function), then we say that this point is the absolute maximum.

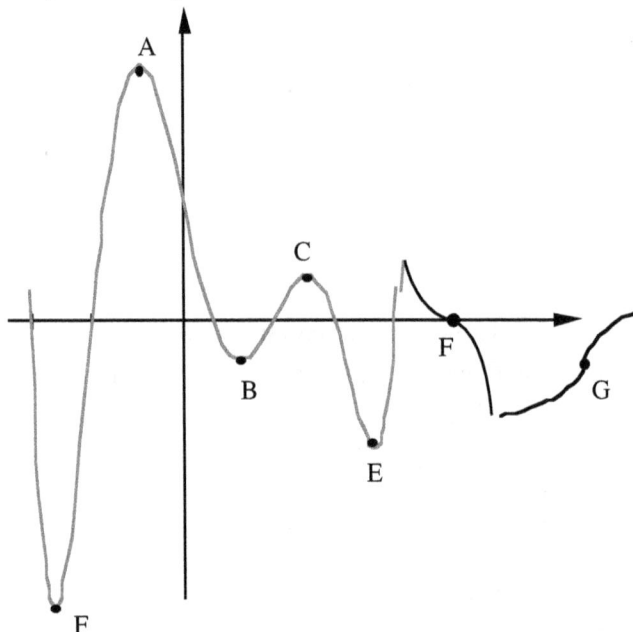

Minimum Point: A given point (turning point) on a curve is a minimum point of the curve if the given point is lower than any other point on the curve in the immediate vicinity (on both sides of the point). At a minimum point, the curve is concave upwards. The y-coordinate of this point is called a minimum value of the curve. Sometimes, a minimum point of a curve may not be the lowest point on the curve. When this happens, this minimum point is lower than only points sufficiently near it and there are other points on the curve which are lower than this point. In this case, we call this minimum point a relative minimum. However, if this minimum point is the lowest point on the curve (irrespective of the domain of the function), then we say that this point is the absolute minimum.

Generally, for a polynomial of degree n, there are at most $n-1$ relative minima or maxima. For example, $y = x^2$ has only one minimum point (and that point is an absolute minimum).

Point of Inflection: A given point (turning point) on a curve is a point of inflection if the curve at this point, changes from being concave up to being concave down or vice versa. At a point of inflection, the curve is either concave down to the left of the point and concave up to the right of the point, or it is concave up to the left of the point and concave down to the right of the point.

Thus, the sense of concavity to the right of the point is opposite to the sense of concavity to the left of the point. Note that a point of inflection is neither a minimum point nor a maximum point.

More formally, for a function f, $f(x_0)$ is a relative maximum of f on an open interval containing x_0 if $f(x_0) > f(x)$ for all $x \neq x_0$ on the interval. Similarly, for a function f, $f(x_0)$ is a relative minimum of f on an open interval containing x_0 if $f(x_0) < f(x)$ for all $x \neq x_0$ on the interval..

Lesson 32 Exercises

1. Identify the concavities of the following graphs:

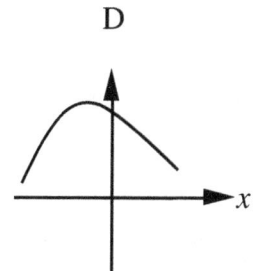

    A   B   C   D

2. Explain the following:

(a) Turning Point, **(b)** Minimum Point, **(c),** Maximum Point,   **(d)** Inflection Point

    Answers: **1.** Concave upwards; **2.** Concave to the right; **3.** Concave to the left; **4.** Concave downwards.

Lesson 33

218

Reflection of Points, Lines and Curves

We will cover reflections about the y-axis, the x-axis, the origin and the line $y = x$.

Let us discuss a few examples of the usefulness of being able to rapidly reflect points, lines and curves. If we know that a curve whose equation we are to find is the reflection of, say, the curve

$y = x^2$ about the x-axis, then we could immediately write the equation of the reflected curve as $y = -x^2$

Similarly, the curves $y = e^x$ and $y = e^{-x}$ are reflections of each other about the y-axis.

Given the graph of a function or a relation, we can find the graph of the inverse of the function or relation by reflecting the given graph about the line $y = x$.

Reflection of a point about the y-axis

Let $P(x,y)$ be a point in an x-y coordinate plane. Then the coordinates of the **point** of reflection of the given point about the y-axis is obtained by replacing the x-coordinate by $-x$, keeping the y-coordinate unchanged. The point of reflection is thus given by $R(-x,y)$.

Example Find the point of reflection of the point $(4, 2)$ about the y-axis.

Solution Change the sign of the x-coordinate and keep the y-coordinate unchanged.
The reflected point is $P(-4, 2)$. (Figure). If the page were folded along the y-axis, the two points $(4, 2)$ and $(-4, 2)$ would coincide (occupy the same space).

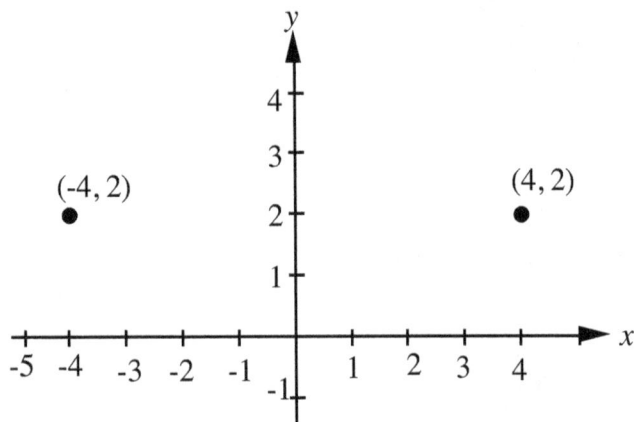

Figure: The graph of the reflection of $(4, 2)$ about the y-axis.

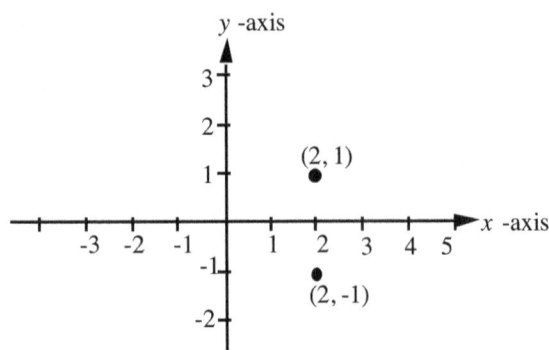

Figure: The graph of the reflection of $(2, 1)$ about the x-axis

Reflection of a point about the x-axis

Let $P_1(x,y)$ be a point in an x-y coordinate plane. Then the coordinates of the point of reflection of the given point about the x-axis is obtained by replacing the y-coordinate by $-y$, keeping the x-coordinate unchanged. The point of reflection is thus given by $P_2(x,-y)$.

Example Find the point of reflection of the point $(2, 1)$ with respect to the x-axis.

Solution Change the sign of the y-coordinate and keep the x-coordinate unchanged.
The point of reflection is $(2,-1)$. (Figure)
If the page were folded along the x-axis the two points $(2 , 1)$ and $(2 -1)$ would coincide.

Reflection of a point about the origin

The point of reflection of the point $P(x,y)$ about the origin is $P(-x,-y)$, obtained by replacing x by $-x$ and y by $-y$, That is, change the signs of both the x- and y-coordinates.

Example Find the point of reflection of $(2,4)$ about the origin.

Solution The point of reflection is given by $P(-x,-y)$ and in the present case it is $Q(-2,-4)$. Practically, if the page were folded first along the positive y-axis (1st quadrant on to 2nd quadrant), and then along the x-axis (1st and 2nd quadrant together on to the 3rd quadrant), the two points $(2\ 4)$, and $(-2\ -4)$, would coincide (Figure); or simply, we would fold the 1st quadrant on to the 3rd quadrant so that the negative x-axis coincides with the positive x-axis; and that the positive y-axis is coincident with the negative y-axis.

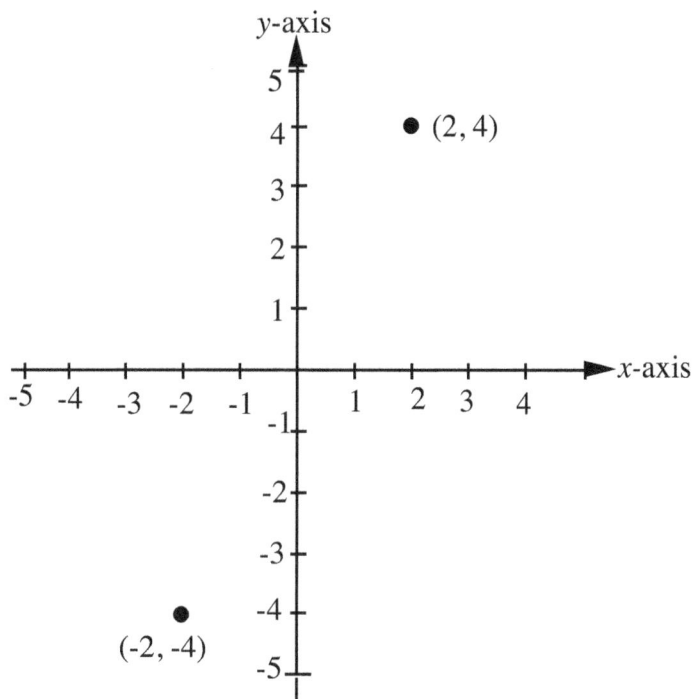

Summary for reflection

1. To reflect about the y-axis, change the sign of the x-coordinate, keeping the y-coordinate unchanged.

2. To reflect about the x-axis, change the sign of the y-coordinate, keeping the x-coordinate unchanged.

3. To reflect about (or through) the origin, change the signs of both the x- and y-coordinates.

Note that changing the sign of the x-coordinate, or the y-coordinate is equivalent to multiplying or dividing the coordinate by -1.

We should note above that the given point and the reflected point are reflections of each other.

Reflection of a point about a vertical line

Geometrically, two points A and B are reflections of each other about a given line if and only if the given line is the perpendicular bisector of AB (i.e., the line divides the segment AB into two equal parts at right angles).

Also: **Reflection of a point about the vertical line** $x = x_m$

Let the point to be reflected be (x_1, y_1).

Let the reflected point be (x_2, y_1)

(The y-coordinate is constant since the reflection is about a vertical line)

Let the midpoint of the line connecting (x_1, y_1) and (x_2, y_1) be (x_m, y_1).

Then $x_m = \dfrac{x_1 + x_2}{2}$ (Note that the equation of the line about which (x_1, y_1) is reflected is $x = x_m$)

Note: For the reflection about the y-axis, $x_m = 0$ (The y-axis is the line $x = 0$.)

Example Find the point of reflection of the point $A(4, 6)$ about
(a) the y-axis ; (b) the line $x = 2$.

Solution
Method 1
(a)

Step 1: Plot the point $(4, 6)$. See Figure.

Step 2: Draw a horizontal line from $(4, 6)$ to intersect the y-axis (the line $x = 0$) at C.
Since the point C is mid-way between the point $(4, 6)$ and the reflected point, count 4 units horizontally to the left, stop and place a dot here. This point is the point of reflection.

The point of reflection is $B(-4, 6)$.

(b) Since the point D is 2 units from the point $(4, 6)$, count 2 units horizontally to the left, stop and place a dot here. This is the point of reflection about the line $x = 2$.

The point of reflection is $C(0, 6)$.

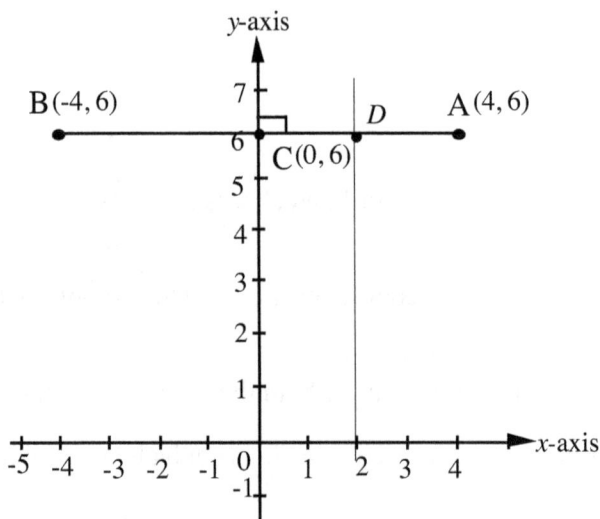

Figure: Graph of the reflection of the point $A(4, 6)$ about the y-axis and the line $x = 2$.

Note above that in (a) $BC = CA$ and BA intersects the y-axis at right angles at C; and in
(b) $CD = DA$ and CA intersects the line $x = 2$ at right angles at D. Note that D is equidistant from A and C.

Method 2

(a)

Apply $x_m = \dfrac{x_1 + x_2}{2}$ (Note that the equation of the line about which (x_1, y_1) is reflected is $x = x_m$.)

$x_1 = 4$, $x_m = 0$ (on the y-axis, $x = 0$). Substituting,

$$0 = \frac{4 + x_2}{2}$$

$$0 = 4 + x_2$$

$$-4 = x_2$$

The x-coordinate of the reflected point -4

Therefore, the point of reflection about the y-axis is $B(-4, 6)$. (the y-coordinate does not change)

(b)

Apply $x_m = \dfrac{x_1 + x_2}{2}$ (Note that the equation of the line about which (x_1, y_1) is reflected is $x = x_m$.)

$x_1 = 4$, $x_m = 2$ (on the line $x = 2$, $x = 2$). Substituting,

$$2 = \frac{4 + x_2}{2}$$

$$4 = 4 + x_2$$

$$0 = x_2$$

The x-coordinate of the reflected point is 0.

Therefore, the point of reflection about the line $x = 2$ is $C(0, 6)$.   (the y-coordinate does not change)

Reflection of a point about the horizontal line $y = y_m$

Let the point to be reflected about the line $y = y_m$ be (x_1, y_1)

Let the reflected point be (x_1, y_2),

(The x-coordinate is constant since the reflection is about a horizontal line)

Let the midpoint of the line connecting (x_1, y_1) and (x_1, y_2) be (x_1, y_{m1}).

$$y_m = \frac{y_1 + y_2}{2}$$ (Note that the equation of the line about which (x_1, y_1) is reflected is $y = y_m$)

Note: For the reflection about the x-axis, $y_m = 0$ (The x-axis is the line $y = 0$.)

Example Find the point of reflection of $(2, 1)$ about the x-axis.

Solution $y_1 = 1$, $y_m = 0$ (On the x-axis, $y = 0$), Substituting in $y_m = \dfrac{y_1 + y_2}{2}$, we obtain

$$0 = \frac{1 + y_2}{2}$$

$$0 = 1 + y_2$$

$$-1 = y_2$$

The y-coordinate of the reflected point is -1.

Therefore, the point of reflection about the x-axis is $(2, -1)$. This problem was solved previously by merely changing the sign of the y-coordinate, keeping the x-coordinate unchanged.

Reflection of a given curve (or line) about the *y*-axis

Step 1: Sketch the given curve using broken or dotted lines.

Step 2: Reflect critical points or important points on the curve about the *y*-axis .(see p.218)

Step 3: Connect the points of reflections from Step l by a smooth solid curve (if it is a curve) or by a straight line (if it is a straight line).

Reflection of a given curve (or line) about the *x*-axis

Method: Repeat steps 1, 2 and 3 above but about the *x*-axis.

Step 1: Sketch the given curve using broken or dotted lines.

Step 2: Reflect critical points or important points on the curve about the *x*-axis .(see p.218)

Step 3: Connect the points of reflections from Step l by a smooth solid curve (if it is a curve) or by a straight line (if it is a straight line).

Reflection of a given curve about (or through) **the origin**

Step 1: Using broken lines rapidly sketch the graph of the given curve.

Step 2: Reflect critical or important points about the origin (as explained on page 219).

Step 3. Connect the points of reflections by a solid curve.

Example 1. Sketch the reflection of the line whose equation is $y = 3x + 2$ about
    (*a*) the *y*-axis; (*b*) the *x*-axis.

(a) Step 1. Sketch the graph of $y = 3x + 2$ using broken lines.

   Step 2: Reflect two points on $y = 3x + 2$ about the *y*-axis and connect the new points by a solid line. (Fig.1)

(b) Step 1: Sketch the graph of $y = 3x + 2$ using broken lines.

   Step 2: Reflect two points on $y = 3x + 2$ about the *x*-axis and connect the new points by a solid line. (Fig. 2)

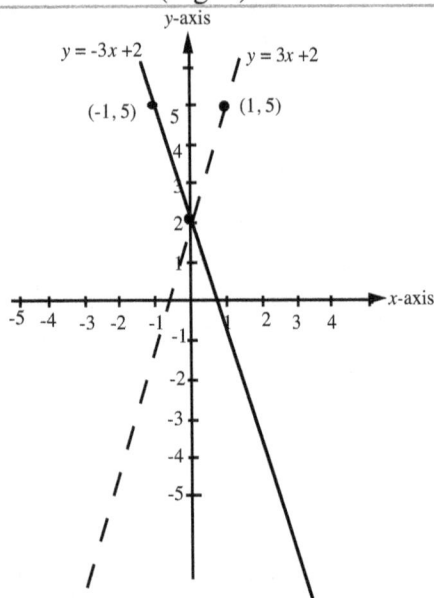

Fig. 1: Graph of the reflection of the line $y = 3x + 2$ about the *y*-axis

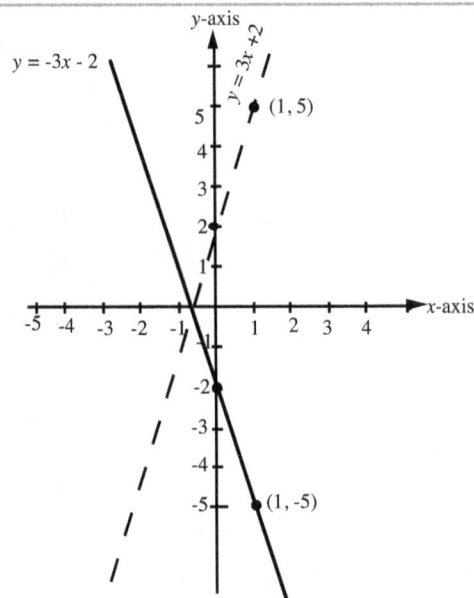

Fig. 2: Graph of the reflection of the line $y = 3x + 2$ about the *x*-axis

Note above: You may also use the slope-intercept method to draw the graph of a straight line.

Reflection of a curve or a line about the line $y = x$

Finding the reflection about the line $y = x$ graphically, is synonymous with finding the inverse of the given graph.

Step 1: Reflect important (critical) points on the given curve about the line $y = x$.

To accomplish the above, **interchange** the ordered pairs (x, y) on the given curve or line and plot the results in a rectangular coordinate system.

Step 2: Connect the points (from Step 1) by a solid curve or line accordingly.

Practically, if the page were folded along the line $y = x$, the given curve (or line) and the reflected curve would coincide.

Example 2 Sketch the reflection of the line whose equation is $y = 3x + 2$ about the line $y = x$

Solution

Step 1. Sketch the graph of $y = 3x + 2$ using broken lines.

Step 2: Reflect two points on $y = 3x + 2$ about the line $y = x$, and connect the new points by a solid line.

Extra: New Theorem by the author
If two lines are inverses of each other, their slopes are reciprocals of each other. If the slope of a line is m, the slope of the inverse of this line is $\frac{1}{m}$.

(see also p.80)

Proof: Let the equation of a line with slope m and y-intercept b be given by $y = mx + b$.

The inverse of this line is

$$x = my + b \text{ or } y = \frac{1}{m}x - \frac{b}{m}$$

Clearly, the slope of the inverse is $\frac{1}{m}$. Q.E.D

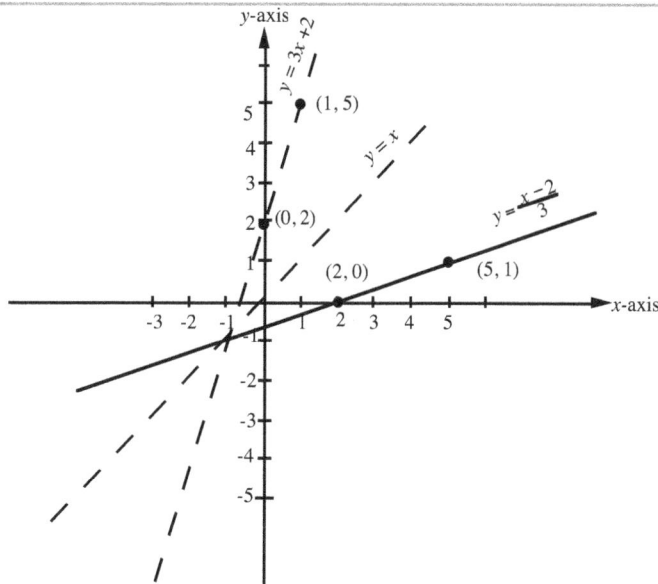

Fig. 3 Graph of the reflection of $y = 3x + 2$ about the line $y = x$.

Equation of the reflection of a curve, given the equation of the curve

(a) **About the y-axis**	(b) **About the x-axis.**
Replace x by $-x$ in the given equation.	Replace y by $-y$ and solve for y in the given equation.
$y = (-x)^3 + 2(-x)^2 + 3(-x) + 1$	$-y = x^3 + 2x^2 + 3x + 1$
$y = -x^3 + 2x^2 - 3x + 1.$	$y = -(x^3 + 2x^2 + 3x + 1) = -x^3 - 2x^2 - 3x - 1$

Example 4 Find the "reflected equation" of $y = 3x + 2$ about the line $y = x$.

The solution is the same as finding the inverse of $y = 3x + 2$. (We repeat a previous solution)

| Step 1 Interchange x and y in the given equation. Then, we obtain $x = 3y + 2$. (2) By tradition, we want to keep the x-axis horizontal and express y as a function of x. | Step 2: Solving for y, $y = \frac{x-2}{3}$ The reflected equation is $y = \frac{x-2}{3}$. |

Lesson 33 Exercises

Find the point of reflection about the y-axis for each of the following:

   1. $(3,2)$;   **2**. $(-4,3)$;   **3**. $(-2,-5)$;   **4**. $(4,-6)$

Exercises 5 – 8 : Repeat Exercises 1 – 4 but with respect to the x–axis.

Exercises 9 – 12. Repeat Exercises 1 – 4 but with respect to the origin.

13. Repeat Exercise 1 – 4 with respect to the line $x = 2$.

Find an equation of the reflection with respect to (a) the x–axis, (b) the y–axis, (c) the origin, (d) the line $y = x$ for the following :

   14. $y = 4x + 3$;   **15**. $y = x^2 - 4x + 3$;   **16**. $y = x^2 - 3x - 2$;   **17**. $-y = x^2 - 3x - 2$.

18. Sketch the graph of $y = 2x + 3$ and reflect the graph with respect to (a) the x–axis, (b) the y–axis.

19. Sketch the graph of $y = -3x + 2$ and reflect the graph with respect to the line $y = x$.

Answers:

1. $(-3, 2)$; **2**. $(4, 3)$; **3**. $(2, -5)$; **4**. $(-4, -6)$; **5**. $(3, -2)$; **6**. $(-4, -3)$; **7**. $(-2, 5)$;

8. $(4, 6)$; **9**. $(-3, -2)$; **10**. $(4, -3)$ **11**. $(2, 5)$; **12**. $(-4, 6)$; **13**. (a) $(1, 2)$;

 (b) $(8, 3)$; (c) $(6, -5)$; (d) $(0, -6)$; **14**. (a) $y = -4x - 3$;

 (b) $y = -4x + 3$; (c) $y = 4x - 3$; (d) $x = 4y + 3$ or $y = \frac{1}{4}x - \frac{3}{4}$; **15**. ($a$) $y = -x^2 + 4x - 3$;

 (b) $y = x^2 + 4x + 3$; (c) $y = -x^2 - 4x - 3$; (d) $x = y^2 - 4y + 3$; **16**. (a) $y = -x^2 + 3x + 2$;

 (b) $y = x^2 + 3x - 2$; (c) $y = -x^2 - 3x + 2$; (d) $x = y^2 - 3y - 2$; **17**. (a) $y = x^2 - 3x + 2$;

 (b) $y = -x^2 - 3x + 2$; (c) $y = x^2 + 3x - 2$; (d) $x = -y^2 + 3y + 2$;

18.

19.

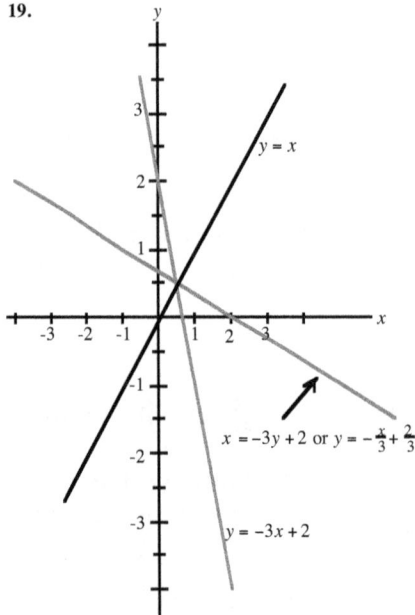

Lesson 34

Introduction to Transformation of Functions and Relations
Translation of Points, Relations, Functions and Axes

There are three main methods for sketching the graphs of functions and relations in an x-y coordinate system of axes, namely, the general method, transformation method, and specific methods for specific types of functions. For example, the slope-intercept method is for linear functions involving two variables, and the "vertex-axis of symmetry" method is for quadratic functions Each method has its relative merits. In the general method,, we choose convenient values for one of the variables and calculate the corresponding values of the other variable to obtain ordered pairs which are then plotted and the points connected. In **transformations**, we are given points (ordered pairs), graphs or equations of known basic functions, and we are required to sketch new graphs using the given information. To sketch the graph of a function or a relation, what we need is to obtain important points on the graph, plot these points and connect them accordingly. Quickly construct a table of values for the given or known points and using the formulas in the box below, determine the new points, plot them and connect them using the appropriate curves or lines. Let $y = f(x)$. Then the general transformation equation is given by

$$y - k = af[b(x - h)] \quad \text{(A)} \quad \text{OR} \quad y = af[b(x - h)] + k \quad \text{(B)}$$

y–modifiers x–modifiers x–modifiers y–modifiers

The parameters k and a are y–modifiers (change the y-coordinates); b and h are x–modifiers (change the x-coordinates).

Some books write the k on the right-hand side of the equation as in (B) above. The form with k on the left-hand side (as in (A) above) makes the signs of k and h consistently easier to recall .

The parameters k and h are for translation (shifting): k is the vertical shift, h is the horizontal shift; a and b are for stretching and shrinking. In constructing tables for the x- and y-coordinates, k and h are addends to the known or given y-coordinates and x-coordinates respectively, while a is a multiplier of the y-coordinates, b is a divisor of the x-coordinates. Note also that the parameters, k, a, b ($b \neq 0$), and h are real.

Let x_0 = original or known x-coordinate, x_n = new x-coordinate. Then $x_n = \frac{x_0}{b} + h$	Let y_0 = original or known y-coordinate, y_n = new y-coordinate. Then $y_n = y_0(a) + k$

Note the following examples:
1. In $y - 2 = 3f[4(x - 5)]$
   $k = 2$, $h = 5$, (set $y - 2 = 0$, $x - 5 = 0$, and solve).
   $a = 3$, $b = 4$ (by comparison with (A) above)
2. In $y + 2 = -3f[-4(x + 5)]$
   $k = -2$, $h = -5$, ($y + 2 = 0$, $x + 5 = 0$, and solve),
   $a = -3$, $b = -4$.

3. For $y = \sqrt{-x + 2}$ rewrite as
   $y = \sqrt{-(x - 2)}$ (basic function: $y = \sqrt{x}$)
   Here, $k = 0$, $a = 1$; $h = 2$, $b = -1$
4. For $y = \sqrt{-2x + 3}$ rewrite as
   $y = \sqrt{-2(x - \frac{3}{2})}$ (basic function: $y = \sqrt{x}$)
   Here, $k = 0$, $a = 1$, $h = \frac{3}{2}$, $b = -2$

To use the usual algebraic **order of operations**, write the equation in the above standard forms (A) or (B). See also Examples 3 and 4 above. **Note:** To divide or multiply by -1, just change the sign. We will first graph a most general case, and later, we will graph less general cases. Two main steps would be involved, namely, **algebraic** action for which we perform the needed calculations; and **geometric** action for which we plot the points and connect them. The **graphical** result, may be a translation (shifting), a stretching, a shrinking, an expansion or a contraction or compression of the given or basic graphs.

Example 1 Given $y = x^2$, draw the graph of $y + 4 = 3\left(\left[-2x + 2\right]^2\right)$

Rewrite $y + 4 = 3\left(\left[-2x + 2\right]^2\right)$ as $y + 4 = 3\left(\left[-2(x - 1)\right]^2\right)$. Here, $k = -4$, $h = 1$, $a = 3$, $b = -2$

$y = x^2$		$y + 4 = 3\left(\left[-2(x-1)\right]^2\right)$		
x_0	y_0	$x_n = \dfrac{x_0}{b} + h$	$y_n = y_0(a) + k$	(x_n, y_n)
0	0	$\frac{0}{-2} + 1 = 0 + 1 = 1$	$0(3) - 4 = 0 - 4 = -4$	$(1, -4)$
1	1	$\frac{1}{-2} + 1 = -\frac{1}{2} + 1 = \frac{1}{2}$	$1(3) - 4 = 3 - 4 = -1$	$(\frac{1}{2}, -1)$
2	4	$\frac{2}{-2} + 1 = -1 + 1 = 0$	$4(3) - 4 = 12 - 4 = 8$	$(0, 8)$
3	9	$\frac{3}{-2} + 1 = -\frac{3}{2} + 1 = -\frac{1}{2}$	$9(3) - 4 = 27 - 4 = 23$	$(-\frac{1}{2}, 23)$
4	16	$\frac{4}{-2} + 1 = -2 + 1 = -1$	$16(3) - 4 = 48 - 4 = 44$	$(-1, 44)$
-1	1	$\frac{-1}{-2} + 1 = \frac{1}{2} + 1 = \frac{3}{2}$	$1(3) - 4 = 3 - 4 = -1$	$(\frac{3}{2}, -1)$
-2	4	$\frac{-2}{-2} + 1 = 1 + 1 = 2$	$4(3) - 4 = 12 - 4 = 8$	$(2, 8)$
-3	9	$\frac{-3}{-2} + 1 = \frac{3}{2} + 1 = \frac{5}{2}$	$9(3) - 4 = 27 - 4 = 23$	$(\frac{5}{2}, 23)$
-4	16	$\frac{-4}{-2} + 1 = 2 + 1 = 3$	$16(3) - 4 = 48 - 4 = 44$	$(3, 44)$

Step 1: Quickly prepare a a table of values for x and y, using the basic parabola equation and convenient x-values.

Step 2: Using the table of values from Step 1, determine the new coordinates x_n and y_n.

Step 3: Plot the new points and connect them by a solid curve (Fig. 1 below)

Students: Try to sketch the same graph using the general method, and compare the two methods.

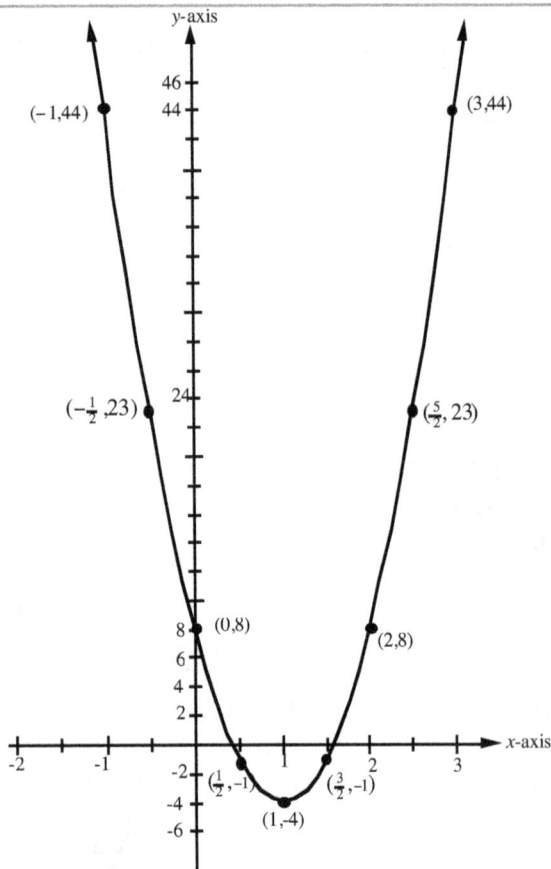

Fig. 1: Graph of $y + 4 = 3\left(\left[-2x + 2\right]^2\right)$

Less General Examples

In Example 1: $y + 4 = 3\left(\left[-2(x - 1)\right]^2\right)$ (A)

$k = -4$ $h = 1$, $a = 3$. $b = -2$

Now, let us observe some **less general examples.**

Case 1: when $k = 0$ (A) becomes
$$y = 3\left(\left[-2(x - 1)\right]^2\right)$$

Case 2: when $k = 0$, $h = 0$ (A) becomes
$$y = 3\left(\left[-2(x)\right]^2\right)$$

Case 3: when $k = 0$, $h = 0$, $a = 1$ (A) becomes $y = \left[-2(x)\right]^2$

Case 4: when $k = 0$, $h = 0$, $a = 1$, $b = 1$, (A) becomes $y = x^2$

Case 5: when $b = 1$, $a = 1$ (A) becomes
$$y + 4 = (x - 1)^2$$

Study the above example very well since you can easily apply the experience gained to trigonometric functions such as $y - 1 = 3\sin(4x + \pi)$.

Note: Apart from the transformation method, if given the equation of the function, you can use the general method where you choose x, and calculate y to obtain ordered pairs for graphing.

Finding a Transformation Equation,   227
Knowing the Graphs of a Function and the Transformed Function

We can write down a specific transformation equation if we know the numerical values of the parameters k, a, b and h of the general transformation equation $y - k = af[b(x - h)]$.

Example 2 We do Example 1, p. 226 backwards. Find the transformation equation, given or knowing the following information in the table below, where

x_0 = original or known x-coordinate, x_n = new x-coordinate,

y_0 = original or known y-coordinate, y_n = new y-coordinate.

$y = f(x)$		(x_n, y_n)
x_0	y_0	
0	0	$(1, -4)$
1	1	$(\frac{1}{2}, -1)$
2	4	$(0, 8)$

Solution:

Step 1: Find b and h.

We will use two x_0-values and two x_n-values together with the equation $x_n = \frac{x_0}{b} + h$ to set up a system of two equations in which the unknowns are b and h.

Substituting $x_0 = 0$, $x_n = 1$ (1st row of table) in

$x_n = \frac{x_0}{b} + h$, we obtain $1 = \frac{0}{b} + h$ (A)

Similarly, using $x_0 = 2$, $x_n = 0$ (3rd row of table)

in $x_n = \frac{x_0}{b} + h$, we obtain $0 = \frac{2}{b} + h$ (B)

We now solve (A) and (B) simultaneously.

From (A) $1 = 0 + h$ and $h = 1$

Substituting $h = 1$ in (B)

$$0 = \frac{2}{b} + 1$$

$$0 = 2 + b \quad (h = 1)$$

$$b = -2$$

Step 2: Find k and a.

We will use two y_0-values and two y_n-values, together with the equation $y_n = y_0(a) + k$ to set up a system of two equations in which the unknowns are k and a.

Substituting $y_0 = 0$, $y_n = -4$ in $y_n = y_0(a) + k$, we obtain $-4 = 0(a) + k$ (C)

Similarly, using $y_0 = 4$, $y_n = 8$ (3rd row of table) in $y_n = y_0(a) + k$, we obtain

$$8 = 4a + k \quad \text{(D)}$$

We now solve (C) and (D) simultaneously.

From (C) $-4 = k$ or

$$k = -4$$

Substituting $k = -4$ in (D)

$$8 = 4a - 4 \quad (k = -4)$$

$$12 = 4a$$

$$a = 3$$

Step 3: Now we substitute $k = -4$, $a = 3$, $b = -2$, and $h = 1$ in
$y - k = af[b(x - h)]$, to obtain
$y - (-4) = 3f[-2(x - 1)]$
$y + 4 = 3f[-2(x - 1)]$ or $y + 4 = 3f[-2x + 2]$

If $f(x)$ is quadratic, we obtain $y + 4 = 3[(-2x + 2)^2]$

Note above that given the graphs of a function and the transformed function, you could read the points $P_0(x_0, y_0)$ and $P_n(x_n, y_n)$ from the graph and construct a table of values as above. You need two P_0 points and two P_n points.. Match the P_0's and P_n's.

Translating (Shifting) the Graphs of Functions, and Relations   228

Given the graph of $y = f(x)$, we can sketch the graph of $y - k = f(x - h)$ by moving (shifting) each point, on $f(x)$, h units horizontally and k units vertically simultaneously.

We agree that if the function is of the form $y - k = f(x - h)$, then the shift is h units to the right and k units up. In the above form, h is positive and k is positive. We agree also that, if the equation is of the form $y + k = f(x + h)$, then the h-shift is to the left and the k-shift is down. In this form, h is negative k is negative.

From the general equation $y - k = af[b(x - h)]$, with $a = 1$, $b = 1$, we obtain $y - k = f(x - h)$.

Example 3 Given $y = x^2$, draw the graph of $y + 4 = (x - 1)^2$ (review p.226)

In $y + 4 = (x - 1)^2$, $k = -4$, $h = 1$, $a = 1$, $b = 1$

$y = x^2$		$y + 4 = (x - 1)^2$		
x_0	y_0	$x_n = x_0 + h$	$y_n = y_0 + k$	(x_n, y_n)
0	0	$0 + 1 = 1$	$0 - 4 = -4$	$(1, -4)$
1	1	$1 + 1 = 2$	$1 - 4 = -3$	$(2, -3)$
2	4	$2 + 1 = 3$	$4 - 4 = 0$	$(3, 0)$
3	9	$3 + 1 = 4$	$9 - 4 = 5$	$(4, 5)$
4	16	$4 + 1 = 5$	$16 - 4 = 12$	$(5, 12)$
-1	1	$-1 + 1 = 0$	$1 - 4 = -3$	$(0, -3)$
-2	4	$-2 + 1 = -1$	$4 - 4 = 0$	$(-1, 0)$
-3	9	$-3 + 1 = -2$	$9 - 4 = 5$	$(-2, 5)$
-4	16	$-4 + 1 = -3$	$16 - 4 = 12$	$(-3, 12)$

Procedure:

Step 1: Quickly prepare a table of values for x and y using the basic parabola equation and convenient x-values.

Step 2: Using the table of values from Step 1, determine the new coordinates x_n and y_n.

Step 3: Plot the new points and connect them by a solid curve., (Fig. 1 below)

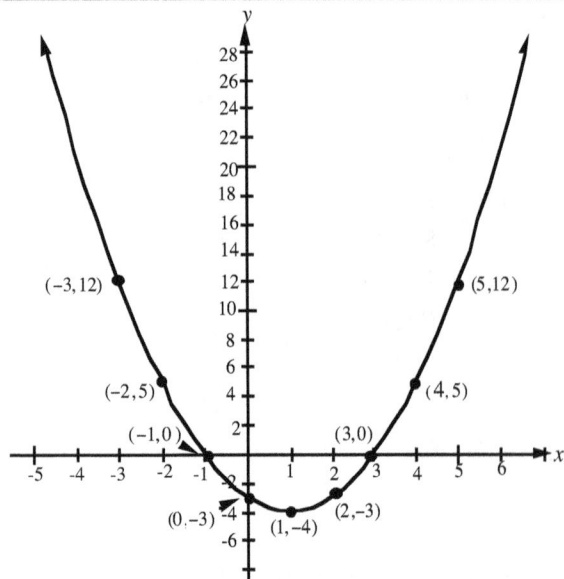

Fig. 1: Graph of $y + 4 = (x - 1)^2$

About Translation of Functions and Relations

Example 2 Given the graph of $y = f(x))$, describe the shifts to be made to obtain the graph of $y - 3 = f(x - 4)$

Solution   Every point on the original graph of $y = f(x)$ is shifted 3 units up and 4 units to the right simultaneously.

Example 3 Given the graph of $y = x^2$, describe the shifts to be made to obtain the graph of $y - 3 = (x - 4)^2$

Solution Same as in Example 2.. Every point on $y = x^2$ is shifted 3 units up and 4 units to the right simultaneously.

Finally, even though descriptions such as "Every point on $y = x^2$ is shifted some units up or down", the practical approach to bring about these translations is to follow the method of Example 1 above in which we construct a table of values for x and y, plot and connect. This approach is also good for "bookkeeping" purposes as well as if $a \neq 1$, $b \neq 1$, in $y - k = af[b(x - h)]$

Study the introduction to transformations of functions (page 225-226).

Example 4 Given the graph of $y = x^2$, describe the shifts to be made to transform it to
the graph   of $y = (x - 3)^2$

In standard form, $y = (x - 3)^2$ becomes
$$y - 0 = (x - 3)^2, \text{ where, } k = 0, h = 3$$
Each point on the graph is shifted 3 units to the right, keeping the y-values unchanged $(k = 0)$.

Translation of Axes

Consider a point P (Figure) with the original coordinates x_0, y_0 and with the coordinate reference axes $O_0 X_0, O_0 Y_0$, where x_0 and y_0 are measured from $O_0 Y_0$ and $O_0 X_0$, respectively .

Suppose that the origin $, O_0(0,0)$, is moved to a new point $O_n(h, k)$, giving rise to a new set of reference coordinate axes $O_n X_n$ and $O_n Y_n$. Let the new coordinates of the point P be x_n, and y_n where x_n and y_n are measured with reference to the new axes, $O_n X_n$ and $O_n Y_n$, respectively.

We must note that the point P does not move but only the origin moves, and that (x_0, y_0) and (x_n, y_n) represent the same point. We also assume that the axes, $O_0 X_0$ and $O_n X_n$, have the same scalar unit and that $O_0 X_0$ and $O_n Y_n$ have the same scalar units. From geometric considerations, we obtain the following relationships:

$$x_0 = x_n + h \qquad (1) \qquad \text{(old coordinate = new coordinate + shift)}$$
$$y_0 = y_n + k \qquad (2)$$

From equations (1) and (2)
$$x_n = x_0 - h \qquad (3)$$
$$y_n = y_0 - k \qquad (4)$$

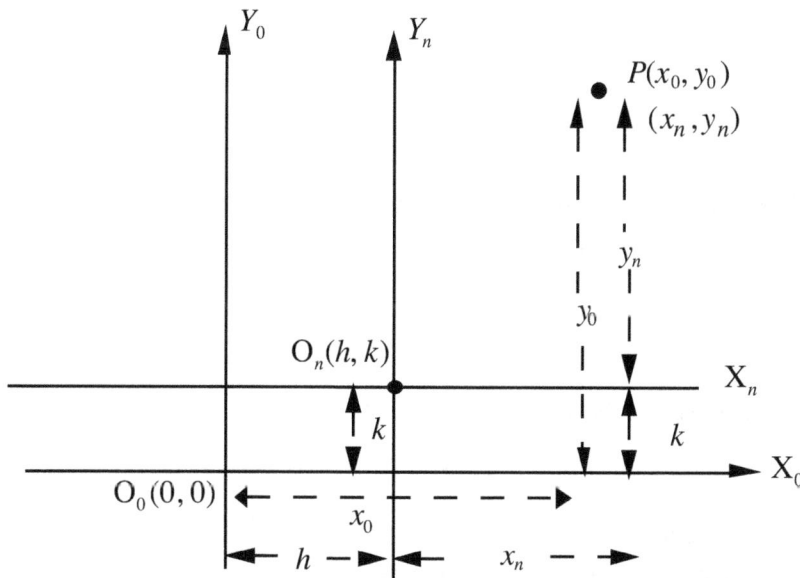

Figure: Graph of translation of axes

Application of Translation of Axes

Example 5 Suppose a point having original coordinates $x = 2$, $y = -3$ has its origin moved to the new origin $O_n(-1, 4)$. Find the new coordinates with reference to the new origin.

Solution: In this example, $x_0 = 2$, $y_0 = -3$, $h = -1$, $k = 4$.

Applying equation (3) and substituting,
$$x_n = x_0 - h$$
(new coordinate = old coordinate - horizontal shift)
$$x_n = 2 - (-1)$$
$$x_n = 3$$
Similarly, substituting in equation (4),
$$y_n = y_0 - k$$
(new coordinate = old coordinate - vertical shift)
$$= -3 - 4$$
$$y_n = -7$$
The new coordinates are $x = 3$, $y = -7$.

Let us say, your old house address was $(2, -3)$ according to the old numbering system. According to the new system, the address of the same house is now $(3, -7)$.

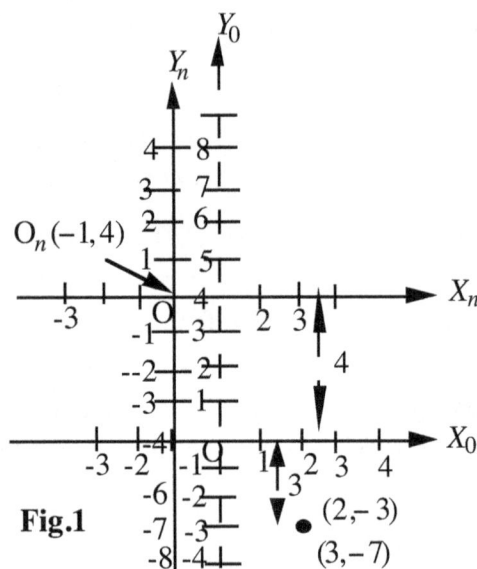

Fig.1

Example 6 Given that the equation of an ellipse is $4x^2 + y^2 + 24x - 2y + 21 = 0$, assuming that the origin of this ellipse is moved to the point $(-2, 3)$, write this equation with reference to the new origin.

Solution: **Approach 1**

We will apply equations (1) and (2) from previous page.

Let $x_0 = x$, $y_0 = y$, $h = -2$, $k = 3$

Then $x = x_n + h$
$$= x_n + (-2)$$
$$x = x_n - 2 \qquad \text{(A)}$$
Similarly, $y = y_n + 3 \qquad \text{(B)}$

We now substitute equations (A) and (B) in the given equation. Then we obtain

$$4(x_n - 2)^2 + (y_n + 3)^2 + 24(x_n - 2) - 2(y_n + 3) + 21 = 0.$$

Simplifying, we obtain

$$4x_n^2 + 8x_n + y_n^2 + 4y_n - 8 = 0$$

If we complete the square, we obtain

$$4(x_n + 1)^2 + (y_n + 2)^2 = 16, \text{ and if we drop the subscript } n, \text{ we obtain}$$

$$4(x + 1)^2 + (y + 2)^2 = 16. \quad \text{<---This is the new equation with reference to the new origin.)}$$

Approach 2

We complete the square first and then substitute $x_n - 2$ and $y_n + 3$ or add -2 and 3 respectively to the x-term and the y-term within the parentheses.

Step 1: $4x^2 + y^2 + 24x - 2y + 21 = 0$,

$\qquad 4(x + 3)^2 + (y - 1)^2 = 16 \qquad$ (by completing the square)

Step 2: $4(x + 3 - 2)^2 + (y - 1 + 3)^2 = 16 \qquad$ (adding -2 and +3 to adjust for the new origin)

$\qquad 4(x + 1)^2 + (y + 2)^2 = 16.$

Motivation for Learning Transformation of Functions and Relations 231

Transformation can be used to sketch new graphs from other/basic graphs as exemplified below.

Polynomials

From $y = x^2$, we can sketch the graphs of:

(a) $y = x^2 + 4$; (b) $y - 5 = 3(x - 2)^2$;

(c) $y = x^2 + 6x + 12$. (hint : complete the square)

From $y = |x|$, we can sketch $y = |x - 2|$

Rational Functions From $y = \dfrac{1}{x}$, we can sketch

(a) $y = \dfrac{1}{x} + 2$; (b) $y = \dfrac{1}{x - 2}$; (c) $y = \dfrac{1}{x + 4}$

Logarithmic Functions

From $y = \log x$, we can sketch

(a) $y = 2\log x + 5$; (b) $y = \log(x - 2)$.

Exponential Functions

From $y = e^x$, we can sketch

(a) $y = e^x - 3$; (b) $y = e^{x+5}$

Trigonometric Functions

From $y = \sin x$, we can sketch (a) $y = \sin x + 2$;

(b) $y = \sin x - 1$; (c) $y = 4\sin(x + 3) + 1$;

Conic Relations

From $x^2 + y^2 = r^2$, $\dfrac{x^2}{a^2} + \dfrac{y^2}{b^2} = 1$, $\dfrac{x^2}{a^2} - \dfrac{y^2}{b^2} = 1$,

we can sketch the graphs of say,

$(x - h)^2 + (y - k)^2 = r^2$, $\dfrac{(x - h)^2}{a^2} + \dfrac{(y - k)^2}{b^2} = 1$,

Lesson 34 Exercises

A Given the graph of $y = f(x)$, describe the shifts to be made to obtain the following:

1. $y - 2 = f(x - 1)$; **2.** $y + 2 = f(x + 1)$; **3.** $y = f(x - 3) + 2$; **4.** $y = f(x + 2) - 4$

5. Sketch the graphs of $y = 2x - 1$ and $y + 4 = 2(x - 3) - 1$ on the same set of rectangular coordinate system of axes.

Answers: 1. Move each point on the graph 2 units up and one unit up simultaneously.

2. Move each point on the graph 2 units down and one unit to the left simultaneously.

3. Move each point on the graph 2 units up and 3 units to the right simultaneously.

4. Move each point on the graph 4 units down and 2 units to the left simultaneously.

B 1. Find the new coordinates if each of the following points is translated 3 units to the right:
(a) $(2, 0)$; (b) $(-5, 0)$; (c) $(4, 1)$

2. The point $(-2, 3)$ is translated 1 unit up and 4 units to the left. Find the coordinates of the new point.

3. The point $(4 -1)$ is translated 2 units down and 3 units to the right.
Find the coordinates of the new point.

4. The point $(3, -2)$ has its origin translated to a new origin $(-4, 1)$. Find the new coordinates with reference to the new origin.

5. The origin of the line $y = 3x - 2$ is moved to a new origin $(-3, 2)$. Find an equation of this line with reference to the new origin..

6. The equation of an ellipse is $4x^2 + 9y^2 - 16x + 18y - 11 = 0$.
Assuming that the origin of this equation is moved (translated) to the new point $(--3, 2)$, transform the above equation in terms of the new origin..

7. From the graph of $y = x^2$, sketch the graph of $y + 4 = (x - 1)^2$

Answers: 1. (a) $(5, 0)$; (b) $(-2, 0)$; (c) $(7, 1)$; **2.** $(-1, -1)$; **3.** $(7, -3)$; **4.** $(7, -3)$;

5. $y_n = 3x_n - 13$, where x_n and y_n are the new coordinates; **6.** $4(x_n - 5)^2 + 9(y_n + 3)^2 = 36$

7. See Example 1, p. 226

Lesson 35   232
Vertical Stretching and Shrinking

Given the sketch of $y = x^2$ and knowing how to stretch and shrink the graphs of functions, we can rapidly sketch the graphs of $y = 3x^2$ and $y = \frac{1}{3}x^2$.

Similarly, from $y = e^x$, we can sketch $y = 4e^x$ and $y = \frac{1}{2}e^x$; from $y = \log x$, we can sketch $y = 5\log x$ and $y = \frac{1}{3}\log x$; from $y = \frac{1}{x}$ we can sketch $y = \frac{2}{x}$ and $y = \frac{1}{2x}$.

Vertical Stretching or Stretching in the y-direction
Action: Keeping the x-coordinates the same while increasing the y-coordinates.
Effect : The curve or line is pulled towards the y-axis.

Example 1   From the graph of $y = x^2$, sketch the graph of $y = 2x^2$.
Solution   (Note here that $a = 2$, $k = 0$, $h = 0$, $b = 1$ in $y - k = af[b(x - h)]$
Step 1: Using broken lines, rapidly sketch the graph of $y = x^2$.
Step 2: Construct a table of values for $(x, 2x^2)$. from a table for (x, x^2) as was done on
    page 226. (Example 1). Each point on $y = x^2$, is shifted to a new point $(x, 2x^2)$.
Step 3: Using a smooth solid curve, connect the new points (**Figure 1**) to obtain the characteristic
    parabolic curve. Pictorially, each half of this curve will be between the curve, $y = x^2$ and
    the y-axis. The characteristic. U-shape is narrower than that of $y = x^2$. The curve is
    contracted in the x-direction. The curve is pulled towards the y-axis.

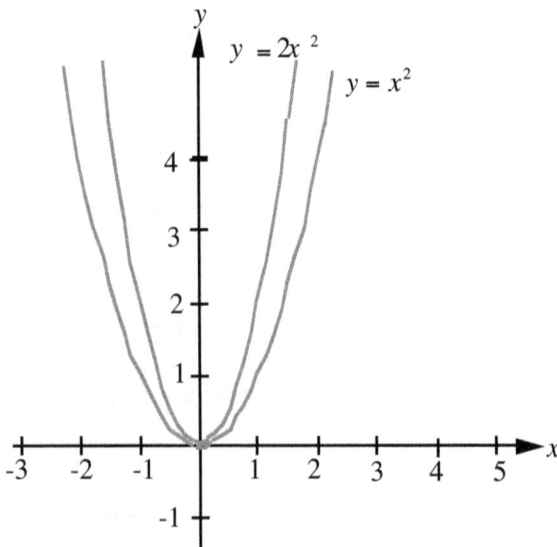

Figure 1: The graphs of $y = x^2$ and $y = 2x^2$.

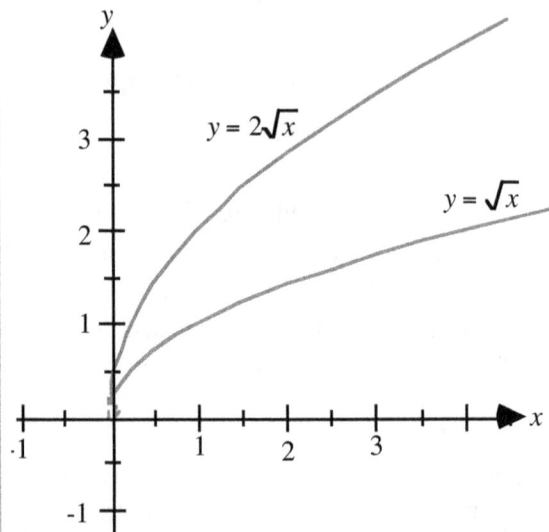

Figure 2: The graphs of $y = \sqrt{x}$, $y = 2\sqrt{x}$.

Example 2 From the graph of $y = \sqrt{x}$ sketch the graph of $y = 2\sqrt{x}$ 233

Solution (Note here that $a = 2$, (stretch factor) $k = 0$, $h = 0$, $b = 1$ in $y - k = af[b(x - h)]$

Step 1: Using broken lines, rapidly sketch the graph of $y = \sqrt{x}$.

Step 2. Construct a table of values for $(x, 2\sqrt{x})$. from a table for $(x, \sqrt{x})$ as was done on page 226. (Example 1).

Step 3: On the same coordinate axes as was used for $y = \sqrt{x}$, plot the new points and connect them (**Figure** 2 above) to obtain the characteristic curve. Note that each point on $y = \sqrt{x}$, is moved to a new point $(x, 2\sqrt{x})$. Pictorially, the curve $y = 2\sqrt{x}$ will be between the curve, $y = \sqrt{x}$ and the y-axis. The curve is pulled towards the y-axis.

Vertical Shrinking or Shrinking in the y-direction

Action: Keeping the x-coordinates the same while decreasing the y-coordinates.

Effect : Curve or line is pushed away from the y-axis and pulled towards the x-axis .

Example 3 From the graph of $y = x^2$, sketch the graph of $y = \frac{1}{3}x^2$,

Solution (Note here that $a = \frac{1}{3}$ (shrink factor); $k = 0$, $h = 0$, $b = 1$ in $y - k = af[b(x - h)]$)

Step 1: Using broken lines, sketch the graph of $y = x^2$.

Step 2: Construct a table of values for $(x, \frac{1}{3}x^2)$. from a table for (x, x^2) as was done on page 226. (Example 1).

Step 3: On the same coordinate axes as was used for $y = x^2$, connect the new points (**Figure 3**) to obtain the characteristic curve. Pictorially, this curve will be between the curve, $y = x^2$ and the x-axis. The characteristic U-shape will be more widely open than that of $y = x^2$. In effect, the curve is pushed away from the y-axis and pulled towards the x-axis. The curve is also expanded in the x-direction. For $x > 0$, $y = \frac{1}{3}x^2$ strictly increases more slowly than $y = x^2$. For $x < 0$, $y = \frac{1}{3}x^2$ strictly decreases more slowly than $y = x^2$.

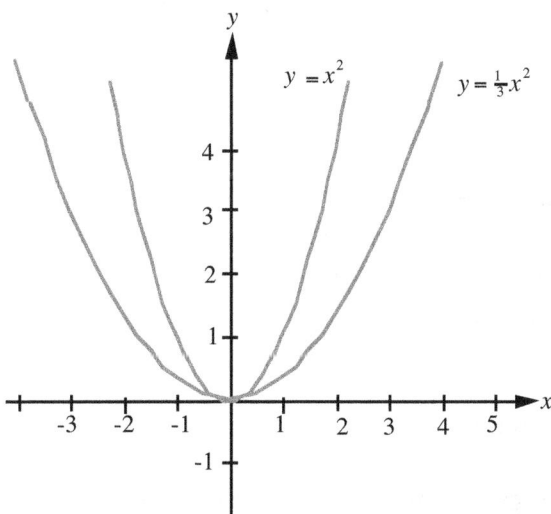

Figure 3 : Graphs of $y = x^2$ and $y = \frac{1}{3}x^2$.

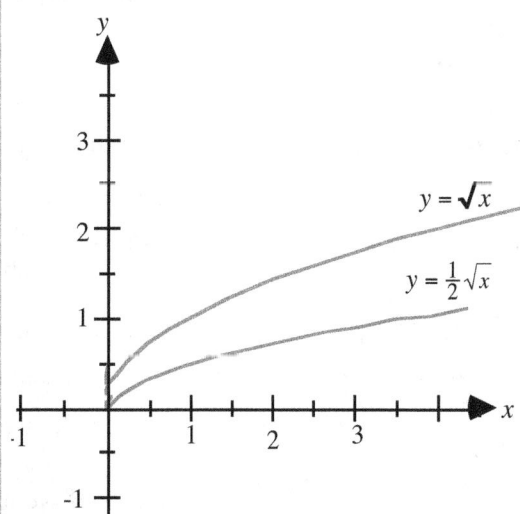

Figure 4: The graphs of $y = \sqrt{x}$, and $y = \frac{1}{2}\sqrt{x}$.

Example 4:   From the graph of $y = \sqrt{x}$ sketch the graph of $y = \frac{1}{2}\sqrt{x}$.

Procedure   (Note here that $a = \frac{1}{2}$ (shrink factor), $k = 0$, $h = 0$, $b = 1$

Step 1: Using broken lines, rapidly sketch the graph of $y = \sqrt{x}$

Step 2: Construct a table of values for $(x, \frac{1}{2}\sqrt{x})$. from a table for $(x, \sqrt{x})$ as was done on
page 226. (Example 1). On the same coordinate axes, plot the new points.

Step 3: Using a solid curve, connect the new points together to obtain the characteristic curve
(**Figure 4** above)

Horizontal Stretching and Shrinking
(Keep the y-coordinates the same and change the x-coordinates)

Example 5 Given $y = x^2$, draw the graph of $y = (-2x)^2$. Note that $y_0 = y_n$

Here, $k = 0$, $h = 0$, $a = 1$, $b = -2$; $y = (-2x)^2 = y - 0 = 1\left(\left[-2(x - 0)\right]^2\right)$

$y = x^2$		$y = (-2x)^2$		
x_0	y_0	$x_n = \frac{x_0}{b}$	y_n	(x_n, y_n)
0	0	$\frac{0}{-2} = 0$	0	$(0, 0)$
1	1	$\frac{1}{-2} = -\frac{1}{2}$	1	$(-\frac{1}{2}, 1)$
2	4	$\frac{2}{-2} = -1$	4	$(-1, 4)$
3	9	$\frac{3}{-2} = -\frac{3}{2}$	9	$(-\frac{3}{2}, 9)$
4	16	$\frac{4}{-2} = -2$	16	$(-2, 16)$
-1	1	$\frac{-1}{-2} = \frac{1}{2}$	1	$(\frac{1}{2}, 1)$
-2	4	$\frac{-2}{-2} = 1$	4	$(1, 4)$
-3	9	$\frac{-3}{-2} = \frac{3}{2}$	9	$(\frac{3}{2}, 9)$
-4	16	$\frac{-4}{-2} = 2$	16	$(2, 16)$

Horizontal Shrinking

Step 1: Quickly prepare a table of values for x and y
using the basic equation $y = x^2$ and convenient
x-values and calculating corresponding y-values

Step 2: Using the table of values from Step 1,
determine the new coordinates x_n and y_n

Note in this example that y_0 from Step 1 is the

same as in Step 2, since $k = 0$, and $a = 1$.

Step 3: Plot the points and connect them by a solid curve. (**Fig. 5** below)

Note above that $y = (-2x)^2 = 4x^2$.

We could therefore graph $y = 4x^2$, with $a = 4$ (stretch factor)
as in Example 1 (p. 232) to obtain the same graph.

Observe the pairs (x_n, y_n): $(0, 0), (-1, 4), (-2, 16), (1, 4), (2, 16)$.

Example 6 Given $y = \sqrt{x}$, draw the graph of $y = \sqrt{2x}$ .Note that $y_0 = y_n$

Here, $k = 0$, $h = 0$, $a = 1$, $b = 2$; ($y = \sqrt{2x} = y - 0 = 1\sqrt{2(x - 0)}$)

$y = \sqrt{x}$		$y = \sqrt{2x}$		
x_0	y_0	$x_n = \frac{x_0}{b}$	y_n	(x_n, y_n)
0	0	$\frac{0}{2} = 0$	0	$(0, 0)$
1	1	$\frac{1}{2} = \frac{1}{2}$	1	$(\frac{1}{2}, 1)$
4	2	$\frac{4}{2} = 2$	2	$(2, 2)$
9	3	$\frac{9}{2} = \frac{9}{2}$	3	$(\frac{9}{2}, 3)$

Step 1: Quickly prepare a table of values for x and y using the basic
parabola equation and convenient x-values.

Step 2: Using the table of values from Step 1, determine the new
coordinates x_n and y_n.

Step 3: Plot the points and connect them by a solid curve. (**Fig. 6** below)

Note: If we are to graph only $y = \sqrt{2x}$ we can choose
convenient x-values such as $0, 2, 8, 32$; calculate y and plot.

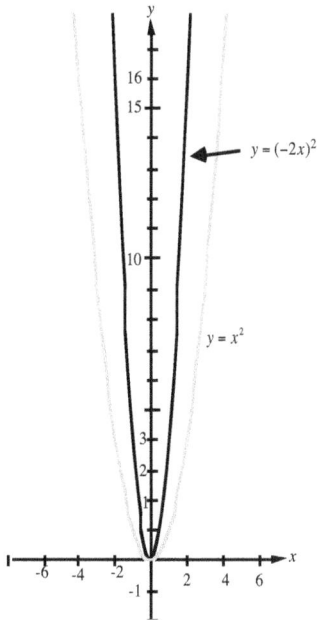

Fig. 5 : Graph of $y = (-2x)^2$

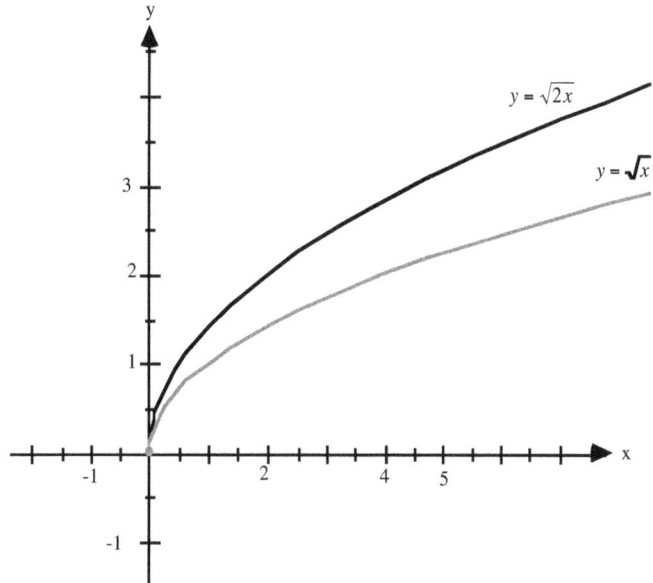

Fig. 6   Graph of $y = \sqrt{2x}$

So far, we have learned how to transform basic graphs to new graphs. In the future we will cover more direct ways of sketching the graphs of some functions such as the quadratic functions (Lesson 38). Remember that it is good practice to construct a table of values for x and y to obtain ordered pairs (points) which are then plotted and connected.

Lesson 35 Exercises

Sketch the graph of $y = x^2$, and on the set of axes, sketch the graphs of the following:

1. $y = \dfrac{1}{2}x^2$;   **2.** $y = \dfrac{x^2}{3}$;   **3.** $y = 3x^2$;   **4.** $y = 12x^2$

From the graph of $y = \pm\sqrt{x}$ sketch the graphs of the following:

5. $x = \frac{1}{2}y^2$;   **6.** $x = \frac{1}{3}y^2$; **7.** $x = 3y^2$;   **8.** $x = 12y^2$.

Answers:

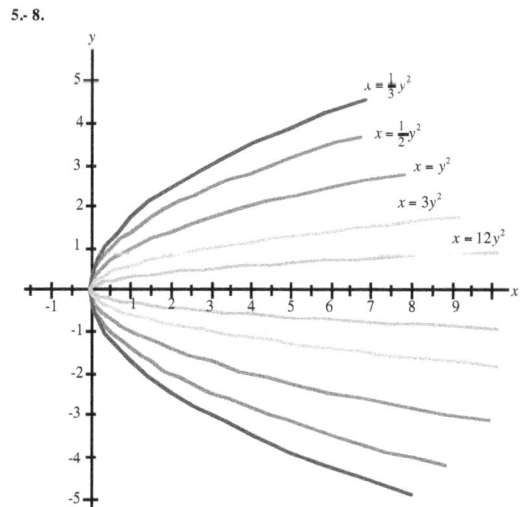

Lesson 36

236

Symmetry; Even and Odd Functions

*The word "symmetric" means having exactly "congruent " parts on either side of a dividing line.

Symmetry about the *y*-axis

A curve is symmetric with respect to the *y*-axis if the equation remains equivalent when the equation is reflected (i.e., replacing *x* by -*x*, keeping *y* unchanged) about the *y*-axis. More formally, a curve is symmetric with respect to the *y*-axis if $f(x) = f(-x)$). If the page were folded along the *y*-axis, the two halves of the curve would coincide. (The equation remaining **equivalent** means that after replacing *x* by -*x* then simplifying the resulting equation, we obtain the original equation}

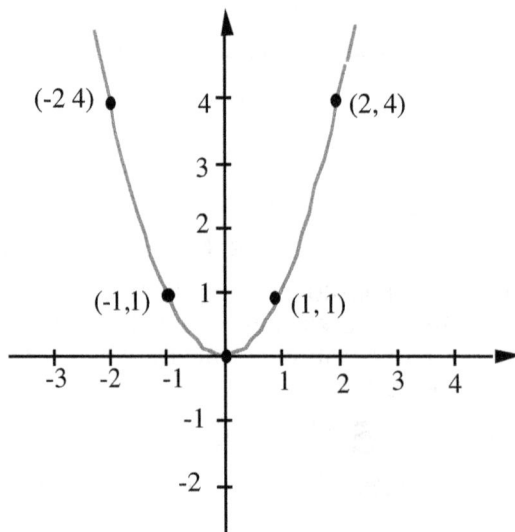

Figure 1 Graph of $y = x^2$ is symmetric about the *y*-axis

Example Is the curve $y = x^2 + 3$ symmetric about the *y*-axis?

Solution: The given curve would be symmetric about the *y*-axis if $f(x) = f(-x)$
Replacing *x* by -*x* in the given equation, we obtain

$$y = (-x)^2 + 3 \qquad \text{(A)}$$

$$y = x^2 + 3 \qquad \text{(B)}$$

On simplifying equation (A), we obtain equation (B) which is the same as the given equation, and therefore, $f(x) = f(-x)$. **Yes**, the given curve is symmetric about the *y*-axis.

Congruent figures are figures which can be made to coincide. Furthermore, the author suggests the following for symmetry about the *y*-axis: " a curve is symmetric about the *y*-axis if the *y*-axis divides the curve into two congruent parts. Perhaps, we can also say that the curve is "congruent" about the *y*-axis: usage of congruency would be consistent with the description that if the page were folded along the *y*-axis the two halves of the curve would coincide.

Outside mathematics, the concept of symmetry is used heavily in structural chemistry, mineralogy and in gemology to identify precious stones such as diamond, ruby and sapphire.

Symmetry about the *x*-axis

A curve is symmetric about the *x*-axis if the equation remains equivalent when the equation is reflected about the *x*-axis . (That is replacing *y* by -*y*, keeping *x* unchanged.). More formally, a curve is symmetric about the *x*-axis if $f(y) = f(-y)$. If the page (Figure 2) were folded along the *x*-axis, the two halves of the curve would coincide.

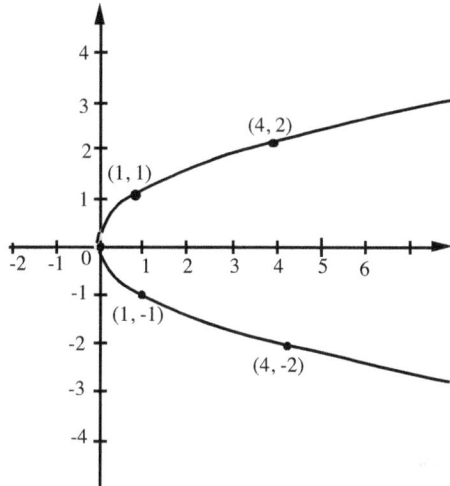

Figure 2

Example 1 Is the curve given by $x = y^2$ symmetric about the *x*-axis?

Solution The given curve would be symmetric about the *x*-axis if $f(y) = f(-y)$
Replacing *y* by -*y* in the given equation, we obtain

$$x = (-y)^2 \qquad \text{(A)}$$
$$x = y^2 \qquad \text{(B)}$$

On simplifying equation (A), we obtain equation (B) which is the same as the given equation. Therefore $f(y) = f(-y)$. **Yes,** the given curve is symmetric about the *x*-axis.

Example 2 Is the curve $y = x^2$ symmetric about the *x*-axis?

Solution The given curve would be symmetric about the *x*-axis if $f(y) = f(-y)$
Replacing *y* by -*y* in the given equation, and solving for *y* we obtain

$$-y = x^2 \qquad \text{(A)}$$
$$y = -x^2 \qquad \text{(B)}$$

On simplifying equation (A), we obtain equation (B) which is **not** equivalent to the original equation. Therefore $f(y) \neq f(-y)$. **No,** the curve is **not** symmetric about the *x*-axis.

Symmetry about the origin 238

 A curve is symmetric about the origin if the equation of the curve remains equivalent when the equation is reflected about the origin . More formally, a curve is symmetric about the origin if on replacing x by $-x$ and y by $-y$ simultaneously, the equation remains equivalent (i.e., on simplifying, the equation remains unchanged)

If the page were folded first along the positive y-axis (1st quadrant on to 2nd quadrant), and then along the x-axis (1st and 2nd quadrants together on to the 3rd quadrant), the two halves of the curve would coincide.
 In Figures 1 and 2 below, each graph is symmetric about the origin.

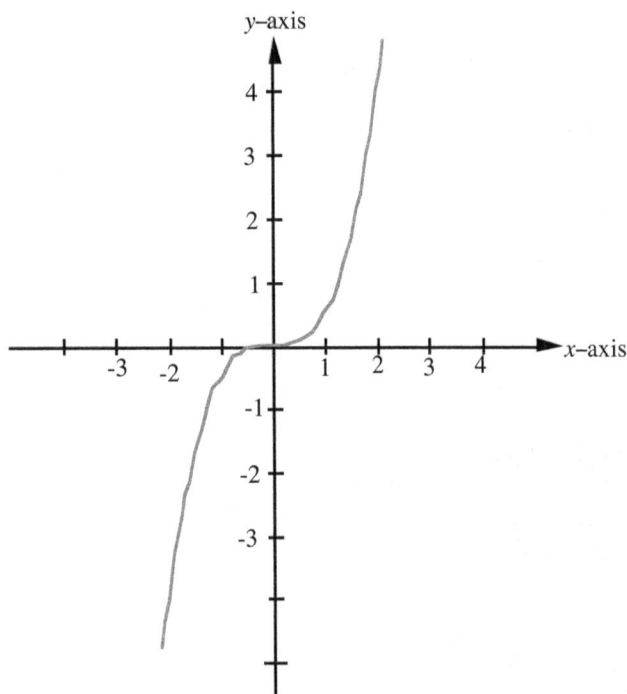

Figure 1 Graph of $y = x^3$ **Figure 2:** Graph of $x^2 + y^2 = 9$

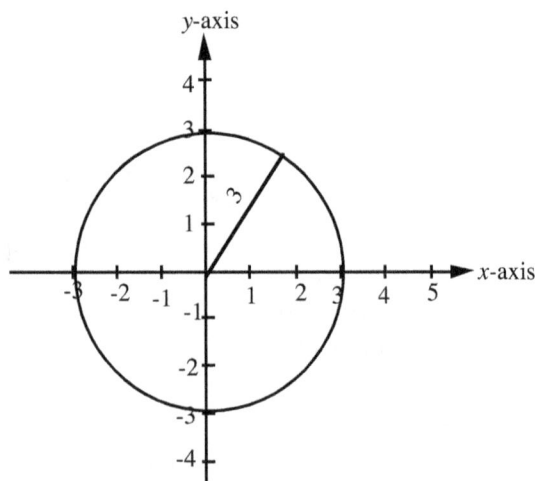

Example Is the curve given by $x^2 + y^2 = 9$ symmetric about the origin.

Solution: Replacing x by $-x$ and y by $-y$ in the given equation,

$$(-x)^2 + (-y)^2 = 9 \qquad \text{(A)}$$
$$x^2 + y^2 = 9 \qquad \text{(B)}$$

 On simplifying equation (A), we obtain equation (B) which is the same as the given equation. .
Yes, the given curve is symmetric about the origin.

From the above discussions on symmetry, we make the following observations:

1. If a curve is symmetric about both the x- and y-axes, then the curve is symmetric with respect to the origin
2. However, if a curve is symmetric about the origin, it does **not** necessarily mean that it is symmetric with respect to both x-and y-axes. Example: In Figure 1 above, even though the curve is symmetric about the origin, it is **not** symmetric about the x-axis or the y-axis
 However Figure **2** is symmetric about both the x- and y-axes.

Even functions

The graph of an even function is symmetric about the y-axis. In sketching the graph of an even function, if we know the part of the graph on one side of the y-axis, then we can easily draw the other part of the curve.

Definition: A function $f(x)$ is **even** in x if $f(x) = f(-x)$ (A)

In words, a function $f(x)$ is even if whenever the negative value of x is substituted in the functional equation, and the resulting equation is simplified, the equation remains unchanged (i.e. the functional value for x is the same as that for -x).

Examples of even functions:

Polynomials: **1.** $f(x) = x^2$;   **2.** $f(x) = 5x^2$;   **3.** $f(x) = 5x^2 + 3$;   **4.** $f(x) = -x^4$

Trigonometric functions: **1.** $f(x) = \cos x$ is an even function of x because $\cos x = \cos(-x)$. So also is $f(x) = \sec x$, the reciprocal of the cosine function.

Exponential functions: $y = 2^{-x^2}$ (an important function in probability and statistics)

Example 1 Is the function $f(x) = x^2 + 5$ an even function?

Solution Substitute the negative of x in the above equation.

Then we obtain $f(-x) = (-x)^2 + 5$

$$= (-x)(-x) + 5$$

$$f(-x) = x^2 + 5 \qquad\qquad \text{(B)}$$

We observe that the right-hand side of $f(x)$ (the given equation) equals the right-hand side of $f(-x)$ (equation (B)), and therefore $f(x) = f(-x)$. **We can also say that each term on the right-hand side did not change sign.** Note that any constant term will not change sign.
Yes, the given function is even.

Testing some numerical values

If $x = 3$ in $f(x) = x^2 + 5$

Then, $f(3) = (3)^2 + 5$; $f(-3) = (-3)^2 + 5$

        $= 9 + 5$ $= 9 + 5$

        $= 14$ $= 14$

We observe above that $f(3) = f(-3)$

Odd functions

The graph of an odd function is symmetric about the origin. In sketching the graph of an odd function, we can use this symmetric property to complete the graph by reflections in the origin.

Definition: A function $f(x)$ is odd in x if

$$f(-x) = -f(x) \qquad\qquad \text{(C)}$$

In words, a function $f(x)$ is odd if whenever the negative value of x is substituted in the functional equation, the new equation obtained is the negative of the original functional equation. (i.e. the functional value for x is the negative of the functional value for -x..). **We can also say that each term on the right-hand side changed sign.** Note that any constant term will not change sign.

Examples of odd functions:

Polynomial functions : **1.** $f(x) = x^3$;   **2.** $f(x) = x^5$;   **3.** $f(x) = x$; Rational functions: $f(x) = \frac{1}{x}$
Trigonometric functions: **1.** $f(x) = \sin x$;   **2.** $f(x) = \tan x$; **3.** $f(x) = \cot x$ and **4.** $f(x) = \csc x$.
We may note that $\cot x$ and $\csc x$ are derived from the sine function which is odd.

Example 1 Is the function $f(x) = 2x$ odd? 240

Solution Substitute the negative of x in the given equation.

Then $f(-x) = 2(-x)$

$f(-x) = -2x$ (D)   (The term on the right-hand side changed sign)

We observe that the right-hand side of $f(-x)$ (equation D) is the negative of the right-hand side of $f(x)$ (the given equation), and therefore $f(-x) = -f(x)$

Yes, the given function is odd.

Testing some numerical values

If $x = 4$ in $f(x) = 2x$

Then $f(4) = 2(4)$ $f(-4) = 2(-4)$

$= 8$ $= -8$

Thus $f(4) = 8$ but $f(-4) = -8$, and therefore $f(-4) = -f(4)$.

Functions which are neither even nor odd

Some functions are neither even nor odd.. **Here, there are sign changes in some of the terms.**

Example The function given by $f(x) = 2x^5 - x^2 + x$ is neither even nor odd. Why?

Lesson 36 Exercises

A 1. Explain what is meant by saying that a curve is symmetric about
(a) the x-axis, (b) the y-axis.

Describe the symmetry of each of the following with respect to the x-axis, the y-axis and the origin.

2. $f(x) = x$;   3. $f(x) = x^2 + 3$;   4. $f(x) = 5x^2 + 1$;   5. $f(x) = 2x^3 + 4$

6. $f(x) = x^3 - 5x - 2$;   7. $f(x) = 4x^6 - x^2 - 5$

Answers: See text..

B Determine which of the following are even functions, odd functions or neither:

1. $f(x) = x$;   2. $f(x) = x^2 + 3$;   3. $f(x) = 5x^2 + 1$;   4. (a) $f(x) = 2x^3 + 4$;

(b) $f(x) = 2x^3 + 4x$

5. (a) $f(x) = x^3 - 5x - 2$; (b) $f(x) = x^3 - 5x$;   6. $f(x) = 4x^6 - x^2 - 5$

7. Name one property of the graph of an even function.

8. Name one property of the graph of an odd function.

9. Can a function specified by a single equation be odd and even simultaneously? If yes, graph an example.

10. Can a function specified by piecewise rule be odd and even simultaneously? If yes, graph an example.

11. Are there any relationships between symmetry about the y-axis and even functions? Explain.

12. Are there any relationships between symmetry about the origin and even functions? Explain.

13. Repeat Exercise 11 for odd functions (instead of even functions).

Answers: 1. Odd;   **2.** Even; **3.** Even; **4.** (a) Neither; (b) Odd; **5.**(a) Neither; (b) Odd; **6.** Even.

CHAPTER 15

POLYNOMIAL FUNCTIONS

Lesson 37: **The x- and y-intercepts of a line or a curve**
   Graphs of Polynomial Functions

Lesson 38A: **Parabola: The Graph of $y = f(x) = ax^2 + bx + c$);**
   Graph of the Inverse Relation of $y = ax^2 + bx + c$

Lesson 38B Applications of Quadratic Functions; Maxima and Minima Problems

Lesson 39: **Graphs of other Polynomial Functions**

Lesson 40: **Graphs of Factorable Polynomial Functions**

Introduction

A number of functions can be approximated by polynomial functions. We use polynomial functions to approximate smooth curves. Polynomial functions can easily be evaluated. The roots of polynomial functions are relatively easy to determine. In calculus, we shall learn that the derivatives and integrals of polynomials are also polynomials.

A polynomial function of degree n is a function which is specified by $f(x) = a_0 x^n + a_1 x^{n-1} + \ldots + a_{n-1} x + a_n$, where n is a whole number, $a_0, a_1, \ldots, a_n$ are real numbers with $a_0 \neq 0$.

Examples of polynomial functions are:

1. $f(x) = 2x + 3$;   **2.** $f(x) = 3x^2 + 8x - 1$;   **3.** $f(x) = x^4 + x^2 + 5$;   **4.** $f(x) = -x^5 + 2$

Lesson 37
The x- and y-intercepts of a Line or a Curve;
Graphs of Polynomial Functions

x-intercepts or zeros

The **x-intercept** of a line (or a curve) is the x-coordinate of the point where the line or curve crosses or meets the x-axis. At this point, $y = 0$. (Technically, we should also say that the x-intercept is the point where the line (or curve) intersects the x-axis, and in which case, we must specify two coordinates, with the y-coordinate always being zero.)

y-intercept

The **y-intercept** of a line (or a curve) is the y-coordinate of the point where the line crosses or meets the y-axis. At this point, $x = 0$. (Geometrically, we should also say that the y-intercept is the point where the line (or curve) intersects the y-axis, and in which case, we must specify two coordinates, with the x-coordinate always being zero.) **Note** that a function cannot have more than one y-intercept, since otherwise, two distinct ordered pairs would have the same first component, namely 0.

Finding the x-intercepts

Step 1: Set y equal to zero in the given equation. .

Step 2: Solve the equation from Step 1 for x. The value or values obtained are the x-intercepts.

Example 1 Find the x-intercepts of the curve given by $y = x^2 + x - 6$

Step 1: Setting $y = 0$, we obtain $0 = x^2 + x - 6$

Step 2: Solving the resulting quadratic equation, $x = -3, x = 2$.
   The x-intercepts are -3 and 2.

Finding the *y*-intercept

Step 1: Set $x = 0$ in the given equation.

Step 2: Solve the equation from Step 1 for *y*. The value of *y* obtained is the *y*-intercept.

Example 2 Find the *y*-intercept of $y = x^2 + x - 6$
Solution

Setting $x = 0$, we obtain
$$y = (0)^2 + (0) - 6$$
$$y = -6$$

The *y*-intercept is -6.

Example 3 In Figure 1 below, find the *x*- and *y*-intercepts.

Solution

The *x*-intercept is -3; and the *y*-intercept is 2.
The *x*- and *y*--intercepts are at (-3, 0) and (0, 2) respectively.

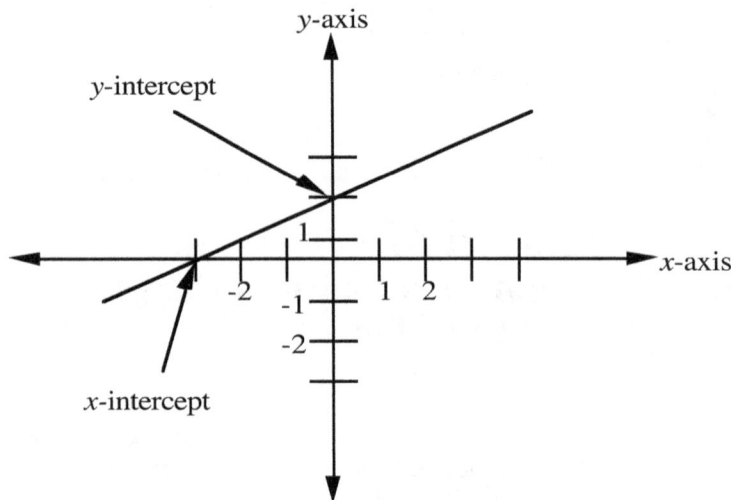

Figure 1

Example 4 Find the *x*- and *y*-intercepts for the graph of $3x + 2y = 12$.

For the *x*-intercept: Let $y = 0$ in $3x + 2y = 12$.
Then $3x + 2(0) = 12$; $3x = 12$; and from which $x = 4$.
The *x*-intercept is 4.

For the *y*-intercept: Let $x = 0$ in $3x + 2y = 12$.
Then $3(0) + 2y = 12$; $2y = 12$; and from which $y = 6$.
The *y*-intercept is 6.

Lesson 37: The x- and y-intercepts of a line or a curve; Graphs of Polynomial Functions

Sketching the Graphs of Polynomial functions

Step 1: Find the x-intercepts by setting $y = 0$ in the given equation and solving for x.

Step 2: Find the y-intercept by setting $x = 0$ in the given equation and solving for y.

Step 3: Plot the intercepts in a rectangular coordinate system.

Step 4:   Additional points may be found by choosing convenient x-values and calculating the corresponding y-values and vice-versa.

Step 5: Connect the points by a straight line if it is a straight line or otherwise, connect the points by a smooth curve. (Note that, for a straight line, two or three points would be sufficient)

Step 6: When uncertain of the direction or location of the curve on an interval, calculate additional points (by choosing x and calculating y and vice-versa) and plot.

Step 7:: Apply the **end behavior** of the graphs of polynomial functions on page 264-265 to help determine the correctness of the end behavior of the graph.

Step 8:: Apply the rule that the number of turning points of a polynomial function of degree n is at most n-1 (page 216) to help determine the correctness of the number of turnings of the graph.

Lesson 37 Exercises

1. Find the x-intercepts of the curve given by $y = x^2 + x - 6$
2. Find the y-intercept of $y = x^2 + x - 6$
4. Find the x- and y-intercepts for the graph of $3x + 2y = 12$.
5. Sketch the graph of the curve given by $y = x^2 + x - 6$

Answers: 1. -3 and 2; **2.** -6; **3.** x-intercept $= 4.$, y-intercept $= 6.$
   5

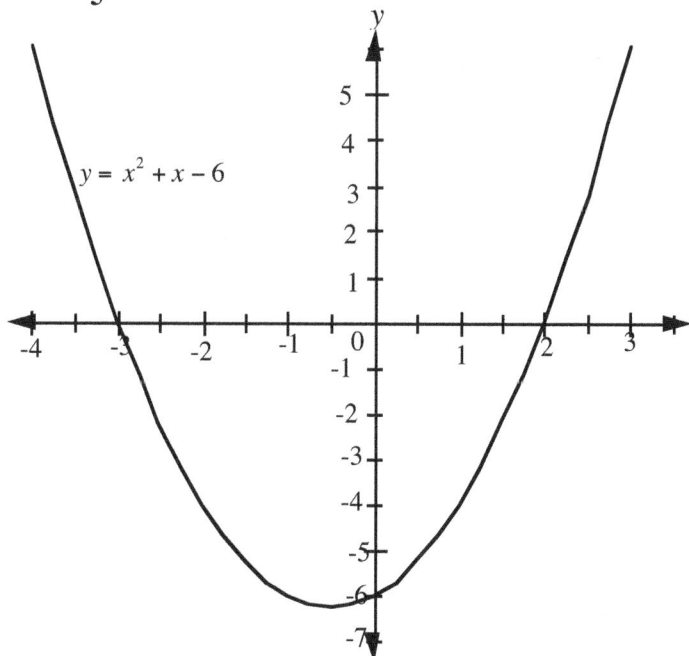

$y = x^2 + x - 6$

Lesson 38A

Parabola: The Graph of $y = f(x) = ax^2 + bx + c$);
Graph of the lnverse Relation of $y = ax^2 + bx + c$

The term parabola is usually reserved for the graphs of equations whose **simple standard** forms are $y = ax^2$. However, curves of forms $y = ax^n$ where n is a positive number, could also be called parabolic curves. For example, $y = ax^3$ could be called a **cubic** parabola; and $y = ax^4$ could be called a **quartic** parabola.

 We must note that the general form of $y = ax^2$ is $y = ax^2 + bx + c$, where a, b and c are constants, and $a \neq 0$. We will learn in the future that knowing the simple form, we can obtain the general form by translation (shifting).

General Properties of the parabola

(1) The parabola is an **U-shaped** curve.

(2) The curve has an axis of symmetry. The axis of symmetry divides the curve into two equal halves.

(3) The axis of symmetry may be the y-axis (Figure). the x-axis (Figure), a vertical line, or an oblique line.
      (Any of the above mentioned axes may be obtained from any of the other axes by the proper translation or rotation of the x- and y-axes)

(4) The axis of symmetry intersects the curve at a point, V, called the vertex of the parabola.

(5) The U-shape of the parabola can "open" in all possible directions. However, we will be interested in the directions either along or parallel to the x- and y-axes. In Figure 1, the parabola opens upwards, and the axis of symmetry is the y-axis. Also, when the parabola opens upwards, we say that it is concave up. When it opens downwards, we say that it is concave down . (Figure).

   The other directions are shown in Figures. We can observe in (Figure) that as x increases indefinitely, y increases indefinitely.

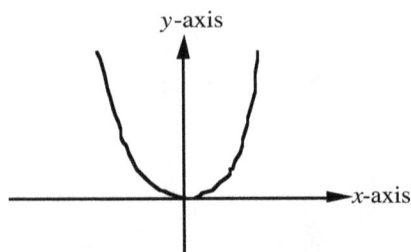

Figure 1: The axis of symmetry
          is the positive y-axis

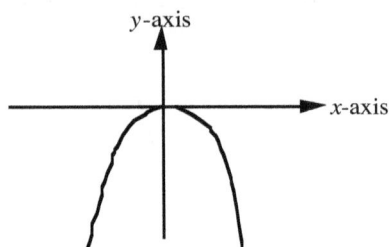

Figure 2: The axis of symmetry
          is the negative y-axis

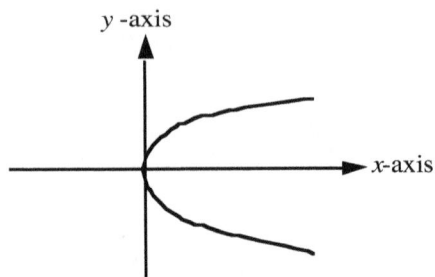

Figure 3: The axis of symmetry
          is the positive x-axis

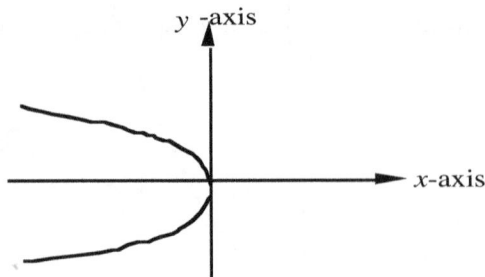

Figure 4: The axis of symmetry
          is the negative x-axis

Note that **Figures 3** and **4** are **not** graphs of functions but relations

Sketching the Parabolas

There are a number of approaches we can use to sketch the parabolas. Each approach depends upon the intended use of the sketch. For instance, if we are interested in the design of searchlight reflectors, reflecting telescopes or radar antennas, we would be interested in properties such as the vertex, axis of symmetry, the focus, focal chord and the directrix. (See chapter 23A).

However, if we are only interested in finding how the shapes of the various parabolas are related to each other, and to other so called parabolic curves, then we will not be interested in finding the location of the focus, the directrix and the focal chord. Nevertheless, if we know the locations of the focus, the focal chord and the directrix, we will of course use these properties to rapidly sketch the parabola. Also, if we are interested in treating the parabola as a function, we will not be interested in the latter three properties.

Parabola as a Function

Under the heading "the parabola as a function", we will be interested mainly in the locations of the vertex, the axis of symmetry, the x-intercepts and the y-intercept if there are any of these intercepts. We may also be interested in some other points on the curve as the occasion demands.

By the so called "vertical line" rule (see p. 50), we can visually determine that the parabolas in Figures 3 and 4 are **not** functions but only relations while Figures 1 and 2 are graphs of functions. Thus in treating the parabola as a function, we will be more interested in Figures 1 and 2 and in which the y-axis or another vertical line is the axis of the parabola. We are not saying that the other parabolas (Figs. 3 and 4) are not important. In fact, we will learn in future that the parabolas in Figure 3 and 4 are the inverse relations of the parabolas in Figures 1 and 2 respectively.

There are two main approaches that we can use to sketch the graphs of the parabolas which are functions. One approach is to start from the simple standard form $y = ax^2$ and then by the necessary translations (shifts) build up to the most general form $y = ax^2 + bx + c$. In using this approach, we may sometimes have to complete the square, plot enough points to obtain the shape.

The other approach (page 251-255) locates the intercepts, the coordinates of the vertex, the axis of symmetry, and one or two calculated values to sketch the curve.

In the equation $y = ax^2$, if $a = 1$, the equation of the parabola becomes $y = x^2$. The axis of symmetry of this parabola is the positive y-axis. The **vertex** is at the origin $(0,0)$, (Figure 5)

We will now proceed to sketch the graphs of the parabolic function $y = ax^2 + bx + c$ by first studying the sketch of $y = x^2$. A sketch of the graph of $y = x^2$ is shown in Figure 5.

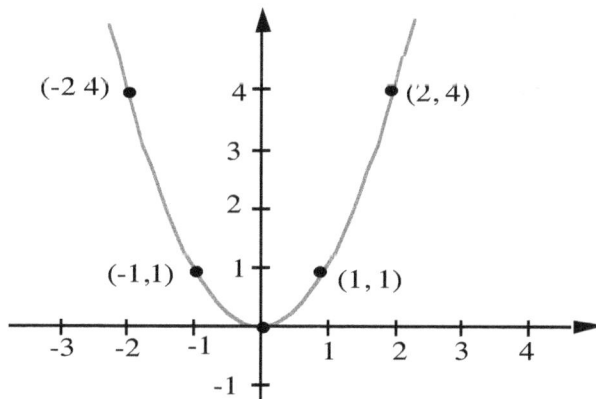

Figure 5   The graph of $y = x^2$

Lesson 38 Graphs of $y = ax^2 + bx + c$); and the Inverse Relation of $y = ax^2 + bx + c$

Now, knowing the properties of the sketch of $y = x^2$ let us sketch the graphs of some more complicated parabolas.

Note below

The transformation method covered on page 225-228 can be applied to graph the following examples (Examples 1-5), even though no tables have been constructed. The student is encouraged to study and imitate the examples on p.224-226 and apply the approach learned to the following examples.

Example 1 Sketch the graph of $y = x^2 + 2$.
In standard form, the equation becomes

$$y - 2 = x^2. \qquad (\text{Recall} \quad y - k = f(x - h))$$

Thus, each point on the curve $y = x^2$ is moved (shifted) up 2 units, keeping the x-values the same. In practice, however, instead of shifting each point, we will move only important points on the curve. Important points on the curve are points such as the turning points (page 216), the x- and y-intercepts.

Step 1: Using dotted lines, rapidly sketch $y = x^2$.

Step 2: Shift important points on the curve 2 units up, keeping the x-values the same.

Step 3: Connect the new points by a solid curve to obtain the characteristic parabolic curve (Figure)

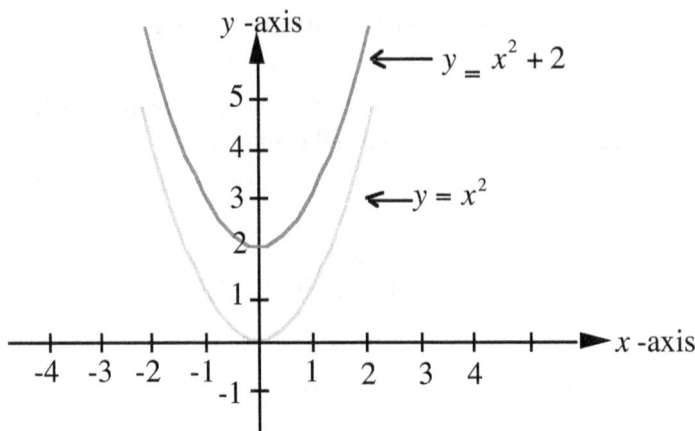

Figure 6: The graph of $y = x^2 + 2$

Example 2   From the graph of $y = x^2$ sketch the graph of $y = -x^2$

Solution

The graph of $y = -x^2$ is the reflection of the graph of $y = x^2$ about the x--axis.

Step 1: Using dotted or broken lines, sketch $y = x^2$

Step 2: Reflect (important) points on $y = x^2$ about the x-axis. Three convenient points on either side
        of the symmetric axis would be sufficient but, by all means, include the vertex and the
        intercepts (if any).
          (Replace each y-coordinate by $-y$, keeping the x-coordinate unchanged and plot.

Step 3: Connect these new points by a smooth solid (darker) curve to obtain the characteristic
        parabola (Figure 7).. The curve is concave down.

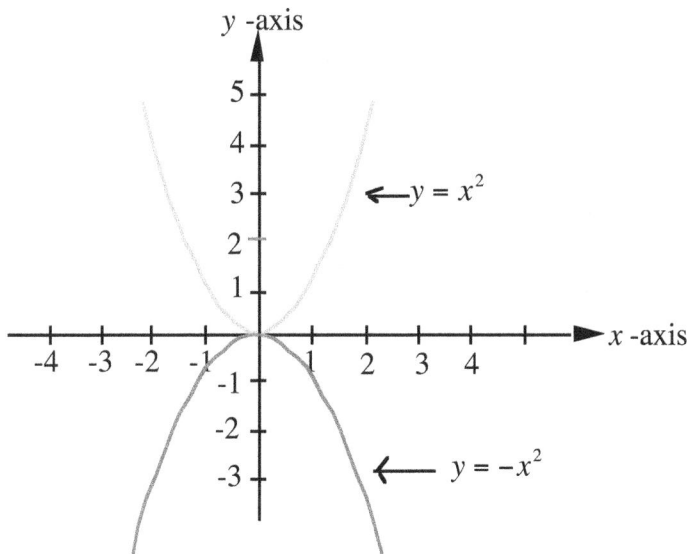

Figure 7: The graph of $y = -x^2$

Lesson 38 Graphs of $y = ax^2 + bx + c$); and the Inverse Relation of $y = ax^2 + bx + c$

Example 3 Sketch the graph of $y = (x + 3)^2$ from the graph of $y = x^2$.

Solution If for a moment, we ignore the 3 in the parentheses, we will observe that we have $y = x^2$

Thus the graph of $y = (x + 3)^2$ is the graph of $y = x^2$ with each point on $y = x^2$ shifted 3 units to the left.

Step 1: Using broken lines quickly sketch the graph of $y = x^2$ (Figure 8).

Step 2: Keeping the y-values the same, move each point 3 units to the left.. Three convenient points on either side of the symmetric axis would be sufficient, but by all means include the vertex and the intercepts (if any).

Step 3: Connect the new set of points from Step 2 by a smooth solid curve to obtain the characteristic parabolic shape.

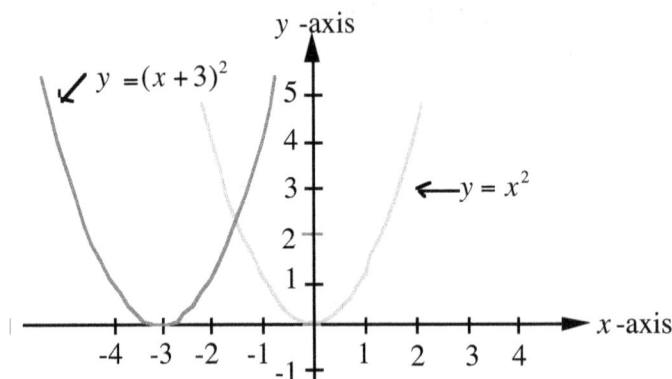

Figure 8: The graph of $y = (x + 3)^2$ or $x^2 + 6x + 9$

Example 4 Sketch the graph of $y - 2 = (x + 3)^2$.

This is the graph of $y = x^2$ with each point shifted 2 units up and 3 units to the left.

Step 1: Using broken lines, quickly sketch $y = x^2$.

Step 2: Move points 2 units up and 3 units to the left (Figure 9). Three convenient points on either side of the symmetric axis would be sufficient, but by all means include the vertex and the intercepts (if any).

Step 3: Connect the new points in Step 2 by a smooth solid curve to obtain the parabola.

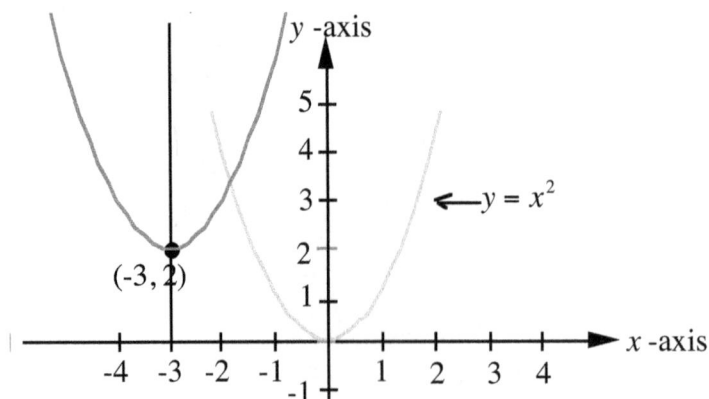

Figure 9: The graph of $y - 2 = (x + 3)^2$

Example 5 Sketch the graph of $y = x^2 + 12x + 36$

Step 1: Complete the square on the right hand-side of the equation. On completing the square, we
obtain $y = (x + 6)^2$.

Step 2: Sketch the graph of $y = x^2$ quickly using dotted lines.

Step 3: Keeping the y-values unchanged, move each point (in practice, we move a few points which
include critical points) 6 units to the left (Figure 10)

Step 4: Connect the points from Step 3 to obtain the curve.

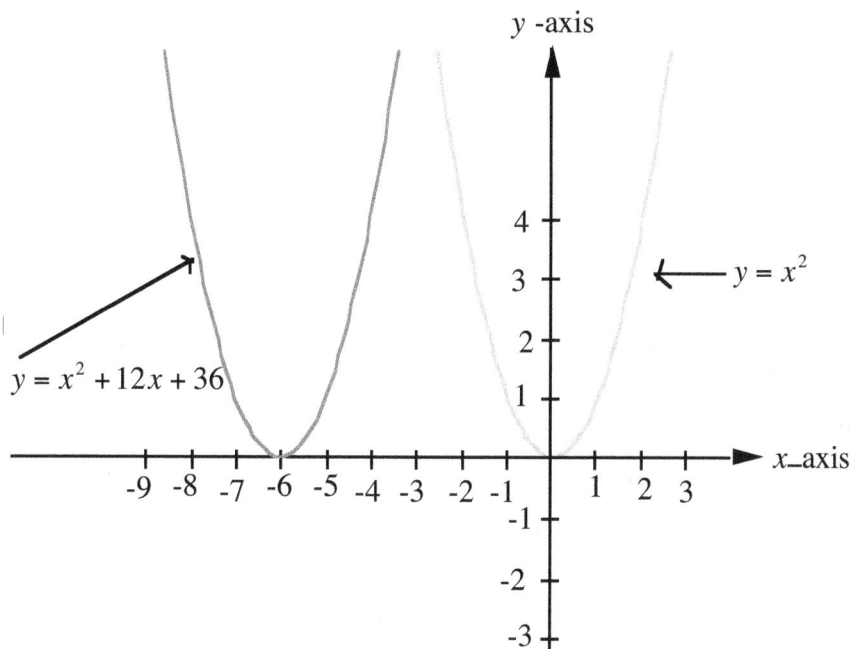

Figure 10: The graph of $y = x^2 + 12x + 36$ or $y = (x + 6)^2$.

Maximum and Minimum Points of the Parabola, $y = f(x) = ax^2 + bx + c$

The graph of the equation $y = f(x) = ax^2 + bx + c$ has either a minimum point or a maximum point. The
minimum point of this curve is an **absolute minimum** and the maximum point is an absolute maximum.
The minimum or maximum point is called the vertex of the parabola.

In particular, if $a > 0$ (i.e., a is positive), then $y = ax^2 + bx + c$ has a minimum point, and the parabola
opens upwards ; but if $a < 0$ (i.e., a is negative), then it has a maximum point.

The coordinates of the **vertex** (minimum or maximum point) are:

x-coordinate $= -\dfrac{b}{2a}$; y-coordinate $= f\left(-\dfrac{b}{2a}\right)$ or $-\dfrac{b^2}{4a} + c$ or $\dfrac{4ac - b^2}{4a}$

The equation of the **axis of symmetry** of the above parabola is $x = -\dfrac{b}{2a}$.

There is another method for sketching the graphs of the quadratic function. The method involves locating the vertex and
symmetric axis either by formula or by completing the square. The method is discussed in the next section.

Lesson 38 Graphs of $y = ax^2 + bx + c$); and the lnverse Relation of $y = ax^2 + bx + c$

Domain and Range of $f(x) = y = ax^2 + bx + c$

250

The above function is a polynomial and as such it is defined for all real values of x.
The domain therefore consists of all real values of x.

The range of $y = ax^2 + bx + c$

When **a is positive**, the parabola opens upwards and the vertex is the minimum point (lowest point) of this curve.

The range is given by the inequality $y \geq \dfrac{4ac - b^2}{4a}$ (Note the sense of the " $\geq$ ")

For example, given that $y = f(x) = x^2 - 6x + 8$, to find the range,   Substitute $a = 1, b = -6, c = 8$, in

$y \geq \dfrac{4ac - b^2}{4a}$ to obtain

$$y \geq \frac{4(1)(8) - (-6)^2}{4(1)}$$

$$y \geq \frac{32 - 36}{4}$$

$$y \geq \frac{-4}{4}$$

$$y \geq -1$$

The range consists of all real numbers greater than or equal to -1. (All real numbers from -1 and up)
When **a is negative**, the parabola opens downwards and the vertex is the maximum point (highest point) of this curve.

The range is given by the inequality $y \leq \dfrac{4ac - b^2}{4a}$ (Note the sense of the " $\leq$ ")

For example, given that $y = f(x) = -x^2 + 6x - 8$, to find the range,

Substitute $a = -1. b = +6, c = -8$, to obtain $y \leq \dfrac{4(-1)(-8) - (+6)^2}{4(-1)}$

$y \leq 1$ (Skipping some steps)

The range consists of all real numbers less than or equal to 1. (All real numbers from 1 and down)

EXTRA: Below, we show how we obtained $-\dfrac{b}{2a}$ and $\dfrac{4ac - b^2}{4a}$, which are the x- and y-coordinates of the vertex

Method 1: By completing the square	**Method 2: For those exposed to Calculus in the past**
$y = ax^2 + bx + c$	If $y = ax^2 + bx + c$; $\dfrac{dy}{dx} = 2ax + b$;
$= a(x^2 + \frac{b}{a}x) + c$	
$= a(x^2 + \frac{b}{a}x + [\frac{b}{2a}]^2 - [\frac{b}{2a}]^2) + c$	$2ax + b = 0$ (at a minimum point, the slope, $\dfrac{dy}{dx} = 0$)
$= a(x^2 + \frac{b}{a}x + [\frac{b}{2a}]^2) - a(\frac{b}{2a})^2 + c$	
$= a(x^2 + \frac{b}{a}x + [\frac{b}{2a}]^2) - a\frac{b^2}{4a^2} + c$	Solving for x, $x = -\dfrac{b}{2a}$;
$y = a(x^2 + \frac{b}{a}x + [\frac{b}{2a}]^2) + c - \frac{b^2}{4a}$	
$y = a(x + \frac{b}{2a})^2 + c - \frac{b^2}{4a}$	Substituting $x = -\dfrac{b}{2a}$ for x in $y = ax^2 + bx + c$, we obtain
$y - (c - \frac{b^2}{4a}) = a(x + \frac{b}{2a})^2$	
(of form $y - k = a(x - h)^2$)	$y = a\left(-\dfrac{b}{2a}\right)^2 + b\left(-\dfrac{b}{2a}\right) + c = \dfrac{4ac - b^2}{4a}$
Setting $x + \frac{b}{2a} = 0$ and solving for x.	
$x = -\frac{b}{2a}$; also, $y = c - \dfrac{b^2}{4a} = \dfrac{4ac - b^2}{4a}$	

Lesson 38 Graphs of $y = ax^2 + bx + c$); and the Inverse Relation of $y = ax^2 + bx + c$

How to Sketch the Graph of $y = f(x) = ax^2 + bx + c$ by the location of the vertex, axis of symmetry and the intercepts

Unlike the method covered on pages 245-249, this method is a direct method

Procedure

Step 1: Find the sign of the coefficient of x^2. If the coefficient is positive, then the curve is concave up (opens up), but if the coefficient is negative the curve is concave down (opens down or is an inverted U).

Step 2: Find the y-intercept by setting $x = 0$ and solving for y. (The y-intercept is at $(0, c)$)

Step 3: Find the x-intercepts by setting $y = 0$ in the given equation and then solving for x. If there are real x-values, the parabola intercepts (crosses or touches) the x-axis. If the x-values are not real, then the parabola does not touch or cross the x-axis. If there are no real x-intercepts, we will choose convenient x-values, say two x-values on both sides of the symmetric axis and calculate the corresponding y-values.

Step 4: Find the coordinates of the vertex:

$$x\text{-coordinate} = -\frac{b}{2a}$$

$$y\text{-coordinate} = f\left(-\frac{b}{2a}\right) \text{ or } -\frac{b^2}{4a} + c$$

(Thus, to calculate the y-coordinate, you can substitute the x-coordinate in the equation and evaluate y; or

use $-\dfrac{b^2}{4a} + c$)

Step 5: Plot the points obtained in Steps 2, 3, and 4.

Step 6: Connect the points by a smooth solid U-shaped curve, noting the direction (from Step 1) in which the parabola opens .

Example 1 Sketch the graph of $f(x) = x^2 - 6x + 8$.

Step 1: Since the coefficient of the x^2-term is positive, the curve opens upwards. The shape is " $\cup$ "

Step 2: Find the x-intercepts.

To find the x-intercept, let $y = 0$ in $y = x^2 - 6x + 8$.

Then, $0 = x^2 - 6x + 8$ or

$x^2 - 6x + 8 = 0$ <-------This is a quadratic equation. Solve by any method

$(x - 2)(x - 4) = 0$ <------We solve by factoring since the factors are easily recognizable.

Solving, $x = 2$, or $x = 4$. The x-intercepts are 2 and 4.

The x-intercepts are at the points $(2, 0)$, and $(4, 0)$.

Step 3: Find the y-intercept.

To find the y-intercept, let $x = 0$ in $y = x^2 - 6x + 8$.

$$\text{Then, } y = (0)^2 - 6(0) + 8$$
$$y = 0 - 0 + 8$$
$$y = 8$$

The y-intercept $= 8$.

The y-intercept is at the point $(0, 8)$. (Note also that the y-intercept is at $(0, c)$)

(In Step 3, we could determine the y-intercept by inspection, if the equation has been solved for y. Here with $c = 8$, we could specify that the y-intercept is 8))

Step 4. Find the axis of symmetry. and the coordinates of the vertex.

$a = 1$ (the coefficient of the x^2-term), $b = -6$ (the coefficient of the x-term).

Axis of symmetry is the line $x = -\dfrac{b}{2a}$ <------formula.

$$= -\dfrac{-6}{2(1)} = 3$$

The axis of symmetry is the line $x = 3$. The x-coordinate of the vertex $= 3$,

The y-coordinate of the vertex is $y = f\left(-\dfrac{b}{2a}\right) = f(3) = 3^2 - 6(3) + 8$ (3 is the x-coordinate of the vertex)

$$= -1.$$

The **vertex** is at $(3, -1)$. This point is also the turning point. (In this problem, it is a minimum point, since a is positive. This minimum point is also the lowest point on this curve.

Step 5: Locate the x- and y-intercepts from (Steps **2** & **3**) above, by plotting the points $(2, 0)$, $(4, 0)$ and $(0, 8)$. (On graph paper and using a rectangular coordinate system of axes)

Step 6: Using a broken line draw, the line $x = 3$, the axis of symmetry (from (**4**) above)

Step 7: Locate the vertex by plotting the point $(3, -1)$. (from Step **4**, above).

Step 8: Locate the point B using symmetry as follows: From the point A (the y-intercept), count horizontally say h units ($h = 3$ in this problem) to the axis of symmetry (the line $x = 3$). Next, continue and count horizontally h units (h = 3) from the axis of symmetry and stop at the point B (this is another point on the curve). You may also locate the point B by the intersection of the lines $y = 8$ and $x = 6$.   (2h = 6)

Step 9: Connect the above points by a smooth U-shaped curve (**Figure 11**), noting that the parabola opens upwards (is concave up), since a is positive ($a = 1$), and also noting that the graph is symmetric about the axis of symmetry (the line $x = 3$, in this example).

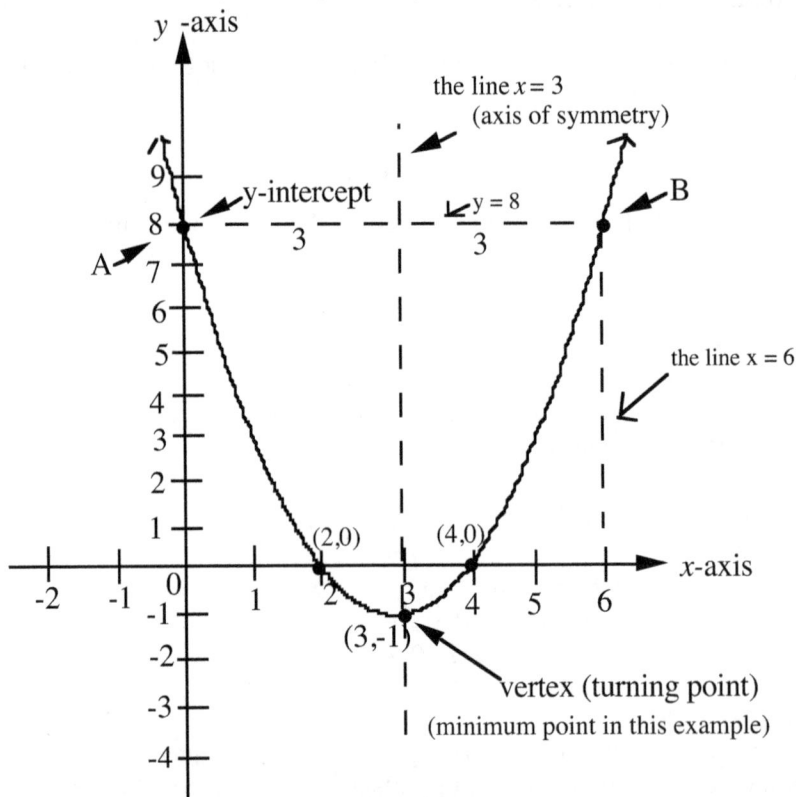

Figure 11: Graph of $y = x^2 - 6x + 8$

Note above that in sketching the graph, you may choose convenient x-values on either side of the axis of symmetry, calculate corresponding y-values to obtain additional points, especially, when there are no (real) x-intercepts.(This may be necessary when in solving the quadratic equation, we obtain solutions involving imaginary numbers.)

Again: When in doubt as to the location of the curve, pick x-values, calculate y, plot the points and connect them.

Example 2 Sketch the graph of $f(x) = -x^2 + 6x - 8$

Step 1: The coefficient of the x^2-term is negative. Therefore, the curve opens downwards.

Step 2. Using the procedure in Example 1, the x-intercepts are 2 and 4.
  The x-intercepts are at the points $(2, 0)$, and $(4, 0)$
  The y-intercept = -8.
  The y-intercept is at $(0, -8)$

Step 3. $a = -1$ (the coefficient of the x^2-term), $b = +6$ (the coefficient of the x-term)
  Axis of symmetry is the line $x = -\dfrac{b}{2a} = -\dfrac{+6}{2(-1)} = 3$.
  The axis of symmetry is the line $x = 3$. The x-coordinate of the vertex = 3,
  and the y-coordinate of the vertex = $y = f(-\dfrac{b}{2a}) = f(3) = -3^2 + 6(3) - 8 = 1$.

Step 4. The vertex is at the point $(3, 1)$.

Step 5 To sketch the graph, plot the points from **Steps 2, 3 & 4,** above; similarly locate the point B (as was done in Example 1) and connect the points by a smooth **inverted** U-shaped curve, noting that the parabola opens downwards (Figure 12), since a is negative ($a = -1$). Also note the symmetry as in Example 1.

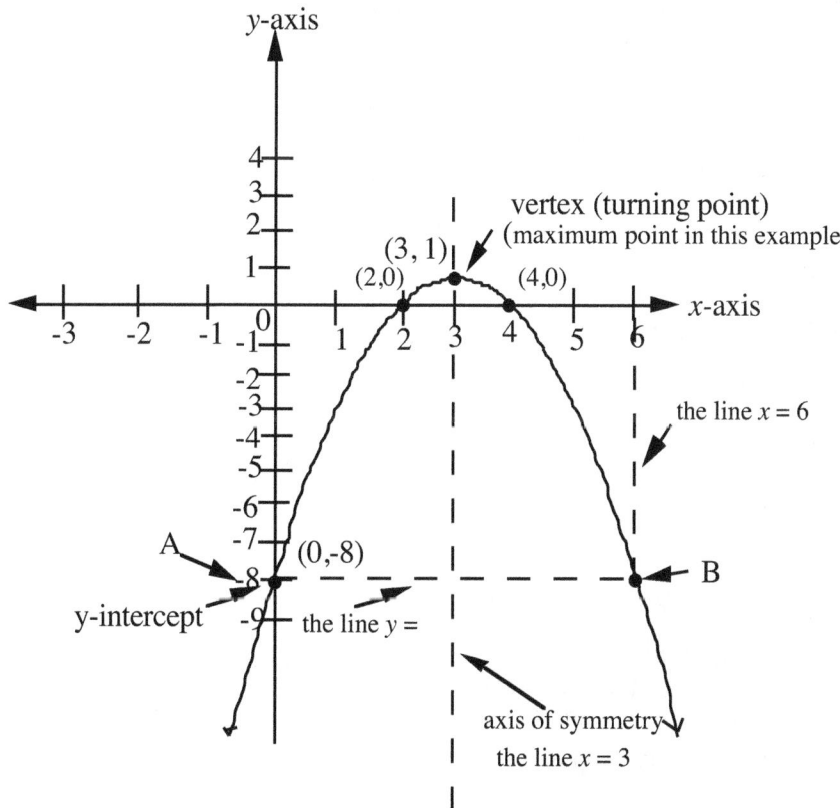

Figure 12: Graph of $y = -x^2 + 6x - 8$

Example 3 Sketch the graph of $y = f(x) = 3x^2 + 2x - 5$

Step 1: The coefficient of the x^2-term is positive. Therefore, the curve opens up.

Step 2: Find the y-intercept. When $x = 0$, $y = 3(0)^2 + 2(0) - 5 = -5$. The y-intercept is $(0, -5)$.

Step 3: Find the x-intercepts by setting $y = 0$..
Then $3x^2 + 2x - 5 = 0$, $a = 3, b = 2, c = -5$.

By the quadratic formula (or otherwise) $x = 1$, or $x = -\dfrac{5}{3}$

The x-intercepts are the points $(1, 0)$ and $(-\frac{5}{3}, 0)$.

Step 4: Find the vertex.. $a = 3, b = 2, c = -5$.

x– coordinate $= -\dfrac{b}{2a} = -\dfrac{2}{2(3)} = -\dfrac{1}{3}$,

y–coordinate $= f\left(-\dfrac{b}{2a}\right) = f\left(-\dfrac{1}{3}\right) = 3\left(-\dfrac{1}{3}\right)^2 + 2\left(-\dfrac{1}{3}\right) - 5 = -\dfrac{16}{3}$

The vertex is at $\left(-\dfrac{1}{3}, -\dfrac{16}{3}\right)$

Step 5: Plot the points $(0, -5)$, $(1, 0)$, $(-\frac{5}{3}, 0)$,

and $\left(-\dfrac{1}{3}, -\dfrac{16}{3}\right)$ from Steps 2, 3 & 4. (Figure 13)

Step 6: Locate the axis of symmetry as done in
Step 6 of Example 1.

Step 7: Connect the points by a smooth solid curve noting that the curve opens up (from Step 1).

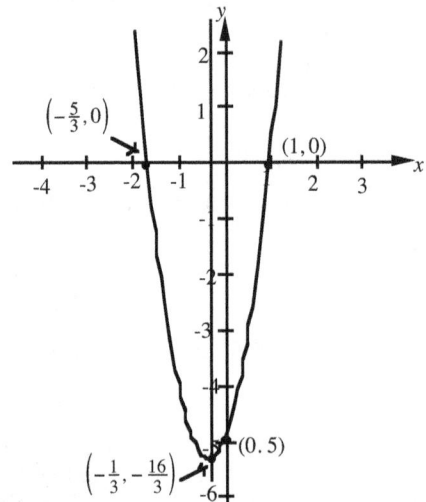

Fig. 13: Graph of $f(x) = 3x^2 + 2x - 5$

Example 4 Sketch the graph of $y = f(x) = x^2 - 2x + 5$.

Step 1: The coefficient of the x^2-term is positive. Therefore, the curve opens up.

Step 2: Find the y-intercept by setting $x = 0$ and solving for y.
Then when $x = 0, y = 5$, The y-intercept is the point $(0, 5)$

Step 3: Find the x-intercepts by setting $y = 0$ and solving for x.
$x^2 - 2x + 5 = 0$; $a = 1$, $b = -2$, $c = 5$

$x = \dfrac{+2 \pm \sqrt{4 - 20}}{2} = \dfrac{+2 \pm \sqrt{-16}}{2}$ which is not real, since $\sqrt{-16}$ is not real

There are no x-intercepts. So, in Step 6, we will choose convenient x-values, and calculate the corresponding y-values.

Step 4: Find the vertex. x–coordinate $= -\dfrac{b}{2a}$, $= -\dfrac{-2}{2(1)} = 1$

y–coordinate $= f(-\dfrac{b}{2a}) = f(1) = (1)^2 - 2(1) + 5 = 4$

The vertex is at $(1, 4)$.

Step 5: Plot the points from the above steps (Fig. 14)

Step 6 Choose convenient points on both sides of the vertex (or the symmetric axis), say $x = -1, x = 3$, and calculate the corresponding y-values $(-1, 8)$ and $(3, 8)$.

Step 7: Locate the axis of symmetry as in Step 8 of Example 1.

Step 8: Plot the points from Step 6 and connect all the points noting that this curve opens up.

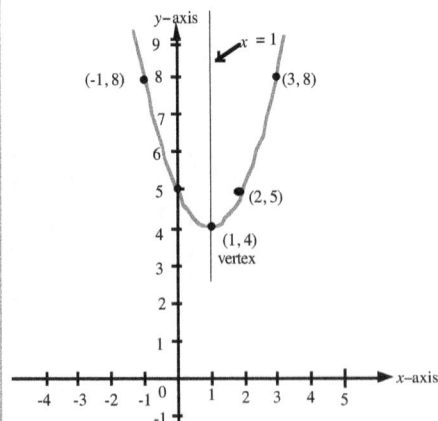

Fig. 14. Graph of $y = f(x) = x^2 - 2x + 5$

EXTRA: Axis of Symmetry , Vertex and Minimum Value

Example 1

Given $f(x) = x^2 - 12x + 8$ (A)

Let $y = x^2 - 12x + 8$ (B)

$= x^2 - 12x + (-6)^2 - (-6)^2 + 8$

$= x^2 - 12x + (-6)^2 - 36 + 8$

$y = (x - 6)^2 - 28$ <---- knowing this

$y + 28 = (x - 6)^2$ <----- or this

Vertex is at $(6, -28)$

Axis of symmetry is the line $x = 6$
 Minimum value is -28.
  Alternatively

From (B), $a = 1$, $b = -12$

Axis of symmetry is given by $x = -\frac{b}{2a}$

$\qquad = -\frac{-12}{2(1)} = \frac{12}{2} = 6$

$f(6) = 6^2 - 12(6) + 8$ (subst.. in (A)

$\qquad = 36 - 72 + 8 = -36 + 8$

$\qquad y = -28$

when $x = 6$. $y = -28$

Vertex is at $(6, -28)$

Axis of symmetry is the line $x = 6$
Minimum value is -28.

Example 2

Given $f(x) = x^2 - 6x + 8$ (A)

Let $y = x^2 - 6x + 8$ (B)

$= x^2 - 6x + (-3)^2 - (-3)^2 + 8$

$= x^2 - 6x + (-3)^2 - 9 + 8$

$y = (x - 3)^2 - 1$ <---- knowing this

$y + 1 = (x - 3)^2$ <----- or this

Vertex is at $(3, -1)$

Axis of symmetry is the line $x = 3$
 Minimum value is -1.
  Alternatively

From (B), $a = 1$, $b = -6$

Axis of symmetry is given by $x = -\frac{b}{2a}$

$\qquad = -\frac{-6}{2(1)} = \frac{6}{2} = 3$

$f(3) = 3^2 - 6(3) + 8$ (subst.. in (A)

$\qquad = 9 - 18 + 8 = -9 + 8$

$\qquad y = -1$

when $x = 3$. $y = -1$

Vertex is at $(3, -1)$

Axis of symmetry is the line $x = 3$
Minimum value is -1.

Example 3

Given $f(x) = -x^2 + 12x - 8$ (A)

Let $y = -x^2 + 12x - 8$ (B)

$y = -(x^2 - 12x + 8)$ (factoring)

$= -[x^2 - 12x + (-6)^2 - (-6)^2 + 8]$

$= -[x^2 - 12x + (-6)^2 - 36 + 8]$

$= -[x^2 - 12x + (-6)^2 - 28]$

$y = -[(x - 6)^2 - 28]$

$y = -(x - 6)^2 + 28$ <---- knowing this

$y - 28 = -(x - 6)^2$ <----- or this

Vertex is at $(6, 28)$

Axis of symmetry is the line $x = 6$
 Maximum value is 28.
  Alternatively

From (A) $a = -1$, $b = 12$

 Axis of symmetry is given by $x = -\frac{b}{2a}$

$x = -\frac{12}{2(-1)} = \frac{12}{2} = 6$; $y = -6^2 + 12(6) - 8 = 28$

Vertex is at $(6, 28)$; Maximum value is 28

Example 4

Given $f(x) = -x^2 + 6x - 8$ (A)

  Let $y = -x^2 + 6x - 8$

 $y = -(x^2 - 6x + 8)$ (factoring)

$= -[x^2 - 6x + (-3)^2 - (-3)^2 + 8]$

$= -[x^2 - 6x + (-3)^2 - 9 + 8]$

$= -[x^2 - 6x + (-3)^2 - 1]$

$y = -[(x - 3)^2 - 1]$

$y = -(x - 3)^2 + 1$ <---- knowing this

$y - 1 = -(x - 3)^2$ <----- or this

Vertex is at $(3, 1)$

Axis of symmetry is the line $x = 3$
 Maximum value is 1
  Alternatively

From (A),: $a = -1$. $b = 6$

Axis of symmetry is given by $x = -\frac{b}{2a}$

$x = -\frac{6}{2(-1)} = 3$; $y = -3^2 + 6(3) - 8 = 1$

Vertex is at $(3, 1)$; Maximum value is 1.

Relative merits (or reconciliation) of the methods of sketching the graphs of quadratic functions

On pages 245-249, we learned how to sketch the graphs of quadratic equations from the simplest form $y = x^2$ by the transformations of translation, reflection. contraction and expansion. This method may be found to be rather involved (especially if we have to complete the square: for example the

equation $y = 3x^2 + 7x - 6$. However, the experience gained in the transformation processes can be extended to other functions such as $y - 3 = 4(x - 1)^3$, $y = 4|x - 3|$, $y + 1 = \log(x + 2)$.

On pages 251-255 , we learned a relatively more direct method of sketching any quadratic function This method involves the location of the vertex (and the symmetric axis). ln this method, we do not have to complete the square and we can directly (by formula) locate the vertex and the symmetric axis.

Of course, if the square has been completed already (i.e., of form $y - k = a(x - h)^2$, we can, by inspection read the coordinates of the vertex with the consequent location of the symmetric axis. The vertex in this case will be at $x = h$, $y = k$. We should note however that the formulas employed to locate the vertex are applicable to only the quadratic equation, but the concepts of locating the vertex and the symmetric axis can be extended to other symmetric relations such as the other conic sections. The method of pages 251-255 is generally recommended especially if the equation is in the expanded form $y = ax^2 + bx + c$. In other cases, it is more convenient to use the method of transformations. In general, experience will dictate which is the preferred method to use.

Inverse Relation of $y = ax^2 + bx + c$ (See also Chapt. 6, Lesson 12)

The inverse relation of $y = ax^2 + bx + c$ is obtained by interchanging the roles of x and y (and keeping the x-axis horizontal). Then, we obtain the inverse relation $x = ay^2 + by + c$ (Figure 1).

We note that the inverse relation $x = ay^2 + by + c$ (Figure) is **generally** not a function , because the graph fails the so called vertical line test (See also page 50). However, by restricting the domain of the original function $y = ax^2 + bx + c$, we can make the inverse relation become a function. We also note that the inverse relation, $x = ay^2 + by + c$ has neither a minimum nor a maximum point.

However, by restricting the domain of a function $y = ax^2 + bx + c$, the inverse relation becomes an inverse function. For example, the inverse relation of $y = x^2 + 6$ is not a function; but by restricting its domain, its inverse relation is a function. (See Figures 2, 3 below.)

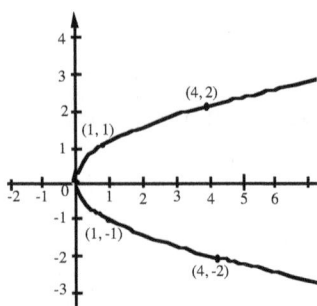

Fig. 1 : The graph of the inverse relation of
$y = ax^2 + bx + c$.
This graph is **not** a function

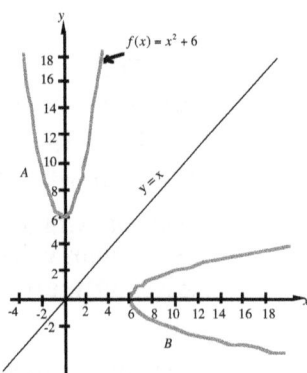

Fig. 2: B, the inverse relation of A, is **not** a function

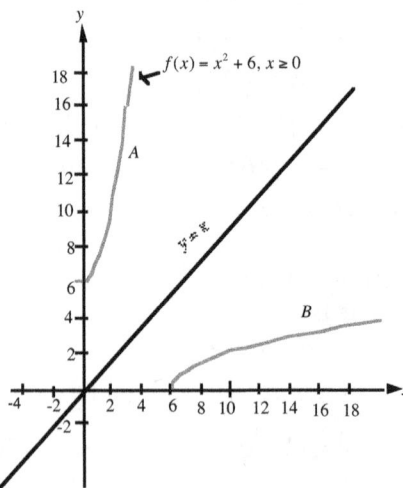

Fig. 3: B, the inverse relation of A, is a function (by the vertical line test.)

Lesson 38A Exercises

Find (a) the x-intercept and (b) the y-intercept of the following.

1. $y = x^2 + 8x + 15$;   **2.** $y = 2x^2 - 2x - 24$;   **3.** $y = (x + 4)^2$;   **4.** $y = -3x - 8$

Given the graph of $y = x^2$ sketch the graphs of the following on the same set of axes:

5. $y = x^2 - 2$;   **6.** $y = -x^2 + 3$;   **7.** $y = (x + 5)^2$;   **8.** $y - 1 = (x + 3)^2$;   **9.** $y = x^2 + 8x + 15$

10. $y = 4x^2$;   **11.** $y + 3 = 2(x - 1)^2$;   **12.** $y = \frac{1}{2}(x + 5)^2$;   **13.** What are the zeros in Problems 1– 4 ?

Find the domain and range of the following: **14.** $y = x^2 - 4x + 4$;   **15.** $y = x^2 - 2$

Problems 16-19: Find the maximum or minimum points in Problems 6-9

20. Can a function have two or more y-intercepts. Give reason for the answer.

21. Can a straight line have two or more x-intercepts. Give reason for the answer.

Answers:
1.(a) x-intercepts: -3, -5; (b) y-intercept: 15; **2.** (a) x-intercepts: -3. 4; (b) y-intercept:-24
3. (a) x-intercept: -4; (b) y-intercept: 16;   **4.**(a) x-intercept: $-\frac{8}{3}$ (b) y-intercept: -8;
13. (a) x-intercepts are the zeros. see solutions above.
14. (a) All real x; (b) range: $y \geq 0$ (all real numbers greater than or equal to 0;
15.(a) All real x; (b) range: $y \geq -2$ (all real numbers greater than or equal to -2;
16. maximum point : $(0, 3)$;   **17.** minimum point : $(-5, 0)$;   **18.** minimum point : $(-3, 1)$;
19. minimum point : $(-4, -1)$.

Lesson 38B
Applications of Quadratic Functions
Maxima and Minima Problems

The main difficulty a student may encounter here is not the application of the quadratic equations but rather being able to obtain the correct relationships between the quantities involved. For quadratic functions, we note the following:

Maximum and Minimum Points of the Parabola, $y = f(x) = ax^2 + bx + c$

The graph of the equation $y = f(x) = ax^2 + bx + c$ has either a minimum point or a maximum point. The minimum point of this curve is an **absolute minimum** and the maximum point is an absolute maximum. The minimum or maximum point is called the vertex of the parabola.

In particular, if $a > 0$ (i.e., a is positive), then $y = ax^2 + bx + c$ has a minimum point, and the parabola opens upwards ; but if $a < 0$ (i.e., a is negative), then it has a maximum point

x-coordinate $= -\dfrac{b}{2a}$; y-coordinate $= f\left(-\dfrac{b}{2a}\right)$ or $-\dfrac{b^2}{4a} + c$ or $\dfrac{4ac - b^2}{4a}$

Guidelines for Solving Maxima and Minima Problems

Step **1.** As in all word problems, read and read the problem, represent the unknowns by letters, and write down what is given or known.

Step **2.** Draw a diagram if helpful, and if geometry is involved,, draw a diagram and identify the quantity that is to be maximized or minimized.

Step **3.** Write down an equation or equations involving the quantity to be maximized or minimized. Try to eliminate any quantity whose value is not given so as to obtain a single independent variable by substitution.

Step **4.** Answer the question using a sentence.

Lesson 38B   Applications of Quadratic Functions; Maxima and Minima Problems

Example 1: Find two numbers whose sum is 48, and whose product is a maximum.

Solution

Step 1: Let one of the numbers $= x$.

Then the other number is $48 - x$

The product, P is then $P(x) = x(48 - x)$

We want to maximize $48x - x^2$ or $-x^2 + 48x$

Step 2: $P(x) = -x^2 + 48x$

$a = -1, b = 48$

(since a is negative, the vertex of the parabola is a maximum point)

$$x = -\frac{48}{2(-1)} = 24 \qquad (x = -\frac{b}{2a})$$

If one number is 24, the other number is $48 - 24 = 24$. The numbers are 24 and 24; and the maximum product is 576.

Example 2 Twice the sum of two numbers 258 is 48. Find these numbers such that their product is a maximum.

Step 1: Let one of the numbers $= x$.

Let the other number $= y$

Twice their sum is 48 translates to:

$$2(x + y) = 48$$

$$x + y = 24 \text{ or } y = (24 - x)$$

The product, P is then $P = xy$

$$P(x) = x(24 - x) = 24x - x^2$$

We want to maximize $P(x) = -x^2 + 24x$,

$a = -1, \ b = 24$

$$x = -\frac{24}{2(-1)} = 12 \qquad (x = -\frac{b}{2a})$$

When $x = 12, y = 24 - 12 = 12$

The numbers are 12 and 12; and the maximum product is 144.

Example 3a The perimeter of a rectangle is 48. Find the length and the width such that the area is a maximum.

Solution

Step 1: Let the width $= x$; and let the length $= y$

The perimeter, P is given by

$$P = 2x + 2y = 48$$

$$x + y = 24 \text{ or } y = (24 - x)$$

The area A is given by $A = xy$

$$A(x) = x(24 - x) = 24x - x^2$$

We want to maximize $A(x) = -x^2 + 24x$ (B)

$a = -1, \ b = 24$

$$x = -\frac{24}{2(-1)} = 12 \qquad (x = -\frac{b}{2a})$$

When $x = 12, y = 24 - 12 = 12$

The length is 12 units; the width is 12 units; and the maximum area is 144 sq. units

Example 3b Find the dimensions of largest rectangular area that can be enclosed using a fencing of 48 units.

Solution   Same as in Example 3a before.

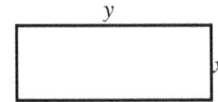

Step 1: Let the width $= x$; and the length $= y$

A rectangular fencing of 48 units implies that the perimeter, P is

(A)
$$P = 2x + 2y = 48 \qquad\qquad (A)$$

$$x + y = 24 \text{ or } y = (24 - x)$$

The area A is given by $A = xy$

$$A(x) = x(24 - x) = 24x - x^2$$

We want to maximize $A(x) = -x^2 + 24x$ (B)

$a = -1, \ b = 24$

$$x = -\frac{24}{2(-1)} = 12 \qquad (x = -\frac{b}{2a})$$

When $x = 12, y = 24 - 12 = 12$

The length is 12 units; the width is 12 units; and the maximum area is 144 sq. units.

Example 4: Find two numbers such that their difference is 12 and their product is a minimum.

Solution

Step 1: Let one of the numbers $= x$.

Let the other number $= y$

$$x - y = 12 \text{ or } y = x - 12$$

The product, P is then $P = xy$

$$P(x) = x(x - 12) = x^2 - 12x$$

(See an extra example on p. 499)

Step 2: We want to minimize $P(x) = x^2 - 12x$,

$a = 1, \ b = -12$

$$x = -\frac{-12}{2(1)} = 6 \qquad (x = -\frac{b}{2a})$$

When $x = 6; \ y = 6 - 12 = -6$

The numbers are 6 and -6; and the minimum product is -36.

Lesson 39   259
Graphs of other Polynomial Functions

The other polynomial functions covered are $y = x^3$, $y = x^4$, $y = x^5$, and $y = x^6$.

Graph of $y = x^3$

Using convenient points, we sketch the graph of $y = x^3$ (Figure 4) and then list its properties.

Properties of $y = x^3$

1. The curve is tangent to the x-axis at the origin and does not cut or cross the x-axis sharply at the origin, but rather, it somehow "rides along" the x-axis for a while before crossing it.

2. The vertex is at the origin.

3. There are two halves of the curve: one half is in the first quadrant, and the other half in the third quadrant.

4. The curve passes through the points $(1, 1)$ and $(-1, -1)$.

5. The curve is symmetric about the origin.

Generally, the curve $y = x^3$ passes through the points $(1, a)$ and $(-1, -a)$

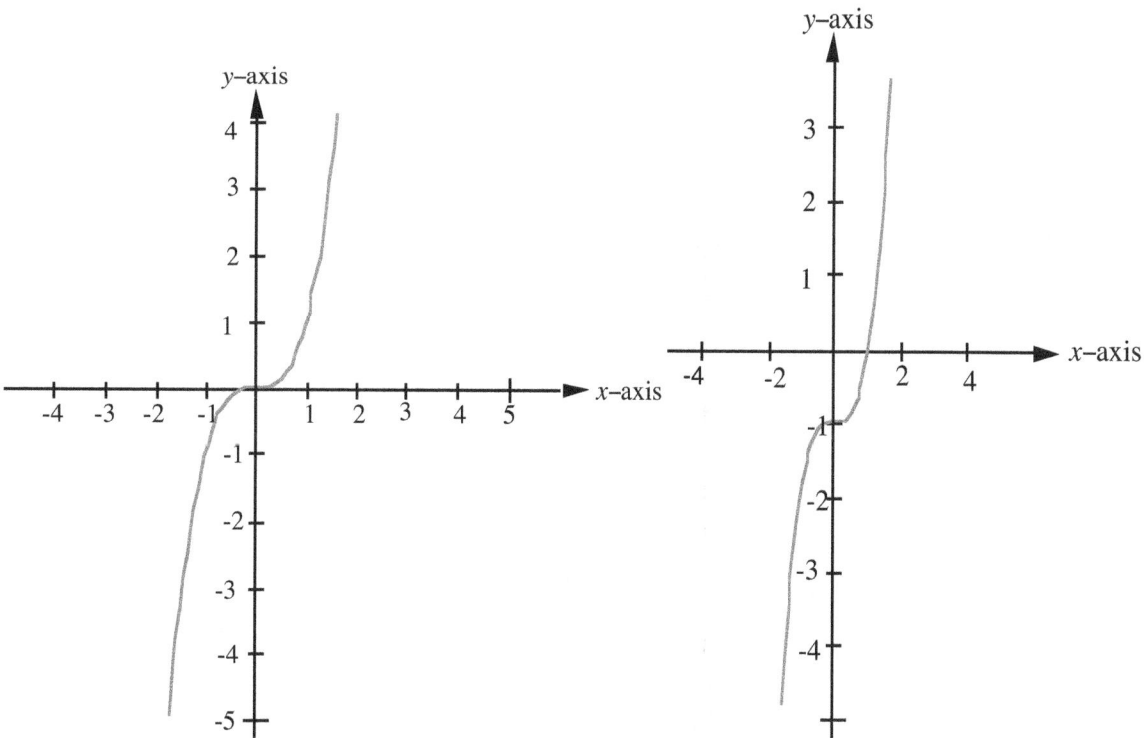

Figure 4 : Graph of $y = x^3$

Figure 5 Graph of $y = x^3 - 1$.

Lesson 39: Graphs of other Polynomial Functions

Examples The graphs of $y = -x^3$, $y = x^3 - 1$, $y = -x^3 - 2$, $y = 2(x - 4)^3 + 1$ are also shown below. Generally, the equation of the "cubic parabola" is given by

$$y - k = a(x - h)^3$$

This curve passes through the point $(x + h, ax^3 + k)$, where if h is positive if the shift is to the right and if negative the shift is to the left. Similarly, k is positive if shift is up and if negative the shift is down.

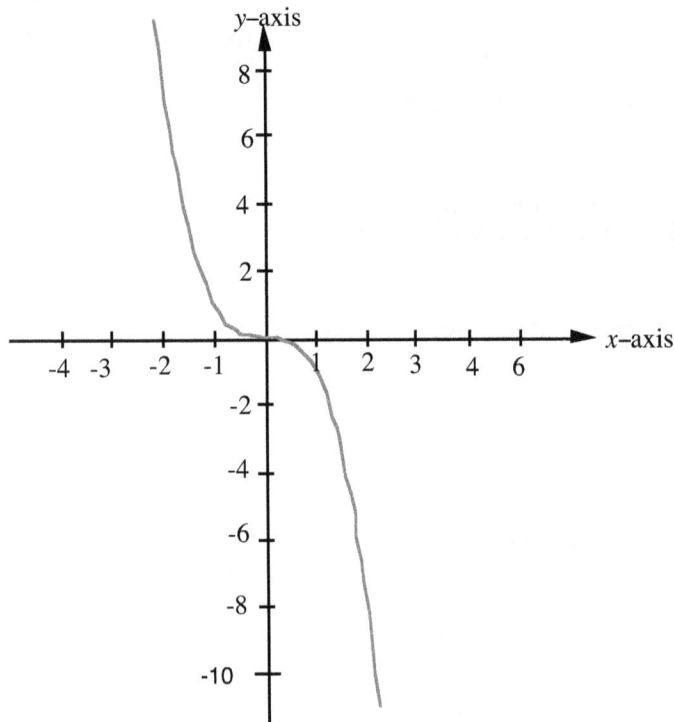

Figure 6 : Graph of $y = -x^3$

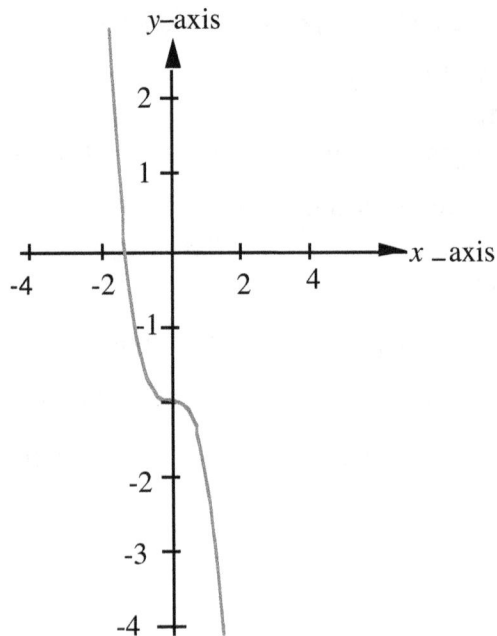

Figure 7 Graph of $y = -x^3 - 2$

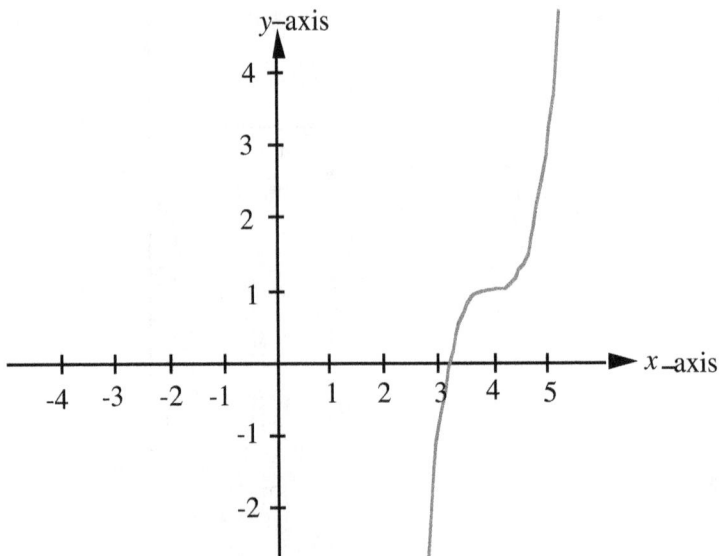

Figure : The graph of $y = 2(x - 4)^3 + 1$

Graphs of $y = x^4$, $y = x^5$, $y = x^6$.   261

The graph of $y = x^4$ is very similar to that of $y = x^2$, being also U-shaped. However, the vertex is much flatter than that of $y = x^2$

Similarly, the graph of $y = x^6$ is U-shaped with the vertex being much flatter than that of $y = x^4$.

On the other hand, $y = x^5$ has the same shape as $y = x^3$, except that the vertex is much flatter than that of $y = x^3$.

Graph of $y = x^4$

The sketching procedure is like that of $y = x^2$.   The graph is shown in Figure

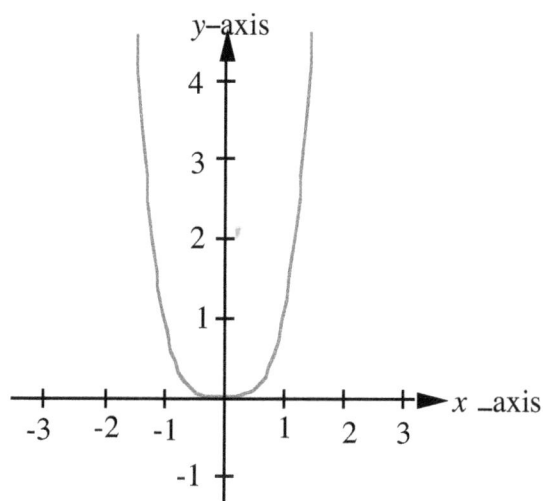

Figure 8 : The graph of $y = x^4$

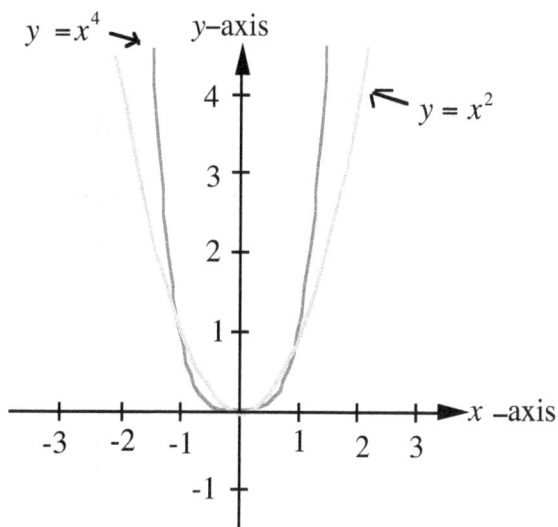

Figure 9 : Graphs of $y = x^4$ and $y = x^2$ compared.

Crossing or Touching of the *X*- and *Y*-axes by Lines and Curves   262

The material covered below (p.262-263) will be useful when sketching the graphs of polynomial functions.

Linear equation. $y = mx + b$

Consider the linear function $f(x) = x + a$. If we let $y = f(x)$, then $y = x + a$

A linear equation cuts or intersects the *x*-axis sharply. It also cuts the *y*-axis sharply. For example, the linear equation $y = 2x - 3$ cuts the *x*-axis sharply at $\frac{3}{2}$ and the *y*-axis sharply at $y = -3$.

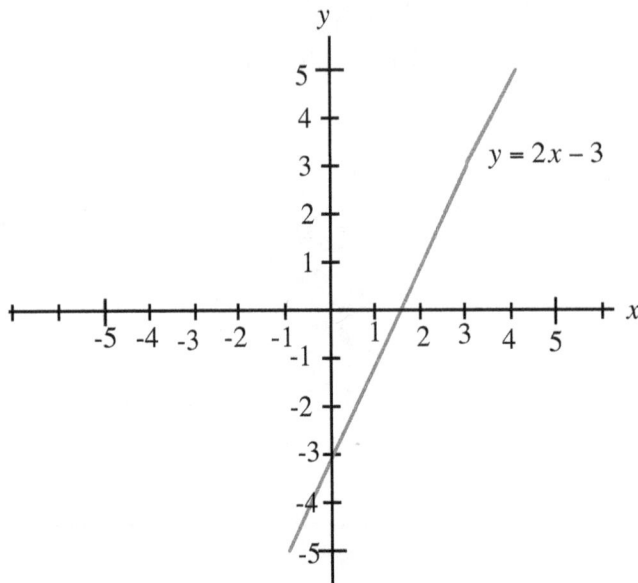

Quadratic equation $y = (x - a)^2$

Example 1 $y = (x - 3)^2$. The vertex of this curve is on one side of the *y*-axis. The vertex is tangent to the *x*-axis at $x = 3$. The curve does not does not cross the *x*-axis, but rather touches it (Figure)

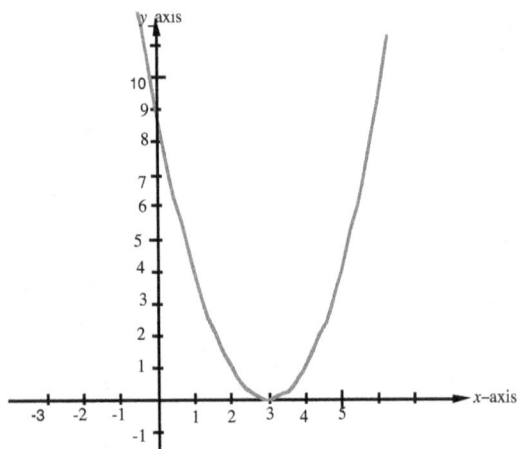

Figure Graph of $y = (x - 3)^2$

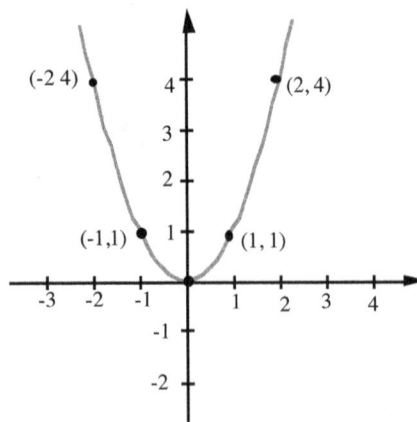

Figure: Graph of f $y = x^2$ is tangent at the origin $(0, 0)$.

Cubic equation $y = (x - a)^3$

Example 2: The vertex of the graph of $y = x^3$ is tangent to the x-axis at the origin. The curve does **not** cut the x-axis sharply (as does the line $y = x$), but rather " rides along the x-axis for a moment before crossing it." The vertex of $y = x^3$ is tangent to the x-axis at the origin $(0, 0)$.

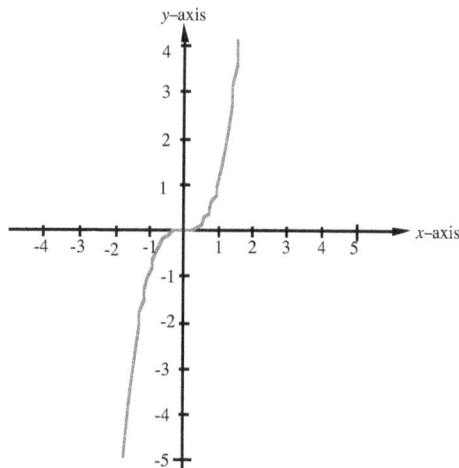

Figure: The graph of $y = x^3$

Quartic equation $y = (x - a)^4$

The behavior of this curve is similar to that of $y = x^2$. See page 261 for comparison of the graphs of $y = x^4$ and $y = x^2$.

Example 3 Discuss how the following curves cross or touch the x-axis:

$$(a) \quad y = (x - 1)^2 (x + 3); \qquad (b) \quad y = (x - 1)(x - 3)(x - 5)^2 (x + 3)^2$$

(a) At $x = 1$, (quadratic behavior) the curve touches the x-axis but does not cross it . The zero $x = 1$ is a double root of the equation. There is an U-shaped curve at the point $(1, 0)$.
   At $x = -3$ (linear behavior) the curve crosses the x-axis sharply.
(b) At $x = -3, x = 5$ (quadratic behavior) the curve touches the x-axis but does not cross it.
   The points $x = -3, x = 5$ are double roots.
   and there are the characteristic U-shapes with the vertices at $(-3, 0)$ and $(5, 0)$ respectively.
   At $x = 1, x = 3$ (linear behavior), the curve crosses the x-axis sharply.

Example 4 Check or confirm the solution to the previous example, Example 3, by sketching the graphs of the equations. Also check (a) the **end behavior** of the graphs: see p.264-265 for help; (b) the number of turning points n-1 (see p.216 for help) of the graph.

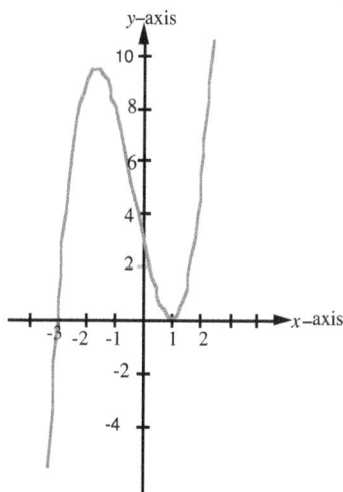

(a)

Fig.: The graph of $y = (x - 1)^2 (x + 3)$

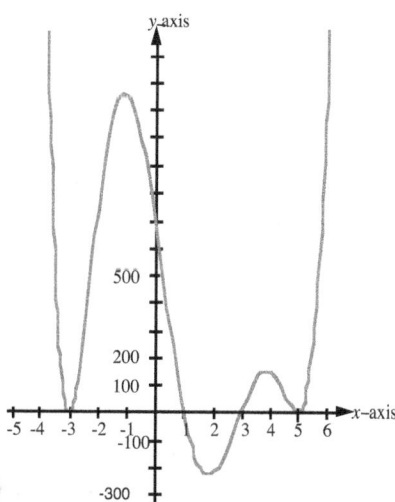

(b)

Fig. The graph of $f(x) = (x - 1)(x - 3)(x - 5)^2 (x + 3)^2$

Sketch the graph of each of the following and state the domain:

1. $y = x^3$; **2.** $y = x^3 + 1$; **3.** $y = x^3 - 1$; **4.** $y = x^5$; **5.** $y = x^4 - 3$; **6.** $y = (x - 5)^4$; **7.** $y = (x + 1)^4 + 2$

Answers:

1.- 3.

4.- 7.

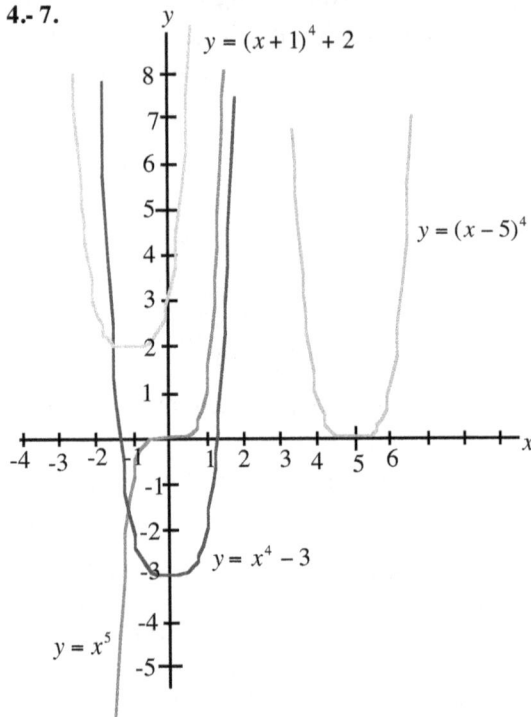

End Behavior of Polynomial Functions

The material covered under the above topic will be useful when sketching the graphs of polynomial functions.

Given the functional equation of a polynomial, it is useful to predict the end behavior of the graph of this function, just as it is useful to predict if the graph of a straight line leans to the left or to the right in the page, given the slope-intercept form of the equation of the line.

The **end behavior** of the graph of a polynomial function, $f(x) = a_n x^n + a_{n-1} x^{n-1} + ... + a_1 x + a_0$, $a_n \neq 0$, is the behavior of the graph as x decreases or increases without bound. (that is, the behavior to the far left or far right of the x–axis). The leading coefficient term, $a_n x^n$, determines the end behavior of the graph of a polynomial function. The **leading coefficient term** of a polynomial is the term with the highest degree. The **leading coefficient** of a polynomial is the coefficient of the term with the highest degree. In $7x^3 + 5x^2 + 2x + 9$, $7x^3$ is the leading coefficient term and 7 is the leading coefficient. The graph of a polynomial function rises or falls as x decreases or increases without bound.

Leading Term Test

The behavior is determined by a_n, and also by n as follows:

n even, $\quad a_n > 0$ Examples:

$y = f(x) = x^2$. $y = f(x) = x^4$; $y = f(x) = x^6$

Memory example

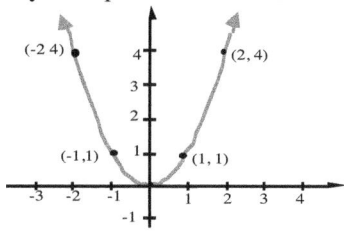

$y = x^2$

Rises to the left and rises to the right
Fig 1

n odd $\quad a_n > 0$ Examples:

$y = f(x) = x^3$. $y = f(x) = x^5$; $y = f(x) = x^7$

Memory example

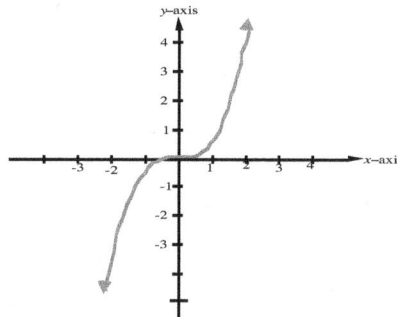

$y = x^3$

Rises to the right and falls to the left

n even, $\quad a_n < 0$. Examples:

$y = f(x) = -x^2$; $y = f(x) = -x^4$; $y = f(x) = -x^6$

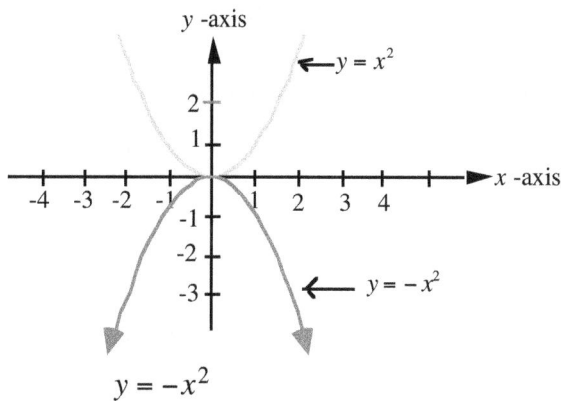

$y = x^2$

$y = -x^2$

$y = -x^2$

Falls to the left and falls to the right

n odd $\quad a_n < 0$. Examples:

$y = f(x) = -x^3$; $y = f(x) = -x^5$; $y = f(x) = -x^7$

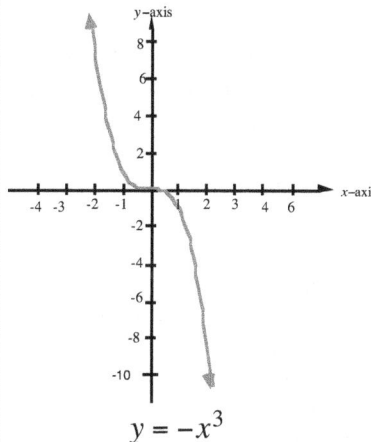

$y = -x^3$

Rises to the left and falls to the right.

n odd, $\quad a_n > 0$ Example:

$y = f(x) = 4x^5 + 3x^2$

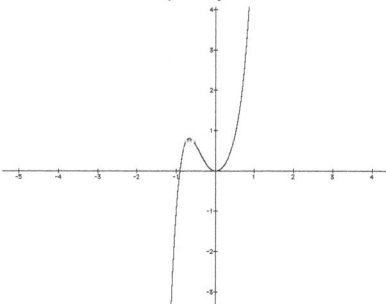

Rises to the right and falls to the left

n odd $\quad a_n < 0$ Example:

$y = f(x) = -4x^5 + 3x^2$

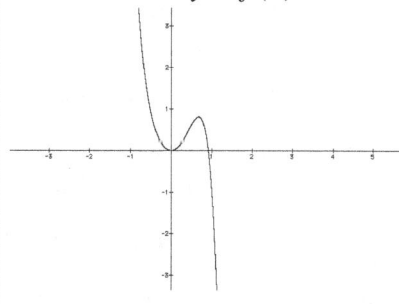

Rises to the left and falls to the right.

Memorize the above graphs. See also page 269 for more graphs.

Lesson 40

Graphs of Factorable Polynomial Functions

Example 1 Sketch the graph of .
Solution

Let $f(x) = y$. Then $y = (x - 1)(x - 2)(x + 3)$

Step 1: Find the x-intercepts by setting $y = 0$ and solving for x.

Apply the zero product rule: That is, set each linear factor to zero and solve for x.

Then from $(x - 1) = 0$, $x = 1$, yielding the point $(1, 0)$.

$(x - 2) = 0$, $x = 2$, yielding the point $(2, 0)$.

$(x + 3) = 0$, $x = -3$, yielding the point $(-3, 0)$.

Step 2: Find the y-intercept by setting $x = 0$ and solving for y.
Then $y = (0 - 1)(0 - 2)(0 + 3)$
$y = (-1)(-2)(+3)$
$y = 6$, yielding the point $(0, 6)$

Step 3: Plot the points (ordered pairs) from Steps 1 and 2.
That is plot $(1, 0)$, $(2, 0)$, $(-3, 0)$, and $(0, 6)$. (Figure)

Step 4: Since there are three x-intercepts, there will be four intervals to consider, namely the intervals
$-\infty$ to -3, -3 to 1, 1 to 2, 2 to $+\infty$.

In particular, we would like to know if the curve is above the x-axis (i.e., positive function), or below the x-axis (negative function) on each of the above intervals. Such information will help us to determine the location and direction of the curve on each interval. We can use a sign diagram (See page 177) to determine the sign of the function (i.e., whether positive or negative).

Alternatively, we can also choose convenient x-values on each interval; calculate the corresponding y-values by substituting the x-values in the given equation, and then plot the ordered pairs. One point on each interval is usually sufficient. The method of calculating values is recommended because it actually yields points which we can plot and connect. The method of sign diagrams determines only whether the curve is below or above the x-axis. In choosing points, whenever convenient, choose x so that it is the mid-point of that interval. When in doubt about the location of the graph on an interval, we will choose more than one x-value on that interval.

Step 5: Continuing, we use the method of calculating values to complete the graphing.

For the interval $-\infty$ to -3, we choose say $x = -4$ and substitute in
$y = (x - 1)(x - 2)(x + 3)$
Then $y = -30$ (i.e., the function is negative and the curve is below the x-axis on this interval
This yields the point $(-4, -30)$.

For the interval -3 to 1 we choose say $x = -1$, substitute in $y = (x - 1)(x - 2)(x + 3)$ to obtain $y = 12$; and the ordered pair $(-1, 12)$. Similarly, we choose $x = -2$, calculate y to obtain the ordered pair $(-2, 12)$, and for $x = -\frac{3}{2}$, $y = \frac{105}{8}$ or $13\frac{1}{8}$ to obtain the point $(-\frac{3}{2}, \frac{105}{8})$.
The curve is above the y axis on this interval, since all the y-values are positive.

Similarly, **for the interval 1 to 2,** we choose say, $x = \frac{3}{2}$, substitute to obtain $y = -\frac{9}{8}$ and the ordered pair $(\frac{3}{2}, -\frac{9}{8})$. The curve is below the x-axis on this interval.

For the interval 2 to $+\infty$, we choose say $x = 3$ to obtain $y = 12$, and the point $(3, 12)$

Lesson 40: Graphs of Factorable Polynomial Functions

Step 6: Plot the points (the ordered pairs) from Step 5., that is, plot (-4, -30), (-1, 12), (−2,12);
$(-\frac{3}{2}, \frac{105}{8})$; $(\frac{3}{2}, -\frac{9}{8})$ and (3, 12),   See Figure.

Step 7: Connect all the plotted points by a smooth solid curve. Usually, this type of curve is drawn
smoothly and continuously as: "up and down", "up and down" or "down and up ", "down and " up alternately.

Step 8:: Apply the **end behavior** of the graphs of polynomial functions on page 264-265 to help
determine the correctness of the end behavior of the graph.

Step 9:: Apply the rule that the number of turning points of a polynomial function of degree n is at
most n-1 (see also 216) to determine the correctness of the number of turnings of the graph.

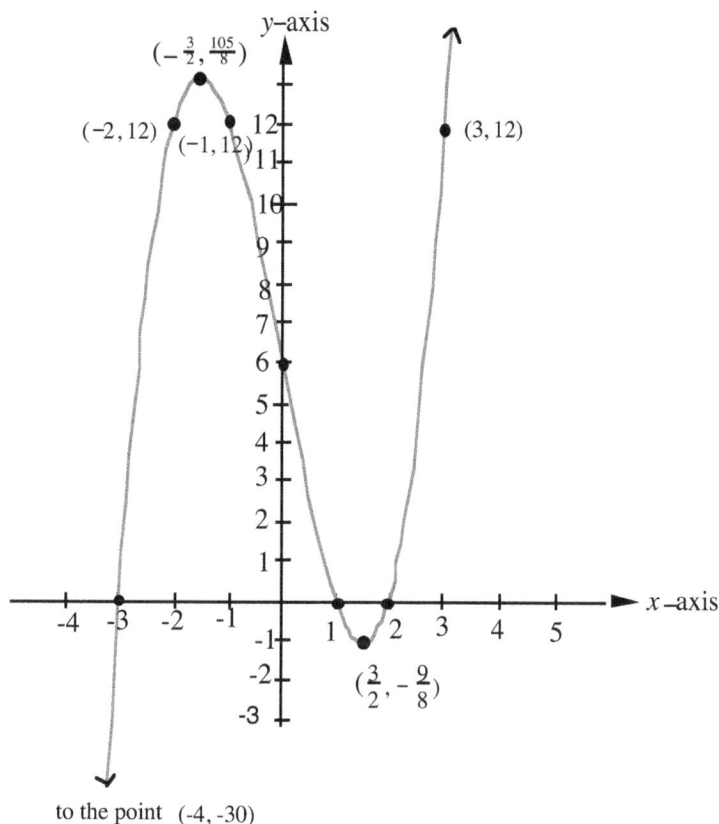

Figure: The graph of $f(x) = (x-1)(x-2)(x+3)$

The above method can also be used to graph inequalities involving polynomials; but in this case, we would like to use sign
diagrams and choice of convenient x-values interchangeably.

Example 2 Sketch the graph of the polynomial given by $f(x) = (x-5)(x+1)(x+4)$.

Solution Let $f(x) = y$. Then $y = (x-5)(x+1)(x+4)$

Step 1: Find the x-intercepts by setting $y = 0$ and solving for x.
Apply the zero product rule: That is, set each linear factor to zero and solve for x.
Then from $(x-5) = 0$, $x = 5$, yielding the point $(5,0)$.

$(x+1) = 0$, $x = -1$, yielding the point $(-1,0)$.

$(x+4) = 0$, $x = -4$, yielding the point $(-4,0)$.

Step 2: Find the y-intercept by setting $x = 0$ and solving for y.
Then $y = (0 - 5)(0 + 1)(0 + 4)$
$y = (-5)(1)(+4)$
$y = -20$, yielding the point $(0, -20)$

Lesson 40: Graphs of Factorable Polynomial Functions

Step 3: Plot the points (ordered pairs) from Steps 1 and 2.
    Thus plot $(5,0)$, $(-1,0)$, $(-4,0)$, and $(0,-20)$. (**Figure**.)

Step 4: Since there are three x-intercepts, there are four intervals to consider, namely the intervals
    $-\infty$ to -4, -4 to -1, -1 to 5, 5 to $+\infty$.

In particular, we will like to know if the curve is above the x-axis (i.e., positive function), or below the x-axis (negative function) on each of the above intervals. Such information will help us to determine the location and direction of the curve on each interval.

We choose convenient x-values on each interval; calculate the corresponding y-values by substituting the x-values in the given equation, and then plot the ordered pairs. One point on each interval is usually sufficient.

 In choosing points, it would be preferable to choose x so that it is the mid-point of that interval. When in doubt about the location of the graph on an interval, we will choose more than one x-value that interval.

Step 5: Continuing, we use the method of calculating values to complete the graphing.
    For the interval $-\infty$ to -4, we choose say $x = -5$ and substitute in $y = (x-5)(x+1)(x+4)$

Then $y = -40$ (i.e., the function is negative and the curve is below the x-axis)
This yields the point $(-5,-40)$.

For the interval -4 to -1 we choose say $x = -3$, substitute to obtain $y = 16$; and
the ordered pair $(-3,16)$ The curve is above the y-axis on this interval, since 16 is positive.
Similarly, for the interval -1 to 5, we choose say $x = 3$, substitute to obtain $y = -56$ and the
ordered pair $(3, -56)$. The curve is below the x-axis.

For the interval 5 to $+\infty$,we choose say $x = 6$ to obtain $y = 70$, and the ordered pair $(6, 70)$

Step 6: Plot the points (the ordered pairs) from Step 5: That is, plot $(-5,-40)$, $(-3, 16)$, $(3, -56)$ and
    $(6, 70)$ See Figure .

Step 7: Connect all the plotted points by a smooth solid curve. Usually, this type of curve is drawn
    smoothly and continuously "up and down" up and down" or down and up, down and up   alternately.

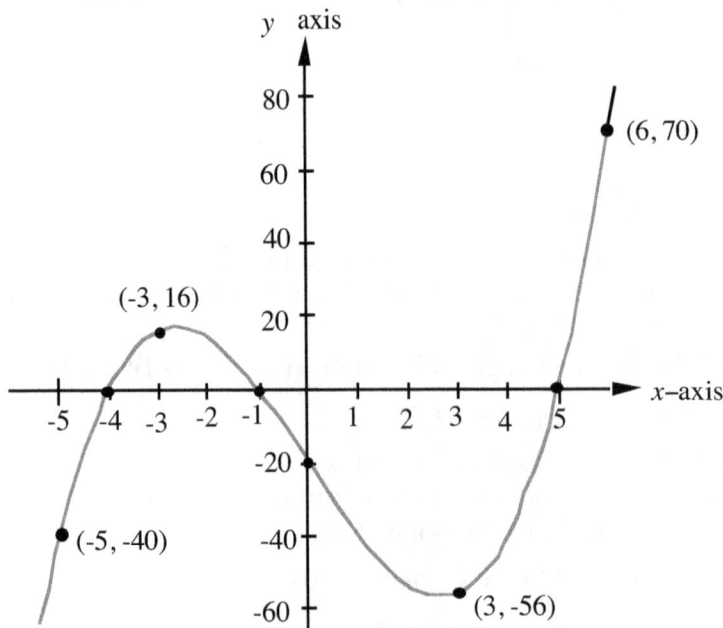

Figure: The graph of $f(x) = (x-5)(x+1)(x+4)$

Lesson 40 Exercises   269

Describe how the following curves cross or touch the x-axis:

1. $y = x - 4$;   **2.** $y = -x$;   **3.** $y = (x+1)^2(x-3)$;   **4.** $y = x^3$;**5.** $y = (x-2)^3$;   **6.**$y = 2x^4$

Sketch the graphs of the following factored polynomials:

7. $y = (x-3)(x-2)$;   **8.** $y = (x-1)(x-4)(x+4)$;   **9.**$y = (x-2)(x+3)(x+5)$

Answers: 1. Cuts the x-axis sharply at $x = 4$: **2.** Cuts the x-axis sharply at $x = 0$.;
3. Touches the x-axis (but does not cross it) at $x = -1$, and cuts the x-axis sharply at $x = 3$;
4.- 6 See page 259-261.

9.
$y = (x-2)(x+3)(x+5)$

8.

$y = (x-1)(x-4)(x+4)$

7.

$y = (x-3)(x-2)$

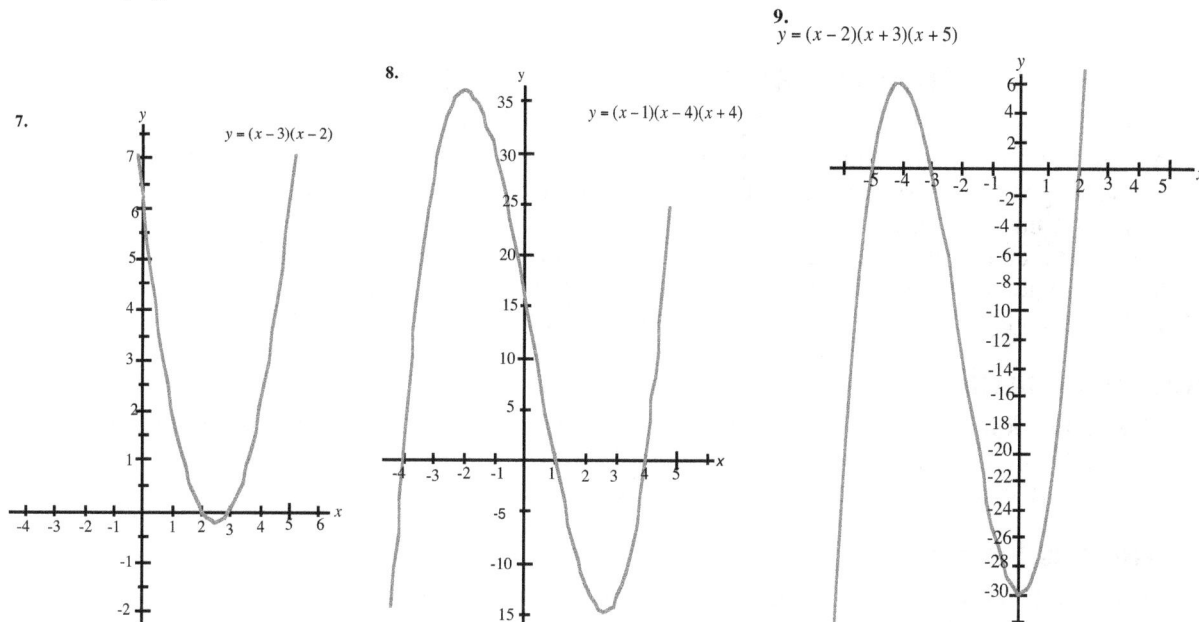

More End Behavior of Polynomials

n even, $a_n > 0$. Example:

$f(x) = 2x^6 - 4x^5 + 5x^3 + 8x$

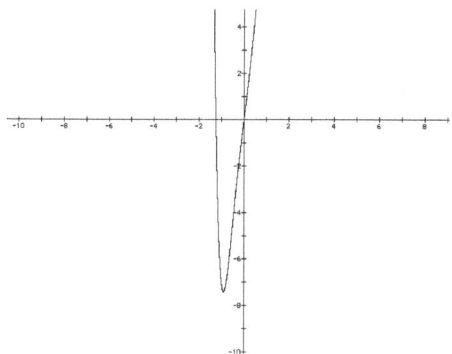

Rises to the left and rises to the right

n odd $a_n < 0$. Example:

$f(x) = -x^3 + 2x^2 + 3x$

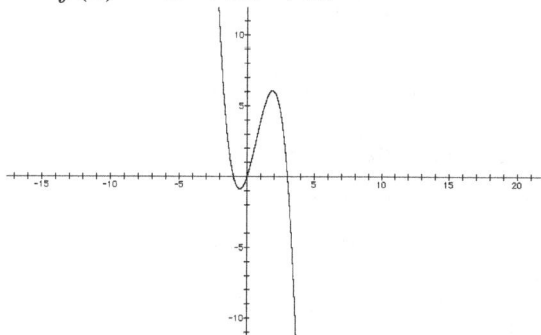

Rises to the left and falls to the right..

CHAPTER 16
RATIONAL FUNCTIONS

Lesson 41: **Continuous and Discontinuous Functions**

Lesson 42: **Asymptotes**

Lesson 43: **Graphs of Rational Functions**

Lesson 41
Continuous and Discontinuous Functions

We use the concepts of continuous and discontinuous functions in sketching the graphs of functions. The graphs of polynomial, sine, and cosine functions are continuous functions. The graphs of rational functions, the tangent ,the cosecant and cotangent functions as well as the graphs of the hyperbolas are generally discontinuous functions.

The meanings given below to continuous and discontinuous functions will be sufficient for our conceptual and qualitative understanding of college algebra and trigonometry. In calculus, we shall learn more formal and complete definitions of continuous and discontinuous functions.

Continuous Functions

Graphically, a function is continuous at a point or on a interval if the graph (curve or line) has no breaks, "jumps" or "holes" in it at that point or on that interval. We can also view a continuous function as one whose line or curve can be drawn without lifting the pencil from the paper.

Discontinuous Functions

Graphically, a function may be discontinuous at a point or on an interval if it has any of the "defective" properties of breaks, jumps, or holes in its curve or line. The breaks represent the excluded values in the domain or range of the function.

There are two main types of discontinuities, namely finite discontinuity and infinite discontinuity.

Finite Discontinuity, Holes

A **finite discontinuity** is a discontinuity which can be removed by redefining the function. Sometimes, we call a finite discontinuity removable or temporary discontinuity. (By redefining the function, in this case, we " bypass or go around the hole".)

Hole: A hole is a circle representing a break in a line or curve at a finite discontinuity. (Figure)

Infinite (Essential) Discontinuity

An **infinite discontinuity** is a discontinuity (a break in a curve) which cannot be removed by redefining the function.

In calculations, we sometimes meet ratios in which certain values when substituted for a variable in the denominator make the denominator zero, and consequently the value of each such ratio (fraction) is undefined.

Examples are **1.** $\frac{3x}{2x+1}$ which is undefined at $x = -\frac{1}{2}$; **2.** $\frac{3x^2 - 7x + 2}{x - 1}$ which is undefined at $x = 1$.

We call such ratios of polynomials **rational expressions**. A rational expression is the ratio of two polynomials.

However, we must note that there are rational expressions which are defined for all real values of the variable in the denominator. For example, $\frac{8}{x^2+4}$ is defined for all real values of the variable in the denominator   The denominator $x^2 + 4$ is positive for all real values of x and never zero, since the square of any nonzero real number is always positive.
A rational expression may be proper or improper.

Proper Rational Expression

ln a proper rational expression, the degree of the numerator polynomial is lower than the degree of the denominator polynomial.

Example $\frac{x+1}{x^2-4}$ (Degree of the numerator polynomial is 1; degree of the denominator polynomial is 2.)

Improper Rational Expression

In an improper fraction, the degree of the numerator polynomial is greater than or equal to the degree of the denominator polynomial.

Examples 1. $\frac{x^2-9}{x-2}$ (Degree of the numerator polynomial is 2; degree of the denominator polynomial is 1.)

2. $\frac{x}{x+2}$ (Degree of the numerator polynomial is 1; degree of the denominator polynomial is 1.)

In working with rational expressions, we shall exclude those values (called **excluded values**) of the variable in the denominator which make the denominator zero. At the excluded values, the rational expressions are undefined.

With polynomial functions, the graph is one smooth unbroken (continuous curve). With rational functions, there may be broken (discontinuous) lines or curves. The curves are broken at the excluded values mentioned above.

Excluded Values and Graphing of Rational Functions

ln sketching the graph of a rational function, we can use either holes or asymptotes (see next section) to indicate the places where there are discontinuities in the curve or line. For finite discontinuities, we shall use holes, but for infinite (essential) discontinuities, we shall use asymptotes.

ln sketching the curves at or near the infinite discontinuities we shall use the asymptotes as "guiding straight lines". In the next section, we cover the definition of an asymptote and then learn how to find equations of asymptotes, as well as sketch selected curves with asymptotes. We shall also justify why so much effort has been devoted to study asymptotes, something that most current text books do not do at this level.

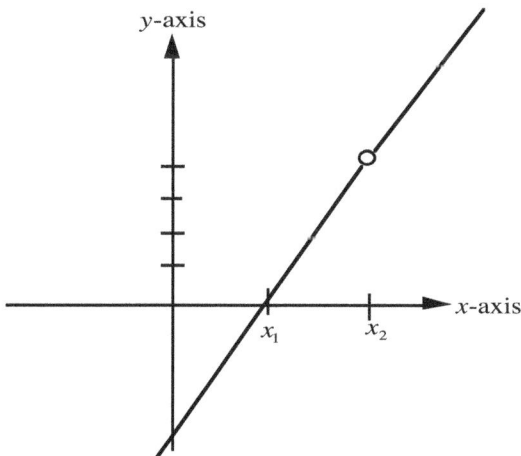

Figure: Function with a finite discontinuity at x_2.

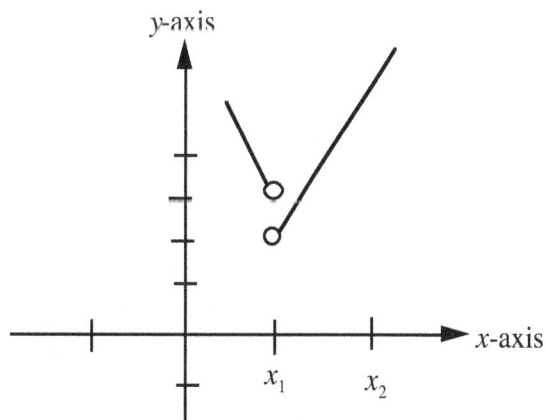

Figure: Function with a hole at x_1

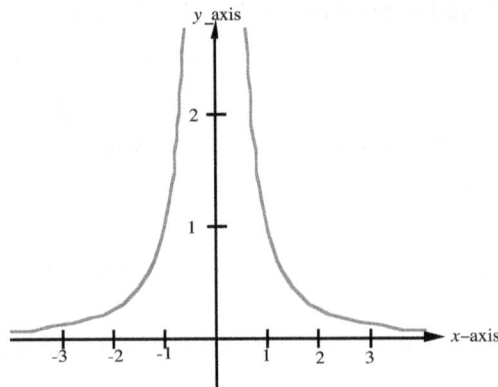

Figure Graph of a function with infinite discontinuity at $x = 0$.

Lesson 41 Exercises

A **1.** Define or explain the following from memory

  (a). Continuous functions, **(b)** Discontinuous functions; **(c)** Finite discontinuity,

  (d) Infinite discontinuity; **(e)** Proper rational Expression; (**f)** Improper rational expression

  (g) Excluded values

2. How do we indicate graphically the discontinuities in a curve or line?

3. Qualitatively, distinguish between the graph of a continuous function and the graph of a discontinuous function. Give an example of each.

4. What is the significance of (a) a hole, (b) an asymptote in sketching the graphs of functions?

Answers: See text.

B State whether the discontinuity in each of the following is finite or infinite,

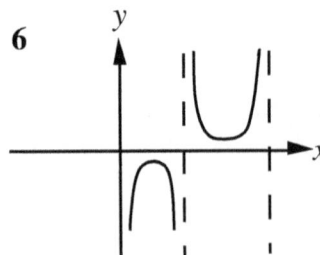

Answers: 1. Finite; **2.** infinite; **3.** finite; **4.** finite;? **5.** infinite; **6.** infinite.

Lesson 42   273
Asymptotes

The word asymptote comes from the Greek word " asymptotos" (Latin: asymptota) which means "not meeting".

Definition: An **asymptote** to a given curve is a straight line which the curve approaches (nearer and nearer) as the curve is produced or extended.

Since an asymptote is a straight line, an asymptote possesses all the properties of straight lines that we learned in the past. An asymptote may be a vertical line (vertical asymptote) with equation of the form $x = a$ (Example: $x = 3$). It may be a horizontal line (horizontal asymptote) with equation of form $y = b$ (Example: $y = 2$); or it may be an oblique line (slant line) with equation of the form $y = mx + b$ (Examples: $y = 3x + 2$, $y = 5x$) where m $\neq 0$.

Vertical Asymptote

A **vertical asympto**te to a curve is the vertical line $x = a$ (where a is the x–intercept) which the equation of the curve approximates as x approaches a and y increases or decreases without bound . Geometrically, the curve smoothly approaches the vertical asymptote nearer and nearer as y increases or decreases without bound.

Horizontal Asymptote

A **horizontal asymptote** to a curve is the horizontal line $y = b$ (where b is the y–intercept) which the equation of the curve approximates as y approaches b and x increases or decreases without bound. Geometrically, the curve smoothly approaches the horizontal asymptote nearer and nearer as x increases or decreases without bound.

Determining the equations of the vertical asymptotes, given the equation of the function

Step 1: Solve the given equation for y in terms of x (if it has not been solved for).

Step 2: Set the denominator from Step 1 to zero and solve for x.

Step 3: The real values of x obtained in Step 2 give the equations of the vertical asymptotes. If the values are **not** real, then there are no vertical asymptotes.

Example 1   Find the equations of the vertical asymptotes of the rational function whose equation is
$$y = \frac{4x^2 + 3}{x^2 - 1}$$

Solution

Step 1: Fortunately, the given equation has been solved for y. We therefore go on to the next step.

Step 2:   Set the denominator to zero and solve for x.
        Then $x^2 - 1 = 0$
        Solving, $x = 1$, or $x = -1$
The equations of the vertical asymptotes are the lines $x = 1$, and $x = -1$.

Example 2 Find the equations of the vertical asymptotes of the rational function

whose equation is $\quad y(x^2 - 1) = 4x^2 + 3$

Solution Step 1: Solve for y

Then $y = \dfrac{4x^2 + 3}{x^2 - 1}$

For the rest of the steps, see Example 1, above.

Example 3 Find the equations of the vertical asymptotes of the rational function whose equation is given by : $y(x^2 + 1) = 4x^2 + 3$

Solution

Step 1: Solve for y

Then $y = \dfrac{4x^2 + 3}{x^2 + 1}$

Step 2: Set the denominator to zero and solve for x.

$x^2 + 1 = 0$

Solving, $x = \sqrt{-1}$, which is not real, There are **no** vertical asymptotes.

Note that the denominator $x^2 + 1$ is positive for all real values of x and never zero, since the square of any nonzero real number is always positive.

Determining the equations of the horizontal asymptotes, given the equation of the function

There are two main methods. One method is similar to the method used in finding the vertical asymptotes. The other method considers the use of limits (by considering large values of x).

Method 1

Step 1: Solve the given equation for x in terms of y.

Step 2: Set the denominator obtained in Step 1 to zero and solve for y. The real values of y obtained give the equations of the horizontal asymptotes. If there are no real values, then there are no horizontal asymptotes.

Method 2

Step 1: Solve the given equation for y.

Step 2: Divide every term in both the numerator and the denominator by the highest power of x in in the **denominator.**

Step 3: Consider large values of x. Note for instance that, for large values of x, $\dfrac{1}{x} \approx 0$, $\dfrac{1}{x^2} \approx 0$.

Step 4: The values of y obtained from Step 3 give the equations of the horizontal asymptotes.

Example 4 Determine the horizontal asymptotes of the curve given by

$$y = \frac{4x^2 + 3}{x^2 - 1}$$

Solution

Method 1

Step 1: Solve the given equation for x in terms of y.

Cross multiplying, $y(x^2 - 1) = 4x^2 + 3$

$x^2 y - y = 4x^2 + 3$

$x^2 y - 4x^2 = y + 3$

$x^2(y - 4) = y + 3$ ⟸ (factoring in order to obtain a single term for x^2. If you have never met this technique before, try to master it. This factoring (monomial factoring) becomes necessary, sometimes, when solving literal equations for one of the variables)

$x^2 = \dfrac{y + 3}{y - 4}$

$x = \pm \dfrac{\sqrt{y + 3}}{\sqrt{y - 4}}$

Step 2: Set the denominator to zero and solve for y.

$$\sqrt{y - 4} = 0$$

$$\left(\sqrt{y - 4}\right)^2 = 0^2$$

$$y - 4 = 0$$

$$y = 4$$

Therefore, the equation of the horizontal asymptote is $y = 4$.

Method 2

Example Find the horizontal asymptotes of $y = \dfrac{4x^2 + 3}{x^2 - 1}$

Solution

Step 1:

$$y = \frac{\frac{4x^2}{x^2} + \frac{3}{x^2}}{\frac{x^2}{x^2} - \frac{1}{x^2}}$$

$$y = \frac{4 + \frac{3}{x^2}}{1 - \frac{1}{x^2}}$$

Step 2: For large values of x, $\frac{3}{x^2} \approx 0$, $\frac{1}{x^2} \approx 0$, and

$$y \approx \frac{4 + 0}{1 - 0}$$

$$y \approx \frac{4}{1}$$

$$y \approx 4$$

The equation of the horizontal asymptote is the line $y = 4$. (We obtain the same result as by the first method.)

Example 5 Find the horizontal asymptotes of

$$y = \frac{\sqrt{x+6}}{x-1}$$

Solution We use Method 2

Step 1: Divide every term by x (That is, divide both the numerator and the denominator by x)

$$y = \frac{\frac{1}{x}\sqrt{x+6}}{\frac{1}{x}(x-1)}$$ (Multiplying by $\frac{1}{x}$ is equivalent to dividing by x)

$$= \frac{\sqrt{\frac{1}{x^2}(x+6)}}{\frac{1}{x}(x-1)}$$ (**Note:** We square $\frac{1}{x}$ before writing it as a factor of the radicand)

$$= \frac{\sqrt{\left(\frac{1}{x}+\frac{6}{x^2}\right)}}{1-\frac{1}{x}}$$

Step 2   For large values of x, $\frac{1}{x} \approx 0$, $\frac{6}{x^2} \approx 0$, and

$$y = \frac{\sqrt{0+0}}{1-0}$$
$$y = 0$$

The equation of the horizontal asymptote is $y = 0$. (The x-axis)

How to determine the equations of oblique (slant) asymptotes

The equation of an oblique asymptote is of the form $y = mx + b$, where m is not zero, and b is any real number. There are a number of methods for finding these equations. However, we shall cover only one method which involves long division.

Example Find the equation of the oblique asymptote of the curve given by
$$y(x-1) = x(x+1)$$

Solution

Step 1: Solve for y.

$$y(x-1) = x(x+1)$$

$$y = \frac{x(x+1)}{x-1}$$

$$y = \frac{x^2+x}{x-1}$$

Step 2: On the right-hand side, divide the numerator by the denominator, using long division to obtain

$$y = x + 2 + \frac{2}{x-1}$$

The polynomial partial quotient $x+2$ is the equation of the oblique asymptote.
Therefore, the equation of the oblique asymptote to the given curve is the line $y = x + 2$.

Non-existence of Vertical and Horizontal Asymptotes

If on solving an equation for x and y to determine vertical or horizontal asymptotes, we obtain non-real or undefined values, then there are no vertical or horizontal asymptotes .

The following generalization about horizontal and oblique asymptotes will be useful: for checking the methods covered

Case 1. If the degree of the numerator polynomial **equals** the degree of the denominator polynomial then the equation of the horizontal asymptote is

$$y = \frac{\text{coefficient of the leading term in the numerator}}{\text{coefficient of the leading term in the denominator}}$$

Case 2. If the degree of the numerator polynomial is **less than** the degree of the denominator polynomial, then the equation of the horizontal asymptote is the line $y = 0$.

Case 3. If the degree of the numerator polynomial **is greater** than the degree of the denominator polynomial then there are no horizontal asymptotes.

Case 4: If the degree of the numerator polynomial **equals** 1 plus the degree of the denominator polynomial, then there are no horizontal asymptotes; but there are **oblique asymptotes**.

Case 5. If the degree of the numerator polynomial is greater than 1 plus the degree of the denominator polynomial, then there are no horizontal asymptotes, and no oblique asymptotes.

Asymptotic Formula for a Given Equation or Expression

The asymptotic formula is the formula which approximates the exact formula when the independent variable (argument) becomes very large (increases indefinitely).

Example 1 Given the exact formula $y = x^2 + x + \frac{5}{x}$. Find the asymptotic formula for y.

Solution For large values of x, $\frac{5}{x} \approx 0$, and the equation becomes

$$y = x^2 + x + 0$$
$$y = x^2 + x$$

The asymptotic formula is $y = x^2 + x$

Example 2 Find the asymptotic expansion of the expression given by $(2x + \frac{1}{x})^2$

Solution   Let $y = (2x + \frac{1}{x})^2$

Then $y = 4x^2 + 4 + \frac{1}{x^2}$

For large values of x, $\frac{1}{x^2} \approx 0$, and

$$y = 4x^2 + 4 + 0$$
$$y = 4x^2 + 4 + 0$$

The asymptotic expansion of $(2x + \frac{1}{x})^2$ is $4x^2 + 4$

How to Draw a Curve and its Asymptote

Example 1 Let us consider the graph of a function (Figure 1) with the following properties:

1. The point A (-7, 0) is an x-intercept of the curve. This point is the nearest known x-intercept on this side of the vertical asymptote, $x = -2$.

2. The function is positive on the interval from $x = -7$ to $x = -2$. (The curve is above the x-axis) We draw this branch curve so that the curve smoothly approaches its asymptote (Figure   )

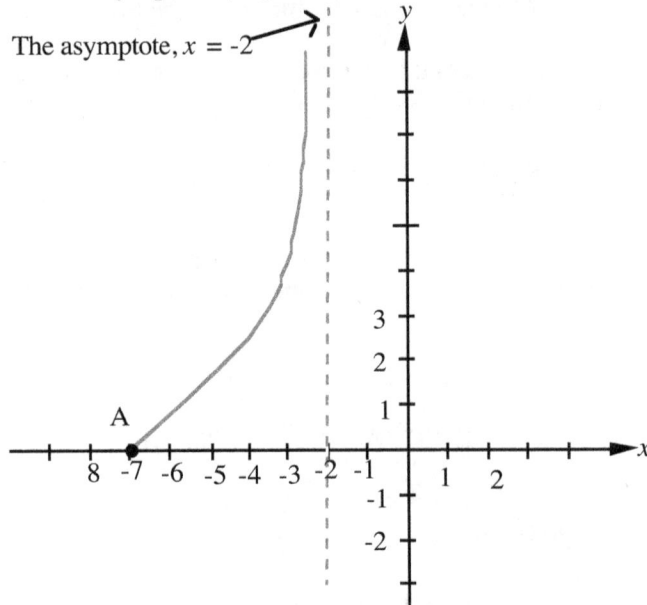

Figure 1: The function is positive between -7 and -2 and has a vertical asymptote at $x = -2$.

Example 2 In Example 1, all the properties remain unchanged except that the curve is below the x-axis (i.e., the function is negative) on the interval from -7 to 2. The graph is shown below.

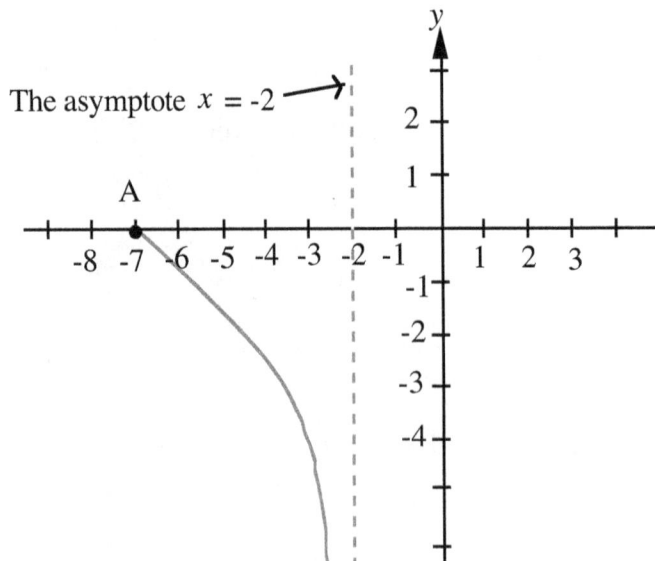

Figure: The function is negative between -7 and -2.

Example 3 Suppose that it is known that A is a point on a curve (either given or calculated); and
that between D and C (Figure) the function is positive; between C and B, the
function is negative. Also, suppose that the curve has a vertical asymptote at $x = -2$,
and an x-intercept at C.
Noting that we are trying to draw the sketch as continuously as permissible, we connect
the points and draw the curve so as to approach its asymptote gradually (Figure)

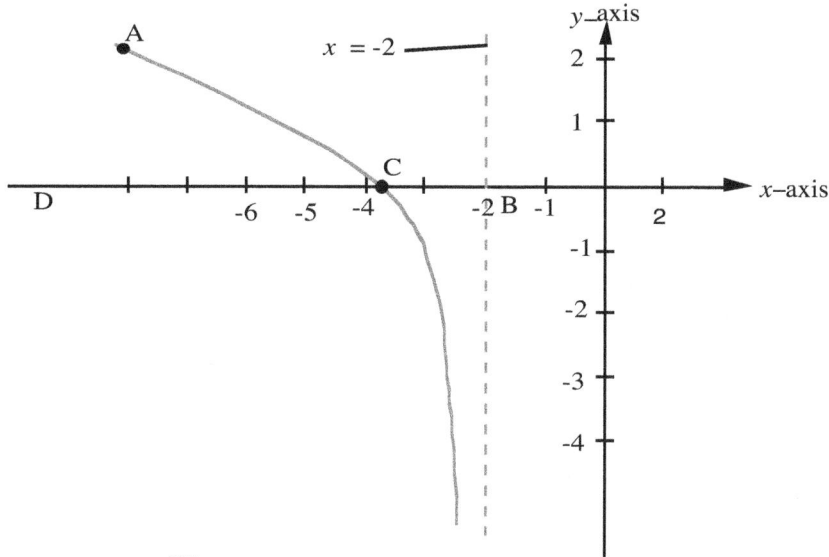

Figure

Example 4 : Everything remains the same as in Example 3 except that there is no x-intercept between A_1
and B; and that the curve is above the x-axis (i.e., the function is positive) between D and B.
The sketch is shown in Figure below.

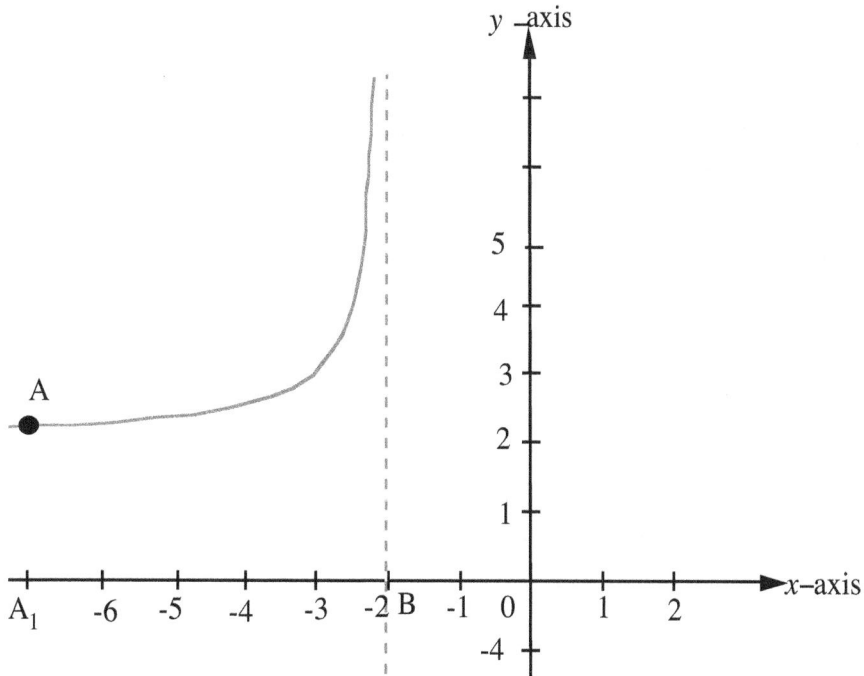

Figure:

Lesson 42: Asymptotes

Example 5 A curve has the following properties: x-intercept at $x = 7$; and no other x-intercept 280
between $x = 3$ and $x = 7$; and a vertical asymptote at $x = 3$. We sketch the curve as shown below.

Example 6 The graph of a function has the following properties:
x-intercepts at $x = -3, x = 0, x = 4$; vertical asymptotes at $x = -1, x = 2$. Curve is above the x-axis between
$x = -3$ and $x = -1$; and between $x = 0$ and $x = 2$. Curve is below the x-axis between
$x = -1$ and $x = 0$; and between $x = 2$ and $x = 3$.

Step 1: Plot the x-intercepts.

Step 2: Draw the vertical asymptotes using dotted or broke lines.

Step 3: Using the experience gained in the previous examples, we draw the various branches of
the curve, noting that the curves are to be drawn through the intercepts and also that the
curves are to be asymptotic to the asymptotes on the respective intervals (Figure below).

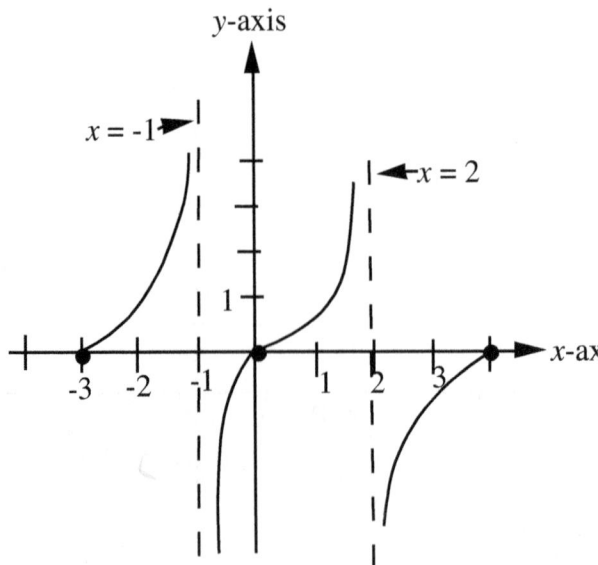

Lesson 42: Asymptotes

Example 7 The graph of a function has the following properties:

281

1. x-intercepts at $x = 0$, $x = 5$;   **2.** Vertical asymptote at $x = 3$;   **3.** Horizontal asymptote at $y = 4$

4. The function is positive between $-\infty$ and 0; and between $x = 5$ and $+\infty$.

5. The function is negative between $x = 0$ and $x = 3$; and between $x = 3$ and $x = 5$.

6. The function is increasing to the right of $x = 3$; **7.** The function is decreasing to the left of $x = 3$.

Solution

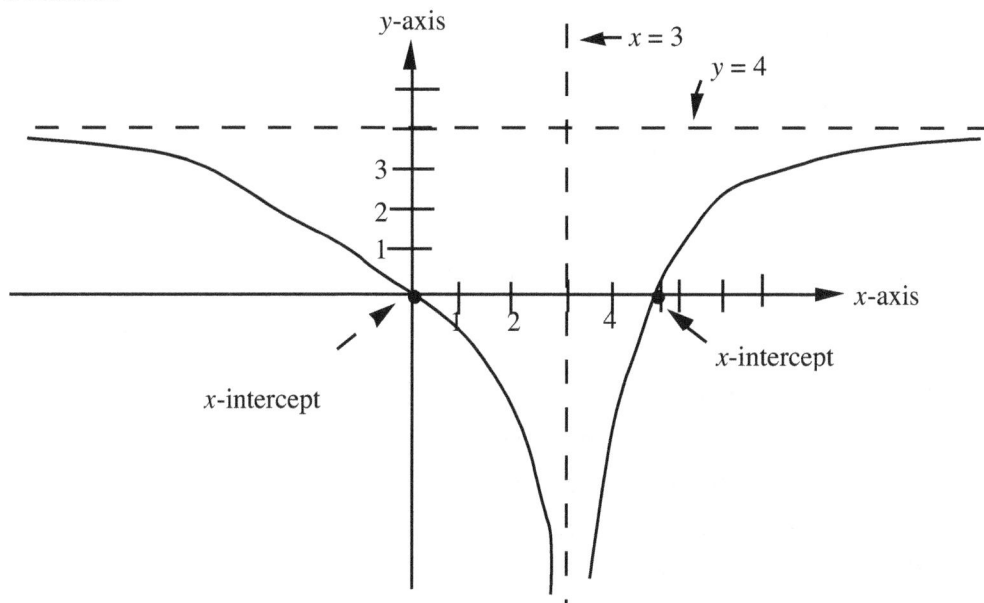

Importance of Devoting Time to Study Asymptotes

Let us justify why we have devoted so much effort and time to study asymptotes.

We do this by presenting examples of the different types of functions or relations that are covered in this book and all of which involve asymptotes.

A **Rational Functions** (Chapter..)

    1. $f(x) = \dfrac{1}{x}$;   **2.** $f(x) = \dfrac{3x-1}{x-4}$;   **3.** $f(x) = \dfrac{5x^2}{(x-1)(x+2)(x+3)}$;   **4.** $f(x) = \dfrac{1}{x^2}$.

B   **Exponential Functions** (Chapter...)

    1. $f(x) = e^x$;   **2.** $f(x) = 2^x$;   **3.** $f(x) = e^x + 3$

C **Logarithmic Functions** (Chapter...)

    1**1.** $f(x) = \log x$;   **2.** $f(x) = \log(x+2)$

D **Trigonometric Functions**   (Chapter...)

    1. $f(x) = \sec x$;   **2.** $f(x) = \csc x$;   **3.** $f(x) = \tan x$;   **4.** $f(x) = \cot x$

E **Others** which are not functions but relations are the hyperbolas (chapter...)

    1. $\dfrac{x^2}{a^2} - \dfrac{y^2}{b^2} = 1$;   **2.** $\dfrac{y^2}{a^2} - \dfrac{x^2}{b^2} = 1$.

We should note that, in the main, it is the asymptotic behavior of these functions which make them different, graphically, from the polynomial and continuous trigonometric functions..

Lesson 42 Exercises

A What is an asymptote to a given curve?

B Find the equations of the vertical and horizontal asymptotes (if any) for each of the following:

1. $y = \dfrac{3x+1}{x+2}$;   2. $y = \dfrac{x+1}{4-x}$;   3. $y = \dfrac{5}{x+1}$;   4. $y = \dfrac{3}{x^2+2}$;   5. $y = \dfrac{4x^2+5}{x^2-9}$

6. $y = \dfrac{5x^2}{x^2-6x}$;   7. $y = \dfrac{2+x^2}{x^2-8x+15}$

8. Find the equation of the oblique asymptote to $y(x-2) = x(x+2)$

9. Given that $A = x^3 + \dfrac{1}{x^2+x+12}$, find the asymptotic formula for A.

10. Find the asymptotic expansion for $(x-\dfrac{1}{x})^2$

Answers: B 1. Vertical asymptote, $x = -2$, horizontal asymptote, $y = 3$.

2. Vertical asymptote, $x = 4$; horizontal asymptote, $y = -1$;

3. Vertical asymptote, $x = -1$; horizontal asymptote, $y = 0$.

4. No Vertical asymptotes, horizontal asymptote, $y = 0$.

5. Vertical asymptotes, $x = -3, x = 3$; horizontal asymptote, $y = 4$.

6. Vertical asymptotes, $x = 0, x = 6$; horizontal asymptote, $y = 5$.

7. Vertical asymptotes, $x = 3, x = 5$; horizontal asymptote, $y = 1$.

8. $y = x+4$;   9. $A = x^3$;   10. $x^2 - 2$

Lesson 43 283

From the Graphs of Polynomial Functions to the Graphs of Rational Functions

A **rational function** is a function which is the ratio of two polynomial functions.

Examples

1. $f(x) = \frac{1}{x}$;   **2.** $f(x) = \frac{6}{x-2}$;   **3.** $f(x) = \frac{3x}{x^2 - 9}$;

4. $f(x) = \frac{1}{(x-2)(x+1)(x+3)}$;   **5.** $f(x) = \frac{x^3 + 5x^2 + 3x - 1}{x^2 - 2x - 1}$

We have already learned how to sketch the graphs of polynomial functions. We have also learned how to find equations of asymptotes as well sketch graphs with asymptotes.
We shall now learn how to sketch the graphs of rational functions.

Graphs of $f(x) = x^n$ and $f(x) = x^{-n}$

The graphs of $f(x) = x^n$ and $f(x) = x^{-n}$ are in the same quadrants, except that the graphs of $f(x) = x^{-n}$ are hyperbolic (see Chapter 23B, Lesson 63) in shape or in character.

Example: The graphs of $f(x) = x$ and $f(x) = x^{-1} = \frac{1}{x}$ are in the same quadrants except that the graph of $f(x) = x^{-1}$ is hyperbolic in shape (Figure 2, below).

Graphs of the Reciprocals of Some Simple Continuous Functions

The author has suggested the following descriptions for relationships between some simple continuous functions and their reciprocals.

For the function $f(x) = \frac{1}{x}$ (the reciprocal of $f(x) = x$)

Whenever the originally continuous line, $y = x$ (Figures) becomes infinitely discontinuous at a point, the line breaks up into two pieces (at this point); a vertical asymptote and a horizontal asymptote are formed at this point; and each piece bends and orientates itself such that one end smoothly becomes asymptotic to the x–axis (the horizontal asymptote) and the other end becomes asymptotic to the y–axis (the vertical asymptote). Similar behavior can be described for

$$f(x) = \frac{1}{mx + b}$$ (the reciprocal of $f(x) = mx + b$)

When $n \geq 2$ for $f(x) = \frac{1}{x^n}$ (e.g., $f(x) = \frac{1}{x^2}$) and reciprocals of trigonometric functions

Whenever a given curve becomes infinitely discontinuous at a point (Figures), the given curve breaks up into two pieces at this point; a vertical asymptote and a horizontal asymptote are formed at this point, and each piece reverses its concavity and orientates itself such that the end of each piece smoothly becomes asymptotic to the asymptotes so formed.

The above described behavior can be applied to the reciprocals of some continuous functions such as the reciprocals of polynomial functions, the reciprocals of some trigonometric functions such as the reciprocals of the sine, cosine and tangent functions. Generally, with the reciprocals of trigonometric functions, only vertical asymptotes are formed at the discontinuous points

Lesson 43: Graphs of Rational Functions

Given the graph of a typical polynomial function, we can readily and by inspection sketch the graph of its reciprocal. Examples of the above behavior are presented below. (Figures)

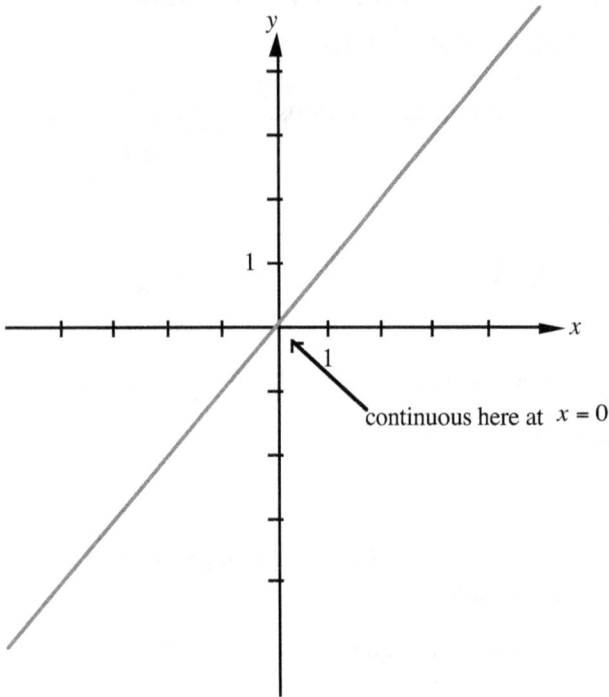

(a) **Figure 1:** The graph of $y = x$ (Polynomial function)

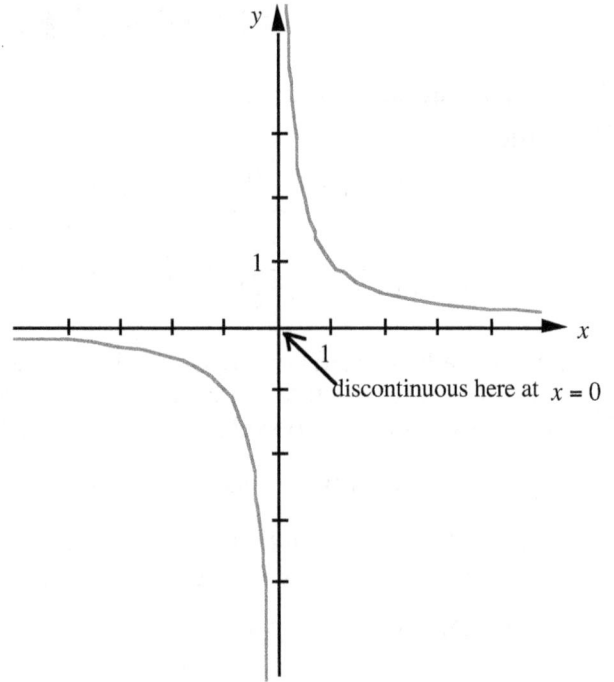

(b) **Figure 2: The** graph of $y = \frac{1}{x}$ (Rational function)

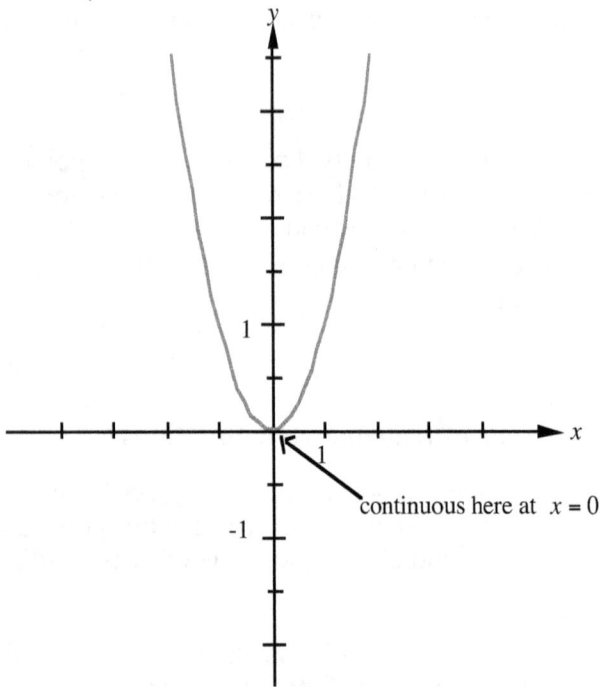

(c) The graph of $y = x^2$ (Polynomial function)

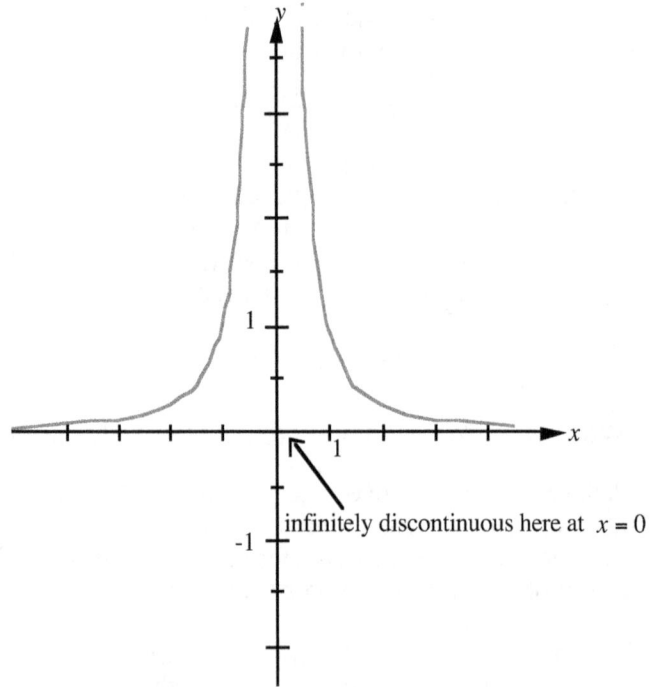

(d) The graph of $y = \frac{1}{x^2}$ (Rational function)

continuous here at $x = 0$

discontinuous here at $x = 0$

continuous here at $x = 0$

infinitely discontinuous here at $x = 0$

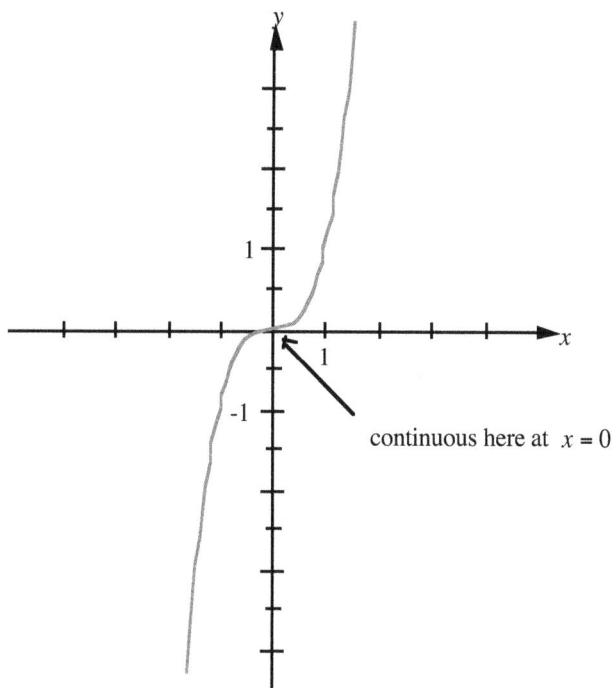

(e) The graph of $y = x^3$ (Polynomial function) .

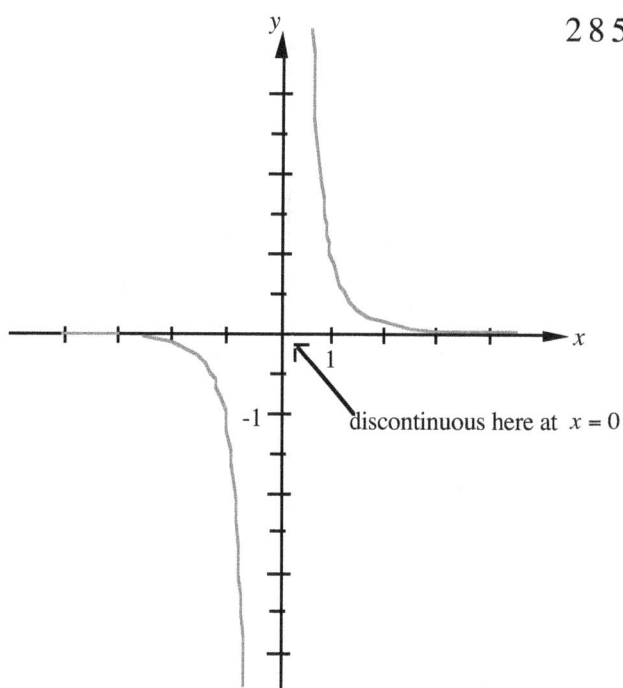

(f) The graph of $y = \dfrac{1}{x^3}$ (Rational function)

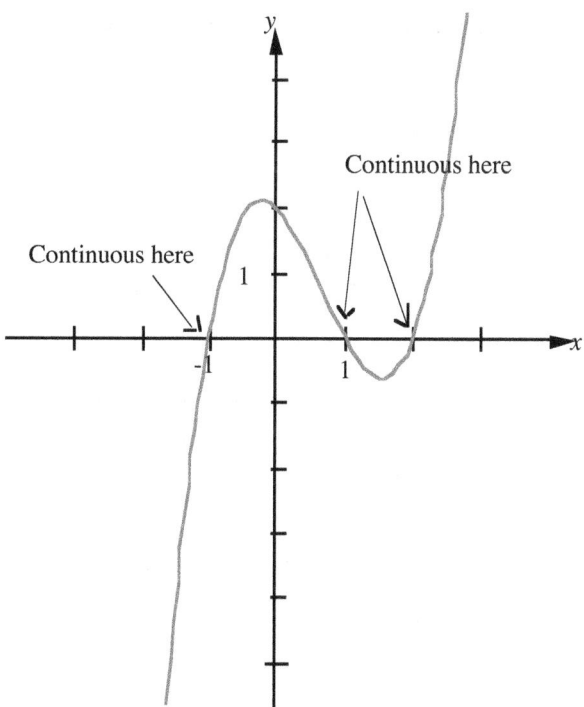

(g) The graph of $y = (x-1)(x-2)(x+1)$.

(Polynomial function)

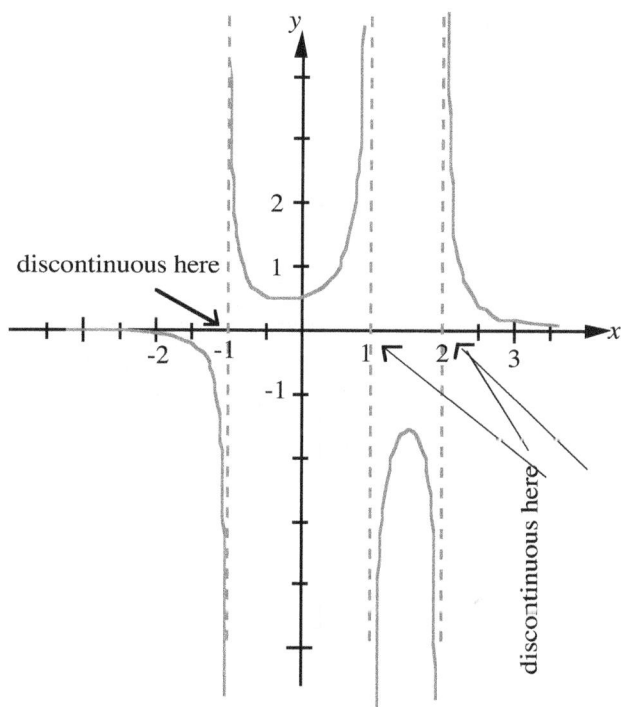

(h) The graph of $y = \dfrac{1}{(x-1)(x-2)(x+1)}$

(Rational function)

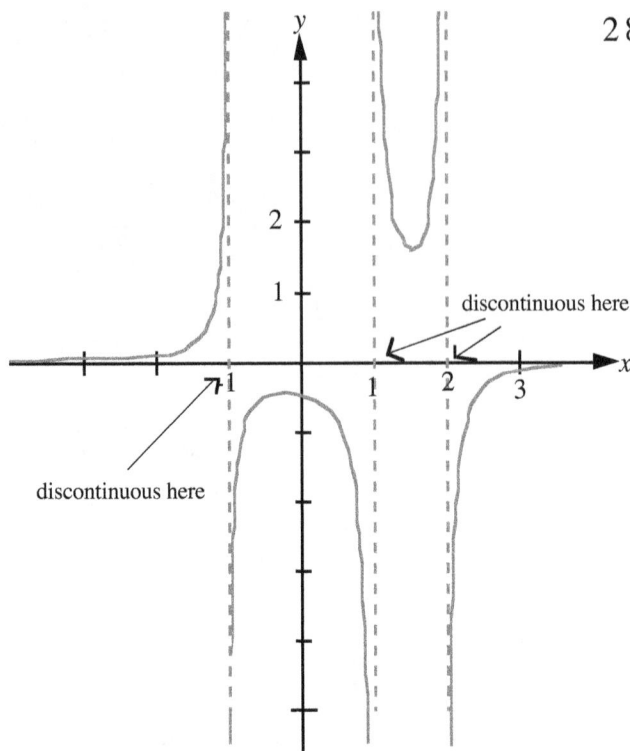

(i) The graph of $y = -(x-1)(x-2)(x+1)$.

(Polynomial function)

(j) The graph of $y = -\dfrac{1}{(x-1)(x-2)(x+1)}$

(Rational function)

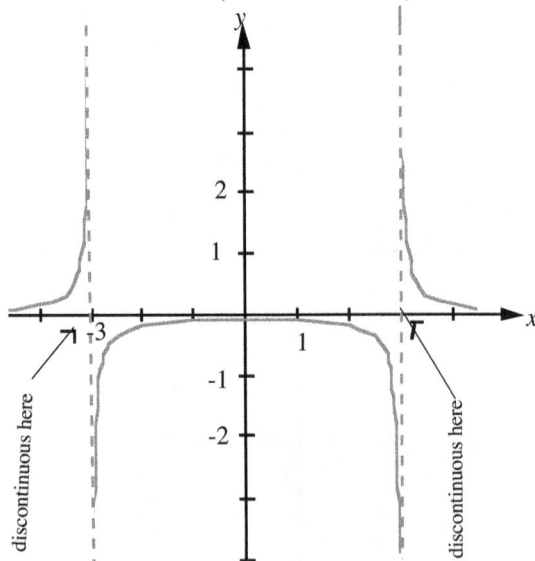

(k) The graph of $y = (x+3)(x-3)$.

Polynomial function)

(l) The graph of $y = \dfrac{1}{(x+3)(x-3)}$

(Rational function)

The following two examples from trigonometry are **not** polynomial or rational functions but exhibit similar behavior. See Chapter 36).

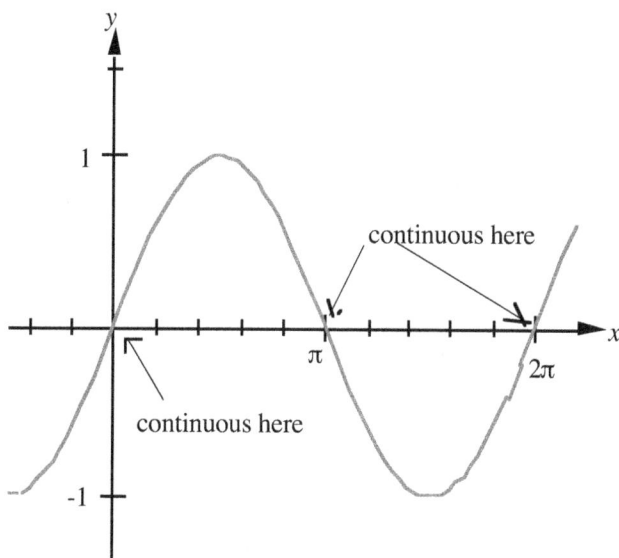

(m) The graph of $y = \sin x$.

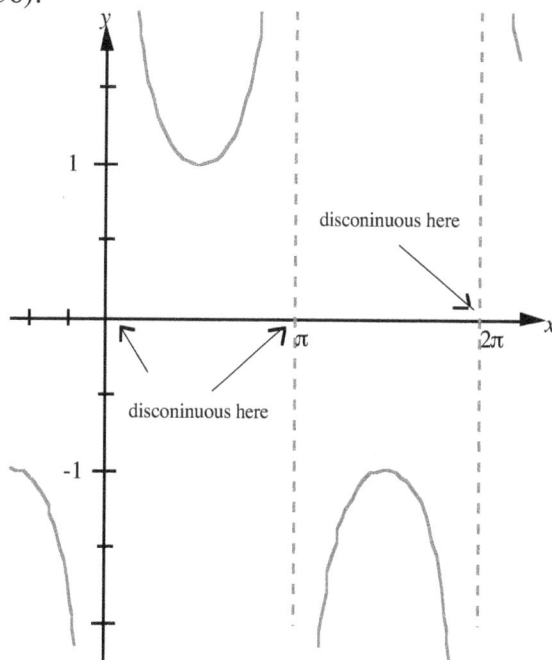

(n) The graph of $y = \dfrac{1}{\sin x} = \csc x$

Sketching the Graphs of $f(x) = \dfrac{1}{x}$ and $f(x) = \dfrac{1}{x-h}$

The function given by $f(x) = \dfrac{1}{x}$ is not defined when $x = 0$.

Let $y = f(x)$. Then, $y = \dfrac{1}{x}$ and $x = \dfrac{1}{y}$. Similarly, $x = \dfrac{1}{y}$ is not defined when $y = 0$.

Thus, the lines $x = 0$ (the y-axis) and $y = 0$ (the x-axis) are vertical and horizontal asymptotes to $y = \dfrac{1}{x}$ and $x = \dfrac{1}{y}$, respectively (Figures)

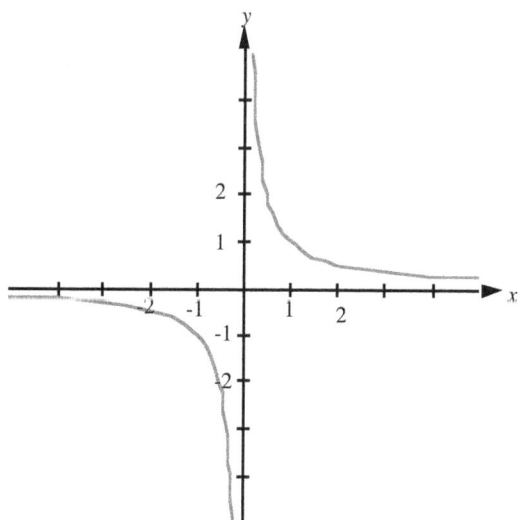

Figure : The graph of $y = \dfrac{1}{x}$

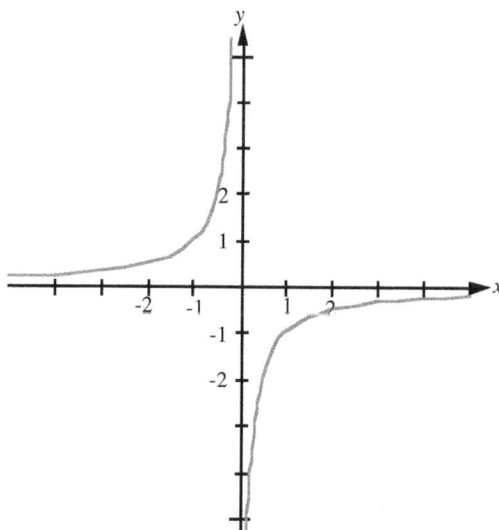

Figure: Graph of $y = -\dfrac{1}{x}$

Lesson 43: Graphs of Rational Functions

Example 1 Sketch the graphs of $y = -\frac{1}{x}, y = \frac{2}{x}$; $y = \frac{1}{x-3}$; and $\frac{2x}{x-3}$ from the graph of $y = \frac{1}{x}$ 288

Solution Reflect the graph of $y = \frac{1}{x}$ about the x-axis. (Figure)

Similarly, the graphs of $y = \frac{2}{x}$, $y = \frac{1}{x-3}$, and $\frac{2x}{x-3}$ are shown in Figure.

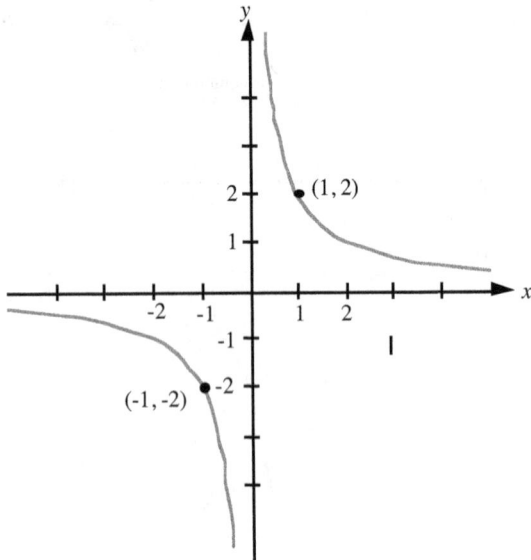

Figure The graph of $y = \frac{2}{x}$

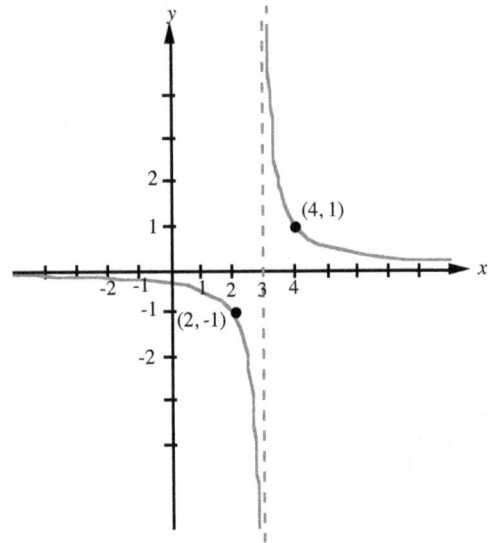

Figure : The graph of $y = \frac{1}{x-3}$

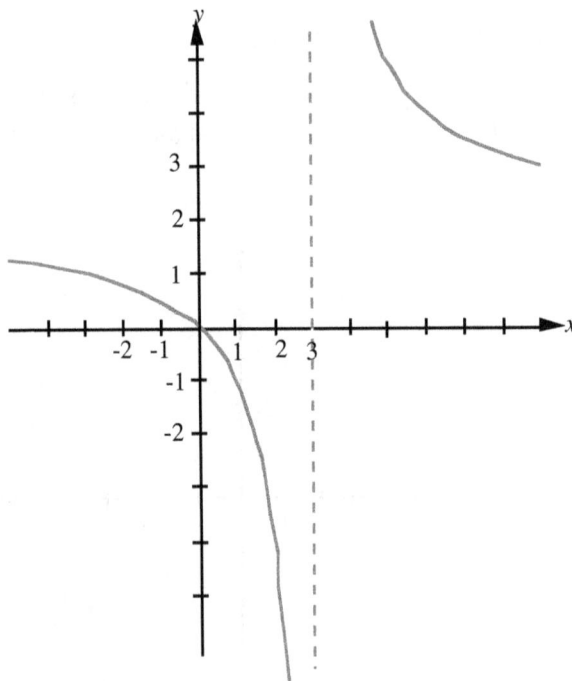

Figure :The graph of $\frac{2x}{x-3}$

General Method of Sketching the Graphs of Rational Functions   289
(Another method)

Before we learn the general method of sketching the graphs of rational functions, we will discuss two types of rational functions with the corresponding discontinuities.

Reducible and Irreducible Rational Functions

Consider the reducible rational function given by

$$f(x) = \frac{(x+1)(x+2)}{x+1}.$$

If we reduce immediately without examining the denominator to note that $x \neq -1$, we will obtain

$$f(x) = x+2.$$

Now, if we sketch the graph of $f(x) = x+2$, we might forget to indicate the discontinuity at $x = -1$ (Figures). Therefore, given a reducible rational function, we must always examine, note and record the excluded values (if any) before proceeding to reduce the fraction, otherwise, we may "lose track" of the excluded values.

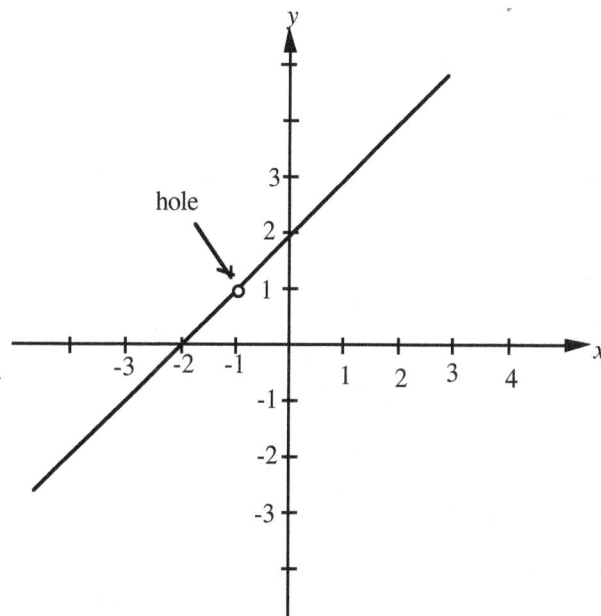

Figure: Graph of $y = x+2$ **Figure**: Graph of $y = \dfrac{(x+1)(x+2)}{x+1}$ (a hole at $x = -1$)

Now, we shall cover properties of examples on irreducible and reducible functions without graphing.

Example 1 $f(x) = \dfrac{1}{x-2}$.

This fraction is proper and irreducible. The graph has an infinite discontinuity and a vertical asymptote at $x = -2$.

Example 2 $f(x) = \dfrac{1}{(x+3(x-5)}$.

This fraction is proper and irreducible. The graph has infinite discontinuities and vertical asymptotes at $x = -3, x = 5$.

Example 3 $f(x) = \dfrac{x+1}{(x+1)(x-2)}$ (reducible)

This fraction is proper but reducible. The common factor is $(x + 1)$. The graph has a finite discontinuity and a hole at $x = -1$. It also has an infinite discontinuity and a vertical asymptote at $x = 2$; and a horizontal asymptote at $y = 0$.

Example 4 $f(x) = \dfrac{x^2+5}{x+1}$

This fraction is improper and irreducible. It has an infinite discontinuity and a vertical asymptote at $x = -1$. By the long division process, the function can be expressed as the sum of a polynomial quotient and a proper, irreducible fraction. In this case, the polynomial quotient is the equation of the oblique asymptote.

$\dfrac{x^2+5}{x+1} = x - 1 + \dfrac{6}{x+1}$. The equation of the oblique asymptote is $y = x - 1$.

Example 5 $f(x) = \dfrac{(x+3)(x-5)}{(x-5)}$

This fraction is improper and reducible. The common factor $(x - 5)$ yields a finite discontinuity (a hole) at $x = 5$. The graph is a straight line ($y = x + 3$) with a hole at $x = 5$. There is **no** vertical asymptote.

Example 6 $f(x) = \dfrac{x}{x}$

The fraction is improper and reducible. The graph is that of the line $y = 1$, with a finite discontinuity (a hole) at $x = 0$.   (Note: On reducing, $\dfrac{x}{x} = 1$)

Example 7 $f(x) = \dfrac{x+2}{x+2}$

The fraction is improper and reducible. The graph is that of the line $y = 1$ with a finite discontinuity (a hole at $x = -2$) .   (Note: On reducing, $\dfrac{x+2}{x+2} = 1$)

We should note that **not** all rational functions have discontinuities and consequently not all rational functions have vertical asymptotes or holes in their graphs.

For example, the function $f(x) = \dfrac{x}{x^2+1}$ is defined for all real values of, since the square of any (non-zero) real number is positive. Moreover, if we attempt to find vertical asymptotes by setting the denominator to zero, we shall obtain non-real solutions. However, the graph of this function has a horizontal asymptote given by the line $y = 0$.

How to sketch the Graphs of Proper Irreducible Rational Functions   291

This is a direct method.

Step 1: Find the x-intercepts by setting $y = 0$, or by setting the numerator to zero and solving for x. If there are no real solutions, then there are no x-intercepts.

Step 2: Find the y-intercept by setting $x = 0$ and solving for y. If the value of y obtained is undefined or non-real, then there is no y-intercept.

Step 3: Find the vertical asymptotes by setting the denominator polynomial to zero and solving for x. If there are no real solutions, then there are no vertical asymptotes.

Step 4: Find the horizontal asymptotes. (See page 274) If there are no real and defined values, then there are no horizontal asymptotes.

Step 5: Using broken or dotted lines, draw the vertical and the horizontal asymptotes as determined in Steps 3 and 4. Plot the x- and y-intercepts from Steps 1 and 2.

Step 6: Using a sign diagram (See page 177), determine the signs of the function on each side of the vertical asymptotes and also the signs of the function between the x-intercepts (or zeros) as was done for the factorable polynomials (page 266) . (That is, determine whether the curve is above or below the x-axis on each side of the vertical asymptotes and between the intervals created by the x-intercepts.) Instead of using sign diagrams, we can use convenient points as was done for factorable polynomials (see page 266).

Step 7: Additional or convenient points may be plotted by choosing convenient x-values and calculating the corresponding y-values, to obtain ordered pairs.

Step 8: Connect the points within each interval by a smooth solid curve noting that the curve does not meet its asymptote but rather approaches it gradually and smoothly as the curve and the asymptote are extended indefinitely (Review examples on page 278)

Example Sketch the graph of $y = \dfrac{3x}{x^2 - 9}$

Step 1: To find the x-intercepts, we set the numerator (or y) to zero.
`Then $3x = 0$ and from which $x = 0$.
The x-intercept $= 0$

Step 2: To find the y-intercept, we set $x = 0$.

Then $y = \dfrac{3(0)}{0 - 9}$

$y = 0$

Therefore, the y-intercept $= 0$

Step 3: To find the vertical asymptotes we set the denominator polynomial to zero and solve for x.
$x^2 - 9 = 0$

$(x + 3)(x - 3) - 0$

Solving, $x = -3$, $x = 3$. (We could use the quadratic formula if the factors are not easily recognizable)

There are vertical asymptotes at $x = -3$ and at $x = 3$.

Step 4: To find the horizontal asymptotes, we can use the method of limits (page 274).

$$y = \dfrac{\dfrac{3x}{x^2}}{\dfrac{x^2}{x^2} - \dfrac{9}{x^2}}$$

$$y = \dfrac{\dfrac{3}{x}}{1 - \dfrac{9}{x^2}}$$

For large values of x, $y \approx \dfrac{0}{1-0} \approx \dfrac{0}{1}$ (Noting that $\dfrac{3}{x} \approx 0, \dfrac{1}{x^2} \approx 0$)

$$y \approx 0$$

Therefore the line $y = 0$ (the x-axis) is a horizontal asymptote.

Step 5: We shall consider the behavior of the function on the intervals
from $-\infty$ to $x = -3$; from $x = -3$ to $x = 0$; from $x = 0$ to $x = 3$; and from $x = 3$ to $+\infty$. We shall use sign a diagram (Figure) to determine the signs of the function on the intervals (See Sign diagram). We can alternatively choose convenient x-values, calculate y-values to determine if the graph is above or below he x-axis on each interval.

Step 6: With the help of the sign diagram, we draw curves (Figure) within each interval taking note of the hyperbolic nature of the curve. We can plot additional convenient points, especially around the turning points.

	Column 1	Column 2	Column 3	Column 4
Number line: $-\infty$ (-3, 0, 3, $+\infty$)				
Factor	Signs of the intervals			
Row 1 $3x$	$-$	$-$	$+$	$+$
Row 2 $x - 3$	$-$	$-$	$-$	$+$
Row 3 $x + 3$	$-$	$+$	$+$	$+$
Row 4 $\dfrac{3x}{(x-3)(x+3)}$	$-$	$+$	$-$	$+$

Sign diagram for $y = \dfrac{3x}{x^2 - 9}$.

Note the following in sketching the curve.

Between $-\infty$ and -3, the function is negative. (The curve is below the x-axis; Row 4 Column 1
Between -3 and 0, the function is positive.   (The curve is above the x-axis; Row 4 Column 2
Between 0 and 3, the function is negative.   (The curve is below the x-axis; Row 4 Column 3
Between 3 and $+\infty$, the function is positive. (The curve is above the x-axis; Row 4 Column 4

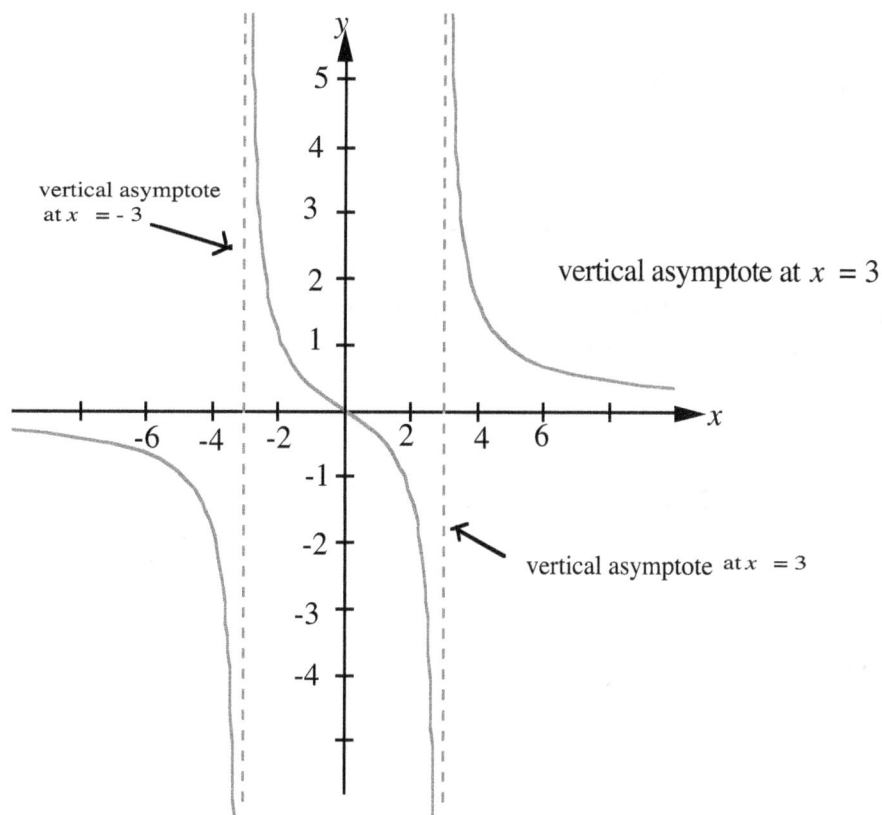

Figure : Graph of $y = \dfrac{3x}{x^2 - 9}$

Note:
Although we have not mentioned (or stressed) anything about the use of oblique asymptotes in sketching the graphs of rational functions, wherever the existence of the oblique asymptote is obvious from the equation, we should not hesitate to determine it and use it to sketch a more accurate graph. An equation has an oblique asymptote if the degree of the numerator polynomial is 1 degree more than the degree of the denominator polynomial.

For example, $f(x) = \dfrac{x^2 + 3}{x + 1}$ has an oblique asymptote which can easily be found by long division (See also page 276)

Lesson 43 Exercises   294

A   Given the graphs of the following, rapidly sketch the graphs of their reciprocals

   1. $y = (x + 1)(x - 3)$;   **2.** $y = (x + 3)(x - 1)(x - 2)$;   **3.** $f(t) = t$

B With examples, discuss the similarities and differences between each the following:
   (a) an improper rational function and an irreducible function.
   (b) a proper rational function and an irreducible function; (c) a reducible
   function and an improper function.

C   Sketch the graphs of the following functions:

   1. $f(x) = \dfrac{3}{x - 4}$;   **2.** $f(x) = \dfrac{3x + 1}{x - 2}$;   **3.** $f(x) = \dfrac{x + 1}{4 - x}$;   **4.** $f(x) = \dfrac{5}{x + 1}$;   **5.** $f(x) = \dfrac{3}{x^2 + 2}$

   6. $f(x) = \dfrac{4x^2 + 3}{x^2 - 9}$;   **7.** $f(x)\dfrac{2x^2}{x^2 - 6x}$;   **8.** $f(x) = \dfrac{2(x - 1)(x - 2)}{x - 2}$;

   9. $f(x) = \dfrac{(x + 3)(x - 1)(x + 2)(x - 2)}{(x + 3)(x - 1)(x + 2)}$.

Answers: C

1.

2.

3.

$$y = \frac{x+1}{4-x}$$

4.

$$y = \frac{5}{x+1}$$

5.

$$y = \frac{3}{x^2+2}$$

$$y = \frac{4x^2+3}{x^2-9}$$

7.

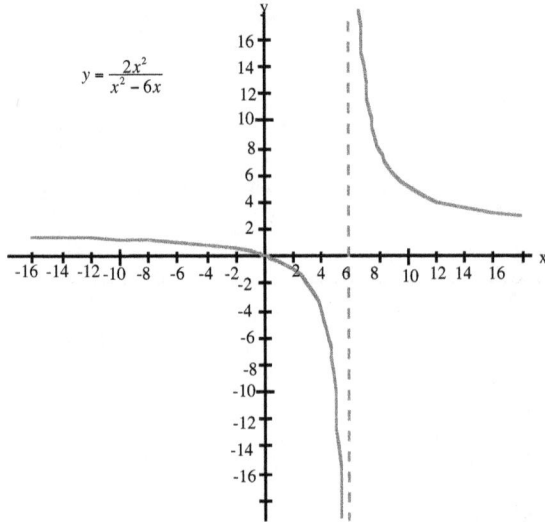

$y = \dfrac{2x^2}{x^2 - 6x}$

8.

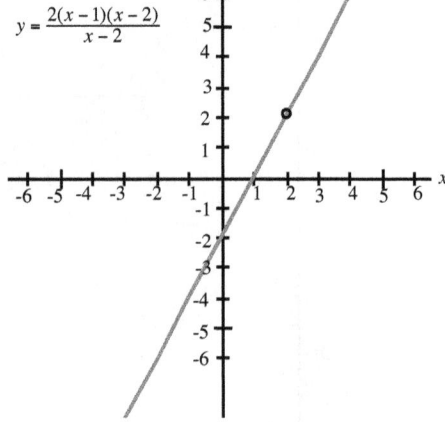

$y = \dfrac{2(x-1)(x-2)}{x-2}$

9.

$y = \dfrac{(x+3)(x-1)(x+2)(x-2)}{(x+3)(x-1)(x+2)}$

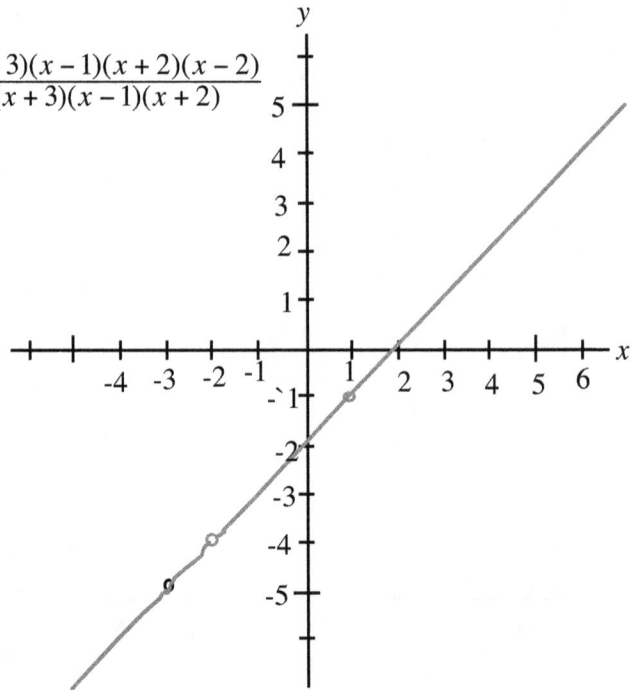

EXTRA

The relative merits (or reconciliation) of the various methods of sketching the graphs of rational functions

The discussion which follows parallels the discussion on quadratic functions (page 256).
In this chapter, we have covered how to relate the graphs of polynomials to the graphs of their reciprocals. This view can be extended to help sketch the graphs of the general factored rational functions by first sketching the corresponding factored polynomial. For instance, we can view the graph of

$$y = \frac{x(x+3)}{(x+1)(x-2)} \text{ as the graph of } y = x(x+3)(x+1)(x-2)$$

in which the linear factors $(x+1)(x-2)$ by being moved into the denominator have brought about discontinuities but still leaving the factors $x, x+3$ in the numerator for the determination of the zeros (x-intercepts) of the rational function. The concepts involved in this method can be extended to the graphs of the continuous trigonometric functions and their corresponding reciprocals. (e.g., $\sin x$ and $\csc x$).

We have also learned how to sketch the graphs of rational equations from the simplest form $y = \frac{1}{x}$ by the transformations of translation, reflection. contraction, and expansion. This method parallels the discussion on $y = x^2$.

A more direct method (p.291) of sketching the graph of any rational function or relation has also been covered.. This method too is in parallelism with the discussion on quadratic functions. This method is the recommended general method especially for more complicated rational fractions.

CHAPTER 17
Logarithms

Lesson 44: **Definitions, Basic Properties and Applications**

Lesson 45: **Evaluation of Logarithms; Common and Natural Logarithms**

Lesson 44
Definitions, Basic Properties and Applications

Definition: The logarithm of a number to a base is the power (or exponent) to which the base must be raised in order to produce that number. A logarithm is therefore an exponent.

For any positive numbers x and b $(b \neq 1)$, and y any real number,

$$\boxed{\log_b x = y \text{ if and only if } b^y = x}$$

(Read " log of x to the base b is y if and only if b to the power y is x.
or log to the base b of x is y if and only if b to the power y is x.)

Note: $\log_b x = y$ is equivalent to $b^y = x$.

Examples: **1.** $\log_{10} 100 = 2$ (because $10^2 = 100$)

           2. $\log_2 16 = 4$ (because $2^4 = 16$)

           3. $\log_{10} \dfrac{1}{100} = -2$ (because $10^{-2} = \dfrac{1}{10^2} = \dfrac{1}{100}$)

If you memorize and understand the definition of the logarithm very well, you will , in some evaluations be able to find the log mentally. For example, to evaluate $\log_2 16$, ask this question: To what power should 2 (the base) be raised in order to produce 16? ; or what is the exponent that must be placed on 2 in order to produce 16?.

Basic Properties of Logarithms

There are three basic rules which are properties of logarithms. Each rule is associated with a fundamental rule of exponents.

Property 1 For any positive numbers M, N, and b $(b \neq 1)$,

$$\log_b (MN) = \log_b M + \log_b N$$

In words, the log of a product equals the sum of the logs of the factors.

Examples   1. $\log_{10} [3(5)] = \log_{10} 3 + \log_{10} 5$.

           2. $\log_{10} [423(785)] = \log_{10} 423 + \log_{10} 785$.

Property 2 $\log_b \dfrac{M}{N} = \log_b M - \log_b N$ $(b \neq 1)$

In words, the log of a fraction equals the log of the numerator minus the log of the denominator.

Example: $\log_5 \dfrac{42}{23} = \log_5 42 - \log_5 23$

Property 3 For any positive numbers M and b and any real number p.

$$\log_b M^P = p \log_b M \qquad (b \neq 1)$$

In words, the log of the pth power of a number equals p times the log of the number.

Example: $\log_2 8^3 = 3 \log_2 8$

Some other useful properties of logarithms

Property 4 $\qquad\qquad \log_b b = 1 \qquad$ (equivalently, $b^1 = b$)

Property 5 $\qquad\qquad \log_b 1 = 0 \qquad$ (equivalently, $b^0 = 1$)

Property 6 $\qquad\qquad b^{\log_b M} = M$

Property 7 $\qquad\quad$ If $M = N$ then

$\qquad\qquad\qquad \log_b M = \log_b N \qquad$ (We may call this step " taking logs of both sides of the equation")

Property 8 $\qquad\quad$ If $\log_b M = \log_b N$, then

$\qquad\qquad\qquad\quad M = N \quad$ (We may call this step "undoing the logs of both sides of the equation";

$\qquad\qquad\qquad\qquad\qquad$ but more formally, " taking antilogs of both sides of the equation)

A very useful property of equality of exponents when the bases are equal:

$$\boxed{\text{If } b^x = b^y \text{ then } x = y}$$

Applications of the above properties:

$$1. \quad C = \frac{t^2 \sqrt[3]{7}}{5} ; \qquad\qquad 2. \quad S = \sqrt{\frac{5A}{\pi R}}$$

Solution

1. $\log c = \log t^2 + \log (7)^{1/3} - \log 5 \quad$ (Applying properties **7, 1, & 2** and changing radical to exponential form)

$\qquad = 2 \log t + \frac{1}{3} \log 7 - \log 5 \quad$ (Applying property **3**)

2. $\qquad S = \sqrt{\frac{5A}{\pi R}}$

$\qquad\quad S = \left(\frac{5A}{\pi R}\right)^{1/2} \qquad\qquad\qquad$ (Changing the radical to exponential form)

$\quad \log S = \log \left(\frac{5A}{\pi R}\right)^{1/2} \qquad\qquad$ (Taking logs)

$\quad \log S = \frac{1}{2} \left(\log \frac{5A}{\pi R}\right) \qquad\qquad$ (Applying property **3**)

$\quad \log S = \frac{1}{2} \left(\log 5 + \log A - \log \pi - \log R\right) \qquad$ (Applying properties **1, & 2**)

Note that $\dfrac{\log_b M}{\log_b N}$ is **not** equal to $\log_b \dfrac{M}{N}$ $\qquad$ ($\log_b \dfrac{M}{N} = \log_b M - \log_b N$)

In the first expression you find the logs first before dividing; in the second expression, you divide first before finding the log.

Antilogarithms

Definition : If $y = \log_a x$, then the antilogarithm (antilog) of y to the base a is x.

The antilog of a given number is, therefore, the number whose logarithm to a given base is the given number. Note that " taking logs" and "taking antilogs" are inverse operations of each other. Each operation reverses the action of the other. Example: $\text{antilog}_2 5 = 2^5 = 32$, since $\log_2 32 = 5$.

Lesson 44 Exercises

Find the following:: **1.** $\log_{10} 100$;   **2.** $\log_2 16$;   **3.** $\log_{10} \dfrac{1}{100}$

4-5: Express log C in terms of the logs of t, 5, and 7; and log S in terms of the logs of A, π, R and 5.

   4. $C = \dfrac{t^2 \sqrt[3]{7}}{5}$;   **5.** $S = \sqrt{\dfrac{5A}{\pi R}}$

Answers: 1. 2;   **2.** 4;   **3.** -2;   **4.** $\log c = 2\log t + \dfrac{1}{3}\log 7 - \log 5$;

   5. $\log S = \dfrac{1}{2}(\log 5 + \log A - \log \pi - \log R)$

Lesson 45

Evaluation of Logarithms; Common and Natural Logarithms

Evaluation of Logarithms

Example 1 Evaluate $\log_2 16$.   (1)

Method 1

Step 1: Use the equivalent definition of logarithms to change (1) to exponential form.

Let $\log_2 16 = y$

Then $2^y = 16$

$2^y = 2^4$ $(16 = 2^4)$

Step 2: Since bases on both sides of the equation are equal, the exponents must be equal.

$\therefore y = 4$

and $\log_2 16 = 4$

Method 2

$\log_2 16 = \log_2 2^4$

$= 4 \log_2 2$ $(\log_2 2 = 1)$

$= 4(1)$

$= 4$

Again, we obtain the same result as by Method 1.

Method 3 Ask yourself this question and answer it : To what power should 2 (the base) be raised in order to produce 16? ; or what is the exponent that must be placed on 2 in order to produce 16?. Of course the answer is 4.

Example 2 Evaluate $\log_3 9^{1/4}$

Method 1

Step 1: Let $\log_3 9^{1/4} = y$.

Then, $3^y = 9^{1/4}$

Step 2: Express the right-hand side as a power of 3.

$3^y = (3^2)^{1/4}$ $(9 = 3^2)$

$3^y = 3^{2/4}$

$3^y = 3^{1/2}$

Step 3: Since the bases are equal, equate the exponents.

then, $y = \dfrac{1}{2}$

$\therefore \log_3 9^{1/4} = \dfrac{1}{2}$

Method 2

$\log_3 9^{1/4} = \log_3 (3^2)^{1/4}$

$= \log_3 3^{2/4}$

$= \log_3 3^{1/2}$

$= \dfrac{1}{2} \log_3 3$ $(\log_3 3 = 1)$

$= \dfrac{1}{2}$

Note: In evaluations and proofs involving logs, whenever you do not know how to proceed, it is good practice to begin by saying " let the given log = say, y, and then apply the equivalent exponential definition and continue.

Common Logarithms

Logarithms to the base 10 are called common logarithms. The base 10 is sometimes omitted. Thus, $\log_{10}$ x may be written as $\log x$. In changing a logarithm to an exponential forms, it helps to indicate the base 10. The common log of a number is the power to which 10 must be raised to produce that number. Thus, the common log of a power of 10 is the exponent on 10.

Examples 1. $\log_{10} 100 = 2$ (because $10^2 = 100$)

     2. $\log_{10} .01 = \log_{10} \dfrac{1}{100} = \log_{10} \dfrac{1}{10^2} = \log_{10} 10^{-2} = -2$ (because $10^{-2} = .01$)

     3. $\log 2 = .3010$ (because $10^{.3010} = 2$)

More evaluations

Example 1: If $\log 2 = .3010$; $\log 3 = .4771$; $\log 7 = .8451$, evaluate the following:

     (a) $\log \dfrac{9}{4}$;   (b) $\log 24$;   (c) $\log 7^{1/2}$.

Solution

Procedure: Express each log in terms of $\log 2, \log 3, \log 7$ or $\log 10$, and simplify.

   (a) $\log \dfrac{9}{4} = \log 9 - \log 4$ $(\log \dfrac{M}{N} = \log M - \log N)$

        $= \log 3^2 - \log 2^2$

        $= 2\log 3 - 2\log 2$ $(\log M^p = p\log M)$

        $= 2(.4771) - 2(.3010)$ (We are given that $\log 3 = .4771$; and $\log 2 = .3010$)

        $= .9542 - .6020$

        $= .3522$

   (b) $\log 24 = \log 8(3)$

        $= \log 8 + \log 3$

        $= \log 2^3 + \log 3$

        $= 3 \log 2 + \log 3$

        $= 3(.3010) + .4771$

        $= .9030 + .4771$

        $= 1.3801$

   (c) $\log 7^{1/2} = \dfrac{1}{2}\log 7$

        $= \dfrac{1}{2}(.8451)$

        $= .4226$

Writing as a single log term

Example   Write as a single log term: $\log 4x + 2\log(x + 1) - \log(x^2 - 1)$

$\log 4x + 2\log(x + 1) - \log(x^2 - 1)$

$= \log 4x + \log(x + 1)^2 - \log(x^2 - 1)$

$= \log \dfrac{4x(x + 1)^2}{x^2 - 1}$

$= \log \dfrac{4x(x + 1)(x + 1)}{(x + 1)(x - 1)}$ (Applying $\log \dfrac{MN}{T} = \log M + \log N - \log T$ backwards)

$= \log \dfrac{4x(x + 1)(\cancel{x + 1})}{\cancel{(x + 1)}(x - 1)}$

$= \log \dfrac{4x(x + 1)}{x - 1}$ or $\log \dfrac{4x^2 + 4x}{x - 1}$

Natural Logarithms

Logarithms to the base e (e $\approx$ 2.72) are called natural logarithms (or Napierian logarithms).

Example The natural log of x is written $\ln x$ or $\log_e x$

Conversions: $\log_b M = \dfrac{\log_c M}{\log_c b}$ <-------**change of base formula** (see also p. 309)

$\ln x = \log_e x \approx 2.3 \log_{10} x$, **or** $\log x = \log_{10} x \approx .43 \log_e x$

Note also: **1.** $\ln 10 \approx 2.3$; **2.** $\log_{10} e = \log e \approx 0.43$. **3.** $\ln 10$ and $\log e$ are reciprocals of each other.

Lesson 45 Exercises

A   1. $\log_3 81$; **2.** $\log_6 6$;   **3.** $\log_2 64$;   **4.** $\log_3 27^2$;   **5.** $\log_5 \dfrac{1}{5}$;   **6.** $\log_2 32$;

   7. $\log_3 81$;   **8.** $\log_3 243$;   **9.** $\log_{10} 1000$;   **10.** $\log_3 \dfrac{1}{27}$;   **11.** $\log 0.001$;   **12.** $\log_{10} 1$

Answers: **1.** 4 ;   **2.** 1 ;   **3.** 6 ;   **4.** 6 ;   **5.** -1 ; **6.** 5 ; **7.** 4 ; **8.** 5 ; **9.** 3 ; **10.** -3 ; **11.** -3 ; **12.** 0

B If $\log 2 = .3010$; $\log 3 = .4771$; $\log 7 = .8451$, evaluate the following:

1. $\log 16$;   **2.** $\log 20$;   **3.** $\log \dfrac{21}{2}$;   **4.** $\log .0072$;   **5.** $\log 50$; **6.** $\log \sqrt{3}$

7. $\log 5^{.01}$;   **8.** $\log \dfrac{1}{600}$;   **9.** $\log \dfrac{1}{2^4}$

Answers: **1.** 1.204 ;   **2.** 1.3010 ;   **3.** 1.0212 ;   **4.** - 2.1428 ;   **5.** 1.699 ;   **6.** 0.2386 ;
          7. 0.007 ;   **8.** -2.7781 ;   **9.** -1.204

C (a) If $f(x) = \log_3 x$, find **1.** $f(9)$; **2.** $f(81)$; **3.** $f(\dfrac{1}{9})$;(b) If $f(x) = \log_4 x$, find **1.** $f(256)$;   **2.** $f(\dfrac{1}{64})$

Answers: **(a) 1.** 2 ;   **2.** 4 ;   **3.** - 2; **(b)**   **1.** 4 ; **2.** -3

D   Write as a single log term:

1. $\log(5x + 10) - \log(x + 2) + \log x$;   **2.** $3\log(x^2 + 7x + 12) - 3\log(x + 3)$;

3. $\log_4 x - \log_4 y + \log_4 z$;   **4.** $2\log_4 5 + 3\log_4 2$;   **5.** $\log(x^2 - 25) - \log(x + 5)$;

Answers: **1.** $\log 5x$; **2.** $\log(x + 4)^3$; **3.** $\log_4 \dfrac{xz}{y}$; **4.** $\log_4 200$; **5.** $\log(x - 5)$

CHAPTER 18

Lesson 46 **Exponential Equations**
Lesson 47: **Logarithmic Equations**

Lesson 46
Exponential Equations

An **exponential equation** is an equation in which the variable appears as an exponent.

Example 1   Solve the exponential equation:

$$15^x = 32$$

Step 1: Take logs of both sides of the equation.

$$\log 15^x = \log 32$$

$$x \log 15 = \log 32 \qquad (\log M^P = p \log M)$$

$$x = \frac{\log 32}{\log 15}$$

$$x = \frac{1.5051}{1.1761} \qquad \text{(From tables or calculator, } \log 32 = 1.5051; \log 15 = 1.1761)$$

$$x = 1.28$$

Example 2   Solve for x: $4^{x+1} = 16$

Method 1	**Method 2:**
$4^{x+1} = (2^2)^{x+1} = 2^4$	$4^{x+1} = 4^2$
$2^{2(x+1)} = 2^4$	$x + 1 = 2$ (Equating exponents, since bases are equal)
$2(x+1) = 4$ (Equating exponents, since bases are equal)	$x = 1$
$2x + 2 = 4$	
$2x = 2$ and from which $x = 1$	**Method 3** Use the method in Example above.

Example 3   Solve for x:

$$7^{x^2 + x} = 34$$

Step 1: Take logs of both sides of the equation:

$$\log 7^{(x^2 + x)} = \log 34$$

$$(x^2 + x)\log 7 = \log 34$$

$$x^2 + x = \frac{\log 34}{\log 7}$$

$$x^2 + x = \frac{1.5315}{.8451} \qquad (\log 34 = 1.5315 \, ; \log 7 = .8451)$$

$$x^2 + x = 1.8122$$

$$x^2 + x - 1.8122 = 0$$

Step 2:   Solve for x by the quadratic formula.

$$x = \frac{-1 \pm \sqrt{1 + 7.2488}}{2} \qquad (a = 1, b = 1, c = -1.8122)$$

$$x = \frac{-1 \pm \sqrt{8.2488}}{2}$$

$$x = \frac{-1 \pm 2.8271}{2}$$

$$x = \frac{-1 + 2.8271}{2} \quad \text{or} \quad x = \frac{-1 - 2.8271}{2}$$

$$x = -1.9360 \text{ or } .9360$$

The solution set is {-1.9360, .9360}

Lesson 46 Exercises

Solve for x: **1**. $12^x = 42$;   **2**. $3^{x+1} = 243$;   **3**. $2^{x^2+1} = 1024$; **4**. $5^{x^2+x} = 42$

Answers: **1**. 1.5041 ;   **2**. 4 ;   **3**. Solution set is {-3, 3} ;   **4**. Solution set is {- 2.1038, 1.1038}

Logarithmic Equations

A **logarithmic equation** is an equation which contains the logarithm of an expression involving the variable. We must check the solutions of logarithmic equations, since the logarithmic function is not defined for negative numbers.

Example 1   Solve for x: $\log(x + 9) - \log x = 1$

Step 1: Write the logarithmic term as a single logarithmic term.

$$\text{Then we obtain } \log_{10}\left(\frac{x + 9}{x}\right) = 1 \qquad \left(\log A - \log B = \log \frac{A}{B}\right)$$

Step 2: $10^1 = \dfrac{x + 9}{x}$ (Applying the equivalent exponential definition)

$$10 = \frac{x + 9}{x}$$

$$10x = x + 9$$

$$x = 1 \qquad (\text{Solving for } x)$$

Checking: Substitute $x = 1$ in the original equation.

$$\log(1 + 9) - \log 1 \overset{?}{=} 1$$

$$\log 10 - \log 1 \overset{?}{=} 1$$

$$1 - 0 = 1 \qquad (\log_{10} 10 = 1)$$

$$1 = 1 \quad \text{True}$$

The solution is 1.

Example 2 Solve for x: $\log x + \log(x - 21) = 2$

Step 1: $\log x + \log(x - 21) = 2$

$\log[x(x - 21)] = 2$ (Writing as a single term)

$\log_{10}[x(x - 21)] = 2$ (Indicating the understood base so as to help transition to exponential form)

$10^2 = x(x - 21)$ (Applying the equivalent exponential definition)

$100 = x^2 - 21x$

Step 2:   Solve the quadratic equation by any method.
We solve by factoring, since the factors are easily recognizable.

$x^2 - 21x - 100 = 0$

$(x + 4)(x - 25) = 0$

$x = -4 \text{ or } x = 25$

Step 3: Since for $x = -4$, $\log(-4) + \log(-4 - 21)$ is not defined (The log of a negative number is not defined)
we reject the solution -4, and check for only 25.

Checking for $x = 25$: $\log 25 + \log(25 - 21) \overset{?}{=} 2$

$$\log 25 + \log 4 \overset{?}{=} 2$$

$$\log 25(4) \overset{?}{=} 2$$

$$\log_{10} 100 \overset{?}{=} 2 \qquad (\log_{10} 100 = 2)$$

$$2 = 2 \quad \text{True} \quad (\text{LHS} = \text{RHS})$$

The solution is 25.

Example 3   Solve for x: $\log \dfrac{3}{x} - 2 = -\log 6x^2$ $(x \neq 0)$

Solution

Step 1: Collect the log terms on the left-hand side.

$$\log \dfrac{3}{x} + \log 6x^2 = 2$$

Step 2: Write the log terms as a single log term.

$$\log \left[\dfrac{3}{x} \cdot 6x^2\right] = 2$$

Step 3: $\log \dfrac{18x^2}{x} = 2$

$\log_{10} 18x = 2$ (simplifying and indicating the base 10)

$10^2 = 18x$ (using the equivalent exponential definition)

$100 = 18x$

$\dfrac{100}{18} = x$, and $x = \dfrac{50}{9}$

After checking (not shown) the solution is $\dfrac{50}{9}$.

Example 4   Solve for x: $\log_x 2 = \dfrac{1}{3}$

Solution

Step 1: Apply the equivalent exponential definition.

If $\log_x 2 = \dfrac{1}{3}$

then $x^{1/3} = 2$

Step 2: $(x^{1/3})^3 = (2)^3$

(Cubing both sides of the equation)

$x = 2^3$

$x = 8$

Checking :

Step 1: $\log_8 2 \overset{?}{=} \dfrac{1}{3}$

Let $\log_8 2 = k$

Then, $8^k = 2$

(Cubing $(2^3)^k = 2$

$2^{3k} = 2^1$

$3k = 1$

$k = \dfrac{1}{3}$

Step 2: $\log_8 2 = \dfrac{1}{3}$

$\dfrac{1}{3} = \dfrac{1}{3}$ True

$\therefore$ the solution is 8.

Equivalent check:

$8^{1/3} \overset{?}{=} 2$

$\sqrt[3]{8} \overset{?}{=} 2$

$2 = 2$ True

Example 5 Solve for x: $\log(x + 2) + \log(x - 5) = \log 8$

Step 1: $\log(x + 2) + \log(x - 5) = \log 8$

$\log(x + 2)(x - 5) = \log 8$ ($\log MN = \log M + \log N$)

$(x + 2)(x - 5) = 8$ (If $\log M = \log N$, then $M = N$, undoing the logs.)

$x^2 - 3x - 10 - 8 = 0$

Step 2: $x^2 - 3x - 18 = 0$ (Solve this quadratic equation)

$(x + 3)(x - 6) = 0$

$x + 3 = 0$ or $x - 6 = 0$

$x = -3$ or $x = 6$

Step 3: We check the values of $x = -3$ and $x = 6$ in the original logarithmic equation.

Checking for $x = -3$: $\log(-3 + 2) + \log(-3 - 5) \overset{?}{=} \log 8$, is not defined , since the logarithm of a negative number is not defined, and we reject the value -3 as a solution.

Checking for $x = 6$: $\log(6 + 2) + \log(6 - 5) \overset{?}{=} \log 8$

$\log 8 + \log 1 \overset{?}{=} \log 8$

$\log 8(1) \overset{?}{=} \log 8$ ($\log MN = \log M + \log N$)   **or** $\log 8 + 0 \overset{?}{=} \log 8$ ($\log 1 = 0$)

$\log 8 \overset{?}{=} \log 8$ True (LHS = RHS) $\log 8 \overset{?}{=} \log 8$

Since $x = 6$ satisfies the original logarithmic equation, the solution is 6.

General discussion about solving exponential and logarithmic equations

1. Given the exponential equation, we (take logs) go through the logarithmic forms to simple algebraic equations (e.g., linear or quadratic equations).

Thus, **exponential** equation---->**logarithmic** equation (and then solve) or ---->**algebraic** equation and then solve. However, sometimes, it may not be necessary to use logarithms. Example: To solve for x, given $2^{x+3} = 8$, we may either do the following : $2^{x+3} = 2^3$; $x + 3 = 3$ (equating the exponents) and from which $x = 0$ or we may do the following: $(x + 3) \log 2 = \log 2^3$

$$x + 3 = \frac{3 \log 2}{\log 2}$$

$$x + 3 = 3 \text{ and from which } x = 0.$$

2. Given the logarithmic equation, we change to the exponential equivalent form, and then to some basic algebraic equations.

Thus, **logarithmic** equation ---->**exponential** equation (and then solve) or---->**algebraic** equation and then solve. The following formulas will be useful, also, in calculus.

Change of base formula for Logarithms	Change of base formula for exponents:
Example Change $\log_b M$ to base c.	(Learn this derivation).
Step 1: Let $\log_b M = y$.	**Example**: Change b^x to base a.
Then, $M = b^y$	Step 1: Let $b^x = a^k$ (1)
(Equivalent exponential definition)	Step 2: Take logs (to base a, the new base) of both sides of the equation.
Step 2: Now, take logs of both sides of the equation and introduce the new base, c.	Then $\log_a b^x = \log_a a^k$
Then, $\log_c M = \log_c b^y$	$x \log_a b = k \log_a a$
Step 3: Solve for y:	$x \log_a b = k$ $(\log_a a = 1)$
$\log_c M = y \log_c b$	and $k = x \log_a b$
$\dfrac{\log_c M}{\log_c b} = y$ (Solving for y)	Step 3: Substitute for k in (1).
Therefore, $\log_b M = \dfrac{\log_c M}{\log_c b}$ <-change of base formula.	Then $\boxed{b^x = a^{x \log_a b}}$.
(Here we changed from base b to base c.)	If $a = e$. $\boxed{b^x = e^{x \log_e b} = e^{x \ln b}}$ (Memorize)

Memory Reminder

Compare the **change of base formulas** for logarithms and exponents, and observe how for the logarithms, the conversion factor, $\ln b$, divides; but for the exponents, $\ln b$ multiplies the exponent.

$$\text{For \textbf{logarithms}: } \log_b x = \frac{\ln x}{\ln b} = \left(\frac{\log_e x}{\log_e b} \right)$$

$$\text{For \textbf{exponents}: } \quad b^x = e^{x \ln b}$$

Some distinctions: Distinguish between the following:

(a) $|\log x|$ and $\log|x|$.

(b) $(\log x)^3$ and $\log(x^3)$.

Solution

(a) For the first expression, we find the log of x before finding the absolute value. For the second expression, we find the absolute value of x before finding the log.

(b) For the first expression, we find the log of x before cubing. For the second expression, we cube x before finding the log.

Lesson 47 Exercises

A Solve for x:

1. $\log(x + 4) - \log x = 1$;   **2.** $\log_2 x + \log_2(x - 7) = 3$;   **3.** $\log(x + 26) = \log(x - 2) + \log 5$

Answers: 1. $\{\frac{4}{9}\}$;   **2.** $\{8\}$;   **3.** 9

B Solve for x: **1.** $7^x = 3$;   **2.** $2^x = 7$;   **3.** $5^{x-2} = 16$;   **4.** $\log x = 4$;

5. $\log(3x + 4) = 2$;   **6.** $\log_2(x - 7) + \log_2 x = 3$;   **7.** $\log_3(x - 2) = 4$;

8. Solve for t: $4^{2t} = \frac{9}{4}$

Answers: **1.** 0.5645 ;   **2.** 2.807 ;   **3.** 3.722 ;   **4.** 10,000 ;   **5.** 32 ; **6.** 8 ; **7.** 83 ;   **8.** 0.2925

CHAPTER 19

Lesson 48: **Graphs of Exponential Functions and Applications of Exponential Functions**
Lesson 49: **Graphs of Logarithmic Functions and Applications of Logarithmic Functions**

Introduction to Exponential and Logarithmic Functions

We use exponential functions in numerical calculations.

1. In physical chemistry, the exponential function is an important function. For example, the exponential equation $C = C_0 e^{-kt}$ relates the decrease in concentration, C, of a reacting material, per unit time, t, where C_0 is the concentration at time $t = 0$. There are similar equations for the decay of a radioactive material.

2. In optics, $I = I_0 e^{-kl}$, where I is the decrease in the intensity of light, l is the depth of the absorbing material.

3. In barometry, $P = P_0 e^{-kh}$, where P_0 is the pressure when h is zero (say at sea level).

4. In physics, the equation of the vibration of a material could be given by $y = Ae^{-kt} \cos 2\pi bt$

5. In chemistry, $\log_e k_p = -\dfrac{\Delta G^0}{RT}$ where k_p is the equilibrium constant, ΔG^0 is the free energy, R is the gas constant, and T is the temperature in degrees Kelvin.

5. In business, we use logarithms, to evaluate the compound interest formula $A = P(1 + \frac{r}{n})^{nt}$ where A is the amount accumulated after t years: P is the principal invested at r % annual interest. rate; and n is the number of times a year the interest is compounded.

Lesson 48
Graphs of Exponential Functions and Applications of Exponential Functions

Let b be a positive constant ($b > 0$), such that b is not equal to 1 ($b \neq 1$). Then the exponential function with a base b is defined by $f(x) = b^x$ (1)
The domain of the exponential function consists of all real numbers and the range consists of all **positive** real numbers.

If we let $f(x) = y,$, then equation (1) above becomes
$$y = b^x \qquad (2)$$
We shall cover two cases of the values of b.

Case 1: $b > 1$ and $y = b^x$. Example: $y = 2^x$

Case 2: $0 < b < 1$ and $y = b^x$. Example: $y = \left(\frac{1}{2}\right)^x$

Lesson 48: Graphs and Applications of Exponential Functions

Case 1: **Properties of** $y = b^x$, $b > 1$

1. The curve crosses the y-axis at $(0, 1)$. That is, the y-intercept $= 1$. (Note: $y = b^0 = 1$).

2. For all values of x, the function is positive (the curve is above the x-axis or all y-values are positive). The function is increasing

3. As x decreases without bound ($x \to -\infty$), y approaches 0 ($y \to 0$.). Thus the negative x-axis is a horizontal asymptote to the curve. Graphically, the curve is concave up to the left.

4. As x increases without bound ($x \to \infty$), the function increases without bound ($y \to \infty$)

5. When $x = 1$, $y = b$.

How to Sketch the Graph of an Exponential Function

Example 1 Sketch the graph of $f(x) = 2^x$.

> Note: Do not be frightened by the type of equation involved in this example. Sketching the graph of an exponential equation is **not** more difficult than sketching the graphs of, say, the line $y = 3x - 5$ or the parabola $y = x^2$. What we need are some points (ordered pairs) on both sides of the y-axis to plot and connect. In this graph, always include $x = 0$ and $x = 1$.

Step 1: Let $f(x) = y$, Then $y = 2^x$.

Calculations   Ordered pairs (points)

⇓   ⇓

Step 2: Find the y-intercept by letting $x = 0$.

$$\left. \begin{array}{l} x = 0, y = 2^0 \\ \quad = 1 \end{array} \right\} \Rightarrow (0,1)$$

Step 3: Choose other convenient x-values (some positive and some negative values) and calculate the corresponding y-values.

$$\left. \begin{array}{l} x = 1, y = 2^1 \\ \quad = 2 \end{array} \right\} \Rightarrow (1,2)$$

$$\left. \begin{array}{l} x = 2, y = 2^2 \\ \quad = 4 \end{array} \right\} \Rightarrow (2,4)$$

$$\left. \begin{array}{l} x = 3, y = 2^3 \\ \quad = 8 \end{array} \right\} \Rightarrow (3,8)$$

$$\left. \begin{array}{l} x = -1, y = 2^{-1} \\ \quad = \dfrac{1}{2} \end{array} \right\} \Rightarrow (-1, \tfrac{1}{2})$$

$$\left. \begin{array}{l} x = -2, y = 2^{-2} \\ \quad = \dfrac{1}{2^2} \\ \quad = \dfrac{1}{4} \end{array} \right\} \Rightarrow (-2, \tfrac{1}{4})$$

Lesson 48: Graphs and Applications of Exponential Functions

Step 4: Plot the points from Steps 2, and 3 (Figure...): That is plot $(0,1)$, $(1,2)$, $(2,4)$, $(3,8)$, $(-1, \frac{1}{2})$ and $(-2, \frac{1}{4})$, noting that as x decreases without bound ($x \to -\infty$), y approaches 0 ($y \to 0$). Thus the negative x-axis is a horizontal asymptote to the curve. Graphically, the curve is concave up to the left. Also, note that as x increases without bound, the function increases without bound.

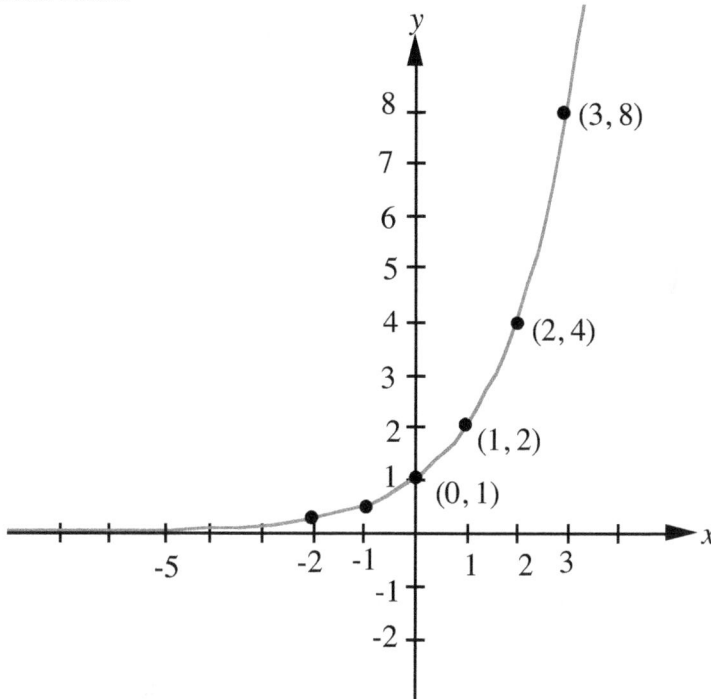

Figure 1: The graph of $y = 2^x$

Case 2: $y = b^x$, $0 < b < 1$

Examples:

1. $y = \left(\frac{1}{2}\right)^x$ or $y = 2^{-x}$ Note: $\left(\frac{1}{2}\right)^x = \left(2^{-1}\right)^x = 2^{-x}$

2. $y = \left(\frac{1}{3}\right)^x$ or $y = 3^{-x}$

Properties of $y = b^x$, $0 < b < 1$

1. The y-intercept is 1. That is, the curve crosses the y-axis at $(0, 1)$.

2. For all values of x the function is positive (All the y-values are positive).

3. The function is decreasing.

4. As x increases without bound ($x \to \infty$), the function approaches zero ($y \to 0$). Thus the positive x-axis is a horizontal asymptote to the curve.

5. As x decreases without bound, ($x \to -\infty$), the function increases without bound ($y \to \infty$).

6. Graphically, the graph is concave up to the right (Figure)

Example 2   Sketch the graph of $f(x) = \left(\dfrac{1}{2}\right)^x$ or $f(x) = 2^{-x}$

Step 1: Let $f(x) = y$, Then $y = \left(\dfrac{1}{2}\right)^x$ or $y = 2^{-x}$

Step 2: Find the y-intercept by letting $x = 0$. $\left.x = 0,\; y = \left(\dfrac{1}{2}\right)^0 = 1\right\} \Rightarrow (0,1)$

Step 3: Choose some other convenient x-values (positive and negative values) ; calculate the corresponding y-values

$\left.x = 1, y = \left(\dfrac{1}{2}\right)^1 = \dfrac{1}{2}\right\} \Rightarrow (1,\tfrac{1}{2})$

$\left.x = 2, y = \left(\dfrac{1}{2}\right)^2 = \dfrac{1}{4}\right\} \Rightarrow (2,\tfrac{1}{4})$

$\left.x = 3, y = \left(\dfrac{1}{2}\right)^3 = \dfrac{1}{8}\right\} \Rightarrow (3,\tfrac{1}{8})$

$x = -2, y = \left(\dfrac{1}{2}\right)^{-2}$
$\left.\begin{aligned} &= \dfrac{1}{2^{-2}} \\ &= 2^2 \\ &= 4 \end{aligned}\right\} \Rightarrow (-2,4)$

$x = -3, y = \left(\dfrac{1}{2}\right)^{-3}$
$\left.\begin{aligned} &= \dfrac{1}{2^{-3}} \\ &= 2^3 \\ &= 8 \end{aligned}\right\} \Rightarrow (-3,8)$

$x = -1, y = \left(\dfrac{1}{2}\right)^{-1} = \dfrac{1}{2^{-1}}$
$\left.\begin{aligned} \\ &= 2^1 \end{aligned}\right\} \Rightarrow (-1,2)$

Step 4: Plot the points from Steps 2, and 3 (Fig.2). That is plot $(0,1)$ $(1,\tfrac{1}{2})$ $(2,\tfrac{1}{4})$ $(3,\tfrac{1}{8})$, $(-1,2)$, $(-2,4))$ and $(-3,8))$, noting that as x increases without bound ($x \to \infty$), y approaches 0 ($y \to 0$). Thus the positive x-axis is a horizontal asymptote to the curve. Graphically, the curve is concave up to the right. Also, note that as x decreases without bound ($x \to -\infty$), the function increases without bound ($y \to \infty$).

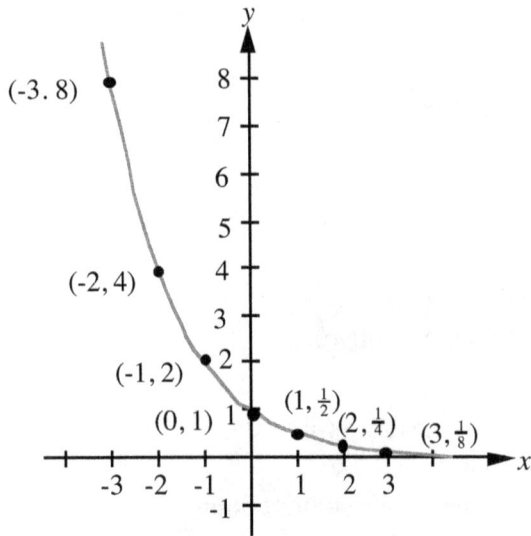

Fig. 2: The graph of $y = \left(\dfrac{1}{2}\right)^x$ or $y = 2^{-x}$

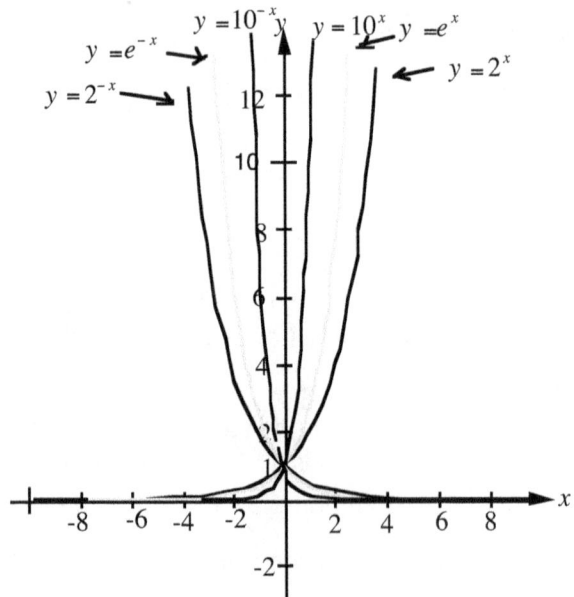

Fig. 3: The graphs of $y = e^x$, $y = 10^x$, $y = 2^x$, $y = e^{-x}$, $y = 10^{-x}$ and $y = 2^{-x}$

Other Exponential Graphs; **The graphs of** $y = e^x$, $y = 10^x$ $y = 2^x$, $y = 2^{-x}$, 315

The graph of $y = e^x$ is a special case of $y = a^x$ in which $a = e$ ($e = 2.72$, and e is the base of the natural logarithms. (see page 304). Similarly, when $a = 10$, we obtain $y = 10^x$. The graphs of $y = 10^x$, $y = 2^x$, $y = e^{-x}$, $y = 10^{-x}$ and $y = 2^{-x}$ on the same set of coordinate axes, are presented in Fig 3, above.

Lesson 48 Exercises

Sketch the graphs of the following pairs on the same set of rectangular axes:

1. $y = 4^x$ and $y = 4^{-x}$;   2. $y = e^x$ and $y = e^{-x}$;   3. $y = 10^x$ and $y = 10^{-x}$

4. What can be said with respect to transformations about the relationship between each pair in Exercises 1-3?

Solution:

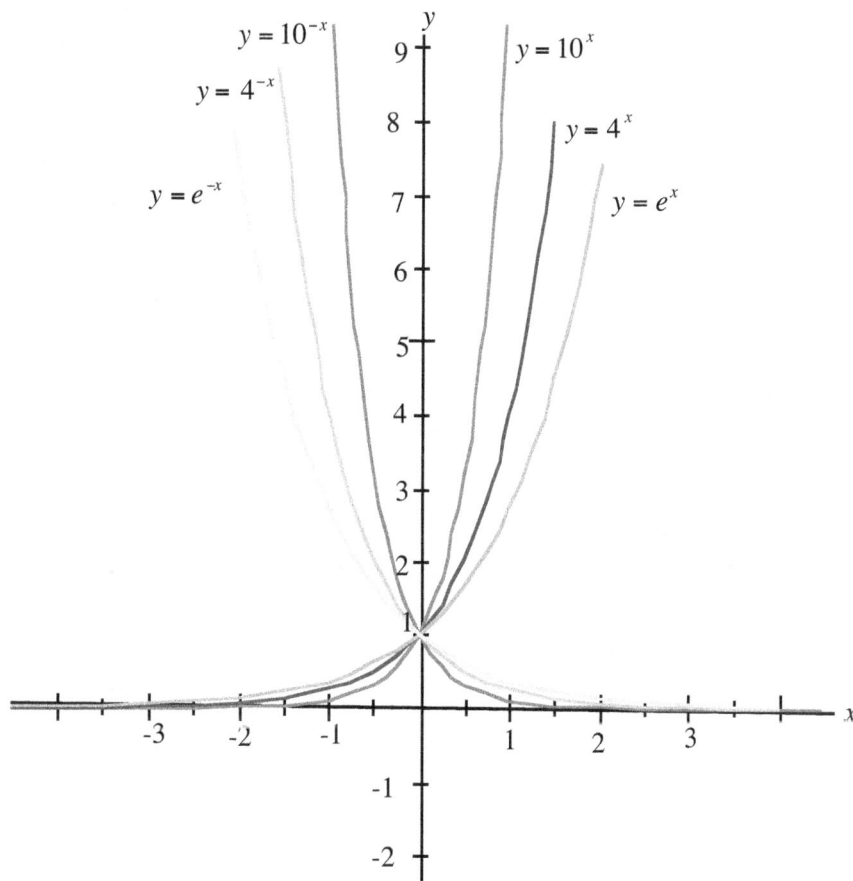

Applications of Exponential Functions 316

Exponential Growth and Decay

The exponential growth or decay model is given by the equation $P(t) = P_0 e^{kt}$, where P_0 is the initial value, $P(t)$ is the value at time t. If $k > 0$, P is said to grow exponentially , and k is then called the growth constant. If $k < 0$, P is said to decay exponentially , and k is then called the decay constant. The applications of this model include population growth, bacterial growth, radioactive decay and compound interest. **Exponential Growth** (For examples on Exponential Decay, see p.500.)

Example 1: At the beginning of an experiment, the population was 3,000. Three hours later, the population was 4500. Assume exponential growth.
(a) Determine the growth constant.   (b) Determine the population after 6 hours.
(c) Determine when the population will reach 15,000.

(a) Step 1: $$P(t) = P_0 e^{kt} \quad (P(3) = 4{,}500)$$ $$4500 = 3000 e^{k(3)}$$ $$\frac{4500}{3000} = e^{3k} \quad (P_0 = 3000)$$ $$\frac{3}{2} = e^{3k}$$ $$\ln\frac{3}{2} = \ln e^{3k}$$ $$= 3k \ln e$$	**Step 2:** $\ln\frac{3}{2} = 3k$ ($\ln e = \log e_e = 1$) $$\left(\ln\frac{3}{2}\right) \div 3 = k \text{ and } k = 0.1352$$ The growth constant = 0.1352 --- **(b)** $P(6) = 3000 e^{[(\ln 1.5)/3](6)}$ $$P(6) = 3000 e^{2(\ln 1.5)}$$ $$= 3000(2.25) \quad (\tfrac{3}{2} = 1.5)$$ $$= 6750$$ After 6 hours, the population is 6750	**(c)** $15{,}000 = 3000 e^{0.1352t}$ $$\frac{15{,}000}{3000} = e^{0.1352t}$$ $$5 = e^{0.1352t}$$ $$\ln 5 = \ln e^{0.1352t}$$ $$\ln 5 = 0.1352t \ln e$$ $$\ln 5 = 0.1352t$$ $$\frac{\ln 5}{0.1352} = t$$ $$11.91 = t$$ The population will reach 15,000 in 11.91 hours

Compound Interest (see also p.477)

The formula relating the principal P, the time, t, taken by P to grow to the amount A is given by
$$A(t) = P(1 + i)^t$$

Example 2: If \$30,000 is invested at 5% interest and compounded annually, in t years, it will grow to an amount , A, given by $A(t) = 30{,}000(1.05)^t$ **(a)** How long will it take to accumulate \$65,000 in the account? **(b)** Find the time required for \$30,000 to double itself. **(a) Step 1:** (a) $A = 30{,}000$ $$65{,}000 = 30{,}000(1.05)^t$$	**Step 2:** $\frac{65{,}000}{30{,}000} = (1.05)^t$ $$2.17 = (1.05)^t$$ $$\log 2.17 = \log 1.05^t$$ $$\log 2.17 = t \log 1.05$$ $$\frac{\log 2.17}{\log 1.05} = t; \ 15.9 = t$$ The time taken by \$30,000 to grow to \$65,000, at the rate of 5%, is 15.9 years.	**Scrapwork** $$\log 2.17 = 0.3365$$ $$\log 1.05 = 0.0212$$ $$\frac{\log 2.17}{\log 1.05} = \frac{0.3365}{0.0212} = 15.9$$

(b) \$30,000 doubles to \$60,000 at the **Step 1:** interest rate of 5% in t years. Therefore, $P = 30{,}000$, $A = 60{,}000$; and substituting in $A(t) = P(1 + i)^t$, we obtain $60{,}000 = 30{,}000(1.05)^t$, and next solve this exponential equation for t. $$\frac{60{,}000}{30{,}000} = (1.05)^t \text{ and } 2 = (1.05)^t$$	**Step 2:** Take logs. $\log 2 = \log(1.05)^t$ $$\log 2 = t \log(1.05)$$ $$\frac{\log 2}{\log 1.05} = t, \text{ and } 14.2 = t$$ At the interest rate of 5%, \$30,000 doubles in 14.2 years.	**Scrapwork** $$\log 2 = 0.3010$$ $$\log 1.05 = 0.0212$$ $$\frac{\log 2}{\log 1.05} = \frac{0.3010}{0.0212} = 14.2$$

Lesson 49   317

Graphs of Logarithmic Functions and Applications of Logarithmic Functions

The logarithmic function is the inverse of the exponential function.

The exponential function is given by $y = b^x$, $b > 0$, $b \neq 1$.. The inverse of this function is obtained by interchanging the x- and y-coordinates of the exponential function. If we interchange the coordinates, we obtain the logarithmic function given by

$$x = b^y \qquad \text{(A)}$$

Now, we solve equation (A) for y using the basic properties of logarithms,

$\log_b x = \log_b b^y$ (Taking logs of both sides of equation (A))

$\log_b x = y \log_b b$

$\log_b x = y$ $(\log_b b = 1)$

We call the function $\log_b x$ the logarithm function of x to the base b. We must note that

$x = b^y$ is equivalent to $\log_b x = y$ (Memorize this equivalent relationship)

Note that the logarithmic function is one-to-one.

Example 1 The graph of $y = \log_2 x$ or $x = 2^y$

The graph of $y = \log_2 x$ can be obtained from the graph of $y = 2^x$ by reflecting $y = 2^x$ about the line $y = x$.

Properties of $y = \log_2 x$ or $x = 2^y$

1. The x-intercept is at $(1, 0)$.

2. As $x \to 0$ (x approaches zero) $\log_2 x \to -\infty$. The curve decreases without bound. The negative y-axis is an asymptote to the curve (in the 4th quadrant).

3. The domain consists of the set of all positive numbers.

4. The range consists of the set of all real numbers.

5. The graph has the negative y-axis as a vertical asymptote (in the 4th quadrant).

6. The graph is concave down to the right (Figure 4)

Figure 4: $y = \log_2 x$ or $x = 2^y$

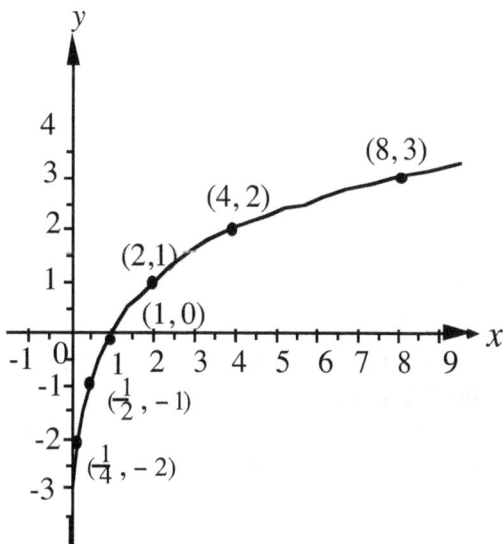

Example Sketch the graph of $f(x) = \log_2 x$.

Step 1: Let $y = f(x)$.

Then $y = \log_2 x$.   (A)

Step 2: Write equation (A) in the equivalent exponential form.

Then $x = 2^y$ <--------- Exponential form.

Calculations   Ordered pairs (points)

$\Downarrow$ $\Downarrow$

Step 3: Choose y and calculate the corresponding values of x. It is more convenient to choose y and calculate x..(This may be the first time you are choosing y and calculating x.)

Find the x-intercept by letting $y = 0$. Note that there is no y-intercept for this function.

$$\left.\begin{array}{l} y = 0, x = 2^0 \\ \quad\quad\quad = 1 \end{array}\right\} \Rightarrow (1,0)$$

Becareful of how you write the ordered pairs, since you are choosing y and calculating x.The first element of each ordered pair is always the x–value irrespective of the order in which the values are obtained.

Step 4: Choose other convenient y-values (positive and negative values) and calculate corresponding the x-values

$$\left.\begin{array}{l} y = 1, x = 2^1 \\ \quad\quad\quad = 2 \end{array}\right\} \Rightarrow (2,1)$$

$$\left.\begin{array}{l} y = 2, x = 2^2 \\ \quad\quad\quad = 4 \end{array}\right\} \Rightarrow (4,2)$$

$$\left.\begin{array}{l} y = 3, x = 2^3 \\ \quad\quad\quad = 8 \end{array}\right\} \Rightarrow (8,3)$$

$$\left.\begin{array}{l} y = -1, x = 2^{-1} \\ \quad\quad\quad = \dfrac{1}{2} \end{array}\right\} \Rightarrow (\tfrac{1}{2},-1)$$

$$\left.\begin{array}{l} y = -2, x = 2^{-2} \\ \quad\quad\quad = \dfrac{1}{2^2} \\ \quad\quad\quad = \dfrac{1}{4} \end{array}\right\} \Rightarrow (\tfrac{1}{4},-2)$$

Step 5: Plot the points from Steps 3, and 4 (Figure 4). (i.e., is plot $(1,0)$, $(2,1)$, $(4,2)$, $(8,3)$, $(\tfrac{1}{2},-1))$, $(\tfrac{1}{4},-2))$ and connect them, noting that as x approaches zero the function decreases without bound, and thus the negative y-axis is a vertical asymptote to this curve. Graphically, the curve is concave down to the right. Also, note that as x increases without bound, the function increases without bound.

Other Logarithmic Graphs

The graphs of $y = \log_a x$ and in particular the graphs of $y = \log_{10} x$ and $y = \log_e x$

The graph of $y = \log_{10} x$ or $x = 10^y$. can be obtained from the graph of $y = 10^x$ by reflecting $y = 10^x$ about the line $y = x$. Similarly, the graph of $y = \log_e x$ or $x = e^y$ can be obtained from the graph of $y = e^x$ by reflecting $y = e^x$ about the line $y = x$.

General Properties of $y = \log_a x$, $y = \log_{10} x$ and $y = \log_e x$

1. The x-intercept is at $(1,0)$.

2. As $x \to 0$, $\log_{10} x \to -\infty$ and $y = \log_e x \to -\infty$

3. The domain consists of the set of all positive numbers.

4. The range consists of the set of all real numbers.

5. The graph has the negative y-axis as a vertical asymptote (in the 4th quadrant).

6. The graph is concave down to the right (Figure)

The graphs of $y = \log_{10} x$ and $y = \log_e x$ are presented below.

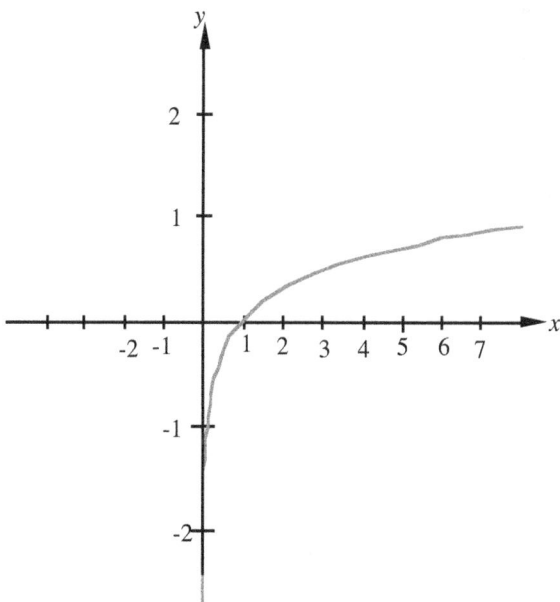

Figure 5: The graph of $y = \log_{10} x$ or $x = 10^y$.

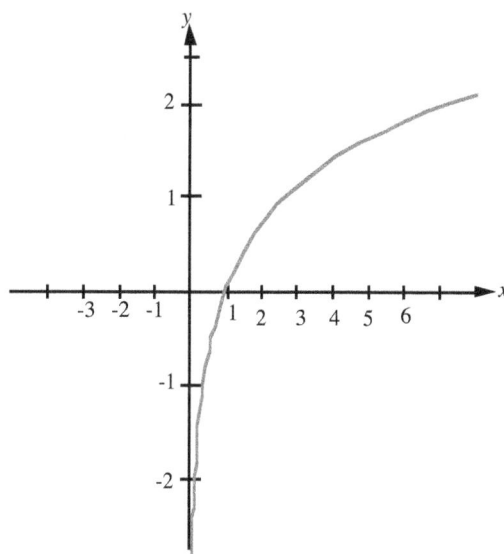

Figure 6: The graph of $y = \log_e x$ or $x = e^y$

We must note the following:

1. For an **exponential equation**, we must be careful not to confuse its equivalent equation in the logarithmic form with its inverse.

Thus $y = e^x$ is equivalent to $x = \log_e y$ (but **not** equivalent to $y = \log_e x$, its inverse).

2. For a **logarithmic equation**, we must be careful not to confuse its equivalent equation in the exponential form with its inverse.

Thus $y = \log_e x$ is equivalent to $x = e^y$, (but **not** equivalent to $y = e^x$, its inverse).
See Figures below.

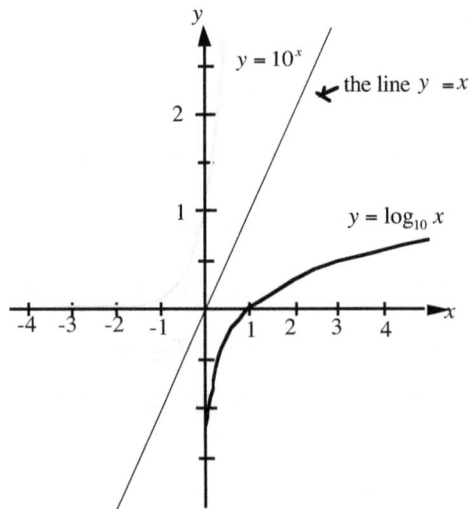

Figure 7: The graph of $y = e^x$ and $\ln x$ **Figure 8** : Graph of $y = 10^x$ and $\log x$

Lesson 49 Exercises

Sketch the graphs of the following:
1. $y = \log x$.;   **2.** $y = \log(x - 1)$;   **3.** $y = \log(x + 2)$;   **4.** $y = \log_3 x$;   **5.** $y = \log_2(x + 3)$.

Answers:

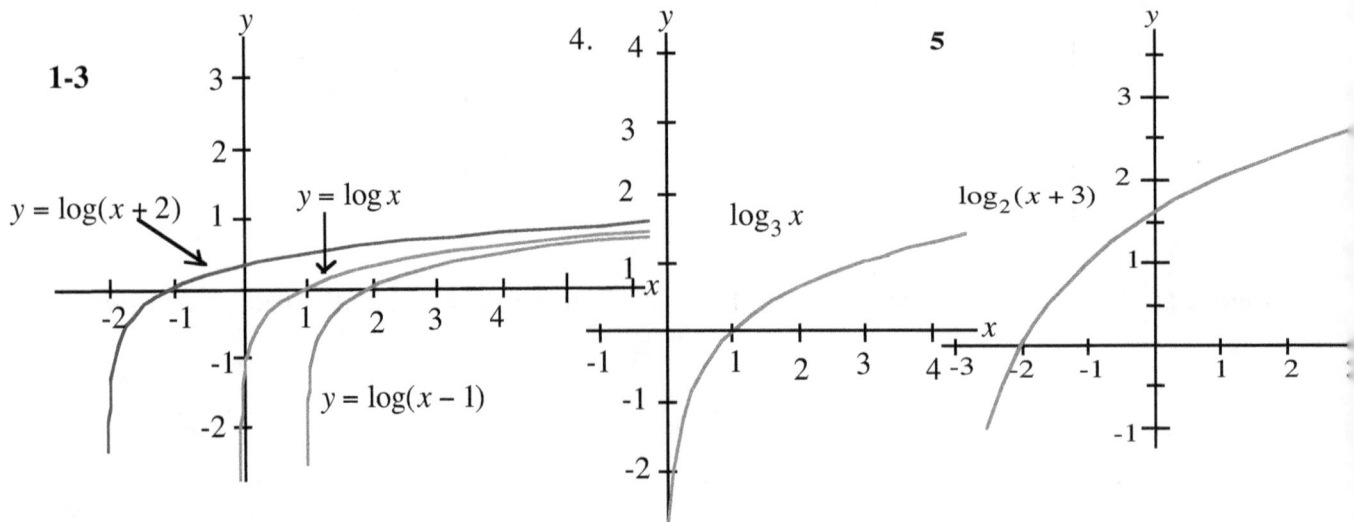

Applications of Logarithmic Functions (see also p.477) 321

Sound Level

Formula for sound level $L = 10\log\dfrac{I}{I_0}$,where L (in decibels, dB) is the loudness of sound, I is the sound intensity in watts per square meter, and $I_0 = 10^{-12}\,W/m^2$

(a) If $I = 10^{-3}\,W/m^2$, find L. (b) If $L = 75$, find I.

Solution

(a) $L = 10\log_{10}\dfrac{10^{-3}}{1}\dfrac{w}{m^2}\cdot\dfrac{m^2}{10^{-12}\,w}$

$= 10\log_{10}\dfrac{10^{-3}}{10^{-12}}$

$= 10\log_{10}10^9$

$= 9(10\log_{10}10)$

$= 10(9)$ (Note: $\log_{10}10 = 1$)

$= 90$ decibels (dB)

(b) $L = 75$

Step 1; Substituting in

$L = 10\log\dfrac{I}{I_0}$

$75 = 10\log\dfrac{I}{10^{-12}}$

$75 = 10\log 10^{12}I$

$\dfrac{75}{10} = \log_{10}(10^{12}I)$

Step 2: $7.5 = \log_{10}10^{12} + \log_{10}I$

$7.5 = 12 + \log_{10}I$

$(\log_{10}10^{12} = 12\log_{10}10 = 12)$

$7.5 - 12 = \log_{10}I$

$-4.5 = \log_{10}I$

$10^{-4.5} = I$

$I = 10^{-4.5}\dfrac{w}{m^2}$

Hydrogen Ion Concentration

The pH of a solution is a measure of the acidity or alkalinity of the liquid. The pH of a solution is defined as follows: $pH = -\log_{10}\left[H^+\right]$, where $\left[H^+\right]$ is the hydrogen ion concentration (concentration per liter).

Example 1 If $\left[H^+\right]$ for a solution is 2.45×10^{-8} moles/liter, find the pH of this solution.

Solution $pH = -\log_{10}\left[2.45\times10^{-6}\right]$

$= -\left(\log_{10}2.45 + \log_{10}10^{-6}\right)$

$= -\left(\log_{10}2.45 + (-6)\log_{10}10\right)$

$= -\left(\log_{10}2.45 + (-6)(1)\right)$ Note: $\log_{10}10 = 1$

$= -(0.389 - 6)$ Note: From a calculator, $\log_{10}2.45 = 0.389$

$= -(-5.61)$

$= 5.61$

The pH of the 2.45×10^{-8} moles/liter solution is 5.61.

Example 2 If the pH of a solution is 7.5, find the $\left[H^+\right]$ of this solution.

Solution

Step 1: $pH = -\log_{10}\left[H^+\right]$

$7.5 = -\log_{10}\left[H^+\right]$ (substituting for $pH = 7..5$)

Step 2: $-7.5 = \log_{10}\left[H^+\right]$

$10^{-7.5} = \left[H^+\right]$

The $\left[H^+\right]$ is $10^{-7.5}$ moles/liter

CHAPTER 20

Lesson 50: **General System of Linear Equations**
Lesson 51: **Matrix and Matrix Methods**
Lesson 52: **Determinants and Applications**

Lesson 50
General System of Linear Equations

Consider a system of three linear equations in three variables x, y, z, as shown below. We call this the standard form of three linear equations in three variables.

$$\begin{cases} a_1x + b_1y + c_1z = d_1 & (1) \\ a_2x + b_2y + c_2z = d_2 & (2) \\ a_3x + b_3y + c_3z = d_3 & (3) \end{cases}$$

A solution of the above system of equations is an ordered **triple** of numbers such that all the equations are satisfied simultaneously by the ordered triple of numbers. We say that a system of equations is **consistent** if it has a solution (i.e., there are values $x, y,$ and z which satisfy the three equations simultaneously). However, if there is no simultaneous solution of the system of equations, we say that the system is **inconsistent**. A consistent system of equations has either one solution (a unique solution) or infinitely many solutions. A system with a unique solution is also said to be independent, and a system with infinitely many solutions is also said to be dependent.

The graph of a linear equation in three variables is that of a plane in space. If the three equations, with a unique solution, were graphed, the three planes would all intersect at only one point. However, lf the three equations, with infinitely many solutions were graphed, the three planes would intersect in a line. If the system is inconsistent, the graph of the system consists of parallel planes.

Triangular form of a system of equations

We say that the above system of equations has been reduced to an equivalent **upper triangular** form if equation (3) contains only the z-term (i.e., $a_3 = 0, b_3 = 0$; equation (2) contains only the y-term and z-term (i.e., $a_2 = 0$); and equation (1) contains the x-term, the y-term and the z-term. We can also say that the system has been reduced to an equivalent **lower triangular** form if equation (1) contains only the x-term, (i.e., $b_1 = 0, c_1 = 0$), equation (2) contains only the x-term and the y-term (i.e., $c_2 = 0$), and equation (3) contains the x-term, the y-term and the z-term.

$a_1x + b_1y + c_1z = d_1$	$a_1x + 0 + 0 = d_1$
$0 + b_2y + c_2z = d_2$	$a_2x + b_2y + 0 = d_2$
$0 + 0 + c_3z = d_3$	$a_3x + b_3y + c_3z = d_3$
Upper triangular form	**Lower triangular form**

If a system of equations cannot be reduced to triangular form, then the system has either an infinite number of solutions or has no solution. However, we must ensure that this irreducibility to triangular form is not due to our inability or careless error in the reduction process.

To reduce a given system of equations to triangular form, we may apply the following operations : 3 2 3

1. Interchange any two equations.
2. Add one equation to another equation.
3. Multiply any equation by a non-zero number.
4. Multiply any equation by a non-zero number and add the resulting equation to another equation.

Let us review how we solved a system of simultaneous equations containing two variables in elementary algebra, by an example.

Example 1: Solve the system of equations simultaneously.

$$\begin{cases} 9x + 4y = 83 & (1) \\ 8x + 3y = 71 & (2) \end{cases}$$

As we learned in elementary algebra, we shall solve the above system of equations by addition
Let us eliminate the variable y, since the y-terms have smaller coefficients (though the size of the coefficient does not matter).

$$\text{Equation (1)} \times 3: \qquad 27x + 12y = 249 \qquad (3)$$
$$\text{Equation (2)} \times -4: \qquad -32x - 12y = -284 \qquad (4)$$
$$\text{Equation (3)} + \text{Equation (4)}: \qquad -5x = -35 \qquad (5)$$
$$\text{solving,} \qquad x = 7$$

Substituting $x = 7$ in equation (2)

$$8(7) + 3y = 71$$
$$56 + 3y = 71$$
$$3y = 15$$
$$y = 5$$

Therefore, $x = 7$, $y = 5$

We could also have eliminated the variable x first, and we would have obtained the same solution.

Gaussian Elimination

Let us retrace the steps used in solving the last system of equations in Example 1, above. We rewrite the system of equations.

$$\text{A}: \quad \begin{cases} 9x + 4y = 83 & (1) \\ 8x + 3y = 71 & (2) \end{cases}$$

$$\text{Equation (1)} \times 3: \qquad 27x + 12y = 249 \qquad (3)$$
$$\text{Equation (2)} \times -4: \qquad -32x - 12y = -284 \qquad (4)$$
$$\text{Equation (3)} + \text{Equation (4)}: \quad -5x = -35 \qquad (5)$$

Now, we replace equation (2) of system (A) by the equivalent equation (5). Then the original system looks like this:

$$\text{B}: \quad \begin{cases} 9x + 4y = 83 & (1) \\ -5x = -35 & (5) \end{cases}$$

We can observe that the left-hand side of the new system (System B) is in upper-triangular form. From equation (5) we can immediately conclude that $x = 7$, substitute $x = 7$ in equation (1) of system (B) to obtain the y-value.

Again we obtain the same solution: $x = 7, y = 5$.

In **Gaussian elimination**, we follow a logical procedure whereby we eliminate the first non-zero terms in a certain sequence until we have reduced the system of equations to triangular form.

Example: Reduce the system of equations to triangular form.

$$\text{A:} \quad \begin{cases} 12x - 15y + 4z = 12 & (1) \\ 3x + 5y + 2z = 0 & (2) \\ 6x + 25y - 8z = -12 & (3) \end{cases}$$

By triangular form, what we want to achieve is something of this form:

$$\begin{cases} 12x - 15y + 4z = 12 \\ 0 + ay + bz = c \\ 0 + 0 + dz = f \end{cases}$$

We shall obtain the above form by the elementary row operations that we used in solving simultaneous equations in elementary algebra. We now proceed to reduce the above system of equations to triangular form.

Step 1: We want to replace equation (2) of system (A) by a new equation whose first term, the x-term is zero. Note that this new equation should be obtained from the above system by elementary row operations on the given equations. We focus on equations (1) and (2). We multiply equation (2) by -12 (12 is the coefficient of the x-term in equation (1)).
Then $-12(3x + 5y + 2z = 0)$ becomes $-36x - 60y - 24z = 0$. $\qquad$ (4)
Next, we multiply equation (1) by 3 (3 is the coefficient of the x-term in equation (2))
Then $3(12x - 15y + 4z = 12)$ becomes $36x - 45y + 12z = 36$ $\qquad$ (5)

Step 2: Add equation (5) and equation (4) to obtain $-105y - 12z = 36$ $\qquad$ (6)
Replace equation (2) of system (A) by the new and equivalent equation (6), keeping the other equations the same. Then the new system of equations looks like as shown, below.

$$\text{B:} \quad \begin{cases} 12x - 15y + 4z = 12 & (1) \\ -105y - 12z = 36 & (6) \\ 6x + 25y - 8z = -12 & (3) \end{cases}$$

(more efficiently, we could multiply eqn.(2) by $- 4$ and add the result to eqn (1))

Step 3: We want to replace equation (3) of system (B) by a new equivalent equation whose first term (the x-term) is zero. (To do this we manipulate equation (1) and equation (3) as was done in Step 1.)
We multiply equation (3) by -12 to obtain

$-72x - 300y + 96z = 144$ $\qquad$ (7)
Next, we multiply equation (1) by 6 to obtain

(More efficiently, we could multiply

$72x - 90y + 24z = 72$ $\qquad$ (8)
eqn (3) by -2 and add the result to eqn (1))
Add equation (8) and equation (7) to obtain
$-390y + 120z = 216$. $\qquad$ (9)
Replace equation (3) in system (B) by equation (9).
Then we obtain the following equivalent system, say C.

$$\text{C:} \quad \begin{cases} 2x - 15y + 4z = 12 & (1) \\ -105y - 12z = 36 & (6) \\ -390y + 120z = 216 & (9) \end{cases}$$

Step 4: We want to replace equation (9) by an equivalent equation whose second term (the y-term) is 0 . Multiply equation (9) by -105 to obtain

Then, $40950y - 12600z = -22680$ (10)

Next, multiply equation (6) by 390 to obtain

$-40950y - 4680z = 14040$ (11)

Add equation (11) and equation (10) to obtain

$-17280z = -8640$ (12)

Now replace equation (9) of system (C) by equation (12). Then, the new system looks like as follows:

$$\text{D:} \quad \begin{cases} 12x - 15y + \quad 4z = 12 & (1) \\ \quad\quad -105y - \quad 12z = 36 & (6) \\ \quad\quad\quad\quad -17280z = -8640 & (12) \end{cases}$$

Observe that system (D) is in triangular form (see page 322) and we can complete the solution as follows:

From equation (12):

$$z = \frac{-8640}{-17280}$$

$$z = \tfrac{1}{2}$$

Substituting $z = \tfrac{1}{2}$ in equation (6) of system (D),

$-105y - 12(\tfrac{1}{2}) = 36$ and solving

$$-105y - 6 = 36$$

$$-105y = 42$$

$$y = \frac{42}{-105} \text{ and } y = -\frac{2}{5}$$

To find x, we substitute $z = \tfrac{1}{2}$ and $y = -\tfrac{2}{5}$ in equation (1) of system (D). Then

$$12x - 15\left(-\tfrac{2}{5}\right) + 4\left(\tfrac{1}{2}\right) = 12$$

$$12x + 6 + 2 = 12$$

$$12x = 4, \text{ and } x = \frac{1}{3}.$$

We now check the solutions, $z = \tfrac{1}{2}$, $y = -\tfrac{2}{5}$ and $x = \tfrac{1}{3}$ in equations (1), (2) and (3) of system (A)

Checking in equation (1): $12(\tfrac{1}{3}) - 15(-\tfrac{2}{5}) + 4(\tfrac{1}{2}) \overset{?}{=} 12$

$$4 + 6 + 2 \overset{?}{=} 12$$

$$12 = 12 \quad \text{(LHS=RHS)}$$

Similarly checking in (2), $3(\tfrac{1}{3}) + 5(-\tfrac{2}{5}) + 2(\tfrac{1}{2}) \overset{?}{=} 0$

$$1 - 2 + 1 \overset{?}{=} 0$$

$$0 = 0 \quad \text{(LHS=RHS)}$$

Similarly checking in (3), $6(\tfrac{1}{3}) + 2\,5(-\tfrac{2}{5}) - 8(\tfrac{1}{2}) \overset{?}{=} -12$

$$2 - 10 - 4 \overset{?}{=} -12$$

$$-12 = -12 \quad (\text{LHS=RHS})$$

Since the values obtained satisfied all three equations of the system, simultaneously, the solutions are $x = \tfrac{1}{3}$, $y = -\tfrac{2}{5}$, and $z = \tfrac{1}{2}$.

Lesson 50 Exercises

A Solve each system of equations by the elimination (addition) method:

1. $\begin{cases} x + y = 2 \\ -x + 2y = -5 \end{cases}$ 2. $\begin{cases} 4x - 7y = 3 \\ 2x + 5y = 1 \end{cases}$ 3. $\begin{cases} 3x + 2y = 5 \\ 7x + 4y = 11 \end{cases}$ 4. $\begin{cases} 3x - 4y = 5 \\ 6x - 8y = 3 \end{cases}$

Answers: **1.** $(3, -1)$; **2.** $(\frac{11}{17}, -\frac{1}{17})$ **3.** $(1, 1)$; **4.** No solution (inconsistent).

B Reduce each of the following system of equations to triangular form and solve:

1. $\begin{cases} 8x + 2y = 20 \\ 4x + y = 10 \end{cases}$ 2. $\begin{cases} x + 2y - z = 0 \\ 2x + 5y + 2z = 0 \\ 3x - y - 4z = 0 \end{cases}$ 3. $\begin{cases} 5x - 3y - z = 16 \\ 2x + y + 2z = 5 \\ 3x - 2y + 2z = 5 \end{cases}$ 4. $\begin{cases} 3x - 2y + z = 0 \\ x + 6y + 3z = 4 \\ 2x + y + 2z = 3 \end{cases}$

Solution:

1.
$$4x + y = 10$$
$$0 = 0$$

Infinite solutions

2.
$$x + 2y - z = 0$$
$$y + 4z = 0$$
$$27z = 0$$
$$x = 0, \ y = 0, \ z = 0$$
or $(0, \ 0, \ 0)$

3.
$$5x - 3y - z = 16$$
$$11y + 12z = -7$$
$$62z = -104$$
$$x = \frac{111}{31}, \ y = \frac{37}{31}, z = -\frac{52}{31}$$
or $\left(\frac{111}{31}, \ \frac{37}{31}, -\frac{52}{31}\right)$

4.
$$3x - 2y + z = 0$$
$$11y + 4z = 5$$
$$8z = 32$$
$$x = -2, \ y = -1, \ z = 4$$
or $(-2, \ -1, \ 4)$

Lesson 51
Matrix and Matrix Methods

We use matrices in solving systems of linear equations, in cost analysis, in transportation, in physics, sociology, psychology and statistics.

An **array** is a grouping of numbers (or things). A **matrix** of numbers is a rectangular array of numbers. We denote a matrix by a bold-faced capital (upper case) letter, for example the letter **A;** We indicate a matrix by enclosing it in brackets.

Example: $\qquad \mathbf{A} = \begin{bmatrix} 3 & 2 & 4 & 7 \\ -1 & 0 & 3 & 9 \\ 11 & 1 & 2 & -6 \end{bmatrix}$

A matrix has rows (across) and columns (down). The rows are horizontal and the columns are vertical. The number of rows m and the number of columns n determine the dimensions of the matrix. If a matrix has m rows and n columns, we say that it is an " m by n" matrix, and we write " $m \times n$" matrix. Thus, m is the number of rows and n is the number of columns. We call each of the numbers in the array an element or an entry of the matrix. In the above matrix, the first row consists of the elements $3, 2, 4$, and 7; the second row consists of the elements $-1, 0, 3$, and 9; and the third row consists of $11, 1, 2$, and -6. The first column consists of $3, -1$, and 11; the second column consists $2, 0$, and 1; and the fourth column consists of $7, 9$, and -6. The matrix **A** has 3 rows and 4 columns. lt is a "3×4" matrix.

If the number of rows equals the number of columns, then we say that the matrix is a square matrix. The **square matrix B** below is a 3×3 matrix. It has 3 rows and 3 columns. We can also give the dimensions of a square matrix by specifying the "order". The order implies either the number of rows or the number of columns.

$$\mathbf{B} = \begin{bmatrix} 2 & -4 & 3 \\ 3 & 5 & 2 \\ 1 & 7 & 9 \end{bmatrix}$$

Square matrices are very important and very convenient to deal with and as such, most of our work involving matrices will be on square matrices.

Special Cases of Matrices: **Row Matrix and Column Matrix**

Sometimes a matrix may have only one row and a number of columns. For example, a 1×3 matrix has only one row and three columns. We call such a matrix a **row matrix** (see matrix **C**, below). However, if a matrix has only one column and a number of rows, we call such a matrix a **column matrix** (see matrix **D**). It is a 4×1 matrix.

Row matrix: $\mathbf{C} = \begin{bmatrix} 7 & 4 & 2 \end{bmatrix}$ $\qquad\qquad$ Column matrix: $\mathbf{D} = \begin{bmatrix} -5 \\ 3 \\ 7 \\ 2 \end{bmatrix}$

Identification of the Entries or the Elements of a Matrix

We usually use double subscripts to identify the elements of a matrix but occasionally, we shall use single subscripts when convenient. Let us consider matrix **A** below.

$$A = \begin{bmatrix} a_{11} & a_{12} & a_{13} \\ a_{21} & a_{22} & a_{23} \\ a_{31} & a_{32} & a_{33} \end{bmatrix}$$

Matrix **A** is a 3 × 3 square matrix. For each element, the first subscript indicates the row in which that element is found, and the second subscript indicates the column in which that element is found. For example, the element a_{23} is identified as being in the second row and the third column. The element a_{12} is identified as being in the first row and the second column. Generally, the element a_{ij} is found in the ith row and the jth column. The elements a_{11}, a_{12}, a_{13} are the first row elements; a_{11}, a_{21}, a_{31} are the first column elements; and a_{11}, a_{22}, a_{33} are the **main diagonal** elements

Matrix in triangular form

A matrix is in **triangular form** if there are only zeros below the main diagonal or if there are zeros above the main diagonal. The matrices **B** and **C** are in triangular form.

$$B = \begin{bmatrix} 3 & 7 & 9 & -3 \\ 0 & 2 & 1 & 4 \\ 0 & 0 & 5 & 6 \\ 0 & 0 & 0 & 4 \end{bmatrix} \qquad C = \begin{bmatrix} 7 & 0 & 0 & 0 \\ 2 & 3 & 0 & 0 \\ 5 & 8 & 2 & 0 \\ 2 & 0 & 1 & 6 \end{bmatrix}$$

Upper triangular form   Lower triangular form

The triangular form of a matrix is similar to the triangular form of a system of equations except that the variables are omitted.

Consider the following system of equations in standard form:

$$\begin{cases} x + 2y + 3z = 3 & (1) \\ 2x + 3y + 8z = 4 & (2) \\ 3x + 2y + 17z = 1 & (3) \end{cases}$$

The coefficients of the variables in the order x, y, z form a 3 × 3 square matrix. The matrix of coefficients of the variables is called the **coefficient matrix**. The coefficient matrix of the above system, say, matrix **A**, is given by:

$$A = \begin{bmatrix} 1 & 2 & 3 \\ 2 & 3 & 8 \\ 3 & 2 & 17 \end{bmatrix}$$

The **column matrix** consisting of the constants of the right-hand sides of the system of equations is called the **column of constants**. In the above case, the column of constants, say **B**, is given by:

$$\mathbf{B} = \begin{bmatrix} 3 \\ 4 \\ 1 \end{bmatrix}$$

If we attach the column matrix to the coefficient matrix, we obtain a 3×4 matrix (matrix **C** below)

$$\mathbf{C} = \left[\begin{array}{ccc|c} 1 & 2 & 3 & 3 \\ 2 & 3 & 8 & 4 \\ 3 & 2 & 17 & 1 \end{array}\right]$$

We call the above 3×4 matrix, the **augmented matrix** of the system of equations under consideration. We may use the same row operations that we used to reduce a system of equations to triangular form, to reduce the augmented matrix to triangular form. Thus, we can perform the following **row operations** on the augmented matrix to obtain an equivalent matrix:

1. We can interchange any two rows.
2. We can add one row to another row.
3. We can multiply any row by a non-zero number
4. We can multiply any row by a non-zero number and add the resulting row to another row.

We should note that before forming the matrix, each of the equations should be put in standard form $ax + by + cz = d$ where $a, b, c,$ and d are constants.

We shall now use these operations to solve a system of equations. Let us rework Example 1 (page 323).

Example 1 Solve the following system of equations using matrix method:

$$\begin{cases} 9x + 4y = 83 & (1) \\ 8x + 3y = 71 & (2) \end{cases}$$

Step 1: Obtain the augmented matrix from the above system. Then we have

$$\mathbf{A} = \left[\begin{array}{cc|c} 9 & 4 & 83 \\ 8 & 3 & 71 \end{array}\right] \qquad \begin{array}{l} \text{Row (1)} \\ \text{Row (2)} \end{array}$$

Step 2: Reduce the augmented matrix to triangular form. We shall indicate the corresponding operations on the original system of equations alongside the matrix manipulations.

$$\begin{array}{llccc|cc} \text{Row (1)} \times 8: & 72 & 32 & | & 664 & \text{Row (3)} \\ \text{Row (2)} \times 9: & 72 & 27 & | & 639 & \text{Row (4)} \\ \text{Row (3)} - \text{Row (4)}: & 0 & 5 & | & 25 & \text{Row (5)} \end{array}$$

Step 3: Now replace second row of matrix **A** by Row (5), keeping the first row of the matrix unchanged. Then we obtain the equivalent matrix, say **B**,

$$\mathbf{B} = \left[\begin{array}{cc|c} 9 & 4 & 83 \\ 0 & 5 & 25 \end{array}\right] \qquad \begin{array}{l} \text{Row (1)} \\ \text{Row (5)} \end{array}$$

Matrix **B** is in triangular form since any element below the main diagonal is zero.

Step 4: To use matrix **B** to solve for x and y, we rewrite matrix **B** in equation form with x and y
variable explicitly indicated. Then we obtain

$$9x + 4y = 83 \qquad (1)$$
$$5y = 25 \qquad (5)$$

Readily from equation (5), $\quad y = \dfrac{25}{5}$

$$y = 5$$

Substituting $y = 5$ in equation (1):

$$9x + 4(5) = 83$$
$$9x + 20 = 83$$
$$9x = 63$$
$$x = 7$$

Therefore, the solution to the system is the ordered pair $(7, 5)$.

Example of a consistent and dependent system of equations

$$\begin{cases} 2x + 4y + 6z = 3 & (1) \\ 2x + 3y + 8z = 4 & (2) \\ 3x + 2y + 17 = 1 & (3) \end{cases}$$

On reducing the above system to triangular form, we obtain

$$\begin{cases} x + 2y + 3z = 3 & (1) \\ \quad\; y - 2z = 2 & (2) \\ \quad\qquad 0 = 0 & (3) \end{cases}$$

Such a system of equations is consistent but has an **infinite number of solutions.** We can therefore assign any value to z and then calculate the corresponding values of y and x.

Lesson 51 Exercises

Solve each system of equations using matrix methods:

1. $\begin{cases} x + y = 2 \\ -x + 2y = -5 \end{cases}$
   2. $\begin{cases} 4x - 7y = 3 \\ 2x + 5y = 1 \end{cases}$
   3. $\begin{cases} 3x + 2y = 5 \\ 7x + 4y = 11 \end{cases}$
   4. $\begin{cases} 3x - 4y = 5 \\ 6x - 8y = 3 \end{cases}$

Answers: **1.** $(3, -1)$; **2.** $(\frac{11}{17}, -\frac{1}{17})$ **3.** $(1, 1)$; **4.** No solution (inconsistent).

Lesson 52   331

Determinants and Applications

Determinants

The **determinant** of a square matrix (n ×n matrix) is the sum of the signed products of all possible combinations of n elements, one element being taken from each row and column. We must note that a determinant is a real number.

Consider a 3 × 3 matrix **A** given by

$$\mathbf{A} = \begin{bmatrix} 4 & 3 & -2 \\ 1 & -1 & 4 \\ -2 & 0 & 5 \end{bmatrix}$$ (Note that the matrix uses brackets)

Then, the determinant of **A** denoted by det **A** is symbolized by using vertical bars.

Thus, det $\mathbf{A} = \begin{vmatrix} 4 & 3 & -2 \\ 1 & -1 & 4 \\ -2 & 0 & 5 \end{vmatrix}$ (Note that the determinant uses vertical bars)

We shall first learn the definition of the determinant of a 2 × 2 matrix, because the determinant of a 3 × 3, 4 × 4 or a higher order matrix is defined in terms of the determinant of a 2 × 2 matrix. Generally, we should know how to find the determinant of a 2 × 2 matrix if we want to able to find the determinants of 3 × 3, or higher order matrices.

Determinant of a 2 × 2 matrix

Let matrix $\mathbf{A} = \begin{bmatrix} a & b \\ c & d \end{bmatrix}$.

Then, det $\mathbf{A} = \begin{vmatrix} a & b \\ c & d \end{vmatrix}$

$$= ad - bc$$

Example   Given that $\mathbf{A} = \begin{bmatrix} 1 & -2 \\ 3 & 4 \end{bmatrix}$, find det **A.**

Solution   By definition,

$$\det \mathbf{A} = \begin{vmatrix} 1 & -2 \\ 3 & 4 \end{vmatrix}$$

$$= (1)(4) - (3)(-2)$$

$$= 4 + 6$$

$$= 10$$

Therefore, det **A** = 10

Example 2 Find det A, given that $\mathbf{A} = \begin{bmatrix} -2 & -3 \\ -5 & -1 \end{bmatrix}$

Solution

$$\det \mathbf{A} = \begin{vmatrix} -2 & -3 \\ -5 & -1 \end{vmatrix}$$

$$= (-2)(-1) - (-3)(-5)$$

$$= 2 - 15$$

$$= -13$$

Therefore, det $\mathbf{A}$ = -13

Minor of an element of a determinant

The **minor** of the element a_{ij} denoted by $\mathbf{M}_{ij}$ is the determinant obtained by deleting the ith row and the jth column. Therefore, to find the minor of an element, delete (ignore) the row and the column the element is in and then find the determinant of the remaining array of elements.

Example Given that $\mathbf{A} = \begin{bmatrix} 3 & 2 & -5 \\ 4 & -2 & 9 \\ 1 & -3 & 10 \end{bmatrix}$

(a) Find the minor of the element in the first row and the third column (that is of the element -5)

(b) Find the minor of the element in the first row and the first column (that is of the element 3)

Solution (a) To find the minor of the -5, we delete (blot out) the first row and the third column (since this element is in the first row and the third column), and then find the determinant of the remaining array of elements.

(a) The minor of the -5 , denoted by $\mathbf{M}_{13}$, is given by

$$\mathbf{M}_{13} = \begin{vmatrix} 4 & -2 \\ 1 & -3 \end{vmatrix}$$

$$= (4)(-3) - (1)(-2)$$

$$= -12 + 2$$

$$= -10$$

(b) To find he minor of the 3, we delete the first row and the first column and find the determinant of the remaining array of elements.

The minor of the 3 , denoted by $\mathbf{M}_{11}$, is given by

$$\mathbf{M}_{11} = \begin{vmatrix} -2 & 9 \\ -3 & 10 \end{vmatrix} = (-2)(10) - (-3)(9) = -20 + 27 = 7$$

Cofactor of an element

The cofactor $\mathbf{C_{ij}}$ of the element a_{ij} , in say det A, is given by $\mathbf{C_{ij}} = (-1)^{i+j}\mathbf{M_{ij}}$

A **cofactor** is a **signed minor** (a minor with a plus or minus sign, irrespective of the sign of the minor)
A cofactor of an element is either the product of +1 and the minor of the element ; or the product of -1 and the minor of the element. The +1 or -1 to use can be determined from the following sign diagram according to location. Example: for a_{11}, we use +1; for a_{12}, we use -1, and for a_{33}, we use +1

Compare the locations of the a's and the +1's and -1's in the two arrays below:

Sign diagram

+1	−1	+	−1	+1	...
−1	+1	−1	+1	−1	...
+1	−1	+1	−1	+1	...
−1	+1	−1	+1	−1	...
...	...	...	...	...	...

Elements and location in array

a_{11}	a_{12}	a_{13}	a_{14}	a_{15}	...
a_{21}	a_{22}	a_{23}	a_{24}	a_{25}	...
a_{31}	a_{32}	a_{33}	a_{34}	a_{35}	...
a_{41}	a_{42}	a_{43}	a_{44}	a_{46}	...
...	...	...	...	...	...

Note that the +1 and -1 alternate in going from the left to the right in any row and in going down in any column. Note also that the main (or principal diagonal) has only plus +1's.

Example: The cofactor of the element a_{11} is the product of +1 and M_{11}.
The cofactor of a_{12} is the product of -1 and M_{12}.

Laplace expansion of a determinant

The determinant of an $n \times n$ matrix **A** is the sum of n products of $a_{ij}\,\mathbf{C_{ij}}$ for any column or row.

The determinant is also the **sum** of n products of $(-1)^{i+j}a_{ij}\mathbf{M_{ij}}$ for any row or column.

To find the value of a determinant by expansion, we expand the determinant about any column or row. The value obtained by expanding the determinant about any column is the same as that obtained by expanding it about any row. Thus, to find the value of a determinant by expansion, we have the option of expanding the determinant about any row or any column. In practice, we prefer to expand the determinant about the row or column with most zeros if there are any zeros in the determinant.

Note: Finding a determinant is finding the sum of some products. The factors involved in each product is the element in the row or column, +1 or -1 , and the minor of the element. Expanding a determinant about a row or column means we are to use the elements of that row or column in finding the products.
If we do this, then the expansions about the zero elements would be zero, saving us time and work.

Example 1   Using cofactors, evaluate the determinant of the matrix given by

$$A = \begin{bmatrix} 5 & 4 & 2 \\ -6 & 0 & -5 \\ 6 & -3 & 4 \end{bmatrix}$$

The **three factors** involved in each product are the **element** (in the array), the **+1 or -1** (from the sign diagram or otherwise) and the **minor** of the element. To find the minor of an element, delete (ignore) the row and the column the element is in and then find the determinant of the remaining array of elements.

Solution

Let us expand the determinant about the second row (-6 0 -5) since it has a zero element.

The first element in the second row
From the sign diagram
The second element in the second row
The third element in the second row

$$\text{Then } \det A = \begin{vmatrix} 5 & 4 & 2 \\ -6 & 0 & -5 \\ 6 & -3 & 4 \end{vmatrix} = -6(-1)\begin{vmatrix} 4 & 2 \\ -3 & 4 \end{vmatrix} + 0\,(+1)\begin{vmatrix} 5 & 2 \\ 6 & 4 \end{vmatrix} + (-5)(-1)\begin{vmatrix} 5 & 4 \\ 6 & -3 \end{vmatrix}$$

The minor of -6
From the sign diagram
The minor of -5

$$= +6[(4)(4) - (-3)(2)] + 0 + 5[(5)(-3) - (6)(4)]$$
$$= 6[16 + 6] + 5[-15 - 24]$$
$$= 6[22] + 5[-39]$$
$$= 132 - 195$$
$$= -63$$

$$\text{(Note: The minor of -6} = \begin{vmatrix} 4 & 2 \\ -3 & 4 \end{vmatrix} = 4(4) - (-3)(2))$$

Expanding a determinant about a row (or about a column):

Step 1: Pick the first element in that row (or column). Write it down.

Step 2: Determine the sign for this element from its location in the sign diagram or from $(-i)^{i+j}$. Write it down next to the element, enclosing it in parentheses

Step 3: Determine the minor of this first element picked.

Step 4: Multiply the factors (numbers) from Steps 1, 2, and 3.

Step 5: Repeat Steps 1, 2, 3, 4 for the other elements in the row (or column).

Step 6: Add all the products from above, noting that some of these products may be negative.

Another method for expanding the determinant of 3×3 matrix.

There is another method (a mnemonic device) for expanding the determinant of a 3×3 matrix. We should note that this method is applicable to **only** 3×3 determinants. It does not apply to a 4×4 or a 5×5 determinant. The method is known as the **rule of Sarrus**, and it illustrated below.

Attach the first and second columns to the right of the original matrix and follow the directions of the arrows. The products going down are positive and the products going up are negative.

Rule of Sarrus

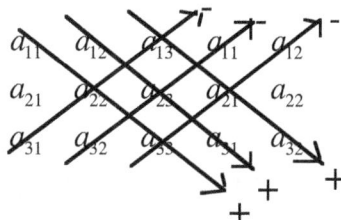

Figure

From the above Figure (Figure), and following the directions of arrows ,we obtain the products.

$$\det \mathbf{A} = a_{11}a_{22}a_{33} + a_{12}a_{23}a_{31} + a_{13}a_{21}a_{32} - a_{31}a_{22}a_{13} - a_{32}a_{23}a_{11} - a_{33}a_{21}a_{12}$$

We apply the rule of Sarrus to Example 1 of the previous page.

Example 2   Using the rule of Sarrus, evaluate the determinant of the matrix given by

$$\mathbf{A} = \begin{bmatrix} 5 & 4 & 2 \\ -6 & 0 & -5 \\ 6 & -3 & 4 \end{bmatrix}$$

$$\det \mathbf{A} = \begin{vmatrix} 5 & 4 & 2 \\ -6 & 0 & -5 \\ 6 & -3 & 4 \end{vmatrix}$$

$$= 5(0)(4) + 4(-5)(6) + 2(-6)(-3) - 6(0)(2) - (-3)(-5)(5) - 4(-6)(4)$$
$$= 0 - 120 + 36 + 0 - 75 + 96$$
$$= -63 .$$

The following **properties** shall be helpful in the evaluation of determinants:
Let **A** be an $n \times n$ square matrix. Then

1. If all the elements of a row (or a column) of det A are zeros, then det **A** is zero.
2. If two rows (or columns) in **A** are interchanged to obtain a new matrix **B**, then det A = - det **B**
3. If any two rows (or columns) of matrix **A** are identical, then det A = 0
4. If the rows and columns of the matrix **A** are interchanged, the value of det **A** remains unchanged. That is, det **A** = det **A**^{T}, where **A**^{T} = the transpose of **A**.
5. If all the elements of a row (or column) of matrix **A** are multiplied (or divided) by a constant k to obtain a new matrix **B** then det A = k det **B** (or 1/k det **B**)
6. 1f a multiple of one row (or column) is added to another rows (or column) then the value of the determinant remains unchanged.

Solution of a Linear System of Equations by Determinants

Consider the following system of equations

$$\begin{cases} ax + by = e & (1) \\ cx + dy = f & (2) \end{cases}$$

Step 1: The coefficient matrix is $\begin{bmatrix} a & b \\ c & d \end{bmatrix}$. The determinant of this matrix is the denominator determinant in Cramer's rule.

The column of constant terms is $\begin{bmatrix} e \\ f \end{bmatrix}$.

The matrix obtained by replacing the column of coefficients of y by the column of constant terms is $\begin{bmatrix} a & e \\ c & f \end{bmatrix}$. The determinant of this matrix is the numerator determinant for calculating y.

The matrix obtained by replacing the column of coefficients of x by the column of constant terms is

$\begin{bmatrix} e & b \\ f & d \end{bmatrix}$. The determinant of this matrix is the numerator determinant for calculating x.

Cramer's Rule for solving a system of linear equations in two variables, with reference to the system of equations, above, is given by

$$y = \frac{\begin{vmatrix} a & e \\ c & f \end{vmatrix}}{\begin{vmatrix} a & b \\ c & d \end{vmatrix}} \quad ; \qquad x = \frac{\begin{vmatrix} e & b \\ f & d \end{vmatrix}}{\begin{vmatrix} a & b \\ c & d \end{vmatrix}}$$

Cramer's rule can be extended to solve a system of three linear equations containing three variables. In general, we can use this rule to solve a system of n equations in n variables. We must note however that Cramer's rule can be applied only to a system with the same number of equations as unknowns.

Cramer's Rule Summarized for two Variables 337

1. The denominator determinant of the formula for calculating x is the same as the denominator determinant of the formula for calculating y. Each denominator determinant is the determinant of the coefficient matrix.

2. The numerator determinant in the formula for calculating x is different from that in the formula for calculating y. Each numerator determinant is obtained from the coefficient matrix by replacing the column of the coefficients of the variable being solved for by the column of constant terms in the augmented matrix of system of equations. The above summary can be extended to a system of n equations in n variables.

Example   Solve the following system of equations using Cramer's rule:

$$\begin{cases} 9x + 4y = 83 \\ 8x + 3y = 71 \end{cases}$$

Step 1: The coefficient matrix is $\begin{bmatrix} 9 & 4 \\ 8 & 3 \end{bmatrix}$.

Step 2: The augmented matrix of the system is $\begin{bmatrix} 9 & 4 & 83 \\ 8 & 3 & 71 \end{bmatrix}$

Step 3: Applying Cramer's rule:

$$x = \frac{\begin{vmatrix} 83 & 4 \\ 71 & 3 \end{vmatrix}}{\begin{vmatrix} 9 & 4 \\ 8 & 3 \end{vmatrix}} = \frac{(83)(3) - (71)(4)}{(9)(3) - (8)(4)} = \frac{-35}{-5} = 7$$

$$y = \frac{\begin{vmatrix} 9 & 83 \\ 8 & 71 \end{vmatrix}}{\begin{vmatrix} 9 & 4 \\ 8 & 3 \end{vmatrix}} = \frac{(9)(71) - (8)(83)}{(9)(3) - (8)(4)} = \frac{-25}{-5} = 5$$

Therefore $x = 7, y = 5$.   The solution is the ordered pair $(7, 5)$

Extension of Cramer's rule to higher order equations 338

Use Cramer's rule to solve the following system of equations:

$$\begin{cases} 2x + 3y - z = 20 \\ 4x - 5y + 2z = 11 \\ 7x - 4y + 3z = 33 \end{cases}$$

Solution

Step 1: Find the coefficient matrix. The coefficient matrix is

$$\begin{bmatrix} 2 & 3 & -1 \\ 4 & -5 & 2 \\ 7 & -4 & 3 \end{bmatrix}$$

Step 2: Find the augmented matrix. The augmented matrix is:

$$\begin{bmatrix} 2 & 3 & -1 & 20 \\ 4 & -5 & 2 & 11 \\ 7 & -4 & 3 & 33 \end{bmatrix}$$

Step 3: Set up the ratios for solving for x, y, and z, and evaluate each determinant. (See pages 334-335)

$$x = \frac{\begin{vmatrix} 20 & 3 & -1 \\ 11 & -5 & 2 \\ 33 & -4 & 3 \end{vmatrix}}{\begin{vmatrix} 2 & 3 & -1 \\ 4 & -5 & 2 \\ 7 & -4 & 3 \end{vmatrix}} = \frac{20(+1)\begin{vmatrix} -5 & 2 \\ -4 & 3 \end{vmatrix} + 11(-1)\begin{vmatrix} 3 & -1 \\ -4 & 3 \end{vmatrix} + 33(+1)\begin{vmatrix} 3 & -1 \\ -5 & 2 \end{vmatrix}}{2(+1)\begin{vmatrix} -5 & 2 \\ -4 & 3 \end{vmatrix} + 4(-)\begin{vmatrix} 3 & -1 \\ -4 & 3 \end{vmatrix} + 7(+1)\begin{vmatrix} 3 & -1 \\ -5 & 2 \end{vmatrix}}$$

$$= \frac{20[-5(3) - (-4(2)] - 11[3(3) - (-4)(-1)] + 33[3(2) - (-5)(-1)]}{2[-5(3) - (-4)(2)] - 4[3(3) - (-4)(-1) + 7[3(2) - (-5)(-1)]} = \frac{-162}{-27} = 6$$

$$y = \frac{\begin{vmatrix} 2 & 20 & -1 \\ 4 & 11 & 2 \\ 7 & 33 & 3 \end{vmatrix}}{\begin{vmatrix} 2 & 3 & -1 \\ 4 & -5 & 2 \\ 7 & -4 & 3 \end{vmatrix}} = \frac{2(+1)\begin{vmatrix} 11 & 2 \\ 33 & 3 \end{vmatrix} + 4(-1)\begin{vmatrix} 20 & -1 \\ 33 & 3 \end{vmatrix} + 7(+1)\begin{vmatrix} 20 & -1 \\ 11 & 2 \end{vmatrix}}{2(+1)\begin{vmatrix} -5 & 2 \\ -4 & 3 \end{vmatrix} + 4(-1)\begin{vmatrix} 3 & -1 \\ -4 & 3 \end{vmatrix} + 7(+1)\begin{vmatrix} 3 & -1 \\ -5 & 2 \end{vmatrix}} = \frac{-81}{-27} = 3$$

$$z = \dfrac{\begin{vmatrix} 2 & 3 & 20 \\ 4 & -5 & 11 \\ 7 & -4 & 33 \end{vmatrix}}{\begin{vmatrix} 2 & 3 & -1 \\ 4 & -5 & 2 \\ 7 & -4 & 3 \end{vmatrix}} = \dfrac{2(+1)\begin{vmatrix} -5 & 11 \\ -4 & 33 \end{vmatrix} + 4(-1)\begin{vmatrix} 3 & 20 \\ -4 & 33 \end{vmatrix} + 7(+1)\begin{vmatrix} 3 & 20 \\ -5 & 11 \end{vmatrix}}{2(+1)\begin{vmatrix} -5 & 2 \\ -4 & 3 \end{vmatrix} + 4(-1)\begin{vmatrix} 3 & -1 \\ -4 & 3 \end{vmatrix} + 7(+1)\begin{vmatrix} 3 & -1 \\ -5 & 2 \end{vmatrix}} = \dfrac{-27}{-27} = 1$$

The solutions are $x = 6, \quad y = 3, \quad z = 1$

Standard forms

In setting up the coefficient matrix and the augmented matrix, the system of equations should be put in standard form if the given system is not already in standard form. By standard form, we mean all the variable terms are on the left-hand sides, the like terms in the same columns and all the corresponding constant terms are on the right-hand sides as shown in system (A) below.

An example of a system in **standard form** is

System (A): $\begin{cases} 4x - 3y + z = 1 \\ 7x + 6y + 2z = 5 \\ -3x + 7y - 4z = -2 \end{cases}$

An example of a system **not** in standard form is

System (B): $\begin{cases} 4x - 3y = -z + 1 \\ 7x + 6y + 2z = 5 \\ 7y - 3x + 2 = 4z \end{cases}$

In standard form, system (B) is identical with System (A)

Using determinants in determining the consistency, inconsistency, dependency and independence of a system of equations.

The following are with reference to Cramer's rule.

Case 1: If the determinant of the coefficient matrix (denominator determinant) is not zero, then the system is **consistent and independent** and there is a unique solution by Cramer's rule.

Case 2: If the determinant of the coefficient matrix is zero and the numerator determinants in Cramer's rule are all zero, then the system is **consistent and dependent**; but there are

infinitely many solutions. (Recall that $\frac{0}{0}$ is indeterminate)

Case 3: If the determinant of the coefficient matrix (denominator determinant) is zero and at least one of the numerator determinants involved in Cramer's rule is not zero, then the system is **inconsistent and independent** and therefore there is no solution.

(Recall for example that $\dfrac{5}{0}$ is undefined)

Relative merits of the different methods of solving systems of linear equations.   340

Cramer's rule: This method is efficient for systems of two equations in two variables. Although, we can extend the method to systems of equations in 3 variables, the method is less efficient when we have systems with four or more variables. Also, for Cramer's rule to work, the number of equations should be the same as the number of variables. In addition, Cramer's rule cannot be used if the denominator determinant is zero.

Reduction to Triangular form: This is a very efficient method for solving linear systems and can be used even if the number of equations is not equal to the number of variables. The method is easily adaptable for use with computers and hence is the most practical method. Other methods we learned previously for solving systems of equations are by addition, substitution. and graphing. However, each of the latter methods is efficient only for systems of two equations in two variables.

Systems of equations with more variables than equations

This system may have an infinite number of solutions or may have no solution at all (i.e., may be inconsistent).

Systems of equations with more equations than variables

Usually, the system may not have any solution but may have a unique solution, if the values obtained in solving say n of the equations satisfy the other $(m - n)$ equations (where m is the number of equations and n is the number of variables), then the system may have a unique solution.

Row operations and the evaluation of determinants

In evaluating determinants of matrices of order more than 2, for example 3×3, or 4×4 matrices, we can use row operations to change a matrix to an equivalent matrix in which, in a particular row or column, all the elements except a few are zero, and then the expansions about the zero elements would be zero. However, we must observe the properties of determinants.(see page 331)

Lesson 52 Exercises

A   Find det **A** of each matrix: **1.** $\begin{bmatrix} 3 & 4 \\ 1 & -5 \end{bmatrix}$ **2.** $\begin{bmatrix} 8 & 2 \\ 4 & 1 \end{bmatrix}$.

Answers: **1.** -19;   **2.** 0

B   Evaluate each of the following determinants:

1. $\begin{vmatrix} 3 & 4 & -2 \\ 1 & -5 & 4 \\ 4 & -4 & -5 \end{vmatrix}$;   **2.** $\begin{vmatrix} -2 & -1 & 3 \\ 4 & -1 & 6 \\ 1 & -1 & 2 \end{vmatrix}$;   **3.** $\begin{vmatrix} 2 & 3 & 5 \\ 1 & 4 & 2 \\ 3 & 1 & 1 \end{vmatrix}$;   **4.** $\begin{vmatrix} 4 & 6 & 2 & -4 \\ 1 & 1 & 2 & -3 \\ -1 & 0 & 3 & -2 \\ 9 & 5 & 0 & -1 \end{vmatrix}$;   **5.** $\begin{vmatrix} 1 & 1 & 1 \\ 0 & 0 & 1 \\ 3 & 4 & -3 \end{vmatrix}$

Answers: **1.** 175; **2.** -15; **3.** -36; **4.** -132; **5.** -1.

C   Solve each system of equations using Cramer's rule:

1. $\begin{cases} x + y = 2 \\ -x + 2y = -5 \end{cases}$ **2.** $\begin{cases} 4x - 7y = 3 \\ 2x + 5y = 1 \end{cases}$ **3.** $\begin{cases} 3x + 2y = 5 \\ 7x + 4y = 11 \end{cases}$ **4.** $\begin{cases} 3x - 4y = 5 \\ 6x - 8y = 3 \end{cases}$

Answers: **1.** $(3, -1)$;   **2.** $(\frac{11}{17}, -\frac{1}{17})$ **3.** $(1, 1)$;   **4.** No solution (inconsistent).

D   Solve the following using matrix methods:

1. $\begin{cases} 5x + 3y = 11 \\ 9x - 2y = 5 \end{cases}$

Answer: $(1, 2)$

E   Solve each of the following system of equations by Cramer's rule:

1. $\begin{cases} 8x + 2y = 20 \\ 4x + y = 10 \end{cases}$ **2.** $\begin{cases} x + 2y - z = 0 \\ 2x + 5y + 2z = 0 \\ 3x - y - 4z = 0 \end{cases}$ **3.** $\begin{cases} 5x - 3y - z = 16 \\ 2x + y + 2z = 5 \\ 3x - 2y + 2z = 5 \end{cases}$ **4.** $\begin{cases} 3x - 2y + z = 0 \\ x + 6y + 3z = 4 \\ 2x + y + 2z = 3 \end{cases}$

Answers

1. Infinite number of solutions;   **2.** $x = 0$, $y = 0$, $z = 0$;   **3.** $x = \frac{111}{31}$, $y = \frac{37}{31}$, $z = -\frac{52}{31}$;   **4.** $x = -2$, $y = -1$, $z = 4$

or $\left(0,\ 0,\ 0 \right)$ or $\left(\frac{111}{31},\ \frac{37}{31}, -\frac{52}{31} \right)$ or $\left(-2,\ -1,\ 4 \right)$

F Solve for x: **1.** $\begin{vmatrix} 2 & 1 & 1 \\ -x & 1 & -3 \\ -2 & x & 1 \end{vmatrix} = 0$;   **2.** $\begin{vmatrix} 2 & 0 & 1 \\ 3 & 0 & 4 \\ 1 & x & 2 \end{vmatrix} = 15$

Answers: **1.** $\{1, -6\}$; **2.** $\{-3\}$;

G   Solve the following by Cramer's rule

1. $\begin{cases} -3x + 3y = 5. \\ 5x + 2y = 1 \end{cases}$ **2.** $\begin{cases} 3x + 4y = -2 \\ 5x - 7y = 1 \end{cases}$ **3.** $\begin{cases} -2x + 4y = 3 \\ 3x - 7y = 1 \end{cases}$ **4.** $\begin{cases} x + 4y - 2z = -4 \\ 3x - y + 3z = 2 \\ -4x + 5y - 4z = 1 \end{cases}$

5. $\begin{cases} x + 3y + 2z = 2 \\ 3x + 2y - z = -1 \\ x - 3y - 3z = 0 \end{cases}$ **6.** $\begin{cases} x + 2y - z = 1 \\ -3x - 5y + 2z = -5 \\ 2x + 6y + 3z = -2 \end{cases}$ **7.** $\begin{cases} 2x + y - z = 9 \\ x - y + z = 0 \\ -x + 3y - 2z = 5 \end{cases}$ **8.** $\begin{cases} x + 3y + 2z = 2 \\ 3x + 2y - z = -1 \\ x - 3y - 3z = 0 \end{cases}$

Solve the following systems using determinants:

9. $\begin{cases} x + y - z = 4 \\ 2x - 2y - 3z = 3 \\ z - 3x - y = -2 \end{cases}$ **10.** $\begin{cases} 5x + 2z = 20 \\ 3x + 2y = 4 \\ 4y - z = -2 \end{cases}$ **11.** $\begin{cases} x + 2y - z = 7 \\ 4x + 3y + 2z = 1 \\ 9x + 8y + 3z = 4 \end{cases}$

Determine the consistency of each of the following systems:

Determine k so that the following system is consistent :

12. $\begin{cases} x + 2y = -1 \\ 2x - y = 3 \\ -3x + 2y = -5 \end{cases}$ **13.** $\begin{cases} x + y - z = 0 \\ x + 2z = 1 \\ y - x = 0 \\ y - 2z - 1 = 1 \end{cases}$ **14.** $\begin{cases} x + 2y - k = 0 \\ y - 3x = 2 \\ x - ky = -12 \end{cases}$

15 A traveler wants to go from a point A to a point B, and to a point C and then to a point D. In going from A to B, the traveler takes a bus which averages 50 m.p.h.. Similarly, from B to C by taxi the average speed is 60 m.p.h. From C to D the traveler went by plane whose average speed was 600 m.p.h. The distance from A to D is 2400 miles and the total time taken from A to D is 14 hr..The distance from C to D is three times as long as the distance from A to C.

Find the time taken to go from   (*a*) A to B;   (*b*) B to C;   (*c*) C to D

Answers: **1.** $x = \frac{1}{3}$, $y = \frac{4}{3}$; **2.** $x = -\frac{10}{41}$, $y = -\frac{13}{41}$; **3.** $x = -\frac{25}{2}$, $y = -\frac{11}{2}$; **4.** $x = -2$, $y = 1$, $z = 3$

5. $x = 3$, $y = -3$; $z = 4$; **6.** $x = \frac{31}{7}$, $y = -\frac{10}{7}$; $z = \frac{4}{7}$, **7.** $x = 3$, $y = 2$; $z = -1$; **8.** $x = 3$, $y = -3$; $z = 4$;

9. $x = -1$, $y = 2$; $z = -3$; **10.** $x = 0$, $y = 2$; $z = 10$, **11.** No solution; **12.** $x = 1$, $y = -1$, consistent;

13. No solution, inconsistent **14.** $k = -\frac{16}{3}$; **15.** (*a*) A to B;: 6 hr;   (*b*) B to C; 5 hr; (*c*) C to D: 3 hr

CHAPTER 21
Complex Numbers

Lesson 53: **Basic Definitions; Basic Operations with Complex Numbers**

Lesson 54: **Equality of Complex Numbers; Roots of Complex Numbers; Equations Involving Complex Numbers**

Lesson 55: **Graphical Representation and Addition of Complex Numbers**

Lesson 56: **Polar (Trigonometric) Form of Complex Numbers**

Lesson 57: **Powers of Complex Numbers; De Moivre's Theorem; Roots of Complex Numbers**

Lesson 53

Definitions; Basic Operations with Complex Numbers

We use complex numbers in the study of electricity and magnetism, in the analysis of feed-back systems especially by the root-locus method, Bode analysis and Nyquist analysis.

Definition, Powers of i, Square Root of Negative Numbers

Sometimes, in attempting to solve certain polynomial equations, we arrive at situations in which we have to find the square roots of negative numbers.

Example: If $x^2 = -1$, then $x = \pm \sqrt{-1}$

Since, for the set of real numbers, there is no provision for the square root of a negative number, we introduce a number "i" which we call the imaginary unit, with the following definition:

Definition: $i = \sqrt{-1}$ or $i^2 = -1$ (We will use both forms of the definition; memorize them)

For example, $(\sqrt{-1})(\sqrt{-1}) = (i)(i) = i^2 = -1$

Powers of i   (Cyclical property of i or i^2) : All powers of i are either equal to ± 1 or $\pm i$.
Examples

    (a) $i^2 = -1$

    (b) $i^3 = (i^2)i$
          $= -1(i)$
          $= -i$

    (c) $i^4 = (i^2)(i^2)$
          $= (-1)(-1)$
          $= 1$

    (d) $i^{10} = (i^4)(i^4)(i^2)$
           $= (1)(1)(-1)$
           $= -1$

Note above: Even powers of i are equal to ± 1. Odd powers of i are equal to $\pm i$.

Note that the imaginary unit "i" is only a tool in mathematics, and that it is not more imaginary (in the literal sense) than the real number 3. The introduction of the imaginary unit allows us to find the square roots of negative numbers.

Example Find the square root: (a) -4 ; (b) -25 ; (c) Simplify: $\sqrt{-15}$.

Solution

(a) $\sqrt{-4} = (\sqrt{-1})(\sqrt{4})$;   (b) $\sqrt{-25} = \sqrt{-1}(\sqrt{25})$;   (c) $\sqrt{-15} = (\sqrt{-1})(\sqrt{15})$

$= i(2)$ $= i(5)$ $= i\sqrt{15} \text{ or} \sqrt{15}i$

$\sqrt{-4} = 2i$ $\sqrt{-25} = 5i$.

Note: In (c) we prefer the first form of the answer; because, sometimes, if we are not careful in writing "i", the "i" may look as if it is under the radical sign . However, if no radical is involved we shall leave the answers as in (a) and (b) above. In some old textbooks, you may find "i" written after the radical.

Generally, $\sqrt{-b} = (\sqrt{-1})(\sqrt{b})$ $(b \geq 0)$

$\qquad\qquad = i\sqrt{b}$

The introduction of the imaginary unit helps us to expand the real number system to a more general system called the complex number system.

If we denote a complex number by z, then $z = a + bi$, where a and b are real numbers; a is called the real part and b is called the imaginary part. If $b = 0$, we have a pure real number (e.g., $z = 3 + 0i = 3$). If $a = 0$, we have a pure imaginary number ($z = 0 + 2i = 2i$). Thus, the product bi is called a pure imaginary number.

Distinction Between Roots of Negative Numbers and Roots of Equations

Roots of numbers (Principal Roots): $\sqrt{-b} = i(\sqrt{b})$ $(b \geq 0)$

Example (a) $\sqrt{-4} = (\sqrt{-1})(\sqrt{4})$
$\qquad\qquad = 2i.$

   (b) $\sqrt{-25} = \sqrt{-1}(\sqrt{25})$
$\qquad\qquad = 5i.$

Roots of equations: Here, we must note that we have more than one root. (two roots.)

Example 1: Consider the solution to the equation $x^2 = -25$.

Solution $x^2 = -25$

$\qquad\qquad x = \pm\sqrt{-25}$

$\qquad\qquad = \pm 5i$, which means $x = +5i$ or $x = -5i$.

Example 2 Solve for x:: $x^2 = -4$

Solution

$\qquad x = \pm\sqrt{-4}$
$\qquad x = \pm 2i$ (or $x = +2i$ or $x = -2i$)

Addition and Subtraction of Complex Numbers

In adding complex numbers, we shall add the real parts, and then add the imaginary parts (in much the same way as we add like terms in polynomial addition).

Perform the indicated operations, leaving the answers in the form $a + bi$.

Example 1 Simplify: $(-3 + 5i) + (-2 + 7i)$

Step 1: Remove the parentheses.

$$(-3 + 5i) + (-2 + 7i)$$

$$= -3 + 5i - 2 + 7i$$

Step 2: Add the real parts, and add the imaginary parts.

$$-3 + 5i - 2 + 7i$$
$$= -5 + 12i \ .$$

Scrapwork:
For the real parts: $-3 - 2 = -5$
For the imaginary parts: $5 + 7 = 12$

Example 2 Simplify: $(6 - 5i) - (-3 + 2i)$

Solution Remove the parentheses and add.

$$(6 - 5i) - (-3 + 2i)$$
$$= 6 - 5i + 3 - 2i$$

$$= 9 - 7i$$

Example 3 Simplify: $(5 + 2i) + (-3 - 6i)$
Solution

Remove the parentheses and add.
$$(5 + 2i) + (-3 - 6i)$$
$$= 5 + 2i - 3 - 6i$$
$$= 5 - 3 + 2i - 6i \ \text{<-------you may skip this step.}$$
$$= 2 - 4i$$

or adding vertically:

$$\begin{array}{r} 5 + 2i \\ \underline{-3 - 6i} \\ 2 - 4i \end{array} \quad (\text{Adding})$$

Example 4 Simplify : $(3 + 4i\} - (2 - 5i)$ (subtraction)
Solution

Remove the parentheses and add.
$$(3 + 4i) - (2 - 5i)$$
$$= 3 + 4i - 2 + 5i$$
$$= 3 - 2 + 4i + 5i \ \text{<-------you may skip this step.}$$
$$= 1 + 9i$$

Example 5 Simplify : $(6 - 3i) + (2 + 3i)$

Solution

$(6 - 3i) + (2 + 3i)$
$= 6 - 3i + 2 + 3i$
$= 8 + 0i$
$= 8$

Multiplication of Complex Numbers

The approach here is multiply, replace i^2 by -1 (or higher powers of i by ± 1 or $\pm i$) , add the real parts and add the imaginary parts.

Example 1 Multiply   -4 - 2i and -5 + i

Solution

Step 1:   Multiply as you multiply binomials
$(-4 - 2i)(-5 + i)$
$= -4(-5 + i) + (-2i)(-5 + i)$ <--- You may skip this step.
$= 20 - 4i + 10i - 2i^2$

Step 2:   (Replace i^2 by -1, since i^2 = -1 by definition)
$= 20 - 4i + 10i - 2(-1)$
$= 20 + 6i + 2$

Step 3:   (Add the like terms: add the real parts; and add the imaginary parts)
$= 22 + 6i$

Example 2   Multiply $3 + 2i$ and $4 + 5i$

Procedure: Multiply, replace i^2 by -1, and simplify.

$(3 + 2i)(4 + 5i)$
$= 12 + 15i + 8i + 10i^2$
$= 12 + 23i + 10(-1)$
$= 12 + 23i - 10$
$= 2 + 23i$ (Adding)

Note: All powers of i are either equal to ± 1 or $\pm i$.   For example $i^3 = -i,\ i^4 = 1,\ i^{10} = -1$

Example 3   Simplify : $4(6 - 3i)$

Solution

$4(6 - 3i)$
$= 24 - 12i$

Example 4   Simplify: $(4 + 2i)(3 - 6i)$

Solution   The above implies multiplication.
Multiply, replace i^2 by -1, and add the like terms.
$(4 + 2i)(3 - 6i)$
$= 12 - 24i + 6i - 12i^2$
$= 12 - 24i + 6i - 12(-1)$
$= 12 - 24i + 6i + 12$
$= 24 - 18i$ (Note: 12 +12 =24 and -24i + 6i = -18i)

Example 5 Simplify: $(7 + 2i)(7 - 2i)$

Solution

$(7 + 2i)(7 - 2i)$
$= 49 - 14i + 14i - 4i^2$
$= 49 - 4(-1)$
$= 49 + 4$
$= 53$

Multiplication of the square roots of negative numbers

Example 1   Find the product of $(\sqrt{-8})$ and $(\sqrt{-2})$

Solution

Step 1: Change to complex number forms.

$\sqrt{-8} = \sqrt{-1}(\sqrt{8})$
$\quad\quad = i\sqrt{8}$

Similarly, $\sqrt{-2} = i\sqrt{2}$

Step 2: Multiply the complex forms now.

$(\sqrt{-8})(\sqrt{-2}) = i\sqrt{8}\ i\sqrt{2}$
$\quad\quad\quad\quad = i^2\sqrt{16}$ $(\sqrt{16} = 4)$
$\quad\quad\quad\quad = i^2\ (4)$
$\quad\quad\quad\quad = (-1)(4)$ $(i^2 = -1)$
$\quad\quad\quad\quad = -4$

In the above problem (Example 1), you may skip Step 1 and show only Step 2.

Example 2   Find the product $(\sqrt{-49})(\sqrt{-25})$

Solution

Step 1: Change to complex number forms.
$\quad\quad\quad$ Then, $\sqrt{-49} = i\sqrt{49} = i(7) = 7i$

$\quad\quad\quad\quad\quad\quad\quad$ <---You may do Step 1 mentally and show only Step 2.

$\quad\quad\quad \sqrt{-25} = i\sqrt{25} = i(5) - 5i$
Step 2: Multiply the complex forms now.
$\quad\quad\quad$ Then, $(\sqrt{-49})(\sqrt{-25}) = (7i)(5i)$
$\quad\quad\quad\quad\quad\quad\quad\quad = (7)(5)i^2$
$\quad\quad\quad\quad\quad\quad\quad\quad = 35(-1)$ $(i^2 = -1)$
$\quad\quad\quad\quad\quad\quad\quad\quad = -35$

We must note in the last two examples that it was necessary first to express the square roots of the negative numbers in terms of i before proceeding to multiply. Failure to do this may result in error such as the following:

Wrong procedure---> $(\sqrt{-8}(\sqrt{-2}) = \sqrt{(-2)(-8)}$
$\quad\quad\quad\quad\quad\quad\quad\quad\quad = \sqrt{16}$
$\quad\quad\quad\quad\quad\quad\quad\quad\quad = 4,$ **which is a wrong answer**.

Generally, it is true that $(\sqrt{a})(\sqrt{b}) = \sqrt{ab}$ if a and b are positive but it is not true if a and b are negative.

Complex Conjugates

Definition : The **conjugate** of a given binomial is another binomial that differs from the given binomial only in the sign of one of the terms. The conjugate of $a + b$ is $a - b$; and the conjugate of $a - b$ is $a + b$.

The **conjugate** of $a + bi$ is $a - bi$. The conjugate of $a - bi$ is $a + bi$. A complex number and its conjugate differ only in the sign of the imaginary part. To find the conjugate of a complex number, change the sign of the imaginary part and keep the sign of the real part unchanged.

Examples: The conjugate of $2 + 3i$ is $2 - 3i$.   The conjugate of $4 - 2i$ is $4 + 2i$.

The conjugate of $7 - 5i$ is $7 + 5i$;   The conjugate of $-3 + 6i$ is $-3 - 6i$.

The conjugate of $-2 -3i$ is $-2 + 3i$.;   The conjugate of $4i$ is $- 4i$.

The conjugate of 7 is 7

The product of a complex number and its conjugate is a real number.

The product of $-2 -3i$ and $-2 + 3i$. $= 4 - 6i + 6i - 9i^2 = 4 - 9(-1) = 4 + 9 = \mathbf{13}$, a real number.

The product of $a + bi$ and $a - bi = a^2 + b^2$

Some uses of complex conjugates: We may use the complex conjugate of a complex number to rationalize the denominator of a fraction or to divide by a complex number.

Division of Complex Numbers

Example 1   Simplify : $\dfrac{2 + 3i}{4 - 5i}$ or divide $2 + 3i$ by $4 - 5i$

To simplify the above complex number is meant we are to write it in the form $z = a + bi$.
The operation here is similar to that of the rationalization of denominators of radical expressions.
To simplify the above expression, we **multiply both the denominator and the numerator by the conjugate of the denominator.**

Solution The conjugate of $4 - 5i$ is $\mathbf{4 + 5i.}$ (See also the note below)
      Multiply both the denominator and the numerator by $4 + 5i$.

Then, we obtain: $\dfrac{2 + 3i}{4 - 5i}$

$$= \frac{(2 + 3i)}{(4 - 5i)} \cdot \frac{\mathbf{(4 + 5i)}}{\mathbf{(4 + 5i)}}$$

$$= \frac{8 + 10i + 12i + 15i^2}{16 + 20i - 20i - 25i^2}$$

$$= \frac{8 + 22i + 15(-1)}{16 + 0 - 25(-1)}$$

$$= \frac{8 + 22i - 15}{16 + 25}$$

$$= \frac{8 + 22i - 15}{41}$$

$$= \frac{-7 + 22i}{41}$$

$$= -\frac{7}{41} + \frac{22}{41}i$$

We may observe above that the **denominator** does **not** contain the imaginary unit "i", even though the numerator contains the imaginary unit.

Note above that we could have multiplied by -4 - 5*i*. (Try it.) It is therefore not critical in this problem which terms should differ in sign, (so far as the denominator rationalization is concerned) provided that either the real parts differ in sign or the imaginary parts differ in sign, but **not** both. We may produce a minus sign in the denominator which we can take care of as usual.

Example 2   Simplify: $\dfrac{5 + 4i}{-3 + 2i}$

Solution Multiply both the denominator and the numerator by -3 - 2i   (The conjugate of -3 + 2i).

$$\frac{5 + 4i}{-3 + 2i}$$

$$= \frac{(5 + 4i)}{(-3 + 2i)} \cdot \frac{(-3 - 2i)}{(-3 - 2i)}$$

$$= \frac{-15 - 10i - 12i - 8i^2}{9 + 6i - 6i - 4i^2}$$

$$= \frac{-15 - 10i - 12i - 8(-1)}{9 + 0 - 4(-1)}$$

$$= \frac{-15 - 22i + 8}{9 + 4}$$

$$= \frac{-7 - 22i}{13}$$

$$= -\frac{7}{13} - \frac{22}{13}i \qquad\qquad (-15 + 8 = -7)$$

Example 3 Simplify: $\dfrac{4 - 2i}{i}$

Solution $\dfrac{4 - 2i}{i}$

$$= \frac{(4 - 2i)}{i} \cdot \frac{(-i)}{(-i)} \qquad \text{(The conjugate of } i \text{ is } -i \text{ ; but you could also use } i)$$

$$= \frac{(4 - 2i)(-i)}{(i)(-i)}$$

$$= \frac{-4i + 2i^2}{-i^2}$$

$$= \frac{-4i + 2(-1)}{-(-1)} \qquad\qquad (i^2 = -1)$$

$$= \frac{-4i - 2}{+1}$$

$$= -2 - 4i$$

Lesson 53 Exercises

A Find the square root or simplify:

1. $\sqrt{-9}$; **2** $\sqrt{-49}$; **3.** $\sqrt{-36}$; **4.** $\sqrt{-18}$; **5.** $\sqrt{-\dfrac{4}{9}}$; **6.** i^{40}; **7.** i^{93} ; **8.** i^{100}

9. $\sqrt{-28}$; **10.** $\sqrt{-32}$

Answers: **1.** $3i$; **2.** $7i$; **3.** $6i$; **4.** $3i\sqrt{2}$; **5.** $\dfrac{2}{3}i$; **6.** 1 ; **7.** i ; **8.** 1 ; **9.** $2i\sqrt{7}$; **10.** $4i\sqrt{2}$

B Simplify: **1.** $4 + 3i - 6 + 5i$; **2.** $(7 - 8i) + (2 + 9i)$ **3.** $(2 - 3i) - (7 - 4i)$;

    4. Subtract $(1 - i)$ from $4 - 2i$; **5.** $4 + 2i - 3(4 - 6i) + i$; **6.** $-i^5 + i^2$; **7.** $4 - i + i^2$

Answers: **1.** $-2 + 8i$; **2.** $9 + i$; **3.** $-5 + i$; **4.** $3 - i$; **5.** $-8 + 21i$; **6.** $-1 - i$; **7.** $3 - i$

C Evaluate the following:

    1. $(-2i)^2$; **2.** $6i^2$; **3.** $(-2i)(3i)$; **4.** $(-5i)(-6i)$; **5.** $(-i)(-2i)$; **6.** $i^3(-4i^3 + 2i^2)$

Answers: **1.** -4; **2.** -6; **3.** 6; **4.** -30; **5.** -2; **6.** $4 + 2i$..

D **Multiply**: $4 + 3i$ and $2 + 5i$; **2.** Multiply $2 - 5i$ and $-2 + 6i$; **3. Simplify**: $(6 - 7i)(2 + 3i)$;

    4. Simplify $(1 - i)(1 + i)$; **5.** $(2 + 3i)^2$; **6.** $(4i^2)(-8i)(i^2)$; **7.** $(5 + 3i)(5 - 3i)$

Answers: **1.** $-7 + 26i$; **2.** $26 + 22i$; **3.** $33 + 4i$; **4.** 2 ; **5.** $-5 + 12i$; **6.** $-32i$; **7.** 34.

E Find the product of each of the following:

    1. $\left(\sqrt{-4}\right)\left(\sqrt{-1}\right)$; **2.** $\left(\sqrt{-16}\right)\left(\sqrt{-4}\right)$; **3.** $\left(\sqrt{-25}\right)\left(\sqrt{49}\right)$; **4.** $\left(\sqrt{-8}\right)\left(\sqrt{-2}\right)$

Answers: **1.** -2; **2.** -8; **3.** $35i$; **4.** -4.

F Find the complex conjugate of each of the following:

    1. $3 + 2i$; **2.** $4 - 5i$; **3.** $6i$; **4.** $-9i$; **5.** $-5 - 3i$; **6.** $-5 + 4i$.

Answers: **1.** $3 - 2i$; **2.** $4 + 5i$; **3.** $-6i$; **4.** $9i$; **5.** $-5 + 3i$; **6.** $-5 - 4i$..

G Divide: **1.** $\dfrac{8 - 6i}{2}$; **2.** $\dfrac{6 + 8i}{2i}$; **3.** $\dfrac{\sqrt{16}}{\sqrt{-16}}$; **4.** $\dfrac{3}{4 - \sqrt{-9}}$; **5.** $\dfrac{4 - 2i}{3 + 5i}$

Answers: **1.** $4 - 3i$; **2.** $4 - 3i$; **3.** $-i$; **4.** $\dfrac{12}{25} + \dfrac{9}{25}i$; **5.** $\dfrac{1}{17} - \dfrac{13}{17}i$

H Simplify: **1.** $\dfrac{2 + 3i}{5 - 2i}$; **2.** $\dfrac{4}{3 - 4i}$; **3.** $\dfrac{5 + 2i}{-i}$ **4.** Divide $4 + 3i$ by $-2 + 3i$;

Simplify: **5.** $\dfrac{2 - 6i}{4 + 3i}$; **6.** $\dfrac{i - 2}{2 - i}$; **7.** $(3 - 2i)(4 + 2i)(i^2)$; **8.** $(6 - 4i)(6 + 4i)$

Answers: **1.** $\dfrac{4}{29} + \dfrac{19}{29}i$; **2.** $\dfrac{12}{25} + \dfrac{16}{25}i$; **3.** $-2 + 5i$; **4.** $\dfrac{1}{13} - \dfrac{18}{13}i$; **5.** $-\dfrac{2}{5} - \dfrac{6}{5}i$; **6.** -1 ; **7.** $-16 + 2i$; **8.** 52

I Find the quotient and simplify: $\dfrac{(2 + 5i)(1 - i)}{(4 - 2i)(3 + i)}$

    Answer: $\dfrac{23 + 14i}{50}$ or $\dfrac{23}{50} + \dfrac{7}{25}i$

Lesson 54

Equality of Complex Numbers; Roots of Complex Numbers; Equations Involving Complex Numbers

Equality of two complex numbers

Two complex numbers $a + bi$ and $c + di$ are equal if and only if $a = c$ (that is the real parts are equal) and $b = d$ (that is the imaginary parts are equal).

Example 4 Find x and y if $2x + 5yi = 8 + 15i$

Solution

Step 1: Set the real parts equal to each other and solve for x.

Then $2x = 8$

$x = 4$

Step 2: Set the imaginary parts equal to each other and solve for y.

Then $5y = 15$ and $y = 3$

Therefore $x = 4, y = 3$.

Roots of Complex Numbers (See also p.372)

Example 5 Find the square root of $5 - 12i$

Step 1 Let $x + yi$ be the square root of $5 - 12i$ (Where x and y are real numbers; Try a nd b instead)

Then $(x + yi)^2 = 5 - 12i$

$$x^2 + 2xyi + y^2i^2 = 5 - 12i$$

$$x^2 - y^2 + 2xyi = 5 - 12i \qquad (i^2 = -1 \text{ and } y^2i^2 = y^2(-1) = -y^2)$$

We apply the equality property of two complex numbers:

For the real parts: $\qquad x^2 - y^2 = 5 \qquad\qquad$ (2)
For the imaginary parts:: $\ 2xy = -12 \qquad\qquad$ (3)

We shall now solve equations (2) and (3) simultaneously.
From equation (3),

$$y = -\frac{12}{2x}$$

$$y = -\frac{6}{x}$$

Substituting for y in equation (2)

$$x^2 - \left(-\frac{6}{x}\right)^2 = 5$$

$$x^2 - \frac{36}{x^2} = 5$$

$$x^4 - 36 = 5x^2$$

$$x^4 - 5x^2 - 36 = 0$$

We shall use the quadratic formula to solve for x. . See page 38 for solving equations quadratic in form.

Step 2:: By the substitution method (Also, see p.38)

Let $x^2 = u$

Then $x^4 - 5x^2 - 36 = 0$ becomes

$$u^2 - 5u - 36 = 0 \qquad (u = x^2)$$

Solving, $u = \dfrac{5 \pm \sqrt{169}}{2}$

$$u = \dfrac{5 \pm 13}{2}$$

Now, converting back to x by letting $u = x^2$

$$x^2 = \dfrac{5 \pm 13}{2}$$

$$x^2 = \dfrac{5 + 13}{2} \qquad \text{or } x^2 = \dfrac{5 - 13}{2}$$

$$x^2 = \dfrac{18}{2} \qquad \text{or } x^2 = \dfrac{-8}{2}$$

$$x^2 = 9 \qquad \text{or } x^2 = -4$$

$$x = \pm\sqrt{9} \qquad \text{or } x = \sqrt{-4}$$

$$x = \pm 3 \qquad \text{or } x = \pm 2i$$

We reject the imaginary solution since x and y must be real.

Therefore, $x = \pm 3$.

Step 3: Substitute $x = \pm 3$ in $y = -\dfrac{6}{x}$ to obtain the corresponding y-values.

When $x = +3$, $y = -\dfrac{6}{3} = -2$, and

when $x = -3$, $y = -\dfrac{6}{-3} = 2$.

Thus, when $x = 3$, $y = -2$, and

when $x = -3$, $y = 2$.

Substituting these values in $x + yi$, the two square roots are $(3 - 2i)$ and $(-3 + 2i)$.

These roots can be verified by squaring each root.

Note above that **in Step 2,** we could have solved the quadratic equation by factoring as follows:

$$x^4 - 5x^2 - 36 = 0$$

$$(x^2 - 9)(x^2 + 4) = 0$$

$$x^2 - 9 = 0; \text{ or } x^2 + 4 = 0$$

$$x^2 = 9; \text{ or } \qquad x^2 = -4$$

$$x = \pm 3 \text{ or } \qquad x = \pm 2i;$$

This approach would be faster than by substitution.

Lesson 54: Complex Numbers: Equality; Roots; and Equations

Example 6   If $z^2 = 8i$, find z satisfying this equation.

Solution: $z = \pm\sqrt{8i}$

Thus, we are to find the square root of $8i$. We can use the same method as in Example 5, above.

Let $a + bi$ be the square root of $8i$.   (where a and b are real)

Then $(a + bi)^2 = 8i$ (By definition, $\sqrt{8i} = a + bi$ if $(a + bi)^2 = 8i$)

$a^2 + 2abi - b^2 = 8i + 0$ (Note: $b^2i^2 = b^2(-1) = -b^2$)

$a^2 - b^2 + 2abi = 8i + 0$ (Also, note that $(a^2 - b^2,\ 2abi) = (0,\ 8i)$)

Equating the real parts, $a^2 - b^2 = 0$ (1)

Equating the imaginary parts, $2ab = 8$ (2)

We shall now solve equations (1) and (2) simultaneously for a and b.

From equation (2), $b = \dfrac{4}{a}$ (3)

Substituting for b from (3) in equation (1),

$$a^2 - \left(\frac{4}{a}\right)^2 = 0$$

$$a^2 - \frac{16}{a^2} = 0$$

$$a^2 \cdot a^2 - \frac{16 \cdot a^2}{a^2} = 0 \cdot a^2 \quad \text{(multiplying each term of the equation by } a^2 \text{ to undo the denominator)}$$

$$a^4 - 16 = 0$$

$$a^4 = 16$$

$$a^2 = \pm 4$$

$$a^2 = +4 \quad \text{or } a^2 = -4$$

$$a = \pm 2 \quad \text{or} \quad a = \pm 2i.$$

Since a must be a real number, we reject the imaginary a-values, $\pm 2i$, and accept the real a-values, ± 2. Therefore $a = +2$ or -2

When $a = +2$, $b = \dfrac{4}{2} = 2$

When $a = -2$, $b = \dfrac{4}{-2} = -2$

Substituting the values of a and b in $a + bi$, the two square roots of $8i$ are $(2 + 2i)$ and $(-2 - 2i)$. Verify these roots by squaring each root.

Note: In the above problem, do not confuse $\pm\sqrt{8i}$ with $\pm\sqrt{-8}$ which is $\pm i\sqrt{8}$ or $\pm 2i\sqrt{2}$.

Equations involving complex numbers

Example 7   Solve for x, given that
$$3ix + 12 - 8i + 2x = 2(6 - i) + 2x$$

Solution We shall get x alone on one side of the equation and the other side should not contain x:

$$3ix + 12 - 8i + 2x = 2(6 - i) + 2x$$
$$3ix - 8i + 12 + 2x = 12 - 2i + 2x$$
$$3ix - 8i = -2i$$
$$(3x - 8)i = -2i$$
$$3x - 8 = -2$$
$$x = 2$$

Example 8   Solve for x, given that $2ix + 3 - 4i = (3 + 2i)x + 5i$

Solution We shall get x alone on one side of the equation and the other side should not contain x:

$$2ix + 3 - 4i = (3 + 2i)x + 5i$$
$$2ix + 3 - 4i = 3x + 2ix + 5i$$
$$2ix - 3x - 2ix = -3 + 9i$$
$$2ix - 3x - 2ix = -3 + 9i$$
$$-3x = -3 + 9i$$
$$x = 1 - 3i$$

Example 9   Solve for x, if x is a real number, given that

Solution

$$2ix + 3 - 4i = (3 + 2i)x + 5i$$
$$2ix + 3 - 4i = 3x + 2ix + 5i$$
$$2ix - 3x - 2ix = -3 + 9i$$
$$2ix - 3x - 2ix = -3 + 9i$$
$$-3x = -3 + 9i$$
$$x = 1 - 3i$$

Since on solving for x, the solution contains the imaginary unit, x is not real.
Therefore, there is **no** solution. (From the original problem, for a solution, x must be real)
.

Lesson 54 Exercises

A Find the square roots of the following 1. $16 - 30i$;   2. $-5 - 12i$

Answers: 1. $5 - 3i$ and $-5 + 3i$;   2. $2 - 3i$ and $-2 + 3i$

B Find z satisfying the given equation 1. $z^2 = 16 - 30i$;   2. $z^2 = -5 - 12i$

Answers: 1. $5 - 3i$ and $-5 + 3i$;   2. $2 - 3i$ and $-2 + 3i$

C   Solve for x: 1. $3ix - 30 + 3i = (12 + 3i)x + 3i$;   2. $x^2 = 16i$

3. Determine the real values of x and y if $4x + 7yi = 12 - 35i$

Answers: 1. $x = -\frac{5}{2}$;   2. $x = 2\sqrt{2} + 2i\sqrt{2}$, $x = -2\sqrt{2} - 2i\sqrt{2}$;   3. $x = 3$, $y = -5$.

Lesson 55

Graphical Representation and Addition of Complex Numbers

A complex number may be considered as an ordered pair of real numbers (a, b). Graphically, we represent a complex number z by a point P whose x-coordinate $= a$, and y-coordinate $= b$. We graph a complex number in a rectangular coordinate system of axes. We call this system the complex plane. The horizontal axis (x-axis) represents the real axis and the vertical axis (y-axis) represents the imaginary axis.

Example: Graph the following in the complex plane:

(a) $2 - 3i$; (b) $3 + 2i$; (c) $-4 - 3i$; (d) $-i$; (e) $5i$; (f) 6.

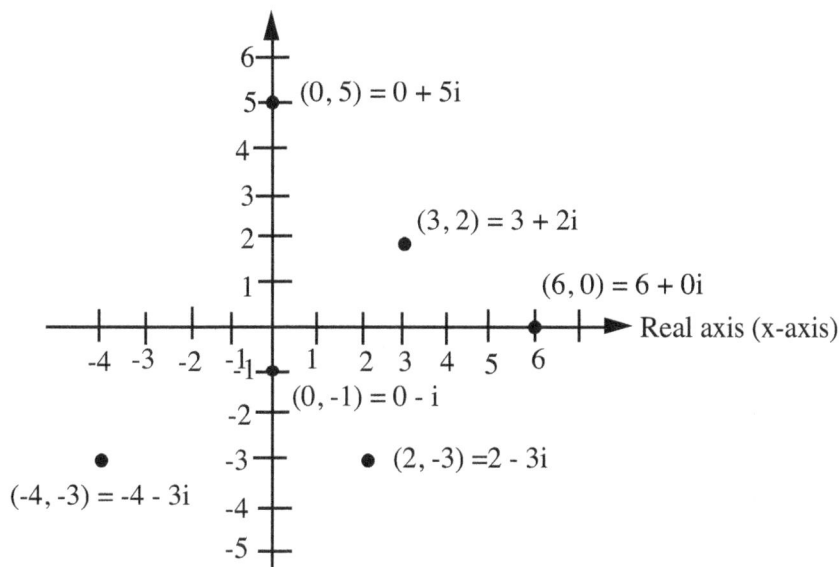

Figure: The graphs of the points, $2 - 3i$, $3 + 2i$, $-4 - 3i$, $-i$, $5i$, 6, in the complex plane.

We may also think of a complex number as a vector. We represent a vector quantity by an arrow drawn to scale. The length of the arrow represents the magnitude of the vector and the direction of the arrow head represents the direction of the vector. In Figure we represent the complex number $a + bi$ as a vector drawn from the origin to the point $(a, b) = a + bi$..

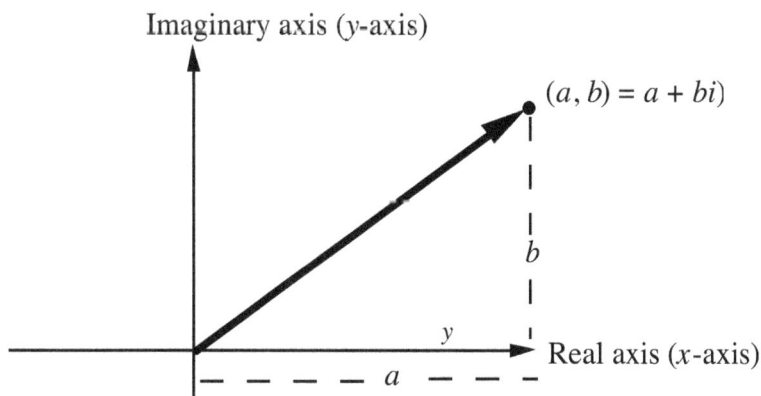

Figure: Graphical representation of a complex number as a vector.

Graphical Addition of Complex Numbers

We may use the parallelogram law of **vector addition** to add complex numbers.

In the complex plane, we draw each of the two vectors $a + bi$ and $c + di$ to scale. The **sum** of the complex numbers is represented by the **diagonal** of the parallelogram (Figure).

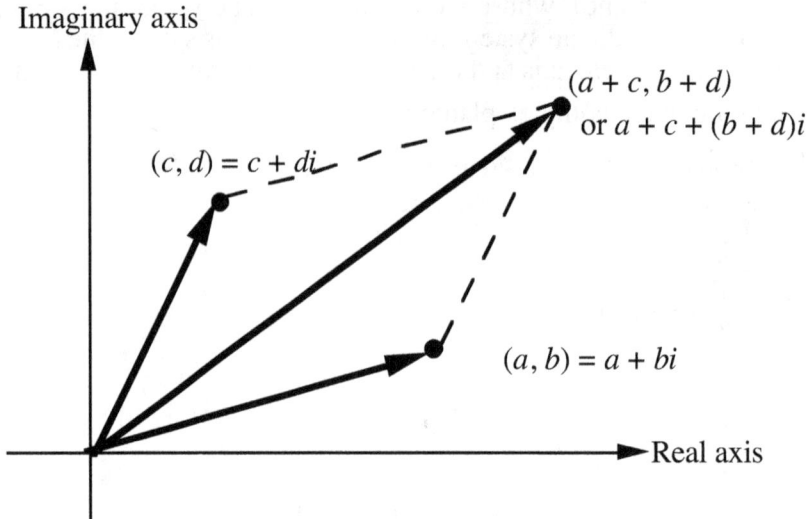

Figure: Graphical addition of complex numbers: $(a + bi) + (c + di)$

Subtraction of Complex Numbers: $(a + bi) - (c + di) = (a + bi) + (-c - di)$

By definition, to subtract $(c + di)$ from $(a + bi)$ means add the negative of $(c + di)$ to $(a + bi)$. Graphically, $(-c - di)$ has the same magnitude as $(c + di)$ except that its direction is opposite to that of $(c + di)$ as shown in Figure

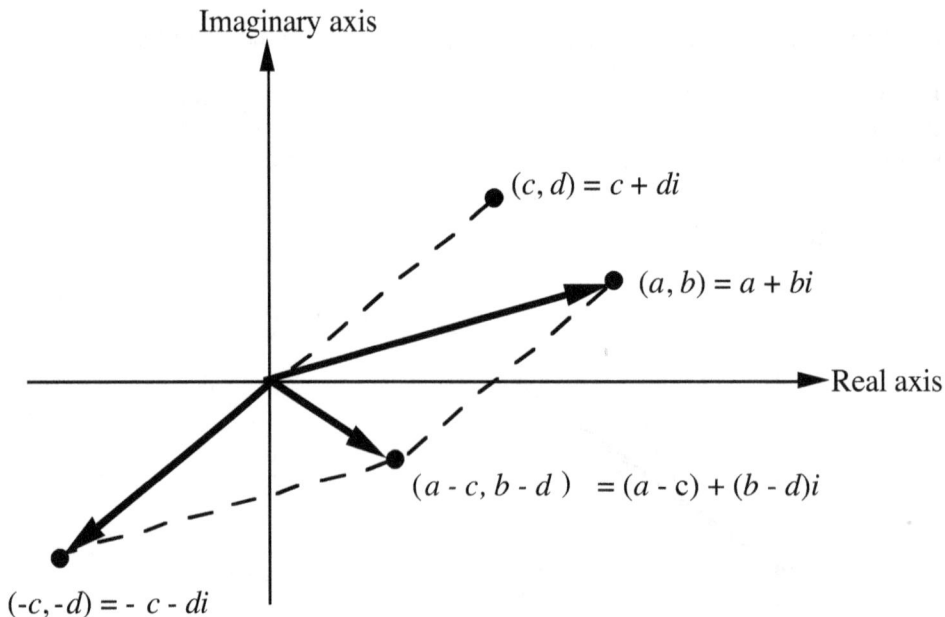

Figure: Graphical subtraction of complex numbers: $(a + bi) - (c + di)$

Example 1 Add algebraically and graphically: $(2 + 3i) + (7 + 5i)$.

Algebraic method

$$(2 + 3i) + (7 + 5i)$$
$$= 2 + 3i + 7 + 5i$$
$$= 9 + 8i$$

Graphical method

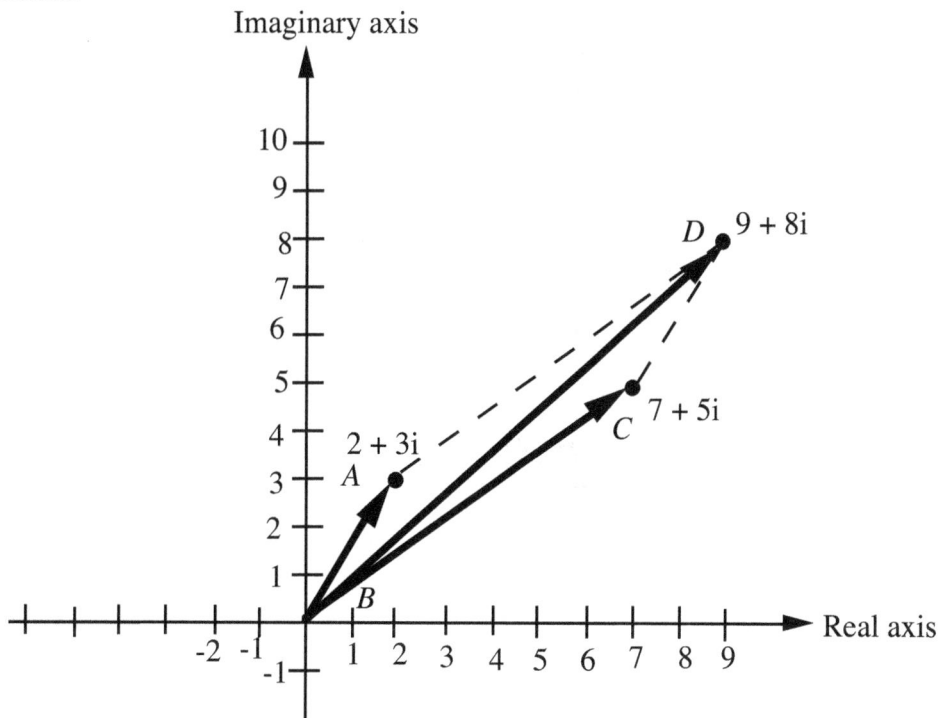

Figure : $(2 + 3i) + (7 + 5i)$ is represented by BD

Graph each of the following in the complex plane:

1. $3 - 2i$;   **2.** $3 + 2i$;   **3.** $-3 - 3i$;   **4.** $-4i$;   **5.** $6i$;   **6.** 5

Solution

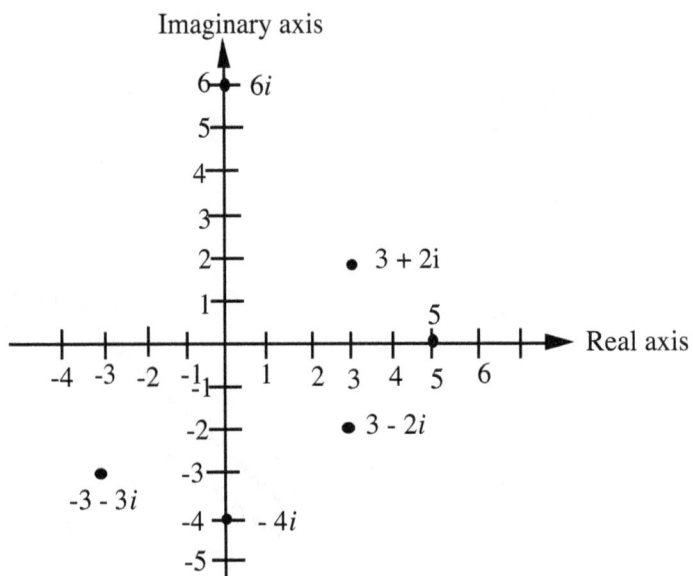

Lesson 56

Polar (Trigonometric) Representation of Complex Numbers.

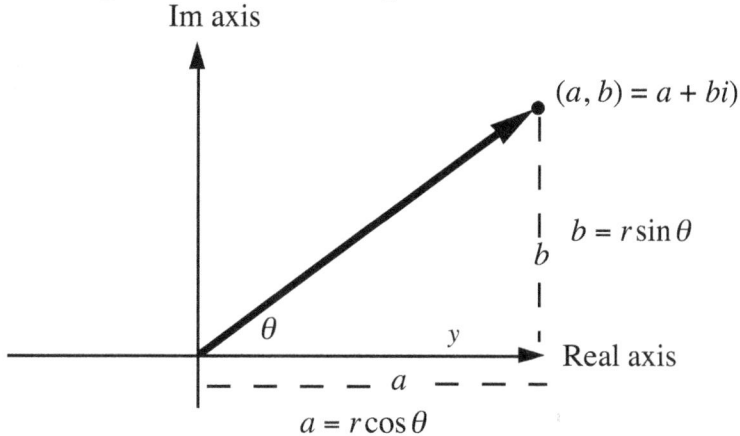

Figure : The graph of polar or trigonometric form of a complex number.

A complex number $z = a + bi$ is represented by a vector drawn from the origin to the point (a, b) and in direction θ with the x-axis. The standard position angle, θ, is called the argument of the complex number.

The **length** r of the vector is given by $r = \sqrt{a^2 + b^2}$
From trigonometric definitions,

$$a = r\cos\theta \qquad (1)$$

$$b = r\sin\theta \qquad (2)$$

$$\theta = \arctan\left(\frac{b}{a}\right) \qquad (3)$$

By substituting for a and b from equations (1) and (2) respectively, in the rectangular form,
$z = a + bi$, we obtain

$z = r\cos\theta + r\sin\theta\, i$ or
$z = r\cos\theta + ir\sin\theta$; and if we factor out the r, we obtain

$$\boxed{z = r(\cos\theta + i\sin\theta)} \qquad (4)$$

Sometimes, the quantity $(\cos\theta + i\sin\theta)$ is abbreviated by cis θ and then equation (4) becomes
$$z = r cis\theta \qquad (5)$$

We can view this abbreviation this way:

$$z = r\,(\,\cos\theta \,+\, i\,\sin\theta\,)$$
$$z = r\ \ c\ \ \ \ \ \ \ i\ s$$

Although it might be cumbersome to write equation (4) repeatedly instead of equation (5), it is recommended that we use the form of equation (4) a number of times before switching to the abbreviated form, $z = r cis\theta$.

Changing from rectangular form to polar form

Rectangular form: $z = a + bi$

Polar form: $z = r(\cos\theta + i\sin\theta)$ or $z = r\,cis\,\theta$

$$r = \sqrt{a^2 + b^2}$$

$$a = r\cos\theta \qquad\qquad (1)$$

$$b = r\sin\theta \qquad\qquad (2)$$

$$\theta_{ref} = \arctan\left(\tfrac{b}{a}\right) \qquad\qquad (3)$$

Example 1 Change the complex number $z = 3 - 4i$ to polar form.

Solution In polar form, we want the form $z = r(\cos\theta + i\sin\theta)$. So we must find r and θ from the rectangular form.

Step 1: Plot $z = 3 - 4i$ as a vector (Figure)

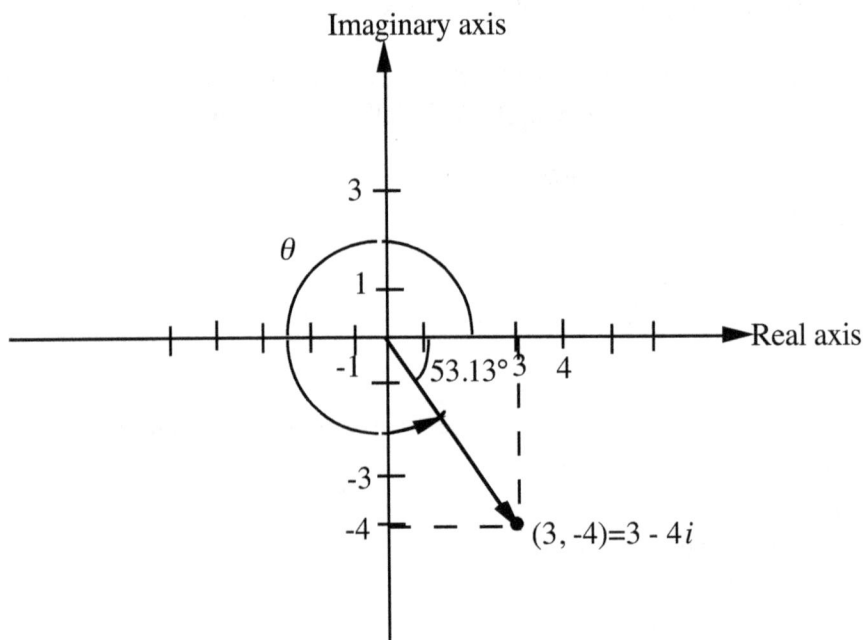

Figure: The graph of $z = 3 - 4i$

Step 2: Find r.

$a = 3, b = $ -4

$$r = \sqrt{3^2 + (-4)^2}$$
$$r = \sqrt{9 + 16}$$
$$r = \sqrt{25}$$
$$r = 5.$$

Step 3: Find θ.

$$\tan \theta_{ref} = \frac{b}{a} \qquad\qquad (\theta_{ref} \text{ is the reference angle})$$

$$= -\frac{4}{3}$$

$$= -1.3$$

Now, from tables or calculator, find the angle whose tangent is 1.3

$$\theta_{ref} = 53.13° \qquad \text{(Terminal side of } \theta \text{ is in the 4th quadrant)}$$

$$\theta = 360° - 53.13°$$

$$\theta = 306.87°$$

(**Note** above that there is a difference between θ and θ_{ref})

Step 4: Substitute $r = 5, \theta = 305.87°$ in $z = r(\cos\theta + i\sin\theta)$.

Then $z = 5(\cos 306.87° + i\sin 306.87°)$

or $z = 5cis306.87°$.

We must note above that the polar form was **not** specified in terms of $53.13°$ but rather in terms of $306.87°$, this angle being in standard position, and as such, it is measured counter-clockwise from the positive x-axis. ($53.13°$ is the reference angle from tables or a calculator.)

Example 2 Change to polar (trigonometric) form: $z = -3 - 5i$

Step 1: Plot $z = -3 - 5i$ as a vector (Figure)

Step 2: Find r.

$$a = -3, b = -5 \quad \text{(from } z = -3 - 5i) \qquad (\textbf{Note}: \text{General form is } z = a + bi)$$

$$r = \sqrt{(-3)^2 + (-5)^2} \qquad\qquad (r = \sqrt{a^2 + b^2})$$

$$r = \sqrt{9 + 25}$$

$$r = \sqrt{34}$$

Step 3: Find θ

$$\tan\theta_{ref} = \frac{-5}{-3} \qquad \left(\frac{b}{a}\right)$$

$$= 1.67$$

$$\theta_{ref} = 59.03° \quad \text{(Now, from tables or calculator, find the angle whose tangent is 1.67)}$$

$$\theta = \theta_{ref} + 180° \quad \text{(The terminal side of } \theta \text{ is in the 3rd quadrant.)}$$

$$\theta = 59.03° + 180°$$

$$\theta = 239.03°$$

Step 4: Substitute $r = \sqrt{34}$, $\theta = 239.03°$ in

$z = r(\cos\theta + i\sin\theta)$ to obtain

$z = \sqrt{34}(\cos 239.03° + i\sin 239.03°)$. <------polar (or trigonometric) form

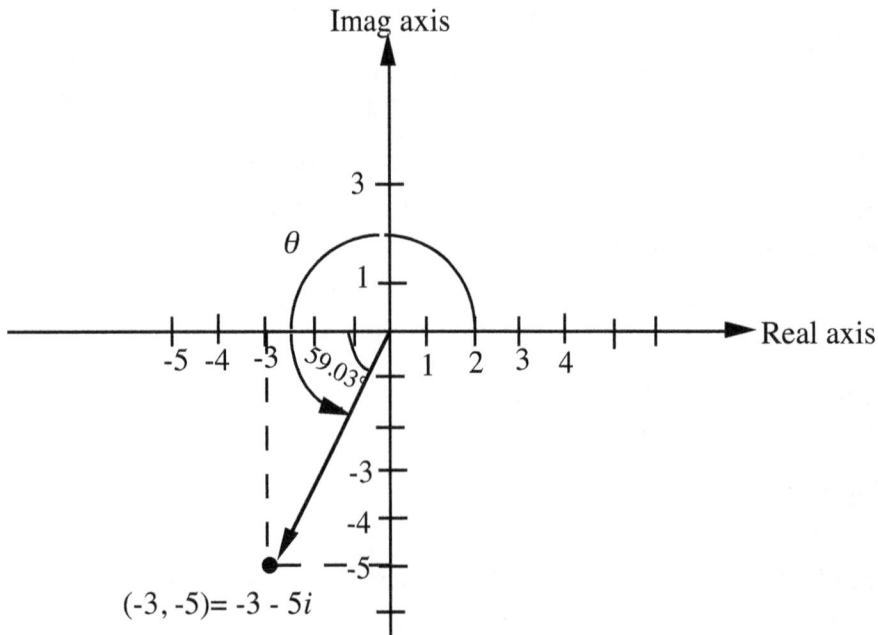

$(-3, -5) = -3 - 5i$

Example 3 Change to polar form.

      (a) 9 ;   (b) -9 ; (c) -3i.

Solution

(a) Step 1: Sketch $z = 9 + 0i$ (Figure)

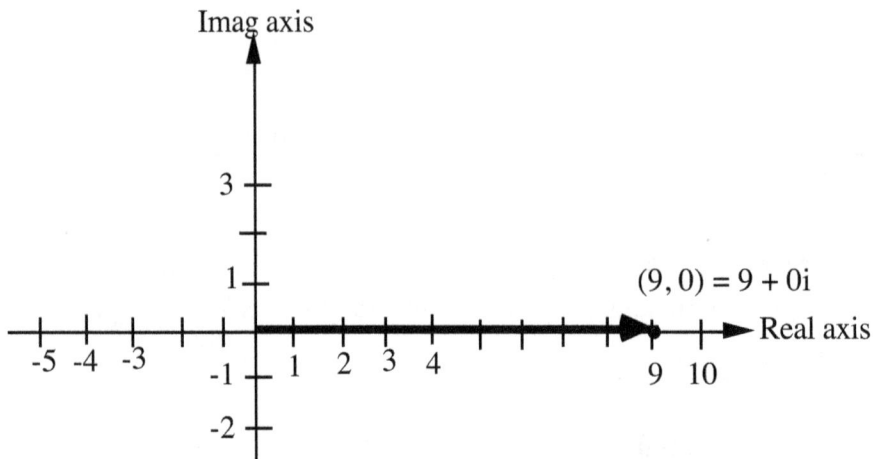

$(9, 0) = 9 + 0i$

Step 2: Find r

     $r = 9$ (since θ is quadrantal; terminal side is along the positive x-axis)
     $a = 9$
     $b = 0$

Step 3: Find θ

     $\tan \theta_{ref} = \dfrac{0}{9} = 0$

$\theta_{ref} = 0°$

In this problem, $\theta = \theta_{ref}$.

$\qquad \theta = 0°$.

Step 4: Substitute $r = 9$, $\theta = 0°$ in the general equation ($z = r(\cos\theta + i\sin\theta)$) to obtain

$$z = 9(\cos 0° + i\sin 0°)$$

$$z = 9\cos 0° \quad \text{or} \quad 9\,cis0° \qquad\qquad (\sin 0° = 0)$$

(b) Step 1: Sketch $z = -9 + 0i$ \qquad (Figure)

Step 2: Find r and θ.

$\qquad r = 9$ \qquad\qquad (Note that r is positive.)

By inspection, since the terminal side falls along the negative x-axis, $\theta_{ref} = 0°$

$\qquad$ Therefore, $\theta = 180°$

Substituting $r = 9$, $\theta = 180°$ in the general equation, we obtain

$z = 9(\cos 180° + i\sin 180°)$

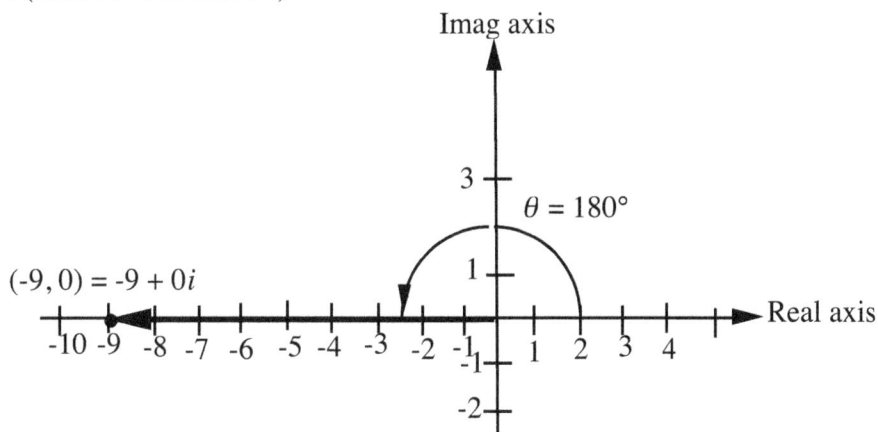

(c) Similarly, for $z = -3i = 0 - 3i$ \quad (Figure)

$\qquad r = 3$, $\theta = 270°$ \quad and

$\qquad z = 3(\cos 270° + i\sin 270°)$ \quad or \quad $z = 3\,cis270°$

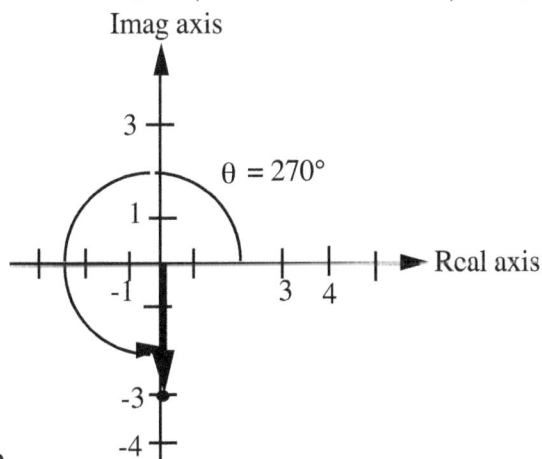

Figure

We must note above that by agreement , we restricted the angular measurement to $0° \le \theta \le 360°$.

Changing from polar form to rectangular (algebraic) form   364

Example 1   Change $z = 10(\cos 220° + i \sin 220°)$ to rectangular form.

The general rectangular form is given by

$$z = a + bi.$$

We must therefore find a and b from the polar form.

By inspection and comparison of $z = r(\cos \theta + i \sin \theta)$, with

$$z = 10(\cos 220° + i \sin 220°)$$
$$r = 10, \quad \theta = 220°$$
$$a = 10 \cos 220° \qquad (a = r \cos \theta)$$
$$a = -7.7 \qquad (\cos 220° = -.77)$$
$$b = 10 \sin 220°$$
$$b = -6.4 \quad (\sin 220° = -.64)$$

Substituting $a = -7.7$, $b = -6.4$ in $z = a + bi$, we obtain

$$z = -7.7 - 6.4i$$
$$10 \, cis \, 220° = -7.7 - 6.4i$$

↑ (Polar form)   ↑ (Rectangular form)

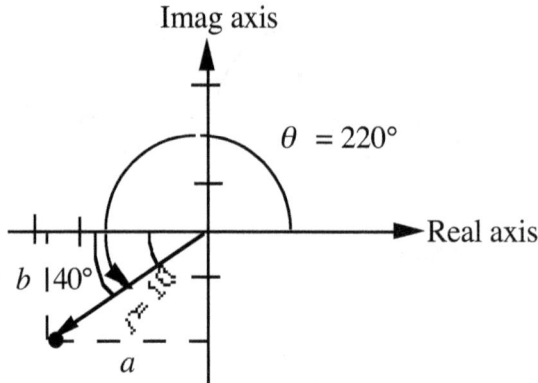

Figure

It may be remarked that it easier to change from the polar form to the rectangular form than vice versa

Exponential form of a complex Number

The exponential form of a complex number, z, is given by $z = re^{i\theta}$, where θ is in radians.

Multiplication of Complex Numbers in Polar Form

Consider the polar (trigonometric) forms of two complex numbers,

$$z_1 = r_1(\cos \theta_1 + i \sin \theta_1) \text{ and}$$
$$z_2 = r_2(\cos_2 + i \sin \theta_2).$$

The product $z_1 z_2$ of the two complex numbers in polar form is given by

$$z_1 z_2 = r_1 r_2 [\cos(\theta_1 + \theta_2) + i \sin(\theta_1 + \theta_2)] \qquad (1)$$

Therefore, to multiply two complex numbers in polar forms, multiply the radii (moduli) and add the angles (arguments).

Example Given that $z_1 = 3(\cos 30° + i \sin 30°$
$$z_2 = 4(\cos 40° + i \sin 40°)$$

(a) Find by analytic means (using equation (1) above) the product of z_1 and z_2.

(b) Show the graphical form of the above method.

Solution

(a) By equation (1) above,

Apply $z_1 z_2 = r_1 r_2 [\cos(\theta_1 + \theta_2) + i \sin(\theta_1 + \theta)]$ with
$r_1 = 3,\ r_2 = 4,\ \theta_1 = 30°,\ \theta_2 = 40°$ to obtain
$z_1 z_2 = (3)(4)[\cos(30° + 40°) + i \sin(30° + 40°)$

$\qquad = 12[\cos 70° + i \sin 70°]$ \hfill (2)

$z_1 z_2 = 12 cis 70°$.

(From tables or calculator we can simplify equation (2) for $z_1 z_2$)

(b) Graphical form:

    Step 1: Find the sum of the angles $30°$ and $40°$
          Then $\theta = 70°$

    Step 2: Find the product $r = r_1 r_2$
          Then $r = (3)(4) = 12$

    Step 3: Draw $\theta = 70°$ in standard position and $r = 12$ (Figure)

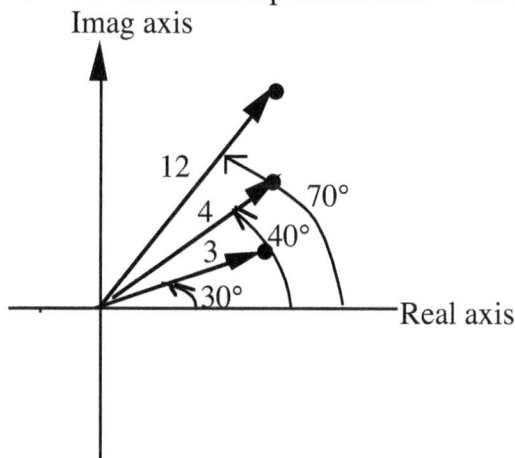

Division of Complex Numbers in Polar Form

Consider the polar (trigonometric) form of two complex numbers, say

$\qquad z_1 = r_1(\cos \theta_1 + i \sin \theta_1)$ and
$\qquad z_2 = r_2(\cos_2 + i \sin \theta_2)$.

The quotient $\dfrac{z_1}{z_2} = \dfrac{r_1}{r_2}[\cos(\theta_1 - \theta_2) + i \sin(\theta_1 - \theta_2)]$

$\qquad$ or $\dfrac{z_1}{z_2} = \dfrac{r_1}{r_2}\ cis(\theta_1 - \theta_2)$.

Therefore, to divide two complex numbers in polar forms, divide the radii (moduli) and subtract the angles (arguments).

Example 1   Given that $z_1 = 15(\cos 50° + i\sin 50°)$ and

$$z_2 = 5(\cos 30° + i\sin 30°), \text{ find the quotient } \frac{z_1}{z_2}$$

Solution $\dfrac{z_1}{z_2} = \dfrac{15}{5}[\cos(50° - 30°) + i\sin(50° - 30°)]$

$$= 3[\cos 20° + i\sin 20°] \text{ or } 3\,cis\,20°$$

Example 2   If $z_1 = 6(\cos 15° + i\sin 15°)$ and $z_2 = 12(\cos 50° + i\sin 50°)$ find the quotient $\dfrac{z_1}{z_2}$.

Solution

$$\frac{z_1}{z_2} = \frac{6}{12}[\cos(15° - 50°) + i\sin(15° - 50°)]$$

$$= \frac{1}{2}[\cos(-35) + i\sin(-35)] \text{ or } \frac{1}{2}cis(-35°)$$

But since we want to specify θ so that $0° \le \theta \le 360°$

$$\frac{z_1}{z_2} = \frac{1}{2}cis(325°)$$

Lesson 56 Exercises

A   Plot each complex number and then change to polar form.

  1. $3 + 4i$;   **2.** $1 + i$;   **3.** $1 - 3i$;   **4.** $3 - i$;   **5.** $1 - i\sqrt{3}$;   **6.** 7;
  7. $-i$;   **8.** $-3 - 4i$;   **9.** $-5i$;   **10.** -3;   **11.** $-.69 + 3.94i$

Answers:

1. $z = 5\,cis\,53.13°$; **2.** $z = \sqrt{2}\,cis\,45°$; **3.** $z = \sqrt{10}\,cis\,288.43°$; **4.** $z = \sqrt{10}\,cis\,341.57°$; **5.** $z = 2\,cis\,300°$;
6. $z = 7\,cis\,0°$; **7.** $z = cis\,270°$; **8.** $z = 5\,cis\,233.13°$; **9.** $z = 5\,cis\,270°$; **10.** $z = 3\,cis\,180°$; **11.** $4\,cis\,100$

B   Plot each complex number and then change to rectangular form.

  1. $4(\cos 30° + i\sin 30°)$;   **2.** $2(\cos 210° + i\sin 210°)$;   **3.** $5\,cis\,30°$;

  4. $3(\cos 0° + i\sin 0°)$;   **5.** $2\,cis\,150°$;   **6.** $9\,cis\,\frac{2\pi}{3}$.

Answers:

  1. $z = 2\sqrt{3} + 2i$; **2.** $z = -\sqrt{3} - i$; **3.** $z = \dfrac{5\sqrt{3}}{2} + \dfrac{5}{2}i$; **4.** $z = 3$; **5.** $z = -\sqrt{3} + i$; **6.** $z = -\dfrac{9}{2} + \dfrac{9i\sqrt{3}}{2}$

C   Perform the indicated operation and leave the answer in polar form.

  1. $(\cos 20° + i\sin 20°)(\cos 10° + i\sin 10°$;   **2.** $3(\cos 25° + i\sin 25) \bullet 4(\cos 125° + i\sin 125°)$;
  3. $[2(\cos 60° + i\sin 60°)]^3$;   **4.** $(3\,cis\,45°)(4\,cis\,360°)$

Answers: 1. $z = cis\,30°$; **2.** $z = 12\,cis\,150°$; **3.** $z = 8\,cis\,180°$; **4.** $z = 12\,cis\,405°$.

D   Perform the indicated operations and leave the answers in polar form.

  1. $\dfrac{14\,cis\,170°}{2\,cis\,60°}$;   **2.** $\dfrac{10\,cis\,60°}{5\,cis\,90°}$ **3.** $\dfrac{15(\cos 100° + i\sin 100°)}{3(\cos 25° + i\sin 25°)}$

Answers: 1. $z = 7\,cis\,110°$; **2.** $z = 2\,cis\,330°$; **3.** $z = 5\,cis\,75°$.

Lesson 57

Powers of Complex Numbers; De Moivre's Theorem; and Roots of Complex Numbers

Powers of Complex Numbers

Let us consider the squaring $z = r \operatorname{cis} \theta$

By the multiplication rule, we multiply the moduli (radii) and add the arguments (angles).

Thus
$$z^2 = (r \operatorname{cis} \theta)^2$$
$$= (r \operatorname{cis} \theta)(r \operatorname{cis} \theta)$$
$$= (r)(r) \operatorname{cis}(\theta + \theta)$$
$$= r^2 \operatorname{cis} 2\theta$$

Similarly,
$$z^3 = r^3 \operatorname{cis} 3\theta$$
$$z^4 = r^4 \operatorname{cis} 4\theta$$

From the above examples, we arrive at a generalization known as **De Moivre's** theorem or formula.

De Moivre's Theorem: If n is a positive integer, then

$$z^n = (r \operatorname{cis}\theta)^n = r^n \operatorname{cis} n\theta$$

or $[r(\cos\theta + i\sin\theta)]^n = r^n(\cos n\theta + i\sin n\theta)$

Example 1   Use De Moivre's theorem to find $(3 - 4i)^6$.

Step 1: Change the complex number to polar (trigonometric) form.
Thus, we want to change $3 - 4i$ to the form $z = r(\cos\theta + i\sin\theta)$

Step 2: From Example 1, page 360, $(3 - 4i) = 5(\cos 306.87° + i\sin 306.87°)$
Applying $z^n = r^n(\cos n\theta + i\sin n\theta)$, we obtain

$$(3 - 4i)^6 = [5(\cos 306.87° + i\sin 306.87°)^6$$
$$= 5^6[\cos(6)(306.87°) + i\sin(6)(306.87°)]$$
$$= 5^6[\cos 1841.22° + i\sin 1841.22°]$$
$$= 15625[\cos 41.22° + i\sin 41.22°]$$
$$= 15625 \operatorname{cis} 41.22° \quad \text{<--- polar form}$$

In exponential form:: $z = r e^{i\theta}$, where θ is in radians

$$41.22° = \frac{41.22°}{180°} \times \frac{\pi \text{ rad}}{1} = 0.72 \text{ rad}$$

$(3 - 4i)^6. = 15625 e^{0.72i}$ <----exponential form ($r = 15625$, $\theta = 0.72$ rad)

$(3 - 4i)^6. = 11753 + 10296i$ **<--- rectangular form**

Comparison

Rectangular form	Polar form	Exponential form
$z = a + bi$ or $z = x + iy$	$z = r(\cos\theta + i\sin\theta) = r\operatorname{cis}\theta$	$z = r e^{i\theta}$

Roots of Complex Numbers (see also p.354)

We may apply De Moivre's theorem to find the roots of complex numbers.

$$\text{if } x^n = A, \qquad \text{(where } n \text{ is a positive integer)}$$

$$\text{then } x = A^{\frac{1}{n}}$$

Example: If $x^3 = 8$, then $x = \sqrt[3]{8} = 8^{\frac{1}{3}}$

Similarly, for a complex number A $\quad$ if $x^n = A$, then $x = A^{\frac{1}{n}}$ $\qquad$ (where n is a positive integer).
There are n distinct roots, where x is an nth root of A.

According to De Moivre's Theorem. (which deals with powers of complex numbers in polar form),

$$[r(\cos\theta + i\sin\theta)]^n = r^n(\cos n\theta + i\sin n\theta) .$$

If we replace n by $\dfrac{1}{n}$ in De Moivre's Theorem, and take into account the periodic nature of the cosine and sine functions, we obtain

$$[r(\cos\theta + i\sin\theta)]^{\frac{1}{n}} = r^{\frac{1}{n}}[\cos\frac{(\theta + 360°k)}{n} + i\sin\frac{(\theta + 360°k)}{n}] \qquad \text{where } k = 0,1,2,...$$

$$= \sqrt[n]{r}[\cos\frac{(\theta + 360°k)}{n} + i\sin\frac{(\theta + 360°k)}{n}]$$

$$(r\operatorname{cis}\theta)^{\frac{1}{n}} = r^{\frac{1}{n}}\operatorname{cis}\frac{(\theta + 360°k)}{n}$$

$$= \sqrt[n]{r}\operatorname{cis}\frac{(\theta + 360°k)}{n}$$

If $n = 2$, there are two distinct roots and we shall use two k-values, namely, $k = 0, k = 1$.
For example, if $k = 0$,

$$\frac{\theta + 360°k}{n} = \frac{\theta + 360°(0)}{n}$$

$$= \frac{\theta + 0}{n}$$

$$= \frac{\theta}{n}.$$

If $k = 1$,

$$\frac{\theta + 360°k}{n} = \frac{\theta + 360°(1)}{n}$$

$$= \frac{\theta + 360°}{n}.$$

Similarly, if $n = 3$, there shall be three distinct roots and we use three k-values, namely, $k = 0, 1$, and 2.

We should note that the k-values, $k = 0, 1, 2,..., n - 1$ are used only in finding the roots of the complex numbers but are **not** used in computing the powers (De Moivre's Theorem).
For $k = n - 1$, the results repeat one of the first n values (previous values).

Applying De Moivre's Theorem to Find the Roots of Numbers $\qquad$

Step 1: Change the given expression to polar (trigonometric) form (if it is already not in polar form).
$\qquad$ (To change to polar form, see page 360)

Step 2: Apply the formula

$$[r(\cos\theta + i\sin\theta)]^{\frac{1}{n}} = \sqrt[n]{r}[\cos\frac{(\theta + 360°k)}{n} + i\sin\frac{(\theta + 360°k)}{n}] \qquad \text{where } k = 0,1,2,\ldots$$

with the appropriate k-values.

Example 1 Find the 4-th roots of $16(\cos120° + i\sin120°)$.

Solution

$\qquad$ The wording of this problem is equivalent to:

$\qquad\qquad$ if $z^4 = 16(\cos120° + i\sin120°)$, solve for z.

$\qquad$ Since the given expression is already in polar form, we go ahead to Step 2 and apply the formula

$$[r(\cos\theta + i\sin\theta)]^{\frac{1}{n}} = \sqrt[n]{r}[\cos\frac{(\theta + 360°k)}{n} + i\sin\frac{(\theta + 360°k)}{n}] \qquad \text{with } k = 0, 1, 2, \text{ and } 3.$$

$$n = 4, \ r = 16, \ \theta = 120°$$

Substituting these values

$$z = \sqrt[4]{16}[\cos\frac{(120° + 360°k)}{4} + i\sin\frac{(120° + 360°k)}{4}]$$

$$z = 2[\cos\frac{(120° + 360°k)}{4} + i\sin\frac{(120° + 360°k)}{4}] \qquad\qquad (A)$$

We can leave the answer in either the trigonometric (polar) form or the rectangular form , $z = a + bi$
Since $n = 4$, we shall use k-values of $k = 0, 1, 2, 3$.
When $k = 0$ in (A), we obtain the root,

$$z = 2[\cos\frac{(120° + 0)}{4} + i\sin\frac{(120° + 0)}{4}]$$

$$= 2[\cos30° + i\sin30°] \quad \text{<-------------------Polar form}$$

$$= 2[\frac{\sqrt{3}}{2} + \frac{1}{2}i]$$

$$= \sqrt{3} + i. \quad \text{<-------------------------------Rectangular form}$$

When $k = 1$ in (A), we obtain the root

$$z = 2[\cos\frac{(120° + 360°(1))}{4} + i\sin\frac{(120° + 360°(1))}{4}]$$

$$= 2[\cos\frac{(120° + 360°}{4} + i\sin\frac{(120° + 360°)}{4}]$$

$$= 2[\cos\frac{480°}{4} + i\sin\frac{480}{4}]$$

$$= 2[\cos120° + i\sin120°] \quad \text{<-----------------polar form}$$

$$= 2[-\cos 60° + i\sin 60°]$$

$$= 2[-\left(\frac{1}{2}\right) + i\left(\frac{\sqrt{3}}{2}\right)]$$

$$= -1 + \sqrt{3}i \ \text{ or } \ -1 + i\sqrt{3} \quad \text{<-------------------rectangular form}$$

For $k = 2$ in (A), we obtain the root.

$$z = 2[\cos\frac{(120° + 360°(2))}{4} + i\sin\frac{(120° + 360°(2))}{4}]$$

$$= 2[\cos\frac{(120° + 720°)}{4} + i\sin\frac{(120° + 720°)}{4}]$$

$$= 2[\cos\frac{840°}{4} + i\sin\frac{840°}{4}]$$

$$= 2[\cos 210° + i\sin 210°] \quad \text{<----------------polar form}$$

$$= 2[-\cos 30° + i\sin 30°]$$

$$= 2[-\left(\frac{\sqrt{3}}{2}\right) + i\left(-\frac{1}{2}\right)]$$

$$= -\sqrt{3} - i \quad \text{<-------------------rectangular form}$$

Similarly, for $k = 3$ in (A) we obtain the root

$$2[\cos 300° + i\sin 300°] \quad \text{<----------------polar form}$$

$$= 2[\cos 60° + i(-\sin 60°)]$$

$$1 - \sqrt{3}i \text{ or } 1 - i\sqrt{3} \quad \text{<-------------------rectangular form}$$

The 4th roots of $16(\cos 120° + i\sin 120°)$ are $\sqrt{3} + i;\ -1 + i\sqrt{3};\ -\sqrt{3} - i;\ \text{and}\ 1 - i\sqrt{3}$

Example 2 If $z^4 = -8 + 8\sqrt{3}i$, solve for z.
Solution

Step 1: Change the right-hand side of the equation to polar form.

$$r = 16, \quad \theta = 120°$$

$$-8 + 8\sqrt{3}i = 16(\cos 120 + i\sin 120)$$

$$\therefore z^4 = 16(\cos 120 + i\sin 120).$$

The rest of the steps are the same as those in Example 1 , above.

Example 3

(Method 1) Find the square root of $5 - 12i$ (This problem was done previously, page 351)

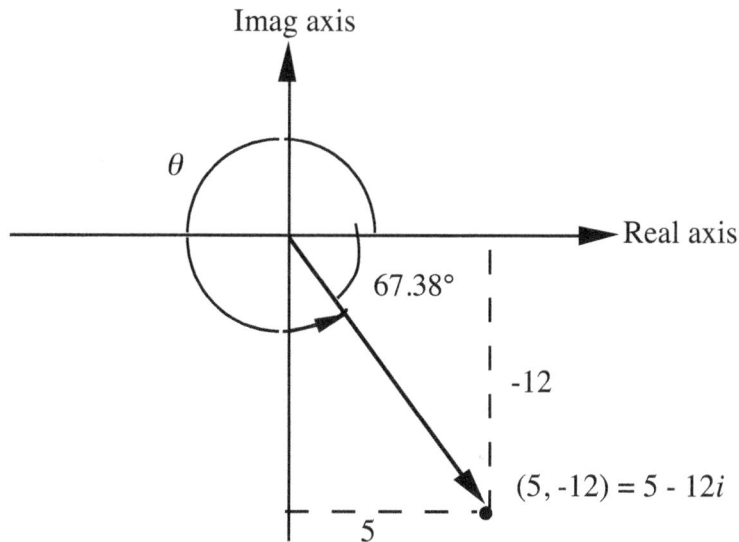

Step 1: Change to polar form.

$a = 5, b = -12$

$r = \sqrt{5^2 + (-12)^2} = 13$

$\tan \theta_{ref} = -\frac{12}{5}$

$\theta_{ref} = 67.38°$

$\theta = 360° - 67.38°$

$\theta = 292.62°$

Now, $r = 13$, $\theta = 292.62°$ and

$5 - 12i = = 13(\cos 292.62° + i \sin 292.62°)$

Step 2: Let $z^2 = 13(\cos 292.62° + i \sin 292.62°)$.

Then $z = \sqrt{13}\left(\cos \frac{292.62° + 360°k}{2} + i \sin \frac{292.62° + 360°k}{2}\right)$, $k = 0, 1$. (Two k-values for $n = 2$)

When $k = 0$, $z = \sqrt{13}\left(\cos \frac{292.62°}{2} + i \sin \frac{292.62°}{2}\right)$

$= \sqrt{13}(\cos 146.31° + i \sin 146.31°)$

$= \sqrt{13}(-.83 + .55i)$

$z = -2.99 + 1.98i$ \qquad (1)

When $k = 1$, $z = \sqrt{13}\left(\cos \frac{292.62° + 360°(1)}{2} + i \sin \frac{292.62° + 360°(1)}{2}\right)$

$z = \sqrt{13}\left(\cos \frac{292.62° + 360°}{2} + i \sin \frac{292.62° + 360°}{2}\right)$

$= \sqrt{13}(\cos 326.31° + i \sin 326.31°)$

$= \sqrt{13}(.83 + i(-.55))$ \qquad ($\cos 326.31° = .83$; \quad $\sin 326.31 = -.55$)

$= \sqrt{13}(.83 - .55i)$

$= 2.99 - 1.98i$ \qquad (2)

The square roots are $-2.99 + 1.98i$ and $2.99 - 1.98i$ (Combining the roots for $k = 0$ and $k = 1$)

Another Method of Finding Square Root (Short-Cut Method)

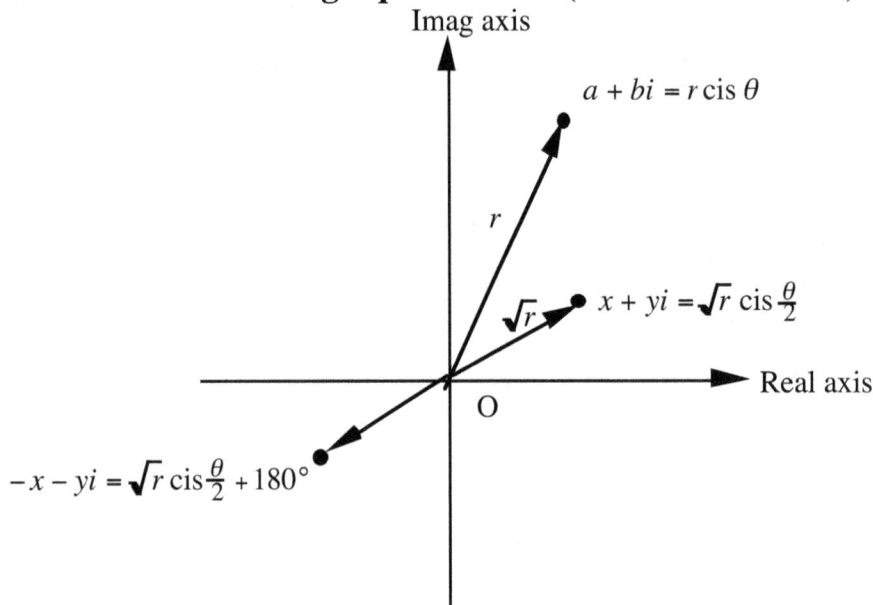

Let $x + yi$ be the square root of $a + bi$

Then $(x + yi)^2 = a + bi$ (1)

$\quad x^2 - y^2 + 2xyi = a + bi$ (2)

Equating the real parts: $x^2 - y^2 = a$ (3)

Equating the imaginary parts: $2xy = b$ (4)

$\quad\quad (\sqrt{r})^2 = x^2 + y^2$ (From figure)

$\quad\quad \therefore \quad r = x^2 + y^2$ (5)

Also, $r^2 = a^2 + b^2$ (From figure)

$\quad\quad\quad r = \sqrt{a^2 + b^2}$ (6)

Equating (5) to (6),

$x^2 + y^2 = \sqrt{a^2 + b^2}$ (7)

We shall apply equations (3), (4), (6), and (7) to find the square root of a complex number, by redoing Example 3, above

Example 3 (Method 2) Find the square root of $5 - 12i$.   (see also p.351-2)

Step 1: Let $x + yi$ be the square root of $5 - 12i$.

$\quad\quad$ From $r^2 = 5^2 + (-12)^2$,

$\quad\quad\quad\quad r = 13$ (1)

$\quad\quad$ Applying $x^2 + y^2 = \sqrt{a^2 + b^2}$

$\quad\quad\quad\quad x^2 + y^2 = 13$ (2)

Step 2: $x^2 - y^2 + 2xy = 5 - 12i$

$\quad\quad$ Equating the real parts: $x^2 - y^2 = 5$ (3)

$\quad\quad$ Equating the imaginary parts: $2xy = -12$ (4)

We now solve for x and y using equations (2), (3) and (4)

Step 3: By adding (2) and (3); we eliminate y^2.

$\quad\quad 2x^2 = 18; \; x^2 = 9$ and $x = +3 \;\; or \;\; -3$

Step 4: We find y by substituting the x-values in equation (4)

When $x = 3$, $2(3)y = -12$ and from which $y = -2$.

When $x = -3$, $2(-3)y = -12$ and from which $y = 2$.

Thus, when $x = 3$, $y = -2$, and when $x = -3$, $y = 2$

Substituting these pairs (of values) in $x + yi$, the square roots are $3 - 2i$ and $3 + 2i$.

Plotting Roots in the Complex Plane

If we graph the roots, all the roots will be found to be equally spaced on a circle of radius, $\sqrt[n]{r}$ (or $r^{\frac{1}{n}}$), and the roots are $\frac{360}{n}$ degrees from one another. The first root is known as the principal nth root. The roots form the vertices of a regular n-sided polygon with its center at the origin.

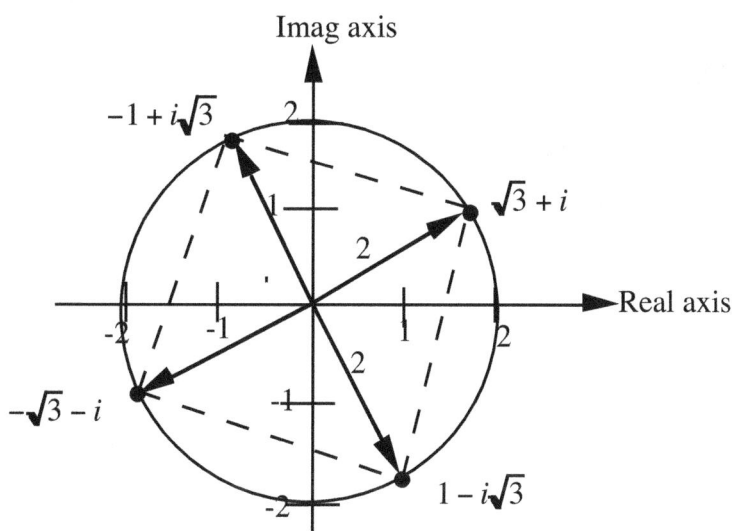

Figure: Graph of the roots for Example 1: $\sqrt{3} + i$; $-1 + i\sqrt{3}$; $-\sqrt{3} - i$; $1 - i\sqrt{3}$.

Lesson 57 Exercises

A Use De Moivre's theorem to evaluate the following:

1. $(2-3i)^4$;   **2.** $(4+4i)^5$;   **3.** $(-3+i)^6$;   **4.** $(4\,cis\,30°)^5$;   **5.** $(-i)^{11}$

Answers: **1.** $-119+120i$;   **2.** $-4096-4096i$;   **3.** $-352-934i$;   **4.** $-886.81+512i$;   **5.** i.

B Find the indicated roots.

1. The cube roots of 8;   **2.** The fifth roots of 1;   **3.** The fifth roots of 3;

4. The square roots of $2-3i$;   **5.** The cube roots of $-1+i$

Find all the roots of the following: **6** $16x^4 = -81$;   **7.** $x^3+8=0$;   **8.** $x^3+8i=0$

Answers:

1. $2,\ -1+i\sqrt{3},\ -1-i\sqrt{3}$;   **2.** $1,\ 0.31+0.95i,\ -0.81+0.59i,\ -0.81-0.59i,\ .31-0.95i$;

3. $\sqrt[5]{3},\ \sqrt[5]{3}(0.31+0.95i),\ \sqrt[5]{3}(-0.81+0.59i),\ \sqrt[5]{3}(-0.81-59i),\sqrt[5]{3}(.31-0.95i)$;

4. $\sqrt[4]{13}(-0.88+0.47i),\sqrt[4]{13}(0.88-0.47i)$;

5. $\sqrt[6]{2}(0.71+0.71i),\ \sqrt[6]{2}(-0.96+0.26i),\ \sqrt[6]{2}(0.26-0.96i)$;

6. $1.061+1.061i,\ -1.061+1.061i,-1.061-1.061i,\ 1.061-1.061i$;

7. $-2,\ 1-i\sqrt{3},\ 1+i\sqrt{3}$;   **8.** $2i,\ -\sqrt{3}-i,\ \sqrt{3}-i$.

CHAPTER 22   375
Lesson 58: **Graphing Polar Coordinates**
Lesson 59: **Graphing Polar Equations**

Introduction to Polar Coordinates

We use polar coordinates in the design of control systems in engineering. Some equations, say, in x and y which are fairly complex in the rectangular forms become relatively simple when changed to the equivalent polar forms. For example, let us consider the equation of the circle

given by $x^2 + y^2 = 9$ <------------rectangular form.

In polar form, $r^2 = 9$ or simply $r = \pm 3$ <------------polar form   (a circle of radius r)

Similarly, in polar form, the equation of the line $y = x$ is given simply by $\theta = 45°$ (for all r)

Lesson 58
Graphing Polar Coordinates

The Polar Coordinate System

As shown in Figures 1 and 2, the **rectangular coordinate system** has two axes, namely the x-axis and the y-axis. These two axes intersect at right angles at the origin which we label $(0, 0)$.
A point P in the plane of this system has just one pair of coordinates (x, y) which is an ordered pair. Also an ordered pair of coordinates (x, y) represents only one point.

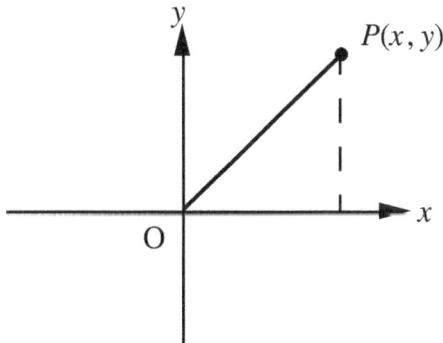

Figure 1: The rectangular coordinate system   **Figure** 2: The polar coordinate System

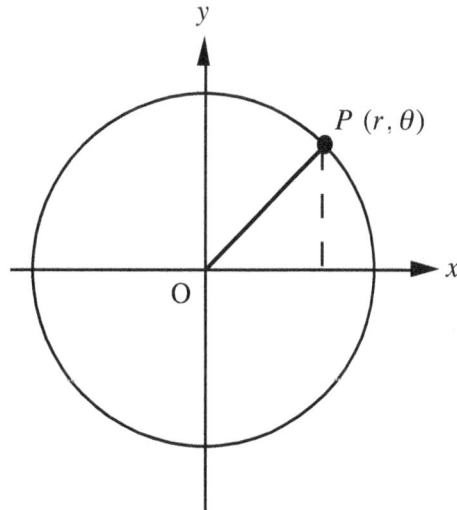

Similarly, the **polar coordinate system** (Figure) has two axes, the x-axis (the **polar-axis** or $0°$-axis) and the y-axis ($\frac{\pi}{2}$ or $90°$ axis) . These two axes intersect at right angles at the origin

(or **the pole**) which we label $(0, \theta)$, where θ is any angle. A **given point** P in the polar coordinate system does **not** have just one pair of coordinates (in contrast to the coordinates in the rectangular coordinate system), but has an unlimited number of polar coordinate pairs. Each polar coordinate pair (r, θ) is ordered. However, we must note that a given **ordered pair** represents only one point.

Thus a given point has many names, but any of the possible names represents only one point, the given point.

A given point in the polar coordinate system has an unlimited number of coordinate pairs because different angular rotations to the same point are possible; and also, the radius can be generated in more than one assumed direction.

For the ordered pair (r, θ), r is the **directed distance** (may be positive or negative) from the origin (the pole) and θ, the **angle of rotation**, may be in degrees or in radians, and may be positive or negative. The angle is positive if measured counterclockwise but negative if measured clockwise. The angles are measured in the same way as done in trigonometry (see page 526).

Although a point has an unlimited polar coordinates, we shall restrict θ so that $0 \le \theta \le 2\pi$. We then obtain what is called the **two primary representations** of the polar coordinates. The primary representations are (r, θ)) and $(-r, \theta + \pi)$.

How to Plot Polar Coordinates

We shall now learn how to **locate a point** in the polar plane given the coordinates (r, θ) of the point. We shall plot on polar graph paper whenever it is available.

Step 1: Assuming we are **standing at the origin** (the pole) and facing in the positive x-direction, we **rotate** through the given angle θ counterclockwise if θ is positive; or clockwise if θ is negative. From this step, we go to Step 2 if r is positive but if r is negative, we go on to Step 3, below.

Step 2: Keeping the direction in which we are facing as a result of our rotation from Step 1, we **count** straight **forwards a distance** r units (r being positive) and then **stop**. We **label** this stopping point (r, θ) if θ is positive or $(r, -\theta)$ if θ is negative.
We then draw r with a **solid line**.

Step 3: If r is negative, then, keeping the direction in which we are facing as a result of rotation from Step 1, we **count** straight **backwards** a distance r units and then **stop**. We **label** this point $(-r, \theta)$ if θ is positive or $(-r, -\theta)$ if θ is negative. Finally, we draw this line segment r with a **broken** (or dotted) **line** to indicate that r is negative.

Example 1   Plot the point having the polar coordinates $(3, 60°)$

Solution

Step 1: Assuming you are at the origin and facing in the positive x-direction (Figure 1.), rotate through $60°$ counterclockwise (draw $60°$ counterclockwise since θ is positive).

Step 2: From the origin count 3 units forwards and stop. Place a dot here and label this stopping point $(3, 60°)$. If θ were in radians, we would label this point $(3, \frac{\pi}{3})$

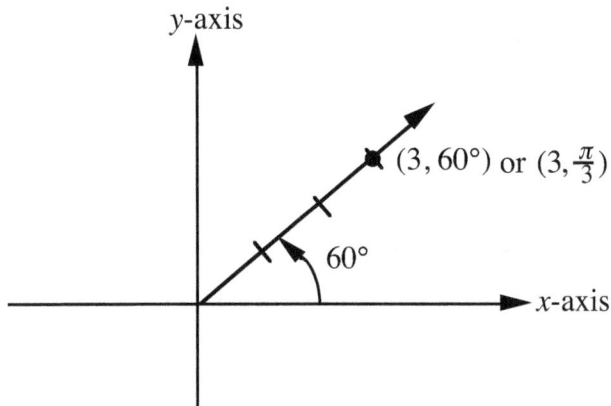

Figure 1 : The graph of the point $P(3, 60°)$

Example 2 Plot the point having the polar coordinates $((3, -60°)$

Solution

Step 1: Since the angle is negative, we draw $60°$ clockwise (Figure 2)

Step 2: From the origin count 3 units forwards and stop. Place a dot here and label this stopping point $(3, -60°)$

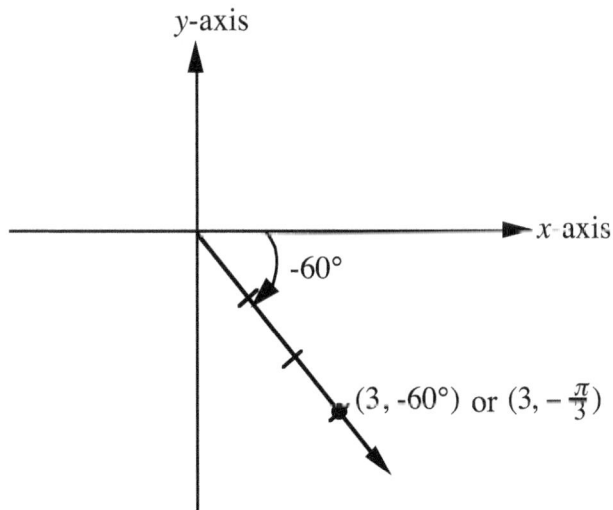

Figure 2 : The graph of the point $P(3, -60°)$

Example 3 Plot the point having the polar coordinates $(-3, 60°)$

Solution

Step 1: Assuming you are at the origin and facing in the positive x-direction (Figure.), rotate through $60°$ counterclockwise (draw $60°$ counterclockwise since θ is positive).

Step 2: Still at the origin and keeping the direction in which you are facing, count 3 units straight backwards along OQ and stop. Place a dot here, and label this point $(-3, 60°)$. Using a broken line draw the radius by connecting OQ.

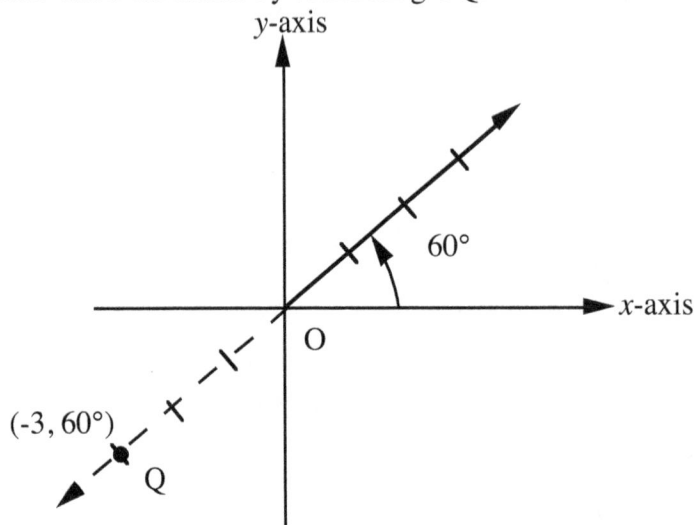

Figure :The graph of the point $Q(-3, 60°)$

Example 4 Plot the point having the polar coordinates $(-4, -60°)$

Step 1: Assuming you are at the origin and facing in the positive x-direction, rotate through $60°$ clockwise (since θ is negative). See Figure 3

Step 2: Still at the origin and keeping the direction in which you are facing as a result of rotation from Step 1, count 4 units straight backwards along OQ and stop. Place a dot here, and label this point $(-4, -60°)$.

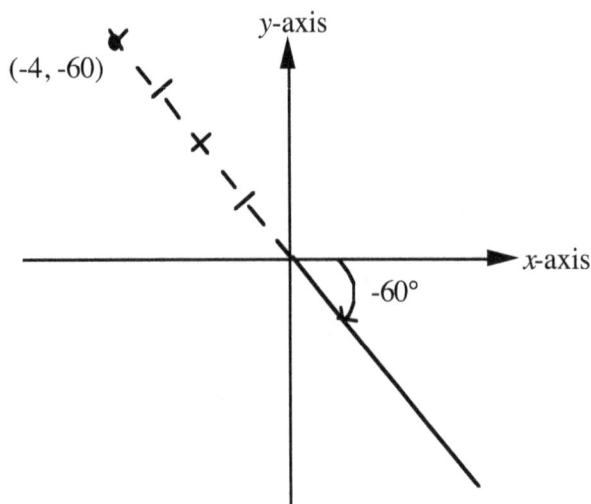

Figure 3 : The graph of the point $(-4, -60°)$

Example 5 Plot the point having the polar coordinates $(-5, -300°)$

Solution

Step 1: Rotate through $300°$ clockwise (since θ is negative).

Step 2: Count 5 units backwards and stop. Place a dot here and label this point $(-5, -300°)$.

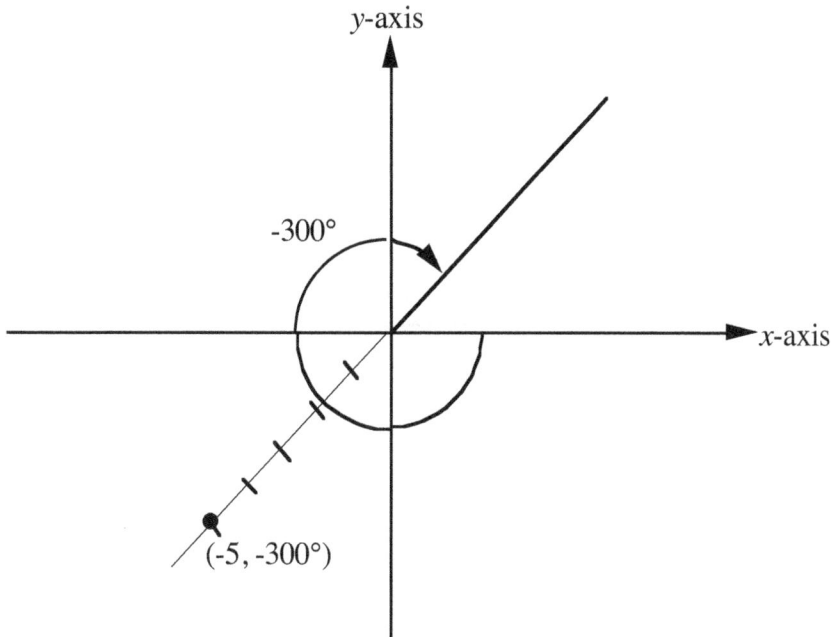

Note: In plotting polar coordinates, we must note that the signs of the ordered pairs (r, θ) depend solely on **how the angles and distances are generated** and do **not** depend on the **quadrant** in which the point happens to be located, unlike the case of the rectangular coordinate system, in which the quadrant determines the signs of the coordinates.

The coordinates (r, θ) and, $(-r, \theta + \pi)$ represent the same point. These two coordinate pairs are the two primary representations of the same given point. There are infinitely many other representations each of which is obtained by adding 2π or (or -2π) successively to the primary representation.

Other representations of the same point: are

$(r, \theta + 2\pi)$ and $(-r, \theta + 3\pi)$.

$(r, \theta - 2\pi)$ and $(-r, \theta - \pi)$.

Determining if a given point is on a curve, given the polar equation

Definition: A given point (except the pole) is on a polar form curve if and only if at least one of the primary representations of the point satisfies the given equation.

Thus if we check for one primary representation and it does not satisfy the given equation, we must also check for the other primary representation.

To check for a point, we substitute the given coordinates in the equation to determine if the left-hand side equals the right-hand side of the equation. If the LHS equals the RHS the point is on the curve, otherwise, it is not on the curve

Lesson 58 Exercises

Plot the following polar coordinates:

1. $(4, 30°)$;   **2.** $(-3, 90°)$;   **3.** $(-3, \frac{\pi}{3})$;   **4.** $(-6, -30°)$;   **5.** $(-4, -400°)$

2.

1.

3.

4.

5.

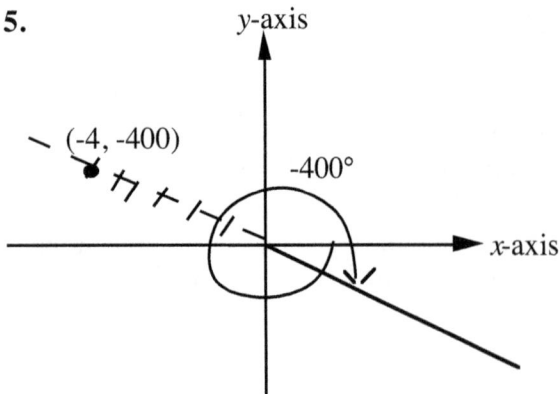

Lesson 59
Graphing Polar Equations

In general, we can sketch the graphs of polar equations by constructing tables for r and θ and plotting points. Consider a table for the polar equation $r = \sin\theta$ (See Table 1). It can be observed from the table that after π radians or 180° the values of r begin to duplicate. We can therefore take advantage of such duplication when sketching polar graphs. We can do this by testing to determine if the graph is symmetric about (a) the x-axis, (b) the y-axis, and (c) the origin.

Table 1

θ	0°	30°	45°	60°	90°	120°	135°	150°	180°	210°	240°	225°	270°	300°	315°	330°	360°
$r = \sin\theta$	0	.5	.71	.87	-1	.87	.71	.5	0	-.5	-.87	-.71	-1	-.87	-.71	-.5	0

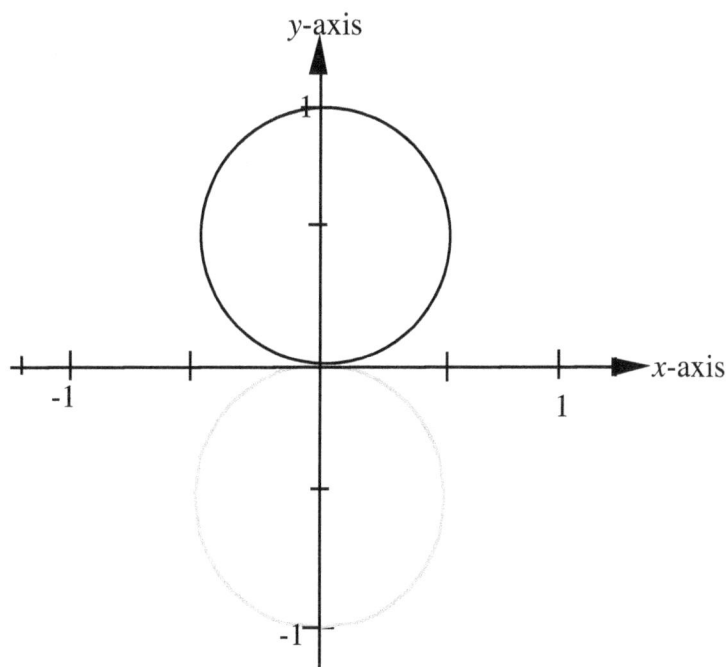

Rules for testing for symmetry about the x-axis

Step 1: Replace (r, θ) by either $(r, -\theta)$, or $(-r, -\theta + \pi)$ and simplify If the equation remains unchanged on simplifying then the curve is symmetric about the x-axis.

Step 2: if $(r, -\theta)$ fails the test, we must check for $(-r, -\theta + \pi)$. If both $(r, -\theta)$ and $(-r, -\theta + \pi)$ fail the test, then the curve is **not** symmetric about the x-axis However if either $(r, -\theta)$ or $(-r, -\theta + \pi)$ passes the test first, then we do not have to test for the other.

Although we can test for symmetry before sketching polar curves , for common practical purposes and to save time, we may not test for symmetry for the common polar graphs. We shall familiarize ourselves with or memorize the properties of the common or simple polar curves (the same as we familiarized ourselves with the simple trigonometric functions such as $y = \sin x$.). When we meet an unfamiliar or a more complicated polar equation, then we shall test for symmetry We shall now

discuss and sketch some common polar graphs. In the future, we should be able to identify $\qquad$
these graphs with the corresponding equations and vice versa.

The following trigonometric identities will be useful when testing for symmetry and graphing polar equations:

1. $\cos(-\theta) = \cos\theta$

2. $\sin(-\theta) = -\sin\theta$

3. $\tan(-\theta) = -\tan\theta$

4. $\cos(\pi - \theta) = -\cos\theta$ **6.** $\cos(\pi + \theta) = -\cos\theta$

5. $\sin(\pi - \theta) = \sin\theta$ **7.** $\sin(\pi + \theta) = -\sin\theta$

Also, $\cos(\theta \pm 2\pi) = \cos\theta$; $\quad \sin(\theta \pm 2\pi) = \sin\theta$

The Rose Curves: The equations of the rose curves are

$$r = a\sin n\theta \quad \text{(where } n \text{ ia a positive integer)}$$
$$r = a\cos n\theta$$

If n is odd each curve has n leaves but has $2n$ leaves if n is even.

Example 1 If $n = 1$ (odd number), the sine equation, $r = a\sin n\theta$ becomes $r = a\sin\theta$ and the curve has one leaf (petal). The leaf is circular and has the y-axis as the axis of symmetry (Figure 1).
Example: $r = 2\sin\theta \qquad$ (where $a = 2,$)

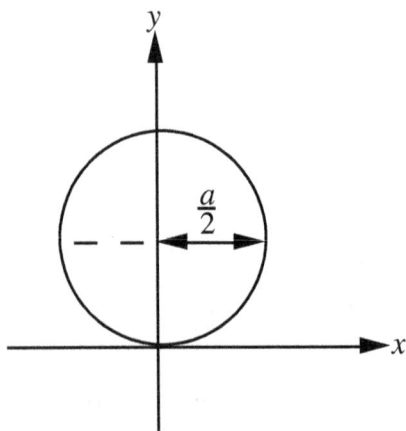

Figure 1 : $r = a\sin\theta$ $\qquad$ **Figure 2 :** Graph of $r = 2\sin\theta$

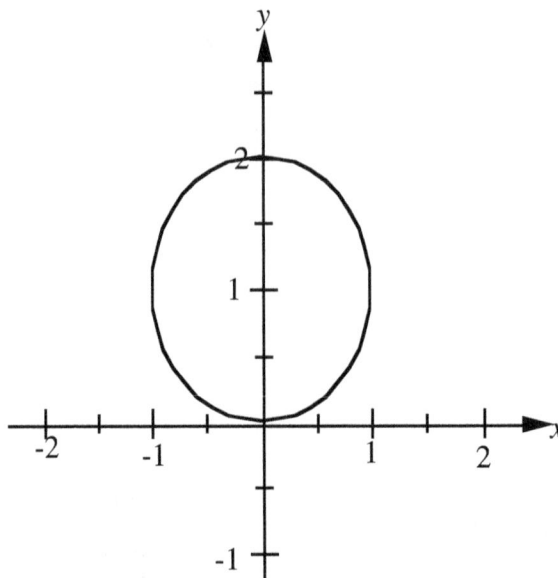

Example 2 If $n = 1$, the cosine equation becomes $r = a\cos\theta$. This curve also like that of the sine has one circular leaf, but the axis of symmetry is the x-axis.
Example: $r = 2\cos\theta$ (where $a = 2$,)

-

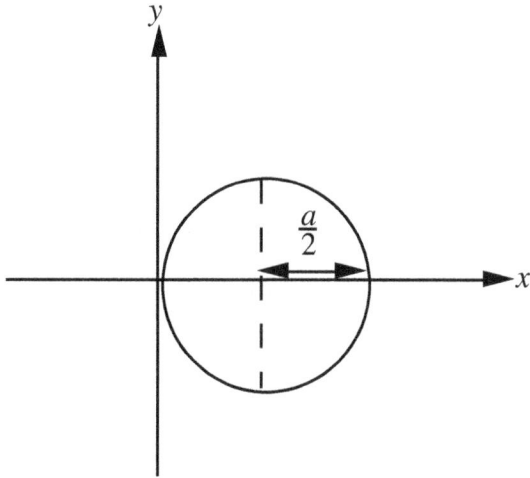

Figure Graph of $r = a\cos\theta$

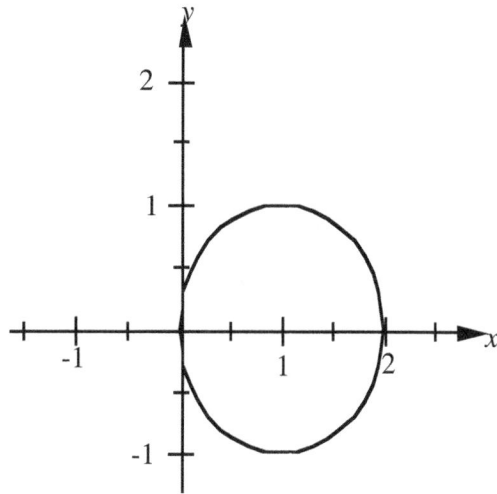

Figure: Graph of $r = 2\cos\theta$

Example 3 Sketch the polar graph for $r = -a\sin\theta$

Solution The graph is the graph of $r = -a\sin\theta$ reflected about the y-axis.
Compare Figure and

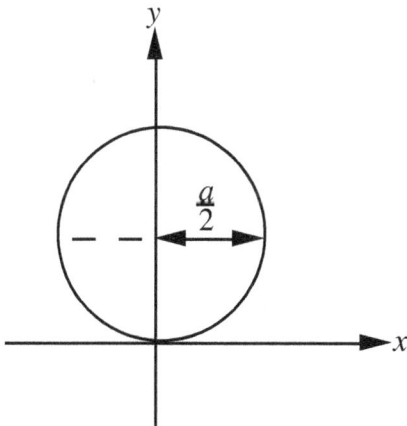

Figure : Graph of $r = a\sin\theta$

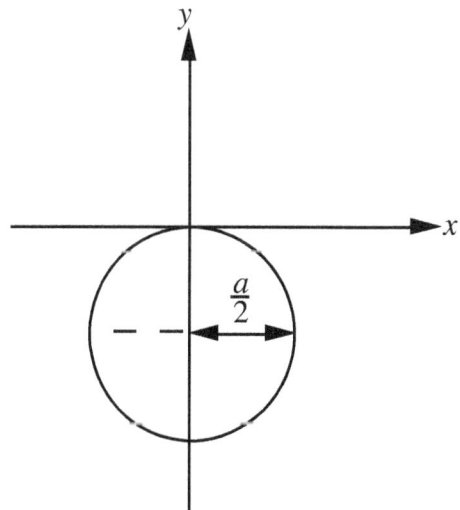

Figure : Graph of $r = -a\sin\theta$

Example 4 Sketch the polar graph for $r = -a\cos\theta$.

Solution The graph of $r = -a\cos\theta$ is the graph of $r = a\cos\theta$ reflected reflected about the y-axis. Compare Figure and

Figure: $r = -a\cos\theta$

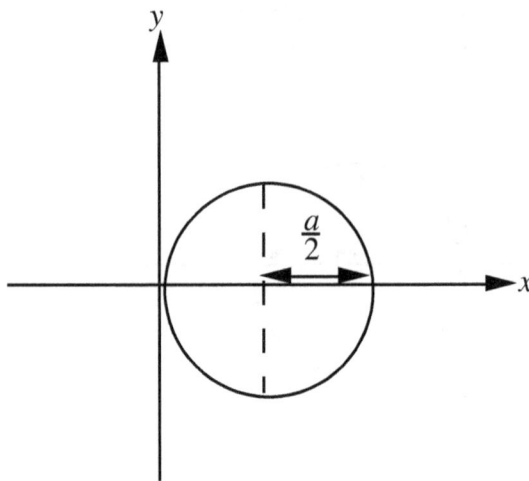

Figure: $r = a\cos\theta$

Example 5 Sketch the graph of $r = 3\sin 2\theta$
Solution

Since n is even ($n = 2$), there are 2(2) or 4 leaves or petals. There are $\frac{360°}{4} = 90°$ between the axes of any two leaves. The first leave is in the first quadrant. The leaves are symmetric about the lines $y = x$, and $y = -x$ (Figure 203)

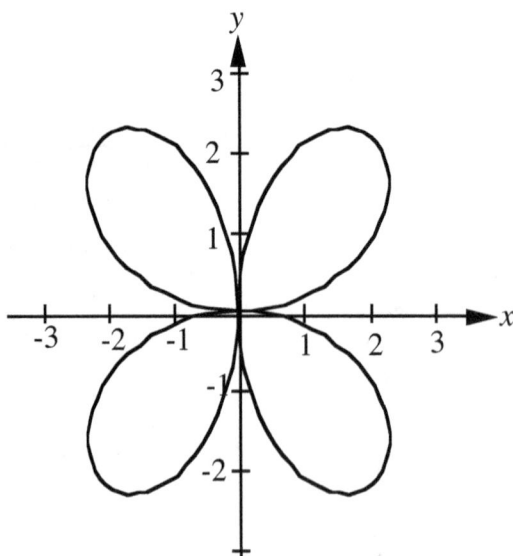

Example 6   Sketch the graph of $r = 3\cos 2\theta$

Solution

Since $n = 2$, there are 2(2) or 4 leaves. The are $\frac{360°}{4} = 90°$ between the axes of any two leaves (as in Example 5 above). However, the axes of symmetry are the x- and y-axes (Figure).

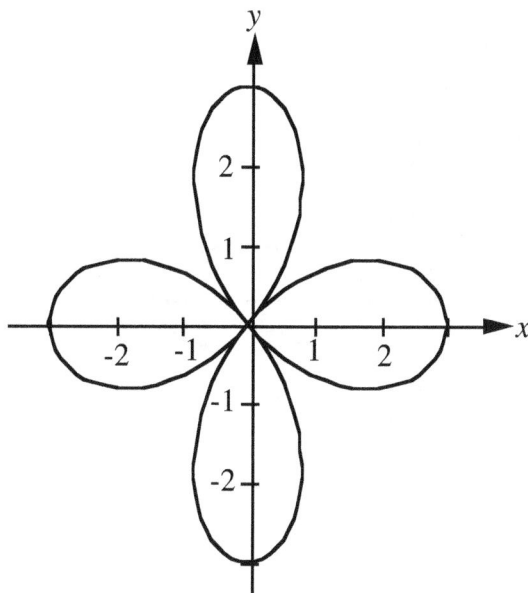

Example 7 Sketch the graph of $r = 4\sin 3\theta$

Solution   Since n is odd and $n = 3$, there are 3 leaves. There are also $\frac{360°}{3} = 120°$ between the axes of any two leaves. The first leave is in the first quadrant and it is symmetric about the line $\frac{\pi}{6}$ (Figure).

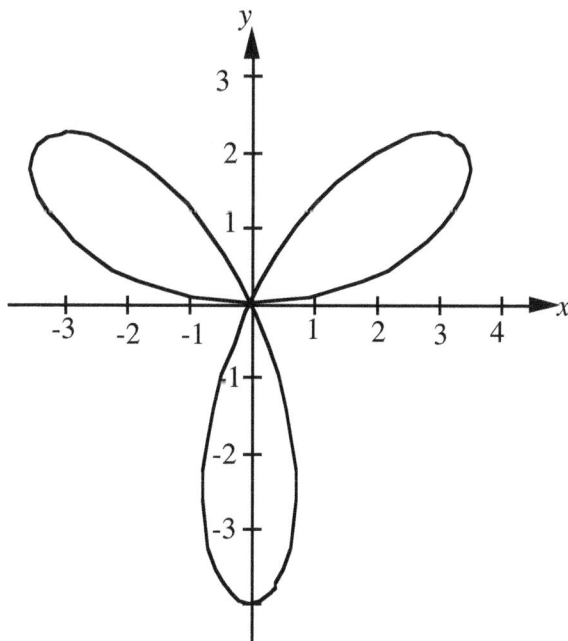

Example 8 Sketch the graph of $r = 4\cos 3\theta$

Solution The graph is similar to that in Example 7, except that there is a difference in the location of the petal (leaf) axes (Figure)

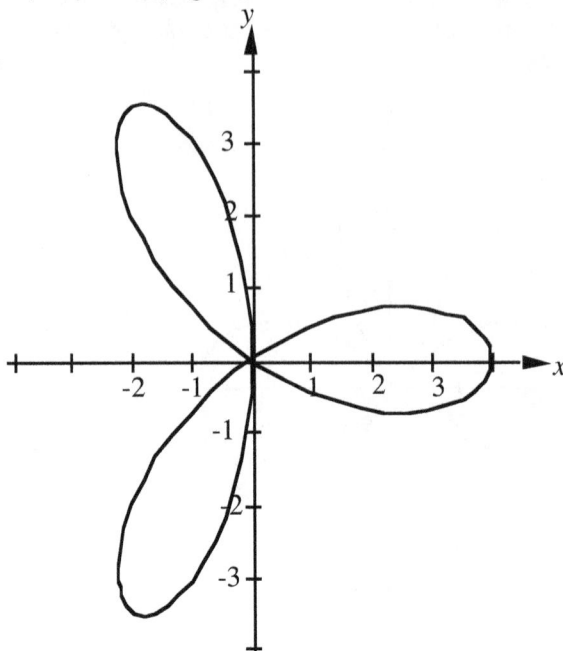

Other polar Graphs

The graphs of the Limacon $r = a \pm b\sin\theta$; $r = a \pm b\cos\theta$; $r^2 = a^2 \sin 2\theta$; $r^2 = a^2 \cos 2\theta$

Lesson 59 Exercises

Identify (e.g., rose curve) and sketch the graph of each equation.

1. $r = 2\cos\theta$;   **2.** $r = 5\sin\theta$;   **3.** $r = -4\sin\theta$;   **4.** $r = 4\sin 3\theta$;   **5.** $r = 5\cos 3\theta$;

6. $r = 4$;   **7.** $r = 2\theta$;   **8.** $r = 4 - 3\sin\theta$;   **9.** $r^2 = 5\cos 2\theta$;   **10.** $r = 3 - 3\cos 3\theta$;

11. $r = e^{2\theta}$;   **12.** $\theta = \frac{\pi}{6}$.

2.

1.

3.

4.

5.

6.

7.

8.

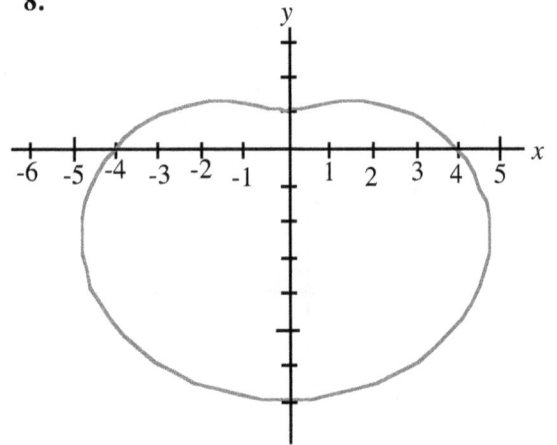

Lesson 59: Graphing Polar Equations

9.

10.

11.

12

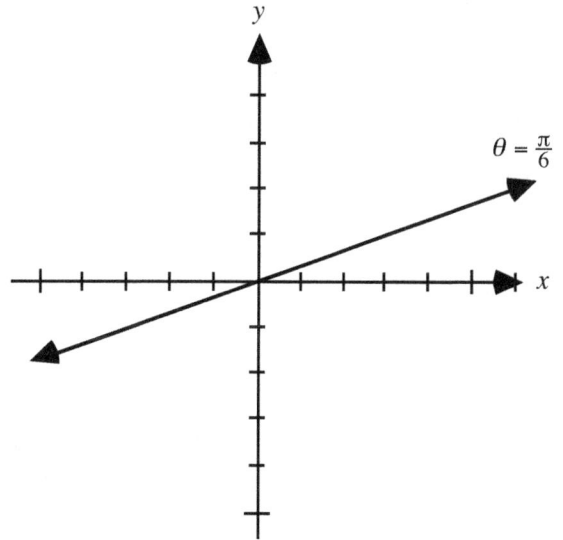

$\theta = \frac{\pi}{6}$

CHAPTER 23A

Conic Sections
Lesson 60: **Circles**
Lesson 61: **Parabolas**
Ellipses (See Chapter 23B)
Hyperbolas (See Chapter 23B)

Some Preliminaries

The conics include the circle, the parabola, the ellipse and the hyperbola

Before we begin with the conics, we shall discuss some terms which are useful in describing some conic properties.

Definitions:

Edge: An edge is a line of intersection of surfaces.

Vertex. A vertex is a point of intersection of edges.

Locus The locus of a point is the collection of points each of which satisfies given conditions.

Vertex of a parabola: The vertex of a parabola is the point where the parabola intersects its axis of symmetry

Projection of a point on a line: The projection of a point P on a line L is the foot of the perpendicular from P to L.

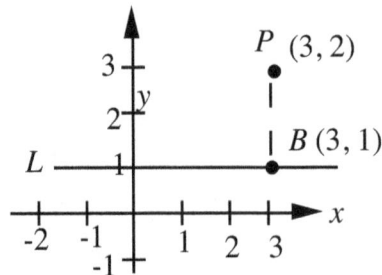

Figure: The projection of $(3, 2)$ on line L is $(3, 1)$

Projection of a line segment on another line: The projection of a line segment AB on (another) line segment CD is the segment of CD that is cut off by the perpendiculars drawn from the ends of AB to CD. (Recall the trajectory of the rays from a projector on to the screen at a movie theater)

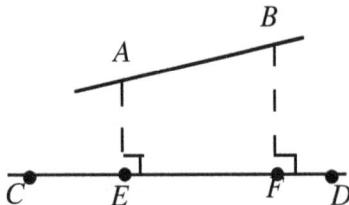

Figure The projection of AB on line CD is EF.   **Figure** The projection of AB on line CD is AF.

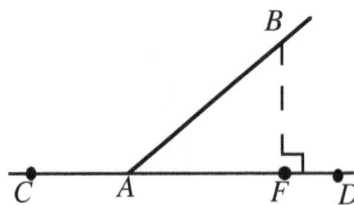

Lesson 60

Circles

Definition: A circle is the set of points in a plane that are equidistant from a fixed point (called the center of the circle). The distance from the center to any point on the circle is called the radius of the circle.

Equation and Graph of a Circle

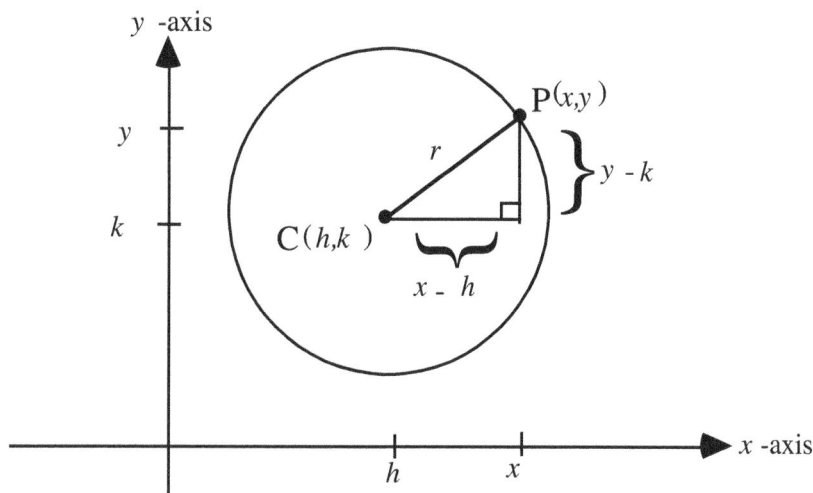

Figure 1

We shall derive an equation of a circle with its center at the point $C(h, k)$ (Figure 1, above) and having radius, r, by applying the Pythagorean theorem.

Let the distance between the center $C(h, k)$ of the circle and a point $P(x,y)$ on the circle be r.
Applying the Pythagorean theorem (since we have a right triangle),

$$r^2 = (x - h)^2 + (y - k)^2 \text{ or rewriting,}$$

center-radius form----> $(x - h)^2 + (y - k)^2 = r^2$ (1)

Equation (1) is the center-radius form of the equation of a circle, because in this form, we can read the radius and the coordinates of the center of the circle by inspection.

Note: We can also obtain the above equation from the distance formula as follows:
Replace d by r; x_2 by x; x_1 by h; y_2 by y , and y_1 by k in

$$d = \sqrt{(x_2 - x_1)^2 + (y_2 - y_1)^2} \quad \text{<--------- distance formula.}$$

Then we obtain $r = \sqrt{(x - h)^2 + (y - k)^2}$

Squaring both sides of the equation, $r^2 = (x - h)^2 + (y - k)^2$
Rewriting, $(x - h)^2 + (y - k)^2 = r^2$ <--------- center-radius form of the equation of a circle.

If we expand equation (1), $(x - h)^2 + (y - k)^2 = r^2$, we obtain
$$x^2 + y^2 - 2hx - 2ky + (h^2 + k^2 - r^2) = 0 \qquad (2)$$
If we let $C = -2h$, $D = -2k$, $E = h^2 + k^2 - r^2$ in equation (2), we obtain
$$x^2 + y^2 + Cx + Dy + E = 0 \qquad (3)$$

Equation (3) is the **general** expanded **form** of the equation of a circle, and every circle has an equation of this type.

Lesson 60: Circles

Equation of the circle with center at the origin $(0,0)$

If we let $h=0, k=0$ in equation (1), we obtain

$$(x-0)^2+(y-0)^2=r^2$$
$$x^2+y^2=r^2 \qquad (4)$$

Equation (4) is the equation a circle with the center at the origin $(0,0)$ (Figure 2).

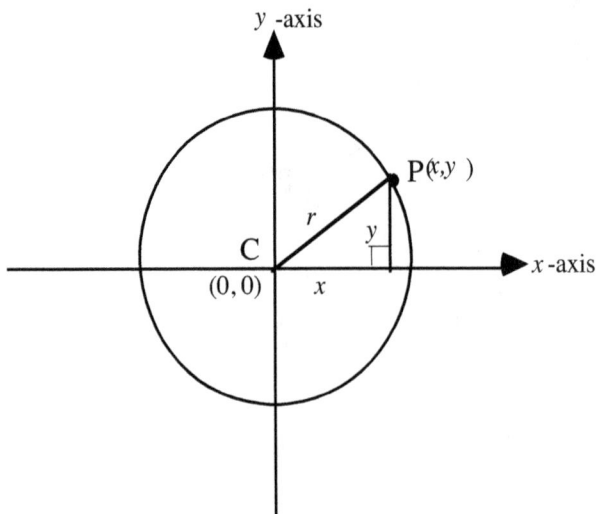

Figure 2

Example 1 (a) Write an equation for the circle with radius 5 and center at $(2,-3)$.
(b) Sketch the graph of the circle.

(a) The center-radius form of the equation of a circle is given by
$(x-h)^2+(y-k)=r^2$, where r is the radius of the circle, $h=$ the x- coordinate of the center, and $k=$ the y-coordinate of the center of the circle.

Substituting $r=5, h=2, k=-3$ in
$(x-h)^2+(y-k)^2=r^2$, we obtain
$(x-2)^2+(y-(-3))^2=5^2$
$(x-2)^2+(y+3)^2=5^2$ <-------center-radius form.
or $(x-2)^2+(y+3)^2=25$.

Expanding this equation: $x^2-4x+4+y^2+6y+9=25$
$x^2+y^2-4x+6y+4+9-25=0$
$x^2+y^2-4x+6y-12=0$ <-------------general form.

(b) Graph

Step 1: Plot the coordinates (2, -3) of the center of the circle.

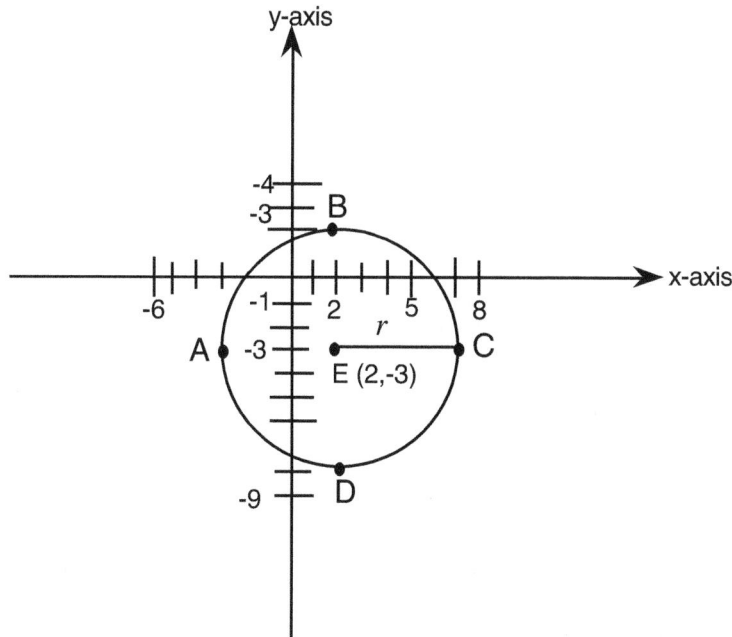

Step 2: Since the radius is 5, count 5 units each, horizontally, from the center (2,-3) to the points A, and C; and also count 5 units each, vertically, from the center (2, -3) to the points B and D. Connect these four points by a smooth "circular " curve to obtain the circle.

Note: Use the same scale for the x- and y-axes otherwise the circle would be distorted and would resemble an ellipse. (see Chapter 23B)

Note also that for a circle, the distance from the center of the circle to any point on the circumference is the same as the distance from the center to any other point on the circumference. Keep this in mind when drawing a circle.

Example 2 Write an equation for the circle with radius 5 and center at (0, 0).
Solution
The center-radius form of the equation of a circle is given by
$$(x - h)^2 + (y - k) = r^2,$$ where r is the radius of the circle.
In this problem $h = 0, k = 0, r = 5$ and substituting these values in

$(x - h)^2 + (y - k)^2 = r^2$ wc obtain

$(x - 0)^2 + (y - 0)^2 = 5^2$

$x^2 + y^2 = 5^2$ <---- Equation of a circle of radius 5 units and with center at the origin (0, 0)

Example 3 Find an equation of the circle with center at the point (4, -2) and passing through the point (-3, 6).

If we know the radius, r, and the coordinates h, k of the center of a circle, we can immediately write down an equation for the circle by applying
$$(x - h)^2 + (y - k)^2 = r^2. \qquad (1)$$

We are given $h = 4, k = -2$, but we do not know r yet; however, we can find it.

Finding r:

Since the point $(-3, 6)$ is given to be on the circle, we may substitute this ordered pair in equation (1).

Step 1: Substitute $h = 4, k = -2, x = -3, y = 6$

in equation (1) and solve for r.

$$(-3 - 4)^2 + (6 - (-2))^2 = r^2$$

$$(-7)^2 + (6 + 2)^2 = r^2$$
$$(-7)^2 + 8^2 = r^2$$
$$49 + 64 = r^2$$
$$113 = r^2$$
$$\sqrt{113} = r$$
$$\text{and } r = \sqrt{113}$$

Step 2: Substitute $h = 4, k = -2, r = \sqrt{113}$ (or $r^2 = 113$) in equation (1) above.

$\therefore$ an equation for the given circle is $(x - 4)^2 + (y + 2)^2 = 113$ <-------center-radius form.

Finding the center and radius of a circle (Given the center-radius form of the equation of a circle)

Example 1 Find the center and radius of the circle whose equation is given by

$$(x + 4)^2 + (y - 3)^2 = 36$$

Procedure: The center-radius form of the equation of the circle with center at the point (h, k) and with radius r is given by:
$$(x - h)^2 + (y - k)^2 = r^2 \qquad\qquad (1)$$

Compare $(x + 4)^2 + (y - 3)^2 = 36$ with equation (1) above.

Then $(x - h)^2 = (x + 4)^2$; $(y - k)^2 = (y - 3)^2$; and $r^2 = 36$
$$x - h = x + 4 ; \qquad y - k = y - 3 \qquad r = \sqrt{36} = 6$$
$$-h = 4 ; \qquad\qquad -k = -3$$
$$h = -4 \qquad\qquad k = 3$$
Therefore $h = -4$; $k = 3$; and $r = 6$.

The center of the circle is at $(-4, 3)$; and its radius is 6 units.

Shortcut method: Step 1: Set $x + 4 = 0$ and solve to obtain $x = -4$ (Thus $h = -4$
Step 2: Set $y - 3 = 0$ and solve to obtain $y = 3$ (Thus $k = 3$
The center of the circle is at $(-4, 3)$;

With some practice, we shall be able to read, by comparison with equation (1) above, the coordinates of the center of the circle without performing any calculations.

Finding the center and radius of a circle given the general form of the equation of a circle.

Example 2 Find the center and radius of a circle whose equation is given by
$$x^2 + y^2 + 8x - 6y - 11 = 0$$

Procedure: We shall group the x-terms, group the y-terms, and complete the square on the x-terms, and on the y-terms. (see page 30 and review how to complete the square.)

Step 1: $x^2 + y^2 + 8x - 6y - 11 = 0$
$$x^2 + 8x + y^2 - 6y = 11 \text{..............................} (1)$$

Step 2: Complete the squares:

* $x^2 + 8x + 16 + y^2 - 6y + 9 = 11 + 16 + 9$ ....(2)

(For the x-terms: $x^2 + 8x + (4)^2$)

$x^2 + 8x + 16 + y^2 - 6y + 9 = 36$ (For the y-terms : $y^2 - 6y + (-3)^2$)

$(x + 4)^2 + (y - 3)^2 = 36$ <---From this center-radius form, follow the steps exactly as in Example 1 above.

The center of the circle is at (-4, 3); and its radius is 6 units.

* Whenever we add any quantity to the left-hand side of equation (1), we must add the same quantity to the right-hand side of the equation: First, we added 16 (i.e.,4^2) for the x-terms, to both sides of the equation; then we added 9 (i.e.,$(-3)^2$) for the y-terms to both sides of the equation.

The Imaginary Circle

Consider the center-radius form of the equation the circle $(x - 1)^2 + (y - 3)^2 = -16$.

The radius, $\sqrt{-16}$, which is imaginary. We therefore conclude that we do not have a real circle.

Moreover, since the left-hand side is the sum of squares (which is never negative), the above equation is false.

Lesson 60 Exercises

A   1. Find the projection of $(3, 5)$ on (a) the x-axis; (b) the y-axis

2. Indicate the projection of AB on CD in Figure

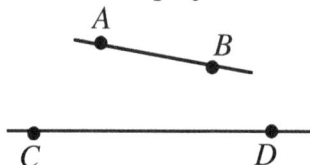

Answers: 1. (a) $(3, 0)$, (b) $(0, 5)$

B (a) Write an equation for the circle with radius 4 and center at (-2, 3).

(b) Sketch the graph of the circle.

Solution: $(x + 2)^2 + (y - 3)^2 = 16$<------- center-radius form.

C Find an equation of the circle with center at the point $(2, 3)$ and passing through the point $(-4, 1)$.

Answer: $(x - 2)^2 + (y - 3)^2 = 40$

D Find the center and radius of the circle whose equation is given by

$$(x - 5)^2 + (y + 2)^2 = 16$$

Answer: Center at (5, -2); and radius, 4 units

E Find an equation for each circle with the given properties.

1. Radius 4 units with center at the origin.

2. Radius 3 units and center at (-3, 2).

3. Radius 5 units and center at (-2, -4)

Problems 4-6: Find the center and radius of each of the given circle

4. $x^2 - 2x + y^2 - 4y - 11 = 0$;   **5.** $x^2 + y^2 + 4x - 6y = 12$;   **6.** $x^2 + y^2 - x - \dfrac{y}{2} = \dfrac{63}{16}$.

Answers: **1.** $x^2 + y^2 = 16$;   **2.** $(x + 3)^2 + (y - 2)^2 = 9$;   **3.** $(x + 2)^2 + (y + 4)^2 = 25$

      4. Center at (1, 2), radius 4 units;   **5.** Center at (-2, 3), radius 5 units;

      6. Center at $(\tfrac{1}{2}, \tfrac{1}{4})$, radius $\dfrac{\sqrt{17}}{2}$ units.

Lesson 61   397
Parabolas

In chapter15, we covered parabolas as functions. In this chapter, we shall cover other aspects such as the geometry and applications in physics and engineering design.

Definition: A parabola is the collection of points in a plane such that each point is at the same distance from a fixed point F (called the focus) as it is from a fixed line D, (called the directrix) with the focus F, not lying on the directrix

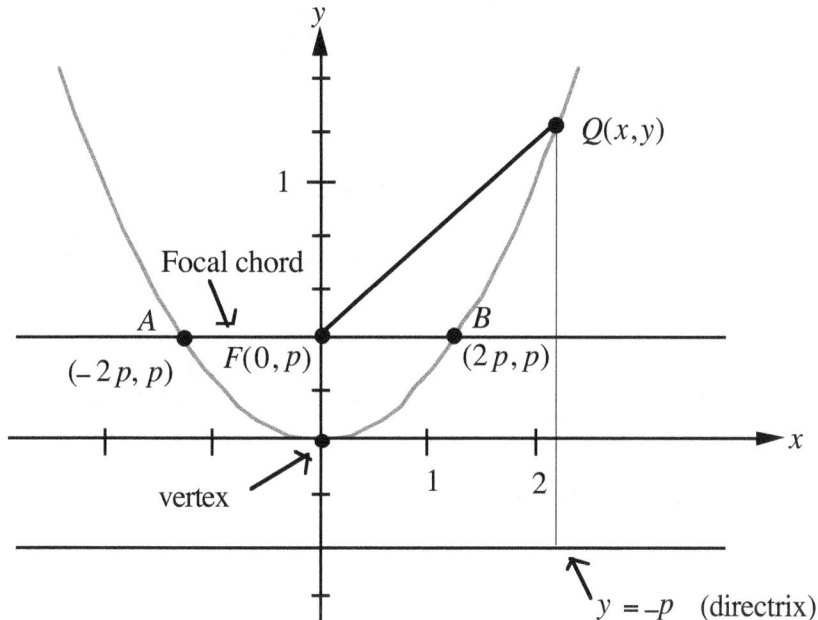

Figure 1 :The graph of $x^2 = 4py$

We repeat some of the properties of a parabola, covered in Chapter 15 , plus additional properties relevant to the present discussion

General Properties of a Parabola

(1) The parabola is an **U-shaped** curve.

(2) The curve has an axis of symmetry. The axis of symmetry divides the curve into two equal halves.

(3) The axis of symmetry may be the y-axis (Figures 2 & 3). the x-axis (Figures 4 & 5), a vertical line, or an oblique line. (Any of the above mentioned axes may be obtained from any other axes by the proper translation or rotation of the x- and y-axes)

(4). The axis of symmetry intersects the curve at a point, V, called the vertex of the parabola.

(5) The axis of the parabola passes through an important point F inside the U shape. This point is called the focus (the eye). The focus is an important point in practical applications. The focus is always inside the U-shape, a fact to remember when sketching the graphs of parabolas.

(6) A line AB intersects the axis of the parabola at right angles at the focus. This line segment is called the focal   chord or latus rectum of the parabola.

(7)   A line called the directrix intersects the axis of symmetry at right angles outside the U-shape. The distance from the focus to the vertex equals the distance from the vertex to the directrix, an important fact to remember when sketching the directrix.

(8). The U-shape of the parabola can "open" in all possible directions. In Figures 2, .3, .4, and 5, some of the possible directions are shown.

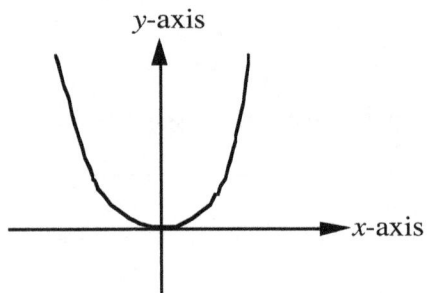

Figure 2 : The axis of symmetry
is the positive *y*-axis

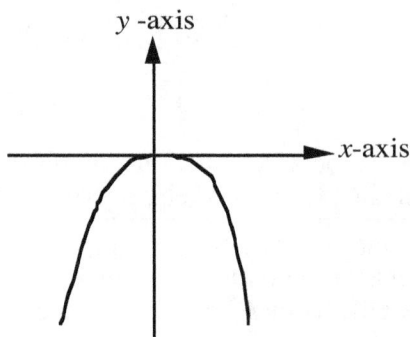

Figure 3 The axis of symmetry
is the negative *y*-axis

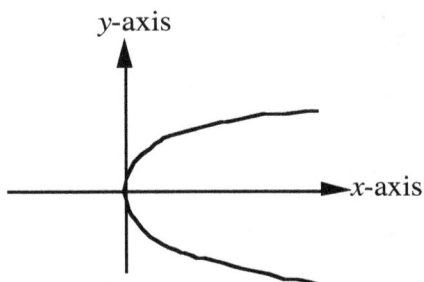

Figure 4 : The axis of symmetry
is the positive *x*-axis

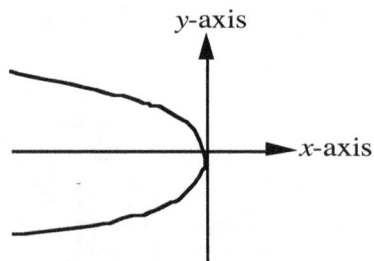

Figure 5 The axis of symmetry
is the negative *x*-axis

Note that Figures 4 and 5 are not graphs of functions

In Chapter 15, we learned how to sketch an identify the graph of the parabola $y = x^2$. In that chapter, we did not derive this equation. In this section, we will derive this equation by identifying additional properties of the parabola

Derivation of equations of parabolas

Case 1: The equation $4py = x^2$

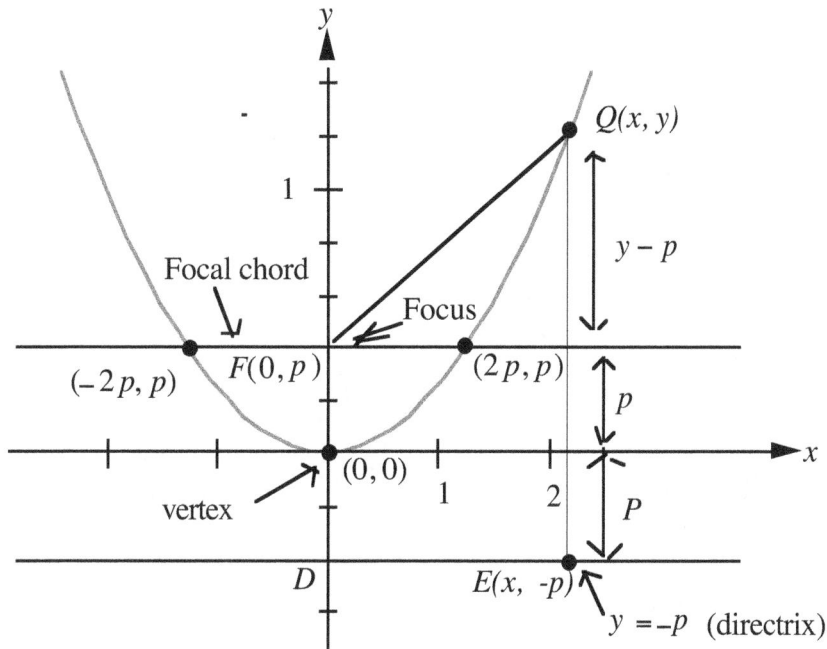

Figure 6

Let the directrix be the line $y = -p$ where $p > 0$. Let the coordinates of the focus F be $(0, p)$, and the coordinates of a point Q on the parabola be (x, y). Then, the coordinates of E are $(x, -p)$, where QE is perpendicular to the directrix. From the definition of the parabola, $|FQ| = |QE| = y + p$.
$|FQ|$ is the distance from $F(0, p)$ to $Q(x, y)$ which, by the distance formula is

$$\sqrt{x^2 + (y - p)^2}.$$

Since we have shown that $|FQ|$ is also equal to $y + p$, we obtain

$$(y + p)^2 = x^2 + y^2 - 2py + p^2$$
$$y^2 + 2py + p^2 = x^2 + y^2 - 2py + p^2$$
$$2py = x^2 - 2py$$
$$4py = x^2 \text{ or}$$
$$x^2 = 4py$$

If $p = \frac{1}{4}$ (the focus is at $(0, \frac{1}{4})$) and the directrix is $y = -\frac{1}{4}$, then $y = x^2$ which is the very familiar equation of the parabola with the vertex at the origin and with the y-axis (the line $x = 0$) as the axis of symmetry. The curve is concave upwards.

Lesson 61: Parabolas

Case 2: The Equation of $x^2 = -4py$ 400

If we reflect (p.219) $x^2 = 4py$ (from Case 1) about the x-axis, we obtain
$$x^2 = -4py.\qquad \text{(Figure 7)}$$
If $p = \frac{1}{4}$, $x^2 = -4py$ becomes $x^2 = -y$ or
$y = -x^2$ which is the equation of the parabola which is concave downwards.

The graph of $x^2 = -4py$ is symmetric about the y-axis and has its focus at $(0, -p)$ and its directrix is the line $y = p$.

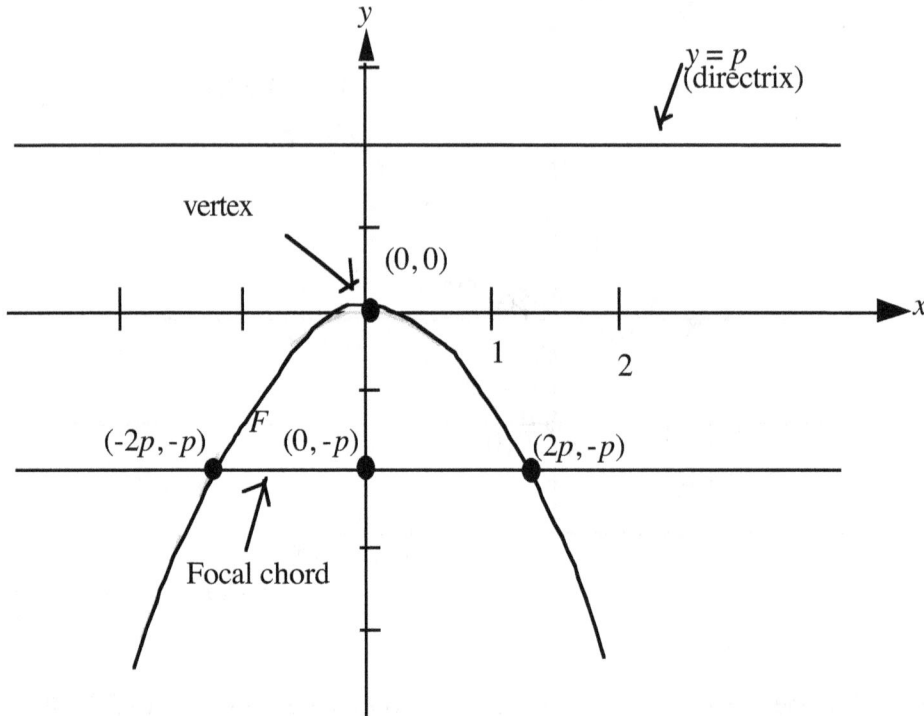

Figure 7 The graph of $x^2 = -4py$

Case 3: The Equation $y^2 = 4px$

By a similar process as in Case 1 , we can derive this equation, if the focus is at $(p,0)$ where $p > 0$ and the directrix is $x = -p$, and obtain

$$y^2 = 4px$$

For real values of y, if p is positive, x must be positive and therefore negative values of x are excluded.

We may **also** derive $y^2 = 4px$ from $x^2 = 4py$ by interchanging x and y.

The graph of this equation is concave to the right and the vertex is at $(0,0)$.

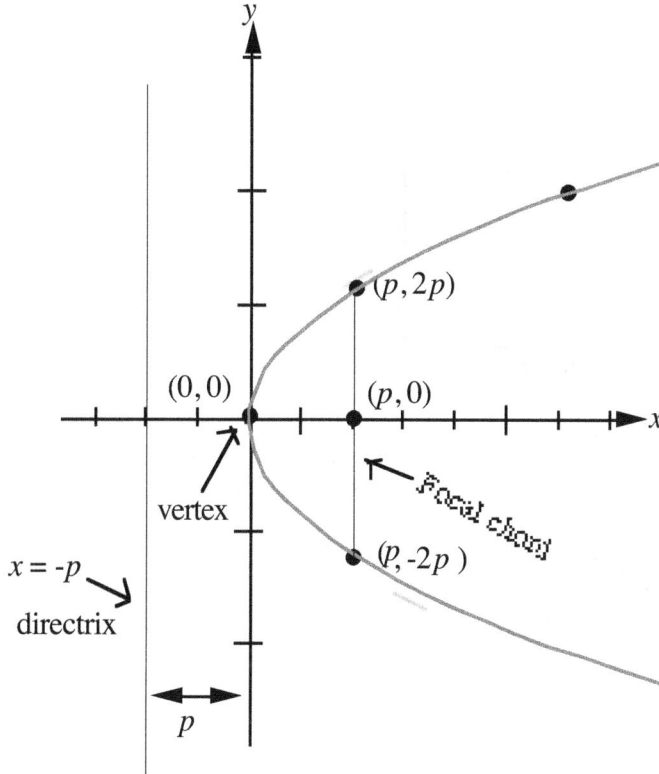

Figure 8: The Graph of $y^2 = 4px$

Case 4: The Equation of $y^2 = -4px$ 402

By reflecting the equation of $y^2 = 4px$ about the y-axis (replacing x by $-x$), we obtain

$$y^2 = -4px$$

From $y^2 = -4px$, $y = \pm 2\sqrt{-px}$ $x \le 0,\ p > 0$.

The graph of $y^2 = -4px$ (Figure) is symmetric about the x-axis. The vertex is at $(0,0)$. The parabola is concave to the left. The directrix is the line $x = p$, and the focus F is at the point $(-p, 0)$

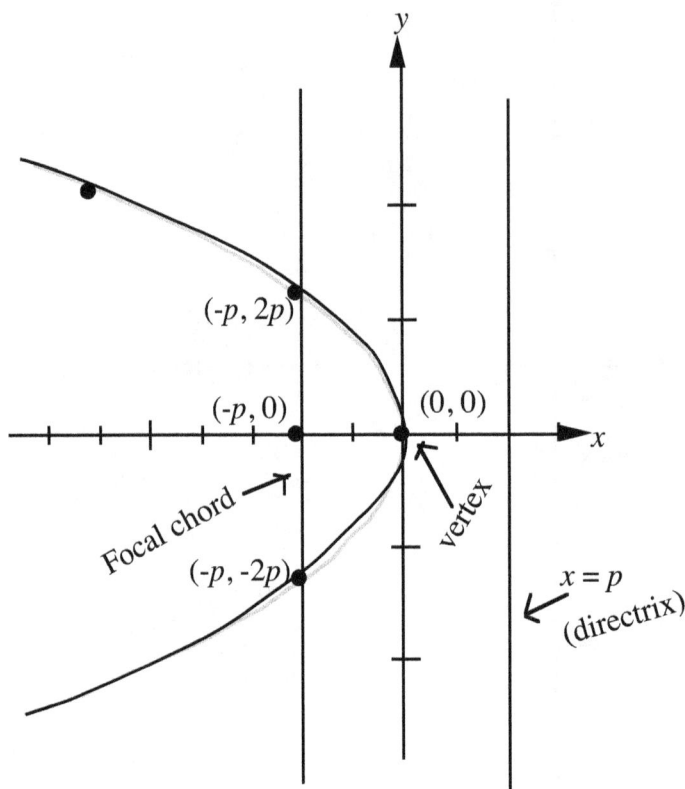

Figure 9: The graph of $y^2 = -4px$

How to Sketch the graph of a parabola when the location of the foci
and the specification of the directrix are of interest.

General forms of the parabola

$(x - h)^2 = 4p(y - k)$ <-------Function

$(y - k)^2 = 4p(x - h)$ <--------Not a function

Step 1: Find the axis of symmetry., (Obtain the axis from the quadratic term)

For $(x - h)^2 = 4p(y - k)$, the axis of symmetry is the line vertical line $x = h$

For $(y - k)^2 = 4p(x - h)$, the axis of symmetry is the line horizontal line $y = k$

For $y = 2x^2$ of general form, $2(x - 0)^2 = y - 0$ „ $x - h = x - 0$, the axis of symmetry is the vertical line $x = 0$ (the y-axis)   (From $x - 0 = 0$)

For $y^2 = 5x$ of general form, $(y - 0)^2 = 5(x - 0)$, the axis of symmetry is the horizontal line $y = 0$ (the x-axis)   (From $y - 0 = 0$)

Step 2: Find the vertex.

For $x^2 = 4py$, the vertex is at $(0, 0)$, and the axis of symmetry is the line, $x = 0$ (y-axis.).

For $y^2 = 4px$, the vertex is at $(0, 0)$ and the axis of symmetry is the line, $y = 0$ (x-axis.)

Step 3: Find the focus.

For $x^2 = 4py$, the focus is at $(0, p)$.

For . $y^2 = 4px$ the focus is at $(p, 0)$.

Finding p

To calculate p, we compare the given equation with the standard equation (compare the coefficients).

Examples: **1**. Find p, given that $y^2 = 8x$.   (1)

Standard form: $y^2 = 4px$ (2)

comparing coefficients, $4p = 8$, and   (or $8x = 4px \Rightarrow p = 2$)

$p = 2$ and focus is at $(2, 0)$.   (The parabola is concave to the right)

The directrix is $x = -p = -2$ (i,e., $x = -2$)

2. Similarly, for $x^2 = -6y$.   (or $y = -\frac{1}{6}x^2$, Parabola is concave downwards)

standard form: $x^2 = 4py$ and $4p = -6$ (or $-6y = 4py \Rightarrow p = -\frac{6}{4} = -\frac{3}{2}$)

Solving for p, $p = -\frac{3}{2}$ and the focus is at $(0, -\frac{3}{2})$

The directrix is $y = -p = -(-\frac{3}{2}) = \frac{3}{2}$ (i.e., $y = \frac{3}{2}$)

(Note the difference between p and $|p|$)

Step 4: Plot the vertex $(0, 0)$, the focus $(p, 0)$, $(-p, 0)$,

$(0, p)$, or $(0, -p)$ depending on the form of the given equation. (Note the difference between p and $|p|$)

Step 5: The length of the focal chord is $4p$. (The focal chord is the line segment through the focus, perpendicular to the axis of symmetry, with both end points on the parabola)

The coordinates of the end points would be $(2p, p)$, $(-2p , p)$ or $(p, 2p)$, $(p, -2p)$, depending on direction in which the parabola 'opens'.

Step 6: Plot the endpoints of the focal chord.

Step 7: Draw the appropriate parabola by connecting the vertex and the end points of the focal chord. If the focal chord is of interest, connect the end points of the focal chord by a straight line.

Lesson 61: Parabolas

Example 1 Sketch the graph of $x^2 = 12y$ and indicate the vertex, focus, focal chord
, and directrix

Step 1: Determine the concavity and the axis of symmetry.

The curve is concave up. The axis of symmetry is the y-axis (Recall: $y = x^2$ or $x^2 = y$)

By equating the right-hand side of $x^2 = 12y$ to the right-hand side of

$$x^2 = 4py \qquad \text{(standard simple form), we obtain}$$
$$4p = 12, \text{ and from which } p = 3.$$

The positive value of p means that the curve is either concave up or concave to the right. In this problem, it is concave up (since the axis of symmetry is a vertical line, given by $x = h$ from the quadratic term..

Step 2: Find the vertex.

Equation (1) can be written as

$(x - 0)^2 = 12(y - 0)$ and which when compared with the general form $(x - h)^2 = 4p(y - k)$ yields $h = 0$, $k = 0$
The vertex is at $(0, 0)$.

Step 3: Find the focus, directrix, and the length of the focal chord.

From Step 1, $p = 3$ and the focus is at $(0, 3)$.   (Coordinates of the focus: $(0, p)$),
The equation of the directrix is $y = -3$ (formula for the directrix: $y = -p$)

Note that the distance from the vertex to the focus equals the distance from the vertex to the directrix.
The length of the focal chord is $4|p| = 4|3| = 12$ (with $p = 3$).

Step 4: Plot the vertex $(0, 0)$, focus $(0, 3)$. Draw the focal chord of length 12 units through the focus $(0, 3)$ and parallel to (in this example) the x-axis.

Step 5: Connect the vertex and the ends of the focal chord by a smooth solid curve to obtain the characteristic parabola.

Step 6: Draw the directrix $y = -3$ (parallel to the focal chord).

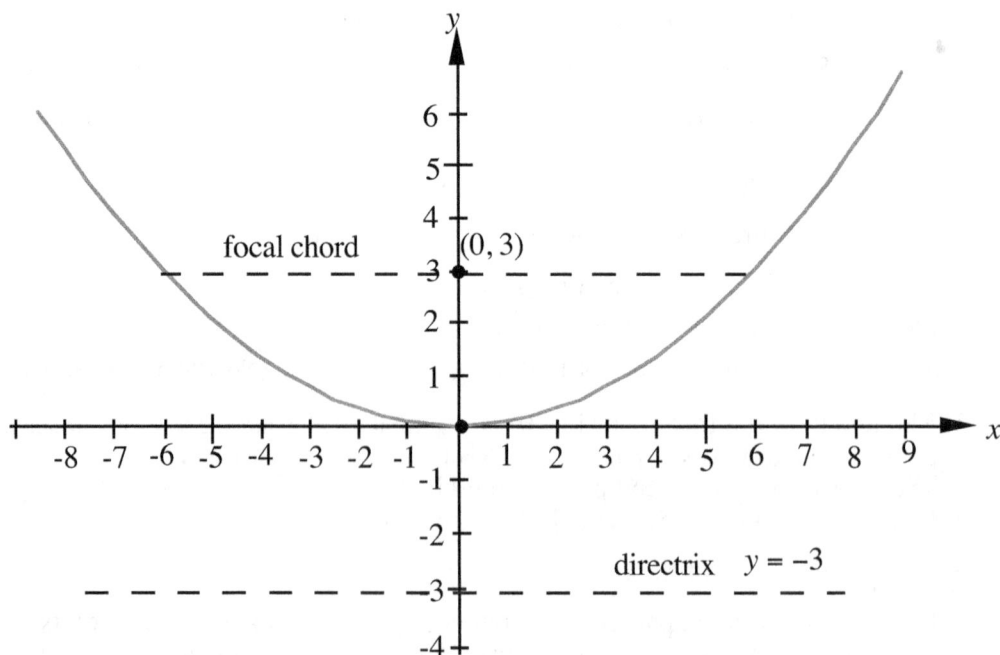

Example 2 Sketch the graph of $(x-1)^2 = 20(y-3)$ (Indicate the vertex, focus, focal chord, and directrix.)

Step 1: Determine the concavity, the axis of symmetry, and the vertex.

The curve is concave up (from the quadratic term, as in Example 1).

By comparison of $(x-1)^2 = 20(y-3)$ with

$$(x-h)^2 = 4p(y-k), \quad h=1, \quad k=3$$

(We may also solve $x-1=0$, and $y-3=0$).

The vertex is at $(1,3)$

The axis of symmetry is the vertical line $x=1$ (parallel to the y-axis)

Step 2: Find the focus, focal chord, focal length, and directrix.

By comparison, $4p = 20$ and from which $p = 5$

(The positive value of p means that the curve is either concave up or concave to the right. It is concave up)

The focus is 5 units from the vertex. The focus is at $(1,8)$. [formula: $(h, k+p)$}

Focal length = $4|5| = 20$ units.

The directrix is 5 units ($p = 5$) from the vertex and the equation of the directrix is
$y = -2$ ($y = k - p = 3 - 5 = -2$)

Note that the distance from the vertex to the focus equals the distance from the vertex to the directrix.

Step 3: If the vertex is not on the coordinate axes, find the x- and y-intercepts.

Note that if you obtain non-real values, then there are no intercepts

There are no x-intercepts, since we obtain imaginary values when $y=0$ in $(x-1)^2 = 20(y-3)$.

By letting $x=0$ in $(x-1)^2 = 20(y-3)$, we obtain the y-intercept = 3.2

The y-intercept is at $(3.2, 0)$

Step 4: Plot the vertex $(1,3)$), focus $(1,8)$. Draw the focal chord of length 20 units through the focus $(1,8)$ and parallel to (in this example) the x-axis.

Step 5: Connect the vertex and the ends of the focal chord by a smooth solid curve to obtain the characteristic U-shaped curve.

Step 6: Draw the directrix, the line, $y=-2$ (parallel to the focal chord). This should pass through $(1,-2)$.

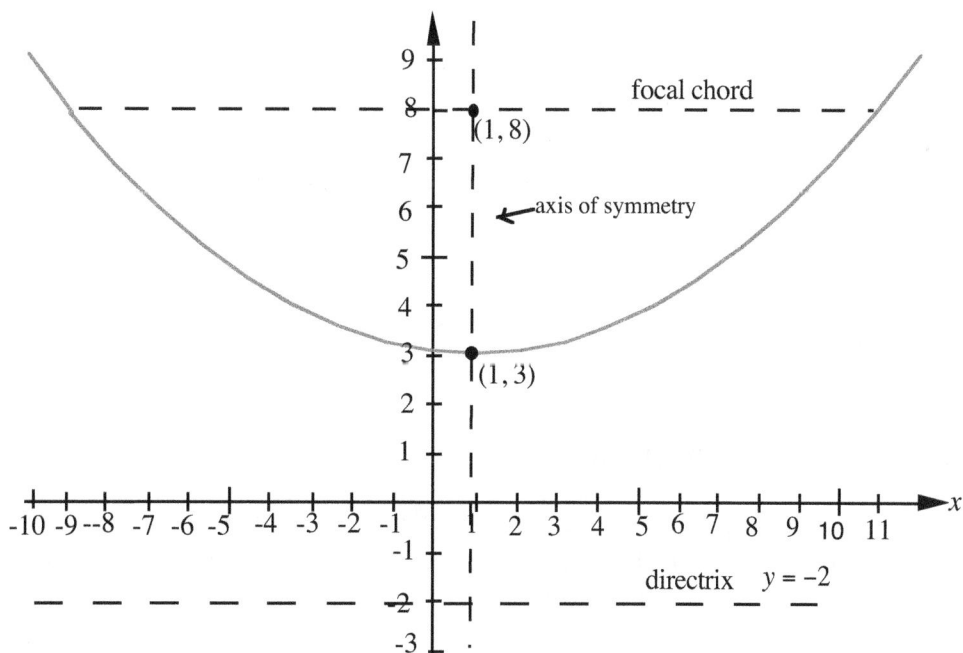

Example 3 Sketch the graph of $x^2 - 8x + 8y - 8 = 0$ and indicate the vertex, focus,

focal chord, and directrix

Step 1: Write the equation in the form $(x - h)^2 = 4p(y - k)$ by completing the square on the x-term and factoring the remaining terms.

$$x^2 - 8x + 16 - 16 + 8y - 8 = 0$$

$$x^2 - 8x + 16 + 8y - 24 = 0$$

$$(x - 4)^2 = -8y + 24$$

$$(x - 4)^2 = -8(y - 3) \qquad (2)$$

Step 2 Determine the concavity, vertex, and the axis of symmetry.

The curve is concave down.(Compare equation (2) with $x^2 = y$)

For the vertex: $h = 4$, $k = 3$, by inspection   (or by solving $(x - 4 = 0, \ y - 3 = 0)$

The vertex is at $(4, 3)$

The axis of symmetry is the line $x = 4$ (parallel to the y-axis)

Step 3: Find the focus, focal chord, focal length, and directrix.

By comparison of right-hand side of equation (2) with the general standard form

$(x - h)^2 = 4p(y - k)$, we obtain

$4p = -8$ and from which $p = $ -2

The negative value of p means that the curve is either concave down or concave to the left

It is concave down since the vertical line $x = 4$ is the axis of symmetry.

The focus is 2 units down from the vertex. The focus is at $(4, 1)$.   [formula: $(h, k + p)$]

Focal length = $4|2| = 8$ units.

The directrix is 2 units ($p = $-2) from the vertex and the equation of the directrix is

$y = 5$ $(y = k - p = 3 - (-2) = 5)$.

(Note that the distance from the vertex to the focus equals the distance from the vertex to the directrix)

Step 4: Since the vertex is not on the coordinate axes, we find the x- and y-intercepts.

(Note that if you obtain non-real values, then there are no intercepts)

For the y-intercept: When $x = 0$, in $x^2 - 8x + 8y - 8 = 0$, we obtain

$(0)^2 - 8(0) + 8y - 8 = 0$ and from which $y = 1$.

The y-intercept is at $(0, 1)$

For the x-intercept: When $y = 0$, in $x^2 - 8x + 8y - 8 = 0$, we obtain

$$x^2 - 8x + 8(0) - 8 = 0$$

$$x^2 - 8x - 8 = 0$$

Solving by the quadratic formula,

$$x = \frac{+8 \pm \sqrt{64 - 4(1)(-8)}}{2}$$

$x \approx -0.9$ or 8.9

The x-intercepts are at $(-0.9, 0)$ and $(8.9, 0)$.

Step 5: Plot the vertex $(4, 3))$, focus $(4, 1)$, the y-intercept $(0, 1)$, the x-intercepts $(-0.9, 0)$ and $(8.9, 0)$. Draw the focal chord of length 8 units through the focus $(4, 1)$ and parallel to (in this example) the x-axis.

Lesson 61: Parabolas

Step 6: Connect the vertex and the ends of the focal chord, the *x*- and *y*-intercepts by a smooth solid curve to obtain the characteristic U-shaped curve.

Step 7: Draw the directrix , the line, $y = 5$ (2 units from the vertex).

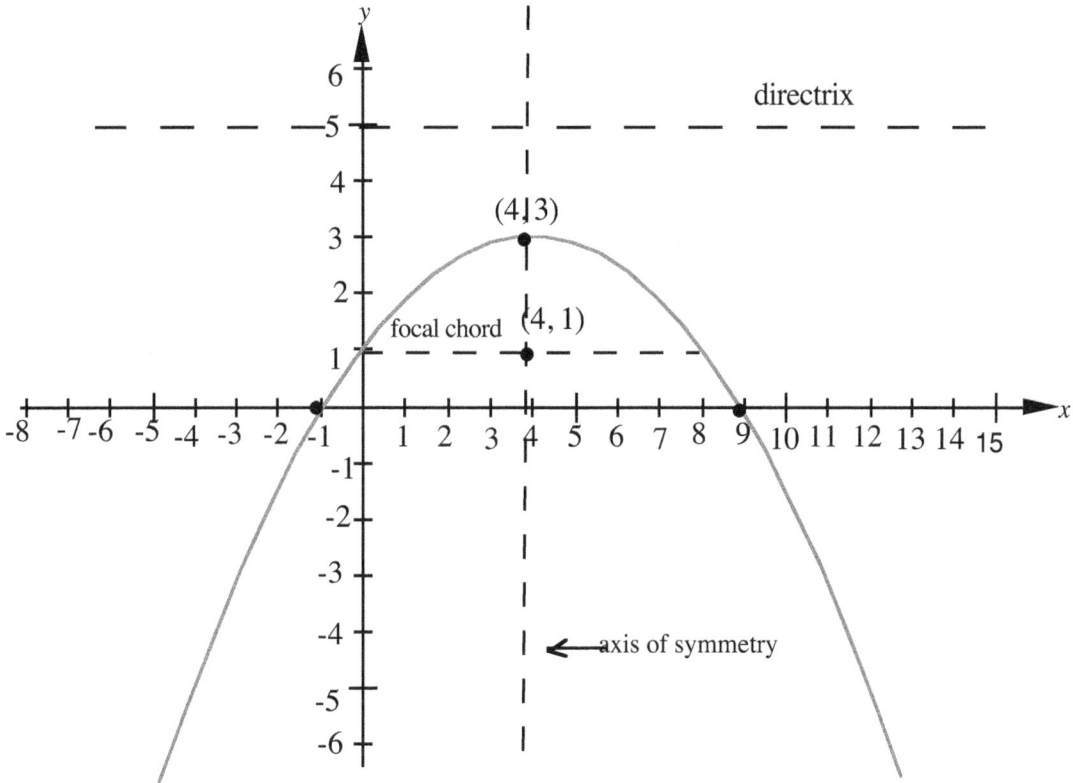

Lesson 61: Parabolas

Example 4 Sketch the graph of $y^2 = -24x$ and indicate the vertex, focus, focal chord, and directrix.

Step 1: Determine the concavity, vertex, and the axis of symmetry.

The curve is concave to the left. (since the axis of symmetry from the quadratic term is of form y = k)

For the vertex: $h = 0$, $k = 0$ (from $(y - 0)^2 = -24(x - 0)$)
The vertex is at $(0, 0)$
The axis of symmetry is the negative x-axis (direction of the linear variable)

Step 2: Find the focus, focal chord, focal length, and directrix.

By comparison, $4p = -24$ and from which $p = -6$

(The negative value of p means that the curve is either concave down or concave to the left.
It is concave to the left here.

The focus is 6 units to the left of the vertex. The focus is at $(-6, 0)$. (formula: $(h + p, k)$

Focal chord length $= 4|-6| = 24$ units.
The directrix is 6 units ($p = -6$) from the vertex and the equation of the directrix is
$x = 6$ (i.e., $x = -p$).
(Note that the distance from the vertex to the focus equals the distance from the vertex to the directrix.)

Step 3: Plot the vertex $(0, 0))$, focus $(-6, 0)$, Draw the focal chord of length 24 units through the   focus $(-6, 0)$ and parallel to the y-axis.

Step 4: Connect the vertex and the ends of the focal chord by a smooth solid curve to obtain the characteristic U-shaped curve.

Step 5: Draw the directrix, the line, $x = 6$. (parallel to the focal chord and also parallel to the y-axis.

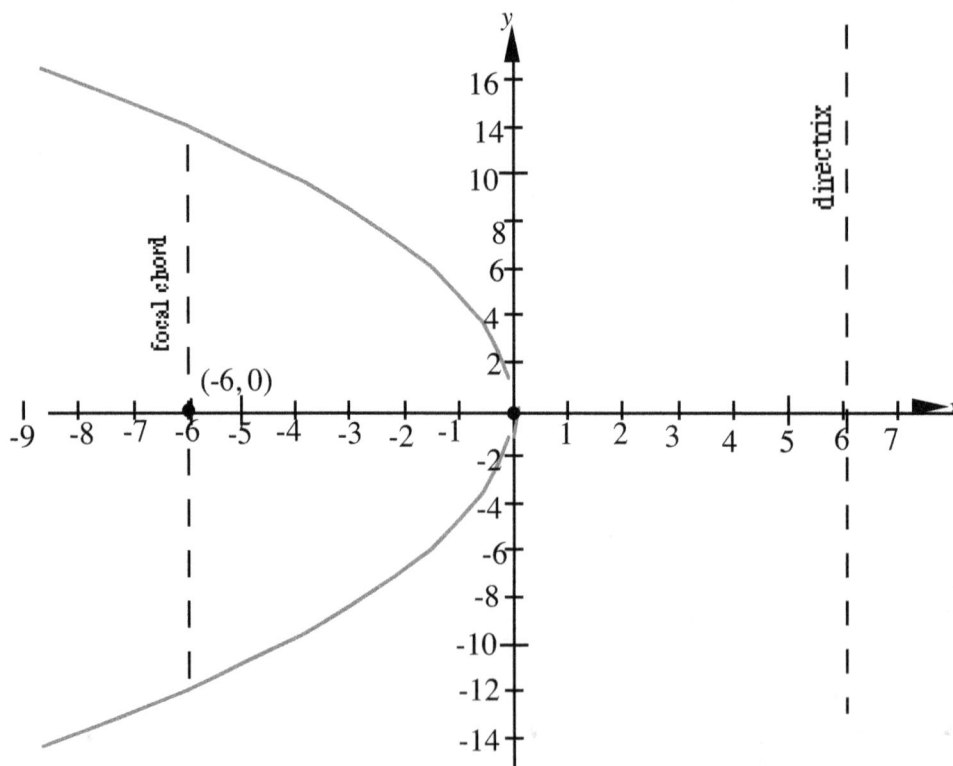

Example 5 Sketch the graph of $(y-5)^2 = 12(x+2)$ indicating the vertex, focus, focal chord, and directrix.

Step 1: Determine the concavity, axis of symmetry, and vertex.

The curve is concave to the right (Recalling $y^2 = x$, the axis of symmetry is parallel to the x-axis)

By comparison of equation $(y-5)^2 = 12(x+2)$ with
$$(y-k)^2 = 4p(x-h)$$

$h = -2$, $k = 5$ by inspection (or by solving $(x+2=0, \ y-5=0)$

The vertex is at (-2, 5).

The axis of symmetry is the line $y = 5$ (parallel to the x-axis)

Step 2: Find the focus, focal chord, focal length, and directrix.

By comparison with the standard equation, $4p = 12$, and from which $p = 3$.

(The positive value of p means that the curve is either concave up or concave to the right. It is concave to the right in this problem)

The focus is 3 units to the right of the vertex. The focus is at $(1,5)$. [(formula: $(h + p, \ k)$}

Focal length $= 4|3| = 12$ units.

The directrix is 3 units ($|p|=3$) to the left of the vertex and the equation of the directrix is $x = -5$ $(x=h-p = -2-3= -5)$

Note that the distance from the vertex to the focus equals the distance from the vertex to the directrix

Step 3: Since the vertex is not on the coordinate axes, we shall find the x- and y-intercepts.

(Note that if we obtain non-real values, then there are no x-intercepts)

By letting $x = 0$, followed by $y = 0$ in $(y-5)^2 = 12(x+2)$, and solving, we obtain:
$y \approx 9.9$ or ≈ 0.1; $x \approx 0.8$

The y-intercepts are at $(0, \ 9.9)$ and $(0, \ 0.1)$; the x-intercept is at $(0.8, 0)$

Step 4: Plot the vertex $(-2, 5)$, focus $(1, 5)$, the intercepts Draw the focal chord of length 12 units through the focus $(1, 5)$ and parallel to (in this example) the y-axis.

Step 5: Connect the vertex and the ends of the focal chord by a smooth solid curve to obtain the characteristic U-shaped curve.

Step 6: Draw the directrix, the line, $x = -5$ (parallel to the focal chord).

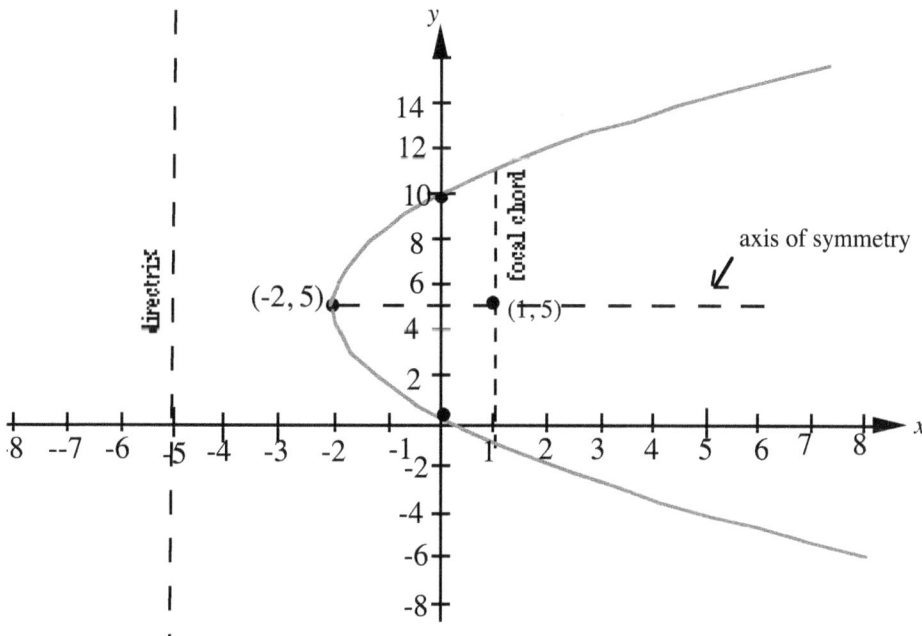

Example 6 Sketch the graph of $y^2 - 10y - 12x + 1 = 0$ and indicate the vertex, focus, $\qquad$ 4 1 0
focal chord, and directrix

Step 1: Write the equation in the form $(y - k)^2 = 4p(x - h)$ by completing the square on the y-term
and factoring the remaining terms.

$$y^2 - 10y - 12x + 1 = 0$$

$$y^2 - 10y + 25 - 25 - 12x + 1 = 0$$

$$(y - 5)^2 - 12x - 24 = 0$$

$$(y - 5)^2 = 12x + 24$$

$$(y - 5)^2 = 12(x + 2) \quad <--------(1)$$

Equation (2) is identical with the last example (Example 5), and therefore, the remaining steps are the
same as those of Example 5.

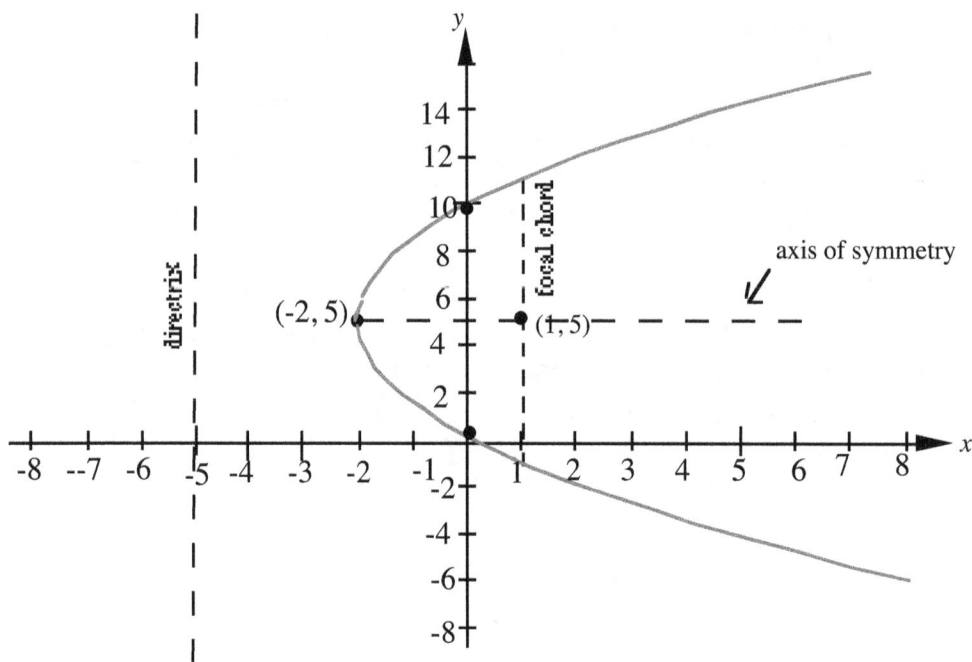

Lesson 61 Exercises 411

Find an equation for each parabola with the specified properties, and sketch its graph:

1. Focus at $(4, 0)$; directrix $x = -4$;   **2.** Focus at $(0, -2)$; directrix $x = -4$;

3. Focus at $(3, 2)$; directrix $y = -2$;   **4.** Focus at $(2, 0)$; vertex at the origin.

5. Focus at $(-3, 0)$; directrix $x = 3$.

   Determine the vertex, focus, directrix and sketch the graph of each of the following

6. $x^2 = 12y$;   **7.** $x^2 = 22y$;   **8.** $y^2 = 12x$;   **9.** $y^2 = -12x$;

10. $y^2 + 2x = 0$;   **11.** $y^2 + 10x = 0$;   **12.** $y^2 + 8x = 0$;   **13.** $x^2 = 4y$;;

14. $x^2 + 3y = 0$;   **15.** $x^2 + 2x = -2y - 7$;   **16.** $4y^2 - 4y = 4x - 24$.

1. $y^2 = 16x$;   **2.** $(y+2)^2 = 8(x+2)$;   **3.** $(x-3)^2 = 8y$;   **4.** $y^2 = 8x$;   **5.** $y^2 = -12x$;
6. Vertex at $(0,0)$, focus at $(0,3)$, directrix $y = -3$;   **7.** Vertex at $(0,0)$, focus at $(0, \frac{11}{2})$,
directrix $y = -\frac{11}{2}$;   **8.** Vertex at $(0,0)$, focus at $(3,0)$, directrix $x = -3$;
9. Vertex at $(0,0)$, focus at $(-3,0)$, directrix $x = 3$;   **10.** Vertex at $(0,0)$, focus at $(-\frac{1}{2},0)$,
directrix $x = \frac{1}{2}$; **11.** Vertex at $(0,0)$, focus at $(-\frac{5}{2},0)$, directrix $x = \frac{5}{2}$;   **12.** Vertex at $(0,0)$,
focus at $(-2,0)$, directrix $x = 2$; **13.** Vertex at $(0,0)$, focus at $(0,1)$, directrix $y = -1$;
14. Vertex at $(0,0)$, focus at $(0, -\frac{3}{4})$, directrix $y = \frac{3}{4}$;   **15.** Vertex at $(-1, -3)$, focus at $(-1, -\frac{7}{2})$,
directrix $y = -\frac{5}{2}$; **16.** Vertex at $(\frac{23}{4}, \frac{1}{2})$, focus at $(6, \frac{1}{2})$, directrix $x = \frac{11}{2}$.

CHAPTER 23B
Conic Sections

Lesson 62: Ellipse
Lesson 63: Hyperbola
Circle and Parabola (see Chapter 23A)

Lesson 62
Ellipse

Definition: An **ellipse** is the set of points such that the sum of the distances of each point from two fixed points, F_1, F_2 (called the foci) is a given constant.

The Ellipse with the major axis along the *x*-axis (center at the origin)

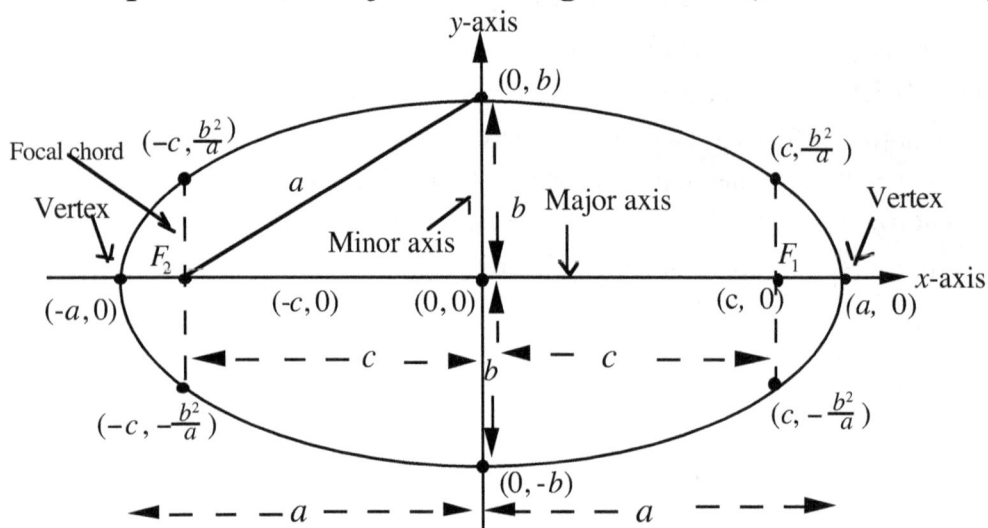

Figure 1: Graph of the ellipse: $\dfrac{x^2}{a^2} + \dfrac{y^2}{b^2} = 1$ $a > b$ (Major axis along the *x*-axis)

Properties of the ellipse with the major axis along the *x*-axis

1. The ellipse has two axes of symmetry: a longer axis called the major axis on which are located the foci and the vertices; and a relatively shorter axis called the minor axis. In the above figure, the major axis is the line segment (also part of the *x*-axis) from $(-a, 0)$ to $(a, 0)$. The minor axis is the line segment (also part of the *y*-axis) which is the perpendicular bisector of the major axis, and it is from $(0, b)$ to $(0, -b)$.

2. The length of the major axis is $2a$ and the length of the minor axis is $2b$. The above lengths were chosen for convenience in deriving the equation of the ellipse. There are a semi-major axis whose length is a (half of the length of the major axis)) and a semi-minor axis whose length is b (half of the length of the minor axis).

3. Each end point of the major-axis is called a vertex. There are thus two vertices. The vertices are equidistant from the center of the ellipse and so also are the foci F_1 and F_2. The foci are at $(-c, 0)$ and $(c, 0)$. The distance between the foci is $2c$ and the distance between the vertices is $2a$.
  There are two focal chords. Each chord is a line segment from a point on the ellipse through a focus to anothe point on the ellipse.

4. The relationship between a, b, and c is given by $a^2 = b^2 + c^2$.

In recalling this relationship, be careful not to confuse the c here with the c in the Pythagorean relationship. Here, c is **not** the hypotenuse.

The standard form of the equation of the ellipse in above figure is given by

$$\frac{x^2}{a^2} + \frac{y^2}{b^2} = 1 \quad a > b \qquad \textbf{(1)} \qquad \text{(Note also that this equation is not a function)}$$

From above: c = the distance from the center to a focus; a = the distance from the center to a vertex. Also, a is the distance between an end of the minor axis and a focus of the ellipse; b = the length of the line segment from the center and perpendicular to the major axis to a point on the ellipse

Properties of the ellipse with the major axis along the *y*-axis

This ellipse can be obtained from the previous ellipse by interchanging the x and y coordinates. Then the

equation of the ellipse becomes $\qquad \frac{y^2}{a^2} + \frac{x^2}{b^2} = 1 \qquad a > b \qquad \textbf{(2)}$

The foci are at $F_1(0, c)$ and $F_2(0, -c)$, where $a^2 = b^2 + c^2$

We observe that equation (2) is the **inverse relation** of equation (1) in the previous case, since either equation can be obtained from the other by interchanging the roles of x and y. See also page 73. Note, however, that these equations are **not** functions.

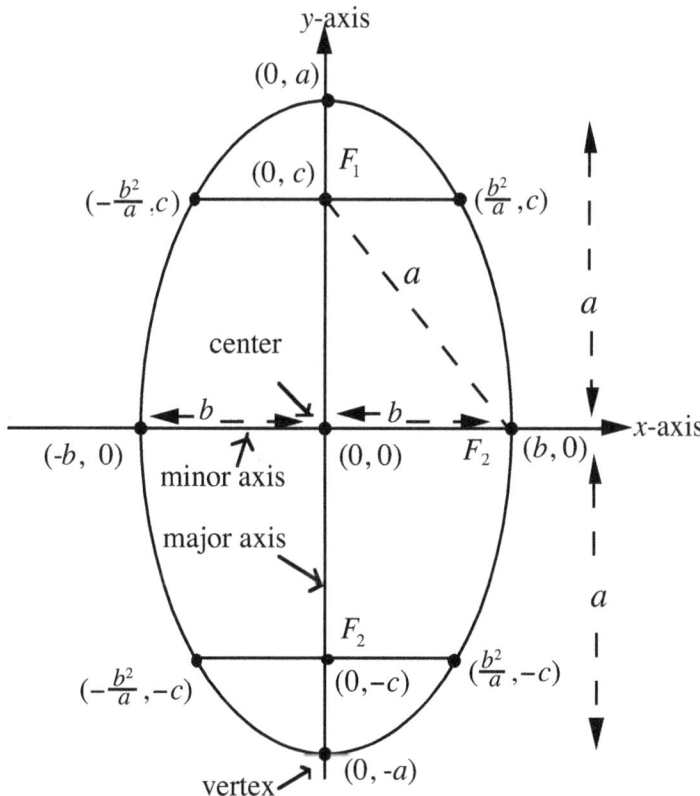

Figure 2 Graph of the ellipse: $\dfrac{y^2}{a^2} + \dfrac{x^2}{b^2} = 1 \qquad a > b$ (Major axis along the y-axis)

Graphically, we can obtain Figure 2 from Figure 1 by reflecting figure about line $y = x$. The major axis is along the y-axis, with the vertices on the y-axis. The minor axis is along the x-axis.

In summary, $\dfrac{x^2}{a^2} + \dfrac{y^2}{b^2} = 1$ and $\dfrac{y^2}{a^2} + \dfrac{x^2}{b^2} = 1$ are inverse relations of each other

How to sketch the graph of an ellipse with center at the origin when location of the foci and the specification of the major and minor axes are of interest

Step 1: Assume the equation is of the form $\dfrac{x^2}{a^2} + \dfrac{y^2}{b^2} = 1$, where $a > 0$, $b > 0$

Step 2: Find the x-intercepts by setting $y = 0$ in the given equation of the ellipse and solving for x. This step shall yield two x-values and therefore the coordinates of two points.

Step 3: Find the y-intercepts by setting $x = 0$ in the given equation and solving for y. This step shall also yield the coordinates of two points.

Step 4: Find the location of the foci.

For the equation $\dfrac{x^2}{a^2} + \dfrac{y^2}{b^2} = 1$, (where $a > b$) the major axis is along x-axis; the foci are at $(c, 0)$ and $(-c, 0)$, where $c^2 = a^2 - b^2$ (from $a^2 = b^2 + c^2$, the coordinates of the focal chord ends are $(c, \frac{b^2}{a})$ and $(-c, \frac{b^2}{a})$.

For the equation $\dfrac{y^2}{a^2} + \dfrac{x^2}{b^2} = 1$, (where $a > b$) the major axis along the y-axis; the foci are at $(0, c)$ and $(0, -c)$ where $c^2 = a^2 - b^2$; the coordinates of the ends of the focal chord are $(\frac{b^2}{a}, -c)$ and $(-\frac{b^2}{a}, -c)$.

Step 5: Using the results from Steps 2 and 3, and **with equal units** for both x- and y-axes, plot the four points. Connect these points by a smooth curve to obtain the ellipse. Note that the center of the ellipse is at $(0, 0)$.

Step 6: Using the results from Step 3, plot the foci points, the focal chord end points and connect these points by straight lines. To improve upon any of the above graphs of the ellipse, we can choose convenient x-values and calculate the corresponding y-values, or choose convenient y-values and calculate the corresponding x-values to obtain additional points. These additional points are then plotted and connected; but if the curve is drawn already, then the curve can be " smoothed" over using pencil and eraser.

General forms of the equations of the ellipse with center at (h, k)

$$\frac{(x-h)^2}{a^2} + \frac{(y-k)^2}{b^2} = 1, \qquad \text{where } a > b$$

$$\frac{(y-k)^2}{a^2} + \frac{(x-h)^2}{b^2} = 1, \qquad \text{where } a > b$$

Extra:

Finding the coordinates of the **center** (h, k) of an ellipse given that the foci are at the points $F_1(x_1, y_1)$ and $F_2(x_2, y_2)$

Since the **foci** are equidistant from the center of the ellipse, by applying the **midpoint** formula.

$$h = \frac{x_1 + x_2}{2}; \quad k = \frac{y_1 + y_2}{2}$$

Similarly, since the **vertices** are also equidistant from the center of the ellipse, if the vertices are at $V_1(x_1, y_1)$; $V_2(x_2, y_2)$, then $h = \dfrac{x_1 + x_2}{2}$; $\quad k = \dfrac{y_1 + y_2}{2}$.

Case 1:   Center of the ellipse at the origin

Example 1 Sketch the graph of the ellipse given by

$$\frac{x^2}{25} + \frac{y^2}{16} = 1 \qquad\qquad (1)$$

Step 1: Determine the center of the ellipse.

Equation (1) can be written as

$$\frac{(x-0)^2}{25} + \frac{(y-0)^2}{16} = 1 \qquad\qquad (2)$$

Comparing equation (2) with the general form,

$$\frac{(x-h)^2}{a^2} + \frac{(y-k)^2}{b^2} = 1. \qquad\qquad (3)$$

$h = 0, k = 0$. Thus, the center is at the origin, $(0, 0)$.

Step 2: Find the constants a, b and c. By comparing the denominators of equations (2) and (3),

$$a^2 = 25; \qquad b^2 = 16$$

$$a = 5. \qquad b = 4$$

$$c^2 = a^2 - b^2 \qquad\qquad \text{(from } a^2 = b^2 + c^2 : \text{ See figure.)}$$

$c^2 = 25 - 16$ and from which $c = 3$.

Step 3: Now, sketch the graph of the ellipse with center at the origin $(0, 0)$; $a = 5$, $b = 4$; and $c = 3$.

The major axis is in the x-direction, since **the denominator of the term containing x is larger than the denominator of the term containing y.** Plot the center $(0, 0)$. From the center, and on the major axis (the x-axis), first count 5 units ($a = 5$) to the right, and again from the center count 5 units to the left to obtain the endpoints of the major axis (Figure....). Similarly, from the center, and on the minor axis (the y-axis), first count 4 units up ($b = 4$), and again from the center count 4 units down to obtain the endpoints of the minor axis.

Step 4: Locate the foci on the major axis by counting 3 units ($c = 3$) from the center in each direction.

Step 5: Connect the endpoints of the major and the minor axes by a smooth curve to obtain the characteristic shape of the ellipse. For a smoother curve, find the endpoints of the focal chords by using

$(c, \frac{b^2}{a})$ and $(-c, \frac{b^2}{a})$.

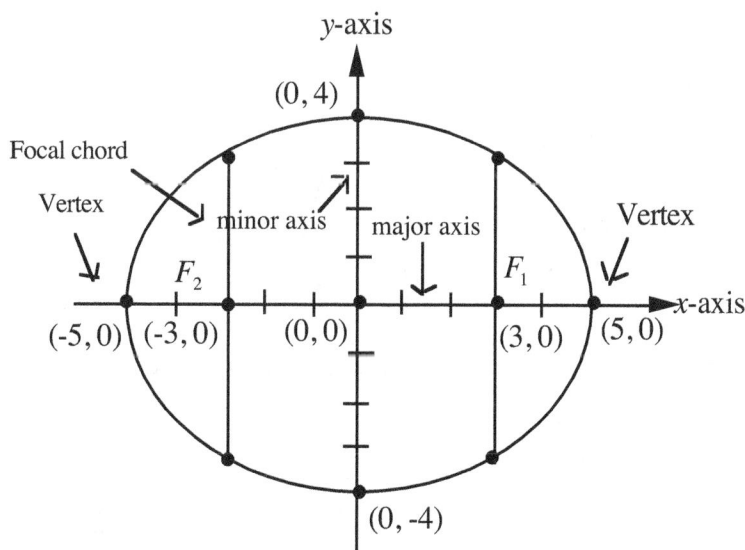

Figure: Graph of $\frac{x^2}{25} + \frac{y^2}{16} = 1$

Case 2: Center of the ellipse not at the origin 416

By the method of locating the center, axes and the endpoints of the axes

Example 2 Sketch the graph of the ellipse specified by

$$\frac{(x-1)^2}{25} + \frac{(y-3)^2}{16} = 1$$

Step 1: Find the center of the ellipse.

Set each numerator to zero and solve: $x - 1 = 0$ or $x = -1$ (that is, $h = 1$)
$y - 3 = 0$, or $y = 3$ (that is, $k = 3$)

Thus, the center is at $(1, 3)$.

Step 2: Find the constants a, b, and c.

By comparing the denominators of the given equation with the general and standard form (See Example 1. Step 2)

$a^2 = 25;$ $b^2 = 16$ (a^2 = the larger denominator; the major axis is parallel to the x-axis)
$a = 5.$ $b = 4$

$c^2 = a^2 - b^2$ (Property relationship between the constants of the equation of the ellipse)

$c^2 = 25 - 16$

$c = 3.$

Step 3: Now, sketch the graph of the ellipse with center at $(1, 3)$, $a = 5, b = 4, c = 3$

First, plot the center $(1, 3)$. The major axis is parallel to the x-axis (that is in the x-direction but not along the x-axis), **since the denominator of the x-term is larger than the denominator of the y-term**. From the center and on the major axis, count 5 units ($a = 5$) to the right, and again from the center, count 5 units to the left to obtain the endpoints of the major axis. Similarly, from the center, count 4 units up ($b = 4$) on the minor axis and 4 units down to obtain the endpoints of the minor axis.

Step 4: Locate the foci on the major axis by counting 3 units ($c = 3$) from the center in each direction.

Step 5: Connect the endpoints of the axes by a smooth curve to obtain the characteristic shape of the ellipse.

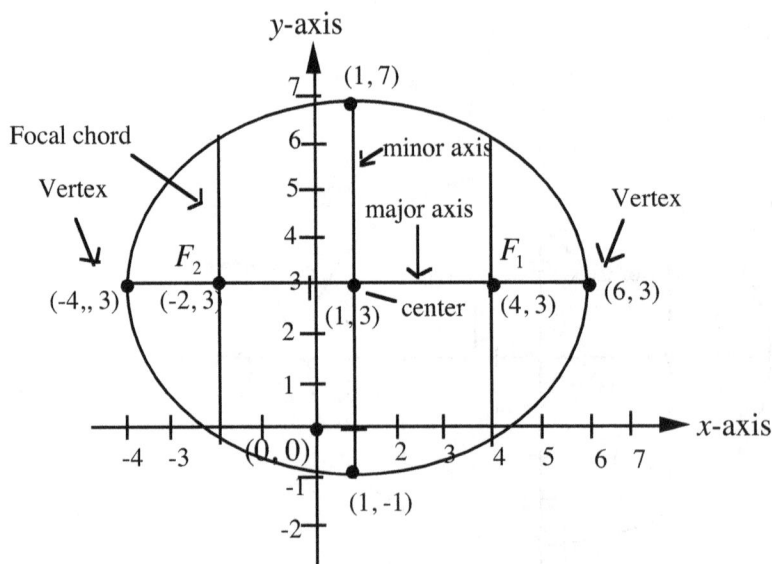

Figure : The graph of $\dfrac{(x-1)^2}{25} + \dfrac{(y-3)^2}{16} = 1$

Example 3 Sketch the graph of $\dfrac{x^2}{5} + \dfrac{5y^2}{16} = 5$ (1)

Step 1: Write the equation in the form $\dfrac{x^2}{a^2} + \dfrac{y^2}{b^2} = 1$ before using the methods outlined in the example .

In standard form, the right-hand side of equation (1) must be 1.

Divide the right-hand side of equation (1) by 5 followed by division of the left-hand side by 5 so that the right-hand side is 1.

$$\frac{x^2}{5(5)} + \frac{5y^2}{16(5)} = \frac{5}{5}$$

$$\frac{x^2}{25} + \frac{y^2}{16} = 1 \quad \text{or} \quad \frac{x^2}{5^2} + \frac{y^2}{4^2} = 1 \; \text{<------(1)}$$

Equation (1) is identical with the Example 1, above. Therefore, see Example 1 for the remaining steps.

Lesson 62: Ellipse

Example 4 Sketch the graph of $100x^2 + 36y^2 = 3600$ (1)

Step 1: Write the given equation in standard form. (i.e., $\dfrac{x^2}{a^2} + \dfrac{y^2}{b^2} = 1$ or $\dfrac{y^2}{a^2} + \dfrac{x^2}{b^2} = 1$).

In standard form, the right-hand side of equation (1) must be 1.

Divide the right-hand side of equation (1) by 3600 followed by division of the left-hand side by 3600 so that the right-hand side is 1.

$$\frac{100x^2}{3600} + \frac{36y^2}{3600} = \frac{3600}{3600}$$

$\dfrac{x^2}{36} + \dfrac{y^2}{100} = 1$ or $\dfrac{y^2}{100} + \dfrac{x^2}{36} = 1$

$a^2 = 100$ and $a = 10$ (a^2 = the larger denominator; the major axis along the y-axis)

$b^2 = 36$ and $b = 6$ (b^2 = the smaller denominator)

$a = 10$, $b = 6$

$c^2 = a^2 - b^2$

$c^2 = 100 - 36$

$c^2 = 64$

$c = 8$

Steps 2, 3, 4, and 5. See Example 1 (page 415). The graph is shown below.

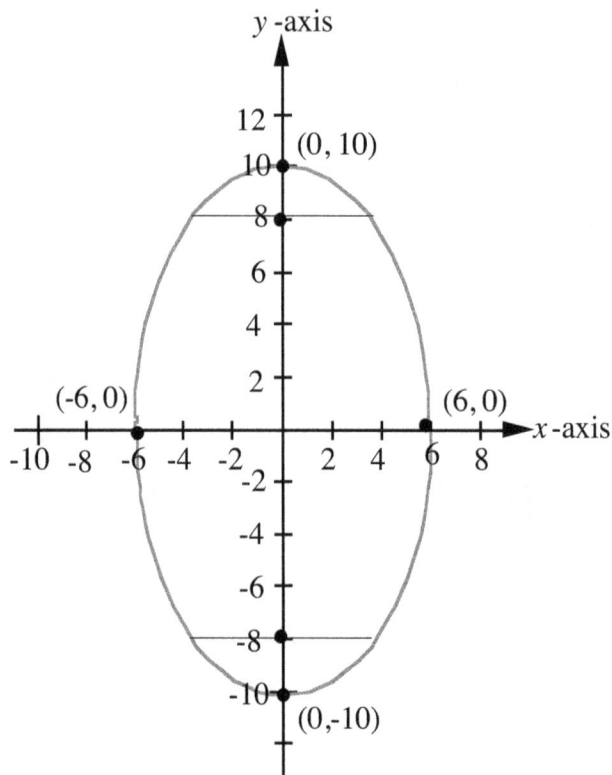

Figure: Graph of $100x^2 + 36y^2 = 3600$

How to sketch the graph of an ellipse with center at the origin **when**
**the location of the foci and the specification of the major and minor
axes are of no interest. (Rapid sketching)**

Method

Step 1: Find the x-intercepts by setting $y = 0$ and solving for x. Assuming the center of the
ellipse is at the origin, this step shall yield the coordinates of two points.
Example: The coordinates $(a, 0)$ and $(-a, 0)$ or $(b, 0)$ and $(-b, 0)$.

Step 2: Find the y-intercepts by setting $x = 0$ in the given equation and solving for y. Similarly,
assuming the center of the ellipse is at the origin, this step yields the coordinates of two
points.

Step 3: Plot the above four points on an x-y plane of rectangular axes using suitably chosen
scalar units. (It is assumed that the x- and y-axes have the same scalar units).

Step 4: Connect the above four points by a smooth curve to obtain the characteristic curve of
the ellipse.

Center not at the origin

If the center is not at the origin, finding the x- and y-intercepts by the intercept method results in having to
deal with tedious quadratic equations; and in this case, use other methods discussed. Also, if we obtain
complex solutions when we try to find the x-intercepts (or y-intercepts), then there are no x-intercepts (or
y-intercepts) , and the intercepts method will not work.

Determining whether a given equation of an ellipse is of the form

$$\frac{x^2}{a^2} + \frac{y^2}{b^2} = 1, \qquad a > b; \qquad \text{or} \qquad \frac{y^2}{a^2} + \frac{x^2}{b^2} = 1 \qquad a > b$$

We use the **relative sizes** of a^2 and b^2 in this determination.

The larger denominator is that fraction containing a^2, and the smaller denominator contains b^2 .

Examples 1. $\frac{x^2}{16} + \frac{y^2}{9} = 1$ ($a^2 = 16$, $b^2 = 9$; major axis is along the x-axis)

2. $\frac{y^2}{25} + \frac{x^2}{16} = 1$ ($a^2 = 25$, $b^2 = 16$, major axis is along the y-axis)

If a and b are nearly equal, the ellipse becomes nearly circular, and if $a = b$, the equation is that of a circle.

The Circle as a Special Ellipse

The equation of the circle can be considered as a special case of the equation of the ellipse in which the
length of the major axis is equal to the length of the minor axis.

Consider the equation of the ellipse given by $\frac{x^2}{a^2} + \frac{y^2}{b^2} = 1$ $\qquad\qquad$ (1)

If we let $a = b$ in equation (1), then equation (1) becomes $\frac{x^2}{a^2} + \frac{y^2}{a^2} = 1$

Multiplying through the above equation by a^2 respectively,

$$\frac{a^2 \bullet x^2}{a^2} + \frac{a^2 \bullet y^2}{a^2} = a^2 \bullet 1$$

$$x^2 + y^2 = a^2 \qquad\qquad (2)$$

In equation (2,) a is usually known as the radius of the circle. Thus $x^2 + y^2 = r^2$ is the equation of the
circle with radius r and center at the origin $(0,0)$. In drawing the circle the scalar unit on the x-axis should
be the same as the scalar unit on the y-axis, otherwise there would be a distortion and the circle would
resemble an ellipse.

Lesson 62 Exercises

A In each of the following, first determine if the principal axis is parallel to the x-axis or the y-axis, and then sketch the graph of each ellipse:

1. $\dfrac{x^2}{9} + \dfrac{y^2}{4} = 1$;   **2.** $\dfrac{x^2}{4} + \dfrac{y^2}{9} = 1$;   **3.** $\dfrac{(x-2)^2}{9} + \dfrac{(y-3)^2}{4} = 1$;   **4.** $\dfrac{(x+2)^2}{9} + \dfrac{(y+3)^2}{25} = 1$;

5. $3x^2 + 4y^2 - 12 = 0$;   **6.** $\dfrac{x^2}{36} + \dfrac{y^2}{16} = 1$;   **7.** $\dfrac{(x-2)^2}{9} + \dfrac{(y-1)^2}{16} = 1$;

8. $\dfrac{(x-1)^2}{49} + \dfrac{(y+2)^2}{81} = 1$;   **9.** $\dfrac{(x-2)^2}{25} + \dfrac{(y+1)^2}{9} = 1$;   **10.** $100x^2 + 25y^2 = 2500$

Answers

1.

$\dfrac{x^2}{9} + \dfrac{y^2}{4} = 1$

2.

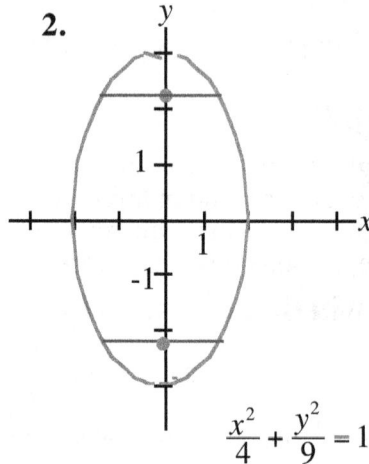

$\dfrac{x^2}{4} + \dfrac{y^2}{9} = 1$

3.

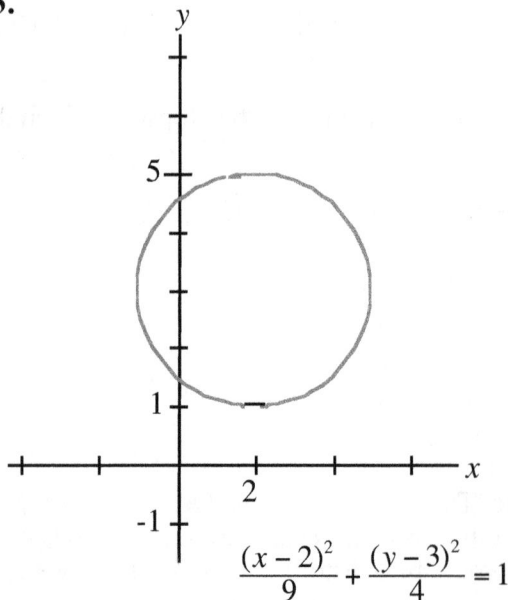

$\dfrac{(x-2)^2}{9} + \dfrac{(y-3)^2}{4} = 1$

4.

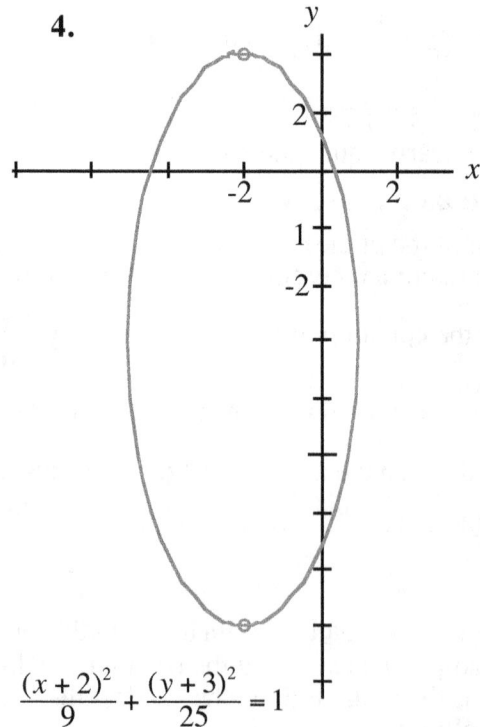

$\dfrac{(x+2)^2}{9} + \dfrac{(y+3)^2}{25} = 1$

5.

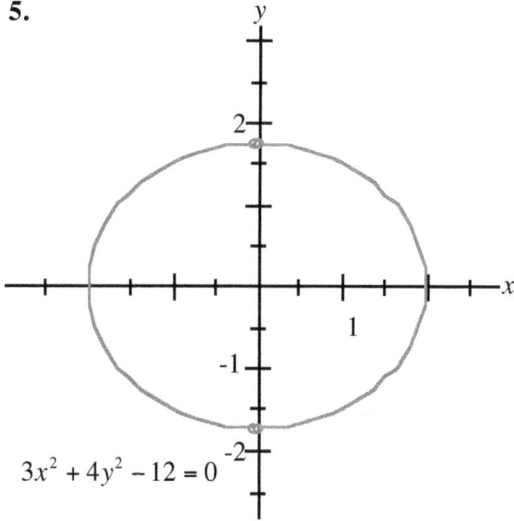

$3x^2 + 4y^2 - 12 = 0$

6.

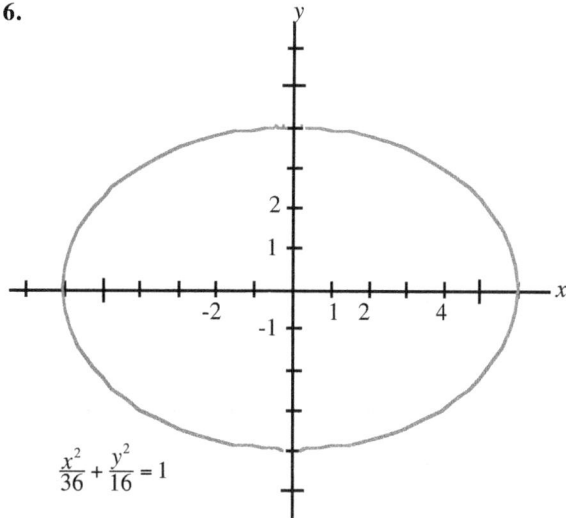

$\frac{x^2}{36} + \frac{y^2}{16} = 1$

7.

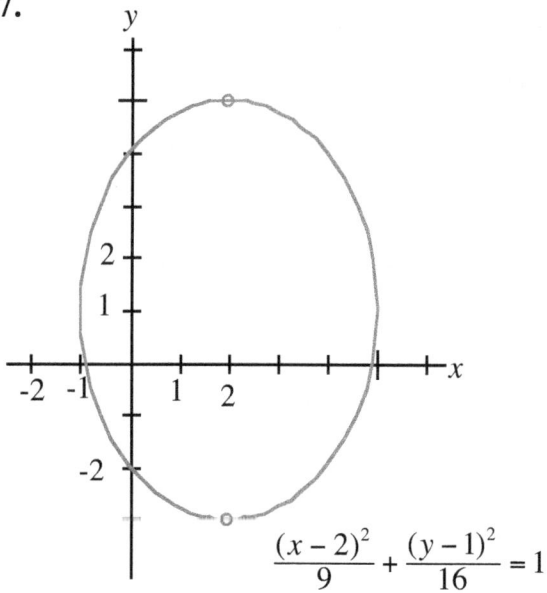

$\frac{(x-2)^2}{9} + \frac{(y-1)^2}{16} = 1$

8.

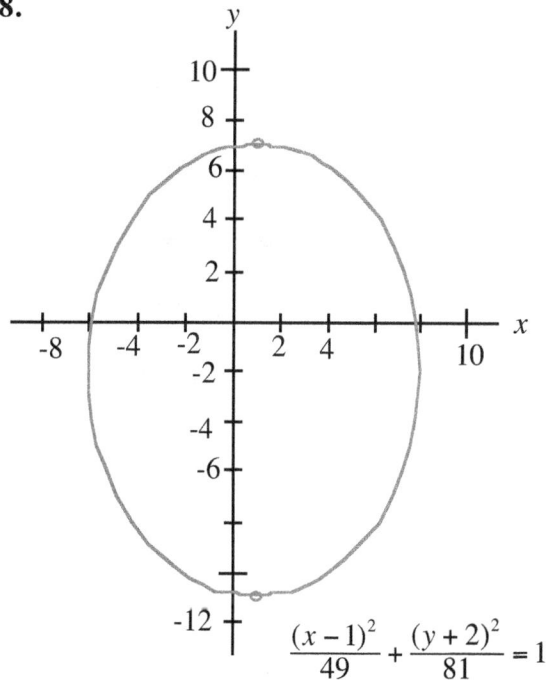

$\frac{(x-1)^2}{49} + \frac{(y+2)^2}{81} = 1$

9.

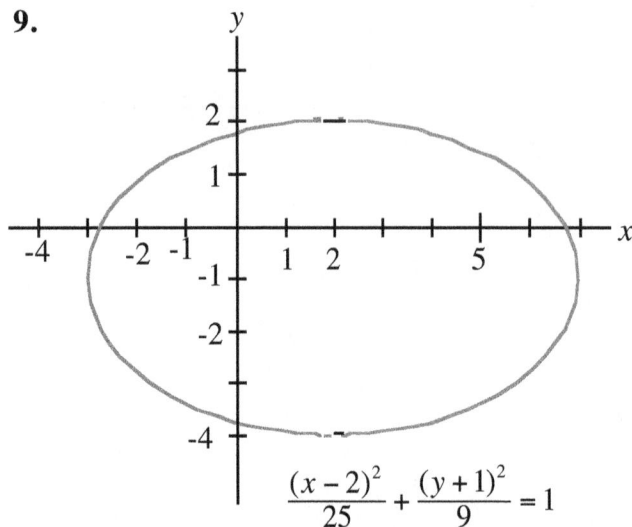

$$\frac{(x-2)^2}{25} + \frac{(y+1)^2}{9} = 1$$

10.

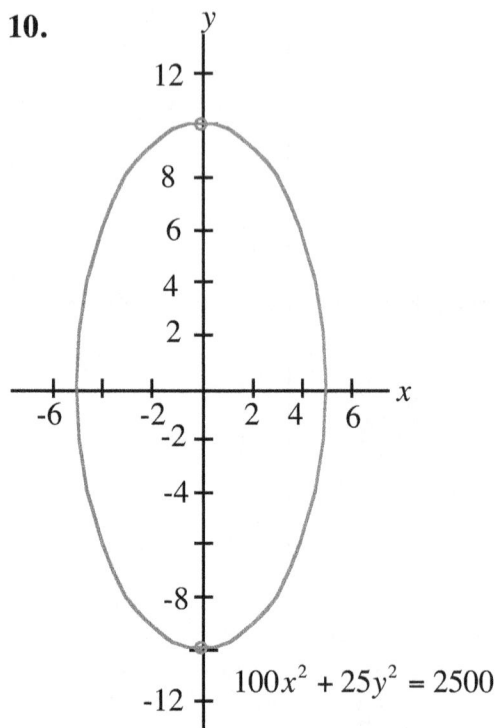

$$100x^2 + 25y^2 = 2500$$

B Find an equation of the ellipse with the specified properties

1. Foci at $(3, 0)$ and $(-3, 0)$, $a = 4$;

2. Foci at $(0. 6)$ and $(0, -6)$, $b = 12$.

3. Foci at $(4, 0)$ and $(-4, 0)$, major axis = 10 units;

4. Foci at $(5, 0)$ and $(-5, 0)$, minor axis = 6 units; **5.** Foci at $(-2, 4)$ and $(4, 4)$, with $a = 5$.

Answers:

1. $\frac{x^2}{16} + \frac{y^2}{7} = 1$; **2.** $\frac{y^2}{180} + \frac{x^2}{144} = 1$; **3.** $\frac{x^2}{25} + \frac{y^2}{9} = 1$; **4.** $\frac{x^2}{34} + \frac{y^2}{9} = 1$; **5.** $\frac{(x-1)^2}{25} + \frac{(y-4)^2}{16} = 1$

Lesson 63
Hyperbolas

Definition: A **hyperbola** is the set of points such that the absolute value of the difference of the distances of each point from two fixed points, called the foci, is a constant (which is always less than the distance between the foci).

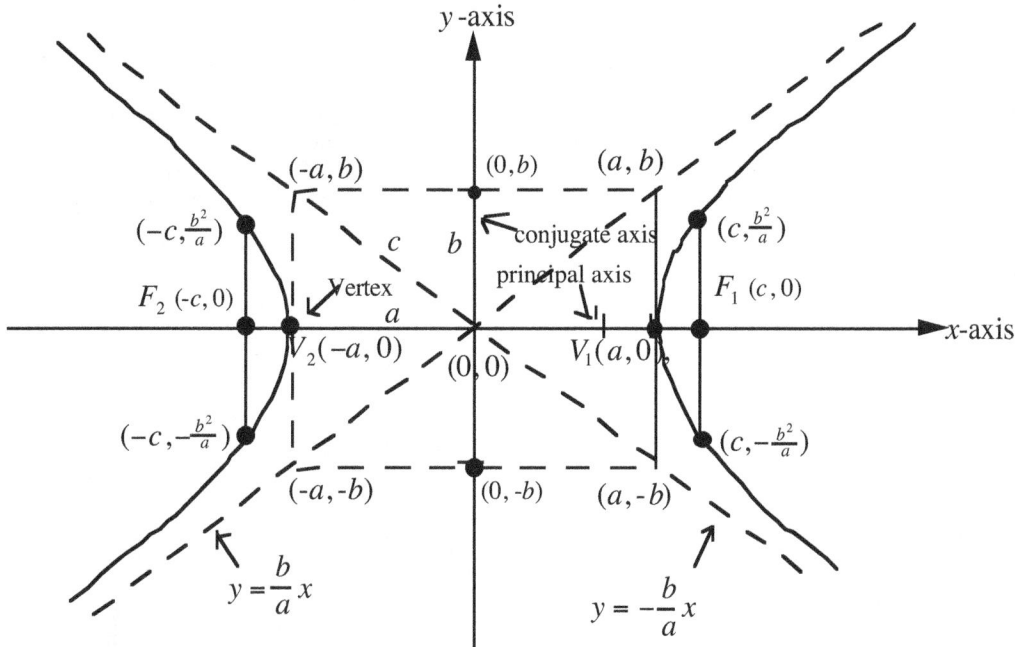

Figure 1: Graph of the hyperbola $\dfrac{x^2}{a^2} - \dfrac{y^2}{b^2} = 1$ $\qquad a > b$ $\qquad$ (Major axis along the x-axis)

Properties of the hyperbola with the transverse axis along the x-axis

1. The graph of a hyperbola consists of two branches. One branch is concave to the left and the other branch is concave to the right. The hyperbola has two axes of symmetry, namely the transverse (or principal) axis, and the conjugate axis. On the transverse axis are located the foci $F_1(-c, 0)$ and $F_2(c, 0)$ and the vertices $V_1(a, 0)$, and $V_2(-a, 0)$, c being the distance from the center of the hyperbola to a focus. Also, c is the length of the semidiagonal of the auxiliary rectangle.
   In the above figure, the transverse axis is also part of the x-axis because of the choice of the location of the origin The conjugate axis (also part of the y-axis) is the perpendicular bisector of the transverse axis, and it passes through $(0, b)$ and $(0, -b)$. We must however note that the transverse axis or the conjugate axis is **not** always part of either of the coordinate axes; the location of the origin will determine whether or not this is the case.

2. The length of the transverse axis is $2a$ and the length of the conjugate axis is $2b$. The above lengths were chosen for convenience in deriving the equation of the hyperbola. There is a semi-transverse axis of length a as well as a semi-conjugate axis of length b .

3. The transverse axis intersects each branch of the hyperbola at a point is called a vertex. There are thus two vertices. The vertices are equidistant from the center of the hyperbola and so also are the foci F_1 and F_2 . The foci are at $(-c, 0)$ and $(c, 0)$. The distance between the foci is $2c$ and the distance between the vertices is $2a$. There are two focal chords. Each chord is a line segment from a point on the hyperbola through a focus to another point on the hyperbola. The above hyperbola has x-intercepts at $V_1(a, 0)$, and $V_2(-a, 0)$ but has no (real) y-intercepts. Thus the above curve does not intersect the y-axis. Note however that generally, the curve may intersect the y-axis due to the location of the center of the hyperbola

Lesson 63: Hyperbola

4. The relationship between a, b, and c is given by $c^2 = a^2 + b^2$.

   Note that this relationship is different from that for the ellipse. By coincidence, this relationship is the same as the Pythagorean theorem. Here, c is the hypotenuse.

The standard form of the equation of the hyperbola in above figure is given by

$$\frac{x^2}{a^2} - \frac{y^2}{b^2} = 1 \qquad \textbf{(1)} \qquad \text{(Note also that this equation is not a function)}$$

5. The curve has oblique asymptotes whose equations are given by

$$y = \frac{b}{a}x \ \text{ and } \ y = -\frac{b}{a}x. \ \text{ (see also p,276-277)}$$

   There is an associated or auxiliary rectangle of length $2a$ and width $2b$., with a semidiagonal of length c. The extensions of the diagonals of the auxiliary rectangle coincide with the asymptotes.

   A mnemonic device for obtaining the equations of the asymptotes is to set the left hand-side of the equation of the standard equation to zero. Then factor, set each factor to zero, and solve for y as shown below.

$$\frac{x^2}{a^2} - \frac{y^2}{b^2} = 0$$

$$\left(\frac{x}{a} + \frac{y}{b}\right)\left(\frac{x}{a} - \frac{y}{b}\right) = 0 \quad \text{(Factoring by the difference between two squares.)}$$

$$\frac{x}{a} + \frac{y}{b} = 0; \qquad \text{or} \quad \frac{x}{a} - \frac{y}{b} = 0 \quad \text{(applying the principle of zero products)}$$

$$\frac{y}{b} = -\frac{x}{a}; \quad \text{or} \qquad -\frac{y}{b} = -\frac{x}{a}$$

$$y = -\frac{b}{a}x; \quad \text{or} \qquad y = \frac{b}{a}x$$

The general form the above hyperbola with center at (h, k) is...

$$\frac{(x - h)^2}{a^2} - \frac{(y - k)^2}{b^2} = 1 \qquad \text{(The transverse axis is parallel to the x-axis)}$$

Note above that c = the distance from the center to a focus = the length of the semi-diagonal. a = the distance from the center to a vertex .

Properties of the hyperbola with the transverse axis along the y-axis   425

This hyperbola can be obtained from the previous hyperbola by interchanging the roles of x and y Then the equation of the hyperbola becomes

$$\frac{y^2}{a^2} - \frac{x^2}{b^2} = 1 \qquad (2) \quad \text{(transverse axis along the y-axis)}$$

The foci are at $F_1(0, c)$ and $F_2(0, -c)$, where $c^2 = a^2 + b^2$. The vertices are at $V_2(0, a)$ and $V_1(0, -a)$ The curve has two branches. One branch is concave up and the other branch is concave down. The curve has real y-intercepts and **no** x-intercepts. The equations of the oblique asymptotes can be obtained by

setting $\dfrac{y^2}{a^2} - \dfrac{x^2}{b^2} = 0$, factoring and solving for y.

Then the equations of the asymptotes are $y = -\dfrac{a}{b}x$ and $y = \dfrac{a}{b}x$

We observe that equation (2) is the **inverse relation** of equation (1) in the previous case, since either equation can be obtained from the other by interchanging the roles of x and y. See also page 73. Note, however, that these equations are **not** functions.

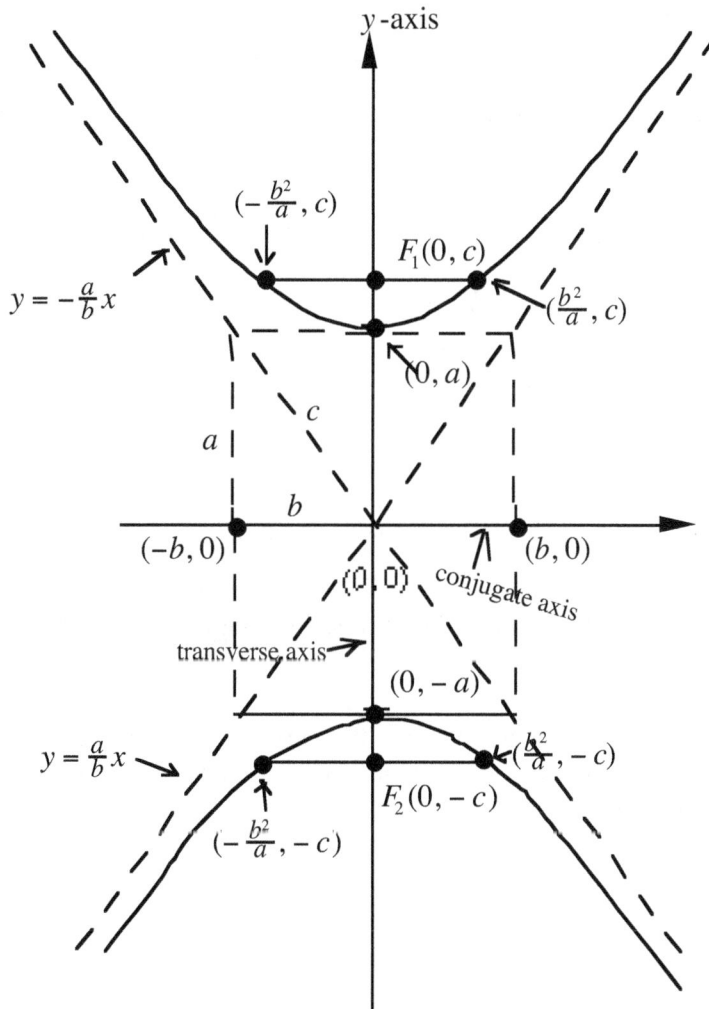

Figure 2: Graph of the hyperbola: $\dfrac{y^2}{a^2} - \dfrac{x^2}{b^2} = 1$ (Transverse axis along the y-axis)

Graphically, we can obtain Figure 2 from Figure 1 by reflecting figure about line $y = x$.

The general form of the equation of this hyperbola equation with center at (h, k) is

$$\frac{(y-k)^2}{a^2} - \frac{(x-h)^2}{b^2} = 1 \qquad \text{(The transverse axis is parallel to the y-axis)}$$

We note again that, $\frac{x^2}{a^2} - \frac{y^2}{b^2} = 1$ and $\frac{y^2}{a^2} - \frac{x^2}{b^2} = 1$ are **inverse relations** of each other but **not** inverse functions

Determining whether a given equation of a hyperbola is of the form

$$\frac{x^2}{a^2} - \frac{y^2}{b^2} = 1 \quad \textbf{or} \quad \frac{y^2}{a^2} - \frac{x^2}{b^2} = 1$$

Here, we use the **signs of the terms** in this determination.

If the term containing x^2 is positive, then the transverse (or principal) axis is along the x-axis, and a^2 is the denominator of this term; while b^2 is the denominator of the negative term .

However, if the term containing y^2 is positive, then the transverse (or principal) axis is along the y-axis, and a^2 is the denominator of this term; while b^2 is the denominator of the negative term .

Examples: **1.** $\frac{x^2}{16} - \frac{y^2}{9} = 1$ ($a^2 = 16$, $b^2 = 9$; transverse axis is along the x-axis)

          2. $\frac{x^2}{9} - \frac{y^2}{16} = 1$ ($a^2 = 9$, $b^2 = 16$; transverse axis is along the x-axis)

          3. $\frac{y^2}{16} - \frac{x^2}{25} = 1$ ($a^2 = 16$, $b^2 = 25$; transverse axis is along the y-axis)

          4. $\frac{y^2}{25} - \frac{x^2}{16} = 1$ ($a^2 = 25$, $b^2 = 16$; transverse axis is along the y-axis)

Note that in the case of the **hyperbola,** the relative sizes of a^2 and b^2 are **not** used in determining which equation we are dealing with (unlike the case for the ellipse). Therefore, a^2 may be larger or smaller than b^2. If $a = b$, the asymptotes of the hyperbola meet at right angles, and the hyperbola is then called a rectangular hyperbola.

Note above: In finding the x- or y-intercepts only the variable in the positive term will yield real values. The variable in the negative term will yield non-real values, and therefore no intercepts.

How to sketch the graph of a hyperbola (General)

There are generally two different methods which we may use to sketch the graphs of hyperbolas.

Method 1   Sketch the graph of $\dfrac{x^2}{a^2} - \dfrac{y^2}{b^2} = 1$ (1)

The principal axis is along the x-axis (along the direction of the variable in the positive term)

Step 1: Find the x-intercepts by setting $y = 0$ in the given equation and solving for x. This solution yields the coordinates $(a, 0)$ and $(-a, 0)$. Plot these points letting the center of symmetry be at the origin.

Step 2: Find the equations of the asymptotes as explained before (page 273).

The equations for the given case are $y = \dfrac{b}{a}x$, and $y = -\dfrac{b}{a}x$.

Step 3: Graph the equations of these asymptotes (straight lines) by choosing two convenient x-values and computing corresponding y-values for each asymptote.

Step 4: Draw vertical lines through the vertices $(a, 0)$ and $(-a, 0)$ to intercept the asymptotes

Step 5: Draw the two branches of the hyperbola, each branch through a vertex. Draw the curves so that they approach the diagonals (the asymptotes) smoothly. One branch is concave to the left and the other branch is concave to the right.

Method 2

Step 1: Read the values a and b from the standard equation by inspection. They are the square roots of the denominators in the standard equation. Plot the points $(a, b), (a, -b), (-a, b), (-a, -b)$. With these points as vertices, draw the rectangle whose center is at the origin $(0, 0)$ and of length $2a$, and width $2b$.

Step 2: Draw the diagonals of the rectangles and extend the diagonals beyond the vertices.

Step 3: Draw the two branches of the hyperbola through the vertices $(a, 0)$, $(-a, 0)$ of the hyperbola. (Note that the vertices of the rectangle and the vertices of the hyperbola are not the same). Each branch is through a vertex and smoothly approaches the extensions of the diagonals (the asymptotes). One branch is concave to the left and the other branch is concave to the

Step 4: Location of the foci. Calculate c from $c^2 = a^2 + b^2$. The foci would be at $(c, 0)$ and $(-c, 0)$.

The end-points of the focal chords are at $(c, \frac{b^2}{a})$ and $(c, -\frac{b^2}{a})$.; $(-c, \frac{b^2}{a})$ and $(-c, -\frac{b^2}{a})$.

Calculate $\frac{b^2}{a}$, and then from each focus, count vertically upwards $\frac{b^2}{a}$ units; and again from

the focus, count vertically downwards $\frac{b^2}{a}$ units to locate the endpoints of the focal chords.

Sketching the graph of the inverse relation

We can similarly sketch the " inverse relation " of the hyperbola in equation (1), above.
The inverse relation is given by

$$\frac{y^2}{a^2} - \frac{x^2}{b^2} = 1 \qquad\qquad (2)$$

The principal axis is along the y-axis. One branch of the hyperbola is concave up and the other branch is concave down. The foci are at $(0, c)$ and $(0, -c)$, where $c^2 = a^2 + b^2$. If we want a more accurate sketch, we can choose convenient x-values, calculate corresponding y-values and then plot the points. Alternatively, we can choose convenient y-values and then compute the corresponding x-values. We should note that for each choice, there would be two x- or two y-values. The double values are due to the quadratic nature of the equations. There will be $\pm x$–values or $\pm y$–values.

Lesson 63: Hyperbola

We must note that the foci are always on the principal axis and inside the concavities of the curve. The principal axis is always intercepts the curves but the conjugate axis does not.

The **hyperbola** $y = \dfrac{1}{x}$ can be considered as a special case of $\dfrac{x^2}{a^2} - \dfrac{y^2}{b^2} = 1$ in which the x- and y-axes have been rotated $+45°$ (or the transverse axis has been rotated $+45°$).

Example 1 Sketch the graph of the hyperbola given by

$$\frac{x^2}{16} - \frac{y^2}{9} = 1$$ (the minus sign indicates that the equation is that of a hyperbola)

Procedure: We shall use a direct method, an analytical method.

Step 1: Find the transverse (or principal) axis. The direction of the principal axis is the direction of the variable in the positive term. In this problem, it is the x-term. Therefore, the transverse axis is in the x-direction, and it is also on the x-axis. We should note that the determination of the axis here is unlike that for the ellipse. It does **not** depend on the relative sizes of the denominators.

Step 2: Find the coordinates of the center of the hyperbola.
By comparison with the general equation, the center is at the origin $(0, 0)$ or $h = 0, k = 0$. or set each numerator to zero and solve.

Step 3: Find the constants $a, b,$ and c.

$$a^2 = 16; \qquad b^2 = 9$$
$$a = 4. \qquad b = 3$$
$$c^2 = a^2 + b^2 \quad \text{(This property relationship is different from that for the ellipse.)}$$
$$c^2 = 16 + 9$$
$$c = 5.$$

Step 4: We shall now draw the auxiliary rectangle (Figure) of length $2a$, or 8 units ($a = 4$); and width $2b$ or 6 units ($b = 3$), and center at $(0, 0)$. Locate the center on the transverse axis. The center is on the x-axis. From the center, count 4 units ($a = 4$) to the right, and again from the center, count 4 units to the left to obtain the vertices of the hyperbola. Similarly, from the center and on the conjugate axis (the y-axis) count 3 units up; and again, from the center, count 3 units down to obtain two points on the opposite sides of the auxiliary rectangle.

Step 5: Using dotted lines, complete the rectangle.

Step 6: Draw the diagonals of the rectangle and extend the diagonals beyond the corners of the rectangle. The extensions of the diagonals are the asymptotes to the two branches of the curve.

Step 7: Locate the foci F_1, F_2 on the principal axis (transverse axis) by counting 5 units ($c = 5$) in each direction from the center.

Step 8: Find $\dfrac{b^2}{a}$ and pair it with c to locate the endpoints of the focal chords. $\dfrac{b^2}{a} = \dfrac{9}{4} = 2\dfrac{1}{4}$.
From each focus, count $2\dfrac{1}{4}$ units vertically upwards, stop, place a dot here, and label it; and again from the focus, count $2\dfrac{1}{4}$ units vertically downwards, stop, place a dot here, and label it. (The end-points of the focal chords are at $(5, \frac{9}{4})$ and $(5, -\frac{9}{4})$ $(-5, \frac{9}{4})$ and $(-5, -\frac{9}{4})$)

Step 9: Draw the two branches of the hyperbola: for each branch, draw a smooth curve beginning from one end of the focal chord through the vertex and to the other end of the focal chord.
Each curve smoothly approaches the extensions of the diagonals (asymptotes). One branch is concave to the right and the other is concave to the left.

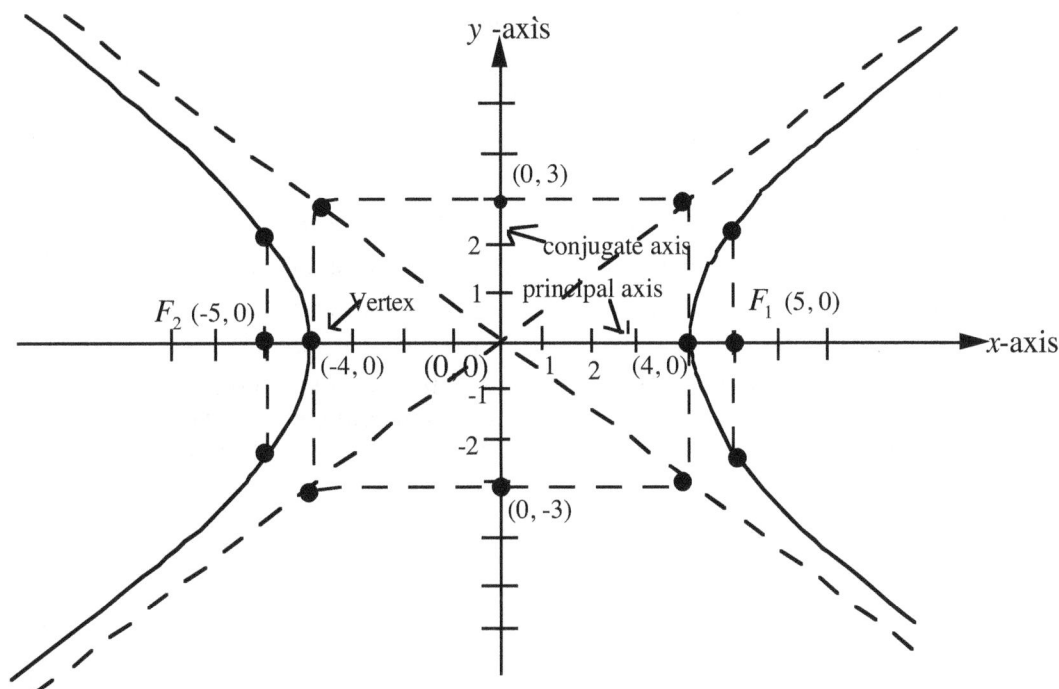

Figure 3: Graph of $\frac{x^2}{16} - \frac{y^2}{9} = 1$

Example 2 Sketch the graph of the hyperbola given by

$$\frac{(x-3)^2}{36} - \frac{(y+4)^2}{64} = 1 \quad \text{(the minus sign indicates that the equation is that of a hyperbola)}$$

Procedure: We use a direct method (an analytic method).

Step 1: Find the principal (or transverse axis) The direction of the principal axis is in the x-direction, but it is not on the x-axis. (That is, it does not coincide with the x-axis.)

Step 2: Find the coordinates of the center of the hyperbola: Set each numerator to zero and solve.
$x - 3 = 0$ and from which $x = 0$. Therefore $h = 0$.
$y + 4 = 0$, and from which $y = -4$. Therefore $k = -4$
The center is at $(3, -4)$

Step 3: Find the constants a, b, and c.

$$a^2 = 36 \text{ and from which } a = 6; \quad b^2 = 64 \text{ and from which } b = 8$$
$$c^2 = 36 + 64 \quad (c^2 = a^2 + b^2; \text{this property relationship is different from that for the ellipse.})$$
$$c = 10$$

Step 4: Plot the center $(3, -4)$ from Step 2.

Step 5: Locate the vertices on the transverse axis by counting 6 units horizontally ($a = 6$) from the center to the right and labeling this point (-9, -4); followed by counting 6 units from the center to the left, stopping, and labeling this point (-3, -4). Connect these two points to obtain the transverse axis.

Step 6: From the center, count 8 units vertically up, stop , place a dot here; (This is one end of the conjugate axis); followed by counting 8 units from the center vertically downwards and stop. (This the other end of the conjugate axis). Connect these two endpoints to obtain the conjugate axis This conjugate axis is parallel to the y-axis.

Lesson 63: Hyperbola

Step 7 Using dotted lines draw the auxiliary rectangle (Figure) of length $2a$, or 12 units ($a = 6$); and width $2b$ or 16 units ($b = 8$), and center at $(3, -4)$.

Step 8: Draw the diagonals of the rectangle and extend the diagonals beyond the corners of the rectangle. The extensions of the diagonals are the asymptotes to the two branches of the curve.

Step 9: Locate the foci F_1, F_2 on the principal axis (transverse axis) by counting 10 units ($c = 10$) each from the center in both directions.

Step 10: Find $\frac{b^2}{a}$ and pair it with c to locate the endpoints of the focal chords. $\frac{b^2}{a} = \frac{64}{6} = 10\frac{2}{3}$.

From each focus, count $10\frac{2}{3}$ units vertically upwards, stop, place a dot here, and label it; and again from the focus, count $10\frac{2}{3}$ units vertically downwards, stop, place a dot here, and label it. .The end-points of the focal chords are at $(13, \frac{20}{3})$ and $(13, -\frac{44}{3})$ $(-7, \frac{20}{3})$ and $(-7, -\frac{44}{3})$.

Step 11: Draw the two branches of the hyperbola: for each branch, draw a smooth curve beginning from one end of the focal chord through the vertex and to the other end of the focal chord. Each curve smoothly approaches the extensions of the diagonals (asymptotes). One branch is concave to the right and the other is concave to the left.

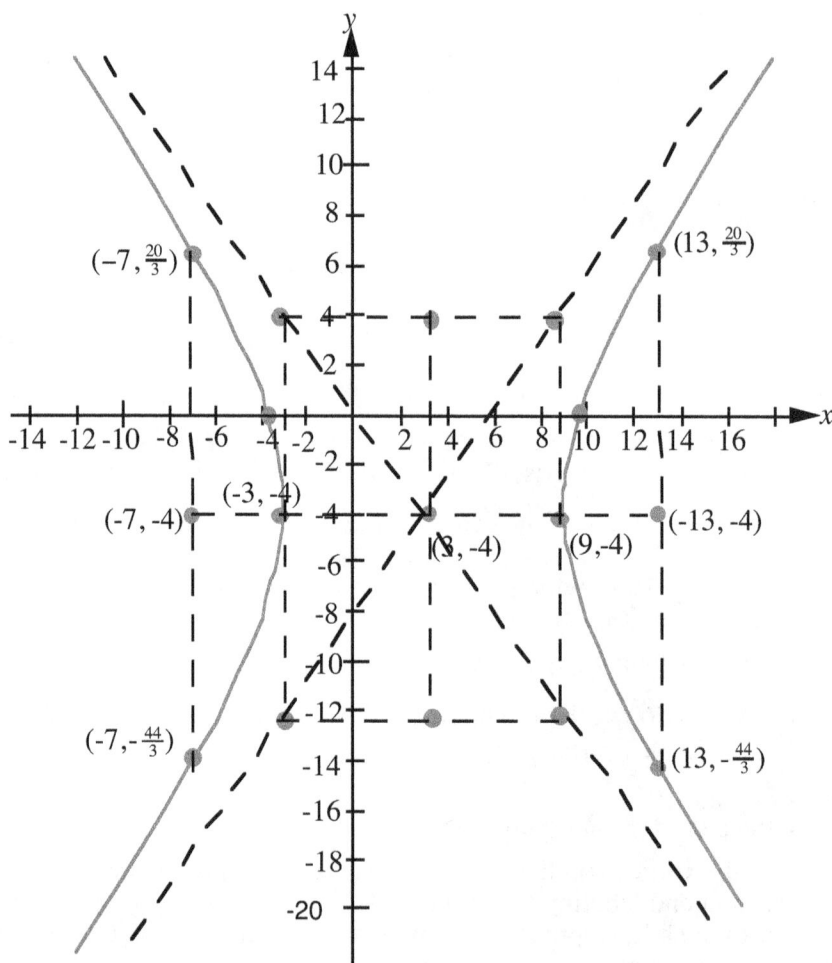

Figure 4: Graph of $\dfrac{(x-3)^2}{36} - \dfrac{(y+4)^2}{64} = 1$

Another Hyperbola, the hyperbola $xy = c$, where c is a constant
We have met this equation under rational functions (Chapter),. For example, the reciprocal of the linear function $y = x$ is $y = \dfrac{1}{x}$ which on cross-multiplying becomes $xy = 1$.

Identifying Conic Equations

(Axes of the Conics are along or parallel to the x- and y-axes)

Each of the conic equations (in the rectangular coordinate system) that we have covered, so far, is a second degree equation of the form

$$Ax^2 + Cy^2 + Dx + Ey + F = 0.$$

The coefficients of the terms determine the type of conic curve involved, and in particular, the coefficients A and C determine whether the equation is that of a circle, an ellipse, a parabola, or a hyperbola.

Method 1: Below, we discuss the conditions on A and C for identifying these curves.

1. If $A = C \neq 0$, the equation is that of a circle.

(Same as if there are both the x^2-term and the y^2-term, and the coefficients of these terms are the same, then the equation is that of a circle)

2. If $A \neq C$, but have the same sign, then the equation is that of an ellipse.

(Same as if there are both the x^2-term and the y^2-term, and the coefficients of these terms are different but of the same sign, then the equation is that of an ellipse.)

3. If A and C have different signs, the equation is that of a hyperbola.

(Same as if there are both the x^2-term and the y^2-term, and the coefficients of these terms have different signs, then the equation is that of a hyperbola.)

4. If either $A = 0$ or $C = 0$ (but both $\neq 0$) then the equation is that of a parabola.

(Same as if there is either the x^2-term or the y^2-term (but not both) then the equation is that of a parabola.)

5. If either $D \neq 0$, or $E \neq 0$ (or both $\neq 0$), the center of the circle, ellipse, hyperbola, or the vertex of the parabola is not at the origin.

(Same as if there is either an x-term or a y-term (or both terms), then the center of the circle, ellipse, hyperbola, or the vertex of the parabola is **not** at the origin.)

Note: If there is an xy-term, the axes of the curve have been rotated.

Note: The equation $4x^2 + 4y^2 - 9 = 0$ is that of a circle, since $A = C$.

However, the equation $4x^2 + 3xy + 4y^2 - 9 = 0$ is **not** that of circle even though $A = C$, because, there is an xy-term. In fact, the equation is that of an ellipse.

Method 2: Another method of identifying conic sections

By investigating the discriminant, AC, of $Ax^2 + Cy^2 + Dx + Ey + F = 0$, we can recognize the type of conic section, whose equation is given, as follows:

1. If $AC = 0$, the equation is that of a parabola.

2. If $AC < 0$, the equation is that of an ellipse.

3. If $AC > 0$, the equation is that of a hyperbola.

4. If $A = C \neq 0$, the equation is that of a circle.

5. If A and C are all zero, then the equation is that of a straight line.

The General Conic Equation

So far, the axes of the conic sections we have covered are either along or parallel to the coordinate x- and y-axes. The more general form the equation includes an xy-term, and is of the form.

$$Ax^2 + Bxy + Cy^2 + Dx + Ey + F = 0.$$

The coefficients of the terms determine the type of curve involved, and in particular, the coefficients A, B, and C determine whether the equation is that of a circle, an ellipse, a parabola, or a hyperbola. Note that if $B \neq 0$ (that is, there is an xy-term), the axis of the conic has been rotated and is not along or parallel to the x- and y-axes.

Method 1: Below, we discuss the conditions on A, B, and C for identifying these curves.

1. If $A = C \neq 0$ and $B = 0$, the equation is that of a circle.

   (Same as if there are both the x^2-term and the y^2-term, and the coefficients of these terms are the same, and there is no xy-term, then the equation is that of a circle)

2. If $A \neq C$, but have the same sign, and $B = 0$, then the equation is that of an ellipse.

   (Same as if there are both the x^2-term and the y^2-term, and the coefficients of these terms are different but of the same sign, and there is no xy-term, then the equation is that of an ellipse.)

3. If A and C have different signs, and $B = 0$, the equation is that of a hyperbola.

   (Same as if there are both the x^2-term and the y^2-term, and the coefficients of these terms have different signs, and there is no xy-term, then the equation is that of a hyperbola.)

Note: If $A = 0$, $C = 0$ but $B \neq 0$, then the equation is that of a hyperbola.

(Same as if there is **no** x^2-term and **no** y^2-term, but **there is** an xy-term then the equation is that of a hyperbola)

4. If either $A = 0$ or $C = 0$ (but both $\neq 0$) and $B = 0$, then the equation is that of a parabola.

   (Same as if there is either the x^2-term or the y^2-term (but not both), and there is no xy-term, then the equation is that of a parabola.)

5. If either $D \neq 0$, or $E \neq 0$ (or both $\neq 0$), the center of the circle, ellipse, hyperbola, or the vertex of the parabola is **not** at the origin.

   (Same as if there is either an x-term or a y-term (or both terms), then the center of the circle, ellipse, hyperbola, or the vertex of the parabola is **not** at the origin.)

6. If $B \neq 0$ (i.e., there is an xy-term), the axis of the curve has been rotated. If $B = 0$, the axis of the curve is either along the coordinate axis or parallel to the coordinate axis.

Note: The equation $4x^2 + 4y^2 - 9 = 0$ is that of a circle, since $A = C$, and $B = 0$.

However, the equation $4x^2 + 3xy + 4y^2 - 9 = 0$ is **not** that of circle even though $A = C$, because $B \neq 0$ (i.e., there is an xy-term). In fact, the equation is that of an ellipse.

Method 2: Another method of identifying conic sections

By investigating the discriminant $B^2 - 4AC$ of $Ax^2 + Bxy + Cy^2 + Dx + Ey + F = 0$, we can recognize the type of conic section , whose equation is given, as follows:

1. If $B^2 - 4AC = 0$, the equation is that of a parabola.

2. If $B^2 - 4AC < 0$, the equation is that of an ellipse.

3. If $B^2 - 4AC > 0$, the equation is that of a hyperbola.

4. If $A = C \neq 0$, and $B = 0$, the equation is that of a circle.

5. If A, B, and C are all zero, then the equation is that of a straight line.

Note: Do not confuse the above discriminant $B^2 - 4AC$ with that of the quadratic equation.

Method 3 Notwithstanding the above methods, one can write the given general equation in standard form, and then easily compare with the standard forms to identify the conic. This approach will relieve one from having to remember the conditions in the above methods

Standard Forms of the Equations of Conics

We have discussed the general conic equation, $Ax^2 + Bxy + Cy^2 + Dx + Ey + F = 0$.
When given the general equation, we can transform it to the standard form. In the standard form, we can readily identify the vertex and the axis of symmetry of a parabola; the center and the symmetric axes of a circle, an ellipse, and a hyperbola. Also, the standard form is the form we usually use in graphing the conics.

Standard Forms Summarized

For a **circle:** $(x - h)^2 + (y - k)^2 = r^2$ with center at (h, k) and of radius r.
(Also called center-radius form of the equation of a circle)

For a **parabola:** $(x - h)^2 = 4p(y - k)$. with vertex at (h, k) and axis parallel to the y-axis

$$(y - k)^2 = 4p(x - h); \text{ with vertex at } (h, k) \text{ and axis parallel to the } x\text{-axis}$$

For an **ellipse**: $\dfrac{(x - h)^2}{a^2} + \dfrac{(y - k)^2}{b^2} = 1$; with center at (h, k) and major axis parallel to the x-axis.

$\dfrac{(y - k)^2}{a^2} + \dfrac{(x - h)^2}{b^2} = 1$; with center at (h, k) and major axis parallel to the y-axis.

For a **hyperbola**: $\dfrac{(x - h)^2}{a^2} - \dfrac{(y - k)^2}{b^2} = 1$, with center at (h, k) and transverse axis parallel to the x-axis.

$\dfrac{(y - k)^2}{a^2} - \dfrac{(x - h)^2}{b^2} = 1$; with center at (h, k) and transverse axis parallel to the y-axis.

Another Method of Defining Conic Sections

Definition: A **circular cone** is a three-dimensional figure formed by a straight line moving around the circumference of a circle and always passing through a fixed point which is not in the plane of the circle. The circle forms the base of the cone; the moving straight line forms the slant side of the cone; and the fixed point forms the vertex.
The **axis** of the circular cone is a line from the vertex to the center of the base.
If the axis is perpendicular to the base, the cone is called a **right circular cone**.

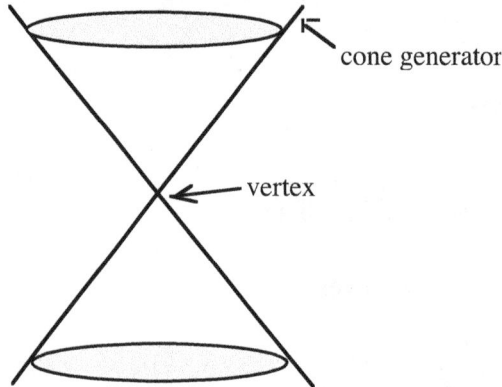

cone generator

vertex

We have previously covered the geometric definitions of conics in terms of the locus of points. We shall now define conics in terms of the intersection of a plane with a right circular cone.

Definition: A **conic section** is a curve formed by the intersection of a plane with a right-circular cone.

The conic section is a **circle** if the intersecting plane is parallel to the base of the cone; the conic is an **ellipse** if the intersecting plane passes through only one part of the cone and not parallel to a slant side nor to the base; the conic is a **parabola** if the intersecting plane passes through the cone and is parallel to the slant side; the conic is a **hyperbola** if the intersecting plane passes through both parts of the cone at an angle greater than the slant angle, the angles being measured from the base.

Lines and **points** are considered **degenerate conics** formed if the intersecting plane includes the vertex of the cone.

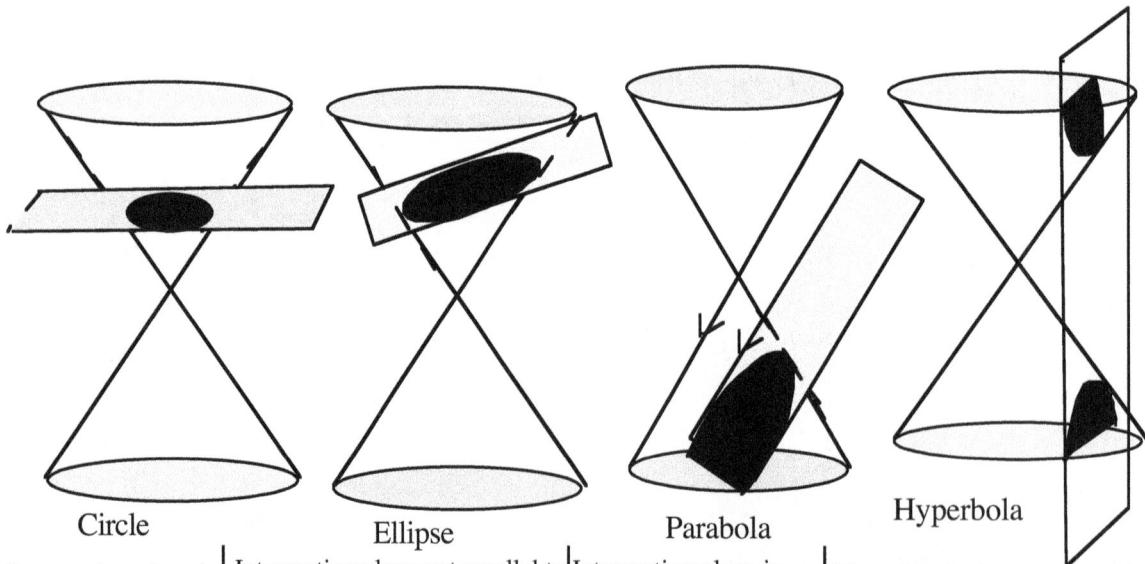

Circle	Ellipse	Parabola	Hyperbola
Intersecting plane is parallel to the base.	Intersecting plane not parallel to the base nor to the slant side.	Intersecting plane is parallel to the slant side.	Plane cuts both cones at angle greater than the slant angle.

Eccentricity

Before defining the eccentricity of conics, we repeat the geometric definitions of conics covered previously.

Definition: A **circle** is the set of points in a plane that are equidistant from a fixed point (called the center of the circle). The distance from the center to any point on the circle is called the radius of the circle.

Definition: A **parabola** is the set of points in a plane such that each point is at the same distance from a fixed point F (called the focus) as it is from a fixed line D, (called the directrix) with the focus F, not lying on the directrix

Definition: An **ellipse** is the set of points such that the sum of the distances of each point from two fixed points, F_1, F_2 (called the foci) is a given constant.

Definition: A **hyperbola** is the set of points such that the absolute value of the difference of the distances of each point from two fixed points, called the foci, is a constant (which is always less than the distance between the foci).

Definition: The **eccentricity** of a conic section, denoted by e, is the ratio,

$$\frac{\text{distance between a point on the curve and the focus.}}{\text{distance between the point and the directrix}} = \frac{c}{a}, \text{ That is } e = \frac{c}{a}$$

This ratio identifies the shape of a conic section. For a circle, $e = 0$; for a parabola, $e = 1$; for an ellipse, $0 < e < 1$; and for a hyperbola, e > 1. We may view the eccentricity of an ellipse as a measure of how much an ellipse differs from a circle, considering that for a circle, $e = 0$.

Note: The word eccentricity pertains to the word eccentric which means not having a common center.

Lesson 63 Exercises

A Find an equation of the hyperbola with the specified properties

1. Foci at $\left(\sqrt{13}\ \ 0\right)$ and $\left(-\sqrt{13}\ \ 0\right)$, and $a = 3$.

2. Foci at $\left(0\ \ 10\right)$ and $\left(0\ -10\right)$, and $a = 6$.

3. Foci at $\left(0\ \ 5\right)$ and $\left(0\ -5\right)$, and $2a = 8$; **4.** Foci at $\left(6\ \ 0\right)$ and $\left(-6\ \ 0\right)$, and $a = 4$.

Answers: **1.** $\dfrac{x^2}{9} - \dfrac{y^2}{4} = 1$; **2.** $\dfrac{y^2}{36} - \dfrac{x^2}{64} = 1$; **3.** $\dfrac{y^2}{16} - \dfrac{x^2}{9} = 1$; **4.** $\dfrac{x^2}{16} - \dfrac{y^2}{20} = 1$.

B Sketch the graph of the hyperbola with the given equation:

1. $\dfrac{x^2}{9} - \dfrac{y^2}{4} = 1$; **2.** $\dfrac{x^2}{4} - \dfrac{y^2}{9} = 1$; **3.** $\dfrac{(x-2)^2}{9} - \dfrac{(y-3)^2}{4} = 1$;

Questions **4- 8**: Draw the graphs of the following using a graphing calculator.

4. $\dfrac{(x+2)^2}{9} - \dfrac{(y+3)^2}{25} = 1$; **5.** $4x^2 - 9y^2 - 144 = 0$; **6.** $\dfrac{(x-2)^2}{9} - \dfrac{(y-1)^2}{16} = 1$;

7. $\dfrac{(x-1)^2}{49} - \dfrac{(y+2)^2}{81} = 1$; **8.** $100x^2 - 25y^2 = 2500$

Solution:

2.

$\dfrac{x^2}{4} - \dfrac{y^2}{9} = 1$

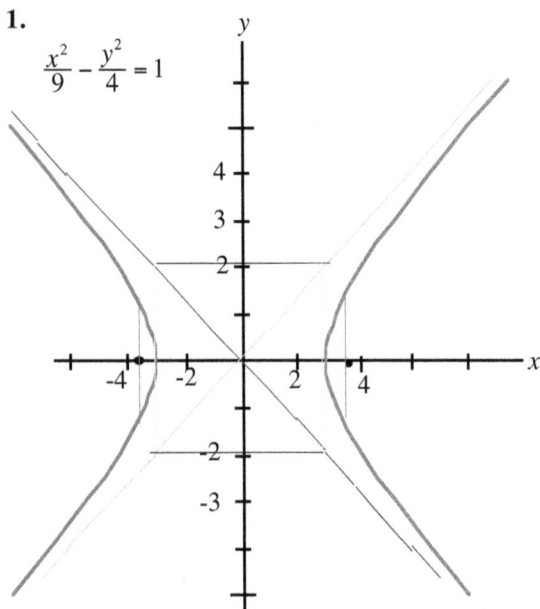

1.

$\dfrac{x^2}{9} - \dfrac{y^2}{4} = 1$

3.

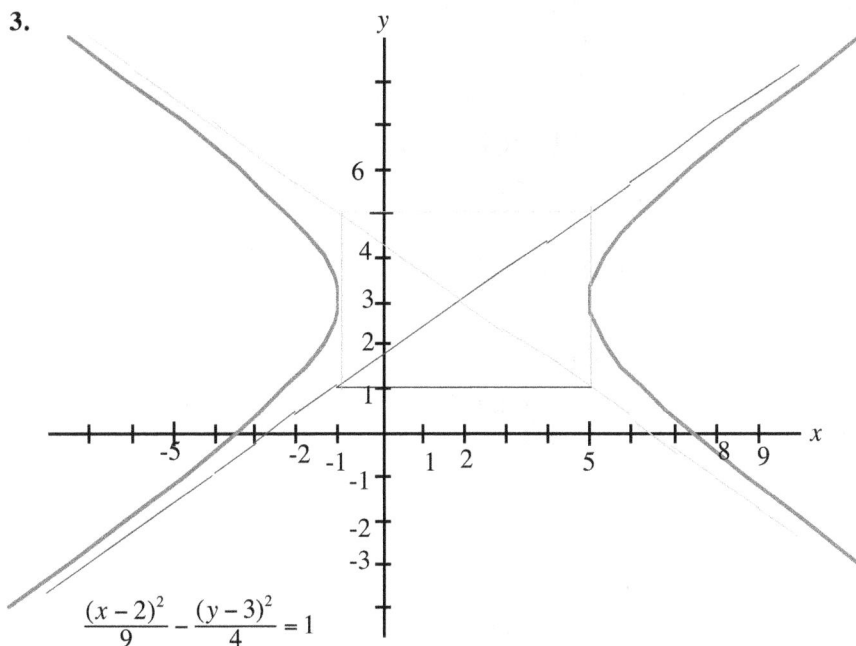

$$\frac{(x-2)^2}{9} - \frac{(y-3)^2}{4} = 1$$

C Determine the type of conic whose equation is given , and sketch its graph.

1. $x^2 - 2x + y^2 - 4y - 11 = 0$; **2.** $x^2 + 4x + y^2 - 6y - 12 = 0$; **3.** $16x^2 - 16x + 16y^2 - 8y - 63 = 0$

4. $x^2 + 2x + 2y + 7 = 0$; **5.** $y^2 - y = x - 6$; **6.** $y^2 + 10x = 0$; **7.** $4x^2 + 9y^2 - 36 = 0$

8. $4x^2 - 16x + 9y^2 - 54y + 61 = 0$; **9.** $100x^2 + 25y^2 - 2500 = 0$

10. $4x^2 - 9y^2 - 36 = 0$; **11.** $4x^2 - 16x - 9y^2 + 54y - 29 = 0$; **12.** $100x^2 - 25y^2 - 2500 = 0$.

Answers: 1. Circle; **2.** Circle; **3.** Circle; **4.** parabola; **5.** parabola; **6. parabola**; **7.** ellipse; **8.** ellipse; **9.** ellipse; **10.** hyperbola; **11.** hyperbola; **12.** hyperbola. For the graphs, see previous problems in this chapter.

CHAPTER 24

Lesson 64: **Synthetic Division, Remainder and Factor Theorems**
Lesson 65: **Descartes' Rules; Rational Roots of Polynomial Equations**

Lesson 64

Synthetic Division, Remainder and Factor Theorems

Synthetic Division

Example 1 Use synthetic division to divide $2x^3 - 14x^2 - x + 6$ by $x - 3$.

Step 1: Arrange the dividend and the divisor in descending powers of the variable (If this has not been done) and write zeros to hold places for missing powers.

Step 2: Write the numerical coefficients of the dividend.

For $2x^3 - 14x^2 - x + 6$. we write

$\qquad 2 \ - \ 14 \ - \ 1 \ + \ 6 \qquad\qquad$ (A)

Step 3: Set the divisor equal to zero and solve for x.

$\qquad x - 3 = 0$, and Solving, $x = 3 \qquad\qquad$ (B)

Step 4: Set up the division process by using expression (A) and solution to equation (B) as dividend and divisor respectively; and leave one blank row space below the dividend and also underline as shown below.

Step 5: Bring down the 2, the first coefficient as indicated by the direction of the arrow and follow the arrows as explained in Step 6 - Step 9.

Step 6: Multiply this 2 by the divisor, 3, and write this product , 6 , in the blank space directly under the -14.

Step 7: Add the -14 and the +6 to obtain -8, and write this sum in that column.

Step 8: Multiply the -8 by the divisor, 3, to obtain -24 and write this product under the -1 and again add in this column to obtain -25.

Step 9: Multiply the -25 by the 3 to obtain -75 and write this product directly below the +6 and add in this column to obtain -69.

The last term in the last row (3rd row) is the remainder, and in this problem, it is -69.
The remaining numbers 2 - 8 -25 are the coefficients of the partial quotient.

Thus $\dfrac{2x^3 - 14x^2 - x + 6}{x - 3} = 2x^2 - 8x - 25$ remainder $- 69$

Note that the highest power of x in the quotient is 1 less than the highest power of x in the dividend.

Note also that if there were more terms in the dividend, we would have repeated the process until the last column is completed; and that if the remainder were zero, then x - 3 would be a factor of $2x^3 - 14x^2 - x + 6$. Note further that unlike ordinary long division, synthetic division requires only three rows irrespective of the number of terms in the dividend.

Example 2 Divide $4x^3 + 2x^2 - 8x + 5$ by $2x + 3$

Solution: Since the coefficient of the x-term of the divisor is not 1, we use the following approach:

Step 1: Factor out the 2 in $2x + 3$ so that the coefficient of the x in the parentheses is 1.

Then we obtain $2(x + \frac{3}{2})$ (multiply the outside by 2 and divide each of $2x$ and 3 by 2)

Step 2: Solve $x + \frac{3}{2} = 0$ for x to obtain $x = -\frac{3}{2}$

Step 3: Use $-\frac{3}{2}$ as a divisor in the synthetic division

The quotient on dividing $4x^3 + 2x^2 - 8x + 5$ by $x + \frac{3}{2}$ is

$4x^2 - 4x - 2$ and the remainder is 8.

Step 4: To obtain the quotient on dividing $4x^3 + 2x^2 - 8x + 5$ by $2x + 3$, we divide the quotient in Step 3 by 2 (the 2 that was factored out in Step 1. "It is time now for it do its part").

Then $\frac{4x^2 - 4x - 2}{2} = 2x^2 - 2x - 1$

Therefore, $\frac{4x^3 + 2x^2 - 8x + 5}{2x + 3} = 2x^2 - 2x - 1$, remainder 8.

Note that the remainder is **not** divided by 2.

Justification for the above method:

$$\frac{4x^3 + 2x^2 - 8x + 5}{2x + 3} = \frac{4x^3 + 2x^2 - 8x + 5}{2\left(x + \frac{3}{2}\right)} = \frac{1}{2}\left\{\frac{4x^3 + 2x^2 - 8x + 5}{x + \frac{3}{2}}\right\} = \frac{1}{2}\left\{4x^2 - 4x - 2 + \frac{8}{x + \frac{3}{2}}\right\}$$

$$= 2x^2 - 2x - 1 + \frac{8}{2x + 3}$$

Comparison of Long Division and Synthetic Division:

In **long division**, we divide, multiply, subtract and repeat the process but in **synthetic division**, we multiply, then add and repeat the process.

Applications of Synthetic Division   440

1. **We** can use synthetic division to find the quotient and remainder when we divide a polynomial by a polynomial (See Examples 1 and 2 above.)

2. We can also use synthetic division to find the functional values of polynomials, especially for cases in which high powers of the variable are involved, Thus we can use synthetic division as an aid in setting up tables for graphing.

Finding Functional Values of Polynomials

Example 3 If $f(x) = 4x^3 - x^2 + 7x + 10$, find $f(-2)$

Solution Set up the synthetic division and carry out the process as shown below.

$f(-2)$ = the last term in the last row (3rd row)

$$-2) \quad 4 \qquad -1 \qquad +7 \qquad +10$$
$$ -8 \qquad +18 \qquad -50$$
$$ 4 \qquad -9 \qquad 25 \qquad -40$$

(arrows labeled $+\cdot -2$ between terms)

Therefore, $f(-2) = -40$.

Note: Above is the same as evaluate $4x^3 - x^2 + 7x + 10$ if $x = -2$

Extra
If $x = 2$, evaluate the following
using (a) normal substitution; and
(b) using synthetic division.

$2x^6 + 3x^5 + 2x^4 + 7x^3 + 3x^2 + 4x + 8$

Ans 340
Which is more efficient, (a) or (b)?

Remainder Theorem (for polynomials)

Let D = dividend, q = quotient, d = divisor. r = remainder

Dividend = quotient × divisor + remainder	**Example** in arithmetic
Then $D = qd + r$	$\frac{14}{3} = 4 + \frac{2}{3} \quad (=4\frac{2}{3})$
$D(x) = q(x)d(x) + r(x)$ (degree of $r(x) <$ degree of $d(x)$)	$\frac{14(3)}{3} = 4(3) + \frac{2(3)}{3}$
If $d(x) = x - c$, then $D(x) = q(x)(x - c) + r(x)$	(undoing the 3 in the denominator))
(degree of $x - c = 1$, degree of $r(x) = 0$)	$14 = 4(3) + 2$
If $x = c$, then $D(c) = q(c)(c - c) + r$	
$ D(c) = q(c)(0) + r$	dividend quotient divisor remainder
$ D(c) = r$	

The **remainder theorem** states that if a polynomial $P(x)$ is divided by $x - c$, (where c is a constant) then the remainder $r = P(c)$.
The remainder theorem allows us to find the remainder in a polynomial division without actually carrying out the division process.

Example 4: Find the remainder if the polynomial $P(x) = x^2 + x - 10$ is divided by $x - 3$.
(a) Use long division and (b) Apply the remainder theorem.

(a) **Long division**

$$\begin{array}{r} x + 4 \\ x - 3 \overline{)\, x^2 + x - 10} \\ \underline{x^2 - 3x} \\ 4x - 10 \\ \underline{4x - 12} \\ 2 \end{array}$$

remainder ->

(b) **Applying the remainder theorem**
Set $x - 3 = 0$ and solve to obtain $x = 3$
$P(x) = x^2 + x - 10$
$P(3) = (3)^2 + (3) - 10 \qquad (x = 3)$
$P(3) = 9 + 3 - 10$
$ = 12 - 10 = 2$ <---remainder

Let us apply this theorem to a previous synthetic division problem (see page 438)

Example 5 Find the remainder when $2x^3 - 14x^2 - x + 6$ is divided by $x - 3$
Solution

Step 1: Set $x - 3 = 0$ and solve for x.
 Then $x = 3$
 Step 2: Substitute $x = 3$ in $2x^3 - 14x^2 - x + 6$
 Then, $P(3) = 2(3)^3 - 14(3)^2 - (3) + 6$
	Step 3: $= 2(27) - 14(9) - 3 + 6$
	  $= 54 - 126 - 3 + 6$
	  $P(3) = -69$
	 By the remainder theorem, $r = P(3)$

Therefore, the remainder is - 69 (the same remainder as by the synthetic division method).

Factor Theorem
Given a polynomial $P(x)$, if $P(c) = 0$, then $x - c$ is a factor of $P(x)$,
That is, $x - c$ divides $P(x)$ with remainder zero..

Example 6: Determine if $x - 3$ is a factor of $x^2 + x - 12$.

Long division

$$x - 3 \overline{) x^2 + x - 12}$$ with quotient $x + 4$

$$\underline{x^2 - 3x}$$
$$4x - 12$$
$$\underline{4x - 12}$$
remainder -> 0

Applying the factor theorem

Set $x - 3 = 0$ and solve to obtain $x = 3$
$P(x) = x^2 + x - 12$
$P(3) = (3)^2 + (3) - 12$ (setting $x - 3 = 0$, $x = 3$)
$P(3) = 9 + 3 - 12$
$= 12 - 12 = 0$

Since $P(3) = 0, x - 3$ is a factor of $x^2 + x - 12$.
(Recall in **arithmetic that** saying that 3 is a factor (divisor) of 12 means that when 12 is divided
by 3, the remainder is zero,)

Example 7 Determine if $x + 4$ is a factor of $x^2 + x - 12$.
Solution

Step 1: Set $x + 4 = 0$, and solve for x.
     Then $x = -4$
Step 2: Substitute for $x = -4$ in $x^2 + x - 12$
          Then $P(-4) = (-4)^2 + (-4) - 12$
                    $= 16 - 4 - 12$
          $P(-4) = 0$
Since $P(-4) = 0$, $x + 4$ is a factor of $x^2 + x - 12$.

Example 8 Determine if $x + 3$ is a factor of $4x^2 + 7x + 5$.
Solution

Step 1: Let $P(x) = 4x^2 + 7x + 5$
 Set $x + 3 = 0$ and solve for x
 Then, $x = -3$.
 Step 2: Substitute $x = -3$ in $P(x)$.
 Then $P(-3) = 4(-3)^2 + 7(-3) + 5$
	Step 3: $P(-3) = 4(9) - 21 + 5$
	  $= 36 - 21 + 5$
	  $P(-3) = 20$ <--- remainder
	 Since $P(-3)$ is **not** zero, $x + 3$ is **not** a factor of
	 $4x^2 + 7x - 5$. (In fact, the remainder is 20)

Converse of the Factor Theorem

If $x - c$ is factor of $P(x)$, then $P(c) = 0$

Question: If $P(x) = -2x^5 + 9x^4 - 3x^3 - 5x^2 + 5x - 4$, and $x - 4$ is a factor of $P(x)$, find $P(4)$.

Solution Since $x - 4$ is a factor of $P(x)$, $P(4) = 0$.

About long division, synthetic division. remainder theorem, and factor theorem.

1. Long division can be used to divide a polynomial $P(x)$ by the linear divisor $x - c$, and produce the quotient and the remainder, which may or may not be zero. If the remainder is zero, then $x - c$ is a factor of $P(x)$.

2. Synthetic division can be used to divide a polynomial $P(x)$ by $x - c$, and produce the quotient and the remainder, which may or may not be zero. If the remainder is zero, then $x - c$ is a factor of $P(x)$.

3. The remainder theorem can be used to find the remainder when a polynomial $P(x)$ is divided by $x - c$. The remainder $r = P(c)$. If the remainder is zero, then $x - c$ is a factor of $P(x)$.

4. Synthetic division can be used to evaluate a polynomial $P(x)$ knowing the numerical value of the variable. The divisor here is the value of the variable, and the remainder is the value of $P(c)$.

5. Long division can be used to evaluate a polynomial, $P(x)$ given $x = c$. Here, the divisor is $x - c$, and the remainder is the value of $P(c)$. However, this is only an exercise since direct substitution and synthetic division are faster. Each of the above methods has relative merits.

In **arithmetic,** saying that 3 is a factor (divisor) of 12 means that when 12 is divided by 3, the remainder is zero, Similarly, saying that $x - 3$ is a factor (divisor) of $x^2 + x - 12$ means when $x^2 + x - 12$ is divided by $x - 3$, the remainder is zero. However, saying that $x + 2$ is **not** a factor (divisor) of $x^2 + x - 12$ means when $x^2 + x - 12$ is divided by $x + 2$, the remainder is **not** zero,

Lesson 64 Exercises

A Use synthetic division in the following problems:

1. Divide $5x^4 - 3x^2 + 3x - 7$ by $x + 1$;   2. Divide $8x^4 + 9x^2 + 14x^3 + 19x + 12$ by $3 + 2x$

3. $(x^4 + 4x^2 + 5x + 2) \div (x + 3)$;   4. Divide $4x^4 - 2x^2 - 3$ by $x - 3$;

5. Divide $x^3 + 2x^2 - 5x - 6$ by $x - 2$;   6. Divide $x^3 + 10x^2 + 33x + 36$ by $x + 4$

7.. Divide $4x^3 - 11x^2 + x - 2$ by $x - 1$;   8. Divide $5x^4 - 2x^3 + 6x^2 + 8$ by $x + 1$

Answer: 1. $5x^3 - 5x^2 + 2x + 1 - \dfrac{8}{x+1}$;   2. $4x^3 + x^2 + 3x + 5 - \dfrac{3}{3 + 2x}$;

3. $x^3 - 3x^2 + 13x - 34 + \dfrac{104}{x + 3}$;   4. $4x^3 + 12x^2 + 34x + 102 + \dfrac{303}{x - 3}$;   5. $x^2 + 4x + 3$

6. $x^2 + 6x + 9$;   7. $4x^2 - 7x - 6 - \frac{8}{x-1}$;   8. $5x^3 - 7x^2 + 13x - 13 + \frac{21}{x+1}$.

B Find the remainder as specified

1. When $x^2 + 6x + 2$ is divided by $x - 1$;   2. When $2x^2 - x + 1$ is divided by $x + 2$

3. When $5x^2 - 3x - 8$ is divided by $x - 2$;   4. When $x^3 - 2x^2 + 3x + 4$ is divided by $x + 2$

Answers: 1. Remainder = 9;   2. Remainder = 11;   3. 6;   4. Remainder = -18

C Determine if $x - 2$ is a factor in each of the following: 1. $4x^2 - 7x + 2$;   2. $-x^2 + 4x - 1$;

3. $\frac{1}{2}x^2 + x$; 4. $3x^5 - 8x^4 + x^2 + 12x + 4$;   5. Is $2x + 3$ a factor of $x^2 + 5x - 1$?

Answers: 1. No;   2. No;   3. No;   4. Yes;   5. No;.

Lesson 65

Rational Roots of Polynomial Equations; Descartes' Rules;

Zeros (roots or solutions) of Polynomial Equations

A **zero** of a function f is a value of x for which $f(x) = 0$

For example, 2 is a zero of $f(x) = 4x - 8$, because $f(2) = 4(2) - 8 = 0$

Note that saying that $x - 2$ is a factor of $f(x) = 4x - 8$ implies that 2 (from $x - 2 = 0$) is a zero of f. Thus, if $x - c$ is a factor of $f(x)$, then c is a zero of f.

In Example 1 above, the question could have been posed as:
Determine if 3 is a zero of $f(x) = x^2 + x - 12$.

Solution $f(3) = 3^2 + 3 - 12 = 12 - 12 = 0$

Since $f(3) = 0$, 3 is zero of $f(x) = x^2 + x - 12$.

Similarly, in a previous example, the question could have been posed as:
Determine if -3 is a zero of $f(x) = 4x^2 + 7x + 5$ (see Example 8, p. 441)

Solution Since $f(-3) = 20 \neq 0$, -3 is **not** a zero of $f(x) = 4x^2 + 7x + 5$.

Determining the number of positive roots and negative roots of a Polynomial

Descartes' Rule of Signs

The number of **positive real roots** of a polynomial $f(x)$ is either equal to the number of changes in sign in successive terms in $f(x)$ or is less than this number by a positive even integer. A root of multiplicity m is counted as m roots. The number of **negative real roots** of $f(x)$ equals the number sign changes occurring in successive terms in $f(-x)$, or is less than this number by a positive even integer.

Example 9 Find (a) the maximum number of positive roots, (b) the maximum number of negative roots of $24x^4 - 26x^3 - 49x^2 + 4x + 12 = 0$.

Solution
(a)   Step 1: Find the number of changes in sign.

$$24x^4 - 26x^3 - 49x^2 + 4x + 12 = 0$$

The coefficient of first term, $24x^4$, is positive and the coefficient of second term, $-26x^3$, is negative. There is, therefore, a change in sign from plus (+) to minus (-). The coefficient of third term, $-49x^2$, is negative, and therefore, there is no sign change from the second   term to the third term. The coefficient of fourth term, $4x$, is positive, and therefore, there is a   change in sign from the third term to the fourth term. The fifth term, 12, is positive, and therefore, there is no sign change from the fourth term to the fifth term. The total number of sign changes is 2 (from the first term to the second term, and from the third term to the fourth term). By Descartes' rule of signs, there are at most either 2 positive roots or 0 (2-2 = 0) positive root.

Step 2: There are at most 2 positive roots.

(*b*)

Step 1: Replace x by $-x$ in $24x^4 - 26x^3 - 49x^2 + 4x + 12$ to obtain $f(-x)$.

Then $f(-x) = 24x^4 + 26x^3 - 49x^2 - 4x + 12$

Step 2: $f(-x) = 24x^4 + 26x^3 - 49x^2 - 4x + 12$

$\underbrace{}_{1} \qquad \underbrace{}_{2}$

The coefficient of first term, $24x^4$, is positive and the coefficient of second term, $26x^3$, is positive. There is, therefore, no change in sign. The coefficient of the third term, $-49x^2$, is negative and therefore, there is a sign change from the second term to the third term. The coefficient of the fourth term, $-4x$, is negative and therefore there no sign change from the third term to the fourth term. The fifth term, 12, is positive, and therefore, there is a sign change from the fourth term to the fifth term.
The total number of sign changes is 2.

Step 3: By Descartes' rule of signs, there are at most 2 negative roots.

Example 10 Find (a) the maximum number of positive roots, (b) the maximum number of negative roots of $6x^3 - 23x^2 + 29x - 12 = 0$.

Solution
(a)

$6x^3 - 23x^2 + 29x - 12 = 0$

$\underbrace{}_{1} \underbrace{}_{2} \underbrace{}_{3}$

We use a similar reasoning as in Example 3. There is a sign change from the first term to the second term. There is a sign change from the second term to the third term. There is also a sign change from the third term to the fourth term. The total number of sign changes is 3. By Descartes' rule of signs, there are at most either 3 positive roots or $3 - 2 = 1$ positive root.
There are at most 3 positive roots.

(b) Step 1: Replace x by $-x$ in $6x^3 - 23x^2 + 29x - 12.$
$-6x^3 - 23x^2 - 29x - 12 = 0$.

Step 2: There are no sign changes, and therefore, there are no negative roots.
Since there are 3 roots ($n = 3$), and there are no negative roots, there are 3 positive roots.

Finding all Positive Integral Factors of a Positive Integer

Example 11 Find all the positive integral factors of 24.

Solution We use positive integers as divisors, beginning from 1.

$$\underbrace{\text{Factors}}$$
$$\Downarrow$$

$$24 \div 1 = 24 \Rightarrow 1 \text{ and } 24$$

$$24 \div 2 = 12 \Rightarrow 2 \text{ and } 12$$

$$24 \div 3 = 8 \Rightarrow 3 \text{ and } 8$$

$$24 \div 4 = 6 \Rightarrow 4 \text{ and } 6$$

$$24 \div 6 = 4 \Rightarrow 6 \text{ and } 4 \quad \text{(We skip 5 since it is not a divisor of 24.)}$$

We stop the division process since there is a repetition of a pair of factors (Here, 4 and 6)
The factors of 24 are 1, 2, 3 , 4. 6, 8, 12 and 24.

Example 12 Find all the positive integral factors of 12.

Solution We use positive integers as divisors, beginning from 1.

$$\underbrace{\text{Factors}}$$
$$\Downarrow$$

$$12 \div 1 = 12 \Rightarrow 1 \text{ and } 12$$

$$12 \div 2 = 6 \Rightarrow 2 \text{ and } 6$$

$$12 \div 3 = 4 \Rightarrow 3 \text{ and } 4$$

$$12 \div 4 = 3 \Rightarrow 4 \text{ and } 3$$

We stop the division process since there is a repetition of a pair of factors (Here, 3 and 4)
The factors of 12 are 1, 2, 3, 4. 6 and 12. (Writing each factor once and in increasing order)

Example 13 Find all the positive integral factors of 6.

Solution We use positive integers as divisors, beginning from 1.

$$\underbrace{\text{Factors}}$$
$$\Downarrow$$

$$6 \div 1 = 6 \Rightarrow 1 \text{ and } 6$$

$$6 \div 2 = 3 \Rightarrow 2 \text{ and } 3$$

$$6 \div 3 = 2 \Rightarrow 3 \text{ and } 2$$

We stop the division since there is a repetition of the factors 2 and 3
The factors of 6 are 1, 2, 3, and 6. (Writing each factor once and in increasing order.)

Rational Zero (or Rational Root) Theorem or Rule

Theorem:

Consider a polynomial function $f(x) = a_n x^n + a_{n-1} x^{n-1} + ...\ ax + a_0$ whose coefficients are all integers.

If $\dfrac{c}{d}$ is a rational zero (root in lowest terms) of $f(x)$, then c is a factor of a_0. and d is a factor of a_n.

Note:

a_n = the coefficient of the highest power of x; a_0 = the constant term (term with no x)

Example: For the polynomial $3x^4 - 4x^3 + x^2 + 6x - 2$.

$$a_n = 3; \qquad a_0 = -2$$

Mnemonic device: Divide through the right-hand side by a_n to obtain for the last term $\dfrac{a_0}{a_n}$ and

then, $\dfrac{c}{d} = \dfrac{a_0}{a_n}$

Note: If $a_n = 1$, then $\frac{c}{d} = a_0$ and $\frac{c}{d}$ is an integer and therefore, the roots are integers.

We must **note** how the polynomial function is specified regarding a_0 and a_n.

In some books, a_n and a_0 have been interchanged as in equation (2) below. In this book we use the form in equation (1) below. The author suggests that form for easy recall

$$f(x) = a_n x^n + a_{n-1} x^{n-1} + ...\ + ax + a_0 \qquad (1)$$

$$f(x) = a_0 x^n + a_1 x^{n-1} + ...\ + a_{n-1} x + a_n \qquad (2)$$

Application of the Rational Root Theorem 447

Finding all possible rational roots of a polynomial equation

We can obtain the possible roots by dividing each factor of the constant term by each factor of the coefficient of the highest power of x in the polynomial. In doing so, there will be repetitions of the quotients. To speed-up the process. we use the following approach:

Example 14 Find all the possible rational roots of $24x^4 - 26x^3 - 49x^2 + 4x + 12 = 0$.

Step 1: The factors of the constant term, 12, say, $a_0's = \pm1, \pm2, \pm3, \pm4, \pm6, \pm12$.

All the factors of the constant term are the possible integral roots, since each factor of the constant is to be divided by the factors ±1 of the highest power of x.. These are therefore, $\pm1, \pm2, \pm3, \pm4, \pm6, \pm12$.

Step 2: The factors of the coefficient of the highest power of x , 24,

say, $a_n's = \pm1, \pm2, \pm3, \pm4, \pm6, \pm8, \pm12, \pm24$.

To obtain the possible fractional roots, divide each factor of a_0 by each factor of a_n, ignoring the results of the division in which the dividend and divisor have a common factor (other than 1) or the result is an integer.

Then, we obtain $\pm\frac{1}{2}, \pm\frac{1}{3}, \pm\frac{1}{4}, \pm\frac{1}{6}, \pm\frac{1}{8}, \pm\frac{1}{12}, \pm\frac{1}{24}, \pm\frac{2}{3}, \pm\frac{3}{2}, \pm\frac{3}{4}, \pm\frac{3}{8}, \pm\frac{4}{3}$.

Step 3: Combining the results from Step 1 with those from Step 2 we obtain the possible rational roots,

$\pm1, \pm2, \pm3, \pm4, \pm6, \pm12, \pm\frac{1}{2}, \pm\frac{1}{3}, \pm\frac{1}{4}, \pm\frac{1}{6}, \pm\frac{1}{8}, \pm\frac{1}{12}, \pm\frac{1}{24}, \pm\frac{2}{3}, \pm\frac{3}{2}, \pm\frac{3}{4},$

$\pm\frac{3}{8}, \pm\frac{4}{3}$.

Example 15: Find all the possible rational roots of $6x^3 - 23x^2 + 29x - 12 = 0$

Step 1: The factors of the constant term, -12, say, $a_0's = \pm1, \pm2, \pm3, \pm4, \pm6, \pm12$

The possible integral roots are all the factors of constant term. These are $\pm1, \pm2, \pm3, \pm4, \pm6, \pm12$.

Step 2: The factors of the coefficient of the highest power of x , 6,

say, $a_n's = \pm1, \pm2, \pm3, \pm6$.

To obtain the possible fractional roots , divide each factor of a_0 by each factor of a_n, ignoring the results of the division in which the dividend and divisor have a common factor (other than 1) or the result is an integer.

Then, we obtain $\pm\frac{1}{2}, \pm\frac{1}{3}, \pm\frac{1}{6}, \pm\frac{2}{3}, \pm\frac{3}{2}, \pm\frac{4}{3}$

Step 3: Combining the results from Step 1 with those from Step 2 we obtain the possible rational roots.

$\pm1, \pm2, \pm3, \pm4, \pm6, \pm12, \pm\frac{1}{2}, \pm\frac{1}{3}, \pm\frac{1}{6}, \pm\frac{2}{3}, \pm\frac{3}{2}, \pm\frac{4}{3}$.

Determining the Roots of a Polynomial Equation $f(x) = 0$ with real coefficients and of degree n

In solving polynomial equations with real coefficients, if the degree of the equation is 3 or higher, we use a trial and error approach to obtain some of the roots until there are only two more roots to be found (in which case the depressed equation is a quadratic equation, and then we can use the quadratic formula or factoring to solve the quadratic equation.

The following guidelines would be helpful in determining the roots of polynomial equations.

1. There are n roots.

2. If there are complex roots, they occur in conjugate pairs

3. Any rational roots are factors of the constant term divided by the coefficient of the leading term.

4. The maximum number of positive roots equals the number of sign changes, and the maximum number of negative roots equals the number of sign changes in $f(-x)$. (From Descartes' Rule of signs)

5. If the degree of the equation is odd, then there is at least one real root.

6. When the depressed equation is a quadratic equation (or if there are only two remaining roots), we can use the quadratic formula or factoring to determine the remaining roots.

7. If the coefficient of the highest-power term is 1, then the rational roots are integers, and are factors of the constant term.

8. If there is only one negative root, then we shall try to find this negative root first.

9. In using synthetic division, if on trying a positive root, there are only positive numbers in the bottom row, then there are no roots larger than the positive root tried and we do not try any value larger than this positive value tried.

10. In using synthetic division, if on trying a negative root, the signs of the numbers in the bottom row alternate, then there are no negative roots less than this negative root tried and we do not try any value less than this negative value tried.

11. We can also find the approximate roots graphically using a graphing calculator.

Example 16 Find all the rational roots of $6x^3 - 23x^2 + 29x - 12 = 0$.

Step 1: Determine the number of roots of the polynomial equation.
There are 3 roots ($n = 3$).

From Example 15, above, the possible rational roots are
$$\pm 1, \pm 2, \pm 3, \pm 4, \pm 6, \pm 12, \pm \tfrac{1}{2}, \pm \tfrac{1}{3}, \pm \tfrac{1}{6}, \pm \tfrac{2}{3}, \pm \tfrac{3}{2}, \pm \tfrac{4}{3}.$$
There are therefore 24 possible rational roots.

Step 2: Find the maximum number of positive roots and the maximum number of negative roots.

$$6x^3 - 23x^2 + 29x - 12 = 0$$

$$\underbrace{\quad}_{1} \underbrace{\quad}_{2} \underbrace{\quad}_{3}$$

From Example 10, there are 3 positive roots, and no negative roots. This implies that we shall test synthetically for only positive roots.

Step 3: Synthetically, test for the positive values from Step 1. After a rational root has been found, the corresponding factor is removed. We test for the following values 1, 2, 3, 4, 6, 12,
$\tfrac{1}{2}, \tfrac{1}{3}, \tfrac{1}{6}, \tfrac{2}{3}, \tfrac{3}{2}, \tfrac{4}{3}.$

Testing for +1:

The zero remainder indicates that $x - 1$ is a factor of $6x^3 - 23x^2 + 29x - 12$.

The depressed equation from the synthetic division is $6x^2 - 17x + 12 = 0$. We can solve this equation by the quadratic formula, but since this quadratic equation is factorable, we solve by factoring:

$6x^2 - 17x + 12 = 0$; $(3x - 4)(2x - 3) = 0$ and from which $x = \tfrac{4}{3}, x = \tfrac{3}{2}$.

The roots are 1, $\tfrac{4}{3}$, and $\tfrac{3}{2}$. (These belong to the prediction set in Step 1.)

Example 17 Find all the rational roots of $24x^4 - 26x^3 - 49x^2 + 4x + 12 = 0$.

Step 1: Determine the number of roots of the polynomial equation.
There are 4 roots ($n = 4$).

Step 2: From Example 14, the possible rational roots are $\pm 1, \pm 2, \pm 3, \pm 4, \pm 6, \pm 12$,
$\pm \frac{1}{2}, \ \pm \frac{1}{3}, \ \pm \frac{1}{4}, \ \pm \frac{1}{6}, \ \pm \frac{1}{8}, \ \pm \frac{1}{12}, \ \pm \frac{1}{24}, \pm \frac{2}{3}, \ \pm \frac{3}{2}, \ \pm \frac{3}{4}, \ \pm \frac{3}{8}, \ \pm \frac{4}{3}, \ \pm \frac{8}{3}$.

Step 3: Find the maximum number of positive roots and the maximum number of negative roots.

$$24x^4 - 26x^3 - 49x^2 + 4x + 12 = 0$$

There are at most 2 positive roots. (From Example 9)

Step 4: $\quad f(-x) = 24x^2 + 26x^3 - 49x^2 - 4x + 12$

There are, at most, 2 negative roots. (From Example 9)

Step 5: Synthetically, test for the possible roots from Step 2. After a rational root has been found, we remove the corresponding factor. We test for the following values: $\pm 1, \pm 2, \pm 3, \pm 4, \pm 6,$
$\pm 12, \ \pm \frac{1}{2}, \ \pm \frac{1}{3}, \ \pm \frac{1}{4}, \ \pm \frac{1}{6}, \ \pm \frac{1}{8}, \ \pm \frac{1}{12}, \ \pm \frac{1}{24}, \pm \frac{2}{3}, \ \pm \frac{3}{2}, \ \pm \frac{3}{4}, \ \pm \frac{3}{8}, \ \pm \frac{4}{3}, \ \pm \frac{8}{3}$.
On testing , ± 1 , -2 are not solutions. We show the test for 2 which is a root.
Testing for 2 :

$$
\begin{array}{r|rrrrr}
2) & 24 & -26 & -49 & 4 & 12 \\
   &   & 48 & 44 & -10 & -12 \\
\hline
   & 24 & 22 & -5 & -6 & 0
\end{array}
$$

The zero remainder indicates that $x - 2$ is a factor of $24x^4 - 26x^3 - 49x^2 + 4x + 12 = 0$.

The depressed equation from the synthetic division is $24x^3 + 22x^2 - 5x - 6 = 0$.

Step 6: We test synthetically for the other values in the depressed equation.

We can also test in the original equation. We show the test for $\frac{1}{2}$:

$$
\begin{array}{r|rrrr}
\frac{1}{2}) & 24 & 22 & -5 & -6 \\
             &   & 12 & 17 & 6 \\
\hline
             & 24 & 34 & 12 & 0
\end{array}
$$

The zero remainder indicates that $x - \frac{1}{2}$ is a factor of $24x^3 + 22x^2 - 5x - 6 = 0$.

The depressed equation from the synthetic division is $24x^2 + 34x - 6 = 0$.
We solve this equation by factoring .(We can also use the quadratic formula.)

$24x^3 + 22x^2 - 5x - 6 = 0$; --> $2(12x^2 + 17x + 6) = 0$; -->$2(4x + 3)(3x + 2) = 0$

Solving, $x = -\frac{3}{4}$, $x = -\frac{2}{3}$. Combining all the solutions, we obtain the solution set $\left\{ 2, \ \frac{1}{2}, \ -\frac{3}{4}, \ -\frac{2}{3} \right\}$.

Lesson 65: Descartes' Rules; Rational Roots of Polynomial Equations

Example 18 Find all rational zeros of $f(x) = x^3 - 4x^2 + 5x - 2$.

Solution Since $n = 3$, there are 3 roots.

By Descartes' rule, there are at most 3 positive roots

In $f(-x) = -x^3 - 4x^2 - 5x - 2$, there are no sign changes and therefore, there are no negative roots. There are therefore 3 positive roots.

Step 1: Since $a_n = 1$, we consider only the constant term -2 for possible roots. The possible roots are ± 1 and ± 2.

Step 2: We test for 1 in $x^3 - 4x^2 + 5x - 2$ to see if 1 is a root.
We determine $f(1)$:
$$f(1) = (1)^3 - 4(1)^2 + 5(1) - 2$$
$$f(1) = 1 - 4 + 5 - 2$$
$$f(1) = 0$$
Since $f(1) = 0$, 1 is a root or a zero.

Note that we could also have used synthetic division to check for possible roots.

Step 3: Since 1 is a root of $x^3 - 4x^2 + 5x - 2$, $x - 1$ is a factor of $x^3 - 4x^2 + 5x - 2$.

Step 4: To find another factor, we divide $x^3 - 4x^2 + 5x - 2$ by $x - 1$. We shall use synthetic division. (Note that we could also use long division.)

```
1)1  - 4    5    - 2
        1   - 3    2
   1  - 3    2    0
```

$$\downarrow \quad \downarrow \quad \downarrow$$

$$x^2 - 3x + 2$$
$$\therefore f(x) = (x - 1)(x^2 - 3x + 2)$$

Step 5: Now, factor the quadratic trinomial $x^2 - 3x + 2$ to obtain
$$x^2 - 3x + 2 = (x - 2)(x - 1)$$
Then $f(x) = (x - 1)(x - 2)(x - 1)$ (putting everything together)
$$f(x) = (x - 1)^2(x - 2)$$

By setting $(x - 1)^2(x - 2) = 0$ and solving,
$x - 1 = 0$, or $x - 1 = 0$, or $x - 2 = 0$
$x = 1, x = 2$.
The roots (or zeros) are 1 with a multiplicity of 2; and 2 with a multiplicity of 1.
Thus, 1 is a repeated or double root.

Example 19 Find a polynomial of degree 4, with -2 as a root of multiplicity of 1, 3 as root of a multiplicity of 2, and -1 as a root of multiplicity of 1.

Solution
$$f(x) = (x - (-2))^1(x - 3)^2(x - (-1))^1$$
$$f(x) = (x + 2)^1(x - 3)^2(x + 1)^1$$
$$f(x) = x^4 - 3x^3 - 7x^2 + 15x + 18.$$

Lesson 65 Exercises

1. Find the rational roots of $x^3 - 6x^2 + 11x - 6 = 0$

2. If $f(x) = x^3 - 2x^2 - x + 3$, and the roots of $f(-x)$ are the negatives of the roots of $f(x)$, then find the equation for $f(-x)$.

3. Find the rational roots of $4x^4 - 25x^2 + 30x - 9 = 0$

Answers: 1. $\{1, 2, 3\}$; **2.** $f(-x) = -x^3 - 2x^2 + x + 3$. **3.** $\left\{-3, \quad 1, \quad \frac{1}{2}, \quad \frac{3}{2}\right\}$

CHAPTER 25
Partial Fractions

Lesson 66

Decomposition of Fractions into Partial Fractions

Consider the addition operations: $\frac{2}{5} + \frac{1}{4} = \frac{13}{20}$

$$\frac{3}{x+2} + \frac{2}{x-1} = \frac{5x+1}{(x+2)(x-1)}$$

In each of the above equations, the terms of left-hand side are the **partial fractions** of the right-hand sides. Note that each left-hand side consists of terms whereas each right-hand side is a single fraction. In integral calculus, there are occasions when we like to break up a single fraction into its equivalent sum of partial fractions.

The procedure for breaking up a given single fraction into a sum of terms is called the **partial fraction decomposition of** the given fraction. We shall assume that the rational functions to be decomposed are proper rational functions

Note that a rational function is **proper** if the degree of the numerator polynomial is less than the degree of the denominator polynomial. If the given fraction is not proper, we will use long division to obtain a polynomial and a proper rational fraction, and then decompose the proper fraction part.

The form of the decomposition depends on the form of the denominators. We shall discuss the various cases.

Case 1. The denominator has unrepeated (distinct) linear factors of form $ax + b$.

In this case, there is one partial fraction of form $\frac{A}{ax+b}$ for each linear factor, where A is a constant to be determined.

Example: $\frac{1}{(x-1)(x+2)} = \frac{A}{x-1} + \frac{B}{x+2}$, where A and B are constants to be determined.

Case 2 The denominator has repeated linear factors, each linear factor $(ax + b)$ repeated n times.

In this case, there will be n partial fractions of form $\frac{A_1}{ax+b} + \frac{A_2}{(ax+b)^2} + \frac{A_3}{(ax+b)^3} + \ldots \frac{A_n}{(ax+b)^n}$

Example 1: $\frac{x^2+5}{(x+4)^3} = \frac{A_1}{(x+4)} + \frac{A_2}{(x+4)^2} + \frac{A_3}{(x+4)^3}$

Example 2: $\frac{4x+1}{x^2} = \frac{A}{x} + \frac{B}{x^2}$

Case 3 The denominator has one unrepeated (distinct) and irreducible quadratic factor $(ax^2 + bx + c)$. In this case, the partial fraction decomposition is of the form $\frac{Ax+b}{ax^2+bx+c}$

where A and B are constants to be determined.

Note: If $b^2 - 4ac$ is negative, then the quadratic factor $ax^2 + bx + c$ is **irreducible**.

For example, $x^2 + 1$ is irreducible since $b^2 - 4ac = 0^2 - 4(1)(1) = -4$

Another way to determine the above irreducibility is if the quadratic factor cannot be decomposed into linear factors without introducing imaginary numbers, then is factor is irreducible

Example $\frac{2x}{(x^2+3)(x^2+2)} = \frac{Ax+B}{x^2+3} + \frac{Cx+D}{x^2+2}$

Lesson 66: Decomposition of Fractions into Partial Fractions

Case 4   The denominator has repeated irreducible quadratic factors repeated n times. For each quadratic factor, there will be n corresponding quadratic factors of form

$$\frac{(A_1 x + B_1)}{ax^2 + bx + c} + \frac{A_2 x + B_2}{(ax^2 + bx + c)^2} + \ldots + \frac{A_n x + B_n}{(ax^2 + bx + c)^n}$$

Example $\dfrac{x+6}{(x^2+4)^3} = \dfrac{(A_1 x + B_1)}{x^2 + 4} + \dfrac{A_2 x + B_2}{(x^2 + 4)^2} + \dfrac{A_3 x + B_3}{(x^2 + 4)^3}$

If we can recall the various forms of the partial fraction, then the only other work we have to do is to determine the constants A, B, C, D, etc, in the numerators.
 Below, we present the process of determining the constants and finding the partial fraction decomposition using examples.

Example 1 Find the partial fraction decomposition of

$$\frac{x-5}{(x-1)(x+2)} \qquad (1)$$

Solution   The partial fraction is the form as in Case 1

Step 1: $\dfrac{x-5}{(x-1)(x+2)} = \dfrac{A}{x-1} + \dfrac{B}{x+2}$

Then $\dfrac{x-5}{(x-1)(x+2)} = \dfrac{A(x+2) + B(x-1)}{(x-1)(x+2)} \qquad (2)$

Since the denominators are identical, we equate the numerators.
(or multiply both sides by $(x-1)(x+2)$)
Then, $x - 5 = A(x+2) + B(x-1) \qquad (3)$
Now, we shall determine the values of A and B.

Step 3: We shall cover two methods for determining A and B, namely, a general method and a substitution method

Method 1: General method

Expand the right-hand side of equation (3) and factor out the x.

$x - 5 = Ax + 2A + Bx - B$

$x - 5 = (A + B)x + 2A - B \qquad (4)$

Equate the coefficients of the like power of x on the left -hand side to the coefficients of like powers of x on the right hand side.

Then for the coefficients,

$1 = A + B \qquad (5)$

$-5 = 2A - B \qquad (6)$

(Note that we could write -5 as $-5x^0$ and $(2A-B)$ as $(2A-B)x^0$)

We shall now solve equations (5) and (6) simultaneously for A and B.

$\qquad$ Solving, $A = -\dfrac{4}{3}$, and $B = \dfrac{7}{3}$

Substituting $A = -\dfrac{4}{3}$, and $B = \dfrac{7}{3}$ in equation (1), we obtain

$$: \quad \frac{x-5}{(x-1)(x+2)} = \frac{-\frac{4}{3}}{x-1} + \frac{\frac{7}{3}}{x+2}$$

$$\frac{x-5}{(x-1)(x+2)} = -\frac{4}{3(x-1)} + \frac{7}{3(x+2)} \quad \text{or} \quad -\frac{4}{3}\left(\frac{1}{x-1}\right) + \frac{7}{3}\left(\frac{1}{x+2}\right)$$

Method 2: Substitution method

This method involves properly choosing values for x so as to eliminate some of the constants, and then solving the resulting equations accordingly. However, this method may not work if all the unknowns drop out (this may occur if there are repeated factors).

We now apply this method to equation (3)

Rewriting equation (3): $\quad x - 5 = A(x+2) + B(x-1)$ $\qquad\qquad$ (3)

If we let $x = -2$ in equation (3) so that the A-term drops out, then

$\qquad -2 - 5 = A(-2+2) + B(-2-1)$

$\qquad\qquad -7 = A(0) + B(-3)$

$\qquad\qquad -7 = 0 - 3B$

$\qquad\qquad \dfrac{7}{3} = B$

If we let $x = 1$ in equation (3) so that the B term drops out, then

$1 - 5 = A(1+2) + B(1-1)$

$\quad -4 = 3A + 0$

$\quad -\dfrac{4}{3} = A$

Therefore $A = -\dfrac{4}{3}$, and $B = \dfrac{7}{3}$.

Again, we obtain the same results as by the general method.

In the determination involving quadratic factors, we may use a combination of the general method and the substitution method. We should also note that if the given rational expression is not proper (i.e. degree of the numerator polynomial is greater than or equal to that of the denominator polynomial), then we shall use the long division process to reduce the improper rational expression to a polynomial and a proper rational expression (preferably in its lowest terms), and then proceed to apply the decomposition process to the fractional part.

Lesson 66: Decomposition of Fractions into Partial Fractions

Example 2: Find the partial fraction decomposition of

$$\frac{x^2 + 3x + 6}{x + 2}$$

Since the expression is improper, we use the long division process.

$$
\begin{array}{r}
x + 1 \\
x + 2 \overline{\smash{)}\ x^2 + 3x + 6} \\
\underline{x^2 + 2x} \\
x + 6 \\
\underline{x + 2} \\
4
\end{array}
$$

Thus $\dfrac{x^2 + 3x + 6}{x + 2} = x + 1 + \dfrac{4}{x + 2}$

In the above case, we do not decompose the fractional part any more. The next example requires further decomposition.

Example 3 Find the partial fraction decomposition of $\dfrac{x^3 + 2x^2 + 3x + 4}{x^2 - 4}$

By the long division process we obtain the following: $\dfrac{x^3 + 2x^2 + 3x + 4}{x^2 - 4} = x + 2 + \dfrac{7x + 12}{x^2 - 4}$

We shall now apply the decomposition process to $\dfrac{7x + 12}{x^2 - 4}$

$$\frac{7x + 12}{x^2 - 4} = \frac{7x + 12}{(x + 2)(x - 2)}$$

Let $\dfrac{7x + 12}{(x + 2)(x - 2)} = \dfrac{A}{x + 2} + \dfrac{B}{x - 2}$ (1)

$$\frac{7x + 12}{(x + 2)(x - 2)} = \frac{A(x - 2) + B(x + 2)}{(x - 2)(x + 2)}$$

Equating the numerators to each other,

$$7x + 12 = A(x - 2) + B(x + 2) \qquad\qquad (2)$$

We now proceed to determine the constants A and B, say by the substitution method.
Letting $x = 2$ in equation (2), the A-term drops out.

$$7(2) + 12 = A(2 - 2) + B(2 + 2)$$
$$26 = A(0) + 4B$$
$$26 = 4B \text{ and } \frac{13}{2} = B \quad \text{(solving)}$$

Letting $x = $ -2 in equation (2) the B-term drops out. $(-14 + 12 = -4A + 0$; and $-2 = -4A; \frac{2}{4} = A)$

Therefore, $A = \dfrac{1}{2}$, and $B = \dfrac{13}{2}$.

We now substitute $A = \dfrac{1}{2}$, and $B = \dfrac{13}{2}$ in equation (1), to obtain

$$\frac{7x + 12}{(x + 2)(x - 2)} = \frac{1}{2}\left(\frac{1}{x + 2}\right) + \frac{13}{2(x - 2)}$$

Substituting in the original given rational expression,

$$\frac{x^3 + 2x^2 + 3x + 4}{x^2 - 4} = x + 2 + \frac{1}{2}\left(\frac{1}{x + 2}\right) + \frac{13}{2(x - 2)}.$$

Lesson 66 Exercises

Find the partial fraction decomposition of each of the following:

1. $\dfrac{2}{(x-2)(x+1)}$;

2. $\dfrac{x-3}{(x-1)(x+2)}$;

3. $\dfrac{3x+11}{x^2-4}$;

4. $\dfrac{7x-21}{x^2-6x+8}$

5. $\dfrac{2x^3-x^2+3x-1}{x^2-25}$;

6. $\dfrac{2+x^2}{x(x-1)(x+3)}$

Answers: **1.** $\dfrac{2}{3(x-2)} - \dfrac{2}{3(x+1)}$; **2.** $\dfrac{5}{3(x+2)} - \dfrac{2}{3(x-1)}$; **3.** $\dfrac{17}{4(x-2)} - \dfrac{5}{4(x+2)}$

4. $\dfrac{7}{2(x-4)} + \dfrac{7}{2(x-2)}$; **5.** $2x-1 + \dfrac{291}{10(x+5)} + \dfrac{239}{10(x-5)}$; **6.** $-\dfrac{2}{3x} + \dfrac{3}{4(x-1)} + \dfrac{11}{12(x+3)}$

CHAPTER 26
Sequences and Series

Lesson 67: **Specification of a Sequence; Finite and Infinite Sequence**
  Finding a Formula the general term of a Sequence

Lesson 68: **The Summation Notation, Σ, Properties and Shifting of the**
  Summation Index

Lesson 69: **Arithmetic Sequence & Arithmetic Series**

Lesson 70: **Geometric Sequence, Geometric Series and Applications**

We use sequences in calculus, topology, actuarial work, and queuing theory.

Lesson 67

Specification of a Sequence; Finite and Infinite Sequence
Finding a Formula the General term of a Sequence

Definition 1:
 A **sequence** is an ordered set of numbers.

Examples : (a) $\{1, \quad 4, \quad 9, \quad 16,... \quad \}$

  (b) $\left\{ 1, \quad \frac{1}{2}, \quad \frac{1}{3}, \quad \frac{1}{4},... \right\}$

  (c) $\{8, \quad 11, \quad 14,... \quad \}$

Each number of a sequence is called a **term** of the sequence.

Definition 2:

A **sequence** is also defined as a function whose domain consists of a set of consecutive positive integers. The **range** of the sequence then consists of the terms of the sequence, and the domain consists of the term numbers. A sequence has a first term, a second term, a third term and so forth. In the sequence 1, 4, 9, 16,... the first term is 1, the second term is 4, the 3rd term is 9.

Specification of sequences

A sequence may be specified in a number of ways:

(1) By listing its terms.
(2) By giving a formula for the nth term (general term).
(3) By stating the first term and a recursive formula for calculating the other terms of the sequence.

The nth term (also called the general term) is denoted by a_n. In this case, we can denote the sequence by the set notation as $\{a_n\}$. We can also write a_n as $a(n)$, but the subscript notation is preferred.

Given a general term of a sequence

Example 1   Suppose that the general term $a_n = n^2$ (where n is the term-number.). Then

the 1st term (i.e., for $n = 1$) is $a_1 = (1)^2 = 1$;

the 2nd term (i.e. for $n = 2$) is $a_2 = (2)^2 = 4$;

the 3rd term (i.e. for $n = 3$) is $a_3 = (3)^2 = 9$;

the 4th term (i.e. for $n = 4$) is $a_4 = (4)^2 = 16$.

Thus, the first four terms of $\{a_n\}$ are 1, 4, 9, and 16.

Given a recursive formula of a sequence

Example 2   Using a recursive formula

$$\begin{cases} a_1 = \mathbf{4} \\ a_{k+1} = 5a_k - 1 \qquad k \geq 1 \end{cases}$$

then $a_{1+1} = 5a_1 - 1$ when $k = 1$

$a_2 = 5a_1 - 1$

$\qquad = 5(4) - 1 \qquad (a_1 = 4)$

$a_2 = \mathbf{19}$

$a_3 = 5a_2 - 1 \qquad\qquad (k = 2)$

$a_3 = 5(19) - 1$

$a_3 = 95 - 1$

$a_3 = \mathbf{94}$

The first three terms of the given sequence are 4, 19, and 94.

Finite and Infinite Sequence

A sequence which has a last term or contains a finite number of terms is called a **finite sequence**.

Examples: (a) **1, 4, 9,** 16.   (The last term is 16.)

$\qquad\qquad\quad (b)$ 8, 11, 14, 17, ..., 35.   (The last term is 35.)

$\qquad\qquad\quad (c)$ 1, $\frac{1}{2}$, $\frac{1}{3}$, $\frac{1}{4}$.   (The last term is $\frac{1}{4}$.)

A sequence which does **not** have a last term (has an infinite number of terms) is called an **infinite sequence**. We usually indicate that a sequence is infinite by placing three dots at the end of the list.

Examples : (a) 8,   11,   14,   17,   ... ; (b) 1, $\frac{1}{2}$, $\frac{1}{3}$, $\frac{1}{4}$,   ...

Finite sequence

If the nth term is given by $a_n = 2n + 1$ and $n = 1, 2,..., 8$ then the sequence $\{a_n\}$ has 8 terms.

$a_1 = 2(1) + 1 = 3$

$a_2 = 2(2) + 1 = 5$

$a_3 = 2(3) + 1 = 7$

....

$a_8 = 2(8) + 1 = 17$

We could write the above sequence as $3, 5, 7,..., 2n + 1$. The above sequence is a finite sequence.

Infinite sequence

Example: If the nth term is $a_{n1} = 2n + 1$ and $n = 1, 2, 3,...$,then we could specify the sequence as

$$3, 5, 7,..., 2n + 1,...$$

Note that this is an infinite sequence, and in this case, the nth term is specified.

Finding a formula for the general term of a sequence

Sometimes, we are given a list of a few successive terms of a sequence without the general term being specified. This is often the case in practical problems and we would like to find a formula for the nth or general term. We should note, however, that any such formula found may not be unique. There may be other formulas which may yield the listed terms. It is also possible that although such a formula found works for the listed terms, the formula may not be the correct one, since we do not know the rest of the terms (the unlisted terms). Therefore, in finding formulas for the nth or general term, we should bear in mind that any such formulas found should be regarded as the **best guesses**.

Example 3: Derive a formula for the general term of the sequence $1, 4, 9, 16,...$
Solution

The approach here is one of trial-and-error. The main principle is to relate each term number (n) to its term by means of addition, subtraction, multiplication, division, roots and powers until a fitting pattern is obtained.

On the 1st term, 1:

$n = 1$. We ask the question: " what shall we do to "1", the term number to get 1, the term ?

Some possibilities are: (a) $1 + 0 = 1$; (b) $1^2 = \mathbf{1}$; (c) $\sqrt{1} = 1$; (d) $\frac{1}{1} = 1$.

On the 2nd term, 4:

$n = 2$. We ask the question: " what shall we do to '2' the term number (by algebraic operations) to obtain 4, the second term.

Some of the possibilities are (a) $2 + 2 = 4$; (b) $2^2 = \mathbf{4}$; (d) $\sqrt{16} = 4$

On the 3rd term, 9

$n = 3$. We ask: "what shall we do to "3" to get 9?

Some possibilities are (a) $3 + 3 \times (3 - 1)$; (b) 3^2.

On the 4th term, 16

$n = 4$. We ask what shall we do to '4' to obtain 16.

The possibilities are (a) $4 + 4 \times 3$; (b) 4^2.

Now, we inspect and compare the various possibilities from each trial. Inspecting the (a)'s for $n = 1, 2, 3, 4$, we do not recognize any pattern. The (b)'s for $n = 1, 2, 3, 4$, show some pattern. We guess that the pattern is n^2. We now test this general term for the given sequence.

Testing our general term n^2:

For $n = 1$, $n^2 = 1$, and $a_1 = 1$

For $n = 2$, $n^2 = 4$, and $a_2 = 4$

For $n = 3$, $n^2 = 9$, and $a_3 = 9$

For $n = 4$, $n^2 = 16$, and $a_4 = 16$

Since the general term n^2 produces all the listed terms of the sequence, we conclude that a formula for the general term $a_n = n^2$. The above example was intended to give the student the kind of process that one may use to find a general term.

Example 4   Find a formula for the general or nth term for the following sequence:

$$1, \frac{1}{4}, \frac{1}{9}, \frac{1}{16}, \ldots$$

Solution

If the first term is written as $\frac{1}{1}$. the sequence then becomes $\frac{1}{1}, \frac{1}{4}, \frac{1}{9}, \frac{1}{16}, \ldots$

We observe that each numerator is 1. In the denominators, we observe that the denominators are the same as the sequence of the last example. Therefore the nth or general term for the above given sequence is

$$a_n = \frac{1}{n^2}$$

Example 5   Find a formula for the general term of the sequence $4, 7, 10, 13,\ldots$

Solution

Considering the first term, $4, n = 1$ and as was explained in Example 1,
Some possibilities are $1 + (2 \times 1 + 1)$, that is, $n + (2n + 1) = 3n + 1$

Considering the 2nd term, $7, n = 2$, we have:
$$2 + (4 + 1)$$
$$n + (2 \times 2 + 1)$$
$$n + (2n + 1) = 3n + 1$$

Considering the 3rd term, 10, where $n = 3$, we have
$$3 + (6 + 1)$$
$$3 + (2 \times 3 + 1)$$
$$n + (2n + 1) = 3n + 1$$

Therefore a formula for the general term is $3n + 1$. We can use this formula to check for the given sequence.

Note that the above formulas were arrived at by mentally or otherwise trying out some patterns until we found a fitting one. So do not be discouraged if the first and second trials do not indicate any pattern.

We present additional examples without much explanation.

Example 6: Find a general formula for the sequence $4, 7, 12, 19,\ldots$

$$
\begin{array}{lll}
n = 1, & 4 = 1 + 3 & i.e., \ n^2 + 3 \\
n = 2, & 7 = 4 + 3 & n^2 + 3 \\
n = 3, & 12 = 9 + 3 & n^2 + 3 \\
n = 4, & 19 = 16 + 3 & n^2 + 3
\end{array}
$$

A formula for the general term is $n^2 + 3$.

Lesson 67: Sequence; Specification, Finite and Infinite Sequence, Formula the general term

Example 7: Find a formula for the general term of the sequence $3, 15, 35, 63,...$

Term :	3	15	35	63
	↓	↓	↓	↓
	$4-1$	$16-1$	$36-1$	$64-1$
	$4 \times 1^2 - 1;$	$4 \times 2^2 - 1;$	$4 \times 3^2 - 1$	$4 \times 4^2 - 1.$

A general term is $4n^2 - 1$

Example 8 Find a formula for the general term of the sequence:

$$\frac{3}{2}, \frac{5}{4}, \frac{7}{8}, \frac{9}{16}, \cdots$$

First, we consider the numerators, $3, 5, 7, 9,..$ and after some trials and errors we obtain $3n - (n-1) = \mathbf{2n + 1}$ as the general term of the numerators.

Secondly ,we consider the denominators, $2, 4, 8, 16,..$ and after some trials, we obtain $\mathbf{2^n}$ as the general term for the denominator. By combining the numerator term, $\mathbf{2n + 1}$ with the denominator term $\mathbf{2^n}$ we obtain the general term $\frac{2n+1}{2^n}$.

Lesson 67 Exercises

A What is a sequence?

Find the first four terms of each sequence whose general term is given.

1. $a_n = 3n + 2;$ 2. $a_n = 3^n;$ 3. $a_n = (-1)(n+5);$ 4. $a_n = \frac{n(n-1)}{2};$ 5. $a_n = (-1)^{n+2};$

6. $a_n = \frac{(-1)^n(n-3)}{n}.$

What conditions must be stated in giving a recursive definition of a sequence.

Find the first four terms of each sequence whose recursive formula is given.

7. $a_1 = 3$ 8. $8a_1 = 3$ 9. $a_1 = -3$

$a_{n+1} = a_n - 2$ $a_{n+1} + 1 = a_n + 4$ $a_{n+1} = a_n + 7$

10. $a_1 = 2$

$a_2 = 2$

$a_n = a_{n-1} + a_{n-2}$ $(n \geq 3)$

11. $a_1 = 1$

$a_2 = 2$

$a_{n+1} = a_n + a_{n-1}$ $(n \geq 2)$

Answers: **1.** $5, 8, 11, 14;$ **2.** $3, 9, 27, 81;$ **3.** $-6, -7, -8, -9;$ **4** $0, 1, 3, 6;$

5. $-1, 1, -1, 1;$ **6.** $2, -\frac{1}{2}, 0, \frac{1}{4};$ **7.** $3, 1, -1, -3;$ **8.** $\frac{3}{8}, \frac{27}{8}, \frac{51}{8}, \frac{75}{8};$ **9.** $-3, 4, 11, 18;$

10. $2, 2, 4, 6;$ **11.** $1, 2, 3, 5.$

B Find a formula for the general term of the sequence whose first four terms are given

1. $2, 2, 4, 6,...$;   **2.** $4, 8, 12, 16, ...$; **3.** $1, 3, 6, 10, ...$

4. $-2, 4, -8, 16,...$;   **5.** $\dfrac{1}{2}, \dfrac{2}{3}, \dfrac{3}{4}, \dfrac{4}{5},...$

Answers: 1. $a_1 = 2,\ a_2 = 2,\ a_n = a_{n-1} + a_{n-2}, n \geq 3$; **2.** $4n$; **3.** $a_n = \dfrac{n(n+1)}{2}$

4. $a_n = (-1)^n 2^n$;   **5.** $a_n = \dfrac{n}{n+1}$

Lesson 68
The Summation Notation, Σ, Properties and Shifting of the Summation Index

Summation Notation

We use the Greek capital letter, Σ, (pronounced sigma) as a summation notation. This notation means the sum of the indicated terms.

Example The series $1 + 4 + 9 + 16$ is written compactly as $\sum_{k=1}^{4} k^2$ and it is read the sum of k^2 as

k goes from 1 to 4. In this example,

$$\text{when } k = 1, \quad k^2 = 1$$
$$k = 2, \quad k^2 = 4$$
$$k = 3, \quad k^2 = 9$$
$$k = 4, \quad k^2 = 16$$

and then $\sum_{k=1}^{4} k^2 = 1 + 4 + 9 + 16 = 30$

Example Write the following in expanded form (1) $\sum_{k=1}^{5} 2K + 1$ (2) $\sum_{k=0}^{3} 3^k + k$

Solution

1. $\displaystyle\sum_{k=1}^{5} 2k + 1 = [2(1) + 1] + [2(2) + 1] + [2(3) + 1] + [2(4) + 1] + [2(5) + 1]$

$$= 3 + 5 + 7 + 9 + 11$$
$$= 35$$

2. $\displaystyle\sum_{k=0}^{3} 3^k + k = (3^0 + 0) + (3^1 + 1) + (3^2 + 2) + (3^3 + 3)$

$$= (1 + 0) + (3 + 1) + (9 + 2) + (27 + 3)$$
$$= 1 + 4 + 11 + 30$$
$$= 46$$

The summation index k above is known as a "dummy" variable in that we can replace it by any other letters or symbols.

As we learned before, associated with each finite sequence is a finite series and corresponding to each infinite sequence is an infinite series.

Application of the summation notation

Example: Associated with the **finite sequence** $1, \frac{1}{2}, \frac{1}{3}, \frac{1}{4}$ is the **finite series**

$$\sum_{n=1}^{4} \frac{1}{n}$$

and associated with the **infinite sequence** $1, \frac{1}{2}, \frac{1}{3}, \frac{1}{4}, \ldots \frac{1}{n}, \ldots$ is the **infinite series**

$$1 + \frac{1}{2} + \frac{1}{3} + \frac{1}{4} + \ldots = \sum_{n=1}^{\infty} \frac{1}{n}$$ which we read as " the sum of $\frac{1}{n}$ as n goes from 1 to infinity.

Some properties of the summation notation

1. $$\sum_{k=1}^{n} ca_k = c \sum_{k=1}^{n} a_k$$

2. $$\sum_{k=1}^{n} a_k + b_k = \sum_{k=1}^{n} a_k + \sum_{k=1}^{n} b_k$$

Shifting (changing) the Summation Index
This shifting will be useful in calculus especially when solving differential equations by power series method. We shall use the shifting of summation index to combine two summations into one. This shifting may be used to make summations run over the same summation index.

Short-cut method for shifting the summation index

Case 1: Infinite upper summation index

Example Write $\displaystyle\sum_{k=1}^{\infty} x^k$ with the index k starting from (a) $k = 0$; (b) $k = 2$.

The operation on the lower summation index, k, and that on the summand x^k are inverses of each other. In (a), as we **decrease** the k in the lower index by 1, we **increase** the k the summand by 1. Similarly, with regards to the solution in (b), as we **increase** the k in the lower summation index by 1, we **decrease** the k in the summand by 1.

(a) To obtain $k = 0$, we **subtract** 1 from the summation index and **add** 1 to the k on the x

Then, $\displaystyle\sum_{k=1}^{\infty} x^k$ becomes $\displaystyle\sum_{k=0}^{\infty} x^{k+1}$ (To change the exponent $k + 1$ back to k, reverse the steps)

Note that $\displaystyle\sum_{k=1}^{\infty} x^k$ and $\displaystyle\sum_{k=0}^{\infty} x^{k+1}$ are equivalent and each will generate the same terms as the other.

(b) To obtain $k = 2$, we **add** 1 to the summation index and subtract 1 from the k on the x

Then, $\displaystyle\sum_{k=1}^{\infty} x^k$ becomes $\displaystyle\sum_{k=2}^{\infty} x^{k-1}$ (To change the exponent $k - 1$ back to k, reverse the steps)

Note also that $\displaystyle\sum_{k=1}^{\infty} x^k$ and $\displaystyle\sum_{k=2}^{\infty} x^{k-1}$ are equivalent and each will generate the same terms as the other.

Detailed Method for shifting the summation index

We repeat the above example. Write $\displaystyle\sum_{k=1}^{\infty} x^k$ with the index k starting from (a) $k = 0$;   (b) $k = 2$.

(a)   Step 1: We want to change $k = 1$ to $k = 0$. To achieve this objective, we subtract 1 from both sides of the equation $k = 1$ to obtain $k - 1 = 0$ (A)

Then $\displaystyle\sum_{k=1}^{\infty} x^k$ becomes $\displaystyle\sum_{k-1=0}^{\infty} x^k$

Step 2: Let $k - 1 = j$.   (B)

Then $k - 1 = 0$ becomes $j = 0$ and

$\displaystyle\sum_{k-1=0}^{\infty} x^k$ becomes $\displaystyle\sum_{j=0}^{\infty} x^k$

Step 3: To compensate for the exponent k on the x, we express k in terms of j.

Solve for k in equation (B) to obtain $k = j + 1$.

Step 4: In $\displaystyle\sum_{j=0}^{\infty} x^k$, replace the k on the x by $j + 1$ to obtain $\displaystyle\sum_{j=0}^{\infty} x^{j+1}$

Step 5: Since j is a dummy variable, we can revert back to k, and then $\displaystyle\sum_{j=0}^{\infty} x^{j+1}$ becomes

$\displaystyle\sum_{k=0}^{\infty} x^{k+1}$, which is what we were asked to write.

Note that $\displaystyle\sum_{k=1}^{\infty} x^k$ and $\displaystyle\sum_{k=0}^{\infty} x^{k+1}$ are equivalent and each will generate the same terms as the other.

(b) Step 1: We want to change $k = 1$ to $k = 2$. To achieve this objective, we add 1 to both sides of the equation $k = 1$ to obtain $k + 1 = 1 + 1$ or $k + 1 = 2$ (A)

Then $\displaystyle\sum_{k=1}^{\infty} x^k$ becomes $\displaystyle\sum_{k+1=2}^{\infty} x^k$

Step 2: Let $k + 1 = j$.   (B)

Then $k + 1 = 2$ becomes $j = 2$, and

$\displaystyle\sum_{k+1=2}^{\infty} x^k$ becomes $\displaystyle\sum_{j=2}^{\infty} x^k$

Step 3: To compensate for the exponent k on the x, we express k in terms of j.

Solving for k in equation (B), $k = j - 1$

Step 4: In $\displaystyle\sum_{j=2}^{\infty} x^k$, replace the k on the x by $j - 1$ to obtain $\displaystyle\sum_{j=2}^{\infty} x^{j-1}$.

Step 5: Since j is a dummy variable, we can revert back to k, and then $\displaystyle\sum_{j=2}^{\infty} x^{j-1}$ becomes

$\displaystyle\sum_{k=2}^{\infty} x^{k-1}$, which is what we were asked to write.

Case 2: Finite upper summation index

Example Write $\displaystyle\sum_{k=2}^{5} x^k$ with the index k starting from (a) $k = 0$;   (b) $k = 3$.

The operation on the lower summation index, k, and the upper index n are the same, while the operation on the summand x^k is the inverse of those on the lower and upper indices.
 In (a), as we **decrease** the k in the lower index by 2 to obtain 0, and decrease the upper index by 2 to obtain 3, we **increase** the k on the summand by 2.

Similarly, in (b), as we **increase** the k in the lower summation index by 1 to obtain 3, and increase the upper summation index by 1 to obtain 6, we **decrease** the k in the summand by 1.

Solution

(a) To obtain $k = 0$, we **subtract** 2 from the summation indices and **add** 2 to the k on the x

Then, $\displaystyle\sum_{k=2}^{5} x^k$ becomes $\displaystyle\sum_{k=0}^{3} x^{k+2}$ (To change the exponent $k + 2$ back to k, reverse the steps)

Note that $\displaystyle\sum_{k=2}^{5} x^k$ and $\displaystyle\sum_{k=0}^{3} x^{k+2}$ are equivalent and each will generate the same terms as the other.

(b) To obtain $k = 3$, we add 1 to the summation indices and subtract. 1 from the k on the x.

Then, $\displaystyle\sum_{k=2}^{5} x^k$ becomes $\displaystyle\sum_{k=3}^{6} x^{k-1}$ (To change the exponent $k - 1$ back to k, reverse the steps)

Note also that $\displaystyle\sum_{k=2}^{5} x^k$ and $\displaystyle\sum_{k=3}^{6} x^{k-1}$ are equivalent and each will generate the same terms as the other.

Lesson 68 Exercises

Write in expanded form and then evaluate:

1. $\displaystyle\sum_{k=0}^{4} (k + 3);$ **2.** $\displaystyle\sum_{k=0}^{3} (2k^2 - 3);$ **3.** $\displaystyle\sum_{n=0}^{4} (-1)^n \frac{1}{n+2};$ **4.** $\displaystyle\sum_{k=2}^{5} \frac{(-1)^k (k+2)}{2^k}$

Write using sigma notation: **5.** $8 + 11 + 14 + 17$;   **6.** $\dfrac{5}{7}, \dfrac{7}{9}, \dfrac{9}{11}, \dfrac{11}{13}, \dots$

7. Write $\displaystyle\sum_{k=2}^{\infty} x^{k+1}$ with the index k starting from (a) $k = 0$; (b) $k = 3$, and verify the results by

reversing the steps.

8. Write $\displaystyle\sum_{k=2}^{5} x^k$ with the index k starting from (a) $k = 0$;   (b) $k = 3$.

Answers: **1.** 25; **2.** 16; **3.** $\dfrac{23}{60}$; **4.** $\dfrac{17}{32}$; **5.** $\displaystyle\sum_{k=0}^{3} (3k + 8)$; **6.** $\displaystyle\sum_{k=3}^{6} \frac{2k-1}{2k+1}$; **7.**(a) $\displaystyle\sum_{k=0}^{\infty} x^{k+3}$, (b) $\displaystyle\sum_{k=3}^{\infty} x^k$

8. (a) $\displaystyle\sum_{k=0}^{3} x^{k+2}$; (b) $\displaystyle\sum_{k=3}^{6} x^{k-1}$

Lesson 69

Arithmetic Sequence and Arithmetic Series

Arithmetic Sequence (or Arithmetic Progression)

We assume that the arithmetic sequence is finite unless otherwise specified.

An **arithmetic sequence** is a sequence of numbers, each of which after the first number is obtained from the preceding one by adding to it a constant number (fixed number). This constant number is called the **common difference** of the sequence and is usually denoted by the letter d; d can be either positive or negative. The common difference is obtained by subtracting any term from the next term. We will specify an arithmetic sequence by a **recursive formula.**

$$\begin{cases} a_1 = \text{the given first number} \\ a_n = a_{n-1} + d \qquad n = 2,3,...k. \end{cases} \Big\} < -\text{recursive formula}$$

Finding an explicit formula for the nth term

Example 1a:	Example 1b:
If the first term $= a_1$.	If the first term $a_1 + (1-1)d = a_1$.
Then the 2nd term $a_2 = \boldsymbol{a_1 + d}$	Then the 2nd term $a_2 = a_1 + (2-1)d = a_1 + d$
The 3rd term $a_3 = a_2 + d = \boldsymbol{a_1 + 2d}$	The 3rd term $a_3 = a_1 + (3-1)d = \boldsymbol{a_1} + 2d$
The 4th term $a_4 = a_3 + d = \boldsymbol{a_1 + 3d}$	The 4th term $a_4 = a_1 + (4-1)d = a_1 + \boldsymbol{3d}$

By inspecting the terms in Examples 1a &1b , we arrive at a **formula** for the general (or nth) term as

$$\boxed{a_n = a_1 + (n-1)d} \qquad <\text{----- Explicit formula}$$

where a_n is the nth term, a_1 is the first term, n is the term number and d is the common difference.

Example 2 Find the 80th term of the arithmetic sequence whose first term is 4 and whose common difference is 5.

Solution: $n = 80$, $a_1 = 4$, $d = 5$
Substituting these values in
$$a_n = a_1 + (n-1)d$$
$$a_{80} = 4 + (80-1)(5)$$
$$= 4 + 79(5)$$
$$= 399$$

The 80th term is 399.

Arithmetic Mean

An arithmetic mean between two given terms is the term between the given terms. A single arithmetic mean is also the average of the two given terms Arithmetic means between two given terms in an arithmetic sequence are the terms between the given terms.

.**Example 3**   The arithmetic mean between 4 and 8 is $\frac{4+8}{2} = \frac{12}{2} = 6$.

Series

Every sequence has an associated series.
A **series** is the **indicated sum** of the terms of a sequence (not the sum itself).

For example, the associated series of the sequence

$4, 7, 10, 13, \dots, 3n + 1$ is the i**ndicated sum**

$4 + 7 + 10 + 13 + \dots + (3n + 1)$.

We can observe that the terms of each series are the terms of the associated sequence. Also with any sequence $\{a_n\}$, there is another associated sequence $\{s_n\}$ called the sequence of partial sums. Each partial sum $\{s_n\}$ is the sum of the first n terms.

If $a_1, a_2, a_3, \dots, a_n \dots$ are the terms of a given sequence, then

$$s_n = a_1 + a_2 + a_3 + \dots + a_n$$

In particular, for $n = 1$, $\quad s_1 = a_1$

$$n = 2, \quad s_2 = a_1 + a_2$$
$$n = 3, \quad s_3 = a_1 + a_2 + a_3$$

Arithmetic Series

For the arithmetic sequence $a_1, (a_1 + d), \dots, (a_n - 2d), (a_n - d), a_n$, there is an associated sequence of the sum of the first n terms, (also known as partial sum).

The sequence of partial sums is $a_1, 2a_1 + d, 3a_1 + 3d,$, and so forth, noting that

$$s_1 = a_1, \quad s_2 = a_1 + a_2, \quad s_3 = a_1 + a_2 + a_3$$

Sum of an arithmetic sequence

Let us derive a formula for the sum of the first n terms of an arithmetic sequence.
Let $s_n =$ the sum of an arithmetic sequence as defined by $a_n = a_1 + (n - 1)d$. We observe that the first term is a_1 the common difference is d, and if the last term is l ($l = a_n$); then next to the last term is $(l - d)$. We write the sum of n terms first forwards and then backwards.

$s_n = a_1 + (a_1 + d) + (a_1 + 2d) + \dots + [a_1 + (n - 1)d]$ (1)
$s_n = l + (l - d) + (l - 2d) + \dots + [l - (n - 1)d]$ (2)
Adding (1) and (2), we obtain
$2s_n = (a_1 + l) + (a_1 + l) + (a_1 + l) \dots + (a_1 + l)$ (3)
$2s_n = n(a_1 + l)$ ($(a_1 + l)$ occurs n times)

$s_n = \dfrac{n(a_1 + l)}{2}$ or $s_n = \dfrac{n(u_1 + u_n)}{2}$ where $l = a_n$ (5)

$$\boxed{s_n = \dfrac{n}{2}(a_1 + l)}$$ <-- n times the mean of the first and last terms.

Scrapwork

$= a_1 + a_1 + d + a_1 + 2d + \dots + a_1 + nd - d$ (3)
$= l + l - d + l - 2d + \dots + l - dn +$ (4)
Adding (3) and (4), we obtain
$2s_n = (a_1 + l) + (a_1 + l) + (a_1 + l) \dots + (a_1 + l)$
$2s_n = n(a_1 + l)$

If we replace l in (5) by $a_1 + (n - 1)d$, we obtain another formula $\boxed{s_n = \dfrac{n}{2}(2a_1 + (n - 1)d)}$ (6)

We can memorize only (5) and then find l by applying $l = a_n = a_1 + (n - 1)d$.

Examples: Sum of the first n terms of an arithmetic sequence

Case 1: If we know the first term a_1, and the last term, l, of an arithmetic sequence, we can find the sum s_n of the first n terms of this sequence by using the formula

$$s_n = \frac{n}{2}(a_1 + l) \text{ or } s_n = \frac{n(a_1 + a_n)}{2}, \text{where } l = a_n$$

Case 2: If we know the first term a_1, the common difference, d, then we can find the sum, s_n, of the first n terms of this sequence by using the formula

$$s_n = \frac{n}{2}[2a_1 + (n-1)d]$$

As noted before, you can memorize Case 1 formula only and then first find l by applying $l = a_n = a_1 + (n-1)d$, and then substitute for l. (see Example 4, Method 2)

Example 4 Find the sum of the first 12 terms of the arithmetic sequence 4, 9, 14, 19,...

Method 1:	**Method 2:**
Step 1 $s_n = \frac{n}{2}[2a_1 + (n-1)d]$ (1)	**Step 1:** We apply $s_n = \frac{n}{2}(a_1 + l)$; (2)

Method 1:

Substitute $a_1 = 4$,, $d = 5$, and $n = 12$ in equation (1),

Step 2: $s_{12} = \frac{12}{2}[2(4) + (12-1)5] = 6[8 + 55]$

$$= 6[63] = 378$$

The sum of the first 12 terms is 378.

Method 2:

but we first find l (last term).

$l = a_n = a_1 + (n-1)d$

$l = a_{12} = 4 + (12-1)5$ ($a_1 = 4$; $n = 12$, $d = 5$)

$= 4 + (11)5$

$l = 4 + 55 = 59$

Step 2: Substitute $a_1 = 4$. and $l = 59$ in (2)

Then $s_{12} = \frac{12}{2}(4 + 59) = 6(63) = 378$

The sum of the first 12 terms is 378.

Example 5 Find the sum of the arithmetic series $\sum\limits_{k=1}^{16} 3k + 7$

Method 1:

$s_n = \frac{n}{2}(a_1 + l)$ (l = last term; a_1 = first term)

$l = 3(16) + 7$ ($l = 3k + 7$; $k = 16$)

$= 48 + 7 = 55$

$a_1 = 3(1) + 7 = 10$ ($a_1 = 3k + 7$; $k = 1$)

$s_{16} = \frac{16}{2}(10 + 55)$ ($a_1 = 10, l = 55$)

$= 8(65)$

$= 520$

The sum of the given series is 520.

Method 3: We can tediously calculate and list the 16 terms and add them.
    With $d = 3$, as in Method 2, keep adding 3.

$S_{16} = 10 + 13 + 16 + 19 + 22 + 25 + 28 + 31 +$

$34 + 37 + 40 + 43 + 46 + 49 + 52 + 55 = 520$

Method 2:

$$s_n = \frac{n}{2}[2a_1 + (n-1)d]$$

To use this formula for s_n, we need to know n, a_1 and d. We know that $n = 16$ from the sigma notation. To find a_1 and d ,we find a few terms from the formula:

when $k = 1$, the first term is $3(1) + 7 = 10$
when $k = 2$, the second term is $3(2) + 7 = 13$
when $k = 3$, the third term is $3(3) + 7 = 16$

The first three terms are 10, 13, and 16.
The first term, $a_1 = 10$, $d = 13 - 10 = 3 = 3$
Substituting $a_1 = 10$, $d = 3$, and $n = 16$ in

$$s_n = \frac{n}{2}[2a_1 + (n-1)d]$$

$$s_{16} = \frac{16}{2}[2(10) + (16-1)3] = 8[20 + 45] = 520$$

Arithmetic sequence as a linear function

Consider the arithmetic sequence given by $a_n = a_1 + (n-1)d$, where n is a variable.

If we let $a_n = y$, and $n = x$, we obtain

$$y = a_1 + (x-1)d$$
$$= (x-1)d + a_1$$
$$= xd - d + a_1$$
$$= dx - d + a_1$$
$$y = xd + (a_1 - d) \qquad (1)$$

Comparing equation (1) with the slope-intercept form of the equation of a straight line given by
$$y = mx + b \qquad (2)$$
the slope of (1) is d and the y-intercept $= (a_1 - d)$
Thus, the common difference, d, is the slope and $(a_1 - d)$ is the y-intercept.

Example 6 Graphically, illustrate that the arithmetic sequence $4, 7, 10, 13,...$ can be considered as a linear function.

Solution $a_1 = 4, d = 3$
$$y = dx + (a_1 - d)$$
$$= 3x + (4 - 3)$$
$$y = 3x + 1$$
$$x = 1, \quad y = 4$$
$$n = 1, \quad a_1 = 4$$

Some ordered pairs for the given sequence are $(1, 4), (2, 7), (3, 10), (4, 13)$
Ordered pairs (x, y) or (n, a_n) $(1, 4),$ $(2, 7),$ $(3, 10)$
$$(n = 1, a_1 = y_1), (n = 2, a_2 = y_2), (n = 3, a_3 = y_3)$$

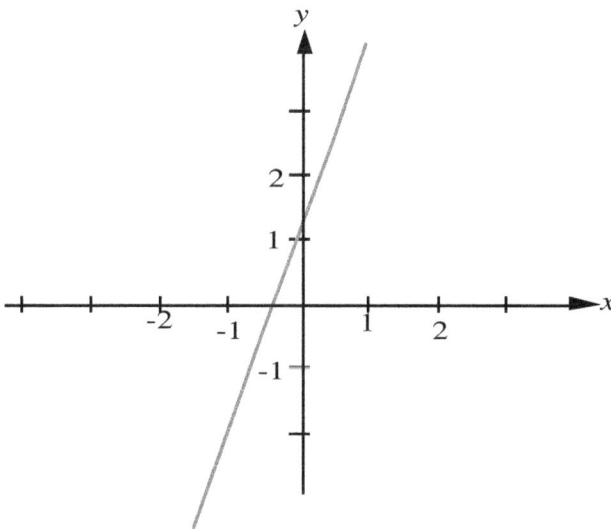

Note: We may also obtain the ordered pairs by noting that a sequence is a set of ordered pairs whose domain (x-values) consists of the consecutive positive integers and pair 1, 2, 3, and 4 with 4, 7, 10. and 13 respectively.

Lesson 69 Exercises

Find the first four terms for each arithmetic sequence, where a_1 is the first term and d is the common difference

1. $a_1 = 2, d = 3$;   **2.** $a_1 = 5, d = -2$;   **3.** $a_1 = \frac{1}{2}, d = \frac{3}{2}$;   **4.** $a_1 = \sqrt{3}, d = \sqrt{3}$.

5. Find the sum of an arithmetic sequence whose first term is 3 and whose last term is 17.

6. Write the first four terms of (a) $5, 9,...$ (b) $k, k - 2$.

7. Determine if the sequence given is arithmetic: $20, 17, 14, 11,...$

8.   Find the 18th term and the sum of the first 18 terms of the arithmetic sequence
    $7, 18, 29, 40,...$

9. Find the 18th term and sun of the first 18 terms of the series given by $7 + 18 + 29 + 40,...$

10. Find he sum of the first 50 positive integers.

11. Graphically, illustrate that the arithmetic sequence $2, 8, 14, 20,...$may be considered as a   linear function.

Answers: **1.** 2, 5, 8, 11; **2.** 5, 3, 1, -1; **3.** $\frac{1}{2}$, 2, $\frac{7}{2}$, 5; **4.** $\sqrt{3}$, $2\sqrt{3}$, $3\sqrt{3}$, $4\sqrt{3}$
5. 80; **6. (a)** 5, 9, 13, 17; (b) $k, k - 2, k - 4, k - 6$; **7.** Yes $(d = -3)$
8. $a_{18} \approx 194$, $s_{18} = 1809$; **9.** $a_{18} \approx 194$, $s_{18} = 1809$; **10.** 1275.

Lesson 70

Geometric Sequence, Geometric Series
Applications of Geometric Series

Geometric Sequence (Geometric Progression);

A **geometric sequence** is a sequence of numbers, each of which after the first number is obtained from the preceding number by multiplying the preceding number by a constant number. The constant number is called the **common ratio** and it is usually denoted by the letter r. We obtain the common ratio by dividing each term by the preceding term. This ratio may be positive or negative.

We specify a geometric sequence by a recursive formula. Thus

$$\begin{cases} a_1 \text{ is the first term} \\ a_n = ra_{n-1} \qquad n = 2, 3, 4, \ldots \end{cases}$$

Applying this recursive formula,

a_1 is the first term

$a_2 = ra_1$

$a_3 = ra_2 = r(ra_1) = r^2 a_1$

$a_4 = ra_2 = r(r^2 a_1) = r^3 a_1$

From the above terms, and by trial and error, we determine the **nth** or **general term** is given by

$$\boxed{a_n = a_1 r^{n-1}}$$

Thus, if we know the first term and the common ratio, we can find any other term.

Example 1: Find the 4th term of a geometric sequence whose first term is 3 and whose common ratio is 2.

Step 1: $a_n = a_1 r^{n-1}$ $a_1 = 3, \ r = 2, \ n = 4$ $a_4 = 3(2)^{4-1}$ $= 3(2)^3$	**Step 2** $= 3(8)$ $= 24$ The fourth term is 24.

Example 2 Find the recursive formula for the geometric sequence whose formula for the general term is $a_n = 6(\frac{1}{2})^{n-1}$.

To write the recursive definition, we must find the first term a_1 and the common ratio, r

Step 1: Finding a_1: $a_1 = 6(\frac{1}{2})^{1-1}$ $(n=1)$ $= 6(\frac{1}{2})^0$ $a_1 = 6$. **Step 2**: Finding r: Divide a term by the preceding term.	**Step 2 continued** $a_n = ra_{n-1}$ (recursive formula for a geometric sequence for $n = 2, 3, 4, \ldots$) a term $= 6(\frac{1}{2})^{n-1}$, preceding term $= 6(\frac{1}{2})^{n-1-1}$ $r = \dfrac{(\frac{1}{2})^{n-1}}{(\frac{1}{2})^{n-2}}$	**Step 3** $r = (\frac{1}{2})^{n-1-n+2}$ $r = \frac{1}{2}$ The recursive formula for the above is: $\begin{cases} a_1 = 6 \\ a_n = \frac{1}{2}a_{n-1} \qquad n = 2, 3, 4, \ldots \end{cases}$

Geometric Mean

Let m be the geometric mean between (or of) a and b. Then a, m, b form a geometric sequence.

Then $\frac{m}{a} = \frac{b}{m}$ (common ratio = common ratio)

$m^2 = ab$

$m = \pm\sqrt{ab}$ (i.e., $+\sqrt{ab}$ or $-\sqrt{ab}$)

Generally, the geometric mean of a set of n numbers is the nth root of the product of the numbers..

Example Find the geometric mean of 4 and 6.

Let $a = 4$, $b = 6$,

Then the mean $m = \pm\sqrt{(4)(6)}$

$m = \pm 2\sqrt{6}$

The geometric mean is $+2\sqrt{6}$ or $-2\sqrt{6}$.

Usually, we look for the positive root, and in which case, the geometric mean would be $2\sqrt{6}$.

Finite Geometric Series

Sum of the First n Terms of a Geometric Sequence

Consider the geometric sequence a_1, $a_1 r$, $a_1 r^2$,..., $a_1 r^{n-1}$

Let $s_n =$ the sum of the first n terms of the sequence

$$s_n = a_1 + a_1 r + a_1 r^2 + ... + a_1 r^{n-2} + a_1 r^{n-1} \qquad (1)$$

Multiplying (1) by r we obtain:

$$r s_n = a_1 r + a_1 r^2 + a_1 r^3 + ... + a_1 r^{n-1} + a_1 r^n \qquad (2)$$

$(1) - (2)$:

$$s_n - r s_n = a_1 - a_1 r^n$$

$$s_n (1 - r) = a_1 - a_1 r^n$$

$$s_n = \frac{a_1 - a_1 r^n}{1 - r}$$

$$\boxed{s_n = \frac{a_1(1 - r^n)}{1 - r}} \qquad (3)$$

Thus knowing the first term and the common ratio we can find the sum of a geometric series, and this formula is the one that we usually use, since we may not know the last term. However, if we know the last term also, then we may use an alternative formula. We derive this formula below.

Step 1: Recalling that $a_n = a_1 r^{n-1}$

$$a_n = a_1 r^n r^{-1}$$

$$\frac{r a_n}{a_1} = r^n$$

$$r^n = \frac{r a_n}{a_1}$$

Step 2: Substituting for r^n in (3) we obtain

$$s_n = \frac{a_1\left(1 - \dfrac{r a_n}{a_1}\right)}{1 - r}$$

$$= \frac{\dfrac{a_1(a_1 - r a_n)}{a_1}}{1 - r}$$

$$\boxed{s_n = \frac{a_1 - r a_n}{1 - r} \qquad r \neq 1}$$

Example 1 Find the sum of the first seven terms of the geometric sequence $4, 8, 16, 32,...$

$$a_1 = 4, \quad r = \tfrac{8}{4} = 2, \quad n = 7 \qquad (\text{Also, } r = \tfrac{16}{8} = \tfrac{32}{16} = 2)$$

$$S_n = \frac{a_1(1 - r^n)}{1 - r}$$

$$S_7 = \frac{4(1 - 2^7)}{1 - 2}$$

$$S_7 = \frac{4(1 - 128)}{-1}$$

$$S_7 = \frac{4(-127)}{-1}$$

$$S_7 = 508$$

The sum of the first seven terns is 508.

Example 2 Find the sum of the following geometric series: $\displaystyle\sum_{k=1}^{4} 3^k$

Solution
Method 1
Step 1: The usual method is faster method, but for illustrative purposes, we use a rather longer method.
We find a_1 and r

$$a_1 = 3^1 = \mathbf{3}; \quad a_2 = 3^2 = \mathbf{9}; \quad a_3 = 3^3 = \mathbf{27}$$

The first three terms are $3, 9, 27$. The common ratio is 3

$$a_1 = 3, \quad r = 3, \quad n = 4$$

$$S_n = \frac{a_1(1 - r^n)}{1 - r}$$

$$S_4 = \frac{3(1 - 3^4)}{1 - 3}$$

Step 2:
$$= \frac{3(1 - 81)}{-2}$$
$$= \frac{3(-80)}{-2}$$
$$S_4 = 120$$
$$\sum_{k=1}^{4} 3^k = 120$$

Usual method--> **Method 2** $\displaystyle\sum_{k=1}^{4} 3^k = 3^1 + 3^2 + 3^3 + 3^4$

$$= 3 + 9 + 27 + 81 = 120$$

Geometric Sequence as an Exponential Function

Previously, we showed that an arithmetic sequence may be considered as a linear function. Here, we consider the geometric sequence as an **exponential function.**

Consider the general term $a_n = a_1 r^{n-1}$ of the geometric sequence

$$\text{Let } a_n = y; \quad n = x$$
$$\text{Then } y = a_1 r^x r^{-1}$$
$$= a_1 r^{-1} r^x$$
$$y = cr^x \qquad \text{where } c = a_1 r^{-1}$$

If x is a positive integer, then the sketch of the graph of the geometric sequence is of the same form as that of the exponential function $f(x) = ca^x$ See also Chapter 19.

Infinite Geometric Series

Consider the geometric series $a_1 + a_1 r + a_1 r^2 + ... + a_1 r^{n-2} + a_1 r^{n-1} + ...$ (1)

The sum of the first n terms is given by $s_n = \dfrac{a_1(1 - r^n)}{1 - r}$ (2)

If $|r| < 1$ or $-1 < r < 1$, and as $n \to \infty$ (i.e., n gets large), $r^n \to 0$ and (2) becomes

$$\boxed{s_n = \dfrac{a_1}{1 - r}}$$ < ---sum of the infinite geometric series when $|r| < 1$ or $-1 < r < 1$

Note that this formula is **not** valid when $|r| > 1$

Example 1 Given the geometric series $6 + 2 + \frac{2}{3} + \frac{2}{9} + ...$, find the sum s_n

Solution $a_1 = 6$. We need to find r , To find r, divide a term of the series by the preceding term.

$$r = \frac{2}{6} = \frac{1}{3} \text{ or } r = \frac{2}{3} \div 2 = \frac{2}{3} \times \frac{1}{2} = \frac{1}{3} \text{ or } r = \frac{2}{9} \div \frac{2}{3} = \frac{2}{9} \times \frac{3}{2} = \frac{1}{3}$$

Now, $a_1 = 6$, $r = \frac{1}{3}$, and $\left|\frac{1}{3}\right| < 1$; and therefore, the formula $s_n = \dfrac{a_1}{1 - r}$ is applicable.

$$s_n = \frac{a_1}{1 - r} = \frac{6}{1 - \frac{1}{3}} = \frac{6}{\frac{2}{3}} = 9$$

Applications:

Example 2: Convert the repeating decimal 0.333...to a common fraction.

$0.333 = \dfrac{3}{10} + \dfrac{3}{100} + \dfrac{3}{1000} + ...$

$a_1 = \dfrac{3}{10}$, $r = \dfrac{1}{10}$. $|r| = \left|\dfrac{1}{10}\right| < 1$ and the forrmula is applicable.

Substituting in $s_n = \dfrac{a_1}{1 - r}$, we obtain

$$\frac{\frac{3}{10}}{1 - \frac{1}{10}} = \frac{\frac{3}{10}}{\frac{9}{10}} = \frac{3}{10} \cdot \frac{10}{9} = \frac{1}{3}$$

$0.333... = \frac{1}{3}$.

Example 3: Convert the repeating decimal 0.1666...to a common fraction.

$0.1666... = \dfrac{1}{10} + \dfrac{6}{100} + \dfrac{6}{1000} + \dfrac{6}{10000} + ...$

$a_1 = \dfrac{6}{100}$, $r = \dfrac{1}{10}$. $|r| = \left|\dfrac{1}{10}\right| < 1$

(formula applies to the repeated part only).

Substituting in $s_n = \dfrac{a_1}{1 - r}$, we obtain

$$\frac{\frac{6}{100}}{1 - \frac{1}{10}} = \frac{\frac{6}{100}}{\frac{9}{10}} = \frac{6}{100} \cdot \frac{10}{9} = \frac{2}{30} = \frac{1}{15}$$

$0.1666... = \frac{1}{10} + \frac{1}{15} = \frac{1}{6}$

There is another method for converting a repeating decimal to a rational number, See Appendix. p.655

General comments on arithmetic and geometric sequence

If we are given a sequence (or series) and we suspect that it is either an arithmetic or a geometric sequence (or series), we should apply the appropriate definitions to test if the sequence or series is arithmetic or geometric and then proceed accordingly.

Applications of Geometric Series

The practical applications of geometric series include the computation of compound interest, and prediction of population growth. Computation of interest: The money that one pays for use of another person's money (the principal) is called interest . If the calculation of the interested is always based on the original amount of money borrowed, the interest is called **simple interest** . However, if the computation of the interest is based on the accumulated amount (the principal plus the interest at the end of a period), the interest is called **compound interest**.

Compound Interest

Compound interest depends on upon four factors, namely, the principal, the interest rate per conversion period, and the duration of the money invested.

An interest rate of 7% percent means 7% per year. If the compounding period is not one year, then the period is specified. For example, an interest rate of 7 % compounded quarterly (4 times a year) means the rate per each quarter is $\frac{7\%}{4} = 1\frac{3}{4}\%$ per quarter. An interest rate of 8% compounded quarterly (that is four times a year) means the interest rate per each quarter is $\frac{8\%}{4} = 2\%$ per quarter.

We cover three main cases.
Case 1: The interest is compounded n times a year;
Case 2: The interest is compounded once a year ($n = 1$);
Case 3: The interest is compounded continuously.

Case 1: The interest is compounded n times a year;

If A = amount at the end of t years.
P = the principal invested initially
r = annual interest rate (as a decimal)
n = number of times per year the interest is computed
t = the number of years, then amount A is given by

$$A = P(1 + \tfrac{r}{n})^{nt}$$

Case 2: The interest is compounded (once) at the end of the year ($n = 1$. that is annually)

If the interest is compounded annually (that is, $n = 1$) for t years, the amount A is given by

(a) $A = P(1 + r)^t$ and

(b)   If $P = 1$, then $A = P(1 + r)^t$ becomes

   $A = (1 + r)^t$ and then we can use the binomial theorem or tables in computing the interest.

Case 3: The interest is compounded continuously.

Here, the formula is $A = Pe^{rt}$

Example for Case 2

If $30,000 is invested at 5% interest and compounded annually, in t years, it will grow to an amount , A, given by $A(t) = 30,00(1.05)^t$

(a) How long will it take to accumulate $65,000 in the account?

(b) Find the time required for $30,000 to double itself.

Solution

a) $A = 30,000$

$65,000 = 30,000(1.05)^t$

$\frac{65,000}{30,000} = (1.05)^t$

$2.17 = (1.05)^t$

$\log 2.17 = \log 1.05^t$

$\log 2.17 = t \log 1.05$

$\frac{\log 2.17}{\log 1.05} = t$

$15.9 = t$

Scrapwork

$\log 2.17 = 0.3365$

$\log 1.05 = 0.0212$

$\frac{\log 2.17}{\log 1.05} = \frac{0.3365}{0.0212} = 15.9$

The time taken by $30,000 to grow to $65,000, at the interest rate of 5%, is 15.9 years.

(b) $30,000 doubles to $60,000 at the interest rate of 5% in t years.

Therefore, $P = 30,000$, $A = 60,000$; and substituting in $A(t) = P(1 + i)^t$, we obtain

$60,000 = 30,000(1.05)^t$, and next solve this exponential equation for t.

$$\frac{60,000}{30,000} = (1.05)^t$$

$$2 = (1.05)^t$$

$$\log 2 = \log(1.05)^t$$

$$\log 2 = t\log(1.05)$$

$$\frac{\log 2}{\log 1.05} = t$$

$$14.2 = t$$

Scrapwork

$\log 2 = 0.3010$

$\log 1.05 = 0.0212$

$$\frac{\log 2}{\log 1.05} = \frac{0.3010}{0.0212} = 14.2$$

Therefore, at the interest rate of 5%, $30,000 doubles in 14.2 years.

For examples on Exponential Growth and Decay, see page 316

Lesson 70 Exercises

A 1. Find the 5th term of a geometric sequence whose first term is 4 and whose common ratio is 3.

  2. Find the 6th term of a geometric sequence whose first terms are given by 3, 12, 48,....

  ,Write the first four terms of each geometric sequence, where a is the first term and r is the common ratio.

  3. $a = 2$, $r = 3$;   4. $a = -3$, $r = 2$;   5. $a = \frac{2}{3}$, $r = \frac{1}{3}$;

  6. Can a term of a geometric sequence be 0? Why?

  7. Find the geometric mean between 5 and 45.

Answers: **1.** $a_5 = 324$; **2.** $a_6 = 3072$; **3.** 2, 6, 18, 54; **4.** -3, -6, -12, -24; **5.** $\frac{2}{3}$, $\frac{2}{9}$, $\frac{2}{27}$, $\frac{2}{81}$

   6. No. **7.** 15

B 1. Find the sum of the first six terms of the geometric series whose first term is 4 and whose common ratio is $\frac{1}{2}$.

  2. Find the sum of the finite geometric series given by $\sum_{k=1}^{5} 2^k$.

Answers: **1.** $\frac{63}{8}$ or $7\frac{7}{8}$; **2.** 62.

C 1. A savings bank pays 5% compounded semiannually. How much will $3,000 amount to after 10 years, assuming that the depositor leaves the account untouched during the 10 years.

  2. How long will it take $5,000 invested at 4% and compounded quarterly to double itself.

  3. A bacterial culture decays according to the equation $A = A_0(10)^t$, where A_0 is the amount of bacterial originally in the culture and t is the time in hours for the decay. If a bacterial culture contains originally 600 bacteria, what is the amount of bacteria in this culture after two hours.

See also, p316

Answers: 1. $4915.85; 2 17.4 yr. 3. 60,000

CHAPTER 27

Lesson 71: **Factorial Notation, *n*!; Operations Involving *n*!**
Lesson 72: **The Binomial Theorem and Applications**

Lesson 71
Factorial Notation, *n*!; Operations Involving *n*!

Factorial Notation *n*!

Definitions
$\begin{cases} \textbf{1.} \ \ n! = n(n-1)(n-2)(n-3)...3 \bullet 2 \bullet 1. \\ \textbf{2.} \ \ 0! = 1 \end{cases}$

Note that $n! = n(n-1)! = n(n-1)(n-2)! = n(n-1)(n-2)(n-3)!$

Examples

$\begin{aligned}
\textbf{1.} \quad 4! &= 4(4-1)(4-2)(4-3)(4-4)! \\
&= 4(3)(2)(1)(0)! \\
&= 4(3)(2)(1)(1) = 24
\end{aligned}$

$\textbf{2.} \quad 6! = 6(5)(4)(3)(2)(1) = 720$

$\textbf{3.} \quad 8! = 8(7)(6)(5)(4)(3)(2)(1) = 40320$

Operations involving the factorial *n*!

Simplify the following

1. $\dfrac{n!}{(n-1)!}$;   **2.** $\dfrac{n!}{(n-2)!}$;   **3.** $\dfrac{(n+1)!}{n(n-2)!}$;   **4.** $\dfrac{(n-3)!}{n!}$

Solution

One approach to simplifying the above fractions is to expand the **least** expanded factorial as far as the **most** expanded factorial in either the numerator or in the denominator, and then divide out the common factors.

Recall that $n! = n(n-1)(n-2)(n-3)...3 \bullet 2 \bullet 1.$

Note also that $n! = n(n-1)! = n(n-1)(n-2)! = n(n-1)(n-2)(n-3)!$

1. $\dfrac{n!}{(n-1)!} = \dfrac{n(n-1)!}{(n-1)!} = n$

$\textbf{2.} \ \dfrac{n!}{(n-2)!} = \dfrac{n(n-1)(n-2)!}{(n-2)!}$

$\qquad\qquad = n(n-1)$

$\qquad\qquad = n^2 - n$

$\textbf{3.} \ \dfrac{(n+1)!}{n(n-2)!} = \dfrac{(n+1)(n+1-1)(n+1-2)(n+1-3)!}{n(n-2)!}$

$\qquad\qquad = \dfrac{(n+1)(n)(n-1)(n-2)!}{n(n-2)!}$

$\qquad\qquad = (n+1)(n-1)$

$\qquad\qquad = n^2 - 1.$

4. $\dfrac{(n-3)!}{n!} = \dfrac{(n-3)!}{n(n-1)(n-2)(n-3)!}$

$\phantom{\dfrac{(n-3)!}{n!}} = \dfrac{1}{n(n-1)(n-2)}$

$\phantom{\dfrac{(n-3)!}{n!}} = \dfrac{1}{n^3 - 3n^2 + 2n}$

Lesson 71 Exercises

Simplify:: **1.** $\dfrac{6!\,4!}{3!}$; **2.** $\dfrac{6!}{(6-2)!}$; **3.** Expand and simplify: $\dfrac{(n-2)!}{(n-3)!}$; **4.** Show that $n + \dfrac{n!}{(n-2)!} = n^2$

Answers: **1.** 2880; **2.** 30; **3.** $n-2$

Lesson 72   481
The Binomial Theorem and Applications

The Binomial Theorem

Binomial : A binomial is a polynomial of two unlike terms.

Expansion of $(a+b)^n$

Let us perform the usual multiplication for $n = 1, 2, 3, 4, 5$

$$n = 1, \quad (a+b)^1 = a+b$$

$$n = 2, \quad (a+b)^2 = a^2 + 2ab + b^2$$

$$n = 3, \quad (a+b)^3 = a^3 + 3a^2b + 3ab^2 + b^3$$

$$n = 4, \quad (a+b)^4 = a^4 + 4a^3b + 6a^2b^2 + 4ab^3 + b^4$$

$$n = 5, \quad (a+b)^5 = a^5 + 5a^4b + 10a^3b^2 + 10a^2b^3 + 5ab^4 + b^5$$

We observe the following properties about the above expansions.

(1) Each expansion contains $(n+1))$ terms.

(2) The first term is a^n and the last term is b^n.

(3) After the first term, as the exponent of a decreases by 1 from one term to the next, the exponent of b increases by 1, and in each term, the sum of the exponents of a and b is n.

(4) The coefficient of the second term is $\frac{n}{1}$, of the 3rd term is $\frac{n(n-1)}{1 \cdot 2}$, of the fourth term is $\frac{n(n-1)(n-2)}{1 \cdot 2 \cdot 3}$.

(5) The coefficient of the next term (after 1st) $= \dfrac{\text{coefficient of the present term} \times \text{ exponent of } a \text{ in the present term}}{\text{number of the present term, counting from the left}}$

(6) The coefficients of the terms are symmetric about the "middle term" if n is even, and about the two middle terms if n is odd.

(7) The number of each term is the exponent of b plus 1.

The applications of the binomial expansion usually do not require the complete expansion. Often, we will be interested only in the coefficients. There are a number of methods for finding the binomial coefficients. One method applies property (5) of the binomial expansion . Another method applies the formula

Binomial coefficient: $\boxed{{}_nC_r = \dfrac{n!}{(n-r)!r!}}$

where ${}_nC_r$, is the binomial coefficient of the $(r+1)$th term in the expansion of $(a+b)^n$. For the first term, $r = 0$, for the second term, $r = 1$, for the third term, $r = 2$.

In the chapter on combinations and permutations, we will learn that ${}_nC_r$ is the symbol for combinations of n objects taken r at a time.

Using the summation notation, the **binomial theorem** or formula says that

Binomial Theorem: $\boxed{(a+b)^n = \sum_{r=0}^{n} {}_nC_r a^{n-r}b^r}$ $({}_nC_0 = 1)$

$$(a+b)^n = {}^nC_0 a^n + {}^nC_1 a^{n-1}b + {}^nC_2 a^{n-2}b^2 + {}^nC_3 a^{n-3}b^3 + {}^nC_4 a^{n-4}b^4 + ... + {}^nC_n b^n$$

Applications of the Binomial Theorem

In the examples that follow, we will distinguish between the **binomial coefficient**, the **coefficient** of a term, and the **term** itself.

Example 1 Find the binomial coefficient of a term in a binomial expansion given that
$n = 4$ and $r = 3$.

Solution Since $r = 3$, the required coefficient is for the $(3+1)$th or the 4th term

$$_4C_3 = \frac{4!}{(4-3)!3!}$$

$$= \frac{(4)(3)!}{(1)!3!}$$

$$= 4$$

Also, for the third term $r = 2$ (since $r + 1 = 3$)

$$_4C_2 = \frac{4!}{(4-2)!2!}$$

$$= \frac{(4)(3)(2)!}{(2)!2!}$$

$$= 6$$

Example 2 Find (a) the term-number, (b) the binomial coefficient, (c) the coefficient of $x^{10}y^3$,

(d) the term whose literal coefficient is $x^{10}y^3$ in the expansion of $(x^2 + y)^8$

Solution The nth term is given by $_nC_r a^{n-r}b^r$:

Step 1: By comparing $(a + b)^n$ with $(x^2 + y)^8$

$$a = x^2, \ b = y, \ n = 8$$

Step 2: Find r. The exponent on y in $x^{10}y^3$ is the value of r if the exponent on y in $(x^2 + y)^8$ is 1.
Therefore, $r = 3$

(a) The term-number is $r + 1 = 3 + 1 = 4$

(b) Finding the binomial coefficient:

Substituting $r = 3$, $n = 8$ in

$$_nC_r = \frac{n!}{(n-r)!r!}$$

$$_8C_3 = \frac{8!}{(8-3)!3!}$$

$$= 56$$

The binomial coefficient is 56.

(c) Since the coefficient of x^2 is $+1$ and the coefficient of y is $+1$, these two terms do not contribute to the numerical coefficient. Therefore, the coefficient of the term is the same as the binomial coefficient. The coefficient of $x^{10}y^3$ is 56

(d) The term is $56x^{10}y^3$.

Example 3 Find (a) the coefficient of $x^4 y^9$ in the expansion of $(2x^2 - y^3)^5$
(b) What is the binomial coefficient of this term?

Step 1: By comparing $(a + b)^n$ with $(2x^2 - y^3)^5$

$$a = 2x^2, \quad b = -y^3, \quad n = 5$$

Step 2: Find r.

Since the exponent on y in $(2x^2 - y^3)^5$ is **not** 1, to find r , we divide the exponent on y in $x^4 y^9$

by the exponent on y in $b = -y^3$.

Therefore, $r = \frac{9}{3} = 3$ (the term involved is the $(r + 1)$th = 4th term)

Step 3: Now, $r = 3$, $a = 2x^2$, $b = -y^3$, $n = 5$.

The nth term is given by $_nC_r a^{n-r} b^r$. Substituting for r, a ,b, and n, we obtain, the fourth term

$$_5C_3 \cdot (2x^2)^{5-3}(-y^3)^3$$

$$= \frac{5!}{(5-3)!3!} \cdot (2x^2)^2(-y^9)$$

$$= \frac{(5)(4)(3)(2)(1)}{(2)(1)(3)(2)(1)} \cdot 4x^4(-y^9)$$

$$= 10 \cdot 4x^4(-y^9)$$

$$= -40x^4 y^9$$

(a) The coefficient of $x^4 y^9$ is -40. (b) The binomial coefficient is 10 (from $_5C_3$)

Note that in the above problem, the binomial coefficient is 10, but the coefficient of the whole term is -40. This is so, because the coefficient of the x^2-term is 2 and **not** 1. The 2 in the x^2-term and the minus sign of $-y^3$ contributed to the overall coefficient.

Note that the binomial formula for the expansion of $(a + b)^n$ as given assumes that the coefficients of a and b are each 1. Therefore, we must be careful to make the necessary adjustments.

Note also that if a is positive when b is negative the expansion contains plus and minus signs which alternate as + - + - .....

Example 4 Find the coefficient of $x^9 y^5$ in the expansion of $(x - 2y)^{14}$

Step 1: By comparing the coefficients of y in $x^9 y^5$ and $(x - 2y)^{14}$, $r = 5$
With $r = 5$, the required coefficient is that of the sixth term. (Term number $= r + 1$)

Step 2: Now, substitute $a = x$, $b = -2y$, $n = 14$, $r = 5$ in $_nC_r a^{n-r} b^r$ to obtain

$$_{14}C_5 x^{14-5}(-2y)^5$$

$$= _{14}C_5 x^9(-32y^5)$$

$$= \frac{14!}{(14-5)!5!} \cdot x^9(-32y^5)$$

$$= \frac{14!}{9!5!} \cdot x^9(-32y^5)$$

$$= -64064x^9 y^5$$

The coefficient $x^9 y^5$ is -64,064.   (Note that the binomial coefficient is $_{14}C_5 = \frac{14!}{9!5!} = 2,002$)

Example 5 Find the 6th term in the expansion of $(x - 2y)^{14}$

Solution

Step 1: Find r:

The term-number $= r + 1 = 6$, and solving, $r = 5$.

Step 2: Now, $r = 5$, $a = x$, $b = -2y$, $n = 14$,. Substituting in $_nC_r a^{n-r} b^r$, we obtain

$$_{14}C_5 x^{14-5}(-2y)^5$$

$$=_{14}C_5 x^9(-32y^5)$$

$$= \frac{14!}{(14-5)!5!} \bullet x^9(-32y^5)$$

$$= \frac{14!}{9!5!} \bullet x^9(-32y^5)$$

$$= -64064 x^9 y^5$$

The sixth term is $-64064 x^9 y^5$. (Same as in Example 4 above)

Compare the approaches used in Examples 4 and 5.

Another method of finding the binomial coefficients of $(a + b)^n$. is by using a triangular array of numbers known as Pascal's triangle. The rows of the triangular array give the coefficients of the terms of the expansion of $(a + b)^n$ The triangle has the following properties. In each row, the first number is 1 and the last number is also 1. To obtain the next row, add the two numbers diagonally above that number, The three numbers involved form a triangle.

$(a + b)^0$	1
$(a + b)^1 = a + b$	1 1
$(a + b)^2 = a^2 + 2ab + b^2$	1 2 1
$(a + b)^3 = a^3 + 3a^2b + 3ab^2 + b^3$	1 3 3 1
$(a + b)^4 = a^4 + 4a^3b + 6a^2b^2 + 4ab^3 + b^4$	1 4 6 4 1
$(a + b)^5 = a^5 + 5a^4b + 10a^3b^2 + 10a^2b^3 + 10ab^4 + b^5$	1 5 10 10 5 1
$(a + b)^6 = a^6 + 6a^5b + 15a^4b^2 + 20a^3b^3 + 15a^2b^4 + 6ab^5 + b^6$	1 6 15 20 15 6 1

More Applications of the Binomial theorem

Example 1 Find $(1.02)^9$ and round-off to the nearest thousand (i.e. 3 decimal places).

Solution

Note that $1.02 = 1 + .02$

Let $a = 1$, $b = .02$

then $(1 + .02)^9 = 1^9 + 9(1)^8(.02) + 36(1)^7(.02)^2 + 84(1)^6(.02)^3$

$$= 1 + .18 + .0144 + .000672$$

$$= 1.195072$$

$$(1.02)^9 = 1.20$$

Note above that we carried out the expansion to four terms before rounding-off to three decimal places.

Example 2   Find $\sqrt{1.03}$ and round-off to the nearest thousand   4 8 5

Solution

Note that $\sqrt{1.03} = (1.03)^{\frac{1}{2}} = (1+.03)^{\frac{1}{2}}$

$$(1 + .03)^{\frac{1}{2}} = 1^{\frac{1}{2}} + \frac{1}{2}(1)^{\frac{1}{2}-1}(.03) + \frac{\frac{1}{2}(\frac{1}{2}-1)}{2}(1)^{\frac{1}{2}-2}(.03)^2 + \frac{\frac{1}{2}(\frac{1}{2}-1)(\frac{1}{2}-2)}{2(3)}(1)^{\frac{1}{2}-3}(.03)^3$$

$$= 1 + .015 - .0001125 + .00001875$$
$$= 1.014906$$

$$(1.03)^{\frac{1}{2}} = 1.01$$

Lesson 72 Exercises

A   **1.** Expand $(x-3y)^4$

   2. Find the fifth term of $(a+2b)^{12}$.

   3. Find the sixth term in the expansion of $(a+2b)^{12}$.

   4. Find the coefficient of $a^5 b^6$ in the expansion of $(a+b)^{11}$.

   5. Find the coefficient of $x^9 y^5$ in the expansion of $(x+2y)^{14}$

Answers: **1.** $x^4 - 12x^3 y + 54x^2 y^2 - 108xy^3 + 81y^4$;   **2.** $7920a^8 b^4$;   **3.** $25344a^7 b^5$;   **4.** 462;   **5.** 64064.

B Find $\sqrt{1.09}$

Answer::   1.04

CHAPTER 28
Combinations and Permutations

Lesson 73: Fundamental Principle of Counting; Introduction to Permutations and Combinations; Tree Diagrams

Lesson 74: **Permutations; Linear Permutations; Cyclical Permutations**

Lesson 75: **Combinations**

Lesson 73

Fundamental Principle of Counting; Introduction to Permutations and Combinations

The formulas we derive here are closely related to the binomial coefficients. We can in fact derive the binomial coefficients in this chapter. The theory of combinations and permutations deals with making a selection (different choices) of a specified nature from a given collection (set). Fundamentally, the theory of combinations and permutations is used in problems involving probability and in pure mathematics in the advanced theory of equations.

Fundamental principle of counting

If an event A (something) can occur in p ways, and another event B (another thing) can occur in q ways either successively or simultaneously, then the two events can occur together in $p \times q$ ways. Similarly if a third event C can occur in r ways, then the three events can occur in $p \times q \times r$ ways. Generally if events E_1, $E_2,...,E_k$ can occur successively or simultaneously in $w_i, w_2,...,w_k$ ways respectively, then all of the events can occur as specified in $w \times w_2 \times ... \times w_k$ ways.

In dealing with problems of selection, we must distinguish between those problems in which the order of selection (i.e. which element is selected first and which is selected second, and so forth), or arrangement of the elements (things) is important and those problems in which the order of selection is immaterial provided the specified elements are selected. The selection in which order is important and must be considered is a **permutation** of the elements while the selection in which the order of selection does not matter is a **combination** of the elements.

Example

Three letters A, B, C are each written on square card. The three cards are to be placed next to each other in a line on a horizontal table. We consider two cases:

Case 1: The order of placement of the cards is important.
Case 2: The order is not important (does not matter)

Case 1: Permutations

In picking the cards and placing them as specified above, the following are the possibilities:

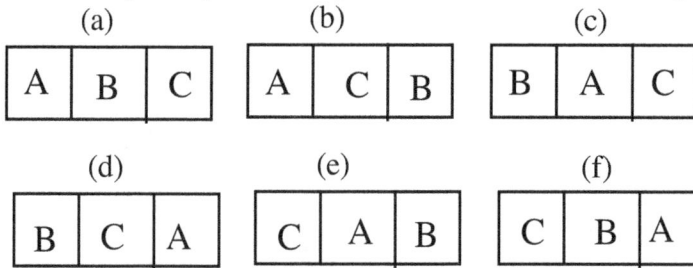

The permutations of 3 letters taking 3 letters at a time.

There are **six** different permutations. We can observe above that there are six permutations of the letters A, B, C. For example, (a) means that the letter A is selected and placed first, followed by the letter B and then the letter C. Similarly, in (b), A is selected and placed first, then C secondly, and thirdly B, and so forth. By trying to pronounce each letter we can readily find the differences especially if each arrangement represented a name.

Case (2): Combination

In this case order is not important and by this, we mean that all we want to do is to place the cards next to each other and that we do not care which letters come first, second or third. In this case, there is only **one combination** of the three letters taking all three at a time. We can add also that so far as combinations are concerned, the arrangements (a), (b), (c), (d), (e), below are all the same.

(a) = (b) = (c) = (d) = (e)

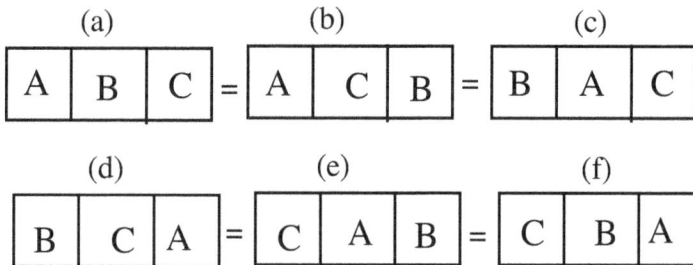

Example 2

Now, as in example (1), suppose we want to place only two of the cards at a time from the three lettered cards.

Case (1) **Permutation** -- (order **important**)

Here are again there are 6 permutations taken 2 letters at a time.

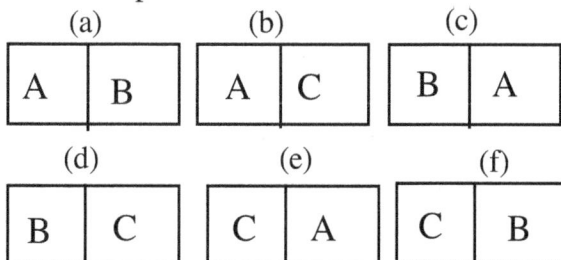

Case (2) **Combination** (order **not** important)   488

There are 3 combinations of 3 letters taken 2 letters at a time.

(a)	(b)	(c)
A B	A C	B C

In the above case of combinations,

A B and B A are the same; B C and C B   are the same; C A   and A C   are the same

Tree Diagram

There is a convenient way of finding the permutations of *n* elements taken *r* at a time. The method is called tree-diagram. A tree diagram consists of branches that show the possible outcomes for two or more experiments. The number of branches at every node equals the number of possible outcomes. The number of experiments equals the number of times the branching operation is performed.

Example
Let us redo Case 1 example using a tree diagram.
Find the permutation of the three letters A, B, and C taking 3 letters at a time
The nodes are at 1, 2, and 3, From node 1, there are 3 possible outcomes, A, or B or C,
From node 2, there are 2 possible outcomes, and from node 3, there is only one possible outcome. since two of the cards have already been used. By counting, there are 6 permutations.

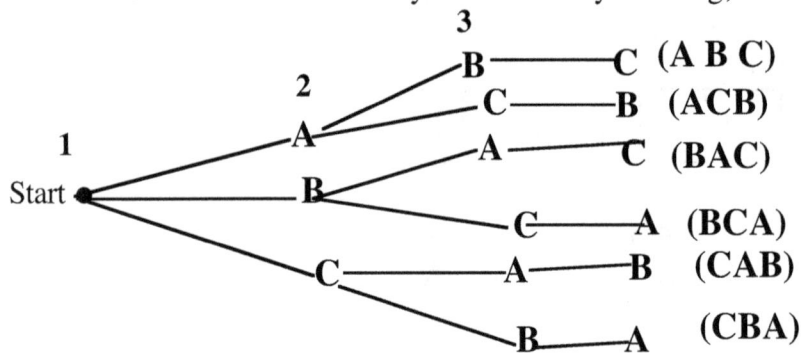

Lesson 73 Exercises

1.. State the fundamental principle of counting.

2. Distinguish between permutations and combinations.. Give an example.

3. What is the number of permutations using the three letters E, F, and G taking 3 letters at a time?

Lesson 74

Permutations: Linear Permutations; Cyclical Permutations

When we want to select from a small number of elements, we can count all the possible arrangements. However, if the number of elements is large, then counting them becomes very tedious and then we resort to simple formulas which we present below.

Let $_nP_r$ represent the permutations of n different things (objects) taken r things at a time. Other notations for $P_{n,r}$ are ^nP_r, P_r^n, $_nP_r$, $P_{(n,r)}$

Example 1 Find the number of linear permutations of 4 objects $a, b, c, d,$ taken 3 at a time.
As in a previous example, we want to arrange them in a line in 3 spaces.

$$\boxed{4}\ \boxed{3}\ \boxed{2}$$

We can fill the first place (box) in 4 ways since we can choose from among 4 objects.
We can fill the second place (box) in 3 ways since we can choose now from 3 remaining objects.
Finally, we can fill the third place (box) in 2 ways since we can choose from among the 2 remaining objects. By the fundamental counting principle. the number of ways of filling the three places (boxes) successively is given by multiplying the numbers of the ways. Thus there are $4 \times 3 \times 2$ ways or 24 ways. Therefore, there are 24 permutations of 4 objects taken 3 at a time.
Symbolically, $_4P_3 = 24$.

From the above example, we can say that we can fill the first place by any of the n objects, the second place by any of the $(n-1)$ objects the third place by any of the $(n-2)$ remaining objects and so forth and then finally the rth place (the last place) by any one of $n - (r-1) = n - r + 1$ objects.

By multiplying the ways of filling the places, the permutations of n objects taken r at a time is
If $_nP_r = n(n-1)(n-2)...(n-r+1)$ \qquad (1) If we denote the permutations of n different objects taken all at a time by $P_{n,n}$, then.

$$_nP_n = n(n-1)(n-2)...(n-r+1)$$
Thus $\qquad = n(n-1)(n-2)...3 \bullet 2 \bullet 1$
$\qquad = n!$
Also, $n(n-1)(n-2)...(n-r+1)(n-r)! = n!$

$$_nP_r = n(n-1)(n-2)...(n-r+1)\frac{(n-r)!}{(n-r)!} \qquad (2)$$

$$\boxed{_nP_r = \frac{n!}{(n-r)!} = P(n,r)} \qquad (3)$$

Equation (3) is for **linear permutation** (straight line arrangement) of different objects taken r at a time
Also, the number of permutations of n objects in a line if r objects must be together is given by

$$\boxed{(n-r+1)!r!}$$

Example 2 Find the number of permutations of 4 objects, taken 3 at a time.

Solution

$$n = 4, r = 3.$$

$$_nP_r = \frac{n!}{(n-r)!}$$

$$_4P_3 = \frac{4!}{(4-3)!}$$

$$= \frac{4!}{1!}$$

$$= \frac{4(3)(2)(1)}{1}$$

$$= 24$$

Cyclical permutations (arrangements in a ring)

The number of cyclical permutations taken all at a time is $(n-1)!$
One can fix one of the "circular places" and then consider the rest as a linear arrangement.

Permutations of n objects, taken all n at a time, some alike (i.e. with repetitions).
The number of permutations of n things with p alike, q alike, taken all at a time, is

$$\frac{n!}{p!q!...}$$

Example: Find the permutations of the letters in the word "Mississippi"

$$n = 11, \quad p = 4, \quad q = 4, \quad r = 2$$

The permutation is $\dfrac{11!}{4!\times 4!\times 2!} = \dfrac{(11)(10)(9)(8)(7)(6)(5)(4)(3)(2)(1)}{(4)(3)(2)(1)\times(4)(3)(2)(1)\times 2(1)}$

$$= 34,650$$

The number of **permutations of n** objects **in a circle** if r objects must sit together is

$$\boxed{(n-r)!r!}$$

Lesson 74 Exercises

1. Find the number of permutations of 6 objects taken 3 objects at a time

2. Find the number of permutations of 6 objects taken all the objects at a time

3. How many three-letter words can be formed from the letters a, b, c, d, using each letter once?

4. Compute: $_6P_6;;.$ 5. Compute: $_6P_5.$ 6. Compute: $_5P_3$.

7. In how many ways can 7 people be seated in a row of 6 seats?

8. In how many ways may 7 people be seated in a row of 4 seats?

9. How many different words of at most 4 letters can be formed from the letters a, b, c, d, e, f?

10. Using the digits $1, 2, 3, 4, 5, 6,,$ how many 3-digit numbers between 200 and 500 can be formed if no repetitions of the digits are allowed?

11. Is $_6P_6 = {_6P_5}$

12. In how many ways can 6 women, 5 men, 4 boys be seated in a row of 15 seats so that the women sit together, the men sit together, and the boys sit together.

Answers: 1. 120; **2.** 720; **3.** 24; **4.** 720; **5.** 720; **6.** 60; **7.** 5040; **8.** 840; **9.** 510; **10.** 60 ; **11.** Yes; **12.** 12,441,600

Lesson 75
Combinations

The combination of n things taken r at a time is denoted by $_nC_r$

The permutation of n things taken r at a time is denoted by $_nP_r$

The number of combinations and the number of permutations are related by

$$_nC_r = \frac{P_{n,r}}{r!}$$

$$_nC_r = \frac{n!}{(n-r)!r!} = C(n,r) \text{ or } \binom{n}{r}$$

Total combination $(x+1)^n = x^n + {}_nC_1 x^{n-1} + {}_nC_2 x^{n-2} + ... + {}_nC_n$

If $x = 1$, we obtain

$$(1+1)^n = 2^n = 1 + {}_nC_1 + {}_nC_2 + ... + {}_nC_n \text{ or}$$

$$2^n = 1 + {}_nC_1 + {}_nC_2 + ... + {}_nC_n$$

$$2^n - 1 = {}_nC_1 + {}_nC_2 + ... + {}_nC_n$$

$$_nC_1 + {}_nC_2 + ... + {}_nC_n = \boxed{2^n - 1} \qquad \textbf{(A)}$$

(A) is a formula for the total number of combinations of n objects taken any number at a time. (that is, taken one or more objects, up to n objects at a time.)

Lesson 75 Exercises

1. Compute: (a) $_6C_6$; (b). $_6C_2$; (c). $_6C_4$; (d). $_6C_1$; (e). $_6C_0$.

2. How many subsets can be formed from a set of 6 elements.

3. How many different triangles can be formed by connecting 6 points, assuming that no three points lie in a line.

4. How many different committees of 3 people can be formed from a group of 8 people?

5. How many sets of five cards can be selected from a deck of 52 cards?

6. In how many ways may one or more people be selected from a group of 6 people?

7. In how many ways can a coach choose a soccer team of 11 from 20 players?

8. Is $_6C_6 = {}_6C_5$

Answers: **1.** (a) 1; (b) 15; (c) 15; (d) 6; (e) 1; **2.** 64; **3.** 20; **4.** 56; **5.** 2,598,960;
6. 63 <-- $2^6 - 1$; **7.** 167,960; **8.** No.

CHAPTER 29

Lesson 76
Mathematical Induction

Mathematical induction deals with the natural numbers (or positive integers).

Axiom: An axiom is a statement in mathematics that we willingly accept or assume to be true so as to deduce other statements.

Deductive reasoning: This is a process of drawing a particular conclusion based on a general statement.

Inductive reasoning: This is a process of drawing a general conclusion based on a number of specific cases. Such conclusions are probable. Also, such a general conclusion can be made to become more probable if we consider more and better choice of cases.

In mathematical induction, we follow a logical procedure which allows us to prove the validity of a general conclusion.

The axiom of mathematical induction

Let P be a set of positive integers (natural numbers). Then

(1)   If the first positive integer, 1, belongs to the set P and

(2)   If the positive integer k belongs to P, then all positive integers belongs to P.

In using the axiom of mathematical induction to construct proofs, we must always verify the above two properties of the axiom to complete the proof. We must thus verify that:

(1) The integer 1 satisfies the statement to be proved and

(2) Assuming that the positive integer k belongs to P , the positive integer $(k + 1)$ satisfies the statement to be proved.

The proofs we shall cover in this chapter will involve only the six basic operations of addition, subtraction, multiplication, division, powers and roots.

Example 1: Prove that the sum of the first n positive even integers is $n(n + 1)$. That is
prove that $2 + 4 + 6 + ... + 2n = n(n + 1)$ (1)

Requirement: We must show that

(a) when $n = 1, 2, ...$ LHS of (1) = RHS of (1) and

(b) when $n = n + 1$ or $k + 1$, LHS of (1) = RHS. of (1)

(LHS and RHS mean Left-Hand Side and Right-Hand Side of the equation, respectively)

Proof

The general term is $2n$, where n = term-number.

Step 1: Verify that when $n = 1, 2$, both LHS and RHS are equal.   ($n = 2$ extra)
$\quad\quad$ ($n = 1$ means considering only the **first** term on the LHS and on the RHS replace n by 1)
$\quad\quad$ Substituting $n = 1$ in (1); that is, in $2n = n(n + 1)$), we obtain

:

$$2(1) \overset{?}{=} 1(1+1)$$
$$2 = 2 \quad \text{True}$$

For $n = 2$, $\underset{n=1}{2\underbrace{(1)}} + \underset{n=2}{2\underbrace{(2)}} \overset{?}{=} \underset{n=2}{2} (\underset{n=2}{2} + 1)$

(**Note** that $n = 2$ means that on the LHS consider the sum of only the first **two** terms and on the right hand side replace n by 2.)

$$2 + 4 \overset{?}{=} 3(3)$$
$$6 = 6 \quad \text{True}$$

Note: We do not have to check for $n = 2$.

Step 2: We prove that when $n = n + 1$, $2 + 4 + 6 + ... + 2n = n(n+1)$ (1)

Let $n = k$ in (1): $2 + 4 + 6 + ... + 2k = k(k+1)$ (2)

$$\boxed{2 + 4 + 6 + ... + 2k = k^2 + k} \quad (3)$$

Also, let $n = (k + 1)$ in (1) i.e. substitute $(k + 1)$ for n in (1).

Then $2 + 4 + 6 + ... + 2(k + 1) = (k+1)(k+1+1)$ (4)

$$\boxed{2 + 4 + 6 + ... + 2(k + 1) = k^2 + 3k + 2} \quad (5)$$

Now, we will show that LHS and RHS of (5) are equal.

Equation (3) is accepted as a true statement since we have shown that this is true when $n = k$
(We want to manipulate equation (3) by some basic algebraic operations so that both sides of the equation are equal.)

In this proof, we add equals to both sides of the equation until the LHS of equation (3) is identical with the LHS of equation (5) and also that the RHS of equation (3) is identical with the RHS of (5).

We write the LHS of equation (5) in another form to indicate explicitly the general term $2k$.

Then $2 + 4 + 6 + ... + 2(k + 1) = 2 + 4 + 6 + ...2k + 2(k + 1)$ (6)

So now replace LHS. of equation (5) by RHS. of equation (6) and equation (5) becomes

$$2 + 4 + 6 + ...+ 2k + 2(k + 1) = k^2 + 3k + 2 \quad (7)$$

So now we prove equation (7) instead of equation (5).

Observe the LHS of equation (3) and the LHS of equation (7), and notice that if we add $2(k+ 1)$ to LHS of equation (3), that side will be identical with LHS. of equation (7), but we should also add $2(k+ 1)$ to RHS of equation (3), that is adding equals to both sides of an equation.

Adding $2(k+ 1)$ to both sides of equation (3), we obtain:

$$2 + 4 + 6 + ... + 2k + 2(k + 1) = k^2 + k + 2(k + 1) \quad (8)$$
$$2 + 4 + 6 + ... + 2k + 2(k + 1) = k^2 + 3k + 2 \quad (9)$$

Now, replacing LHS of equation (9) by LHS of equation (6) (since LHS of (9) = RHS of (6))

$$2 + 4 + 6 + ...+ 2(k + 1) = k^2 + 3k + 2$$

Hence, we have shown that equation (5) is true.

We have therefore shown that equation (1) holds for $n = 1$ and $n = k + 1$,

$$\therefore \ 2 + 4 + 6 + ... + 2n = n(n + 1) \text{ and the proof is complete.}$$

Note: Some of the comments and explanations in the above proof are usually not part of the proof, and may therefore be omitted from the proof. They were added to aid the student understand the principles involved in the proof.

Example 2 Prove that $1 \times 2 + 2 \times 3 + ... + n(n+1) = \dfrac{n(n+1)(n+2)}{3}$ (1)

Proof: For $n = 1$,

       LHS of (1) $= 1(1+1) = 1(2) = 2$

RHS of (1) $= \dfrac{1(1+1)(1+2)}{3} = \dfrac{1(2)(3)}{3} = 2$ (Quantities equal to he same quantity are equal to each other)

Therefore, LHS of (1) = RHS of (1).

Let $n = k$ in (1). Then

$1 \times 2 + 2 \times 3 + ... + k(k+1) = \dfrac{k(k+1)(k+2)}{3}$ true   (2)

Also, let $n = k+1$ in (1): $1 \times 2 + 2 \times 3 + ... + (k+1)(k+2) = \dfrac{(k+1)(k+2)(k+3)}{3}$ (3)

Now, we want to prove that the LHS of (3) = RHS of (3).

We rewrite LHS of (3) so that the general term is explicitly indicated. Then equation (3) becomes

$1 \times 2 + 2 \times 3 + ... + \underbrace{k(K+1)}_{\text{general term}} + (k+1)(k+2) = \dfrac{(k+1)(k+2)(k+3)}{3}$ (4)

Now, instead of proving (3), we prove (4).

Expanding equation (4):

$1 \times 2 + 2 \times 3 + ... + 2k^2 + 4k + 2 \overset{?}{=} \dfrac{k^3 + 6k^2 + 11k + 6}{3}$ true   (5)

Similarly, expanding (2), we obtain

$1 \times 2 + 2 \times 3 + ... + k^2 + k = \dfrac{k^3 + 3k^2 + 2k}{3}$ true   (6)

Since, equation (6) is true (why?) we can add equal quantities to both sides of the equation to obtain an equivalent. It is up to us now to obtain equation (5) from equation (6) by algebraic operations.

Thus, we add $\dfrac{3k^2 + 9k + 6}{3}$ to both sides of (6). (RHS of (5) minus RHS of (6) yields this addend)

Then we obtain $1 \times 2 + 2 \times 3 + \dfrac{3k^2 + 9k + 6}{3} + ... + k^2 + k = \dfrac{3k^2 + 9k + 6}{3} + \dfrac{k^3 + 3k^2 + 2k}{3}$ (7)

Simplifying (7): $1 \times 2 + 2 \times 3 + ... + 2k^{2i} + 4k + 2 = \dfrac{k^3 + 6k^2 + 11k + 6}{3}$ (8)

Comparing equations (5) and (8), we observe that the LHS's are identical and their RHS's are also identical. We have therefore proved (4) and also the original equation (3).

Since we have shown that equation (1) holds for $n = 1$ and $n = k+1$,

$1 \times 2 + 2 \times 3 + ... + n(n+1) = \dfrac{n(n+1)(n+2)}{3}$, and the proof is complete.

Note above: In deciding on adding $\dfrac{3k^2 + 9k + 6}{3}$ to both sides of (6), we asked: "what can be done to the RHS of (6) so that it is the same as the RHS of (5)? " .Practice this technique.

Example 3: Use mathematical induction to prove that : $\dfrac{n+1}{n!} < \dfrac{8}{2^n}$ (1)

Proof

For $n = 1$: $\dfrac{2}{1} \overset{?}{<} \dfrac{8}{2}$

$2 < 4$ True

For $n = 2$: $\dfrac{3}{2} \overset{?}{<} \dfrac{8}{4}$

$\dfrac{3}{2} < 2$ True

Thus, inequality (1) above holds for $n = 1$, and 2.

Now, assume that inequality (1) holds for $n = k$ (where k is an integer)

If it can be shown that inequality (1) holds for $n = k + 1$, the proof would be complete.

Replacing n by $k + 1$ in inequality (1), we obtain

$$\dfrac{k+2}{(k+1)!} \overset{?}{<} \dfrac{8}{2^{k+1}} \qquad (2)$$

Rewriting inequality (2) in a different form: $\dfrac{k+2}{(k+1)!} \overset{?}{<} 2^{2-k}$ (3)

Similarly, rewriting inequality (1): $\dfrac{k+1}{k!} \overset{?}{<} 2^{3-k}$ (1)

Assuming that inequality (1) is true, we would show that inequality (3) is true.

Multiplying inequality (1) by 2^{-1}: $\dfrac{k+1}{2k!} \overset{?}{<} 2^{2-k}$ (4)

We will now show that left-hand side of inequality (3) < left-hand side of inequality (4):

$$\dfrac{k+2}{(k+1)k!} \overset{?}{<} \dfrac{k+1}{2k!} \qquad \text{(Note: } (k+1)! = (k+1)(K+1-1)! = (k+1)k!$$

$$2k + 4 \overset{?}{<} k^2 + 2k + 1 \quad \text{(dividing out the } k! \text{ and undoing the denominators)}$$

$$4 < k^2 + 1 \qquad (5)$$

Clearly, inequality (5) is true for $k \geq 2$.

Now, $\dfrac{k+2}{(k+1)k!} < \dfrac{k+1}{2k!} < 2^{2-k}$

$\therefore$ $\dfrac{k+2}{(k+1)k!} < 2^{2-k}$ and inequality (3) above holds.

Consequently, inequality (1) holds for $n = k + 1$.

It has also been shown that inequality (1) holds for $n = 1$.

We have therefore shown that inequality (1) holds for $n = 1$ and $n = k + 1$.

$$\dfrac{n+1}{n!} < \dfrac{8}{2^n}$$

QED

Lesson 76 Exercises

Prove each of the following using mathematical induction:

1. $1 + 3 + 5 + \ldots + 2n - 1 = n^2$;   **2.** $2 + 4 + 6 + \ldots + 2n = n(n+1)$ **3.** $1 + 2 + 3 \ldots + n = \dfrac{n(n+1)}{2}$;

4. $1^3 + 2^3 + 3^3 + \ldots + n^3 = \dfrac{n^2(n+1)^2}{4}$;   **5.** $4 + 7 + 10 + \ldots + 3n + 1 = \dfrac{n(3n+5)}{2}$;   **6.** : $\dfrac{n+1}{n!} < \dfrac{8}{2^n}$

Extra: Constructing Direct Mathematical Proofs
I want to use basic things to do big things.
The best things in life are free.

In constructing a mathematical proof, we show that a given mathematical statement is true. If the statement is in the "If-then" form, the **hypothesis** is the clause that follows the word "if" and the **conclusion** is the clause that follows the word "then". The hypothesis is what is given, and the conclusion is what is to be proved. Usually, we begin with the hypothesis and proceed to the conclusion by logically combining axioms, definitions, and already proved statements.

There are a number of strategies that can be used to construct a proof.
Strategies
As shown in the diagram below, we can view the construction of a proof as drawing a line between two points A and B, where we can start from A and proceed to B (This is the usual approach.) or start from B and proceed to A; or start from A to a point C between A and B, and then complete the line by starting from B and proceeding to C.

Let **A** represent the **hypothesis** and let **B** represent the **conclusion**.

Approach 1. Begin from A and proceed continuously and logically to B.
Approach 2: Begin from B and proceed continuously and logically to A.
Approach 3: Begin from A and proceed continuously and logically to any point C between A and B, followed by beginning from B and proceeding to C.
Approach 4: Begin from B and proceed continuously and logically to any point C between A and B, followed by beginning from A and proceeding to C. There are other strategies such as start from C.

1. A •————————————→• B
2. A •←———————————• B
3. A •————→←———• B
        C
4. A •——————→←——• B
              C

Finally, check to make sure all the statements flow logically from **A**, the hypothesis to **B**, the conclusion.
Note: From any step to the next step, first determine what you want for the next step, and perform a mathematical operation to get to the next step.

If you have experience in constructing proofs (e.g., proving theorems) in geometry, then you are ready, with some minor adjustments, to construct proofs in calculus and in mathematics in general.
In constructing proofs, we usually operate on expressions, on equations and on inequalities.

Case 1: Performing Operations on Expressions
On an expression, we perform two operations which are inverses of each other so that the resulting expression is equivalent to the original expression.
Always, think of the properties of the basic operations in elementary mathematics such as the **multiplicative property of 1**, and the **additive property of zero.** These properties are applied to expressions.
Multiplicative property of 1: For any real number n, $n \times 1 = 1 \times n = n$

Example 1. Forming equivalent fractions: Given the fraction $\frac{1}{2}$, we can do the

following: $\frac{1 \times 5}{2 \times 5} = \frac{5}{10}$ (We multiplied by $\frac{5}{5}$ which equals 1.

Example 2. Rationalizing the denominator of a radical expression.

Given the radical $\frac{1}{\sqrt{2}}$, we can do the following

$$\frac{1}{\sqrt{2}} \cdot \frac{\sqrt{2}}{\sqrt{2}} = \frac{\sqrt{2}}{2} \quad (\frac{\sqrt{2}}{\sqrt{2}} = 1)$$

Example 3. Dividing a complex number. Divide: $\frac{4}{3-2i}$

$$\frac{4}{3-2i} = \frac{4}{(3-2i)} \cdot \frac{(3+2i)}{(3+2i)} \qquad \left(\text{Note}: \frac{3+2i}{3+2i} = 1\right)$$

$$= \frac{12}{13} + \frac{8}{13}i$$

Additive property of zero: for any real number n, $n + 0 = 0 + n = n$

Example 1 Complete the square: $x^2 - 12x + 8 = 0$

Solution: On the left side of the equation, **add** and **subtract** the square of $\frac{b}{2}$.

$$\boxed{x^2 - 12x + (-6)^2 - (-6)^2 + 8 = 0} \quad (b = -12, \ \frac{b}{2} = -6, \ \left(\frac{b}{2}\right)^2 = (-6)^2)$$

$$(x-6)^2 - (-6)^2 + 8 = 0$$

Adding $(-6)^2$ and subtracting $(-6)^2$ on the left side of the equation is equivalent to adding **zero** to the left side of the equation.

Example 2: Divide $\frac{x}{x+2}$

Solution In the numerator, add and subtract 2.

$$\frac{x+2-2}{x+2} \qquad ((2-2=0) \ \textbf{<--} \text{ Our interest is in this step}$$

$$= \frac{x+2}{x+2} - \frac{2}{x+2} \quad \text{(splitting the numerators)}$$

$$= 1 - \frac{2}{x+2} \quad \text{(Note that we could obtain the same result by long division)}.$$

Case 2: Operating on equations and inequalities

On equations and inequalities, we perform the same operation on both sides of an equation or an inequality.

Example: Completing the square: $x^2 - 12x = -8$

Step 1: Add the square of half the coefficient of the x-term to both sides of the

equation. (i.e., add the square of $\frac{b}{2}$ to both sides of the equation)

$$x^2 - 12x + \left(\frac{-12}{2}\right)^2 = -8 + \left(\frac{-12}{2}\right)^2 \quad (\ b = -12, \ \left(\frac{b}{2}\right)^2 = (-6)^2$$

$$x^2 - 12x + (-6)^2 = -8 + (-6)^2$$

Step 2:   Complete the square on the left-hand side of the equation

$$(x-6)^2 = -8 + 36$$

$$(x-6)^2 = 28$$

Our interest here is in **Step 1**, where we added $(-6)^2$ to both sides of the equation.

Exaggeration

To exaggerate the substitution axiom:

If the cost of 2 books = \$100, then, the cost of $\sin\frac{\pi}{2}$ book **= \$ 50**

(same as the cost of one book = \$50; since $\sin\frac{\pi}{2} = 1$)

We could also say that the cost of $\tan\frac{\pi}{4}$ book = \$50, since $\tan\frac{\pi}{4} = 1.$

Similarly, the cost of x^0 book = \$50.. Also $x^0 = \sin\frac{\pi}{2} = \tan\frac{\pi}{4} = 7 - 6 = \frac{5}{5} = 1., \ (x \neq 0)$

Extra: Similarly, $(\sin^2\theta + \cos^2\theta)$ book costs \$50., since $\sin^2\theta + \cos^2\theta = 1.$

CHAPTER 30

Introduction to Trigonometry, Right Triangle Trigonometry

Lesson 77: **Basic Review for Geometry**

Lesson 78: **Right Triangle Trigonometry**

Lesson 79: **Angles of Elevation and Depression; Bearing; Linear Interpolation**

Introduction to Trigonometry

Trigonometry has applications in the study of physics, astronomy, engineering, medicine and so on. We use trigonometry in surveying, in the study of alternating electrical currents, periodic motion and stretched strings. The sine curve can be used in the study of alternating currents, periodic motion, vibration of stretched strings, mechanical vibrations, and motion of electrons in atoms. The laws of sines and cosines can be used in determining the location of an object relative to a given point. The law of cosines is also used in air traffic control as well as in osteopathy (in Medicine).

To those students who might have been exposed previously to trigonometry and have developed some phobia towards trigonometry, now is the time to undo this phobia. The topics in trigonometry, are **not** more difficult to learn than the topics in algebra that we have previously covered. What we have to do is to "learn the rules of the game very well and play the game".

In this book, we will structure the coverage of trigonometry as follows:

1. Beginning with some basic geometric review, we will cover the trigonometry of the right triangle where we restrict the measures of the angles involved to between $0°$ and $90°$. (Chapter..)

2. We will then cover the trigonometry of any angle, where the measures of the angles involved may be positive or negative and of any measure. (Chapter..)

3. In Chapter.., we will extend the trigonometry of any angle to the trigonometry of real numbers by letting the radian measure, x , of any angle equal to a real number x.

Lesson 77
Basic Review for Geometry

Triangles

A triangle is a closed geometric figure bounded by three straight line segments.
The point at which any two sides of the triangle meet is called a vertex. A triangle has three sides and three vertices. We shall call the three sides and the three angles the six parts of the triangle.
We may name a triangle by using three capital letters in any order, where the letters represent the vertices.

Example Triangle ABC is symbolized $\triangle ABC$ (Fig.10)

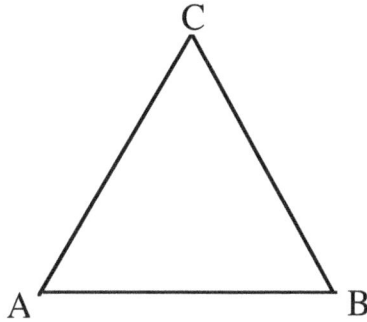

Fig.10: The points A, B, and C are the vertices of $\triangle ABC$.

Standard notation for labeling the sides of a triangle

Consider triangle ABC , below, in which the capital (or upper case) letters represent the vertices. By agreement, we shall label the sides opposite to these angles with small letters (lower case letters). These lower case letters represent the lengths of the sides of the triangle: c is the length of side opposite $\angle C, b$ is the length of side opposite $\angle B$ and a is the length of the side opposite $\angle A$. We shall call this notation the **standard notation** for labeling a triangle.

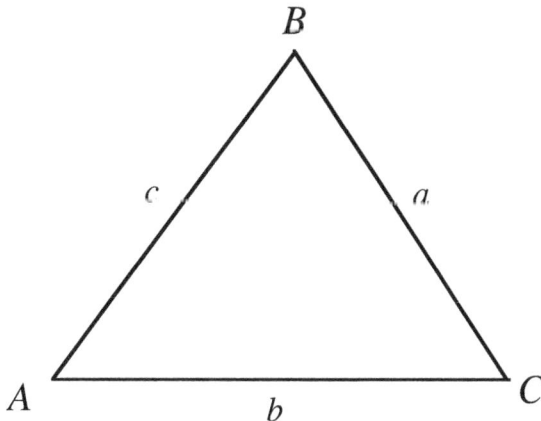

Classification of Triangles

A triangle may be classified either by its sides or by its angles.

Classification by Angles

By its angles a triangle may be classified as being acute, obtuse or right.

Acute triangle: A triangle in which all the angles are acute. The measure of each angle is between $0°$ and $90°$.

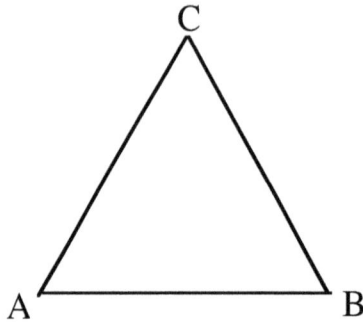

Obtuse triangle: A triangle in which there is an obtuse angle. **Note** that a triangle cannot have more than one obtuse angle. The measure of one of the angles is between $90°$ and $180°$.

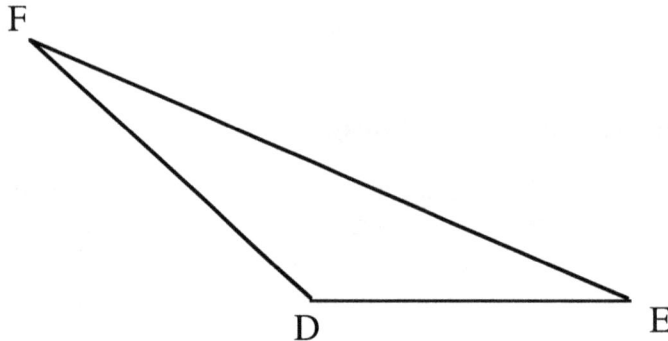

Right triangle (or right-angled triangle): A triangle which has one right angle ($90°$). **Note** that a triangle cannot have more than one right angle.

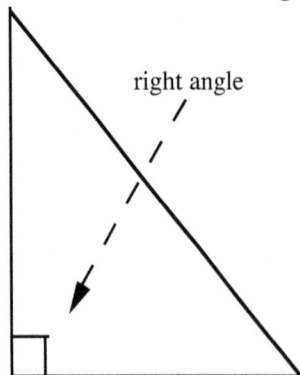

right angle

Classification by sides

We can also classify a triangle by sides. By its sides, a triangle may be classified as being scalene, isosceles or equilateral.

Equilateral triangle: A triangle in which all three sides are congruent.
(The lengths of all three sides are equal)

All the interior angles of an equilateral triangle are congruent.
(Each angle has a measure of 60°).

Isosceles triangle: A triangle in which two sides are congruent. In an isosceles triangle, the   angles opposite to the congruent sides are congruent.

Scalene triangle

This is a triangle in which the lengths of all the three sides are different from one another (Fig.11).

How to determine if a triangle is acute, obtuse, or right, knowing the lengths of the sides

Let the longest side of a triangle be of length c units, and let the lengths of the other two sides be of lengths a units and b units.

Then, (1) if $c^2 = a^2 + b^2$ (the Pythagorean Theorem) then the triangle is a right triangle
Example: A triangle whose sides are of lengths $5, 4$, and 3.

(2) if $c^2 > a^2 + b^2$, then the triangle is obtuse.
Example: A triangle whose sides are of lengths $8, 6$, and 5.

(3) if $c^2 < a^2 + b^2$ then the triangle is acute.
Example: A triangle whose sides are of lengths $6, 5$, and 4.

Theorem: The sum of the measures of the interior angles of a triangle is 180°.
In the figure below, $m \angle A + m \angle B + m \angle C = 180°$

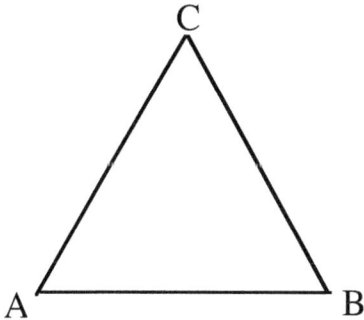

Right Triangle

A **right triangle** (Fig.18) has two acute angles and one right angle. In Fig.18, the right angle is at the vertex C. The side opposite to $\angle A$. is $\overline{BC}$; and the side adjacent to angle $\angle A$. is $\overline{AC}$. $\overline{AB}$ is the side opposite to the right angle. The side opposite the right angle is called the hypotenuse.

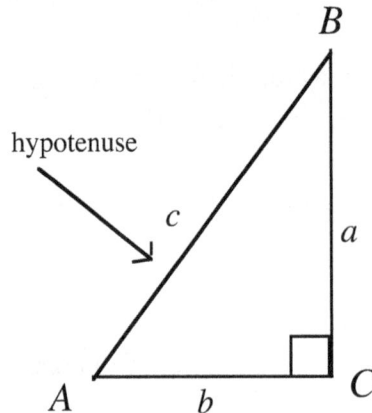

Similarly, if we refer to $\angle B$, $\overline{AC}$ is the side opposite $\angle B$ (or simply the opposite side); $\overline{BC}$ is then the adjacent side but $\overline{AB}$ is still the hypotenuse. Note that the hypotenuse remains unchanged whether we refer to angle A or angle B but the adjacent and opposite sides change depending upon which angle we are referring to.

The Pythagorean Theorem

The Pythagorean theorem states that in a right triangle, the square of the length of the hypotenuse equals the sum of the squares of the lengths of the other two sides.

Symbolically, $c^2 = a^2 + b^2$ (using standard notation)

Useful relationships concerning the relative measures of the sides and angles of a triangle

1. In any triangle, the **sum** of the **lengths** of any **two sides** is larger than the length of the **third** side.

2. In any triangle, the **longest side** is opposite to the **largest angle**, and the **shortest side** is opposite to the **smallest angle**.

3. In an **acute triangle, the** square of the length of **longest side is smaller than** the sum of the squares of the lengths of the other two sides.

4. In an **obtuse triangle**, the **square** of the length of the **longest side is larger than** the sum of the **squares** of the lengths of the other two sides.

5. In a **right triangle**, the **square** of the length of the **longest side** (the hypotenuse) **is equal** to the sum of the **squares** of the lengths of the other two sides. **(The Pythagorean theorem)**

Lesson 77 Exercises

A   Define the following:
   1. Triangle; 2. Acute triangle, 3. Obtuse triangle,;
   4. Right triangle,; 5. Equilateral triangle,
   6. Isosceles triangle, 7. scalene triangle,

B. State the Pythagorean Theorem.

Lesson 78

Right Triangle Trigonometry

Units of measuring angles in Trigonometry

As it is in the case of geometry, we may measure angles in degrees. However, in addition, we may also measure angles in another unit called the "radian". In calculus and scientific work, we will usually measure angles in radians.

Trigonometric Functions (Trigonometric ratios)

These ratios are sine (**sin**); cosine (**cos**); tangent (**tan**); cosecant (**csc**); secant (**sec**); and cotangent (**cot**).

We can derive these six ratios by comparing two similar right triangles. Each ratio is independent of the lengths of the sides involved. Each ratio is not a function of the lengths of the sides involved. However, each ratio of corresponding sides is a function of the angles involved. As the angle we refer to changes, the ratios of corresponding sides change and hence we call these relations trigonometric functions. We summarize the trigonometric functions below,.

Definitions: The following definitions are applicable to **right** triangles only.

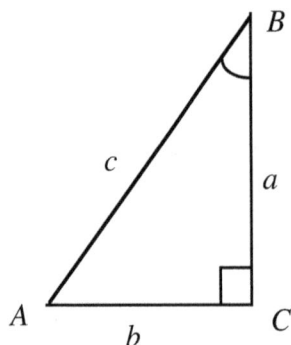

1. $\sin A = \dfrac{\text{opposite side}}{\text{hypotenuse}} = \dfrac{a}{c}$ (sin A is abbreviation for the sine of A)

2. $\cos A = \dfrac{\text{adjacent side}}{\text{hypotenuse}} = \dfrac{b}{c}$ (cos A is abbreviation for the cosine of A)

3. $\tan A = \dfrac{\text{opposite side}}{\text{adjacent side}} = \dfrac{a}{b}$ (tan A is abbreviation for the tangent of A)

4. $\csc A = \dfrac{1}{\sin A} = \dfrac{\text{hypotenuse}}{\text{opposite side}} = \dfrac{c}{a}$ (csc A is abbreviation for the cosecant of A)

5. $\sec A = \dfrac{1}{\cos A} = \dfrac{\text{hypotenuse}}{\text{adjacent side}} = \dfrac{c}{b}$ (sec A is abbreviation for the secant of A)

6. $\cot A = \dfrac{1}{\tan A} = \dfrac{\text{adjacent side}}{\text{opposite side}} = \dfrac{b}{a}$ (cot A is abbreviation for the cotangent of A)

If instead of referring to $\angle A$, we refer to $\angle B$, then we have the following

1. $\cos B = \dfrac{a}{c}$; $\sin A = \dfrac{a}{c}$

2. $\sin B = \dfrac{b}{c}$; $\cos A = \dfrac{b}{c}$ (Note also that $\cos B = \dfrac{a}{c} = \sin A = \dfrac{a}{c}$)

3. $\cot B = \dfrac{a}{b}$; $\tan A = \dfrac{a}{b}$.

Note above that for $\angle B$, the opposite side is $\overline{AC}$, and the adjacent side is $\overline{BC}$. For $\angle A$, the opposite side is $\overline{BC}$ and the adjacent side is $\overline{AC}$

As we have already observed, certain functions of $\angle A$, are equal to certain functions of $\angle B$, We may deduce from the above equations that:

$$\sin A = \cos B$$

$$\sin B = \cos A$$

$$\tan B = \cot A$$

(Note that $\angle A$ and $\angle B$ are complementary angles. See also page 510-511)

To summarize for all the six functions, we have

1. $\sin A = \dfrac{a}{c}$ $\quad = \quad$ $\cos B = \dfrac{a}{c}$

2. $\cos A = \dfrac{b}{c}$ $\quad = \quad$ $\sin B = \dfrac{b}{c}$

3. $\tan A = \dfrac{a}{b}$ $\quad = \quad$ $\cot B = \dfrac{a}{b}$

4. $\csc A = \dfrac{c}{a}$ $\quad = \quad$ $\sec B = \dfrac{c}{a}$

5. $\sec A = \dfrac{c}{b}$ $\quad = \quad$ $\csc B = \dfrac{c}{b}$

6. $\cot A = \dfrac{b}{a}$ $\quad = \quad$ $\tan B = \dfrac{b}{a}$

A memory **device** for remembering the reciprocal relationships for sine and cosine is as follows. Usually, students remember that the secant and the cosecant are reciprocals of some functions. The recall problem is: Is the secant for the sine or for the cosine, and is the cosecant for the sine or for the cosine?

You might think that because each of **s**ine and **s**ecant begins with the letter " **s** " the reciprocal of the sine function (sin) is the secant function (sec) , but this is **not** so, they are **not** reciprocals of each other. Similarly, you might also think that because each of **c**osine and **c**osecant begins with the letter " **c** " the reciprocal of the cosine function (cos) is the cosecant function (csc) , but this also is **not** so, they are **not** reciprocals of each other. Therefore, note that, the name of each given function and its reciprocal do not rhyme with each other. As such, always think the opposite: The " reciprocal of sine is **c**osecant " and the " reciprocal of **c**osine is **s**ecant". It is rather unfortunate that mathematicians did not choose the names to facilitate recall.. Perhaps, for recall purposes, we can use" S-C " for sine and cosecant; and " C-S "for cosine and secant.

Important: Remember that all the above trigonometric definitions for the sine, cosine, tangent, cosecant, secant and cotangent apply only to a triangle which has a right angle. The ratios do **not** apply to a triangle which does not have a right angle.

How to use the electronic calculator as well as trigonometric tables

We use trigonometric tables or the trigonometric functions on an electronic calculator for two main tasks. One task is to find the trigonometric functional value for an angle when the measure of the angle is known (or given). The other task (the opposite task) is to find the measure of an angle when given the functional value of the angle.

Many scientific calculators have built-in trigonometric functions. To use the calculator, we push or depress the appropriate buttons (or keys).

Given the measure of an angle, how to find the value of the trigonometric 506 function

Most calculators have buttons for the sine, the cosine and the tangent functions. Recall that we measure the angles either in degrees or in radians. As such, we may have to determine which units we are using in operating the calculator. We do this by pushing the button for degree measure (degree mode) or radian measure (radian mode) according to which unit of measurement we want to use. However, some old calculators may not have buttons (keys) for degrees and radians. In this case, we would want to change from one unit using the relation $\boxed{180° = \pi \text{ radians}}$.

Example 1 Find $\sin 60°$
Solution

Step 1: Put the calculator in degree mode by pushing the button for degrees. Some calculators may use the same button for both degrees and radians (R/D). In this case, one push of   button may indicate radian and a second push of button may indicate degrees and vice-   versa. In some calculators a red dot (depending on color of display panel) on the label   "radian" on the display panel indicates radians measure and no dot is understood to mean   degree measure, however, some calculators can display both the degrees and the radians on   the panel.

Step 2: Press the buttons in the following order from left to right.
Thus, press (the button) 6, then 0, and finally press sin.
and then read 0.8660254.

$$\boxed{6} \quad \boxed{0} \quad \boxed{\sin}$$

Then $\sin 60° = 0.8660254$

Note: On some calculators, the order of pressing the buttons may be reversed. For the above example, we press "sin" followed by 6 and 0. This order is sometimes labeled DAL (Direct Algebraic Logic). Therefore, experiment with the order. Before using a calculator on an exam, practice (and review) how to use your calculator, since this will save you time on the exam. Even, if you knew how to use your calculator, in the past, it is good practice to review its usage before the exam.

Example 2 Find $\cos 60°$.
Solution

Step 1: Put calculator in degree mode.
Step 2: Press the buttons for 6, 0, and then cos and read 0.5.

$$\boxed{6} \quad \boxed{0} \quad \boxed{\cos}$$

Thus, $\cos 60° = 0.5$.

Note: For calculators with the label "**D.A.L.**" (Direct Algebraic Logic), the sequence for example, for finding $\cos 60°$ is: Press the button for **cos** first, followed by **60°**
Thus:

$$\boxed{\cos} \quad \boxed{6} \quad \boxed{0}$$

Example 3   Find sec $60°$.

Solution

sec key on calculator

If your calculator has key for the sec function, then press $6, 0$, then sec to obtain sec $60° = 2$.

No sec key on calculator

If your calculator has no key (button) for the sec function, then recalling that sec $60° = \dfrac{1}{\cos 60°}$, we can find $\cos 60°$, and then find the reciprocal of $\cos 60°$.

Step 1: Put calculator in degree mode.

Step 2: Press the buttons in the following order $6, 0$, and cos.
          This gives you 0.5

Step 3: Press on the button for $\dfrac{1}{x}$ to obtain 2.

          Thus sec $60° = 2$.

Similarly for csc $60°$ we can use the sin and $\dfrac{1}{x}$ buttons; for cot $60°$, we can use the tan and $\dfrac{1}{x}$ keys.

Example 4   Find sin $390°$

Solution

          Press the buttons in the following order $3, 9, 0$, and then sin, to obtain
          sin $390° = 0.5$.

Examples using radian measure

Example 5 Find sin 1.5

Solution

          Since no unit is indicated, the unit is understood to be the radian unit.

Step 1: Put calculator in radian mode (Follow operating instructions of calculator.).
Step 2: Push the button in the following order: 1.5, sin and read 0.997495.
          Thus, sin $1.5 = 0.997495$

Example 6   Find sec 1.5

Step 1: Put calculator in radian mode (Since 1.5 is understood to be in radians)
Step 2: Press 1.5, then cos.

Step 3: Then push the $\dfrac{1}{x}$ (Recall sec $1.5 = \dfrac{1}{\cos 1.5}$ and read 14.1368329.
          Therefore sec $1.5 = 14.1368329$. If you have the sec key, then use it directly.

If we are given the radian measure and we want to operate in degree mode, we may change the radian measure to degree measure by multiplying by $\dfrac{180°}{\pi}$. On the other hand, given the degree measure, to change to radian measure, we multiply the degree measure by $\dfrac{\pi}{180°}$. However, if the calculator has an automatic conversion button then use it.

Finding the measure of an angle given the value of the trigonometric function, using a calculator

Here, we shall use the following keys if available: the buttons $\sin^{-1}$, $\cos^{-1}$, $\tan^{-1}$. If these are not available, you may have to use a combination of the keys: "inv or arc together with sin, cos and tan.

Example 7   If sin = .97, find θ in degrees.
Step 1: Since we want θ to be in degrees we put the calculator in degree mode.

Step 2: Press .97.
Step 3: Push the "2nd" button, then the $\sin^{-1}$ button and read 75.93013°

$$\boxed{\;.\;}\quad\boxed{9}\quad\boxed{7}\quad\text{then}\;\boxed{\sin^{-1}}$$

Therefore, $\theta = 75.93013$.   (to the nearest tenth, $\theta \approx 75.9$)

In step 3, if your the calculator does not have $\sin^{-1}$, $\cos^{-1}$, $\tan^{-1}$ keys, press inv (inverse) or arc, then sin as:

$$\boxed{\;.\;}\quad\boxed{9}\quad\boxed{7}\quad\boxed{\text{inv}}\;\text{or}\;\boxed{\text{arc}}\quad\boxed{\text{sin}}$$

How to find the values of trigonometric functions by using trigonometric tables

On occasion, we may not have access to a calculator. When this is the case, then we may resort to using trigonometric tables. Trigonometric tables are usually in the back pages of most books. In most representative trigonometric tables, the trigonometric functions are for angles from 0° to 90° for degree measures and from 0 to 1.57 for radian measures. The angles and the corresponding functional values are not listed continuously in one column from 0° to 90° (0 to 1.57 radian) but the readings for the angles between 0° and 45° (0 to .7854 radians) are in the left-hand column while the readings for the angles between 45° and 90° (.7854 and 1.57) are located in the right-hand column. Note this discontinuous arrangement as it will save you time and frustration. Some books, however, may have separate tables for the angles measured in radians and in this case, the angles may be listed.

Partial Table of values for Trigonometric Functions

↓ Angle θ									
Degrees	Radians	Sin	Cos	Tan	Csc	Sec	Cot		
…………	…………	…………	…………	…………	…………	…………	…………	…………	…………
29° 00'	.5061	.4848	.8746	.5543	2.063	1.143	1.804	1.0647	61° 00'
29° 10'	.5091	.4874	.8732	.5581	2.052	1.145	1.792	1.0617	60° 50'
29° 20'	.5120	.4899	.8718	.5619	2.041	1.147	1.780	1.0588	60° 40'
29° 30'	.5149	.4924	.8704	.5658	2.031	1.149	1.767	1.0559	60° 30'
29° 40'	.5178	.4950	.8689	.5696	2.020	1.151	1.756	1.0530	60° 20'
29° 50'	.5207	.4975	.8675	.5735	2.010	1.153	1.744	1.0501	60° 10'
--------	…………	…………	…………	…………	…………	…………	…………	…………	-------
		Cos	Sin	Cot	Sec	Csc	Tan	Radians	Degrees
								Angle θ ↑	

Example 8 Find the functional value for angles between **0° and 45°**.(table on p.508)

    (a)   sin 29° 50'   (b) cos 29° 50'

    (c)   tan 29° 50'   (d) Sec 29° 50'

(a) Step 1: From the table p.508, we locate the angle 29° 50' in the left column, headed "Degrees". (Since it is between 0° and 45°).

    Step 2: We read across this row to the right until this row meets (or intersects) the column for sin (sin being at the top of the table)

    The value at the intersection of this row and the column for the sin function is the value we are looking for and we would read .4975. Therefore, sin 29° 50' = 0.4975.

(b)   For cos 29° 50', we read .8675 = 0.8675

(c)   For tan 29° 50', we read .5735 = 0.5735

(d)   For sec 29° 50' we read 1.153.

    We must note that if the angle whose functional value we want to find is more than 90° , then we will find the reference angle first and then the functional value as on pages 528-530.

Example 9 Angles between 45° and 90°

    Find the functional value for cos 61° 00' .

    Reading down the right-hand column.

Step 1: We locate the angle 61° 00' in the right-hand (right-most) column.

Step 2: We then read across the row to the left until we intersect the column with the heading $\cos\theta$ (or cos x). The intersection of the row for 61° 00' and the column for $\cos\theta$ (at the bottom of the table) gives the desired functional value. In this case, note the headings cos x, sin x , tan x are at the **bottom** of the table.

Step 3: We read .4848. Therefore cos 61° 00' = 0.4848

    Similarly we shall obtain the following
        sin 61° 00' = 0.8746
        csc 61° 00' = 1.143.

Given the functional values we can also use the trigonometric tables to find the corresponding angles. In this case, we will locate the functional value first ,and then move across the table row-wise either to the left or right to read the corresponding angles. Note that we are reversing the steps in this case.

Complementary Angles and Cofunctions

Complementary angles : If the sum of the measures of two angles is 90°, then the two angles are complementary. The complement of a 40° angle is a 50° angle.

$\angle 1$ and $\angle 2$ are complementary angles.(i.e., m $\angle 1$ + m $\angle 2 = 90°$)

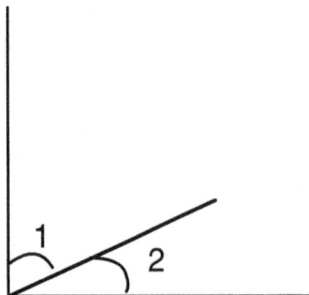

Figure 15

Note that it is **not** necessary that the angles have the same vertex in order to be complementary.

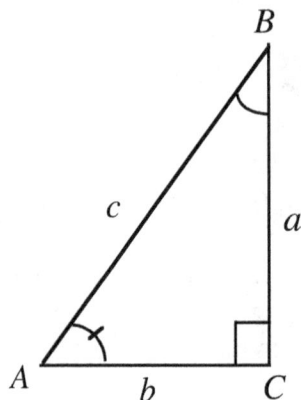

We know from geometry that the sum of the measures of $\angle A$, $\angle B$, and $\angle C$ is 180°. Since $m \angle C = 90°$,

$$m \angle A + m \angle B = 90° \dots\dots\dots (7)$$

Equation (7) shows that angle A and angle B are complementary (i.e. their sum = 90°).

From equation (7) $m\angle A = 90° - m\angle B$, and

$$m\angle B = 90° - m\angle A$$

We know that $\sin A = \cos B$ (since $\sin A = \frac{a}{c}$, and $\cos B = \frac{a}{c}$.)

Therefore, if we replace $m\angle A$ by $90° - m\angle B$

$$\sin(90° - B) = \cos B \dots\dots\dots\dots(8)$$

Similarly, $\sin(90° - A) = \cos A$

$$\cot(90° - A) = \tan A$$

We summarize the above relations which are functions below:

Cofunctions (Complementary functions)

$$\left.\begin{array}{l} \sin A = \cos (90° - A) \\ \cos A = \sin (90° - A) \end{array}\right\} \leftarrow \text{The sine and the \textbf{co}sine are \textbf{co}functions.}$$

$$\left.\begin{array}{l} \tan A = \cot (90° - A) \\ \cot A = \tan (90° - A) \end{array}\right\} \leftarrow \text{The tangent and the \textbf{co}tangent are cofuntions.}$$

$$\left.\begin{array}{l} \sec A = \csc (90° - A) \\ \csc A = \sec (90° - A) \end{array}\right\} \leftarrow \text{The secant and the \textbf{co}secant are cofuntions.}$$

We call the above relations cofunctions and we also say that the **cofunctions** of **complementary angles are equal in value.**

Example 10 Given $\triangle ABC$ in which m $\angle C = 90°$ and m $\angle A = 60°$, we derive the following:

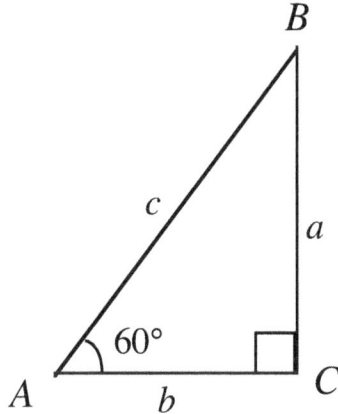

$$\sin 60° = \cos(90° - 60°)$$
$$\sin 60° = \cos 30° \quad (\text{Note}: \; \sin 60° = .866; \; \cos 30° = .866, \text{ from calculator or tables})$$
$$\sin 30° = \cos(90° - 30°)$$
$$\sin 30° = \cos 60° = .5$$
$$\tan 30° = \cot(90° - 30°)$$
$$\tan 30° = \cot 60° = .58$$
$$\sin 10° = \cos 80°.$$

If θ is in radians:

$$60° = \frac{\pi}{3}; \quad 90° = \frac{\pi}{2}, \text{ and}$$

$$\sin \frac{\pi}{3} = \cos\left(\frac{\pi}{2} - \frac{\pi}{3}\right) \qquad \text{Note}: \; \frac{\pi}{2} - \frac{\pi}{3} = \frac{\pi}{6}$$

$$\sin \frac{\pi}{3} = \cos \frac{\pi}{6} \qquad (\text{equivalent to } \sin 60° = \cos 30°)$$

Similarly, $\sin \frac{\pi}{6} = \cos \frac{\pi}{3}$ (equivalent to $\sin 30° = \cos 60°$)

Special triangles: The 30º-60º-90º triangle and the 45º-45º-90º triangle   512

The trigonometric functions for $30°, 45°, 60°$, and $90°$ occur very frequently in mathematics that students are usually required to memorize them.

The 30º-60º-90º triangle

For the $30°-60°-90°$ triangle (Figure 2), the dimensions are obtained by considering an equilateral triangle whose sides are each 2 units long. From geometry, the length of the hypotenuse is twice the length of the shortest side. The third side is calculated by the Pythagorean theorem.

Thus if $AB = 2$, then $AC = 1$ and $2^2 = 1^2 + BC^2$ and from which $BC = \sqrt{3}$. Usually, after having gone through the derivation, using geometric considerations and the Pythagorean theorem, students become familiar with the dimensions $1, 2, \sqrt{3}$. The problem of recall that remains then is how to place these dimensions on the $30°-60°-90°$ triangle, if one does not draw the equilateral triangle, but draws only Δ ABC. The following mnemonic device will be helpful:

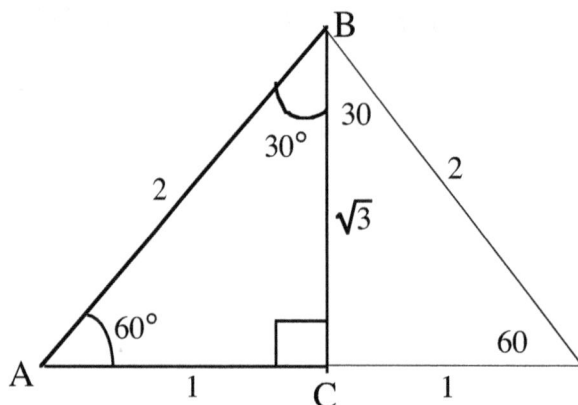

Figure 2

Mnemonic device: If you can remember the numbers $1, 2, \sqrt{3}$ as being the lengths of the sides of this triangle, and also remember that in any triangle, the longest side is opposite the largest angle, and that the shortest side is opposite the smallest angle, you should be able to place these dimensions accordingly on the triangle. Thus, 1 is opposite the $30°$ angle, 2 is opposite the $90°$ angle (the right angle) , and $\sqrt{3}$ is opposite the $60°$ angle. $(\sqrt{3} \approx 1.73)$

By applying the mnemonic devices, SOH , CAH, TOA we can write down some functional values.
Examples

$$\sin 30° = \frac{\text{opposite side}}{\text{hypotenuse}} = \frac{1}{2}$$

$$\cos 30° = \frac{\text{adjacent side}}{\text{hypotenuse}} = \frac{\sqrt{3}}{2}$$

$$\sin 60° = \frac{\text{opposite side}}{\text{hypotenuse}} = \frac{\sqrt{3}}{2}$$

$$\tan 60° = \frac{\text{opposite}}{\text{adjacent side}} = \frac{\sqrt{3}}{1} = \sqrt{3} .$$

The 45º-45º-90º triangle

By sketching an isosceles right triangle with measures of 45º, 45º, and 90º (Figure 1) we will be able, using the basic definitions, write down the trigonometric functional value for 45º.

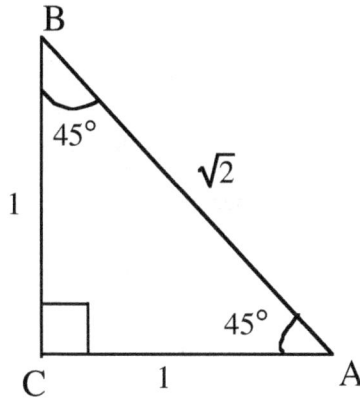

Figure 1

In an isosceles triangle, the sides opposite the congruent angles have the same length. If we let the length of the sides opposite the congruent angles be 1 unit each, we can calculate the length of the hypotenuse using the Pythagorean theorem.

$$AB^2 = AC^2 + BC^2$$
$$AB^2 = 1^2 + 1^2$$
$$AB^2 = 2$$
$$AB = \sqrt{2}.$$

Example 11 $\sin 45º = \dfrac{\text{opposite side}}{\text{hypotenuse}} = \dfrac{1}{\sqrt{2}} = \dfrac{1}{\sqrt{2}} \cdot \dfrac{\sqrt{2}}{\sqrt{2}} = \dfrac{\sqrt{2}}{2}$ (Rationalizing the denominator)

$\cos 45º = \dfrac{\text{adjacent side}}{\text{hypotenuse}} = \dfrac{1}{\sqrt{2}} = \dfrac{1}{\sqrt{2}} \cdot \dfrac{\sqrt{2}}{\sqrt{2}} = \dfrac{\sqrt{2}}{2}$

$\tan 45º = \dfrac{\text{opposite side}}{\text{adjacent side}} \dfrac{1}{1} = 1.$

Solutions of Right Triangles

514

In any triangle, there are six parts, namely, three sides and three angles. When we want to "solve a triangle" what we want to do is to find the measures of all the unknown parts. These parts may be sides or angles or a combination of angles and sides. Usually, the values of three parts are given or known and we are to find the other three parts.

How to determine which function to use in solving right triangles

The determination of which trigonometric relationship to use depends on what we are given and what we want to find. A good practice is to scribble down the following three relationships:

$$\sin \theta = \frac{\text{opposite side}}{\text{hypotenuse}} \qquad \left(\frac{\textbf{O}\text{pposite}}{\textbf{H}\text{ypotenuse}}\right. : \textbf{"SOH"} \text{ <-------- mnemonic device)}$$

$$\cos \theta = \frac{\text{adjacent. side}}{\text{hypotenuse.}} \qquad \left(\frac{\textbf{A}\text{djacent}}{\textbf{H}\text{ypotenuse}}\right. : \textbf{"CAH"} \text{ <-------- mnemonic device)}$$

$$\tan \theta = \frac{\text{opposite side}}{\text{adjacent side}} \qquad \left(\frac{\textbf{O}\text{pposite}}{\textbf{A}\text{djacent}}\right. : \textbf{"TOA"} \text{ <-------- mnemonic device)}$$

and use the relationship in which we know all but one quantity. Sometimes, we may use two relationships simultaneously to find two unknowns.

Example 1 In right triangle ABC, below, find the length of BC by using the appropriate trigonometric function.

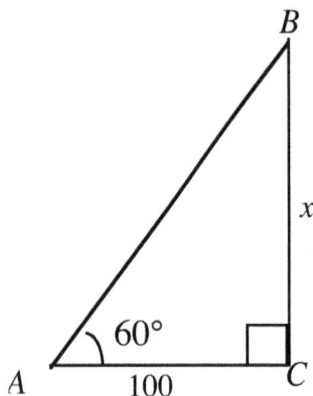

Step 1: Find out which parts are given. We are given m $\angle A = 60°$, length of $AC = 100$ units.

Step 2: Note the part that we want to find. In this problem it is the side $\overline{BC}$ which we represent by x

Step 3: Determine how the side we want to find is related to the given (known) angle, and also how the given side, $\overline{AC}$, is related to the given angle (m $\angle A = 60°$), by using any of the mnemonic devices: **SOH, CAH, TOA.** $\overline{BC}$ is opposite to $\angle A$ and AC is adjacent to $\angle A$. Thus, opposite side (say ,"O") and adjacent side (say, "A") are involved and we have " OA".

Now, "**OA**" is contained only in the device **TOA** which translates to: Tangent $= \dfrac{\textbf{O}\text{pposite}}{\textbf{A}\text{djacent}}$

Hence, the trigonometric function to use in finding BC is that of the tangent. Therefore,

$$\tan A = \frac{x}{100} \qquad \left(\frac{\text{opposite side}}{\text{adjacent side}}\right)$$

$$\tan 60° = \frac{x}{100}$$

Solving for x, we obtain $x = 100 \tan 60°$

From tables, calculator, or from memory tan 60° = 1.73 approximately.

$$x = 100(1.73)$$

$$= 173 \text{ units.}$$

Therefore the length of BC is 173 units..

In the last example, if what we were **given** and what we were to **find** involved the Opposite side and the Hypotenuse, with reference to the given angle, we would then have "OH" and we would have used the relationship

$$\sin A = \frac{BC}{AB} \quad (\textbf{O}pposite \ side \ and \ \textbf{H}ypotenuse \rightarrow \textbf{OH} \rightarrow \textbf{SOH} \rightarrow \sin = \frac{O}{H})$$

Similarly, if the Adjacent side and the Hypotenuse were involved, then we would have "AH" and we would have used the cosine relationship

$$\cos A = \frac{AC}{AB} \quad (\textbf{A}djacent \ side \ and \ \textbf{H}ypotenuse \rightarrow \textbf{AH} \rightarrow \textbf{CAH} \rightarrow \cos = \frac{A}{H})$$

Example 2 Solve triangle ABC.

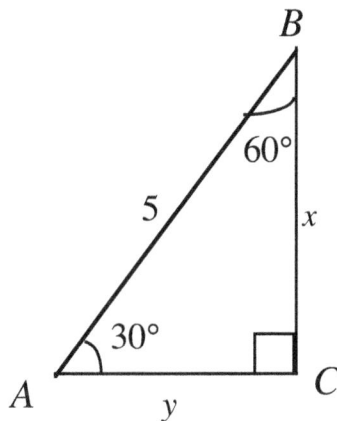

Solution There are only two parts that are unknown, and therefore we have to find only two parts namely, the sides $\overline{AC}$ and $\overline{BC}$.

Step 1: Finding x

The approach here is to find a ratio which involves the known (or given) side and the unknown side. With reference to $\angle A$, we will use the opposite side, $\overline{BC}$ and the hypotenuse, $\overline{AB}$

(Note: **O**pposite side and **H**ypotenuse → **OH** → **SOH** → $\sin = \frac{O}{H}$

$$\sin 30 = \frac{x}{5} \quad \left(\frac{\text{Opposite}}{\text{Hypotenuse}}\right)$$

$$x = 5\sin 30 \quad \text{(solving for } x\text{)}$$

$$= 5\left(\frac{1}{2}\right) \quad (\sin 30 = \frac{1}{2}, \text{from tables or from memory})$$

$$= 2.5$$

Step 2: Finding y

With reference to $\angle A$, the sides involved here are the adjacent side (unknown) and the hypotenuse (known). Note: **A**djacent side and **H**ypotenuse → **AH** → **CAH** → $\cos = \frac{A}{H}$.

$$\cos 30° = \frac{y}{5} \qquad \left(\frac{\text{Adjacent}}{\text{Hypotenuse}}\right)$$

$$y = 5\cos 30°$$

$$y = 5(.866) \qquad (\cos 30° = .866, \text{ from tables or calculator})$$

$$= 4.33$$

We could have found y by referring to $\angle B$; and in which case, the opposite side and hypotenuse would be involved. i.e.

$(\mathbf{O}pposite$ side and $\mathbf{H}ypotenuse \to \mathbf{OH} \to \mathbf{SOH} \to \sin = \frac{O}{H}$

$$\sin 60° = \frac{y}{5}$$

$$y = 5\sin 60°$$

$$= 5(.866) \qquad (\sin 60° = .866)$$

$$= 4.33$$

Again, we obtain the same value as when we referred to $\angle A$.

Example 3   Solve the right triangle ABC , given that $b = 4$, $m \angle B = 30°$

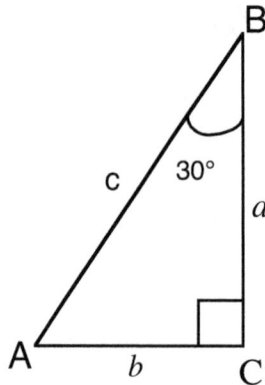

Solution

   (a) $m \angle A = 180° - (30° + 90°)$ $(m\angle C = 90°;$ sum of the measures of the $\angle$'s of a Δ is $180°)$

$$= 180° - 120°$$

$$m \angle A = 60°$$

   (b)   To find c:
         c is the hypotenuse, $b = 4$ is opposite $\angle B$.

$$\sin 30 = \frac{4}{c} \qquad \left(\frac{\text{opposite}}{\text{hypotenuse}}, \text{also } m\angle B = 30°, b = 4\right)$$

$$c \sin 30° = 4$$

$$c = \frac{4}{\sin 30°} = \frac{4}{\frac{1}{2}} = 8 \qquad (\sin 30° = \frac{1}{2})$$

(c)   To find a :

   Now $c = 8$, $b = 4$

Apply the Pythagorean theorem (We may apply this theorem since $\triangle$ ABC is a right triangle)

$8^2 = 4 + a^2$

$64 = 16 + a^2$

$48 = a^2$

$a = \sqrt{48} = \sqrt{16}\sqrt{3} = 4\sqrt{3}$

Note: In (b), we could have found a first using the tangent relationship

$\tan A = \dfrac{a}{4}$ $\left(\dfrac{\textbf{O}\text{pposite side}}{\textbf{A}\text{djacent side}} \text{ "TOA"} \text{ <-------- mnemonic device}\right)$

Solving, $a = 4 \tan A$ **Also**, $\cos 30° = \dfrac{a}{c}$

          $a = 4 \tan 60°$ $\dfrac{\sqrt{3}}{2} = \dfrac{a}{8}$ and $a = \dfrac{8(\sqrt{3})}{2} = 4\sqrt{3}$

          $a = 4\sqrt{3}$

We conclude that, sometimes, we have the option of choosing from two or more of the trigonometric functions.

Finding other trigonometric ratios given one functional value   518

In a right triangle, if we are given one trigonometric ratio for an acute angle we can determine the other trigonometric ratios.

Example: If $\sin\theta = \frac{3}{7}$, (a) find $\cos\theta$; (b) find $\tan\theta$.

Step 1: Sketch the a right triangle ABC.   Let angle $A = \theta$

Step 2: Since $\sin\theta = \frac{3}{7} = \dfrac{\text{Opposite side}}{\text{Hypotenuse}} = \frac{a}{c} = \frac{3}{7}$ and therefore, a $=3$, $c = 7$.

We first calculate $b.$, using the Pythagorean theorem

From $c^2 = a^2 + b^2$

$$7^2 = 3^2 + b^2$$

$$49 = 9 + b^2$$

Calculating b. $40 = b^2$

$$\sqrt{40} = b$$

$$2\sqrt{10} = b$$

$$b = 2\sqrt{10}$$

Step 3: Now, $a = 3$, $b = 2\sqrt{10}$, and, $c = 7$.

$$\cos\theta = \frac{b}{c}$$

$$= \frac{2\sqrt{10}}{7}$$

$$\tan\theta = \frac{a}{b}$$

$$= \frac{3}{2\sqrt{10}}$$

$$= \frac{3\sqrt{10}}{20} \qquad \text{(rationalizing the denominator)}$$

Lesson 78 Exercises

A Without using tables or calculator answer the following questions

1. If $\sin 49° = 0.7547$, then $\cos 41° = ...$?
2. If $\tan 50° = 1.1918$, then $\cot 40° = ...$?
3. If $\cot 60° = 0.5774$, then $\tan 30° = ...$?
4. If $\cot 43° = 1.0724$, then $\tan 47° = ...$?
5. If $\cot 40° = t$, then $\tan 50° = ...$?
6. If $\sec 60° = k$, then $\csc 30° = ...$?

Answers: **1.** 0.7547; **2.** 1.1918; **3.** 0.5774; **4.** 1.0724; **5.** t; **6.** k.

B Solve the following right triangles:

1. $m\angle A = 38°$, $b = 4$; 2. $m\angle A = 30°$, $b = 10$; 3. $a = 8$, $c = 10$; 4. $a = 40$, $b = 32$.
5. $b = 6$, and $m \angle A = 60°$

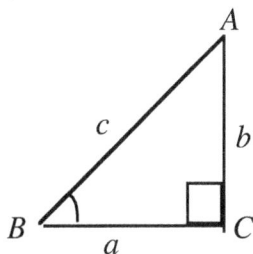

Answers: **1.** $a = 3.13$, $c = 5.08$, $m\angle B = 52°$; **2** $a = 5.77$, $c = 11.54$, $m\angle A = 60°$
3. $b = 6$, $m\angle A = 53.13°$; $m\angle B = 36.87°$; **4.** $c = 51.22$; $m\angle A = 51.34°$; $m\angle B = 38.66°$.
5. $m \angle B = 30°$; $a = 6\sqrt{3}$; $c = 12$.

C If $\sin\theta = \frac{3}{5}$, find (a) $\cos\theta$; (b) $\tan\theta$; (c) $\sec\theta$.

Answers: **(a)** $\frac{4}{5}$(if θ in quad I) or $-\frac{4}{5}$(if θ in quad II); **(b)** $\frac{3}{4}$(if θ in quad I) or $-\frac{3}{4}$(if θ in quad II)

(c) $\frac{5}{4}$(if θ in quad I) or $-\frac{5}{4}$(if θ in quad II).

D **1.** If $\sin\theta = -\frac{3}{5}$, find $\cos\theta$; (b) $\csc\theta$; $\tan\theta$ and $\cot\theta$.

2. If $\cos\theta = \frac{2}{3}$, find (a) $\sin\theta$; (b) $\tan\theta$, and $\cot\theta$.

Answers: **1.** If θ is in the **third** quadrant: $\cos\theta = -\frac{4}{5}$; $\csc\theta = -\frac{5}{3}$; $\tan\theta = \frac{3}{4}$; $\cot\theta = \frac{4}{3}$.

If θ is in the **fourth** quadrant: $\cos\theta = \frac{4}{5}$; $\csc\theta = -\frac{5}{3}$; $\tan\theta = -\frac{3}{4}$; $\cot\theta = -\frac{4}{3}$.

2. If θ is in the **first** quadrant: $\sin\theta = \frac{\sqrt{5}}{3}$; $\tan\theta = \frac{\sqrt{5}}{2}$; $\cot\theta = \frac{2\sqrt{5}}{5}$.

If θ is in **fourth** the quadrant: $\sin\theta = -\frac{\sqrt{5}}{3}$; $\tan\theta = -\frac{\sqrt{5}}{2}$; $\cot\theta = -\frac{2\sqrt{5}}{5}$.

Lesson 79

Application of Trigonometric Ratios; Angles of Elevation and Depression; Bearing; Linear Interpolation

Angle of elevation

Definition: For an observer sighting (looking) at something above the observer, **the angle of elevation** is the angle between the horizontal line (*x*-axis) from the observer's eye and the line of sight to the object, the angle being measured from the horizontal to the line of sight.

Example   Find the height reached by a kite, if the length of the kite is 400 ft, and the angle of elevation of the kite is 60°.

Step 1: Sketch a diagram for the system under consideration.
Assume that the hand holding on to the kite and the eye coincide at one point.

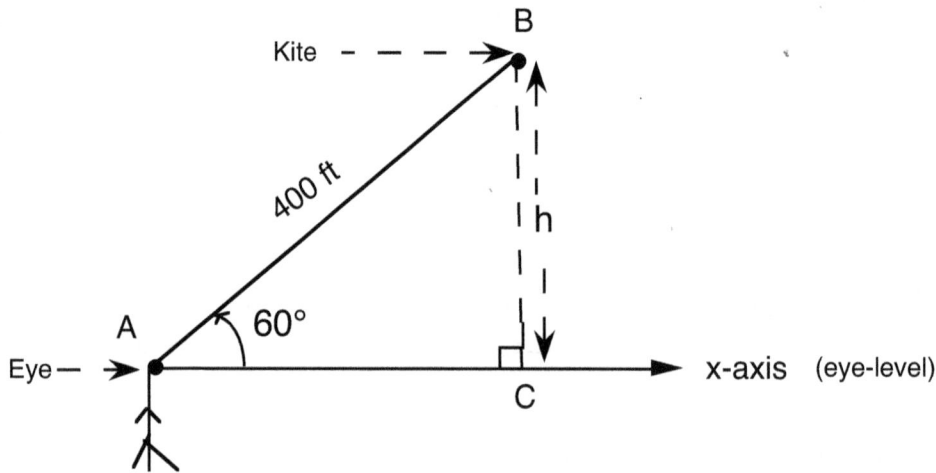

AB = Length of kite = 400 ft.

BC = *h* = height of kite (vertical distance from eye-level to the kite)

Step 2:   Δ ABC is a right triangle, and therefore we may apply the trigonometric relationships.

$$\sin 60° = \frac{h}{400}$$ $\left(\text{SOH} : \frac{\textbf{O}\text{pposite}}{\textbf{H}\text{ypotenuse}}\right)$

$$h = 400 \sin 60°$$

$$= 400 \left(\frac{\sqrt{3}}{2}\right) \text{ ft.}$$

$$= 200 \sqrt{3} \text{ ft.} \qquad\qquad (\sqrt{3} \approx 1.73)$$

$$\approx 346.41 \text{ ft.}$$

The height reached is $200 \sqrt{3}$ ft. ($\approx$346.41 ft.)

Angle of depression   521

Definition: For an observer sighting (looking) at something below the observer, the **angle of depression** is the angle between the horizontal line (x-axis) from the observer's eye and the line of sight to the object, the angle being measured from the horizontal to the line of sight.   In the figure below: $\angle \alpha$ is the angle of elevation of B from A.

$\angle \beta$ is the angle of depression of A from B.

The angle of elevation is congruent to the angle of depression.

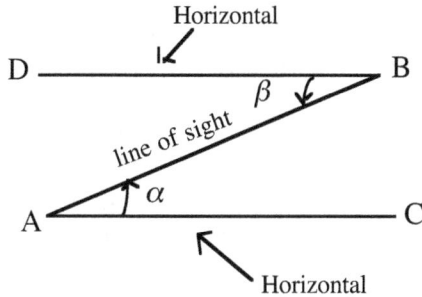

Bearing

The **bearing** of a given line in a horizontal plane is the acute angle between the given line and the North-Side (N-S) line. The angle is measured from a point on the N-S line to a a point on the given line. In finding bearings, the reference line is the North-South line which is a vertical line.

To indicate a bearing, we first specify whether North (N) or South (S), specify the measure of the angle and then specify whether East (E) or West (W). Note that the angle is always between the N-S line and the given line. Note also that the first letter is either N or S and the second letter after the angle is either E or W. (Contrast this angle with the reference angle of an angle in standard position in an x-y coordinate system of axes) Furthermore, note that if the angle is above the E-W line, the first letter is N; and below the E-W line, the first letter is S.

Examples:

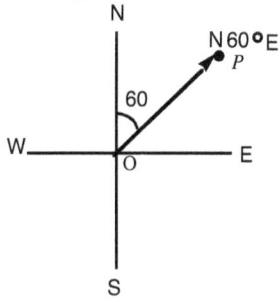

Bearing of P from O is $N60^\circ E$

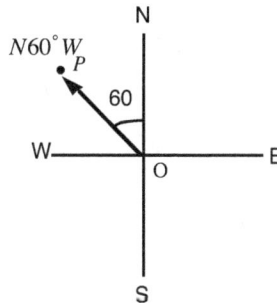

Bearing of P from O is $N60^\circ W$

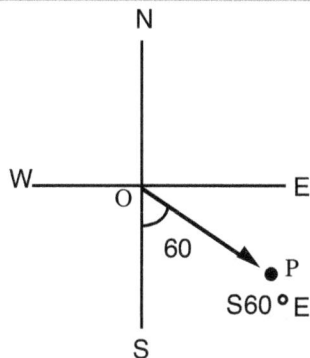

Bearing of P from O is $S60^\circ E$

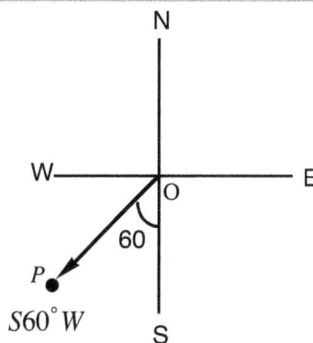

Bearing of P from O is $S60^\circ W$

Linear Interpolation

Sometimes, we may know two data points or corresponding values (x_1,y_1) and (x_2,y_2) and we would like to know either the y-value or the x-value of a point between the two known data points. This situation arises in applications such as below:

(1) In using logarithmic, trigonometric tables (and or similar tables) to approximate an unknown value between two known values.

(2) To approximate an unknown root between two other known roots in finding the real roots of polynomial equations for which there are no exact formulas for calculating the real roots.

The method of linear interpolation assumes that small changes in. say, x results in proportional changes in y , and therefore, the graph between the two points (x_1,y_1) and (x_2,y_2) is a straight line. Geometrically, there are two popular methods for setting up the proportion which will yield the required value:

(1) By similar triangles.
(2) By using the two-point-slope form of the equation of a straight line, which is given by

$$(y - y_1) = \frac{y_2 - y_1}{x_2 - x_1}(x - x_1) \text{ or } \frac{y - y_1}{x - x_1} = \frac{y_2 - y_1}{x_2 - x_1}$$

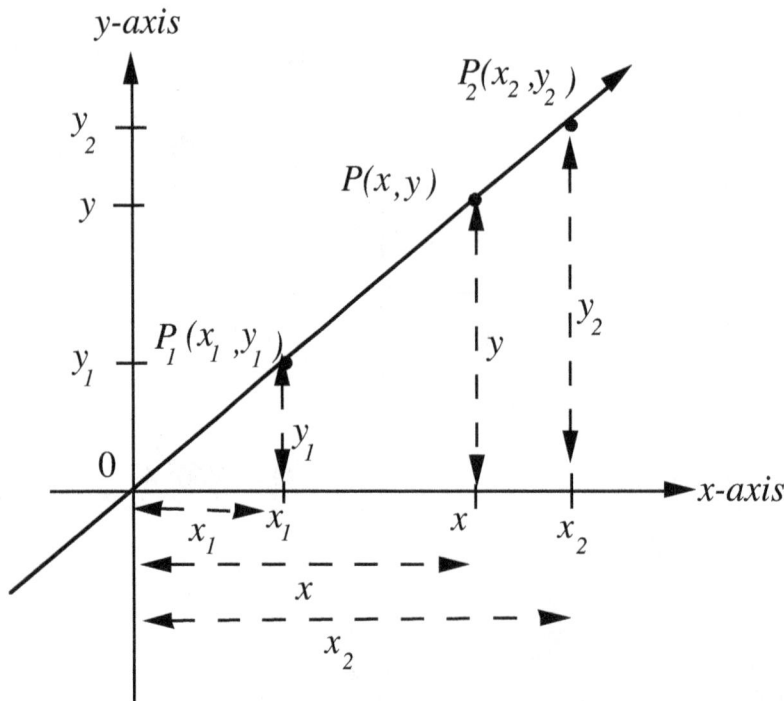

We will use the two-point-slope form, since this can be applied by mere substitution, and moreover, may be easier to remember (from elementary mathematics).
In using the above equation, we would know the values of five quantities in the proportion and we will calculate the sixth quantity.

Lesson 79 : Angles of Elevation and Depression; Bearing; Linear Interpolation

Example Consider the incomplete table, below, and find y.

523

x	y
$x_1 = .42$	$y_1 = .4078$
$x = .43$	$y = ?$
$x_2 = .44$	$y_2 = .4259$

We will calculate the y-coordinate of the point $P(.43, y)$ between $P_1(x_1, y_1)$ and $P_2(x_2, y_2)$ by substituting the known values x_1, y_1, x_2, y_2, and x in $\dfrac{y_2 - y_1}{x_2 - x_1} = \dfrac{y - y_1}{x - x_1}$, we obtain

$$\frac{.4259 - .4078}{.44 - .42} = \frac{y - .4078}{.43 - .42}$$

Solving, $y = \dfrac{(.43 - .42)(.4259 - .4078)}{.44 - .42} + .4078$

$$y = .4169$$

Lesson 79 Exercises

A 1. A 20-meter ladder leans against the wall of a room. If the angle of elevation of the top of the ladder is $30°$, how far above the floor does the top of the ladder reach the wall?

2. From a point on the ground 30 feet from the base of a tree, the angle of elevation of the top of the tree is $60°$. Find the height of the tree.

3. The angle of depression of a boat from a cliff 100 meters above the shore line is $40°$. Find the distance from the boat to a point directly below the cliff.

4. Find the height reached by a kite, if the length of the kite is 600 ft, and the angle of elevation of the kite is $30°$.

Answers: **1.** 10 meters; **2.** height $= 30\sqrt{3}$ ft (≈ 52 ft); **3.** ≈ 119 meters; **4.** 300ft.

B 1. City A is 400 miles due east of City B. A third city, City C, due north of City B, is $N60°E$. How far is City C from City B?

2. Two ships A and B depart a port at 1pm. Ship A travels at 20 mph in direction $N30°W$. Ship B travels in direction $S60°W$ at 15 mph. Find the bearing of ship A from ship B at 3 pm?

3. Give the bearing for A, B, and C.

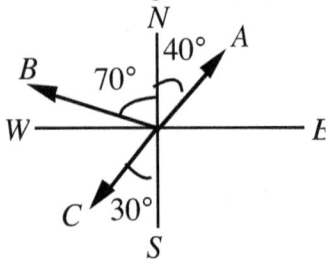

4. Criticize the following specifications for bearings:
   (a) $W15°N$; (b) $E15°N.$; (c) $S40°N$

5. Draw the bearing for each of the following:
   (a) $N15°W$; (b) $N15°E.$; (c) $S40°E$; (d) $S75°E$; (e) $S75°W$

Answers:

1.

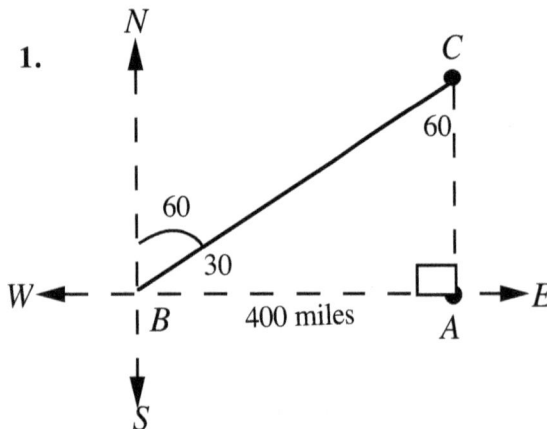

City C is 462 miles from City B.

2.

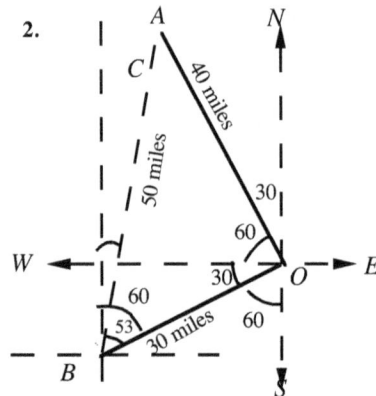

The bearing of A from B at 3 p.m.is $N7°E$.

3. $A:$ $N40°E$; $B:$ $N70°W$; $C:$ $S30°W$;
4. (a) First letter must be N or S and second letter must be E or W.
   (b) First letter must be N or S and second letter must be E or W.
   (c) Second letter must be E or W.

Answers continued.

5.

a

b

c

d

e

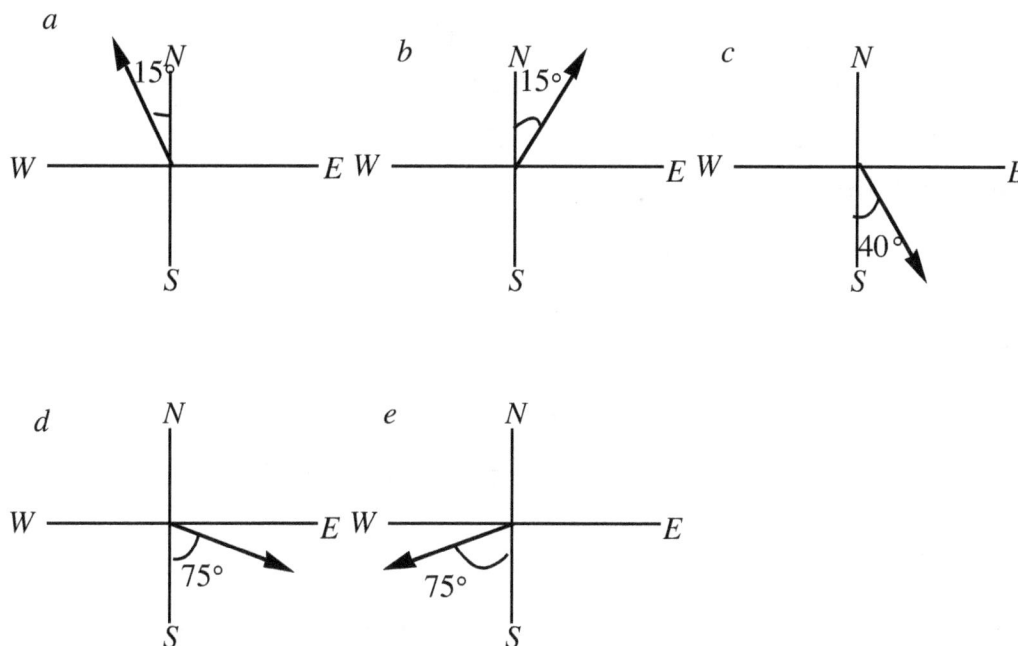

C   **1.** Using linear interpolation, find the *y* -value corresponding to *x* = .6500 in the following table:

x	y
.6421	.4460
.6500	?
.6612	.4613

2. In the following, interpolate if necessary:

(a) Find $\sin 15° \, 18'$;   (b) Find $\cos 60° \, 15'$;

3. Find the measure of θ given that

(a) $\cos \theta - .4000$;   (b) $\sin \theta = .9623$

Answers: **1.** 0.4523; **2.** (a) 0.2639; (b) 0.4962; 66.42° (or 1.16 rad); (b) 74.22° (or 1.30 rad).

CHAPTER 31

Trigonometrical Functional Value of any Angle

Lesson 80: **Definitions; Functional Value Given the Measure of the Angle**
Lesson 81: **Given a Point on the Terminal Side of the Angle**
Lesson 82: **Trigonometric Functional Values of Quadrantal Angles**

So far, we have defined the trigonometric ratios (functions) in terms of the sides of a right triangle and it did not matter where the vertex and the initial side of the angle were located, provided we obtained the required angular rotation. In this chapter, we will define two additional terms so that we can apply the trigonometric functions much more widely and we will not be restricted to angles between 0° and 90°.

Lesson 80

Basic Definitions; Functional Values given the Measure of the Angle

Basic Definitions

Angle

In trigonometry, we define an **angle** as being formed by rotating a line segment (a ray) about one of its end points, from an initial position to a final or terminal position. We call the end point, the **vertex**. We call the initial position, the initial side and the terminal position, the terminal side. The angle formed is measured by the amount of rotation. We shall refer to the angles formed by these rotations as **directed angle**s. By a directed angle we mean if the rotation (used in forming the angle) is in a **counterclockwise direction**, then we shall say that the measure of the angle formed is **positive** (Figure. 7). However, if the rotation is in a **clockwise direction**, then we say that the measure of the angle formed is **negative** (Figure 8). A curved arrow is used to denote the direction and amount of rotation. If no direction is indicated, then it is to be understood that the angle involved is the smallest positive angle.

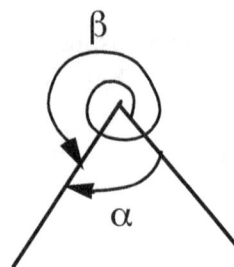

Fig. 7: θ is a positive angle   **Fig. 8**: θ is a negative angle   **Fig. 9:** β is a positive angle

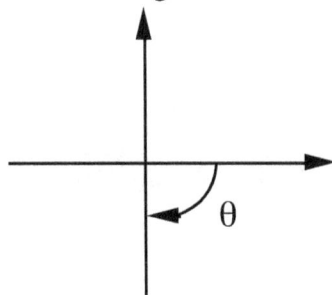

α is a negative angle

A ray or a side can be rotated clockwise or counter-clockwise indefinitely. Consequently, the measure of an angle has an unlimited number of values. However, with some restrictions we can have unique values for the angles.

Angle in standard position

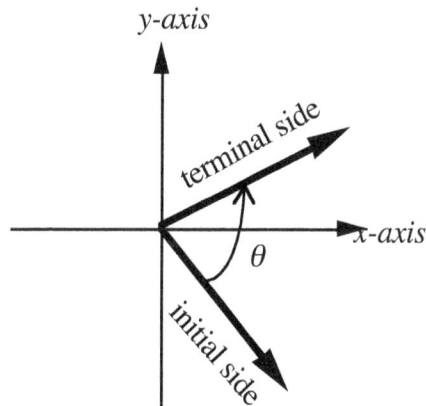

Definition: An angle is in standard position in an *x-y* rectangular coordinate system if it's vertex is at the origin and its initial side is along positive *x* -axis.

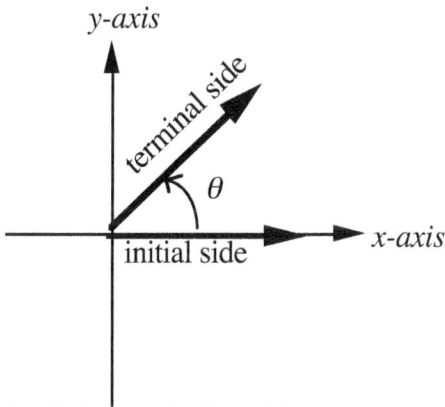

positive y-axis

... negative x-axis   positive x-axis   ...

negative y-axis

y-axis

terminal side

θ

initial side   *x-axis*

Angle in standard position

y-axis

terminal side

θ

x-axis

initial side

Angle **not** in standard position
(Initial side not along the positive *x*-axis)

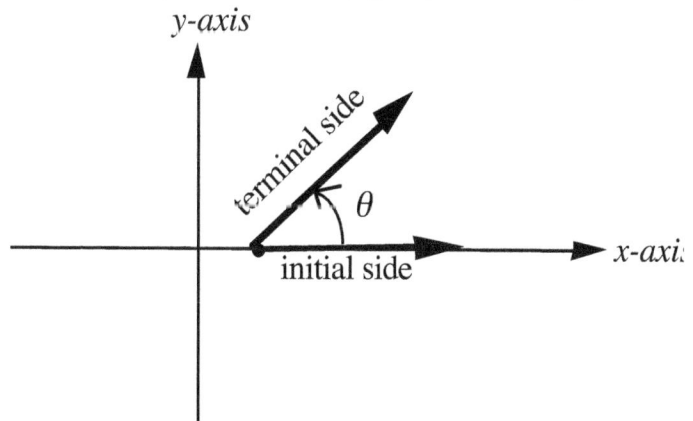

y-axis

initial side

x-axis

θ

terminal side

Angle **not** in standard position
(Initial side not along the positive *x*-axis)

y-axis

terminal side

θ

initial side   *x-axis*

Angle **not** in standard position
(Vertex not at the origin)

Angle in standard position

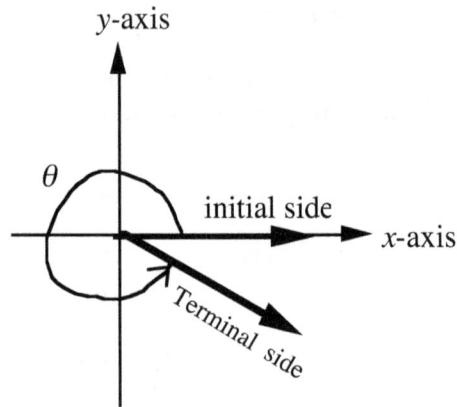

Angle in standard position

Reference Angle

Definition: If an angle, θ, is in standard position, then the reference angle is the positive acute angle between the terminal side of θ and the x- axis (the horizontal axis).

Example 1   Find the reference angle for 230°.

Solution

Step 1: Draw (or sketch) $\theta = 230°$ in standard position.
To draw the arc, begin from the positive x-axis and draw the arc counterclockwise.

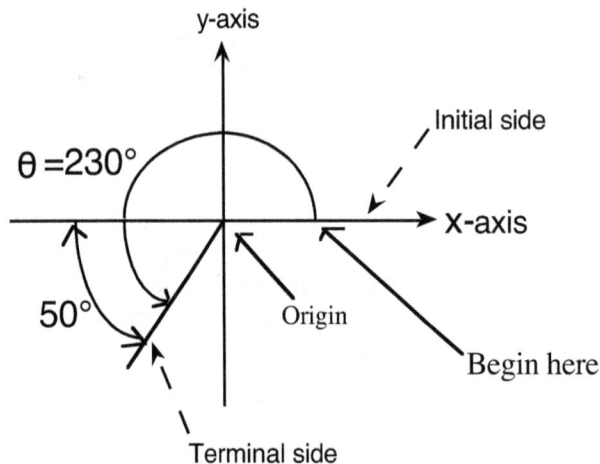

Step 2:   Determine the reference angle by difference. The reference angle for 230° is 230 - 180
i.e. 50°
The reference angle for 230° is 50° .

Note: The reference angle is always between the terminal side and the x-axis.

Example 2 Find the reference angle for 460°

Solution

Step 1: Draw (sketch) θ = 460° in standard position.

Step 2: The reference angle is the difference between 180° and 100°.

$$\text{i.e.,}\quad 180° - 100° = 80°$$

Note that 360° will bring you back to the initial side, leaving 460 - 360 = 100 to consider.

Example 3   Find the reference angle for $\theta = -135°$

Solution

Step 1: Draw (sketch) θ = -135° (Draw this clockwise, since θ is negative.)

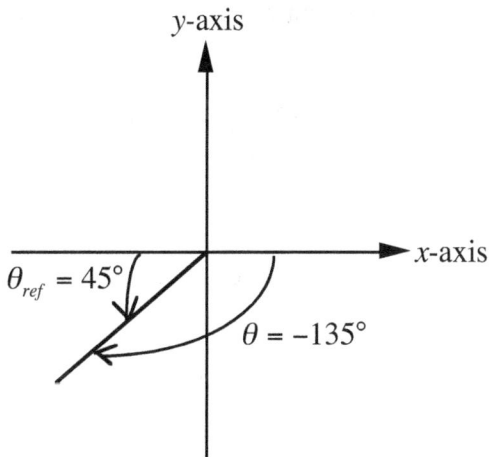

Step 2: The reference angle is $180° - 135° = 45°$, but note that the reference angle is always between the terminal side and the x-axis (never between the y-axis).

Relationship between θ (standard-position angle) and θ_{ref}

1. If θ is in the **first quadrant**, then $\theta_{ref} = \theta$

2. If θ is in the **second quadrant**, then $\theta_{ref} = 180° - \theta$ or equivalently $\theta = 180° - \theta_{ref}$

3. If θ is in the **third quadrant**, then $\theta_{ref} = \theta - 180°$ or equivalently $\theta = \theta_{ref} + 180°$

4. If θ is in the **fourth quadrant**, then $\theta_{ref} = 360° - \theta = \theta$ or equivalently $\theta = 360° - \theta_{ref}$

Note that θ and θ_{ref} are always in the **same** quadrant.

Trigonometric Functional Value of any Angle (Given the measure of the angle)   530

Example   (a) Find the value of sin 210° without using a calculator: use trigonometric tables.
                (b) Verify your answer using a calculator.

Solution (a)

Step 1:   Draw (or sketch) the given angle in standard position.

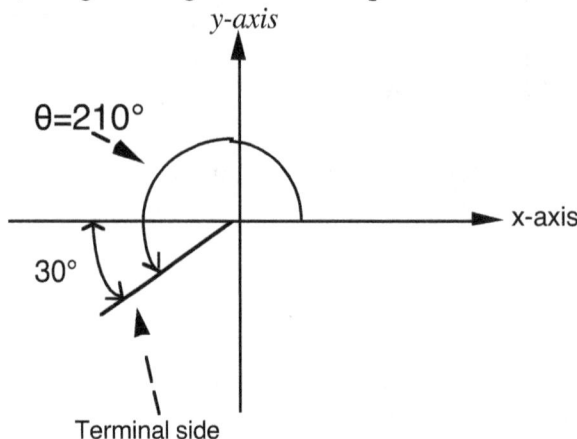

Figure 1

Step 2:   Find the reference angle.
          The reference angle is 210° - 180° = 30°

Step 3: From tables or from memory, find sin 30°. This value will be positive, and we must then
determine the sign of sin 210° according to the quadrant in which the terminal side of this angle lies.
(See Figure 2 below) The terminal side of 210° is in the 3rd quadrant where the sine function is negative.

$$\therefore \; \sin 210 = -\sin 30 = -\frac{1}{2}$$

Signs of Trigonometric Functions According to Quadrants

Mnemonic device " **CAST**" for remembering in which quadrant a trigonometric function is positive or negative

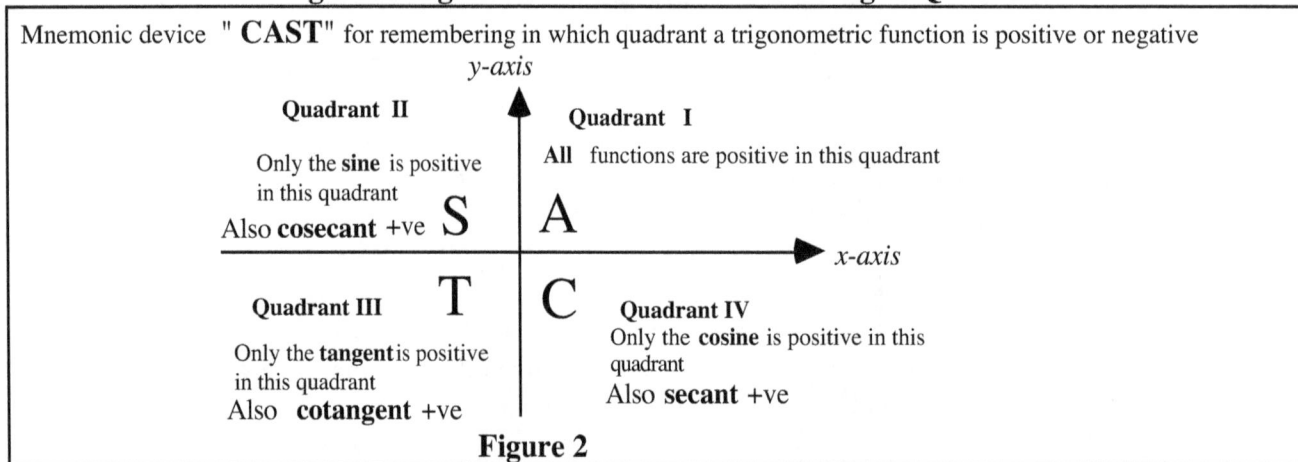

Figure 2

In the above device: The **A** in "**CAST**" means **A**ll functions are positive in this quadrant; **S** for only sine
is positive in this quadrant; and similarly **T** for tangent; and **C** for cosine.
Note: The order of the above "CAST" is counterclockwise. There are other mnemonic devices such as
          "**A**ll **S**tudents **T**ake **C**alculus".

(b) With the calculator in degree mode, press 210; then press "the sin" key and read -0.5

Note above: The sign of a trigonometric function and its reciprocal are the same.

Lesson 80 Exercises

A Find the reference angles for the following: **1.** 250° ; **2.** 295°; **3.** 560°; **4.** 75°; **5.** 360°

6 −35°; **7.** −125°; **8.** −265°

Answers. **1.** 70° ; **2.** 65° ; **3.** 20°; **4.** 75° ; **5.** 0°; 6 35°; **7.** 55°; **8.** 85°.

B In the following problems, first use tables; and then verify your answers using a calculator.

Find the values of the following: **1.** sin 250° ; **2.** cos 250°; **3.** sin 460°; **4.** sin 75°

Answers: **1.** - 0.94 ; **2.** -0.34; **3.** 0.98; **4.** 0.96

C Find the following:

(a) $\sin(-60°)$; (b) $\cos(-60°)$; (c) $\sin(-220°)$; (d) $\cos(-380°)$; (e) $\sin\frac{5\pi}{4}$; (f) $\cos\left(-\frac{\pi}{3}\right)$.

(g) $\sin(-150°)$; (h) $\cos(-250°)$; (i) $\sin(-460°)$; (j) $\sin(-75°)$

Answers: (a) -.86; (b) .5; (c) .64; (d) .94; (e) -.71; (f) .5; (g) -.5; (h) -.34; (i) -.98; (j) -.97

Lesson 81

Trigonometric Functional Value of any Angle
(Given the coordinates of a point on the terminal side of an angle)

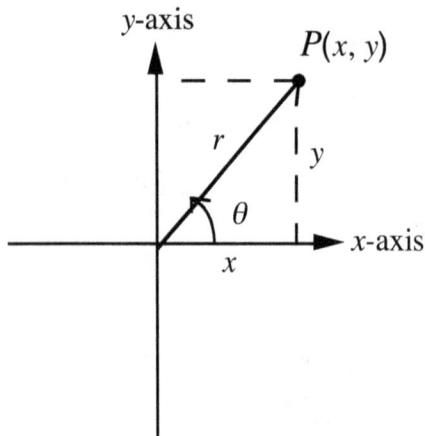

Let θ be an angle in standard position in a rectangular coordinate plane. Let $P(x,y)$ be a point on the terminal side of the angle θ and at a positive distance r from the origin (**Figure**).

Applying Pythagorean theorem to the above figure,

$$r^2 = x^2 + y^2$$

$$r = \sqrt{x^2 + y^2}$$

With reference to θ, r is the hypotenuse, x the adjacent side, y is the opposite side.
We define the six trigonometric functions (see also page 504) for θ in terms of r, x and y.
(Recall: SOH , CAH, TOA)

$$\sin \theta = \frac{y}{r} \; ; \qquad\qquad \csc \theta = \frac{r}{y}.$$

$$\cos \theta = \frac{x}{r} \; ; \qquad\qquad \sec \theta = \frac{r}{x}.$$

$$\tan \theta = \frac{y}{x} \; ; \qquad\qquad \cot \theta = \frac{x}{y}.$$

Note above that r is always positive while x and y may be positive or negative.

Example Find the six trigonometric values of θ , (if θ is an angle in standard position) given
that the terminal side of θ passes through the point $(-3, 2\sqrt{3}\,)$.

Step 1:   Draw (or sketch) θ with terminal side passing through the point $(-3, 2\sqrt{3})$
(x-coordinate = -3; y-coordinate = $2\sqrt{3}$)

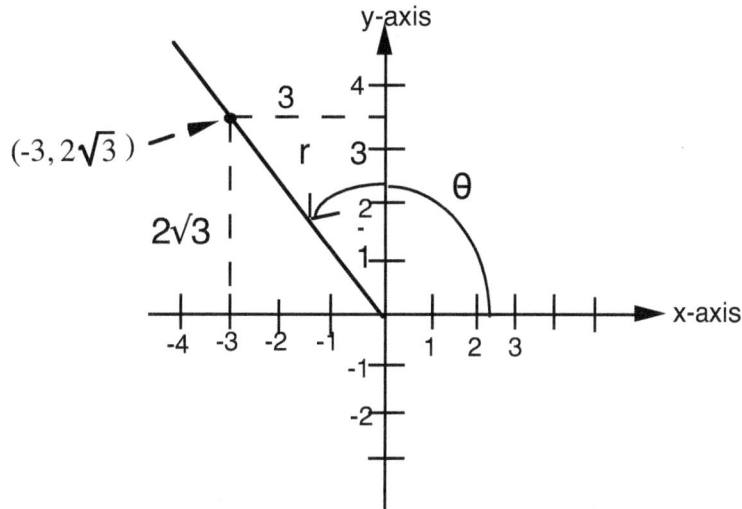

Step 2: Find r using the Pythagorean Theorem.
$$r^2 = (-3)^2 + (\,2\sqrt{3}\,)^2 \quad (r^2 = x^2 + y^2)$$
$$r^2 = 9 + 4\sqrt{9}$$
$$r^2 = 9 + 12 \qquad \text{(Note: } 4\sqrt{9} = 4(3) = 12))$$
$$r^2 = 21$$
$$r = \sqrt{21}$$

Step 3:   Now, $r = \sqrt{21}$, $x = -3$, $y = 2\sqrt{3}$

(a) $\cos\theta = \dfrac{x}{r}$

$$= \frac{-3}{\sqrt{21}}$$

$$= -\frac{3}{\sqrt{21}} \cdot \frac{\sqrt{21}}{\sqrt{21}} \quad \text{(Rationalizing the denominator)}$$

$$\therefore \cos\theta = -\frac{3\sqrt{21}}{21}$$

(b) $\sin\theta = \dfrac{y}{r}$ $(y = 2\sqrt{3},\ r = \sqrt{21})$

$$= \frac{2\sqrt{3}}{\sqrt{21}}$$

$$= \frac{2\sqrt{3}}{\sqrt{21}} \cdot \frac{\sqrt{21}}{\sqrt{21}} \qquad \text{(Rationalizing the denominator)}$$

$$= \frac{2\sqrt{63}}{21}$$

$$= \frac{2\sqrt{9}\sqrt{7}}{21}$$

$$= \frac{2\cancel{(3)}\sqrt{7}}{\cancel{21}_{7}} \qquad\qquad (\sqrt{9} = 3)$$

$$\sin \theta = \frac{2\sqrt{7}}{7}$$

(c) $\quad \tan \theta = \frac{y}{x} \qquad\qquad (y = 2\sqrt{3}, x = -3)$

$$= \frac{2\sqrt{3}}{-3}$$

$$\therefore \tan \theta = -\frac{2\sqrt{3}}{3}$$

(d) $\quad \sec \theta = \frac{1}{\cos \theta}$

$$= \frac{1}{\frac{x}{r}} = \frac{r}{x} = \frac{\sqrt{21}}{-3}$$

$$\therefore \sec \theta = -\frac{\sqrt{21}}{3}$$

(e) $\quad \csc \theta = \frac{1}{\sin \theta}$

$$= \frac{r}{y}$$

$$= \frac{\sqrt{21}}{2\sqrt{3}}$$

$$= \frac{\sqrt{21}}{2\sqrt{3}} \cdot \frac{\sqrt{3}}{\sqrt{3}} \qquad\qquad \text{(Rationalizing the denominator)}$$

$$= \frac{\sqrt{63}}{2\sqrt{9}}$$

$$= \frac{\sqrt{63}}{2(3)} \qquad\qquad (\sqrt{9} = 3)$$

$$= \frac{\sqrt{9}\sqrt{7}}{6}$$

$$= \frac{3\sqrt{7}}{6}$$

$$\therefore \csc \theta = \frac{\sqrt{7}}{2}$$

(f) $\quad \cot \theta = \frac{1}{\tan \theta}$

$$= \frac{1}{\frac{y}{x}}$$

$$= \frac{x}{y}$$

$$= \frac{-3}{2\sqrt{3}} \qquad\qquad (x = -3, y = 2\sqrt{3})$$

$$= -\frac{3}{2\sqrt{3}} \cdot \frac{\sqrt{3}}{\sqrt{3}} \qquad\qquad \text{(Rationalizing the denominator)}.$$

$$= -\frac{3\sqrt{3}}{2\sqrt{9}}$$

$$= -\frac{\cancel{3}\sqrt{3}}{2(\cancel{3})} \qquad\qquad (\sqrt{9} = 3)$$

$$\therefore \cot \theta = -\frac{\sqrt{3}}{2}$$

Comparatively, note in the last two examples that:

A. On page 530, we had to determine the sign of the function according to the quadrant (in which the terminal side of the angle lies) if we obtain the values of the acute angles from tables or from memory.

B. On page 532, we did not have to determine the sign of the function. The signs were obtained solely from the signs of the *x*- and *y*-coordinates; but note also that *r* is always positive.
Note also however that we may use the method in **A** to do the problems referred to in **B**.

Lesson 81 Exercises

A Find the six trigonometric values of θ , if θ is an angle in standard position, given that the terminal side of θ passes through the point $(5, 2)$.

Solutions: $\sin \theta = \frac{2\sqrt{29}}{29}$; $\cos \theta = \frac{5\sqrt{29}}{29}$; $\tan \theta = \frac{2}{5}$; $\csc \theta = \frac{\sqrt{29}}{2}$; $\sec \theta = \frac{\sqrt{29}}{5}$; $\cot \theta = \frac{5}{2}$

B Redo the Example on page 533 by applying the mnemonic devices SOH, CAH, TOA together with "CAST" referred to on page 530

Lesson 82

Trigonometric Functional Values of Quadrantal Angles

Quadrantal angles are angles whose terminal sides lie on either the x-axis or the y-axis. These angles are multiples of 90°. Examples of quadrantal angles are 0° , 90°, 180, -180°, 270°, and 360°.

Trigonometric functional values for the quadrantal angle 0°

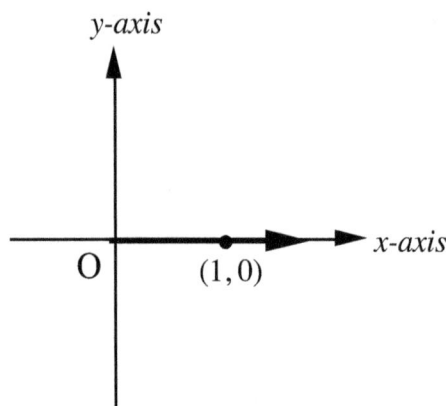

Consider a point $P(1,0)$ on the terminal side of an angle $\theta = 0°$ in standard position and its distance from the origin to this point is 1. From the figure, $x = 1$, $y = 0$, and $r = 1$.

(*a*) $\sin \theta = \dfrac{y}{r}$

       $\sin 0 = \dfrac{0}{1}$ $(y = 0,\ r = 1)$

       $\mathbf{\sin 0° = 0}$

(*b*) $\cos\theta = \dfrac{x}{r}$

       $\cos 0° = \dfrac{1}{1}$ $(x = 1,\ r = 1)$

       $\mathbf{\cos 0° = 1}$

(*c*) $\tan \theta = \dfrac{y}{x}$

       $\tan 0° = \dfrac{0}{1}$ $(y = 0,\ x = 1)$

       $\mathbf{\tan 0° = 0}$

The reciprocal relationships are obtained by inverting (a), (b), b and (c) above.

(*e*) $\csc 0° = \dfrac{r}{y} = \dfrac{1}{0}$ **is undefined.**

(*f*) $\sec 0° = \dfrac{r}{x} = \dfrac{1}{1}$

       $\mathbf{\sec 0° = 1}$

(*g*) $\cot 0° = \dfrac{1}{0}$ **is undefined**

Trigonometric functional values for the quadrantal angle 90°

Consider a point $P(0, 1)$ on the terminal side of an angle $\theta = 90°$ in standard position. Then the distance from origin to P is 1. Thus, $r = 1$, $x = 0$ and $y = 1$.

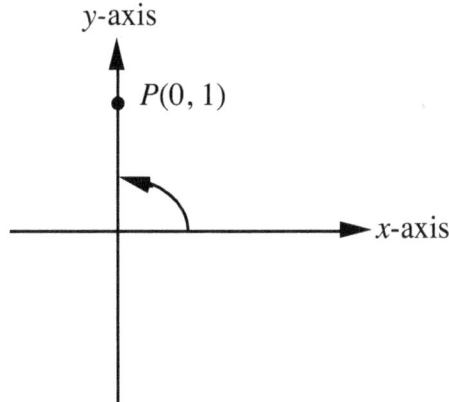

(a) $\sin \theta = \dfrac{y}{r}$

$\sin 90° = \dfrac{1}{1}$ $(y = 1, r = 1)$

$\sin 90° = 1$

(b) $\cos \theta = \dfrac{x}{r}$

$\cos 90° = \dfrac{0}{1}$ $(x = 0, r = 1)$

$\cos 90° = 0$

(c) $\tan \theta = \dfrac{y}{x}$

$\tan 90° = \dfrac{1}{0}$ $(y = 1, x = 0)$

tan 90° is undefined

To find the functional values for the cosecant, the secant, and the cotangent , invert the right-hand sides of the sine, cosine, and tangent functions, respectively, and deduce the results.

Trigonometric functional values for the quadrantal angle 180°   538

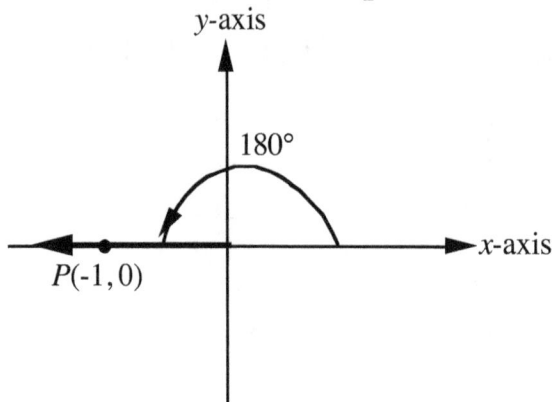

Consider a point $P(-1,0)$ on the terminal side of an angle $\theta = 180°$ in standard position. Then the distance from the origin to P is 1. Thus, $r = 1$, $x = -1$ and $y = 0$.

$$(a) \quad \sin 180° = \frac{y}{r}$$

$$\sin 180° = \frac{0}{1} \qquad (y = 0,\ r = 1)$$

$$\mathbf{\sin 180° = 0}$$

$$(b) \quad \cos 180° = \frac{x}{r}$$

$$= \frac{-1}{1} \qquad (x = -1,\ r = 1)$$

$$\mathbf{\cos 180° = -1}$$

$$(c) \quad \tan 180° = \frac{y}{x}$$

$$= \frac{0}{-1} \qquad (y = 0,\ x = -1)$$

$$\mathbf{\tan 180° = 0}$$

To find the functional values for the cosecant, the secant, and the cotangent , invert the right-hand sides of the sine, cosine, and tangent functions , respectively, and deduce the results.

$$\csc 180° = \text{undefined}$$
$$\sec 180° = -1$$
$$\cot 180° = \text{undefined}$$

Trigonometric functional values for the quadrantal angle 270°   539

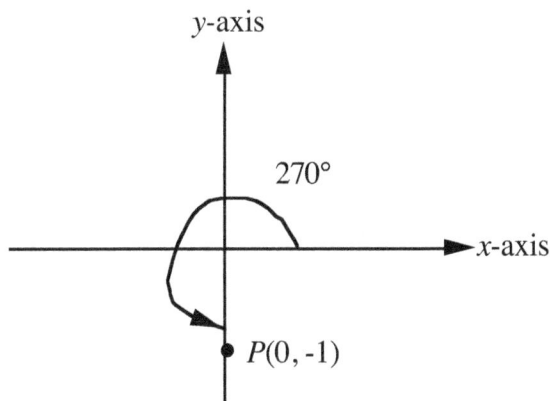

Consider a point $P(0,-1)$ on the terminal side of an angle $\theta = 270°$ in standard position. Then the distance from origin to P is 1. Thus, $r = 1$, $x = 0$ and $y = -1$.

(a) $\sin 270° = \dfrac{y}{r}$

$\sin 270° = \dfrac{-1}{1}$ $(y = -1,\ r = 1)$

$\mathbf{\sin 270° = -1}$

(b) $\cos 270° = \dfrac{x}{r}$

$= \dfrac{0}{1}$ $(x = 0,\ r = 1)$

$\mathbf{\cos 270° = 0}$

(c) $\tan 270° = \dfrac{y}{x}$

$= \dfrac{-1}{0}$ $(y = -1,\ x = 0)$

$\mathbf{\tan 270°}$ is undefined

To find the functional values for the cosecant, the secant, and the cotangent, invert the right-hand sides of the sine, cosine, and tangent functions, respectively, and deduce the results.

Note   In the above derivations, the value of r does not matter.

For example, for $\theta = 0°$, if $r = a$, $x = a$, and $y = 0$ we obtain the following:

$\sin 0° = \dfrac{y}{r} = \dfrac{0}{a} = 0$

$\cos 0° = \dfrac{x}{r} = \dfrac{a}{a} = 1$

$\tan 0° = \dfrac{y}{x} = \dfrac{0}{a} = 0$

The trigonometric functional values for the quadrantal angle 360° are the same as those for 0°.

You may also observe that the values for sine, cosine, tangent, cotangent, secant, and cosecant are either $-1, 0, 1$, or undefined. For sine and cosine, the values are either $-1, 0, 1$. A mnemonic device for remembering these values is to locate the appropriate axis and choose a convenient " r " (say, $r = 1$) and then apply the fundamental definitions.

Functional Values for Coterminal Angles

Coterminal angles are angles which have the same terminal sides when both angles are placed in standard position on the same coordinate system of axes. Since the location of the terminal side of an angle completely determines the trigonometric functional value, **coterminal angles have the same functional values**.

Examples

      (*a*) $270°$ and $-90°$ are coterminal, and therefore have the same functional value.

      (*b*) $-270°$ and $90°$ are coterminal, and therefore have the same functional value.

      (*c*) $180°$ and $-180°$ are coterminal, and therefore have the same functional value.

      (*d*) $30°$ and $390°$ are coterminal, and therefore have the same functional value.

      (*e*) $40°$ and $760°$ are coterminal, and therefore have the same functional value.

Trigonometric functional values for negative quadrantal angles

Since coterminal angles have the same functional values, the functional values for $270°$ and $-90°$ are the same, since these two angles are coterminal.

Therefore for $-90°$ see values for $270°$. Similarly, for $-180°$ see values for $180°$ and for $-270°$ see values for $90°$.

Functional Values for $\sin(180° - A)$, $\cos(180° - A)$, and $\tan(180° - A)$

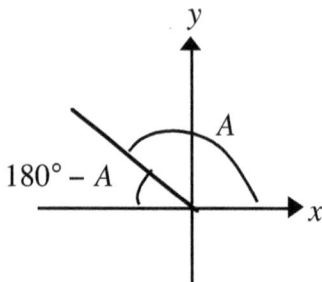

$\sin A = +\sin(180° - A)$ (the sine is positive in the second quadrant)
$\therefore \sin(180° - A) = \sin A$

  $\cos A = -\cos(180° - A)$ (the cosine is negative in the second quadrant)

$\therefore \cos(180° - A) = -\cos A$

$\tan A = -\tan(180° - A)$ (the tangent is negative in the second quadrant)
  $\therefore \tan(180° - A) = -\tan A$.

Derive the trig functional values for the following:

$0°$, $-180°$, $-270°$

Answers: $\sin 0° = 0$, $\cos 0° = 1$; $\tan 0° = 0$; $\csc 0°$ is undefined; $\sec 0° = 1$; $\cot 0°$ is undefined

$\sin(-180°) = 0$; $\cos(-180°) = -1$; $\tan(-180°) = 0$; $\sec(-180°) = -1$; $\csc(-180°)$ is undefined.; $\cot(-180°)$ is undefined.

$\sin(-270°) = 1$; $\cos(-270°) = 0$; $\tan(-270°)$ is undefined; $\csc(-270°) = 1$; $\sec(-270°)$ is undefined; $\cot(-270°) = 0$.

CHAPTER 32

Finding Other Trigonometric Functional Values

Lesson 83: **Given a Functional Value and the Quadrant of the Angle**

Lesson 84: **Given a Functional Value and no Specification of the Quadrant of the Angle**

Lesson 83

Given a Functional Value and the Quadrant of the Angle

See also page 518 for the corresponding topic on the right triangle. This topic will be helpful when we cover inverse functions in Chapter 37.

Example   If $\sin \theta = -\dfrac{4}{5}$ and the terminal side of θ is in the fourth quadrant (θ being in standard position)

        (a) Find $\cos \theta$, (b) $\tan \theta$, (c) $\csc \theta$

        We will determine r, x and y and apply the definitions of the trigonometric functions

Step 1: Since we know that the terminal side is in the 4th quadrant, we will draw θ accordingly.

fig.1

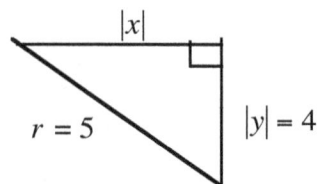

fig.2

Step 2: In this quadrant , the y-coordinate is negative and the x-coordinate is positive.

$$\sin \theta = \frac{y}{r} = -\frac{4}{5}$$

$y = -4$ since r is always positive (We give the minus sign to the "4") .
$r = 5$

Step 3:   To find x, we use the Pythagorean theorem.

$$5^2 = (-4)^2 + x^2$$
$$25 = 16 + x^2$$
$$9 = x^2$$
$$3 = x$$

∴ $x = 3, \ y = -4, \ r = 5$

Step 4:

(a) $\cos \theta = \dfrac{x}{r}$

$\cos \theta = \dfrac{3}{5}$ $(x = 3, r = 5)$

(b) $\tan \theta = \dfrac{y}{x}$

$\quad = \dfrac{-4}{3}$ $(y = -4, x = 3)$

$\quad = -\dfrac{4}{3}$

(c) $\csc \theta = \dfrac{1}{\sin \theta}$

$\quad = \dfrac{r}{y}$

$\quad = \dfrac{5}{-4}$

$\quad = -\dfrac{5}{4}$

Note: We could have obtained $\csc \theta$ from $\sin \theta = -\dfrac{4}{5}$ by inversion.

We may observe from the last two examples that the value of a trigonometric function does not depend upon the direction of rotation used in generating the angle. Thus if we know the coordinates of a point on the terminal side of the angle (and hence the quadrant in which the terminal side is) we can find the values of the trigonometric functions.

Lesson 83 Exercises

1. If $\sin \theta = -\dfrac{3}{7}$ and the terminal side of θ is in the third quadrant (θ being in standard position)

(a) Find $\cos \theta$, (b) $\tan \theta$, (c) $\csc \theta$

2.   If $\sin \theta = -\dfrac{3}{5}$, and the terminal side of θ passes through the point $(-4, -3)$, find the other five trigonometric functional values.

Solutions: **1.** $\cos \theta = \dfrac{-2\sqrt{10}}{7}$; (b) $\tan \theta = \dfrac{3\sqrt{10}}{20}$; (c) $\csc \theta = -\dfrac{7}{3}$

2. $\cos \theta = -\dfrac{4}{5}$; . $\tan \theta = \dfrac{3}{4}$; $\csc \theta = -\dfrac{5}{3}$; $\sec \theta = -\dfrac{5}{4}$; $\cot \theta = \dfrac{4}{3}$.

Lesson 84

Finding Other Trigonometric Functional Values
(Given a functional value and no specification of the quadrant of the angle.)

Example: If $\sin \theta = -\frac{3}{7}$, (a) find $\cos \theta$; (b) find $\tan \theta$.

Step 1: The sine function is negative in the 3rd and 4th quadrants. Since the exact quadrant is not specified in this problem, we shall consider two cases here.

Case 1 Terminal side of θ is in 3rd quadrant.

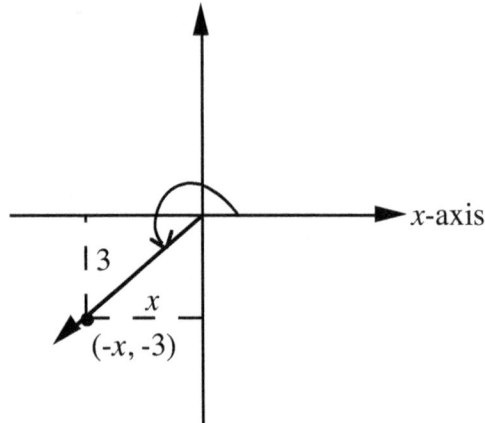

In the third quadrant, the x and y-coordinates are negative.

From the definition $\sin \theta = \frac{y}{r} = \frac{-3}{7}$. $y = $ -3, $r = 7$, since r is always positive,

Plot $(-x, -3)$, where x is the distance from the point (on the terminal side of the angle) to the y-axis.

Step 1: Find x.

$$\text{From } r^2 = x^2 + y^2$$
$$7^2 = (-x)^2 + (-3)^2$$
$$49 = x^2 + 9$$
$$40 = x^2$$
$$\sqrt{40} = x$$
$$2\sqrt{10} = x$$

Step 2:   Now, the x-coordinate is $-2\sqrt{10}$, $y = -3$, $r = 7$.

$$\cos \theta = \frac{x}{r}$$
$$= -\frac{2\sqrt{10}}{7} \qquad \text{(the } x \text{ - coordinate is } -2\sqrt{10})$$

$$\tan \theta = \frac{y}{x}$$

$$= \frac{-3}{-2\sqrt{10}}$$

$$= \frac{3\sqrt{10}}{20}$$

Case 2 Terminal side of θ is in 4th quadrant.

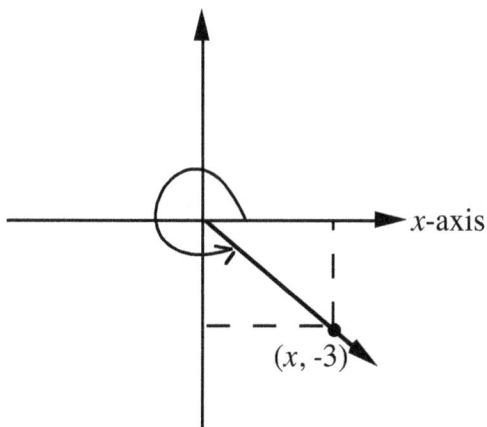

In this quadrant, the x-coordinate is positive and it is $2\sqrt{10}$, $y = -3$, $r = 7$.

$$\cos \theta = \frac{x}{r}$$

$$= \frac{2\sqrt{10}}{7}$$

$$\tan \theta = \frac{y}{x}$$

$$= \frac{-3}{2\sqrt{10}}$$

$$= -\frac{3\sqrt{10}}{20}$$

You can check the signs of the functional values by using the mnemonic device "CAST" see p.530

Lesson 84 Exercises

Redo Step 2 of the last Example by applying the mnemonic devices SOH, CAH, TOA together with "CAST" referred to on page 530.

CHAPTER 33

Trigonometry of Oblique Triangles
Lesson 85: **The Law of Cosine; The Law of Sines**
Lesson 86: **Applications of the Laws of Cosine and Sines**

An oblique triangle is a triangle in which no angle is a right angle. We may consider the right triangle as a special case of the oblique triangle and that the laws which are applicable to oblique triangles are also applicable to the right triangle.

In solving oblique triangles, we are usually given three parts of the oblique triangle and we are required to find the remaining three parts. We will usually seek a unique solution (that is only one solution). There may be a unique solution, two distinct solutions or no solutions at all depending on the relative lengths of the sides and the measures of the angles

Lesson 85

The Law of Cosine; The Law of Sines

The Law of Cosines.

Direct use of the Law of Cosines applies to the following cases:

(1) Given two sides and an included angle (abbreviated SAS), we are therefore to find the remaining side and two angles.

(2) Given all three sides (abbreviated SSS) we are therefore to find the remaining three angles.

The above cases yield unique solutions. A mnemonic device for the applicability of this law of cosines is **Co-SAS -SSS** - pronounced Kosaszz.

Statement of the Law of Cosines

Using standard notation for corresponding angles and sides, the law of cosines states that :

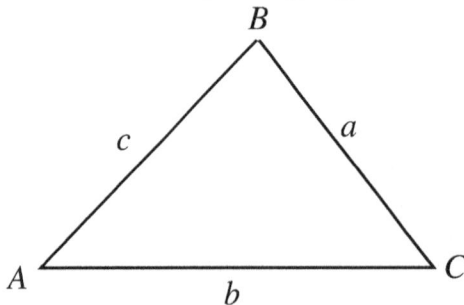

$$a^2 = b^2 + c^2 - 2bc \cos A$$

or

$$b^2 = a^2 + c^2 - 2ac \cos B$$

or

$$c^2 = a^2 + b^2 - 2ab \cos\ C$$

In words, the square of the length of one side is equal to the sum of the squares (of the lengths) of the other two sides minus twice the product (of the lengths) of the other two sides multiplied by the cosine of the angle between these two sides.

Derivation of the law of cosines

We will consider   (1) an acute triangle; (2) an obtuse triangle; (3) a right triangle as a special case.

Case 1 For an acute triangle

Given   Acute triangle ABC with sides a, b, c

Required: To derive $a^2 = b^2 + c^2 - 2bc \cos A$

Derivation

Consider the acute triangle ABC shown labeled with the standard notation for corresponding sides and angles.

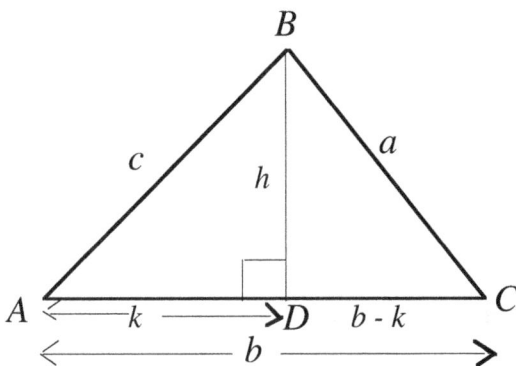

Draw the height h of triangle ABC by drawing $\overline{BD}$ perpendicular to $\overline{AC}$

Let $AD = k$. Then $DC = b - k$ We will consider subtriangles ABD and CBD; apply the Pythagorean theorem to both triangles and then eliminate h and k in terms of the given sides.

$$\text{In rt. } \Delta \ ABD, \ \ c^2 = h^2 + k^2 \ \dots\dots\dots\dots\dots\dots\dots\dots\dots\dots\dots\dots(1)$$

$$\text{In rt. } \Delta \ BCD, \ \ a^2 = h^2 + (b - k)^2 \ \dots\dots\dots\dots\dots\dots\dots\dots\dots(2)$$

$$a^2 = h^2 + b^2 - 2bk + k^2 \ \dots\dots\dots\dots\dots\dots\dots(3)$$

Equation (3) - Equation (1): $a^2 - c^2 = h^2 + b^2 - 2bk + k^2 - h^2 - k^2$

$$a^2 = c^2 + b^2 - 2bk$$

$$a^2 = b^2 + c^2 - 2bc \cos A \quad (k = c \cos A, \text{from } \cos A = \frac{k}{c})$$

Therefore, the derivation is complete.

Case 2 For an for obtuse triangle

Given   Obtuse triangle ABC with sides a, b, c

Required: To derive $a^2 = b^2 + c^2 - 2bc \cos A$

Derivation

Consider the obtuse triangle ABC shown labeled with the standard notation for corresponding sides and angles.

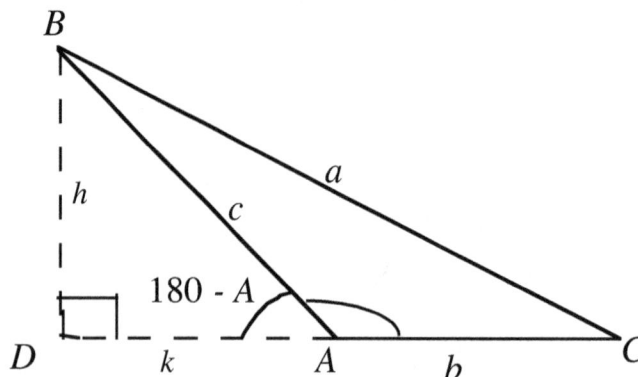

Draw the height h of triangle ABC by drawing $\overline{BD}$ perpendicular to $\overline{CA}$ extended (produced), and note that the height h of the triangle falls outside the given triangle .
Let $AD = k$. Then $DC = (b + k.)$ We will consider right triangles BAD and BCD ; apply the Pythagorean theorem to both triangles and then eliminate h and k in terms of the given sides.

$$\text{In rt. } \Delta\ BAD, \ c^2 = h^2 + k^2 \dots\dots\dots\dots\dots\dots\dots\dots\dots\dots\dots\dots\dots(1)$$

$$\text{In rt. } \Delta\ BCD, \ a^2 = h^2 + (b + k)^2 \dots\dots\dots\dots\dots\dots\dots\dots\dots\dots(2)$$

$$a^2 = h^2 + b^2 + 2bk + k^2 \dots\dots\dots\dots\dots\dots\dots\dots(3)$$

Eqn. (3) - Eqn (1): $a^2 - c^2 = h^2 + b^2 + 2bk + k^2 - h^2 - k^2$

$$a^2 = c^2 + b^2 + 2bk$$

$$a^2 = b^2 + c^2 + 2b(-c \cos A) \quad (k = -c \cos A, \text{ from } \cos (180 - A) = \tfrac{k}{c})$$

$$a^2 = b^2 + c^2 - 2bc \cos A$$

Expressing k in terms of C and A. (Case 2): Since the given angle $\angle A$ is not in right triangle BAD , we have to express $\angle A$ indirectly in terms of its supplement (180 - A). $\cos (180 - A) = -\cos A$.

In triangle BAD, $\cos A = -\cos (180 - A)$ or $\cos (180 - A) = -\cos A = \dfrac{k}{c}$, and from which

$k = -c \cos A$
Again, we obtain the same result as for the acute triangle.

One of the objectives in the above derivation (and subsequent derivations) is to introduce the student to the techniques of deducing statements from other statements as well as in proving statements in mathematics, generally.

Law of Cosine for the Right Triangle (a special case)

For a right triangle , let $m\angle A = 90°$. Then the law of cosines, $a^2 = b^2 + c^2 - 2bc \cos A$ becomes

$$a^2 = b^2 + c^2 - 2bc \cos 90°$$

$$a^2 = b^2 + c^2 - 2bc\,(0) \qquad\qquad (\cos 90° = 0)$$

$$a^2 = b^2 + c^2 \;\leftarrow\text{- - - -}\text{The Pythagorean theorem}$$

We may conclude that the law of cosines is a generalization of the Pythagorean theorem.

The Law of Sines

Use of the Law of Sines.

The direct use of the law of sines will apply in the following cases:

(1) Given **two angles** and an **included side** (ASA) and therefore, we are to find the remaining   sides and angle. This case yields a unique solution.

(2) Given two sides and an angle (ASS or SSA) there may be only one solution (one triangle), two solutions (two triangles) or there may be no solution at all (no triangle) depending upon the relative lengths of the sides and the measures of the angles. This case is commonly known as the ambiguous case of the law of sines.

A mnemonic device for its application is:
(1) sin ASA - pronounced "sinasa" **unique case**
(2) ASS - t**he ambiguous case**

Statement of the Law of Sines

The law of sines states that in any triangle, the sides are proportional to the sines of the angles opposite (facing) these sides. That is, the ratio of the length of a side to the sine of the angle opposite this side is equal to the ratio of the length of any other side to the sine of the angle opposite its side

Using standard notation for angles and sides,

a is to **sin A** as b is to **sin B** as c is to **sin C**

Symbolically, $\dfrac{a}{\sin A} = \dfrac{b}{\sin B} = \dfrac{c}{\sin C}$ (equivalent to $\dfrac{\sin A}{a} = \dfrac{\sin B}{b} = \dfrac{\sin C}{c}$)

Derivation of the Law of Sines

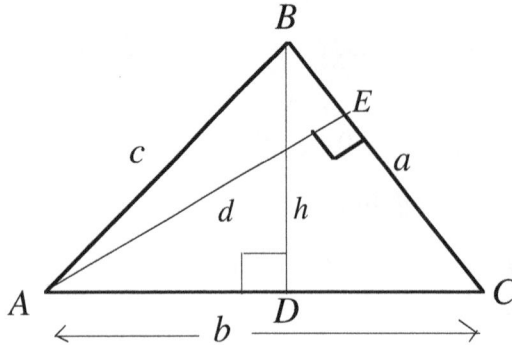

Given: $\triangle ABC$

Derivation   From B, draw $\overline{BD}$ perpendicular to $\overline{AC}$.

In right $\triangle ABD$, $\sin A = \dfrac{h}{c}$

$$h = c \sin A ....(1)$$

Similarly, in right $\triangle CBD$, $\sin C = \dfrac{h}{a}$

$$h = a \sin C ....(2)$$

Set right - hand sides of (1) and (2) equal to each other. Then

$$c \sin A = a \sin C$$

$$\frac{c \sin A}{\sin C \sin A} = \frac{a \sin C}{\sin A \sin C} \quad \text{(Divide both sides of the equation by } \sin A \text{ and by } \sin C;$$

$$\frac{c}{\sin C} = \frac{a}{\sin A} \qquad (3) \quad \text{(or solving for } \frac{c}{\sin C} \text{)}$$

Similarly, we draw $\overline{AE}$ perpendicular to $\overline{BC}$.

In right $\triangle AEC$, $\sin C = \dfrac{d}{b}$

$$d = b \sin C ....(4)$$

and in right $\triangle AEB$, $\sin B = \dfrac{d}{c}$

$$d = c \sin B ....(5)$$

Set right - hand sides of (4) and (5) equal to each other. Then

$$b \sin C = c \sin B$$

$$\frac{b \sin C}{\sin B \sin C} = \frac{c \sin B}{\sin C \sin B} \quad \text{(Divide both sides of the equation by } \sin C \text{ and by } \sin B)$$

$$\frac{b}{\sin B} = \frac{c}{\sin C} \qquad (6) \quad \text{(or solving for } \frac{b}{\sin B} \text{)}$$

From (3), $\qquad \dfrac{a}{\sin A} = \dfrac{c}{\sin C}$

Therefore, $\quad \dfrac{a}{\sin A} = \dfrac{b}{\sin B} = \dfrac{c}{\sin C}$ <-------The law of sines.

Lesson 85 Exercises

1. State the Law of Cosines;   **2.** State the Law of Sines

3. Derive the Law of Cosines;   **4.** Derive the Law of sines.

Lesson 86

Applications of the Laws of Cosine and Sines

Example 1: Solve $\triangle$ ABC given that, $m\angle C = 150°$, $b = 3, c = 8$

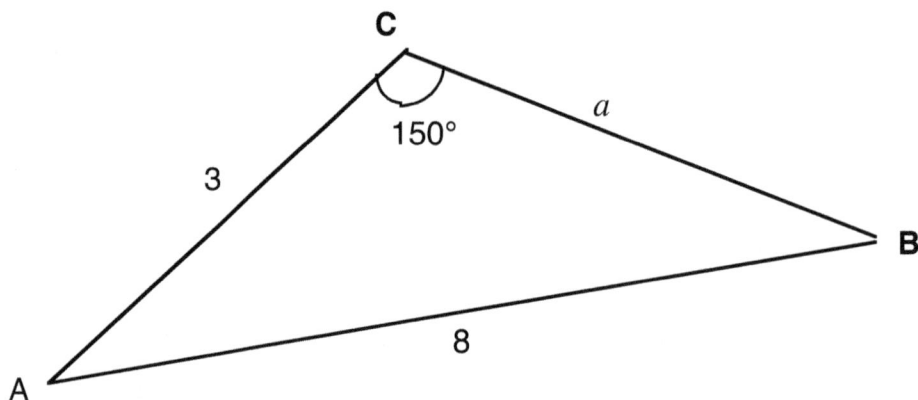

To **solve the triangle**, we will find the remaining three parts, namely the length a, $m \angle A$, and $m \angle B$.

 Solution: We **cannot** apply the Pythagorean theorem, since $\triangle$ ABC is **not** a right triangle.

   We will try the law of sines, or the law of cosines.

We apply the law of sines: $\dfrac{a}{\sin A} = \dfrac{b}{\sin B} = \dfrac{c}{\sin C}$

Step 1:   We apply $\dfrac{c}{\sin C} = \dfrac{b}{\sin B}$

$$\frac{8}{\sin 150°} = \frac{3}{\sin B}$$

$$\frac{8}{\sin 30°} = \frac{3}{\sin B} \qquad\qquad (\sin 150° = \sin 30° = \tfrac{1}{2})$$

$$\frac{8}{\frac{1}{2}} = \frac{3}{\sin B} \qquad\qquad (b = 3;\ c = 8;\ \sin 150° = \sin 30° = \tfrac{1}{2})$$

$$8 \sin B = \frac{3}{2} \qquad\qquad \text{(by cross-multiplication or otherwise)}$$

$$\sin B = \frac{3}{2} \times \frac{1}{8} = \frac{3}{16}$$

$$= .1875$$

Step 2:   Find the angle whose sine is .1875 (From tables or calculator)
      then, $m \angle B = 10.81°$
      $m \angle A = 180° - (150° + 10.8°) = 19.19°.$

Finding a

Method 1 Given: $c = 8$, m $\angle$ C= $150º$, $b = 3$

Using the law of cosines:

$$c^2 = a^2 + b^2 - 2ab \cos C \quad (\angle\ C \text{ is an included angle})$$

$$8^2 = a^2 + (3)^2 - 2a(3) \cos 150º$$

$$64 = a^2 + 9 - 2a(3)\left(-\frac{\sqrt{3}}{2}\right) \qquad (\cos 150° = -\cos 30° = -\frac{\sqrt{3}}{2})$$

$$64 = a^2 + 9 + 2a(3)\frac{\sqrt{3}}{2}$$

$$64 = a^2 + 9 + 3\sqrt{3}\ a$$

$$0 = a^2 + 3\sqrt{3}\ a - 55 \quad \longleftarrow\text{quadratic equation}$$

Solve by the quadratic formula.
$$a^2 + 3\sqrt{3}a - 55 = 0$$

$$a = \frac{-3\sqrt{3} \pm \sqrt{(3\sqrt{3})^2 - 4(1)(-55)}}{2} \qquad (a = 1, b = 3\sqrt{3}, c = -55)$$

$$= \frac{-3\sqrt{3} \pm \sqrt{9(3) + 220}}{2}$$

$$= \frac{-3\sqrt{3} \pm \sqrt{247}}{2}$$

$$= \frac{-3\sqrt{3} \pm 15.71}{2} = \frac{-3\sqrt{3} + 15.71}{2} \quad \text{or} \quad \frac{-3\sqrt{3} - 15.71}{2}$$
$$a = 5.26 \text{ or } -10.45$$

we reject the negative root since the length of a side of a triangle is always positive.
$\therefore\ a = 5.26.$

Finding *a*

Method 2 We could also have found *a*, using the law of sines after having found.

m ∠ A = 19.19°

$$\frac{a}{\sin A} = \frac{c}{\sin C}$$

$$\frac{a}{\sin 19.19°} = \frac{8}{\sin 150°}$$

(sin 19.19 = .3287)

(sin 150° = sin 30° = .5)

$$a = \frac{8 \sin 19.19}{\sin 150}$$

$$= \frac{8(.32870)}{.5}$$

$$a = 5.26$$

Again, we obtain the same value for *a*. Method 2 is less involved than Method 1. However, the disadvantage of Method 2, in this particular problem, is that we used a value, m ∠ A = 19.19°, which we calculated; and if there were error in m ∠ A, there would be error in the value of *a*. The advantage of Method 1 is that we used only given values and no calculated values (*c* = 8, m ∠ C = 150, *b* = 3); but then, we had to solve a quadratic equation.

Example 2 Solve Δ ABC given that m ∠ C = 130° , *a* = 6, *b* = 9.

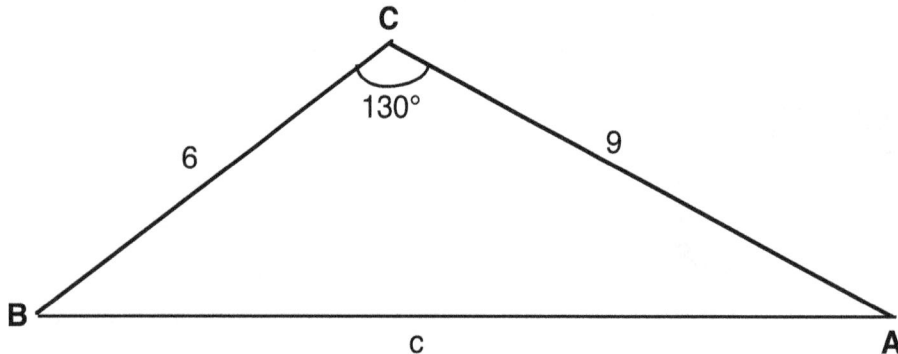

We will find m ∠ B, M ∠A and length *c*.

Finding *c*

$$c^2 = a^2 + b^2 - 2ab \cos C \quad \text{<-------Law of cosines}$$

$$c^2 = 6^2 + 9^2 - 2\ (6)(9) \cos 130°$$

$$= 36 + 81 - 108(-.642) \qquad (\cos 130° = -.642)$$

$$= 36 + 81 + 69.42$$

$$c^2 = 186.42$$

$$c = \sqrt{186.42}$$

$$c = 13.65$$

(Note above that if we had applied the law of sines to find c, we would have had one equation with two unknowns; and we would not be able to obtain a numerical value for c without an additional equation with the same unknowns.)

Finding M ∠ A:

$$\frac{c}{\sin 130^\circ} = \frac{a}{\sin A} \qquad \text{(law of sines)}.$$

$$\frac{13.65}{\sin 130} = \frac{6}{\sin A} \quad (c = 13,65, a = 6)$$

We solve for sine A.

$$\sin A = \frac{6 \sin 130}{13.65}$$

$$= \frac{6(.766)}{13.65}$$

$$\sin A = .3367$$

Find inverse sine of .3367 (from tables or a calculator)

(That is, find the angle whose sine is .3367)

m ∠ A = 19.68

m ∠ B = 180 - (130 + 19.68)

m ∠ B = 30.32

(We could also use the law of sines to find m ∠ B, and then check that the sum of the measures of the angles equals 180°)

∴   c = 13.65 ; m ∠ A = 19.68° ; m ∠ B = 30.32°

When do we apply the Pythagorean theorem, the law of cosines, or the law of sines?

We have learned how to solve triangles using the Pythagorean theorem, the law of sines and the law of cosines. Given a triangle to solve, how do we know which theorem or law to apply? Which theorem or law to apply depends on what we are given and what we want to find.

1. We may apply the Pythagorean theorem only if the triangle is a right triangle. However, if by construction, we are able to obtain right triangles, then we can apply the theorem to any such triangles obtained.

2. For the law of cosines:

(a) Given **two sides and an included angle** (SAS), use of the law of cosines yields a unique solution (i.e. one triangle). In this case, we will find the remaining side and the measures of two angles.

(b) Given all **three sides** (SSS), the application of the law of cosines yields a unique solution. In this case, we will find the measures of all the angles.

3. For the law of sines:

(a) Given **two angles** and **any side** (ASA, AAS, SAA), the use of the law of sines will yield a unique solution (one triangle).

Note that given any two angles, we can easily find the third angle, since the sum of the measures of the angles of a triangle equals 180°.

(b) **Ambiguous Case:** Given **two sides** and **an angle opposite one of the sides** (ASS or SSA), there may be only one solution (one triangle), two solutions (two Δ's), or there may be no solution at all (no triangle) depending on relative lengths of the sides and the measures of the angles

Note also that if the measure of any angle given is 180° or more there is no solution, since the sum of the measures of the three angles of a triangle is 180°.

Also, if the sum of the measures of any two angles given is 180° or more there is no solution.

Note also that if the triangle is a **right triangle**, we can consider applying any of the six trigonometric functions (ratios): sine, cosine, tangent, cosecant, secant, and cotangent.

Question: How do we remember the above cases? Answer: By practice, repetition and finding mnemonic devices.

Lesson 86 Exercises

A Solve the triangle:

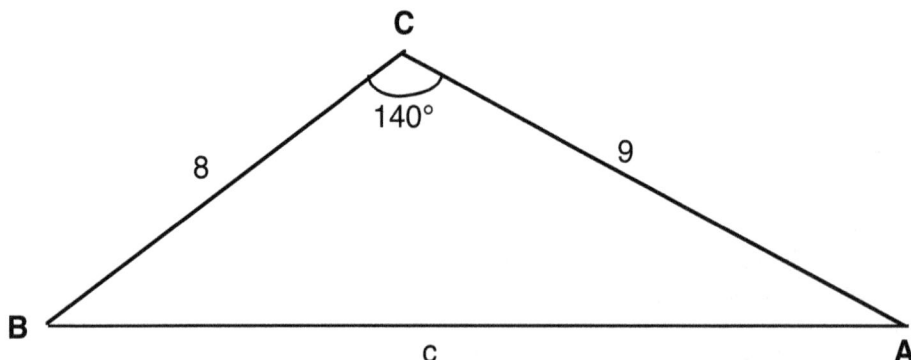

Solutions: c = 15.98;   m ∠ B = 21.2° ;   m ∠ A = 18.8°
≈ 16;   ≈ 21°;   ≈ 19°

B Solve the following triangles:

1.

2.

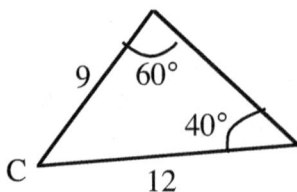

Solve the following triangles using the given dimensions:

3. $a = 15$; $m\angle A = 50°$; $m\angle B = 110°$;   **4.** $b = 8$; $m\angle B = 75°$; $m\angle C = 55°$

Answers: **1.** $c = 17$, $m\angle A = 20°$; $m\angle B = 25°$; **2.** $c = 13.7$, $m\angle C = 80°$
3. $m\angle C = 20°$; $b = 18$; $c = 6.7$; **4.** $m\angle A = 50°$; $a = 6.3$; $c = 6.8$

C   Determine the measure of the largest angle to the nearest degree.

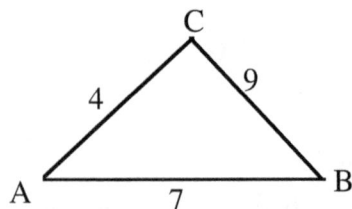

Answer: $m\angle A = 107°$ (Hint: Largest angle is opposite longest side)

CHAPTER 34

Application of Trigonometry to Vectors

Lesson 87: **Basic Definitions; Representation of Vectors**
Lesson 88: **Addition (Sum, Resultant, or Composition) of Vectors**

For anyone who will study physics, mechanics or engineering, this chapter is a very important one. The mastery of the principles and techniques for resolving vectors into components as well as finding the resultant of a number of vectors will relieve you of the mathematical aspects in your first year physics and mechanics courses.

Lesson 87
Basic Definitions; Representation of Vectors

Measurable quantities can be classified either as scalars or as vectors.

Scalar Quantity

Definition   A **scalar quantity** is a measurable quantity which has only magnitude and it is specified by a number and units.

Examples   (1) **Speed**
A car travels at 20 miles per hour.
magnitude

(2) **Mass**
The mass of a block of wood is 6 kilograms.
magnitude

(3) **Distance**
The distance between the points A and B is 8 meters.
magnitude

Addition of Scalar Quantities

We add scalar quantities by ordinary arithmetic addition.

Example   3 kilograms + 5 kilograms = 8 kilograms

Vector Quantity

Definition: A **vector quantity** is a quantity that has both magnitude and direction.
We specify a vector quantity by giving its magnitude and its direction.

Examples (1) **Displacement**.
A car travels 20 miles northwards
magnitude   *direction*

(2) **Force** of 40 lbs acts on a body vertically downwards.
magnitude   *direction*

(3) **Velocity**
A train travels at 50 mph 60° east of north.
magnitude   *direction*

Graphical Representation of Vector Quantities

We represent a vector quantity by a line segment drawn to scale and with an arrow at one end. The length of the arrow represents the magnitude of the vector and the direction of the arrow head represents the direction of the vector. The above drawing, shown below, can be done by using a ruler and a protractor.

We must specify the scalar unit used.

Examples (a)   Represent graphically a force of 10 lbs in a direction 60° with the positive x-axis

Solution See Figure below

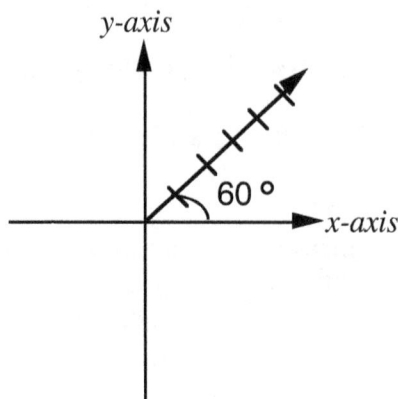

Scale: 1 unit = 2 lb

(b)   Represent graphically a displacement of 40 meters in direction -45° below the horizontal.

Solution See Figure

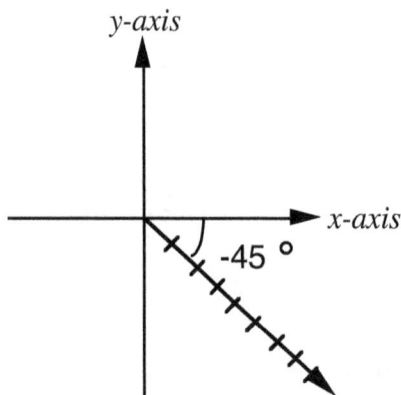

Scale: 1 unit = 5 meters

Symbolic Representation of Vectors   559

We must note the following differences carefully when reading and writing vectors symbolically:

In writing as done on chalkboard or in notebook by hand

We indicate that a quantity is a vector by placing an arrow over the symbol(s) (letter (s)).

Examples: (1) Using the letter F to represent a vector, we would write $\vec{F}$.

(2) Using the letter d to represent a vector we would write $\vec{d}$.

(3) Similarly in using the letter x we write $\vec{x}$.

(4)- Using two upper-case letters O and P , we write $\overrightarrow{OP}$ (Fig. ...)

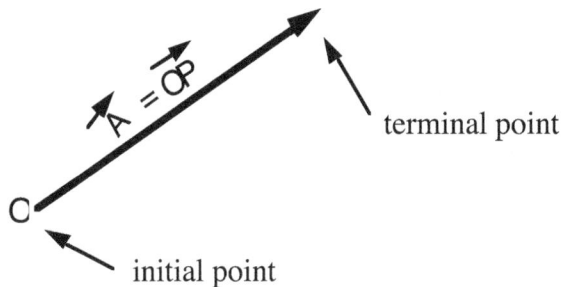

terminal point

initial point

The **magnitude** of the vector, $\vec{F}$, symbolized by $\left|\vec{F}\right| = F$

Written without the arrow, the letter represents the magnitude of the vector.

In printing as it is done in textbooks

A bold face letter is used to indicate that a quantity is a vector.

Examples   (1) The vector $\vec{F}$ would be printed **F** (bold face), with magnitude F .

(2) The vector $\vec{V}$ would be printed **V** (bold face)m with magnitude V .

Lesson 87 Exercises

1. Distinguish between a scalar quantity and a vector quantity.

2. How do we add scalar quantities, and how do we add vector quantities?

Lesson 88 560

Addition (Sum, Resultant, or Composition) of Vectors

We add vectors by geometric methods. However, if the vectors have the same line of action then the magnitude of the resultant can be found algebraically. The geometric methods are either graphical or analytic.

Graphically, we can use:

1. The vector polygon method (for any number of vectors) ;

2. The parallelogram law method (for two vectors at a time).

Analytically, we can use:

1. The law of cosines and the law of sines (for two vectors at a time) and

2. The component method (for any number of vectors). The component method is the one that we shall use most since it can be applied to any number of vectors at a time.

Definition **Resultant**

The resultant or the sum of a number of vectors is that single vector which when acting alone (so far as motion is concerned) has the same effect as the original vectors acting together.

How to Resolve ("break-up") a Vector into Rectangular components

Definition   A **component** of a vector in a specified direction is the effective value of the vector in   the specified direction. By **rectangular components**, we mean the components must   be at right angles (perpendicular) to each other.

Procedure: We will draw a rectangle on the vector being resolved.

Step 1: Draw the rectangle so that the vector being resolved forms the diagonal of the rectangle and that the vector components will form the sides (adjacent sides) of the rectangle.

Step 2:   Use the properties of a right triangle to find the components as illustrated in the following examples.

Example 1 Given a vector $\mathbf{F}$ acting at an angle θ_x with the positive x-axis, resolve $\mathbf{F}$ into rectangular components along the x- and y-axes.

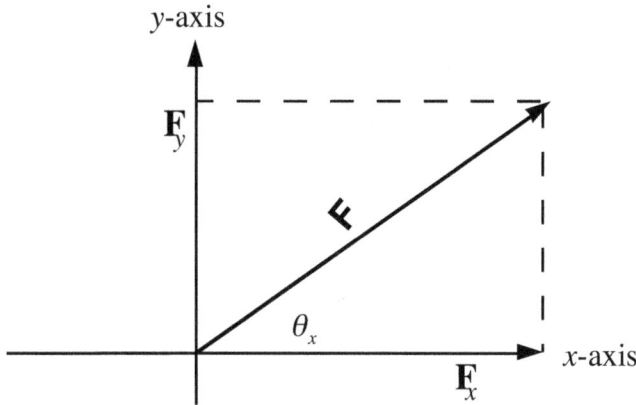

Fig 1

Solution We will call the component in the x-direction (along the x-axis), the x-component and the component in the y-direction (along the y-axis), the y-component (see Fig. 1)

Let the x-component in the positive x-direction (to the right) be positive. Then the x-component in the negative x-direction (to the left) would be negative. Similarly, let the y-component in the positive y-direction (up) be positive. Then the y-component in the negative y-direction (down) is negative. Note that the choice of positive and negative directions is arbitrary and that when convenient we may reverse the positivity or negativity of any of the directions. However we must be consistent with the choice of directions in any particular problem.

Continuing, and applying right-triangle trigonometry:
The horizontal component (x-component) $\mathbf{F}_x = +F\cos\theta_x$ (the plus sign indicates the positive direction)
The vertical component (y-component) $\mathbf{F}_x = +F\sin\theta_x$ (the plus sign indicates the positive direction)

(**Note** that if θ, say θ_y, is between $\mathbf{F}$ and the y-axis, then $\mathbf{F}_x = +F\sin\theta_y$ and $\mathbf{F}_y = +F\cos\theta_y$)

Continuing,

$$F^2 = F_x^2 + F_y^2 \qquad \text{(Applying the Pythagorean Theorem)}$$

$$\tan\theta_x = \frac{F_y}{F_x} \quad \text{or} \quad \theta_x = \tan^{-1}\left(\frac{F_y}{F_x}\right) \quad \text{(We will use these relationships in the next section)}$$

Example 2 Given that $F = 10$ lbs, $\theta_x = 30°$, resolve **F** into horizontal and vertical components.   562

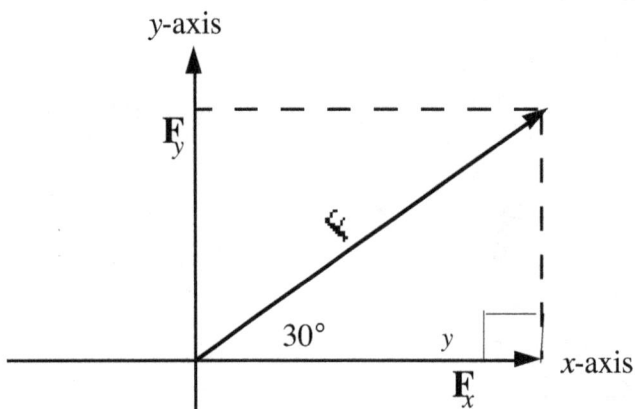

Solution

$$\mathbf{F}_x = +(10 \; lb)\cos 30°$$

$$= +10(.866) \; lb \qquad (\cos 30° = .866)$$

$$\mathbf{F}_x = +8.66 \; lb \quad \text{(the plus sign indicates that the component is in the positive } x \text{ - direction)}$$

$$\mathbf{F}_y = +(10 \; lb)\sin 30°$$

$$= +10(.5) \; lb \qquad (\sin 30° = .5)$$

$$\mathbf{F}_x = +5 \; lb \quad \text{(the plus sign indicates that the component is in the positive } y \text{ - direction)}$$

Therefore, the x-component = 8.66 lbs to the right; and the y-component is 5 lbs up.

Example 2 Resolve into horizontal and vertical components: $F = 40$ units, inclined at
$\theta_x = 60°$ below the horizontal, as shown in **Figure** below.

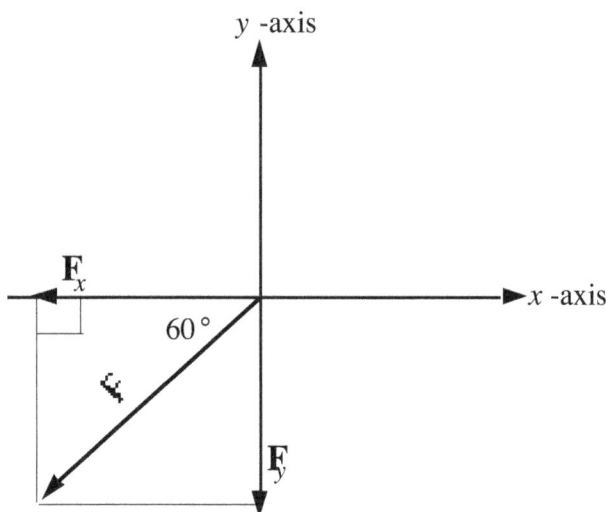

Solution Assume the horizontal components to the right are positive, and vertical components up are positive; then the horizontal components to the left are negative and the vertical components down are negative.

Horizontal component, $\mathbf{F}_x = -F\cos 60°$ units

$$= -(40)\cos 60° \text{ units}$$

$$= -(40)(.5) \text{ units} \qquad (\cos 60° = .5)$$

$$\mathbf{F}_x = -20 \text{ units}$$

(the minus sign shows that this component is in the negative x - direction)

Vertical component, $\mathbf{F}_y = -F\sin 60°$ units

$$= -(40)\sin 60° \text{ units}$$

$$= -(40)(.866) \text{ units} \qquad (\sin 60° = .866)$$

$$= -34.64 \text{ units}$$

$$\mathbf{F}_y = -34.64 \text{ units}$$

(the minus sign shows that this component is in the negative y - direction)

The horizontal component is 20 units to the left; and the vertical component is 34.64 units down.

The main applications of the resolution of vectors into rectangular components are (a) to find the resultant (sum) of a number of vectors (displacements, forces, velocities, accelerations, etc.) and (b) to solve problems in Physics.

Component Method of Vector Addition

Procedure

Step 1: Resolve each vector into rectangular components (say, the horizontal component and the vertical component) along any convenient pair of axes.
(say x-and y-axes). We usually call the horizontal component the x-component and the vertical component the y-component.
Note that the signs of the non-zero components are either positive or negative.

Step 2: Add the x-components, noting their signs. Let the sum of the x-components be R_x.

Step 3: Add the y-components, noting their signs. Let the sum of the y-components be R_y.

Step 4: Draw R_x and R_y and let their resultant (sum) be R. The magnitude of R is given by:

$$R = \sqrt{R_x^2 + R_y^2} \qquad \text{(by the Pythagorean theorem)}$$

Step 5: Find θ, the standard-position angle giving the direction of the resultant.
Let θ_{ref} be the reference angle (the angle between the x-axis and the resultant).

$$\tan\theta_{ref} = \frac{R_y}{R_x} \quad \text{or}$$

$$\theta_{ref} = \tan^{-1}\left(\frac{R_y}{R_x}\right)$$

(θ_{ref} can be found from calculator or tables.)

Note that $\tan\theta$ is positive either in the first quadrant or in the third quadrant, but negative in either the second or fourth quadrant. We consider four cases of the location of θ_{ref} or θ.

Case 1: Both R_x and R_y are positive. In this case, θ_{ref} is in the first quadrant, and $\theta = \theta_{ref}$.

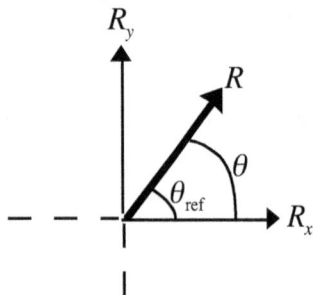

Case 2: Both R_x and R_y are negative. In this case, θ_{ref} is in the third quadrant, and $\theta = \theta_{ref} + 180°$.

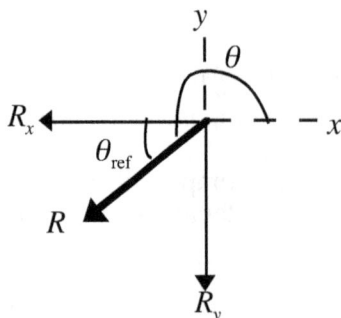

Note that saying that θ is in a specified quadrant means that the terminal side of θ lies in that quadrant.

Case 3: R_x is negative and R_y is positive. In this case, θ_{ref} is in the second quadrant and

$\theta = 180° - \theta_{ref}$

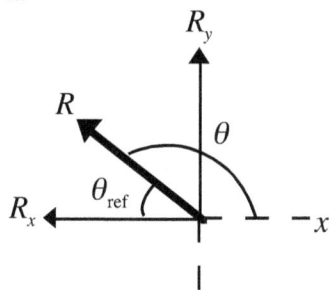

Case 4: R_x is positive and R_y is negative . In this case, θ_{ref} is in the fourth quadrant and

$\theta = 360° - \theta_{ref}$

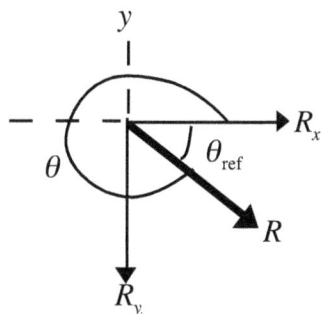

Note: Stating the direction of the resultant

Sometimes, after determining θ_{ref} (instead of stating the standard-position angle), the direction of the resultant may be specified with reference to one of the vectors (involved in the vector addition), or with respect to either the x- or y-axis.

Example 3 Find the resultant of the following forces acting on a particle at the origin.

$\mathbf{F}_1 = 200$ units at $60°$ to the horizontal; $\mathbf{F}_2 = 120$ units along the positive x-axis;

$\mathbf{F}_3 = 150$ units at $45°$ to the horizontal.

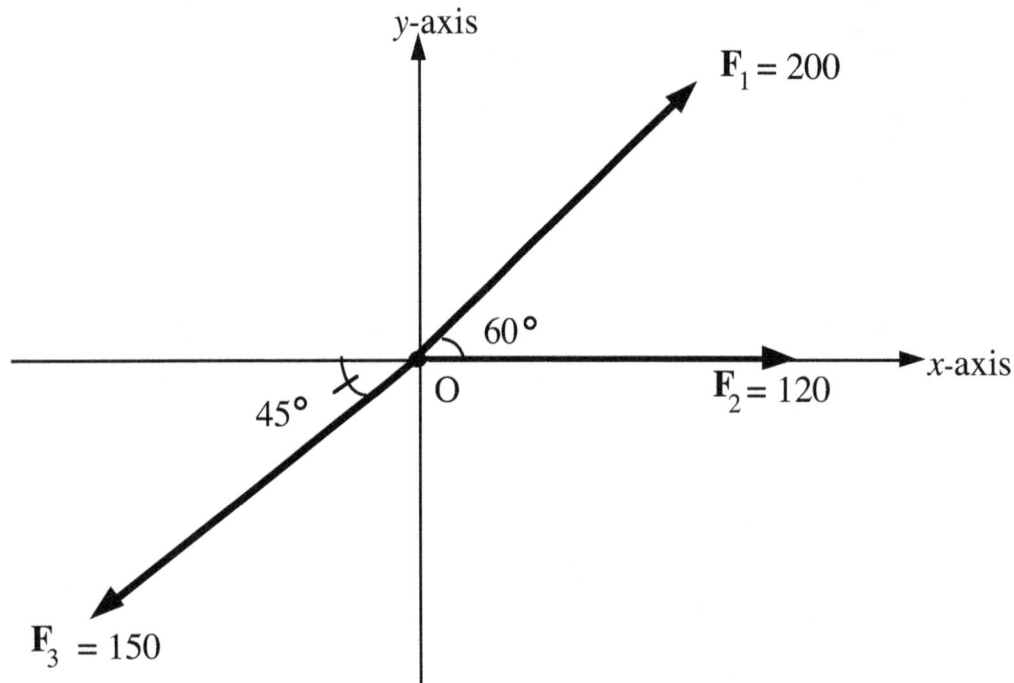

Solution Let the forces in the $+x$-direction be positive and let those in the $+y$-direction be positive.

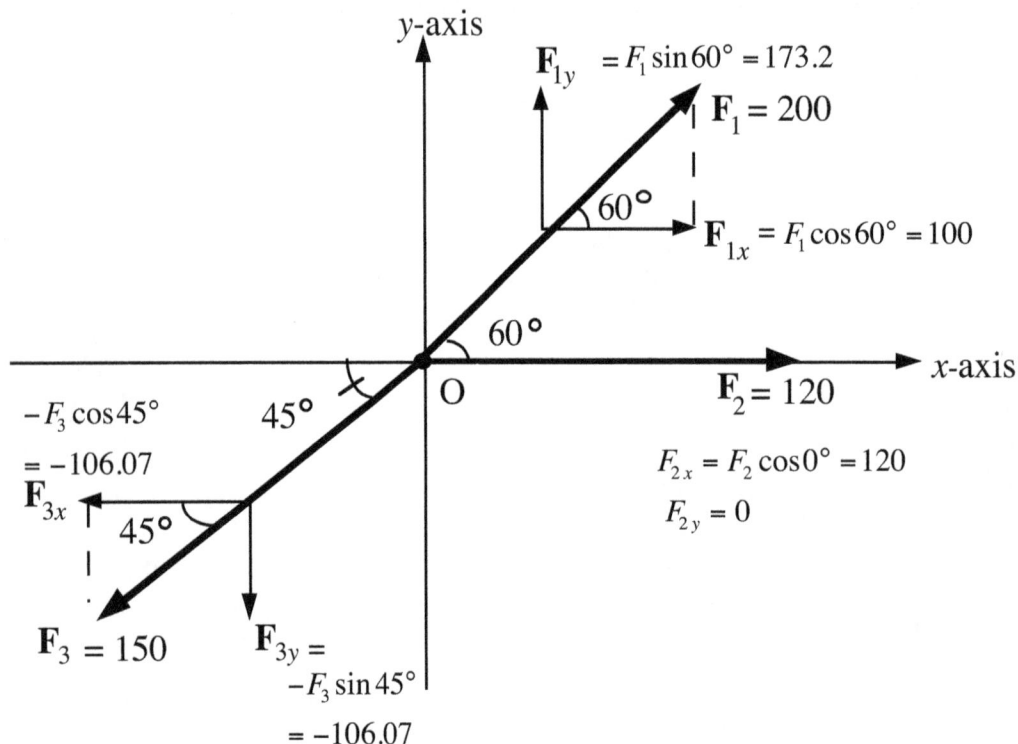

Step 1: Calculations on $\mathbf{F}_1$

$$\mathbf{F}_{1x} = +F_1 \cos 60° \qquad\qquad \mathbf{F}_{1y} = +F_1 \sin 60°$$

$$= +200(.5) \qquad\qquad\qquad = +200(.866)$$

$$\mathbf{F}_{1x} = +100 \text{ units} \qquad\qquad \mathbf{F}_{1y} = +173.2 \text{ units}$$

Step 2: Calculations on $\mathbf{F}_2$

$$\mathbf{F}_{2x} = +F_2 \cos 0° \qquad\qquad \mathbf{F}_{2y} = +F_2 \sin 0°$$

$$= +120(1) \qquad\qquad\qquad = +120(0)$$

$$\mathbf{F}_{2x} = +120 \text{ units} \qquad\qquad \mathbf{F}_{2y} = 0 \text{ units}$$

Note above that if the given force ($\mathbf{F}_{2y}$, in Step 2) is along the x-axis, then the y-component is zero, and we do not have to do any calculations. Similarly, if the given force were along the y-axis, the x-component would be zero.

Step 3: Calculations on $\mathbf{F}_3$

$$\mathbf{F}_{3x} = -F_3 \cos 45° \qquad\qquad \mathbf{F}_{3y} = -F_3 \sin 45°$$

$$= -(150)(.7071) \qquad\qquad = -(150)(.7071)$$

$$\mathbf{F}_{3x} = -106.07 \text{ units} \qquad\quad \mathbf{F}_{3y} = -106.07 \text{ units}$$

Step 4: Add the x-components

$$R_x = F_{1x} + F_{2x} + F_{3x}$$

$$= (+100) + (+120) + (-.106.07)$$

$$= 100 + 120 - 106.07$$

$$R_x = +113.93 \text{ units}$$

Step 5: Add the y-components

$$R_y = F_{1y} + F_{2y} + F_{3y}$$

$$= +(173.2) + 0 + (-106.07)$$

$$= +173.2 - 106.07$$

$$R_y = +67.13 \text{ units}$$

Step 6: Find the resultant R.

$$R = \sqrt{R_x^2 + R_y^2}$$

$$R = \sqrt{(113.93)^2 + (67.13)^2}$$

$$R = 132.24 \text{ units}$$

Step 7: Find the direction, θ., where θ is in standard position.

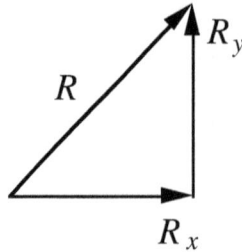

$$\tan \theta_{ref} = \frac{R_y}{R_x}$$

$$= \frac{67.13}{113.93}$$

$$= .589$$

$$\theta_{ref} = 30.5°$$

Since both R_x and R_y. are positive, $\theta = \theta_{ref}$. Therefore, $\theta = 30.5°$
The magnitude of the resultant force is 132.24 units, and is directed at standard-position angle 30.5°.

Finding the direction of the resultant

Let θ be a standard-position angle and let θ_{ref} be the reference angle.

If $\frac{R_y}{R_x}$ is positive, then θ is either in the first or third quadrant , since $\tan \theta$ is positive in these quadrants.

If $\frac{R_y}{R_x}$ is negative, θ is either in the second or fourth quadrant , since $\tan \theta$ is negative in these quadrants.

Example on finding θ, having found R_x and R_y.
Case 1: $R_x = 140$ and $R_y = 50$: For solution, imitate Step 7 of Example 3 to obtain $\theta = 19.65°$.
Case 2: $R_x = -140$ and $R_y = -50$ **Case 3** $R_x = 140$, and $R_y = -50$

    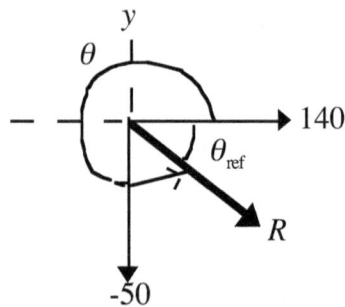

$$\tan \theta_{ref} = \frac{R_y}{R_x} \qquad\qquad \tan \theta_{ref} = \frac{R_y}{R_x}$$

$$= \frac{-50}{-140} \qquad\qquad\qquad = \frac{-50}{140}$$

$$= .36 \qquad\qquad\qquad\qquad = -.36$$

$$\theta_{ref} = 19.65 \qquad\qquad\qquad \theta_{ref} = 19.65$$

$$\theta = 180 + 19.65 = \mathbf{199.65°} \qquad \theta = 360 - 19.65 = \mathbf{340.35°}$$

Case 4: $R_x = -140$ units and $R_y = 50$ units

$$\tan\theta_{ref} = \frac{R_y}{R_x}$$

$$= \frac{50}{-140}$$

$$= -.36$$

$$\theta_{ref} = 19.65$$

$$\theta = 180 - 19.65 = \mathbf{160.35°}$$

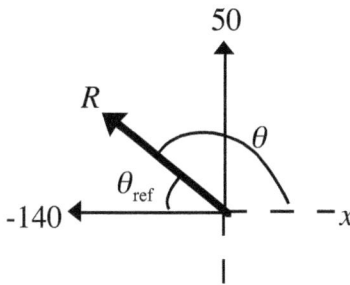

We repeat a previous note below:

Stating the direction of the resultant

Note that sometimes, after determining θ_{ref} (instead of stating the standard-position angle), the direction of the resultant may be specified with reference to one of the vectors (involved in the vector addition), or with respect to either the x- or y-axis.

Lesson 88 Exercises

A   Resolve into horizontal and vertical components:

  Suppose $F = 15$ lbs, $\theta_x = 60°$ resolve $\mathbf{F}$ into horizontal and vertical components.

Answer: $F_x = +7.5 \, \ell b$ (to the right); $F_y = +13 \, \ell b$ (up).

B   Find the resultant of the following forces acting on an object at the origin.

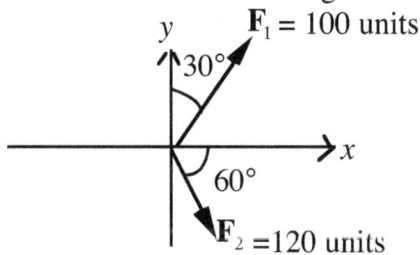

Answer: Resultant = 111 units, at $\theta_x = -9.0°$

CHAPTER 35

Trigonometry of Real Numbers

Lesson 89: **Definitions; Radian Measure; Arc Length; Reference Number**
Lesson 90: **Trig Functional Values of Angles and of Real Numbers**

Lesson 89

Definitions; Radian Measure; Arc Length; Reference Number

Definition of the "radian"

As shown in the figure below, one radian is the measure of an angle with its vertex at the center of a circle and with its sides intersecting an arc of a circle whose length is equal to the radius of the circle.

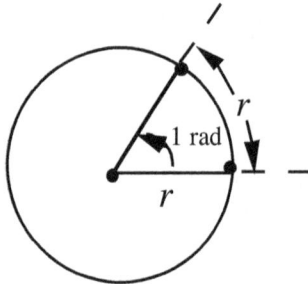

We can therefore say that radian measure = length of arc intersected by the sides of an angle divided by radius of the circle.

i.e., $\dfrac{\text{length of the arc intercepted by the sides of an angle}}{\text{the radius of the circle}}$

The circumference, C, of a circle of radius r is given by $C = 2\pi r$. Also, the circumference generates an angle of $360°$ at a center of the circle.

$$m(\theta) = \frac{\text{arc length}}{\text{radius}}, \text{ where } m(\theta) \text{ is the measure of } \theta.$$

In radians, $m(\theta) = \dfrac{2\pi r}{r}$
$= 2\pi$

Since the circumference subtends an angle of $360°$ at the center (for one revolution)

$2\pi = 360°$ (note that $\pi = 3.14$, approximately)
$\pi = 180°$
$\pi \text{ radians} = 180°$
and $1 \text{ radian} = \dfrac{180°}{\pi}$ degrees
$1 \text{ radian} = 57.3$ degrees approximately

Similarly, since 180 degrees $= \pi$ radians

$1 \text{ degree} = \dfrac{\pi}{180°}$ radians
$1 \text{ degree} = .017$ radians approximately.

Relationship between radian measure and arc length of a circle

Consider an arc (of a circle) of length S which subtends an angle θ at the center of a circle, where θ is in radians. By definition, θ in radians $= \dfrac{\text{arc length}}{\text{radius}}$

$$\theta = \frac{s}{r}$$ (θ must be in radians. If θ is in degrees, it must be

$$s = r\theta$$ converted to radians before substituting in the formula.)

Note also that sometimes we call the angle involved in the above formula, the central angle.

Example 1

Find the arc length intercepted by the central angle of 25 radians in a circle of radius 15 cm.

Solution

$\theta = 25$ radians

$r = 15$ cm

$s = r\theta$

$s = (15)(25)$

$s = 375$ cm

Example 2

Find the arc length intercepted by the central angle of $25°$ in a circle of radius 15 in.

Step 1: Convert the angle in degrees to radians.

$$25° = \frac{25°}{180°}\pi \text{ radians}$$

Step 2: $s = r\theta$

$$= (15)\left(\frac{25°}{180°}\right)\pi \text{ ins.}$$

$$= (15)\left(\frac{25°}{180°}\right)(3.14) \text{ ins.} \qquad (\pi \approx 3.14)$$

$$= 6.54 \text{ in.}$$

Trigonometry of Real Numbers: Circular Functions 572

Some of the uses of trigonometry involve **real numbers** rather than angles. It is therefore important that we obtain definitions in trigonometry in terms of real numbers. Such definitions will reconcile the topics in calculus courses with those in trigonometry. We have already learned trigonometric definitions in terms of angles. We used these definitions in solving triangles.

However, there are problems in science to which we may apply trigonometric definitions even though triangles are not involved in these problems. It has been suggested that, perhaps, if there were no trigonometry of triangles, there would have to be trigonometry of real numbers. By first studying trigonometric definitions in terms of angles, we have become prejudiced with **angles** before we study trigonometry of **real numbers** We shall now make the transition from trigonometry of triangles to trigonometry of real numbers using the **unit circle**.

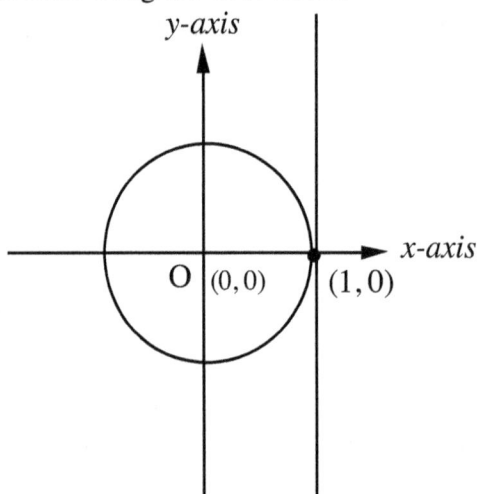

Consider a circle of radius 1 unit drawn in the x-y plane. We will use the same unit for measuring the arc length and the x - and y-axes. Consider also a real number line S with its origin at the point $(1,0)$ on the circle. Assuming that the real number line is flexible we wind the positive half of the line anticlockwise about the circle. Similarly, we wind the negative half of the line in a clockwise direction about the circle. Having completed a circle we withdraw the excess portions of the real number line. Then we obtain the figure below

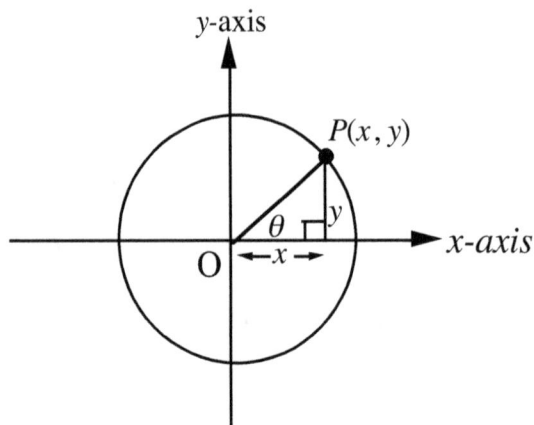

We observe that to every real number on the just wrapped real number line, there corresponds only one point on the circle.

Perhaps we may say that by wrapping the real number line around the unit circle, we obtain" the real number circle". We may thus define "the real number circle as a circle with radius 1 unit and a circumference of 6.28 units.

It is the above relationship which allows us to apply the trigonometry of angles to trigonometry of real numbers directly, when the angles are in radians.

Relationship between the angle θ in radians and the real number S.

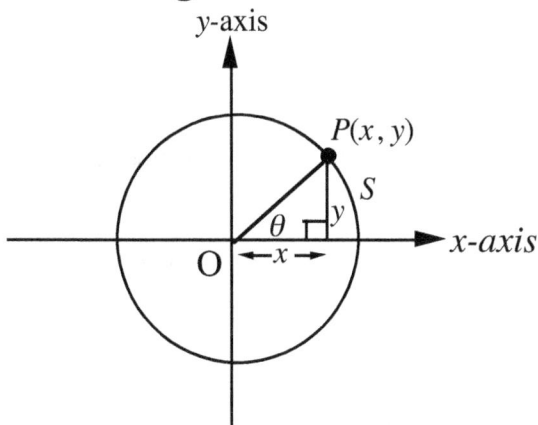

Consider a point $P(x, y)$ on the unit circle of radius r. Let θ radian be an angle in standard position. By definition, arc length $S = r\theta$ (where θ is in radians). For the unit circle, $r = 1$.

$$\text{and } S = (1) \ \theta$$
$$S = \theta \qquad \text{(where } \theta \text{ is in radians)}$$

Since for any real number S, θ (in radians) = S we can replace θ by S in the trigonometric definitions to obtain

$$\sin \theta = \sin S; \quad \cos \theta = \cos S; \quad \tan \theta = \tan S; \quad \csc \theta = \csc S; \quad \sec \theta = \sec S; \quad \cot \theta = \cot S$$

From above and in future if S is any real number, then, for example $\sin S$ may mean either the sine of the real number S or the geometric sine of S (where S in radians). Therefore if S is written without any units it is understood that S is the real number S or the angle S in radians. However if we want to indicate the measure of the angle in degrees we must explicitly indicate the degree measure.

The trigonometric function of the real number S = the trigonometric function of the angle S in radians and the above definitions become

$$\sin S = \sin S \text{ radians}$$

$$\cos S = \cos S \text{ radians}$$

$$\tan S = \tan S \text{ radians}$$

$$\csc S = \csc S \text{ radians}$$

$$\sec S = \sec S \text{ radians}$$

$$\cot S = \cot S \text{ radians}$$

Note above that no conversions are needed if θ is in radians.

Because the circle was used in obtaining these functions, they are sometimes referred to **as circular functions**

The use of the unit circle to approximate values of sin S and cos S for $0 \le S \le 2\pi$

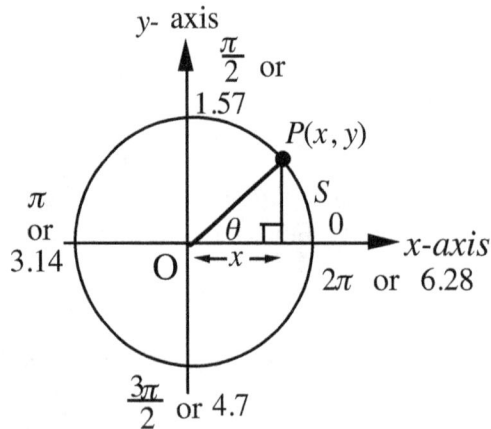

Consider a point $P(x, y)$ on the unit circle, where $x = f(s)$ and $y = f(s)$.

By definition, we have

$$\sin S = \frac{y}{r}$$

With $r = 1$, $\sin S = y.$

Similarly, $\cos \theta = \frac{x}{r}$ and

$$\cos S = x$$

$$\tan S = \frac{y}{x}$$

To find cos S we locate the value of S on the circumference of the unit circle and then read the x-coordinate of this point .(We may draw the perpendicular from this point to the x axis.)

Similarly, to find sin S we locate S on the circumference and then read the corresponding y-value. (We may draw a horizontals to intersect the y-axis)

Reference number:

Let us recall the concept of the reference angle, where $0 \le \theta \le 2\pi$ and θ was in degrees or in radians. We have a similar case in dealing with real numbers but in this case, we have a **reference number** S which we used above in finding the trigonometric values for the cosine and sine functions.

The reference angle in this case is the acute angle in radians between the terminal side and the x-axis. Note that this acute angle, S_{ref}, is between 0 and 1.57 radians or between 0 and $\frac{\pi}{2}$.

Note below that $1.57 = \frac{\pi}{2} = 90°$; $3.14 = \pi = 180°$; $6.28 = 2\pi = 360°$

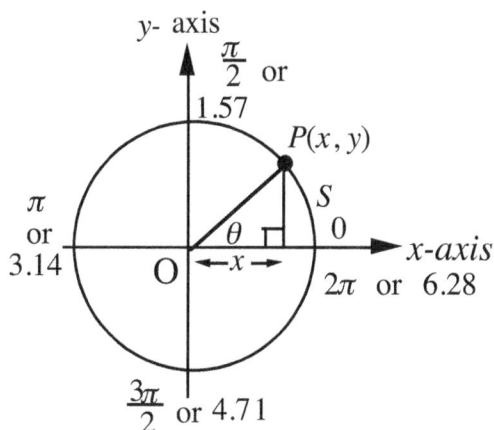

Example 1

Find the reference angle and the reference number for $S = 1.05$

Solution

Step 1:   Sketch the angle 1.05 radian

Step 2: Determine the reference angle which in the above case is the same as the given angle.
The given angle is in the 1st quadrant. (It is acute and therefore less than 1.57 radians)
The reference angle, S_{ref} =1.05.
The reference number is 1.05.

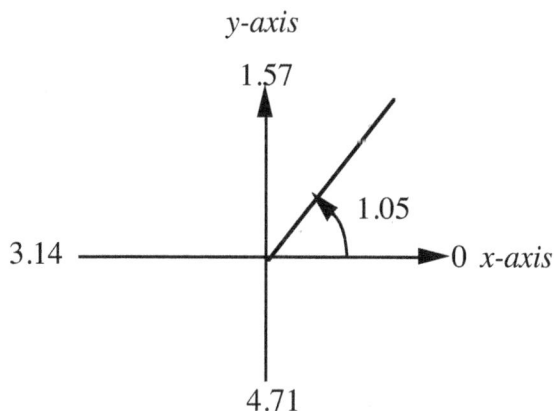

Example 2   Find the reference angle and reference number for $S = 2.09$

Solution

Step 1: Sketch (draw) the angle $S = 2.07$ radian

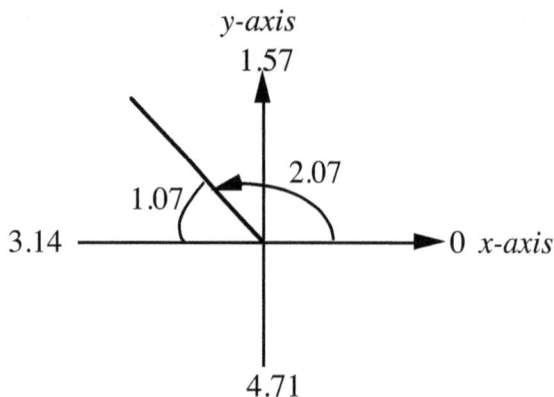

Step 2: The reference angle in this case is found by the difference between 3.14 and 2.07, noting that the reference angle is acute (less than 1.57 radians) and it is always between the terminal side and the horizontal axis. $\theta_{ref} = 3.14 - 2.09 = 1.07$ rads
The reference angle for 2.09 radians is 1.07 rad. The reference number is 1.07.

Also, note above that $\theta_{ref} = \pi - 2.09$ (since $\pi = 3.14$)

Example 3   Find the reference angle and reference number for 4.28.
Solution
Step 1: Draw the angle 4.28 in radians.

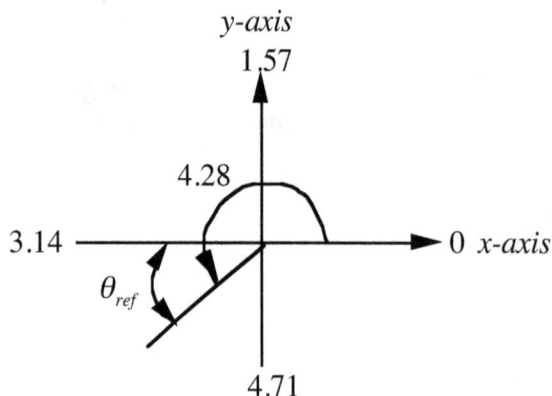

Step 2: Find the reference angle by difference
$$\theta_{ref} = (4.28 - \pi)$$
$$\theta_{ref} = (4.28 - 3.14) \text{ rad} \qquad \text{(Note that } \pi = 3.14 \text{ approximately.)}$$
$$= 1.14 \text{ rad.}$$

The reference angle for 4.28 radians is 1.14 radians
The reference number is 1.14.

Example 4: Find the reference angle and the reference number if $\theta = 5.15$ radians

Solution

Step 1: Sketch $\theta = 5.15$ radians

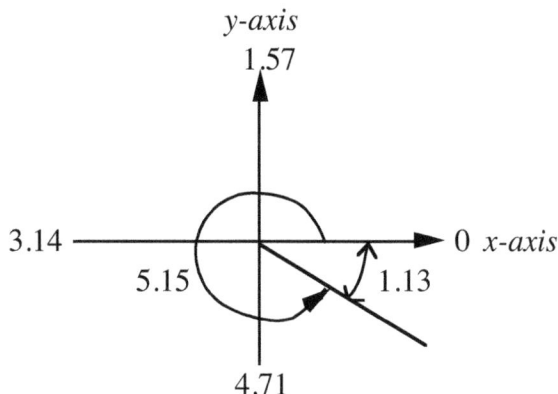

Step 2: $\theta_{ref} = (2\pi - 5.15)$ radians

$= (2(3.14) - 5.15)$ radians

$= (6.28 - 5.15)$ radians

$= 1.13$ radians

The reference angle for 5.15 radians is 1.13 radians.
The reference number is 1.13.

Example 5 Find the reference angle for $\theta = 10.42$.

Solution Refer to both Figures below.

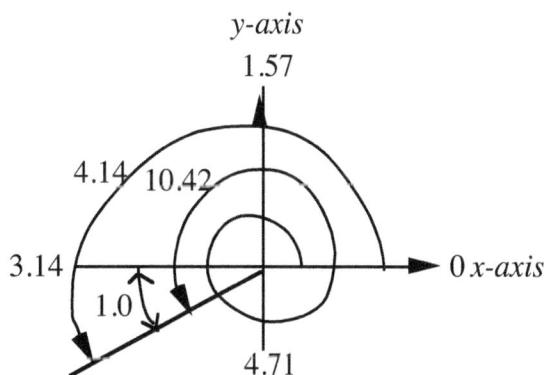

After the first revolution, there are 10.42 - 6.28 = 4.14 radians left over.

The reference angle = 4.14 - 3.14 = 1 rad.
The reference number is 1

Example 6 Find the reference angle in radians and reference number if $\theta = -4.96$

Solution

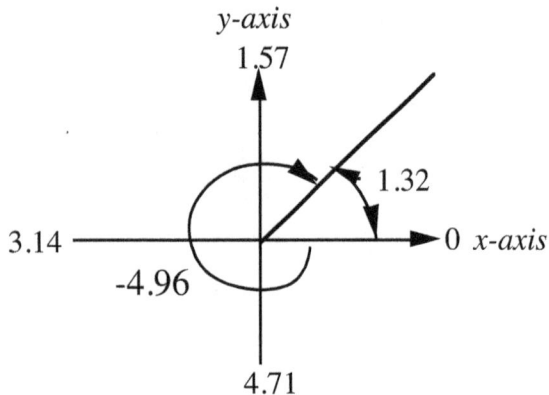

Step 1:   Draw $\theta = -4.96$ in radians (Draw it clockwise.)

Step 2: The reference angle is is 6.28 - 4..96
$$\theta_{ref} = 1.32$$

The reference number = 1.32

Lesson 89 Exercises

Find the reference angle and the reference number for the following:
(a) $\theta = 5.4$ radians;   (b) 11.30;   (c) – 4.92;   (d) – 8.28;   (e) 1.63 radians .

Answers:
(a) 0.88 radians, 0.88;   (b) 1.26 radians, 1.26;   (c) 1.36 radians, 1.36;
(d) 1.14 radians, 1.14;   (e) 1.51 radians, 1.51.

Lesson 90   579

Trig Functional Values of Angles and of Real Numbers

Finding the trigonometric functional value of any angle in radians

Procedure:

Step 1: Sketch the given angle in radians

Step 2: Find the reference angle as described previously (See page 575)

Step 3: From tables or calculator find the value of the trigonometric function.

Step 4: The value from Step 3 is always positive and we must determine the sign of the function according to the quadrant in which the terminal side of the angle lies (see page 530 for how to determine the signs of the functions in various quadrants.)

For some values of θ, using the calculator alone (as in Step 3) would be sufficient.

The following example parallels the Example on page 530 and noting that

$$1.57 = \frac{\pi}{2} = 90°; \quad 3.14 = \pi = 180°; \quad 6.28 = 2\pi = 360°$$

Example Find $\sin \theta$, $\cos \theta$; $\tan \theta$ if (1) if $\theta = 3$; (2) $\theta = 6.8$; (3) $\theta = -4.96$.

Solution

(1) Step 1: Draw $\theta = 3$

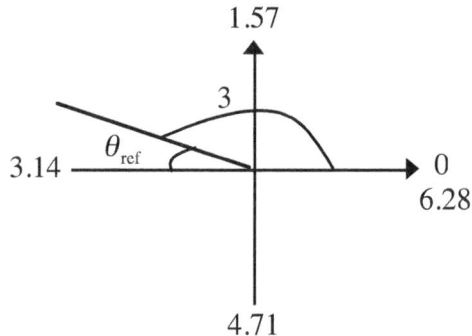

Step 2: Find the reference angle, θ_{ref}.

$$\theta_{ref} = 3.14 - 3 = .14$$

$$\theta_{ref} = .14$$

Step 3: $\sin 3 = 0.14$ ($\sin 0.14 = 0.14$ <---- This result for the sine is interesting. In the future you will learn that for very small values of θ, $\sin \theta = \theta$, a fact we will apply when we cover the derivative of the sine function in calculus)

$$\cos 3 = -0.99; \quad \tan 3 = -0.14$$

(2) For $\theta = 6.8$,

$$\theta_{ref} = 6.8 - 6.28 = 0.52 \text{ and}$$

$$\sin 6.8 = 0.49; \quad \cos 6.8 = 0.87; \quad \tan 6.8 = 0.57$$

(3) For $\theta = -4.96$,

$$\theta_{ref} = 6.28 - 4.96 = 1.32$$

$$\theta_{ref} = 1.32$$

$$\sin (-4.96) = 0.97; \quad \cos (-4.96) = 0.25; \quad \tan (-4.96) = 4.0 .$$

Trigonometric functional value of a real number S 580

Given the real number S, to find its trigonometric functional value, we assume that S is an angle in radians , find the reference angle and then find the value of the corresponding geometrical function of S in radians. (from tables, using a calculator, or from memory).

The domain of trigonometric functions consists of angles but the domain of circular functions consists of real numbers.

Relationship between θ (standard-position angle in radians) and θ_{ref}

1. If θ is in the **first quadrant**, then $\theta_{ref} = \theta$

2. If θ is in the **second quadrant**, then $\theta_{ref} = 3.14 - \theta$ or equivalently $\theta = 3.14 - \theta_{ref}$

3. If θ is in the **third quadrant**, then $\theta_{ref} = \theta - 3.14$ or equivalently $\theta = \theta_{ref} + 3.14$

4. If θ is in the **fourth quadrant**, then $\theta_{ref} = 6.28 - \theta = \theta$ or equivalently $\theta = 6.28 - \theta_{ref}$

Note that θ and θ_{ref} are always in the **same** quadrant. Also note that $\pi \approx 3.14$

Lesson 90 Exercises

A   Find (a) $\sin\theta$; (b) $\cos\theta$ for the following:

(1) $\theta = 5.4$ radians;   (2) $\theta = 11.30$ radians;   (3) $\theta = -4.92$;   (4) $\theta = -8.28$ radians;

(5) $\theta = 1.63$ radians

Answers: 1. (a) -0.77, (b) 0.63;   2. (a) -0.95, (b) 0.30; 3. (a) 0.98, 0.21;

4. (a) -0.91, (b) -0.41;   (5) (a) 1.0, (b) -0.06.

B   Find θ in radians if $0 \le \theta \le 6.28$ given that:

(1) $\sin\theta = -.77$;   (2) $\cos\theta = -0.06$.

Answers: (1) $\theta = 5.4$ rad, 4.02 rad;   (2) $\theta = 1.63$ rad, 4.65 rad.

CHAPTER 36

Graphs of Trigonometric Functions

Lesson 91: Introduction; Labeling the Coordinate Axes;Illustration of the Periodicity of Trigonometric Functions

Lesson 92: Sketching the Graph of $y = \sin x$

Lesson 93: Sketching the Graph of $y = \cos x$

Lesson 94: Graphs of Discontinuous Trigonometric Functions:
$y = \tan x$; $y = \csc x$; $y = \sec x$; $y = \cot x$

Lesson 95: Direct Procedure for Sketching the Graphs of $y - k = a \sin (bx - h)$, and $y - k = a \cos (bx - h)$

Lesson 91

Introduction; Labeling the Coordinate Axes; Illustration of the Periodicity of Trigonometric Functions

Introduction

Having mastered the sketching of polynomial and rational functions, we should **not** develop any phobias when we learn how to sketch trigonometric graphs. The approach and principles are similar to those outlined for graphing polynomial and rational functions.

As in any sketching in an $x-y$ coordinate plane, what we need to do is to obtain important points (ordered pairs) on each curve, plot these points, and connect them to obtain the characteristic shape.

In the main, we should have the following objectives in mind when drawing trigonometric graphs.

1. From which starting point, through which points and to which endpoint do we draw the curve

2. The unit interval for the coordinate axes.

3. The number of the basic curves to draw: This depends on the domain and period of the function.

4. For each basic curve we need **five important points** (critical points) 2 end points (a starting point and a stopping point), 1 mid-point and two quarter-points .
Each point is an ordered pair (x, y). The five points include points at which the curve crosses (intersects) the x- and y-axes; and the highest point and the lowest point on the curve.
The highest and the lowest points are determined by a constant called the amplitude of the function.
Examples: The amplitude of $y = 3\sin x$ is 3 . The amplitude of $y = 2\sin x$ is 2; The amplitude of $y = \sin x$ is 1. The amplitude of $y = \frac{1}{2}\cos x$ is $\frac{1}{2}$.

There are a number of approaches for sketching the graphs of trigonometric functions. Each approach depends on the complexity of the given function.

Case 1: Sketching the graphs of the basic functions $y = \sin x$. $y = \cos x$, $y = \tan x$, $y = \csc x$,
$y = \sec x$, $y = \cot x$. We will first cover the basic curves (one cycle), the sine, cosine, and the tangent functions as well as the basic curves for the cosecant, secant and cotangent functions. We will then cover drawing the graphs given the general equations of the sine and cosine functions. See Lesson 95

Case 2 Sketching the graphs of functions such as $y = 3\sin x$, $y = \sin x + 2$, $y = \sin 3x$ (See Lesson 92)

Case 3: Sketching the graphs of the more general forms such as $y - 2 = 3\sin(4x + \pi)$,
$y - 2 = 3\sin(4x + 1)$; $y - 2 = 3\cos(4x + \pi)$; $y - 2 = 3\cos(4x + 1)$;
$y - 2 = 3\tan(4x + \pi)$; $y - 2 = 3\tan(4x + 1)$,   See Lesson 95

Case 4: Sketching the graphs of the inverse trigonometric functions such as
$y = \text{Arcsin } x$, $y = \arccos x$,, $y = \text{Arctan } x$, $y = \text{arcsec } x$, $y = arc\csc x$, $y = \text{Arccot } x$.

Labeling the Coordinate Axes of Trigonometric Graphs

Sometimes, the difficulty students may have in graphing trigonometric functions is in labeling the coordinate-axes . We will therefore devote sometime to learn how to label these axes.

Labeling the number line both forwards and backwards

Suppose we want to label the number line or the x-axis with the unit interval $\frac{1}{2}$, then the end of the first interval is labeled $1 \cdot \frac{1}{2}$ or $\frac{1}{2}$, the end of the next interval is labeled $2 \cdot \frac{1}{2}$ or 1; the end of the next interval is labeled $3 \cdot \frac{1}{2}$ or $\frac{3}{2}$; the end of the fourth interval is labeled $4 \cdot \frac{1}{2}$ or 2. Notice the sequence of the factors 1, 2, 3, and 4 in determining multiples of the unit interval.
Note that in labeling the line, the convention is that all fractions must be reduced to lowest terms.

For example, $4 \cdot \frac{1}{2}$ is normally labeled 2; $4 \cdot \frac{\pi}{2}$ is normally be labeled 2π; and

similarly $3 \cdot \frac{1}{2}$ and $3 \cdot \frac{\pi}{2}$ are normally labeled $\frac{3}{2}$ and $\frac{3\pi}{2}$ respectively.

(a)   **Using fractions**

Using fractions:

| -2 or $-4 \cdot \frac{1}{2}$ | $-\frac{3}{2}$ or $-3 \cdot \frac{1}{2}$ | -1 or $-2 \cdot \frac{1}{2}$ | $-\frac{1}{2}$ or $-1 \cdot \frac{1}{2}$ | 0 | $\frac{1}{2}$ or $1 \cdot \frac{1}{2}$ | 1 or $2 \cdot \frac{1}{2}$ | $\frac{3}{2}$ or $3 \cdot \frac{1}{2}$ | 2 or $4 \cdot \frac{1}{2}$ | $\frac{5}{2}$ or $5 \cdot \frac{1}{2}$ |

(b)   **Using degrees**
If we use degrees for labeling, with the " unit interval $90°$, we obtain the figure below:

Using degrees

| $-360°$ or $-4 \cdot 90°$ | $-270°$ or $-3 \cdot 90°$ | $-180°$ or $-2 \cdot 90°$ | $-90°$ or $-1 \cdot 90°$ | 0 | $90°$ or $1 \cdot 90°$ | $180°$ or $2 \cdot 90°$ | $270°$ or $3 \cdot 90°$ | $360°$ or $4 \cdot 90°$ | $450°$ or $5 \cdot 90°$ |

(c)   **Using π units**
If we use π units for labeling, with the " unit interval" $\frac{\pi}{2}$, we obtain the figure below:

Using π units:

| -2π or $-4 \cdot \frac{\pi}{2}$ | $-\frac{3\pi}{2}$ or $-3 \cdot \frac{\pi}{2}$ | $-\pi$ or $-2 \cdot \frac{\pi}{2}$ | $-\frac{\pi}{2}$ or $-1 \cdot \frac{\pi}{2}$ | 0 | $\frac{\pi}{2}$ or $1 \cdot \frac{\pi}{2}$ | π or $2 \cdot \frac{\pi}{2}$ | $\frac{3\pi}{2}$ or $3 \cdot \frac{\pi}{2}$ | 2π or $4 \cdot \frac{\pi}{2}$ | $\frac{5\pi}{2}$ or $5 \cdot \frac{\pi}{2}$ |

(d) **Using Decimals**

If we use decimals for labeling, with the "unit interval" 1.57, we obtain the figure below:

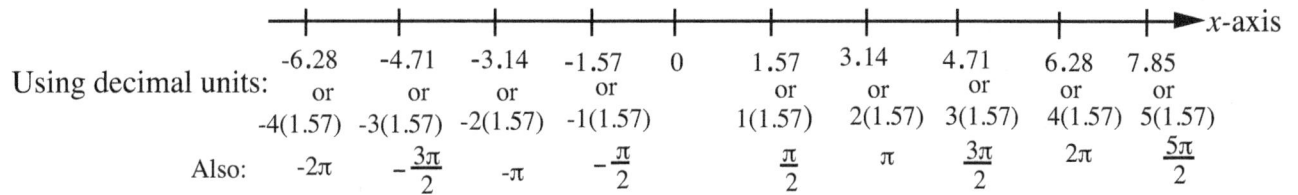

Using decimal units:

-6.28	-4.71	-3.14	-1.57	0	1.57	3.14	4.71	6.28	7.85
or	or	or	or		or	or	or	or	or
-4(1.57)	-3(1.57)	-2(1.57)	-1(1.57)		1(1.57)	2(1.57)	3(1.57)	4(1.57)	5(1.57)

Also: -2π $-\dfrac{3\pi}{2}$ $-\pi$ $-\dfrac{\pi}{2}$ $\dfrac{\pi}{2}$ π $\dfrac{3\pi}{2}$ 2π $\dfrac{5\pi}{2}$

Note also that $1.57 = \dfrac{\pi}{2} = 90°$; $3.14 = \pi = 180°$; $6.28 = 2\pi = 360°$

Illustrating the periodicity of trigonometric functions   584

One common property the graphs of trigonometric functions have is the property of periodicity (regular repetition of the same *y*-values for different *x*-values.).We will use the graph of the sine function to illustrate this repetitive property. The other trigonometric functions also have repetitive properties.

The graph of $y = \sin x$

The graph of $y = \sin x$ (*x* in radians) is drawn below, using a table of values (or a set of ordered pairs). We will learn how to graph this function later, but for the meantime, let us familiarize ourselves with the basic properties. Observe the shape of the curve and its relationship with the axes (*x*- and *y*-axes). The period of $y = \sin x$ is 2π. Thus, the curve drawn

will **repeat itself every** 2π units (Note: $\pi \approx 3.14$, and therefore $2\pi \approx 2(3.14) \approx 6.28$). Algebraically, from time to time, we obtain the same *y*-values when we substitute different *x*-values in $y = \sin x$.

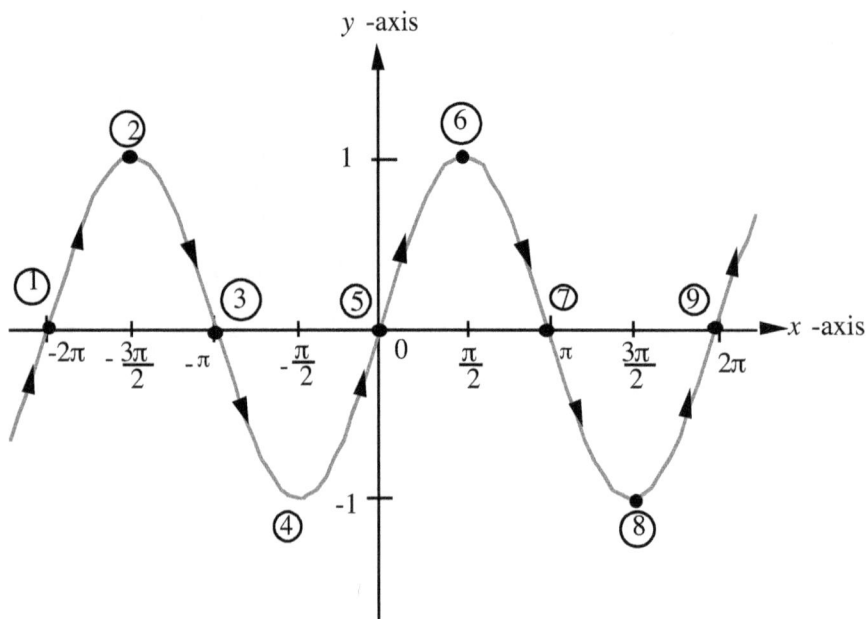

Figure

Graphical meaning of the phrase "repeats itself"

In the above figure, observe the curve at point **1**, with the coordinates $x = -2\pi, y = 0$, and note the arrow and its direction (the arrow is used for illustrative purposes only and we do not show the arrows when we draw these graphs). At point **1,** the curve is increasing (going up) and crossing the *x*- axis. At point **5** (with $x = 0, y = 0$) the curve does the same thing as it did at point **1**, namely it is increasing (going up) and crossing the *x*-axis again. Thus, at point **5**, the curve repeats what it did at point **1**, namely, crossing the *x*-axis and going up. The distance between point **1** and point **5** is the period of this function. In the above case, this period is 2π. Hence we say that the curve repeats itself every 2π (or 6.28 units). Similarly, at point **9**, the curve is increasing (going up) and crossing the *x*-axis. Thus, the curve repeats itself again at point **9**. Again, the distance between the points **5** and **9** is 2π units, the period of this function. Observe that repetition as explained above involves two simultaneous (at the same time) events, namely, crossing the *x*-axis and increasing (going up). Instead of starting from
 point **1** or from point **5**, we could also start from point **3**. Suppose, we start from point **3**, then observe that at this point, the curve is crossing the *x*-axis and decreasing (going down). At point **7**, the

curve repeats the same two events occurring at point **3**, namely, crossing the *x*-axis and decreasing (going down). Thus, starting from point **3**, the curve repeats itself at point **7.**
The distance between point **3** and point **7** is 2π units. Similarly, if the curve starts at point **2** (turning point), it will repeat the pattern (starting from the turning point and decreasing) at point **6.**
If the curve starts at point **4**, it will repeat at point **8.**
Therefore, to determine the period from the graph, we pick a starting point on the curve, observe two simultaneous events at this point and then observe the next time the same "simultaneous events" take place and then determine the distance between these two nearest "simultaneous- events" points.

Another property of the graph of the sine function is the **amplitude**. The amplitude determines the highest and the lowest points on the graph of this function. In the above figure, the amplitude is the vertical distance from any of the points marked **2, 4, 6,** or **8** to the *x*-axis.

Examples of amplitudes: The amplitude of $y = 3\sin x$ is 3 . The amplitude of $y = 2\sin x$ is 2;

The amplitude of $y = \sin x$ is 1.   The amplitude of $y = \frac{1}{2}\cos x$ is $\frac{1}{2}$.

In the next section, we will now learn how to sketch the graph of $y = \sin x$ for $0 \le x \le 2\pi$.

Lesson 91 Exercises

1. Use the following unit intervals to label the number line from $0°$ to $360°$:
   (a) $30°$;   (b) $40°$, (c) $45°$

2. Use the following unit intervals to label the number line from 0 to 2π: (a) $\frac{\pi}{3}$;   (b) $\frac{\pi}{4}$.

3. Use the following unit interval to label the number line from 0 to 6: $\frac{1}{4}$

4. Use the following unit interval to label the number line from 0 to 6.28: **1.57**

1.(a)

1.(b)

1.(c)

2.(a)

2.(b)

3.

4.

Lesson 92

Sketching the Graph of $y = \sin x$

To sketch the graph of $y = \sin x$ we must know the following basic properties of this curve:

1. Its **period** is 2π. Its **amplitude** which is 1.

2. Five important points (**critical points**): 2 end points (a starting point and a stopping point), 1 mid-point and two quarter-points are needed. Each point is an ordered pair (x, y). The five points include the x-intercepts, the y-intercept; and the amplitude. The amplitude determines the highest and the lowest points on the graph of this function

Example 1 Sketch the graph of $y = \sin x$ for $0 \leq x \leq 2\pi$.

Step 1: We construct a table containing these important points, using the formula $y = \sin x$

x	0	$\frac{\pi}{2}$	π	$\frac{3\pi}{2}$	2π
$y = \sin x$	0	1	0	-1	0

Step 2:   From the origin and on the x-axis, mark-off a convenient interval and label its end-point 2π, the period of $y = \sin x$. Divide this interval into four equal intervals and label them with $\frac{\pi}{2}$, π, and $\frac{3\pi}{2}$, using Figure (c) p, 459 as a guide.

Step 3: On the y-axis, choosing a convenient interval, mark and label the points $y = 1$ and $y = -1$.

Step 4: Plot the points $(0,0)$, $(\frac{\pi}{2},1)$, $(\pi,0)$, $(\frac{3\pi}{2}, -1)$ and $(2\pi,0)$. (Points on $y = \sin x$)

Step 5: Connect the points from Step 4 by a smooth curve to obtain the characteristic shape of $y = \sin x$. (Figure)

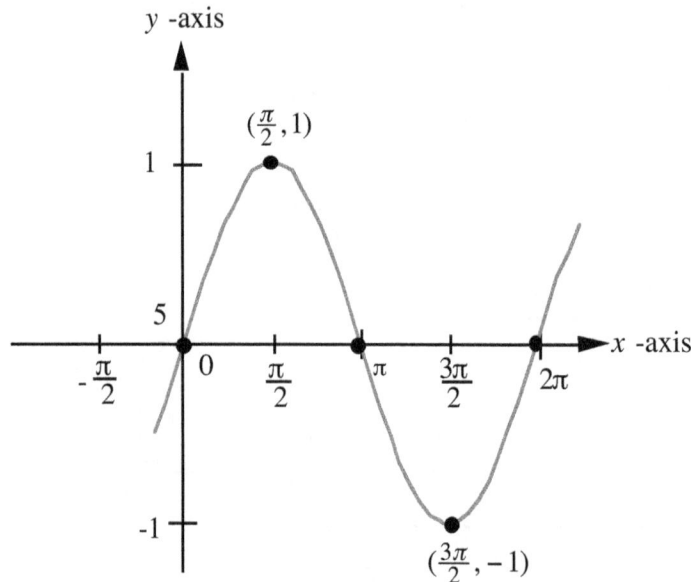

Figure: The graph of $y = \sin x$

Practice sketching this graph a number of times, noting its features, so that this graph (picture) becomes part of your "trigonometric vocabulary", just like the multiplication tables.

Example 2   Sketch the graph of $y = \sin x$ for $-2\pi \le x \le 2\pi$

Solution: The steps are the same as those for Example 1 above, except that we extend the graph to
the left of the y-axis. .

Step 1:   From the origin and on the x-axis, mark-off a convenient interval and label its end-point 2π,
the period of $y = \sin x$.

Step 2:   Divide this interval from 0 to 2π into four equal intervals and label them with
0, $\frac{\pi}{2}$, π, and $\frac{3\pi}{2}$ (See also Figure (c), page 582 as a guide.)

Step 3:   Similarly, to the left of the y-axis, on the x-axis, mark-off another interval 2π units, and
label this endpoint -2π. Divide this interval into four equal intervals, and label this
interval on the negative x-axis with the negatives of the x-values from Step 2.

Step 4:   On the y-axis, using a convenient interval, mark-off and label the points $y = 1$ and $y = -1$.

Step 5:   Plot the points $(0,0)$, $(\frac{\pi}{2},1)$, $(\pi,0)$, $(\frac{3\pi}{2},-1)$, $(2\pi,0)$ and also plot $(-\frac{\pi}{2},-1)$,
$(-\pi,0)$, $(-\frac{3\pi}{2},1)$ and $(-2\pi,0)$.

Step 7:   Connect the plotted points, from left to right, by a smooth curve to
obtain the characteristic shape of $y = \sin x$. (two cycles) (Figure)

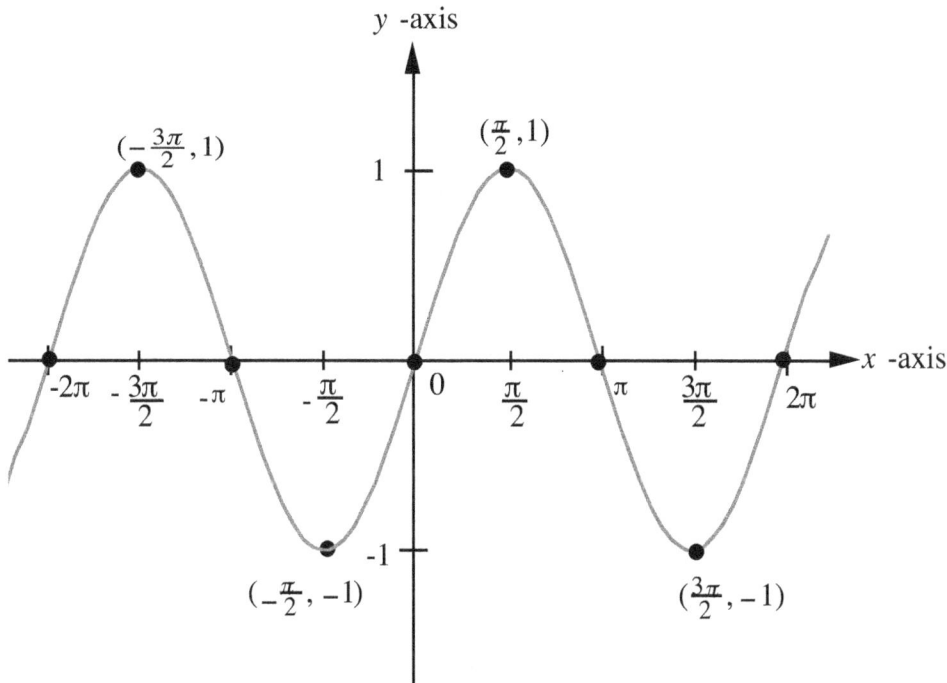

Figure : The graph of $y = \sin x$ for $-2\pi \le x \le 2\pi$

Example 3 Given $y = \sin x$, draw the graph of $y = \sin x + 2$ 588

Rewrite $y = \sin x + 2$ as $y - 2 = \sin x$ and apply the transformation method. See **Figure 3**.

Here, $k = 2, h = 0, a = 1, b = 1$ (also review page 225-226)

$y = \sin x$		$y - 2 = \sin x$		
x_0	y_0	$x_n = x_0$	$y_n = y_0(a) + k$	(x_n, y_n)
0	0	0	$0(1) + 2 = 2$	$(0,2)$
$\dfrac{\pi}{2}$	1	$\dfrac{\pi}{2}$	$1(1) + 2 = 1 + 2 = 3$	$(\dfrac{\pi}{2}, 3)$
π	0	π	$0(1) + 2 = 0 + 2 = 2$	$(\pi, 2)$
$\dfrac{3\pi}{2}$	-1	$\dfrac{3\pi}{2}$	$-1(1) + 2 = -1 + 2 = 1$	$(\dfrac{3\pi}{2}, 1)$
2π	0	2π	$0(1) + 2 = 0 + 2 = 2$	$(2\pi, 2)$

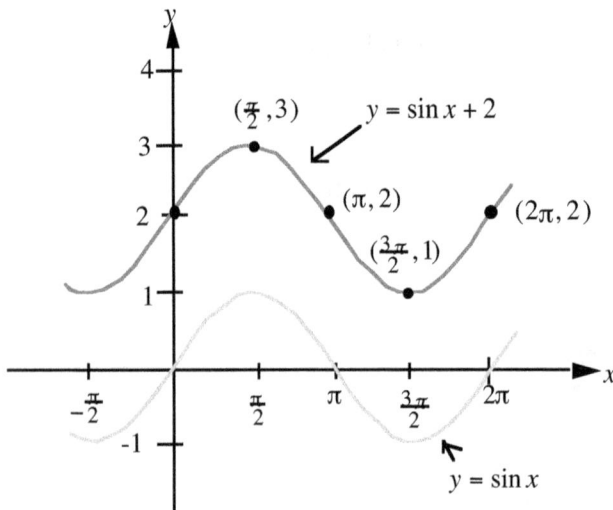

Figure 3:   The graph of $y = \sin x + 2$

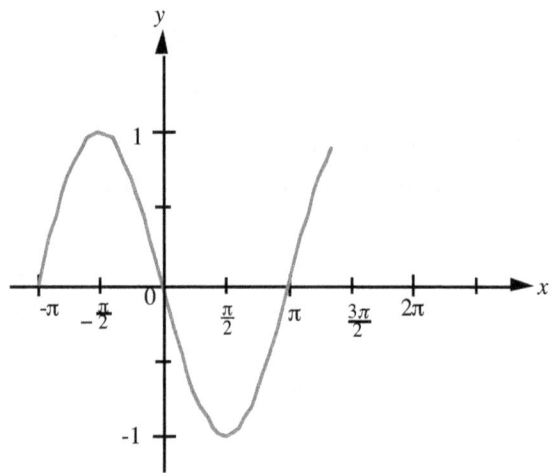

Figure 4: Graph of $y = \sin(x + \pi)$

Example 4 : Sketch the graph of $y = \sin(x + \pi)$ (also review page 225-226)

Solution We can obtain the graph of $y = \sin(x + \pi)$ from the graph of $y = \sin x$ by moving each critical point π units to the left, keeping the y-coordinates unchanged. More formally, we apply the transformation method.

See **Figure 4.** $y = \sin(x + \pi)$. Here, $k = 0, h = -\pi, a = 1, b = 1$

$y = \sin x$		$y = \sin(x + \pi)$		
x_0	y_0	$x_n = \dfrac{x_0}{b} + h$	$y_n = y_0$	(x_n, y_n)
0	0	$\dfrac{0}{1} - \pi = 0 - \pi = -\pi$	0	$(-\pi, 0)$
$\dfrac{\pi}{2}$	1	$\dfrac{\pi}{2 \cdot 1} - \pi = \dfrac{\pi}{2} - \pi = -\dfrac{\pi}{2}$	1	$(-\dfrac{\pi}{2}, 1)$
π	0	$\dfrac{\pi}{1} - \pi = 0$	0	$(0, 0)$
$\dfrac{3\pi}{2}$	-1	$\dfrac{3\pi}{2 \cdot 1} - \pi = \dfrac{3\pi}{2} - \pi = \dfrac{\pi}{2}$	-1	$(\dfrac{\pi}{2}, -1)$
2π	0	$\dfrac{2\pi}{1} - \pi = 2\pi - \pi = \pi$	0	$(\pi, 0)$

Example 5 Sketch the graph of $y = 5\sin(2x - 3)$

Solution See figure below.

We can apply the transformation method as in the previous examples

Rewrite $y = 5\sin(2x - 3)$ as $y = 5\sin[2(x - \frac{3}{2})]$ and apply the transformation method.

Here, $k = 0$, $h = \frac{3}{2}$, $a = 5$, $b = 2$. Construct a table of values and graph. **Figure 5**

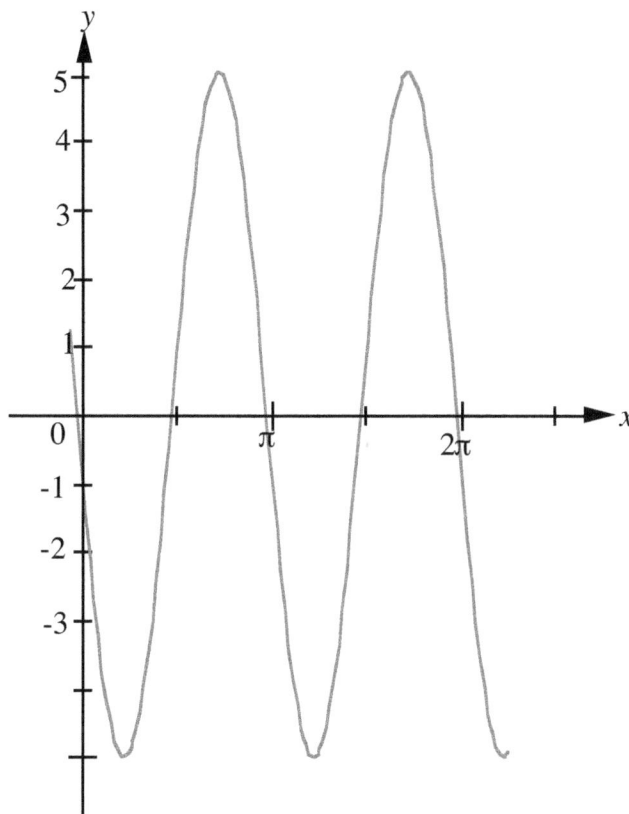

Figure : The Graph of $y = 5\sin(2x - 3)$

Now, having more or less mastered the sketching of the sine curves, let us go on to learn to sketch the cosine, tangent, secant and the cosecant curves. In Lesson 95, we will learn a direct method for sketching the graphs of trigonometric functions such as $y - k = a\sin(bx - h)$ and $y - k = a\cos(bx - h)$. However always keep in mind the transformation method as well as the general method for sketching the graphs of functions.

We will use the same principles that we used in sketching sine curves. The only thing new we have to know are the properties of each curve.

We will begin by first drawing the graph of $y = \cos x$ using values from memory, tables or calculator.

Lesson 92 Exercises

Sketch the graphs of the following:

1. $y = \sin x$ for $0 \le x \le 2\pi$;   **2.** $y = -\sin x$ for $0 \le x \le 2\pi$;;   **3.** $y = 3\sin x$ for $0 \le x \le 2\pi$;;

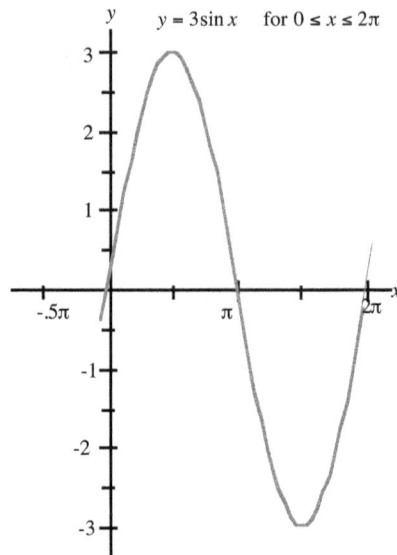

Lesson 93 591

Sketching the Graph of $y = \cos x$

Properties of the cosine curve $y = \cos x$

(1) The period of $y = \cos x$ is 2π, the same period as that of $y = \sin x$.
(2) The amplitude is I.

Observe in **Figure** that if the curve begins from the point $(0, 1)$ marked **2**, on the y-axis, then (in going from left to right) we observe that, at the point $(2\pi, 1)$, marked **5**, the curve repeats.
Thus the curve is again at its maximum point at **5** as it was at **2**. The period is the distance between point **2** and point **5**, and it is 2π, the period of the cosine curve. However, if we had started drawing the curve from the point $(-\frac{\pi}{2}, 0)$ marked **1**, then the next time the curve repeats. would be at point **4**. The period in this case will be the distance between **1** and **4**. This period is also 2π, the same period as before.
In fact, from trigonometric identities, $\cos x = \sin (x + \frac{\pi}{2})$
This means that given the sketch of the $y = \sin x$, to obtain the sketch of $y = \cos x$, we shift the sine curve horizontally to the left by $\frac{\pi}{2}$ units. (We shift each important point on $y = \sin x$).

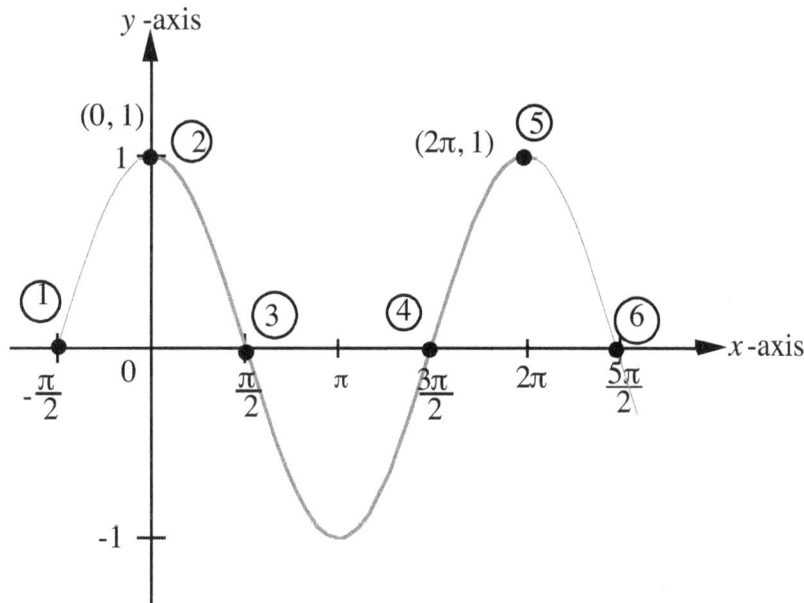

Figure : The graph of $y = \cos x$ for $-\frac{\pi}{2} \leq x \leq \frac{5\pi}{2}$

Observe the shape of the cosine curve and its relationship with the x- and y- axes. Also observe the similarities and differences between the sine and the cosine curves.

How to sketch the graph of $y = \cos x$

To sketch the graph of $y = \cos x$ we must know the following basic properties of this curve:

1. Its **period** is 2π; and its **amplitude** is 1.

2. Five important points (**critical points**) 2 end points (a starting point and a stopping point), 1 mid-point and two quarter-points are needed. Each point is an ordered pair (x, y). The five points include the x-intercepts, the y-intercept; and the amplitude. The amplitude determines the highest and the lowest points on the graph of this function

Example 6   Sketch the graph of $y = \cos x$ for $0 \le x \le 2\pi$.

Step 1: We construct a table containing these important points, using the formula $y = \cos x$

Table for $y = \cos x$

x	0	$\frac{\pi}{2}$	π	$\frac{3\pi}{2}$	2π
$y = \cos x$	1	0	-1	0	1

Step 2:   From the origin and on the x-axis, mark-off a convenient interval and label its end-point 2π, the period of $y = \cos x$. Divide this interval into four equal intervals and label them with $\frac{\pi}{2}$, π, and $\frac{3\pi}{2}$,, using Figure   .. p.466 as a guide.

Step 4: On the y-axis, using a convenient interval, mark-off and label the points $y = 1$ and $y = -1$.

Step 5: Plot the points $(0, 1)$, $(\frac{\pi}{2}, 0)$, $(\pi, -1)$, $(\frac{3\pi}{2}, 0)$ and $(2\pi, 1)$. (obtained from the above table)

Step 6: Connect the points from Step 5 by a smooth curve to obtain the characteristic shape of $y = \cos x$. (Figure)

Practice sketching this graph a number of times, noting its features, so that this graph (picture) becomes part of your "trigonometric vocabulary", just like the multiplication tables.

Sketch the graphs of the following:

1. $y = \cos x$ for $0 \le x \le 2\pi$; **2.** $y = -\cos x$ for $0 \le x \le 2\pi$;   **3.** $y = 3\cos x$ for $0 \le x \le 2\pi$;

4. $y = -\cos x$ for $0 \le x \le 2\pi$;

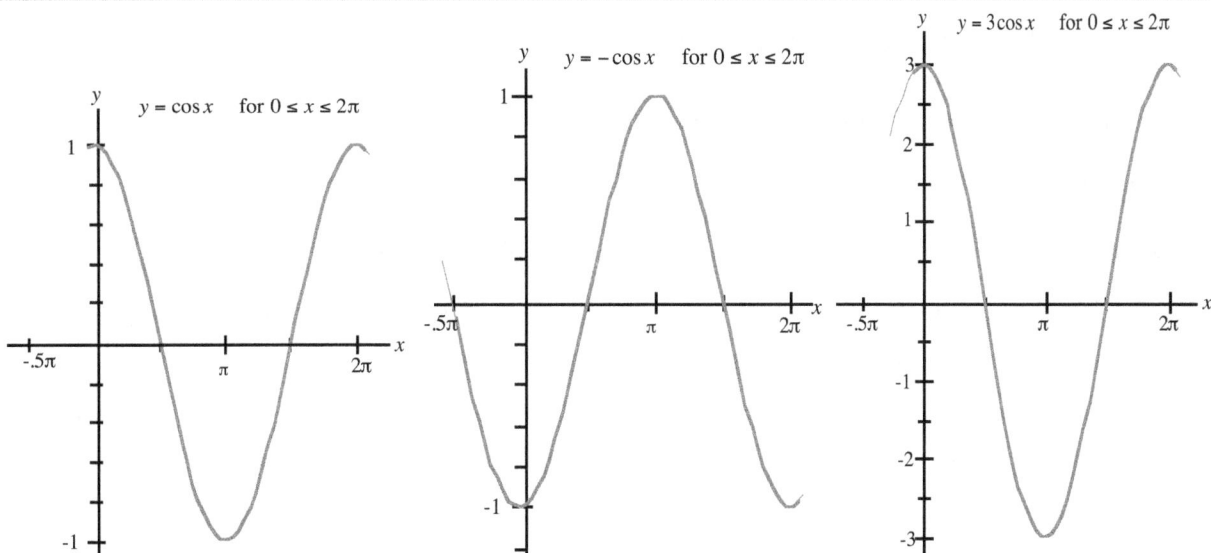

Lesson 94 594

Graphs of Discontinuous Trigonometric Functions:
$$y = \tan x; \quad y = \csc x; \quad y = \sec x; \quad y = \cot x$$

The approach used in sketching the graphs $y = \tan x$ and the reciprocal functions, $y = \csc x$, $y = \sec x$ and $y = \cot x$, is similar to that used in sketching the graphs of rational functions (see Chapter 16)

The Graph of the Tangent Function, $y = \tan x$

Here, we will use the same principles that we used in sketching the sine and cosine curves. However, we must first learn the properties of the tangent curve.

Properties of $y = \tan x$

(1) The period of $y = \tan x$ is π (Thus, the curve repeats itself every π units)

(2) The x-intercepts are at where $\sin x = 0$ and $\cos x \neq 0$, since for x-intercepts, $y = \tan x = \frac{\sin x}{\cos x} = 0$..

(3) The y-intercept is at where $x = 0$. (Here, $\sin x = 0$, $\cos x = 1$)

(4) The curve is not defined where $\cos x$ is zero: That is it is not defined at $x = -\frac{\pi}{2}, \frac{\pi}{2}, \frac{3\pi}{2}, \frac{5\pi}{2}$ etc.

The curve is discontinuous at these points and the curve is asymptotic to the vertical lines $x = -\frac{\pi}{2}, \frac{\pi}{2}, \frac{3\pi}{2}, \frac{5\pi}{2}$ etc

(4) The tangent curve has points of inflection at all x-intercepts.

Table for $y = \tan x$

x	$-\frac{\pi}{2}$	$-\frac{\pi}{4}$	0	$\frac{\pi}{4}$	$\frac{\pi}{2}$
$y = \tan x$	undefined	-1	0	1	undefined

Note that $\tan x$ and $\sec x$ have the same vertical asymptotes $\tan x = \frac{\sin x}{\cos x}$. and $\sec x = \frac{1}{\cos x}$
Note also that the tangent curve is discontinuous (broken up) at the x-intercepts (zeros) of the cosine curve.

How to Sketch the Graph of $y = \tan x$

Example 7 Sketch the graph of $y = \tan x$ for $-\frac{\pi}{2} \le x \le \frac{\pi}{2}$.

Solution

Step 1: Determine the x-intercept. The x-intercept is at $(0, 0)$.

Step 2: Determine the y-intercept. The y-intercept is at $(0, 0)$.

Step 3: Find the vertical asymptotes: The vertical asymptotes are the lines $x = -\frac{\pi}{2}$, and $x = \frac{\pi}{2}$.

(We may use the cosine curve as a guide to help identify the vertical asymptotes since the tangent

($\tan x = \frac{\sin x}{\cos x}$) is not defined at where the cosine is 0)

Step 4: Determine the quarter points for $x = -\frac{\pi}{4}$, and $\frac{\pi}{4}$,

The quarter points are at $(-\frac{\pi}{4}, -1)$ and $(\frac{\pi}{4}, 1)$.

Step 5: Plot the points from Steps 1, 2, and 4, above.

Step 6: Using broken or dotted lines, draw the vertical asymptotes $x = -\frac{\pi}{2}$, and $x = \frac{\pi}{2}$.

Step 7: Connect the points in Step 5 by a smooth solid curve noting that the curve does not meet its asymptotes but rather approaches them smoothly as the curve and its asymptotes are extended indefinitely. (Figure below)

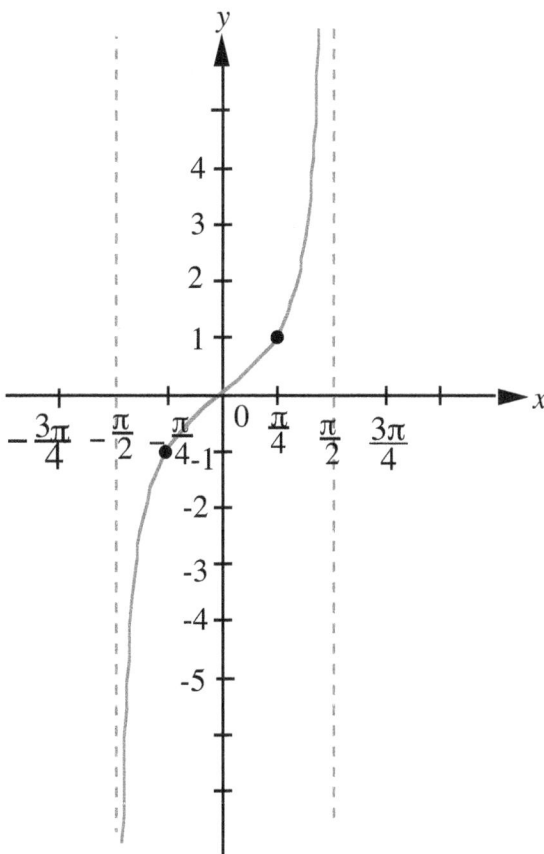

The graph of $y = \tan x$ for $-\frac{\pi}{2} \le x \le \frac{\pi}{2}$

Lesson 94: Graphs of Discontinuous Trig. Functions: $y = \tan x$; $y = \csc x$; $y = \sec x$; $y = \cot x$

Example 8   Sketch the graph of $y = \tan x$ for $-\pi \le x \le 2\pi$

Solution   See also Example 7, above.

Step 1: Determine the x-intercepts. The x-intercepts are at $(-\pi, 0)$, $(0, 0)$, $(\pi, 0)$ and $(2\pi, 0)$

Step 2: Determine the y-intercept. The y-intercept is at $(0, 0)$.

Step 3: Find the vertical asymptotes: The vertical asymptotes are the lines $x = -\frac{\pi}{2}, x = \frac{\pi}{2}$,

and $x = \frac{3\pi}{2}$. (We can use the cosine curve as a guide to help identify the vertical asymptotes since the

tangent ($\tan x = \frac{\sin x}{\cos x}$) is not defined at where the cosine is 0)

Step 4: Determine the quarter points:

The quarter points are: $(-\frac{3\pi}{4}, 1)$, $(-\frac{\pi}{4}, -1)$, $(\frac{\pi}{4}, 1)$, $(\frac{3\pi}{4}, -1)$, $(\frac{5\pi}{4}, 1)$, $(\frac{7\pi}{4}, -1)$.

Step 5: Plot the points from Steps 1, 2, and 4,.

Step 6: Using broken or dotted lined, draw the vertical asymptotes from Step 3.

Step 7: Using solid curves draw the tangent curves (referring to Example 7) as shown in Figure.

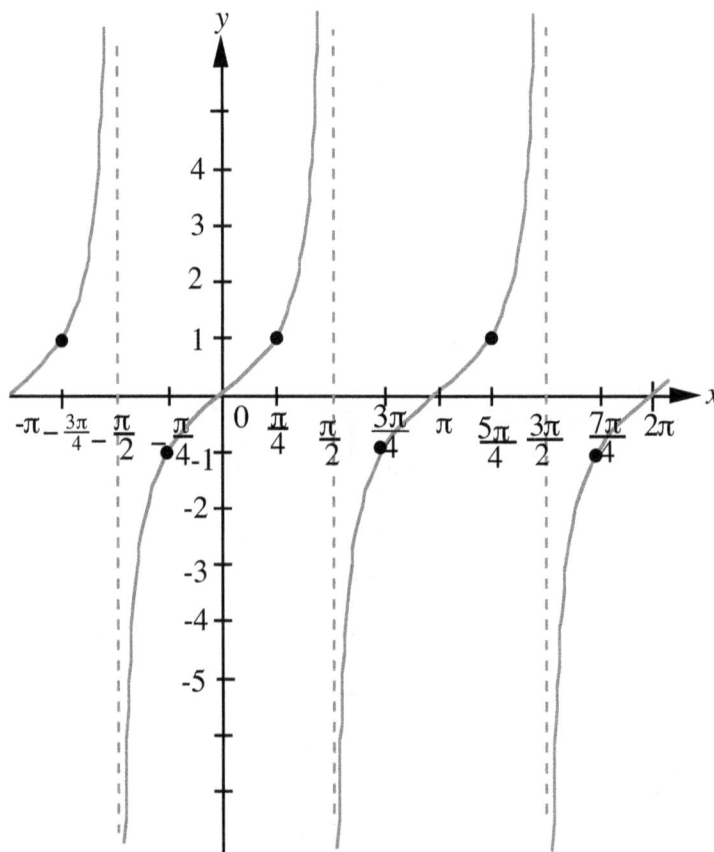

Figure : The graph of $y = \tan x$ $-\pi \le x \le 2\pi$

The Graph of $y = \csc x$

By definition, $\csc x = \dfrac{1}{\sin x}$ and so $\csc x$ is not defined where $\sin x = 0$. The cosecant curve is discontinuous (broken up) at the x-intercepts of the sine curve, and has vertical asymptotes at these intercepts. In addition, the concavities of the cosecant curve are the reversed concavities of the sine curve. The maximum (absolute maximum) points of the sine curve are the minimum (relative minimum) points of the cosecant curve. The period of $\csc x$ is 2π (same as that of $\sin x$).
See also the descriptions of reciprocal functions on page 283-288

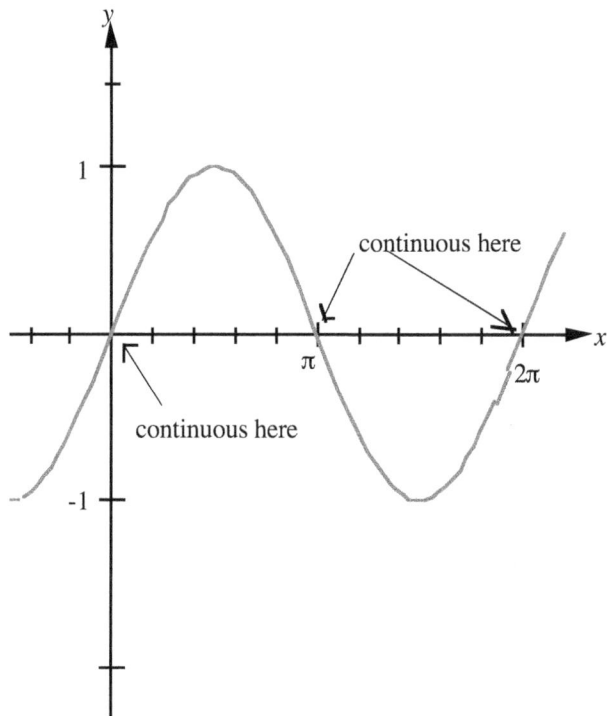

(k) The graph of $y = \sin x$.

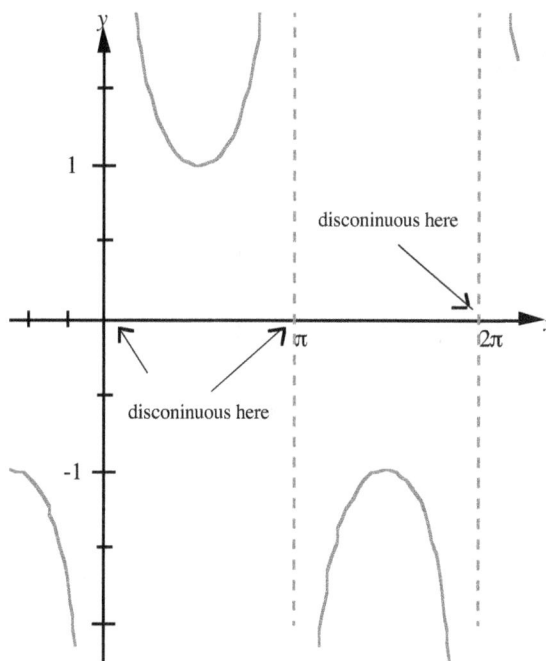

(l) The graph of $y = \dfrac{1}{\sin x} = \csc x$

How to Sketch the Graph of $y = \csc x$; $0 \le x \le 2\pi$ 598

(See also page 283 for the graphs of functions and their reciprocals)

Example Sketch the graph of $y = \csc x$ for $0 \le x \le 2\pi$.

Note: There are no x-intercepts and no y-intercepts.

Step 1: Using dotted lines, sketch the graph of the corresponding sine curve.

Step 2: Draw vertical asymptotes at the x-intercepts of the sine curve.

Step 3: Using solid curved lines, draw the cosecant curve by visually breaking up the sine curve at its x-intercepts, and reversing the concavity each portion of broken curve to obtain the cosecant curve as shown in Figure.

Note: You dot not have to sketch the sine curve before sketching the cosecant curve. Using the sine curve is only a mnemonic device, since by now we are very familiar with the sine curve.

Table for the graph

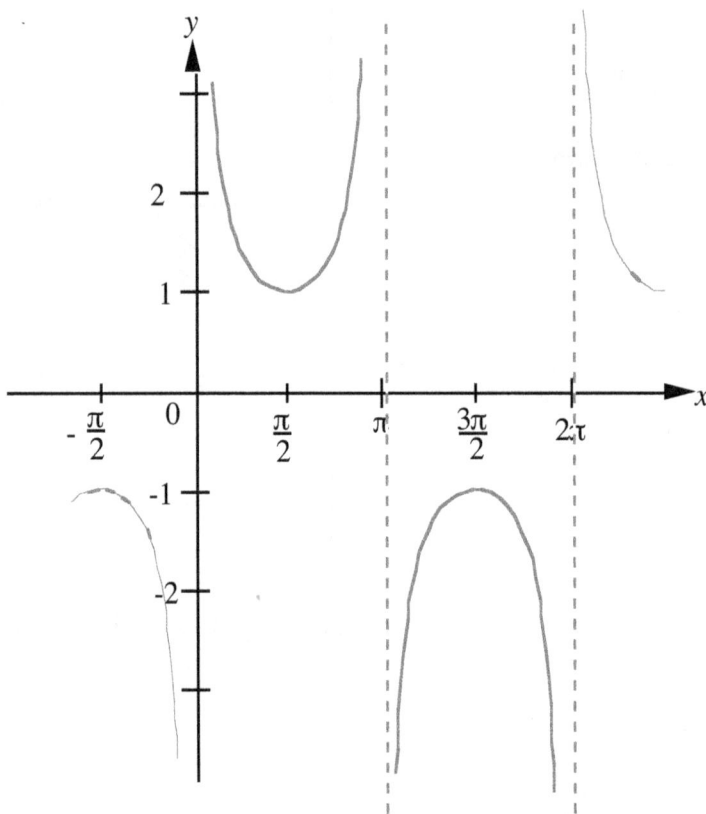

Figure : The graph of $\csc x$

The graph of $y = \sec x$

By definition, $\sec x = \dfrac{1}{\cos x}$ and so $\sec x$ is not defined where $\cos x = 0$. The secant curve is discontinuous (broken up) at where $\cos x = 0$ and the graph of $y = \sec x$ is broken up at the x-intercepts of the cosine curve, and has vertical asymptotes at these intercepts. Thus we can sketch $y = \sec x$ by first sketching the cosine curve (since by now we are used to sketching the cosine curve) and noting the following: The secant curve has no absolute maximum or minimum but has relative maximum and minimum. The concavity of the secant curve is the reversed concavity of the cosine curve. The maximum point of the cosine curve becomes the minimum point of the secant curve and the minimum point of the cosine curve becomes the maximum (relative maximum) point of the secant curve. The period of $\sec x$ is 2π (same as that of $\cos x$)

How to sketch the graph of $y = \sec x$
(See also page 283 for the graphs of functions and their reciprocals)

We can sketch the secant curve in a similar way as we sketched the cosecant curve except that we can obtain it from the cosine curve.

Step 1: Using dotted lines, sketch the graph of the corresponding cosine function.

Step 2: Draw vertical asymptotes, at the x-intercepts of the cosine curve.

Step 3: Using solid curved lines, draw the secant curve by visually breaking up the cosine curve at its x-intercepts, and reversing the concavity each portion of broken curve to obtain the secant curve.

Note: You dot not have to sketch the cosine curve before sketching the secant curve. Using the cosine curve is only a mnemonic device, since by now we are very familiar with the cosine curve.

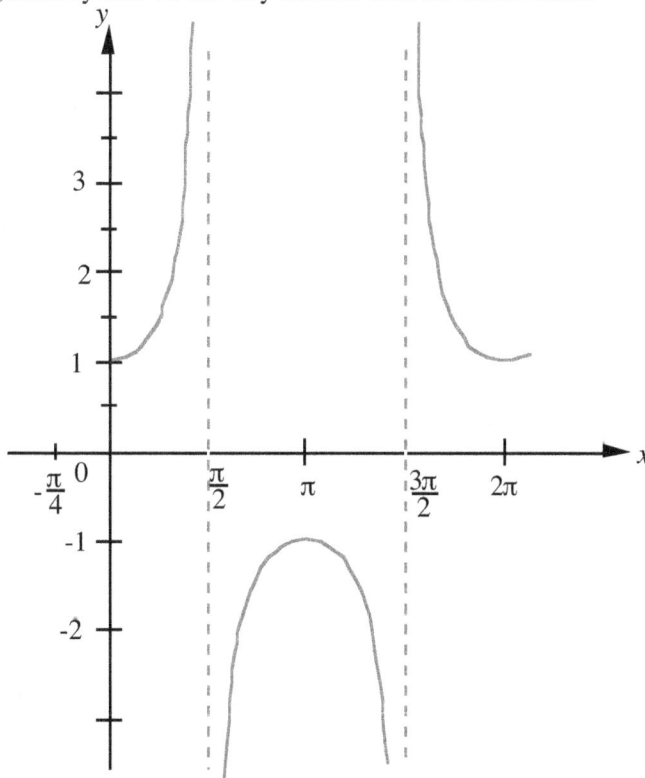

Figure : The graph of $y = \sec x$

The graph of $y = \cot x$ 600

By definition, $\cot x = \dfrac{1}{\tan x} = \dfrac{\cos x}{\sin x}$. Thus $\cot x$ is not defined at the x-intercepts of $\tan x$ or

$\sin x$, (since $\tan x = \frac{\sin x}{\cos x}$). The graph $y = \cot x$ has vertical asymptotes at the x-intercepts of $\tan x$ or $\sin x$.

How to sketch the graph of $y = \cot x$

Step 1: Using broken lines, draw the vertical asymptotes at the intercepts above.

Step 2: Noting that the cotangent curve and the tangent curve have the same shape except that the concavities of the cotangent curve are the reversed concavities of the tangent curve.
Using solid curves draw the cotangent curve as shown in figure.

Table for the graph

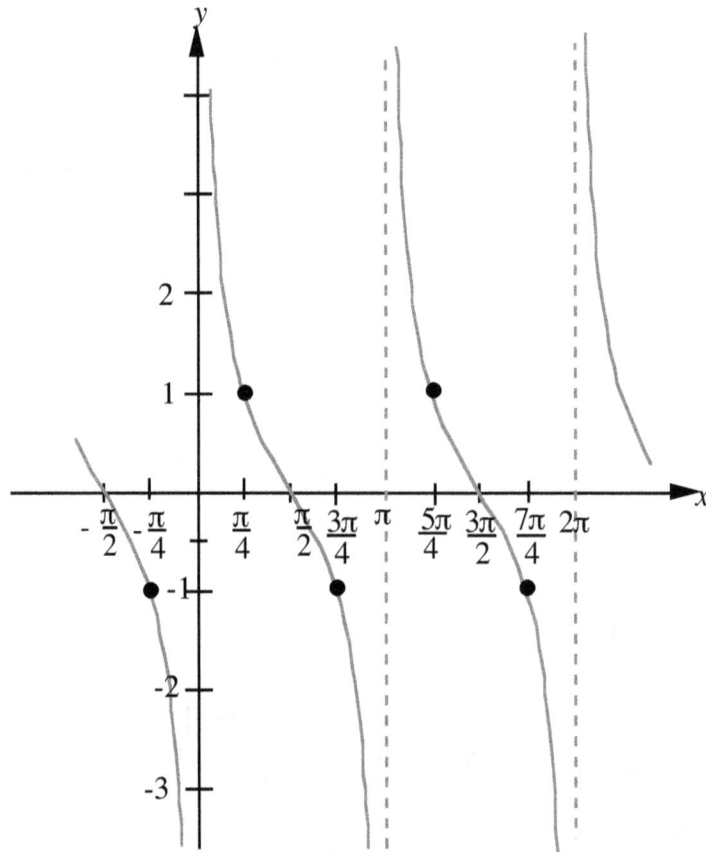

Figure :The graph of $y = \cot x$

The author's previous description (page 283 repeated below) of the relationship between the graphs of continuous functions and their reciprocals applies also to $y = \tan x$ and $y = \cot x$ if we consider the graph of $y = \tan x$ as a collection of continuous pieces of a curve and apply the description to each piece.

"**Whenever a given curve becomes infinitely discontinuous at a point, the given curve breaks up into two pieces at this point; an asymptote is formed at this point, and each piece reverses its concavity and orientates itself such that the end of each piece smoothly becomes asymptotic to the asymptote so formed.** "

Lesson 94 Exercises

Sketch the graphs of $y = 2\tan x$ for $0 \le x \le \pi$

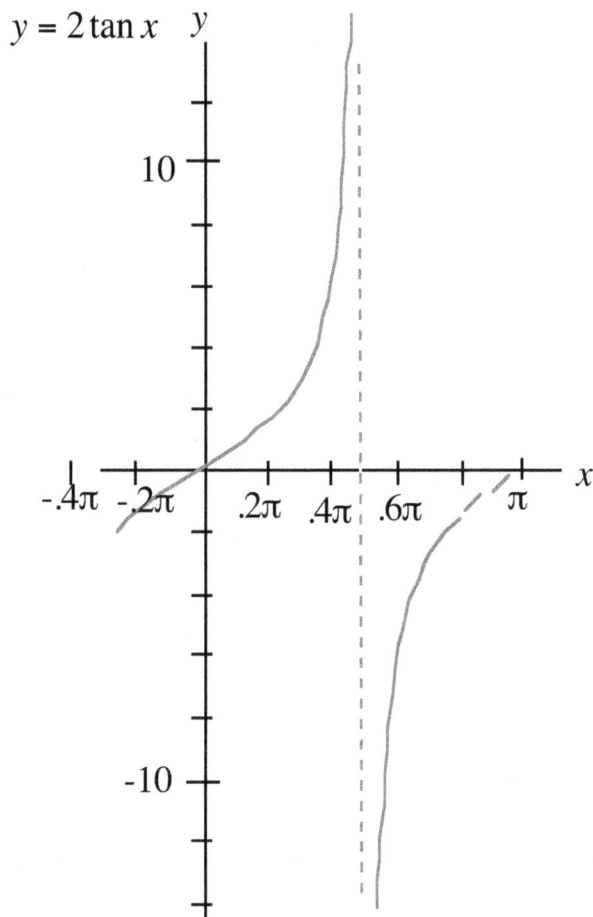

Lesson 95

Direct Procedure for Sketching the Graphs of

$y - k = a \sin(bx - h)$, and $y - k = a \cos(bx - h)$

Note below that in addition to the direct methods, we can use the transformation method to sketch the graphs of the more general functions. See method 2 of Example 2

General Equations of the Sine and Cosine Curves

Generally the equation of the sine curve is given by

$$y - k = a \sin(bx - h) \qquad (1) \quad \text{<-- general form}$$

and that of the cosine is given by

$$y - k = a \cos(bx - h) \qquad (2) \quad \text{<-- general form}$$

where (in each equation) k is the vertical shift, a is the amplitude, b is the constant used in determining the period; b is also the number of cycles in a given interval:

The period, $p = \dfrac{2\pi}{b}$.

If $a > 0$, the graph has the regular shape, but if $a < 0$, the graph has the inverted shape.

When the constant, b, of the right-hand side of $y - k = a \sin(bx - h)$ and $y - k = a \cos(bx - h)$ is factored out, we obtain, respectively,

$$y - k = a \sin b \left(x - \frac{h}{b}\right) \quad \text{<----------- standard form}$$

and

$$y - k = a \cos b \left(x - \frac{h}{b}\right) \quad \text{<----------- standard form}$$

in which $\dfrac{h}{b}$ is the horizontal shift or the phase shift or phase constant.

We may obtain $\dfrac{h}{b}$ from (2) by setting $bx - h = 0$ and solving for x.

If $h = 0$ (no horizontal shift from origin), $k = 0$ (no vertical shift from the origin), $a = 1$, and $b = 1$, we obtain

$$y = \sin x \qquad\qquad (3)$$
$$y = \cos x \qquad\qquad (4)$$

Equations (3) and (4) are the simplest forms of the equations of the sine and cosine curves, respectively.

Its period is $\dfrac{2\pi}{1}$ or 2π..

The graph starts from $x = h$ and ends at $x = 2\pi + h$. The minimum point is at $y = K - A$, and the maximum point, $y = K + A$

Note:

In the main, we should have the following objectives in mind when drawing the trigonometric curves.
1. From which starting point, through which points and to which final point do we draw the curve
2. The unit interval for the coordinate axes.
3. The number of the basic curves to draw: This depends on the domain and period.
4. For each basic curve we need five important points (critical points) 2 end points (a starting point and a stopping point), 1 mid-point and two quarter-points . Each point is an ordered pair (x, y).
   The five points include points at which the curve crosses (intersects) the x- and y-axes; and how far up and how far down (its amplitude) in the y-direction does it go.

Procedure for sketching the graph of the general equation, $-k = a \cos(bx - h)$ 603
$$y - k = a \sin(bx - h) \text{ and}$$

Step 1: Find the period p.
$$p = \frac{2\pi}{b}$$

Step 2: Find the beginning-point x-coordinate, x_1, of the first cycle.

Set $bx - h = 0$ and solve for x. to obtain $x = \frac{h}{b}$. Let this x-value be x_1

Step 3: Find the end-point x-coordinate, x_5, of the first cycle.

Set $bx - h = 2\pi$ and solve for x to obtain $x = \frac{2\pi + h}{b}$ Let this x-value be x_5

(We may combine Steps 1 and 2, by solving the domain inequality: $0 \le bx - h \le 2\pi$ to obtain
$$\frac{h}{b} \le x \le \frac{2\pi + h}{b}$$

Step 4: Determine the domain (x-values) to use to obtain ordered pairs for the points..
Divide the period p by 4 to obtain "the unit interval" for the x-axis.

Add $\frac{p}{4}$ successfully to x_1 until x_5 is reached (to obtain the x-values for the next Step)

Step 5: Using these x-values (from Step 4) calculate the corresponding y-values by substituting   these values in turns in the functional equation to obtain ordered pairs or a table of values
for x and y.

Step 6: Find the amplitude, The amplitude is $|a|$
For this step, we can also solve the range inequality $-a \le y - k \le a$ to obtain
$$k - a \le y \le k + a \quad (\textbf{Note:} \text{ For recall} -a \le y - k \le a \text{ is equivalent to} |y - k| \le a)$$

Step 7: Plot the ordered pairs from Step 5.

Step 8 : Determine the concavities of the basic curve
If $a > 0$ the curve is regular (has the regular concavity). If $a < 0$, the curve is inverted.
(Each portion of the regular curve is reflected about the x-axis))

Step 9: Determine the number of cycles for the specified interval.
The number of cycles $= b$ ($0 \le x \le 2\pi$). If only one cycle is asked for, then draw it for   only one cycle (period).

Step 10: Draw the remaining cycles by repetition and following Step 4 to Step 8.

Note: Skipping Step 5
If you do not want to calculate y but use only the following information to graph:
the x-values, $|a|$, p, $a > 0$ or $a < 0$; then note the following:

1. There are five x-values on the x-axis, There are four equal intervals between the first x-value and the fifth x-value.

2. In starting to draw the curve note that **if** the first point is **on** the x-axis (a zero), the next point , (the second point) is **not** on the x-axis (a minimum or maximum point); the third point is **on** the x-axis (a zero) the fourth point is **not** on the x-axis (a minimum or maximum point) , and the fifth point is **on** the x-axis (a zero)..
However, **if** the first point is **not** on the x-axis (**not** a zero), the second point is **on** the x-axis
(a zero), the third point is **not** on the x-axis (not a zero), the fourth point is **on** the x-axis, and the fifth point is **not** on the x-axis. Using this alternation in the location of points on or off the x-axis, together with the amplitude, we can draw the curves. We will begin with an example in which $k = 0$.

Example 1: Sketch the graph of $y = 3 \sin (4x + \pi)$ $y - k = a \cos (bx - h)$

Solution

Method 1

Step 1: Find the period p: $b = 4$, and $p = \dfrac{2\pi}{b} = \dfrac{2\pi}{4} = \dfrac{\pi}{2}$

Step 2: Find the beginning point x-coordinate, x_1, of the first cycle.

Set $4x + \pi = 0$ and solve for x. to obtain $x = -\dfrac{\pi}{4}$. Let $x_1 = -\dfrac{\pi}{4}$.

Step 3: Find the endpoint x-coordinate, x_5, of the first cycle:

Set $4x + \pi = 2\pi$., and solve to obtain $x = \dfrac{\pi}{4}$. Let $x_5 = \dfrac{\pi}{4}$.

(We may combine Steps 2 and 3, by solving the domain inequality: $0 \le 4x + \pi \le 2\pi$ to obtain

$-\dfrac{\pi}{4} \le x \le \dfrac{\pi}{4}$ (The first cycle of the curve is between $-\dfrac{\pi}{4}$ and $\dfrac{\pi}{4}$.

Step 4: Determine the domain (x-values) to use to obtain ordered pairs for the points.

Divide the period p by 4 to obtain "the quarter -interval", $\dfrac{p}{4}$, for the x-axis.

Add $\dfrac{p}{4}$ successfully to x_1 until x_5 is reached (to obtain the x-values for the next Step)

$$\dfrac{p}{4} = \dfrac{\pi}{2} \cdot \dfrac{1}{4} \text{ or } \dfrac{1}{4}\text{of}\left(\dfrac{\pi}{2}\right) \qquad (p = \dfrac{\pi}{2})$$

$$= \dfrac{\pi}{8} <----- \dfrac{p}{4}$$

$x_1 = -\dfrac{\pi}{4}$ (from Step 2)	$x_4 = x_3 + \dfrac{\pi}{8}$
$x_2 = x_1 + \dfrac{\pi}{8}$	$= 0 + \dfrac{\pi}{8}$
$= -\dfrac{\pi}{4} + \dfrac{\pi}{8}$	$= \dfrac{\pi}{8}$
$= -\dfrac{\pi}{8}$	$x_5 = \dfrac{\pi}{8} + \dfrac{\pi}{8}$
$x_3 = x_2 + \dfrac{\pi}{8}$	$= \dfrac{2\pi}{8}$
$= -\dfrac{\pi}{8} + \dfrac{\pi}{8}$	$= \dfrac{\pi}{4}$ (same value as in Step 3)
$= 0$	

Summarizing, the x-coordinates are $-\dfrac{\pi}{4}$, $-\dfrac{\pi}{8}$, 0, $\dfrac{\pi}{8}$, and $\dfrac{\pi}{4}$.

Step 5: Label the x-axis (using $\dfrac{p}{4} = \dfrac{\pi}{8}$ as the unit interval). See Fig. 1, p.607

Step 6: Find the y-coordinates of the minimum and the maximum points.

The amplitude is $|a| = |-3| = 3$.

For this step, we may also write the range inequality $-3 \le y \le 3$ (noting that $k = 0$)
The y--coordinates of the minimum and the maximum points are -3 and 3 respectively.

Lesson 95: Direct Procedure: Sketching the Graphs of $y - k = a \sin (bx - h)$,

Step 7: Using the x-coordinates (from Step 4) calculate the corresponding y-coordinates by substituting these values in turns in the functional equation to obtain ordered pairs or a table of values for x and y. The ordered pairs are:

$$(-\tfrac{\pi}{4},0); \quad (-\tfrac{\pi}{8},3); \quad (0,0); \quad (\tfrac{\pi}{8},-3); \quad (\tfrac{\pi}{4},0).$$

For the calculations of the y-coordinates, see **Box 1** below.

Step 8: Label the y-axis.

Step 9: Plot the ordered pairs from Step 7.

Step 10: Determine the concavities of the basic curve
   Since $a > 0$, the curve has the regular concavity.

Step 11: Connect the points plotted, noting the concavities and the characteristic shape of the sine curve.

Step 12: Determine the number of cycles for the specified interval.
   The number of cycles $= b$. If asked for only one cycle, draw the curve for one cycle.

Step 13: Draw the remaining cycles by extension and following Step 4 to Step 11..

Step 14: The number of cycles to draw $= b$, and in the present problem, $b = 4$, and therefore we draw 3 more cycles.

Box 1: Calculations for Step 7

$x_1 = -\tfrac{\pi}{4}$, $y = 3 \sin [4(-\tfrac{\pi}{4}) + \pi]$
$= 3 \sin [-\pi + \pi]$
$= 3 \sin [0] \qquad (\sin 0 = 0)$
$= 0$
$\Rightarrow (-\tfrac{\pi}{4}, 0)$

$x_3 = 0$, $y = 3 \sin [4(0) + \pi]$
$= 3 \sin [0 + \pi]$
$= 3 \sin [\pi]$
$= 3(0) \qquad (\sin \pi = 0)$
$= 0$
$\Rightarrow (0,0)$

$x_2 = -\tfrac{\pi}{8}$, $y = 3 \sin [4(-\tfrac{\pi}{8}) + \pi]$
$= 3 \sin [-\tfrac{\pi}{2} + \pi]$
$= 3 \sin [\tfrac{\pi}{2}]$
$= 3(1) = 3 \qquad (\sin \tfrac{\pi}{2} = 1)$
$\Rightarrow (-\tfrac{\pi}{8}, 3)$

$x_4 = \tfrac{\pi}{8}$, $y = 3 \sin [4(\tfrac{\pi}{8}) + \pi]$
$= 3 \sin [\tfrac{\pi}{2} + \pi]$
$= 3 \sin [\tfrac{3\pi}{2}]$
$= 3(-1) = -3$
$\Rightarrow (\tfrac{\pi}{8}, -3)$

$x_5 = \tfrac{\pi}{4}$, $y = 3 \sin [4(\tfrac{\pi}{4}) + \pi] = 3 \sin [\pi + \pi] = 3 \sin [2\pi] = 3(0) = 0 \Rightarrow (\tfrac{\pi}{4}, 0)$

Note : $\sin[2\pi] = 0$

Example 1: Sketch the graph of $y = 3\sin(4x + \pi)$ $\quad y - k = a\cos(bx - h)$

Method 2 (You may also use method 2 (transformation) of Example 2)

Step 1: Find the period p. $b = 4$, and $p = \dfrac{2\pi}{b} = \dfrac{2\pi}{4} = \dfrac{\pi}{2}$

Step 2: Find the beginning-point x-coordinate, x_1, of the first cycle.

   Set $4x + \pi = 0$ and solve for x.: Let $x_1 = -\dfrac{\pi}{4}$

Step 3: Find the end-point x-coordinate, x_5, of the first cycle:

   Set $4x + \pi = 2\pi$. and solve for x to obtain $x = \dfrac{\pi}{4}$. Let $x_5 = \dfrac{\pi}{4}$.

(We may combine Steps 2 and 3, by solving the domain inequality: $0 \le 4x + \pi \le 2\pi$ to obtain

$\qquad -\dfrac{\pi}{4} \le x \le \dfrac{\pi}{4}$ (The first cycle of the curve is between $-\dfrac{\pi}{4}$ and $\dfrac{\pi}{4}$.)

   There are 4 cycles, since $b = 4$.

Step 4: Determine the domain (x-values) to use to obtain points (for the first cycle).

   Divide the period p by 4 to obtain "the quarter-interval", $\dfrac{p}{4}$, for the x-axis.

   Add $\dfrac{p}{4}$ successfully to x_1 until x_5 is reached (to obtain the x-values for the next Step)

$$\dfrac{p}{4} = \dfrac{\pi}{2} \cdot \dfrac{1}{4} \text{ or } \dfrac{1}{4}\text{of}\left(\dfrac{\pi}{2}\right) \qquad (p = \dfrac{\pi}{2})$$

$$= \dfrac{\pi}{8} < - - - - - \dfrac{p}{4}$$

$x_1 = -\dfrac{\pi}{4}$ (from Step 2)	$x_4 = x_3 + \dfrac{\pi}{8}$
$x_2 = x_1 + \dfrac{\pi}{8}$	$= 0 + \dfrac{\pi}{8}$
$= -\dfrac{\pi}{4} + \dfrac{\pi}{8}$	$= \dfrac{\pi}{8}$
$= -\dfrac{\pi}{8}$	$x_5 = \dfrac{\pi}{8} + \dfrac{\pi}{8}$
$x_3 = x_2 + \dfrac{\pi}{8}$	$= \dfrac{2\pi}{8}$
$= -\dfrac{\pi}{8} + \dfrac{\pi}{8}$	$= \dfrac{\pi}{4}$ (same value as in Step 3)
$= 0$	

Summarizing, the x-coordinates are $-\dfrac{\pi}{4}$, $-\dfrac{\pi}{8}$, 0, $\dfrac{\pi}{8}$, and $\dfrac{\pi}{4}$

Step 5: Label the x-axis (using $\dfrac{p}{4} = \dfrac{\pi}{8}$ as the unit interval)

Step 6: Determine the corresponding y-coordinates. Instead of calculating the y-coordinates by substituting the x-coordinates in the functional equation, we shall **alternatively** find the y-coordinates by using the mnemonic device "YoMaxYoMinYo (pronounced yo-max-yo-min-yo) or MaxYoMinYoMax where if $k = 0$, Yo $= 0$, Max $= |a| = 3$, and Min $= -|a| = -3$. Therefore, the y-coordinates are $0, 3, 0, -3$ and 0.

   Pairing these y-values with the corresponding x-values, $-\dfrac{\pi}{4}$, $-\dfrac{\pi}{8}$, 0, $\dfrac{\pi}{8}$, $\dfrac{\pi}{4}$ (from Step 4),

   we obtain the ordered pairs $(-\dfrac{\pi}{4}, 0)$; $(-\dfrac{\pi}{8}, 3)$; $(0, 0)$; $(\dfrac{\pi}{8}, -3)$; $(\dfrac{\pi}{4}, 0)$. for the graph.

Step 7: Label the y-axis

Step 8: Plot the ordered pairs from Step 6.

Step 9: Determine the concavity of the basic curve. Since $a > 0$, the curve has the regular concavity.

Step 10: Connect the five points plotted noting the concavity and applying the following:

> **If** the first point is **on** the x-axis (a zero), the next point (the second point) is **not** on the x-axis (a minimum or maximum point); the third point is **on** the x-axis (a zero); the fourth point is **not** on the x-axis (a minimum or maximum point); and the fifth point is **on** the x-axis (a zero). However, **if** the first point is **not** on the x-axis (**not** a zero), the second point is **on** the x-axis (a zero); the third point is **not** on the x-axis (not a zero), the fourth point is **on** the x-axis, and the fifth point is **not** on the x-axis. Using this alternation in the location of points on or off the x-axis, together with the amplitude, we can draw the curves.

Step 11: The number of cycles to draw $= 4$, since $b = 4$, and therefore we would draw 3 more cycles by repetition.

Step 12: Draw the remaining cycles by extension and following Step 4 to Step 10.

> Figure 1 shows the first cycle, Figure 2 shows all the cycles between 0 and 2π, using a different scale. Since $b = 4$, four cycles are drawn.

Figure 1

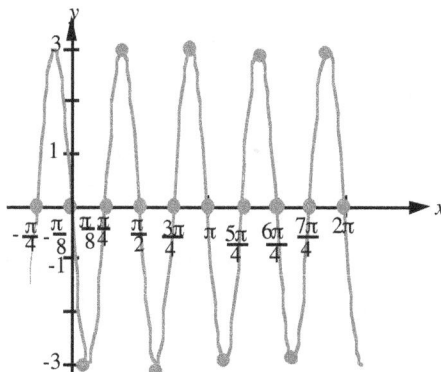

Figure 2

Example 1b   Sketch the graph of $y = 3 \sin(4x + \pi)$ $0 \le x \le 2\pi$

The sketch is the same as that of Example 1a, except that only that part of the curve between 0 and 2π is shown. (You may also use method 2 of Example 2)

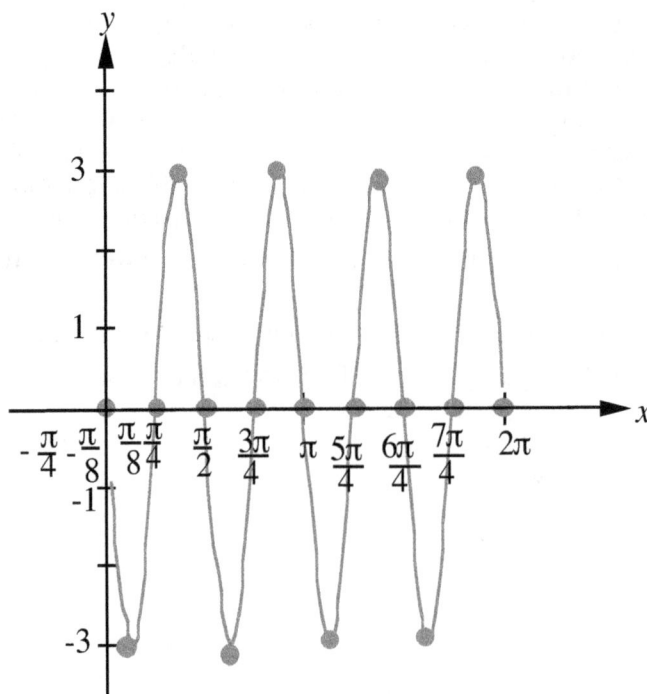

Example 2: Sketch the graph of $y - 1 = 3 \sin(4x + \pi)$

We cover three methods

Method 1

This Example is the same as Example 1, except that here, k = 1. . Each point on the graph in Example 1 is shifted 1 unit up, (a vertical shift). Steps 1 to 4 are the same. The difference in Step 5 is the 1. Redo the calculations for Step 5, and observe later that we could obtain the ordered pairs in Step 5 in Example 2 by increasing the y-value of each ordered pair in Example 1 by 1. We repeat the steps in Example 1 with the appropriate changes for Example 2..

Step 1: Find the period p. $b = 4$, and $p = \dfrac{2\pi}{b} = \dfrac{2\pi}{4} = \dfrac{\pi}{2}$

Step 2: Find the beginning point x-coordinate, x_1 of the first cycle

   Set $4x + \pi = 0$ and solve for x.: Let $x_1 = -\dfrac{\pi}{4}$

Step 3: Find the end-point x-coordinate, x_5, of the first cycle of the curve:

   Set $4x + \pi = 2\pi$.and solve for x   Let $x_5 = \dfrac{\pi}{4}$

(We may combine Steps 2 and 3, by solving the domain inequality: $0 \le 4x + \pi \le 2\pi$ to obtain

   $-\dfrac{\pi}{4} \le x \le \dfrac{\pi}{4}$ (The first cycle of the curve is between $-\dfrac{\pi}{4}$ and $\dfrac{\pi}{4}$.

Step 4: Determine the domain (x-values) to use to obtain ordered pairs for the points..

   Divide the period p by 4 to obtain "the quarter -interval", $\dfrac{p}{4}$, for the x-axis.

Add $\dfrac{p}{4}$ successfully to x_1 until x_5 is reached (to obtain the x-values for the next Step)

$\dfrac{p}{4} = \dfrac{\pi}{2} \cdot \dfrac{1}{4}$ or $\dfrac{1}{4}$ of $\left(\dfrac{\pi}{2}\right)$ $\qquad \left(p = \dfrac{\pi}{2}\right)$

$\qquad = \dfrac{\pi}{8} < ----- \dfrac{p}{4}$

$\begin{bmatrix} x_1 = -\dfrac{\pi}{4} \text{ (from Step 2)} \\[2mm] x_2 = x_1 + \dfrac{\pi}{8} \\[2mm] \quad = -\dfrac{\pi}{4} + \dfrac{\pi}{8} \\[2mm] \quad = -\dfrac{\pi}{8} \\[2mm] x_3 = x_2 + \dfrac{\pi}{8} \\[2mm] \quad = -\dfrac{\pi}{8} + \dfrac{\pi}{8} \\[2mm] \quad = 0 \end{bmatrix}$ $\begin{bmatrix} x_4 = x_3 + \dfrac{\pi}{8} \\[2mm] \quad = 0 + \dfrac{\pi}{8} \\[2mm] \quad = \dfrac{\pi}{8} \\[2mm] x_5 = \dfrac{\pi}{8} + \dfrac{\pi}{8} \\[2mm] \quad = \dfrac{2\pi}{8} \\[2mm] \quad = \dfrac{\pi}{4} \end{bmatrix}$

(same value in Step 3)

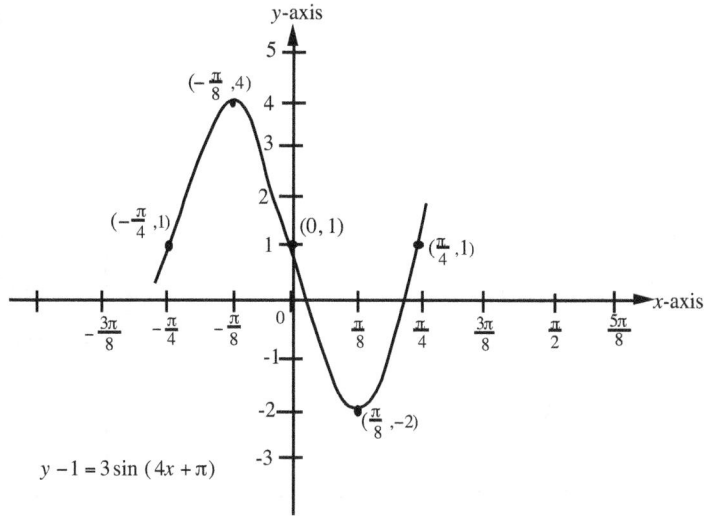

$y - 1 = 3\sin(4x + \pi)$

Figure 3

Summarizing, the x-coordinates are $-\dfrac{\pi}{4}$, $-\dfrac{\pi}{8}$, 0, $\dfrac{\pi}{8}$, and $\dfrac{\pi}{4}$

Step 5: Label the x-axis (using $\dfrac{p}{4} = \dfrac{\pi}{8}$ as the unit interval)

Step 6: Using these x-coordinates (from Step 4) calculate the corresponding y-coordinates by substituting these values in turns in the functional equation to obtain ordered pairs or a table of values for x and y. The ordered pairs are $\left(-\dfrac{\pi}{4}, 1\right)$; $\left(-\dfrac{\pi}{8}, 4\right)$; $(0, 1)$; $\left(\dfrac{\pi}{8}, -2\right)$; $\left(\dfrac{\pi}{4}, 1\right)$. For the calculations of the y-coordinates see Box 2 below.

Step 7: Find the y-coordinates of the minimum and maximum points.
The amplitude is $|a| = 3$, $k = 1$ We can also solve the range inequality $-3 \leq y - 1 \leq 3$ (from $-a \leq y - k \leq a$) to obtain $-2 \leq y \leq 4$.
The y-coordinate of the minimum point is -2 and that of the maximum point is 4.

Step 8: Label the y-axis

Step 9: Plot the ordered pairs from Step 6.

Step 10: Determine the concavities of the basic curve
Since $a > 0$, the curve has the regular concavity.

Step 11: Connect the points plotted, noting the concavities and the characteristic shape of the sine curve.

Step 12: Determine the number of cycles for the specified interval.
The number of cycles $= b$. If only one cycle is required, draw the curve for one cycle.

Step 13: Draw the remaining cycles by extension and following Step 4 to Step 10.
See **Figure 3**, above.

$x_1 = -\frac{\pi}{4}$, $y = 3 \sin[4(-\frac{\pi}{4}) + \pi] + 1$

$= 3 \sin[-\pi + \pi] + 1$

$= 3 \sin[0] + 1$ $(\sin 0 = 0)$

$= 1$

$\Rightarrow (-\frac{\pi}{4}, 1)$

$x_3 = 0$, $y = 3 \sin[4(0) + \pi] + 1$

$= 3 \sin[0 + \pi] + 1$

$= 3 \sin[\pi] + 1$

$= 3(0) + 1$ $(\sin \pi = 0)$

$= 1$

$\Rightarrow (0, 1)$

$x_2 = -\frac{\pi}{8}$, $y = 3 \sin[4(-\frac{\pi}{8}) + \pi] + 1$

$= 3 \sin[-\frac{\pi}{2} + \pi] + 1$

$= 3 \sin[\frac{\pi}{2}] + 1$

$= 3(1)$ $(\sin \frac{\pi}{2} = 1)$

$= 4$

$\Rightarrow (-\frac{\pi}{8}, 4)$

$x_4 = \frac{\pi}{8}$, $y = 3 \sin[4(\frac{\pi}{8}) + \pi] + 1$

$= 3 \sin[\frac{\pi}{2} + \pi] + 1$

$= 3 \sin[\frac{3\pi}{2}] + 1$

$= 3(-1) + 1$

$= -2$

$\Rightarrow (\frac{\pi}{8}, -2)$

$x_5 = \frac{\pi}{4}$, $y = 3 \sin[4(\frac{\pi}{4}) + \pi] + 1 = 3 \sin[\pi + \pi] + 1 = 3 \sin[2\pi] + 1 = 3(0) + 1 = 1 \Rightarrow (\frac{\pi}{4}, 1)$.

Example 2
Method 2 : (Transformation of functions method)   (also review page 225-226)

Given $y = \sin x$, draw the graph of $y - 1 = 3\sin(4x + \pi)$

Rewrite $y - 1 = 3\sin(4x + \pi)$ as $y - 1 = 3\sin[4(x + \frac{\pi}{4})]$. Here, $k = 1$, $h = -\frac{\pi}{4}$, $a = 3$, $b = 4$

$y = \sin x$		\multicolumn{3}{c}{$y - 1 = 3\sin[4(x + \frac{\pi}{4})]$}		
x_0	y_0	$x_n = \dfrac{x_0}{b} + h$	$y_n = y_0(a) + k$	(x_n, y_n)	
0	0	$\frac{0}{4} - \frac{\pi}{4} = 0 + -\frac{\pi}{4} = -\frac{\pi}{4}$	$0(3) + 1 = 0 + 1 = 1$	$(-\frac{\pi}{4}, 1)$	Step 1: Quickly prepare a a table of values for x and y, using the basic $\sin x$ equation and convenient x-values.
$\frac{\pi}{2}$	1	$\frac{\pi}{2 \cdot 4} - \frac{\pi}{4} = \frac{\pi}{8} - \frac{\pi}{4} = -\frac{\pi}{8}$	$1(3) + 1 = 3 + 1 = 4$	$(-\frac{\pi}{8}, 4)$	Step 2: Using the table of values from Step 1, determine the
π	0	$\frac{\pi}{4} - \frac{\pi}{4} = 0$	$0(3) + 1 = 0 + 1 = 1$	$(0, 1)$	new coordinates x_n and y_n.
$\frac{3\pi}{2}$	-1	$\frac{3\pi}{2 \cdot 4} - \frac{\pi}{4} = \frac{3\pi}{8} - \frac{\pi}{4} = \frac{\pi}{8}$	$-1(3) + 1 = -3 + 1 = -2$	$(\frac{\pi}{8}, -2)$	Step 3: Plot the new points and connect them by a solid curve
2π	0	$\frac{2\pi}{4} - \frac{\pi}{4} = \frac{\pi}{2} - \frac{\pi}{4} = \frac{\pi}{4}$	$0(3) + 1 = 0 + 1 = 1$	$(\frac{\pi}{4}, 1)$	(See Figure 3, previous page) or next page.
$-\frac{\pi}{2}$	-1	$-\frac{\pi}{2 \cdot 4} - \frac{\pi}{4} = -\frac{\pi}{8} - \frac{\pi}{4} = -\frac{3\pi}{8}$	$-1(3) + 1 = -3 + 1 = -2$	$(-\frac{3\pi}{8}, -2)$	
$-\pi$	0	$-\frac{\pi}{4} - \frac{\pi}{4} = -\frac{\pi}{2}$	$0(3) + 1 = 0 + 1 = 1$	$(-\frac{\pi}{2}, 1)$	
$-\frac{3\pi}{2}$	1	$-\frac{3\pi}{2 \cdot 4} - \frac{\pi}{4} = -\frac{3\pi}{8} - \frac{\pi}{4} = -\frac{5\pi}{8}$	$1(3) + 1 = 3 + 1 = 4$	$(-\frac{5\pi}{8}, 4)$	
-2π	0	$-\frac{2\pi}{4} - \frac{\pi}{4} = -\frac{3\pi}{4}$	$0(3) + 1 = 0 + 1 = 1$	$(-\frac{3\pi}{4}, 1)$	

In Example 2 above, compare method 1 and method 2, the transformation method. Also compare the calculations in step 6 of Method 1 and the calculations in Method 2. Observe the compactness of the transformation method compared to method 1. The first five rows of method 2 cover all the calculations in method 1. A word to the wise is enough.

Example 2: Sketch the graph of $y - 1 = 3 \sin (4x + \pi)$

Method 3

This Example is the same as Example 1, except that here $k = 1$. . Each point on the graph in Example 1 is shifted 1 unit up, (a vertical shift). Steps 1 to 4 are the same. We repeat the steps with the appropriate changes.

Step 1: Find the period p. $b = 4$, and $p = \dfrac{2\pi}{b} = \dfrac{2\pi}{4} = \dfrac{\pi}{2}$

Step 2: Find the beginning point x-coordinate, x_1, of the first cycle.

   Set $4x + \pi = 0$ and solve for x.: Let $x_1 = -\dfrac{\pi}{4}$

Step 3: Find the end-point x-coordinate of the first cycle:

   Set $4x + \pi = 2\pi$. and solve for x to obtain $x = \dfrac{\pi}{4}$. Let $x_5 = \dfrac{\pi}{4}$.

(We may combine Steps 2 and 3, by solving the domain inequality: $0 \le 4x + \pi \le 2\pi$ to obtain

   $-\dfrac{\pi}{4} \le x \le \dfrac{\pi}{4}$ (The first cycle of the curve is between $-\dfrac{\pi}{4}$ and $\dfrac{\pi}{4}$.
   Four 4 cycles are to be drawn. since $b = 4$.

Step 4: Determine the domain (x-values) to use to obtain points for the (for the first cycle).

   Divide the period p by 4 to obtain "the quarter -interval", $\dfrac{p}{4}$, for the x-axis.

   Add $\dfrac{p}{4}$ successfully to x_1 until x_5 is reached (to obtain the x-values for the next Step)

$$\dfrac{p}{4} = \dfrac{\pi}{2} \cdot \dfrac{1}{4} \text{ or } \dfrac{1}{4} \text{of} \left(\dfrac{\pi}{2}\right) \qquad (p = \tfrac{\pi}{2})$$

$$= \dfrac{\pi}{8} < - - - - - \dfrac{p}{4}$$

$x_1 = -\dfrac{\pi}{4}$ (from Step 2) $x_4 = x_3 + \dfrac{\pi}{8}$

$x_2 = x_1 + \dfrac{\pi}{8}$ $= 0 + \dfrac{\pi}{8}$

$\quad = -\dfrac{\pi}{4} + \dfrac{\pi}{8}$ $= \dfrac{\pi}{8}$

$\quad = -\dfrac{\pi}{8}$ $x_5 = \dfrac{\pi}{8} + \dfrac{\pi}{8}$

$x_3 = x_2 + \dfrac{\pi}{8}$ $= \dfrac{2\pi}{8}$

$\quad = -\dfrac{\pi}{8} + \dfrac{\pi}{8}$ $= \dfrac{\pi}{4}$ (same value as in Step 3)

$\quad = 0$

Summarizing, the x-coordinates are $-\dfrac{\pi}{4}$, $-\dfrac{\pi}{8}$, 0, $\dfrac{\pi}{8}$, and $\dfrac{\pi}{4}$.

Step 5: Label the x-axis (using $\dfrac{p}{4} = \dfrac{\pi}{8}$ as the unit interval)

Step 6: Determine the corresponding y-coordinates. We can obtain the y-coordinates by substituting the x-values in the functional equation, $y - 1 = 3 \sin (4x + \pi)$.

However, we can **alternatively** find these values by using the mnemonic device $(bx - h)$
"YoMaxYoMinYo (pronounced yo-max-yo-min-yo) or MaxYoMinYoMax,
where if $k = 1$, Yo $= k = 1$, Max $= k + |a| = 1 + 3 = 4$, Min $= = 1 - 3 = -2$.
The corresponding y-coordinates are $1, 4, 1, -2$ and 1. Pairing these y-values with the
corresponding x-values, $-\frac{\pi}{4}, -\frac{\pi}{8}, 0, \frac{\pi}{8}, \frac{\pi}{4}$ (from Step 5), we obtain the ordered pairs

$(-\frac{\pi}{4}, 1)$; $(-\frac{\pi}{8}, 4)$; $(0, 1)$; $(\frac{\pi}{8}, -2)$; $(\frac{\pi}{4}, 1)$. for the graph.

Step 7: Label the y-axis.

Step 8: Plot the ordered pairs from Step 6.

Step 9: Determine the concavity of the basic curve. Since $a > 0$, the curve has the regular concavity.

Step 10: Connect the five points plotted, noting the concavity, the characteristic shape of the sine   curve, and applying the following:

Step 11: The number of cycles to draw $= 4$, since $b = 4$; and therefore, we draw 3 more cycles by repetition.

Step 12: Draw the remaining cycles by extension and following Step 4 to Step 10.
Figure shows the first cycle Figure shows all the cycles between 0 and 2π, using a different scale. Since $b = 4$, Four cycles are drawn.

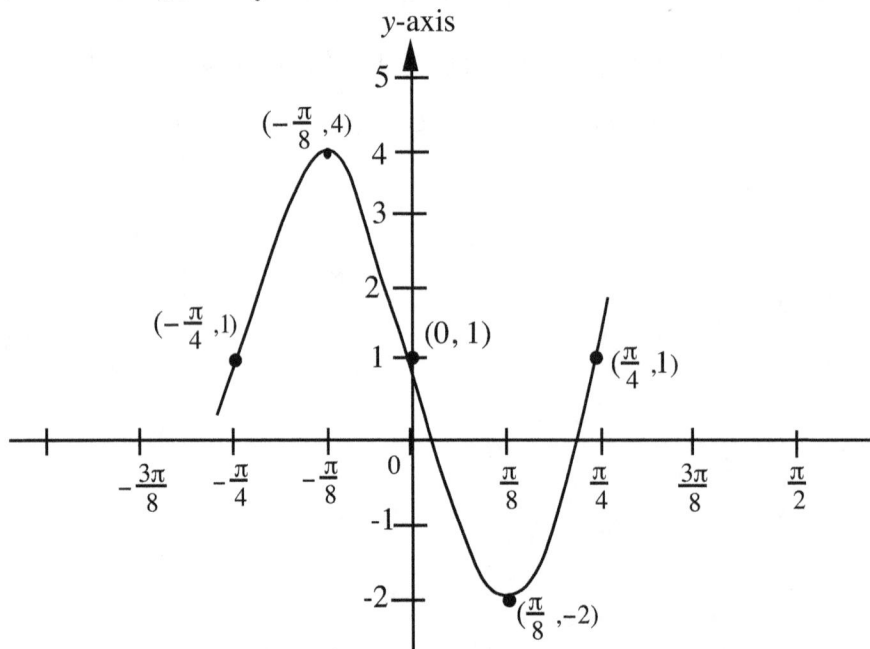

Figure: Graph of $y - 1 = 3 \sin(4x + \pi)$

Example 3: To draw the graph of $y = 4 \cos (2x - \pi)$ or $y - 1 = 3 \cos (4x + \pi)$ imitate the
method of the last two examples for the sine curves, $y = 3 \sin (4x + \pi)$
and $y - 1 = 3 \sin (4x + \pi)$. In particular, imitate the **transformation method**.
The graph of $y = 4 \cos (2x - \pi)$ is shown in Figure below

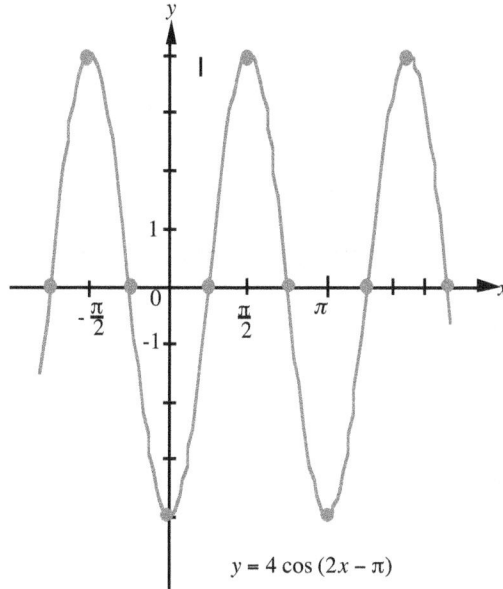

$y = 4 \cos (2x - \pi)$

More Examples on Trigonometric Graphs

Example 4   Sketch the graph of $y - 1 = 3 \sin (4x + \pi)$ $0 \leq x \leq 2\pi$
Solution
The graph is the same as that of Example 2, except that we extend the graph to 2π and delete the part of
the graph for which $x < 0$. Imitate the **transformation method**, and make sure x_n goes up to 2π.

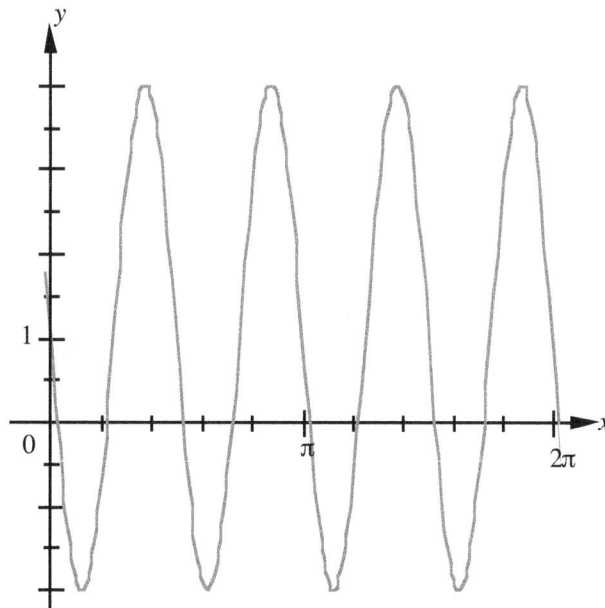

Graph of $y - 1 = 3 \sin (4x + \pi)$ $0 \leq x \leq 2\pi$

Example 5   Sketch the graph of $y = -\sin x$ $(0 \le x \le 2\pi)$.

Solution Imitate the **transformation method,** .and make sure the x_n's satisfy $0 \le x \le 2\pi$.

The graph is the reflection of $y = \sin x$ about the x-axis.

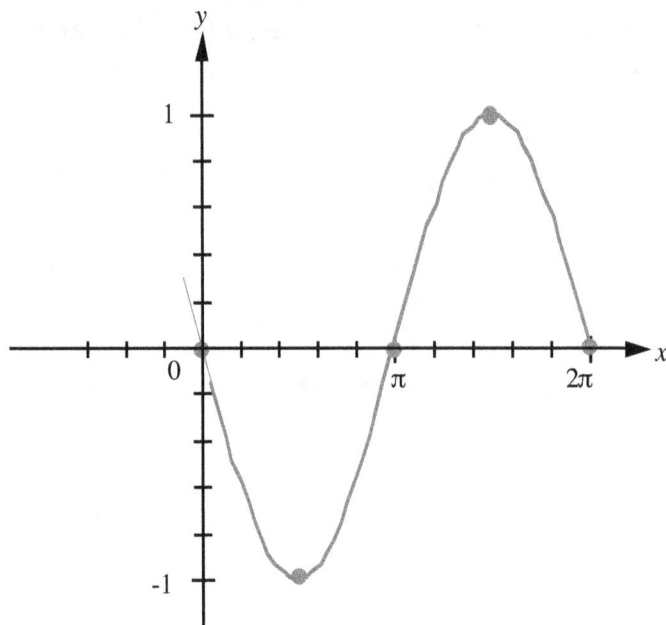

$y = -\sin x$ $(0 \le x \le 2\pi)$.

Example 6   Sketch the graph of $y = -\sin x$ $(-\pi \le x \le 2\pi)$.

Solution: Imitate the **transformation method,** and make sure the x_n's satisfy $-\pi \le x \le 2\pi$ ......

The sketch is similar to that of Example 5 above, except that we extend the curve to -π.

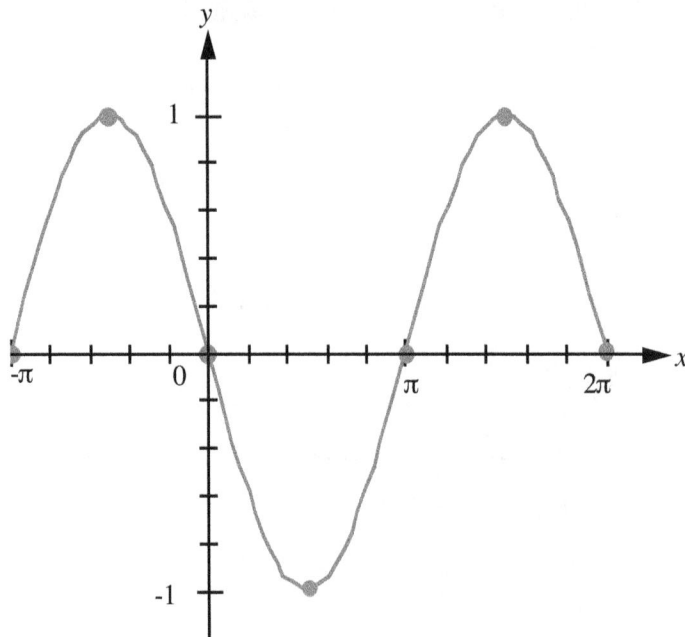

$y = -\sin x$ $(-\pi \le x \le 2\pi)$.

Lesson 95: Direct Procedure: Sketching the Graphs of $y - k = a \sin (bx - h)$, $y - k = a \cos (bx - h)$

Example 7 Sketch the graph of $y = \sin x$ $(0 \le x \le 3\pi)$ 615

Solution   In this case, we sketch one and half sine curve starting from the origin and
terminating at the point $(3\pi, 0)$.

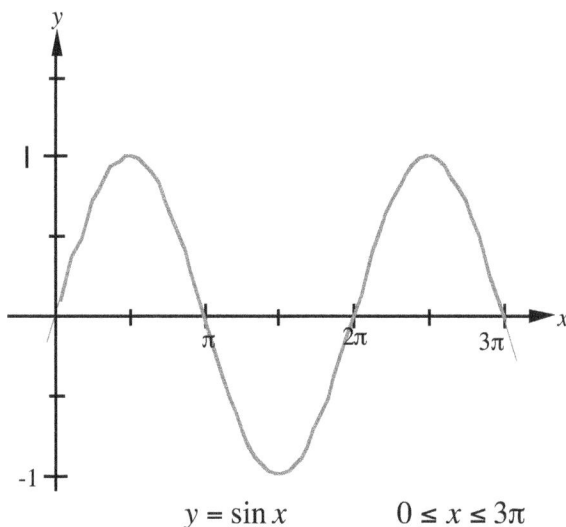

$$y = \sin x \qquad 0 \le x \le 3\pi$$

Example 8 Sketch the graph of $y = 3 \sin x$ $(0 \le x \le 2\pi)$

Given $y = a \sin x$, a is called the amplitude. The amplitude indicates how far up (maximum value or
highest point on this curve) or how far down (minimum value or lowest point) the curve goes from the x-axis.
The maximum y-value for $y = \sin x$ is 1 and the minimum value is -1. The maximum y-value for
$y = 3 \sin x$ is 3 and the minimum y-value is -3.

The graph of $y = 3 \sin x$ and $y = \sin x$ have the same period, 2π.

To sketch $y = 3 \sin x$, repeat Step 1- Step 5 in Example 7 above, but replace $(\frac{\pi}{2}, 1)$ by $(\frac{\pi}{2}, 3)$ and
replace $(\frac{3\pi}{2}, -1)$ by $(\frac{3\pi}{2}, -3)$.

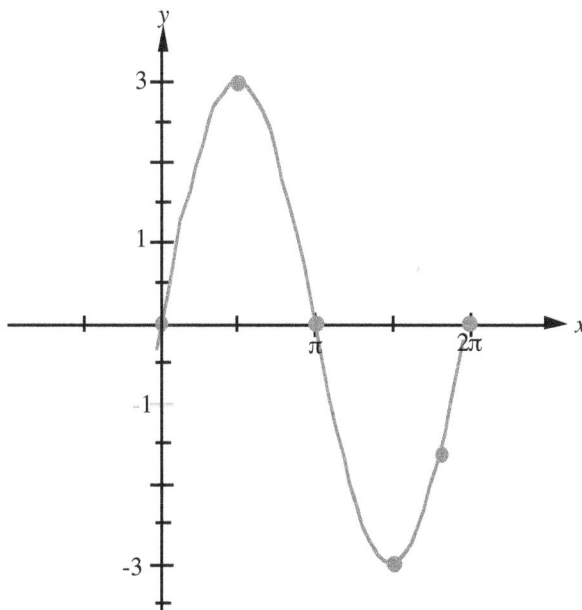

Figure $y = 3 \sin x$ $(0 \le x \le 2\pi)$

Example 9 Sketch the graph of $y = \sin 4x$ (see p.606-612 for the general case) $k = a \cos(bx - h)$
Method 1

Step 1: $b = 4$ the period, $p = \frac{2\pi}{4} = \frac{\pi}{2}$

Since the period is $\frac{\pi}{2}$ the graph the curve must repeat itself every $\frac{\pi}{2}$ units.

Step 2: Find the starting point: Set $4x = 0$ and solve to obtain $x = 0$.

Step 3: Find the endpoint: Set $4x = 2\pi$ and solve for x to obtain $x = \frac{\pi}{2}$.

Step 4: Determine the domain: The unit interval $u = \frac{1}{4}(p) = \frac{1}{4}(\frac{\pi}{2}) = \frac{\pi}{8}$.

$$x_1 = 0$$

$$x_2 = 0 + \frac{\pi}{8} = \frac{\pi}{8}$$

$$x_3 = \frac{\pi}{8} + \frac{\pi}{8} = \frac{2\pi}{8} \text{ or } \frac{\pi}{4}$$

$$x_4 = \frac{2\pi}{8} + \frac{\pi}{8} = \frac{3\pi}{8}$$

$$x_5 = \frac{3\pi}{8} + \frac{\pi}{8} = \frac{4\pi}{8} \text{ or } \frac{\pi}{2}$$

Step 5: The amplitude of $y = \sin 4x$ is 1 (i.e. $a = 1$).
   The range is $-1 \leq y \leq 1$. The minimum point is at $y = -1$ and the maximum point is at $y = 1$.

Step 6: Label the x-axis with the x-values from Step 4 .

Step 7: Label the y-axis with the results from Step 5.

Step 8: Draw a complete sine curve starting from the origin $(0,0)$) so that the curve repeats at the mark, $\frac{\pi}{2}$. (Review the properties of the sine curve).

However, we have not finished sketching yet, because the interval is $0 \leq x \leq 2\pi$. So we must repeat the sketch until we have drawn enough sine curves to complete the interval up to the 2π mark. So, from the point $B(\frac{\pi}{2}, 0)$ we draw three more additional sine (cycles) curves (using the properties of the sine curve). For the interval $0 \leq x \leq 2\pi$, the number of cycles is b..
For $y = \sin 4x$, $0 \leq x \leq 2\pi$ the number of cycles is 4.

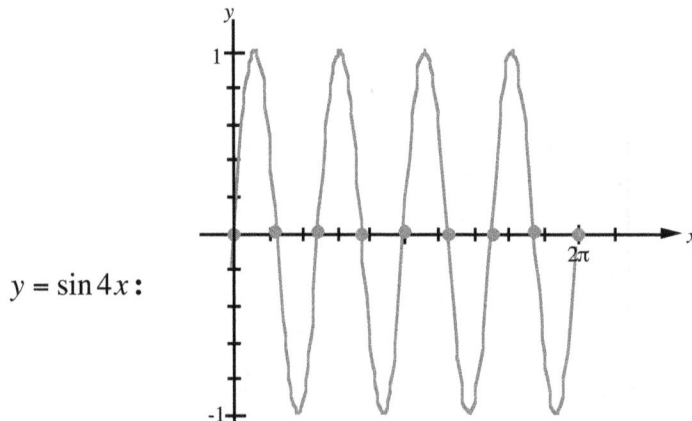

$y = \sin 4x$:

Method 2: Apply the transformation method

Example 10   Sketch the graph of $y = 3\sin 4x$ $0 \leq x \leq 2\pi$

Method 1

Step 1: The amplitude is 3

Step 2:   Find the period: the period, $p = \frac{2\pi}{4} = \frac{\pi}{2}$

Step 3:   Number of cycles = 4

The graph is shown below.

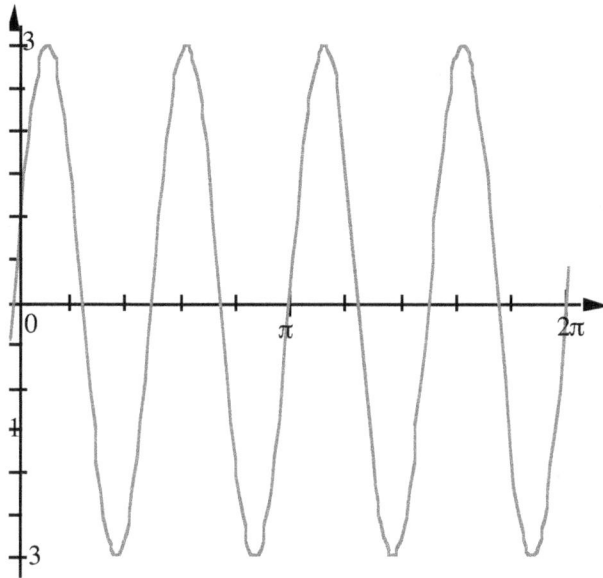

$y = 3\sin 4x$ $0 \leq x \leq 2\pi$

Method 2: Apply the transformation method

Example 11 Sketch $y = \sin x + 2$ $0 \leq x \leq 2\pi$

In standard form $y - 2 = \sin x$ Apply the transformation method)

Each point on the graph of $y = \sin x$ is shifted up 3 units, keeping the x-values unchanged.

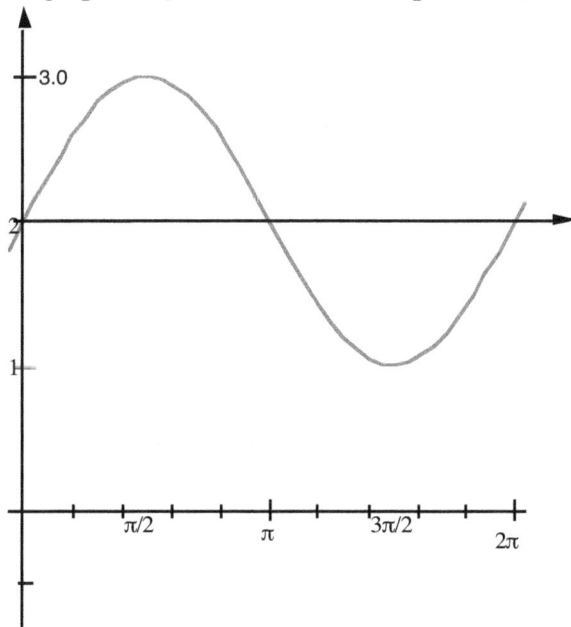

Example 12 Sketch the graph of $y - 2 = \sin(x + 3)$
Method 1

The period is $p = \frac{2\pi}{1} = 2\pi$

Domain: $0 \leq x + 3 \leq 6.28$

$0 - 3 \leq x \leq 6.28 - 3$

$-3 \leq x \leq 3.28$

The graph of the given equation is the graph of $y = \sin x$ with each x-coordinate shifted 3 units to the left and each y-coordinate shifted up 2 units, simultaneously.

Use decimal units to label the x-axis.

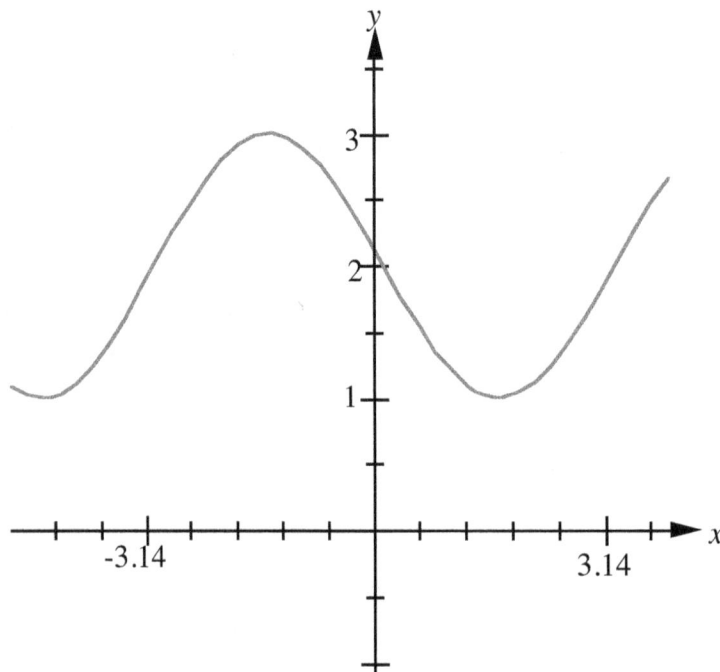

$$y - 2 = \sin(x + 3)$$

Method 2: Apply the transformation method

Sketching graphs of functions by addition of y-coordinates (ordinates)
Example Sketch the graph of $f(x) = \sin x + \cos x$,
Let the functions $f(x)$, $g(x)$ and $h(x)$ have the same domain and that $f(x) = g(x) + h(x)$ (1)

To sketch $f(x)$, we first sketch $g(x)$ and $h(x)$ separately on the same set of rectangular axes, then (at each) for each critical or important x-value, the corresponding y-values of $g(x)$ and $h(x)$) are added together. The sum so obtained gives the (new) y-value of $f(x)$. These summed y-values are plotted. The new points are then connected by a smooth curve. Note above that we add only the y-values keeping the x-values unchanged.

Lesson 95: Direct Procedure: Sketching the Graphs of $y - k = a \sin(bx - h)$, $y - k = a \cos(bx - h)$

Example 13: Sketch the graph of $y = \sin x + \cos x$ 619

Let $y_1 = \sin x$ and $y_2 = \cos x$

Step 1: Sketch the graph of $y_1 = \sin x$

Step 2: Sketch the graph of $y_2 = \cos x$ on the same axes as $y_1 = \sin x$

Step 3: For each x-coordinate of important or critical points add the corresponding y_1 and y_2 to obtain $y = y_1 + y_2$, and ordered pairs (x, y)

Step 4: Plot the ordered pairs from Step 3, noting that the x-coordinates remain the same.

Step 5: Connect the new points with a smooth curve.
  Note that y_1 and y_2 are added algebraically, and we must take into consideration the signs (positive or negative) when we add of y_1 and y_2.

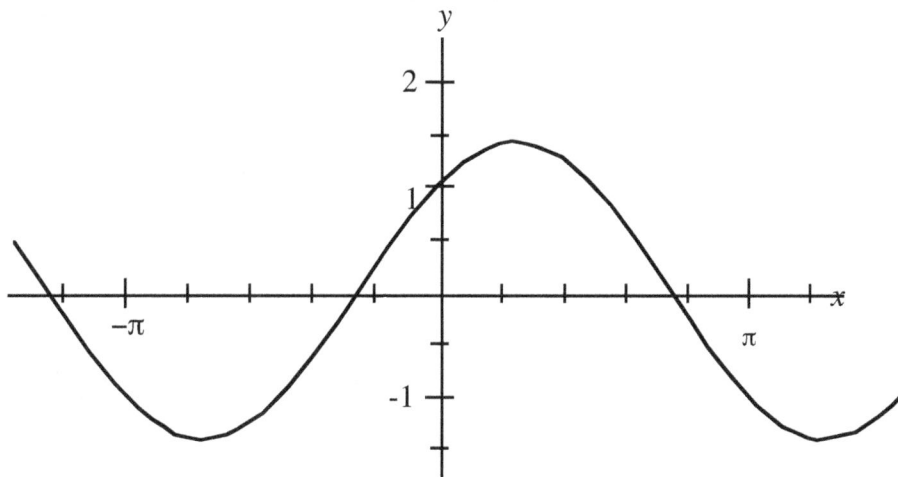

Example 14 Sketch the graph of $y = \sec x$ for $2\pi \le x \le 3\pi$
Solution See graph below

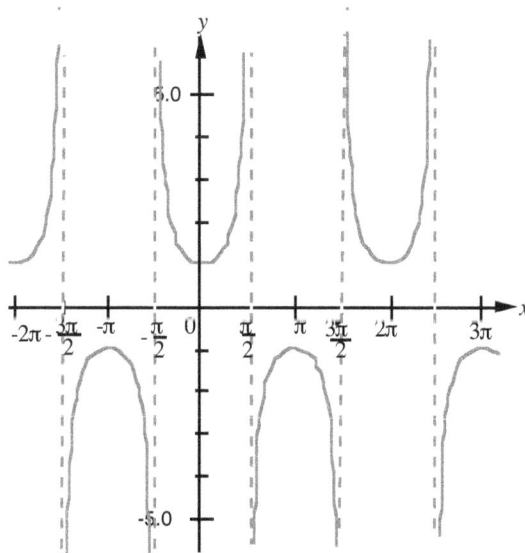

Example 15: Sketch the graph of $y = 2 \tan \left(x + \frac{\pi}{3}\right)$

Method 1

Solution Before proceeding, see page 594 to review the basic properties of the tangent curve.

Step 1: Find the period p. $b = 1$, and $p = \frac{\pi}{b} = \frac{\pi}{1} = \pi$

Step 2: Find the beginning-point x-coordinate, x_1, of the first cycle which locates a vertical asymptote.

Set $x + \frac{\pi}{3} = -\frac{\pi}{2}$, and solve for x to obtain $x_1 = -\frac{5\pi}{6}$<---This is a vertical asymptote.

Step 3: Find the end-point x-coordinate, x_5, of the first cycle which locates a vertical asymptote.

Set $x + \frac{\pi}{3} = \frac{\pi}{2}$ and solve for x to obtain $x_5 = \frac{\pi}{6}$.<---This is a vertical asymptote.

(We may combine Steps 2 and 3, by solving the domain inequality: $-\frac{\pi}{2} \le x + \frac{\pi}{3} \le \frac{\pi}{2}$ to obtain

$-\frac{5\pi}{6} \le x \le \frac{\pi}{6}$ (The first cycle of the curve is between the asymptotes $x = -\frac{5\pi}{6}$ and $x = \frac{\pi}{6}$)

Step 4: Find the x-intercepts: Set $y = 0$ in $y = 2 \tan \left(x + \frac{\pi}{3}\right)$

Then $0 = 2 \tan \left(x + \frac{\pi}{3}\right)$ or $\tan \left(x + \frac{\pi}{3}\right) = 0$

$x + \frac{\pi}{3} = 0$ or π. When $x + \frac{\pi}{3} = 0$, $x = -\frac{\pi}{3}$. When $x + \frac{\pi}{3} = \pi$, $x = \frac{2\pi}{3}$

Since for the cycle of interest , $-\frac{5\pi}{6} \le x \le \frac{\pi}{6}$, we consider only $x = -\frac{\pi}{3}$.

The x-intercept is at $\left(-\frac{\pi}{3}, 0\right)$

Step 5: Find the y-intercept. Let $x = 0$ in $y = 2 \tan \left(x + \frac{\pi}{3}\right)$

Then $y = 2 \tan \left(0 + \frac{\pi}{3}\right)$ and $y = 2 \tan \frac{\pi}{3}$ or $y = 2(1.732)$ $\left(\tan \frac{\pi}{3} = 1.732\right)$

$y = 3.46$. The y-intercept is at $(0, 3.5)$.

Step 6: Determine the domain (x-values) to use to obtain points (for the first cycle).

Divide the period p by 4 to obtain "the quarter -interval", $\frac{p}{4}$, for the x-axis.

Add $\frac{p}{4}$ successfully to x_1 until x_5 is reached (to obtain the x-values for the next Step)

$$\frac{p}{4} = \frac{\pi}{1} \cdot \frac{1}{4} \text{ or } \frac{1}{4} \text{of} \left(\frac{\pi}{1}\right) \qquad (p = \pi)$$

$$= \frac{\pi}{4} <----- \frac{p}{4}$$

$x_1 = -\frac{5\pi}{6}$ (from Step 2, Also a vertical asymptote)

$x_2 = -\frac{5\pi}{6} + \frac{\pi}{4}$

$\quad = -\frac{7\pi}{12}$

$x_3 = x_2 + \frac{\pi}{8} = -\frac{7\pi}{12} + \frac{\pi}{4} = -\frac{4\pi}{12}$

$x_4 = -\frac{4\pi}{12} + \frac{\pi}{4}$

$\quad = -\frac{\pi}{12}$

$x_5 = -\frac{\pi}{12} + \frac{\pi}{4}$

$\quad = \frac{2\pi}{12}$ or $\frac{\pi}{6}$ (From Step 3. A vertical asymptote)

Summarizing, the x-coordinates are $-\frac{5\pi}{6}$ or $-\frac{10\pi}{12}$, $-\frac{7\pi}{12}$, $-\frac{4\pi}{12}$, $-\frac{\pi}{12}$, and $\frac{2\pi}{12}$, noting that the first and the last values locate vertical asymptotes.

Step 7: Label the x-axis with the summary from Step 6. $\qquad y - k = a \cos(bx - h)$

Step 8: Determine the corresponding y-coordinates by substituting

the x-coordinates $-\frac{7\pi}{12}$, $-\frac{4\pi}{12}$, $-\frac{\pi}{12}$ in the equation $y = 2 \tan(x + \frac{\pi}{3})$ to obtain

-2, 0, and 2 respectively. (For calculations, see Box 3, below)

Step 9: Pair these y-values with $-\frac{7\pi}{12}$, $-\frac{4\pi}{12}$, $-\frac{\pi}{12}$ respectively to obtain

$$\left(-\frac{7\pi}{12}, -2\right), \left(-\frac{4\pi}{12}, 0\right), \left(-\frac{\pi}{12}, 2\right)$$

Step 10: Label the y-axis

Step 11: Plot the ordered pairs from Step 9.

Step 12: Draw the vertical asymptotes $x_1 = -\frac{5\pi}{6}$ and $x_5 = \frac{\pi}{6}$

Step 13: Connect the three points plotted noting the concavity , the characteristic shape of the tangent curve and extend the curve so that each end is asymptotic to its asymptote.

Box 3: Calculations for Step 8

$x_2 = -\frac{7\pi}{12}$, $y = 2\tan(-\frac{7\pi}{12} + \frac{\pi}{3})$

$\qquad = 2\tan(-\frac{3\pi}{12})$

$\qquad = 2(-\tan\frac{3\pi}{12}) = 2(-1) = -2$ $\left. \right\rbrace \Rightarrow (-\frac{7\pi}{12}, -2)$

$x_3 = -\frac{4\pi}{12}$, $y = 2\tan(-\frac{4\pi}{12} + \frac{\pi}{3})$

$\qquad = 2\tan(0)$

$\qquad = 0$ $\left. \right\rbrace \Rightarrow (-\frac{4\pi}{12}, 0)$

$x_4 = -\frac{\pi}{12}$, $y = 2\tan(-\frac{\pi}{12} + \frac{\pi}{3})$

$\qquad = 2\tan(\frac{3\pi}{12}) = 2(1) = 2$ $\left. \right\rbrace \Rightarrow (-\frac{\pi}{12}, 2)$

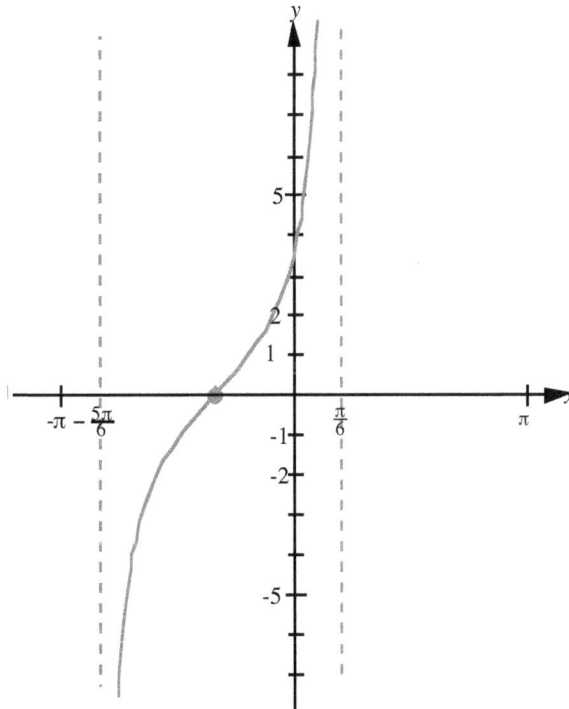

Method 2: Apply the transformation method (imitate page 610.

Sketch the graph of each of the following:

1. $y = 4\sin(2x + \pi)$; $\qquad$ **2.** $y = -2\sin(3x + 4)$; $\qquad$ **3.** $y - 1 = 3\cos(x - \frac{\pi}{4})$;

4. $y = 3\tan\left(x - \frac{\pi}{4}\right)$ $\quad$ for $0 \le x \le 2\pi$; $\qquad$ **5.** $y = -\cos x$ $\quad$ for $0 \le x \le 2\pi$.

Answers:

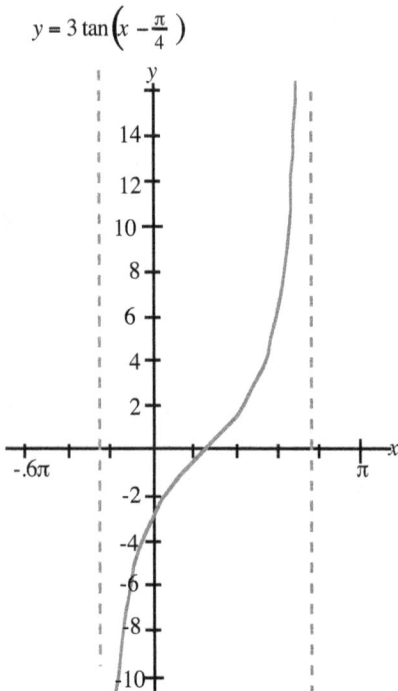

CHAPTER 37

From Trigonometric Functions to
Inverse Trigonometric Functions
("Exchange is no Robbery")

Lesson 96: **Introduction; Definitions; Operations Involving Inverse Trigonometric Functions**

Lesson 97: **How to Sketch the Graph of** $y = \text{Arcsin } x$

Lesson 98: **How to Sketch the Graph of** $y = \text{Arccos } x$

Lesson 99: **How to Sketch the Graph of** $y = \text{Arctan } x$

Lesson 96
Introduction; Definitions; Operations Involving
Inverse Trigonometric Functions

We use inverse trigonometric functions in solving trigonometric equations. In calculus, we use inverse trigonometric functions to obtain formulas for various results involving algebraic functions.

In Chapter 31, we learned how to find the functional values (y-values), given the measures of the angles. There, we obtained the functional values using trigonometric tables, the electronic calculator or from memory. We also learned how to find the measures of angles given the functional values.
With inverse trigonometric functions, we are also given the functional value, say y, and we are asked to find an angle or a number, say x. The process of finding y and the process of finding x are inverse operations (in much the same way as multiplication and division are inverse operations).

In Chapter 6, we also learned how to determine the inverses of algebraic functions, given the various functional forms. If the function was specified by a set form, we obtained its inverse by interchanging the first and second components of each ordered pair. With trigonometric functions, interchanging the first and second components of each ordered pair of numbers generally produce relations which are **not** functions because of the periodicity of the trigonometric functions.

Consider the **trigonometric relation** $y = \sin x$, which is specified by the set of ordered pairs,

$A = \{(-2\pi, 0), (-\frac{3\pi}{2}, 1), (-\pi, 0), (-\frac{\pi}{2}, -1), (0, 0), (\frac{\pi}{2}, 1), (\pi, 0), (\frac{3\pi}{2}, -1), (2\pi, 0)\}.$

This relation is also a **function** since no two distinct ordered pairs have the same first component. (A function is a set of ordered pairs of numbers in which no two distinct ordered pairs have the same first component).
On interchanging the first and second components of each ordered pair, we obtain the inverse relation, B

$= \{(0, -2\pi), (1, -\frac{3\pi}{2}), (0, -\pi), (-1, -\frac{\pi}{2}), (0, 0), (1, \frac{\pi}{2}), (0, \pi), (-1, \frac{3\pi}{2}), (0, 2\pi)\}.$

This new **relation** is **not a function** since the first, the third, the fifth, the seventh and the ninth ordered pairs have the same first component, namely 0, or also, the second, and the sixth ordered pairs have the same first component, namely 1. We observe that the original set, A, does **not** represent a **one-to-one function** and therefore, the inverse relation is not a function. The graph in the figure below fails the horizontal line test.

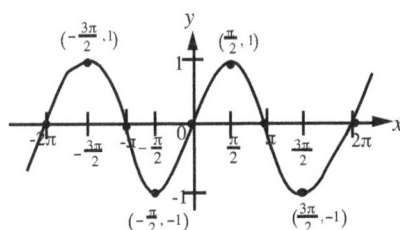

To make the set B a function, we can eliminate some of the ordered pairs so that no two distinct
ordered pairs have the same first component. If we consider the ordered pair $(0, 0)$ and choose
ordered pairs to its right and ordered pairs to its left so that no two distinct ordered pairs have the
same first component, we obtain the following: $C = \{(-1, -\frac{\pi}{2}), (0, 0), (1, \frac{\pi}{2})\}$..This set C is a **function**
since all the first components are different from one another. We observe also that in the original function
given by set A, if we keep the fourth, the fifth and the sixth ordered pairs and drop the other pairs, we
obtain $\{(-\frac{\pi}{2}, -1), (0, 0), (\frac{\pi}{2}, 1)\}$ which is a one-to-one function and therefore, its inverse relation is also a
function.

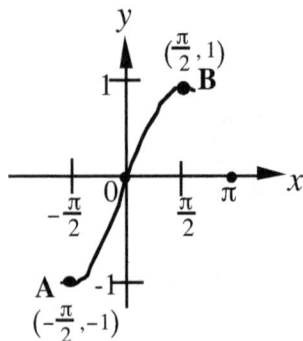

Principal branch

Fig 1. $y = \sin x$ for $-\frac{\pi}{2} \le x \le \frac{\pi}{2}$
$-1 \le y \le 1$

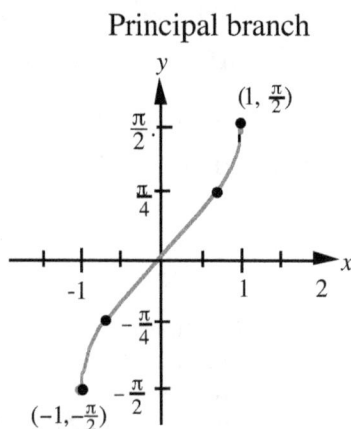

Fig. 2 $x = \sin y, -\frac{\pi}{2} \le y \le \frac{\pi}{2}.$
$-1 \le x \le 1$

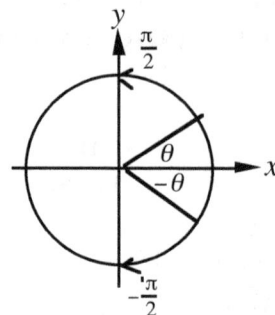

Fig. 3: For $-\frac{\pi}{2} \le y \le \frac{\pi}{2}.$

From above, we notice that the inverse relation for the sine function, $y = \sin x$, is generally **not** a function.
Similarly, the inverse relations for $y = \cos x$, $y = \sec x$, $y = \csc x$, $y = \tan x$, and $y = \tan x$ are generally
not functions.
Some of these inverse relations are many-valued and consist of a collection of single-valued functions
known as branches. Since it is more convenient to work with functions, we will make these inverse
relations become functions by restricting their ranges. By placing restrictions on the domains of those
original functions which are **not** one-to-one, we obtain a particular branch called the **principal branch** of
each inverse relation. Each branch **chosen** is an inverse function and we capitalize the first character of
the functional name or its abbreviation. (Note that some authors do not adhere to this **capitalization** and
in which case the principal branch and its range are to be understood).

Inverse Sine Function

Definition   Given, the equation, $y = \sin x$ $\quad -\frac{\pi}{2} \le x \le \frac{\pi}{2}$, what is its inverse?

Since the inverse is found by **interchanging** the roles of x and y,

Domain ↓   Range ↓

the inverse of $y = \sin x$ for $-\frac{\pi}{2} \le x \le \frac{\pi}{2},$ $\quad -1 \le y \le 1$ is

$\qquad x = \sin y$ for $-\frac{\pi}{2} \le y \le \frac{\pi}{2},$ $\quad -1 \le x \le 1$

Range   Domain

(Observe that the domain
and range are interchanged.)

If we solve $x = \sin y$ for y, and use the notation, $\text{Sin}^{-1} x$, we obtain

$y = \text{Sin}^{-1} x$ $\qquad$ (Recall previously that the inverse of $f(x)$ was symbolized $f^{-1}(x)$)

Another notation for this inverse is $y = \text{Arcsin } x$. We may use either $y = \text{Sin}^{-1} x$ or $y = \text{Arcsin } x.$

Lesson 96: Definitions; Operations Involving Inverse Trigonometric Functions

Example 1 If $\frac{\pi}{6} = \text{Arc}\sin\left(\frac{1}{2}\right)$ or $\text{Sin}^{-1}\left(\frac{1}{2}\right)$,

then $\sin\frac{\pi}{6} = \frac{1}{2}$

(is read " if $\frac{\pi}{6}$ is the angle or number whose sine is $\frac{1}{2}$.")

(Memorize this example to help remember the definitions)

$y = \text{Arcsin } x$ (y is the angle or the number whose sine is x.

Inverse Cosine Function

Definition Given, the equation, $y = \cos x$ $0 \le x \le \pi$, what is its inverse?

The inverse is found by interchanging the roles of x and y.

Domain ↓   Range ↓

The inverse of $y = \cos x$ for $0 \le x \le \pi$, $-1 \le y \le 1$ is

$x = \cos y$ for $0 \le y \le \pi$, $-1 \le x \le 1$

Range   Domain

If we solve $x = \cos y$ for y, and use the notation,

$\text{Cos}^{-1}x$, we obtain $y = \text{Cos}^{-1}x$ or $y = \text{Arccos } x$

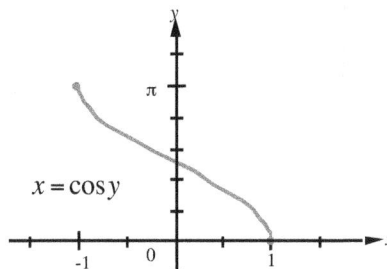

Inverse Tangent Function

Definition Given, the equation, $y = \tan x$, $-\frac{\pi}{2} < x < \frac{\pi}{2}$, what is its inverse?

The inverse is found by interchanging the roles of x and y.

Domain ↓   Range ↓

The inverse of $y = \tan x$ for $-\frac{\pi}{2} < x < \frac{\pi}{2}$, $-\infty < y < \infty$ is

$x = \tan y$ for $-\frac{\pi}{2} < y < \frac{\pi}{2}$, $-\infty < x < \infty$

Range   Domain

If we solve $x = \tan y$ for y, and use the notation,

$\text{Tan}^{-1}x$, we obtain $y = \text{Tan}^{-1}x$ or $y = \text{Arctan } x$

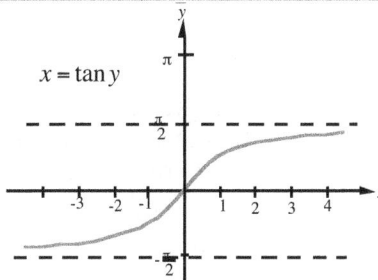

Other Inverse Trigonometric Functions

The inverse functions $\text{Csc}^{-1}x$, $\text{Sec}^{-1}x$, and $\text{Cot}^{-1}x$ can be expressed in terms of $\text{Sin}^{-1}x$, $\text{Cos}^{-1}x$, and $\text{Tan}^{-1}x$ respectively as follows:

$\text{Csc}^{-1}x = \text{Sin}^{-1}\left(\frac{1}{x}\right)$ for $|x| \ge 1$; $\text{Sec}^{-1}x = \text{Cos}^{-1}\left(\frac{1}{x}\right)$ for $|x| \ge 1$; and $\text{Cot}^{-1}x = \text{Tan}^{-1}\left(\frac{1}{x}\right)$.

For example, $\text{Csc}^{-1}2 = \text{Sin}^{-1}\left(\frac{1}{2}\right) = \frac{\pi}{6}$. That is, to find $\text{Csc}^{-1}2$, first find the reciprocal of 2, which is $\frac{1}{2}$ and then find $\text{Sin}^{-1}\left(\frac{1}{2}\right)$, which is $\frac{\pi}{6}$. This approach gives only the reference angle and you have to place the angle in standard position for the original reciprocal inverse function according to the quadrant for its definition. For example, if x is negative and we use $\text{Cot}^{-1}x = \text{Tan}^{-1}\left(\frac{1}{x}\right)$, we must place the reference angle in the second quadrant for $\text{Cot}^{-1}x$, and write the standard position angle.

Choice of the ranges and quadrants for the inverse functions (See also p. 629)

If x is positive, $\text{Sin}^{-1}x$ $\text{Cos}^{-1}x$, $\text{Tan}^{-1}x$, $\text{Csc}^{-1}x$, $\text{Sec}^{-1}x$, and $\text{Cot}^{-1}x$ are all in the first quadrant, or appropriately quadrantal. However, if x is negative, $\text{Sin}^{-1}x$,$\text{Csc}^{-1}x$, or $\text{Tan}^{-1}x$ are in the fourth quadrant or appropriately quadrantal and they are expressed as negative angles; while if x is negative, $\text{Cos}^{-1}x$, $\text{Sec}^{-1}x$, and $\text{Cot}^{-1}x$ are in the second quadrant or appropriately quadrantal.

Note the difference between the quadrants for, say, the angle, $\text{Sin}^{-1}x$ (Fig. 3, previous page) and the quadrants for the graph of $y = \text{Sin}^{-1}x$ (Fig. 2, previous page)..The graph consists of the points (x,y).

We summarize the definitions of the **inverse trigonometric functions** below. The restrictions made to obtain the inverse functions are arbitrary, and other branches may be used. The above branches are the ones that are usually used.

Function	Domain	Range	Inverse	Notation	Domain	Range
$y = \sin x,$	$-\frac{\pi}{2} \leq x \leq \frac{\pi}{2},$	$-1 \leq y \leq 1$	$x = \sin y,$	$y = \text{Arcsin } x$ or $\text{Sin}^{-1}x;$	$-1 \leq x \leq 1;$	$-\frac{\pi}{2} \leq y \leq \frac{\pi}{2}$
$y = \cos x,$	$0 \leq x \leq \pi,$	$-1 \leq y \leq 1$	$x = \cos y,$	$y = \text{Arccos } x$ or $\text{Cos}^{-1}x;$	$-1 \leq x \leq 1;$	$0 \leq y \leq \pi$
$y = \tan x,$	$-\frac{\pi}{2} < x < \frac{\pi}{2},$	$-\infty < y < \infty$	$x = \tan y,$	$y = \text{Arctan } x$ or $\text{Tan}^{-1}x;$	$-\infty < x < \infty;$	$-\frac{\pi}{2} < y < \frac{\pi}{2}$
$y = \cot x,$	$0 < x < \pi,$	$-\infty < y < \infty$	$x = \cot y,$	$y = \text{Arccot } x$ or $\text{Cot}^{-1}x;$	$-\infty < x < \infty;$	$0 < y < \pi$
(Also: $\text{Tan}^{-1}x + \text{Cot}^{-1}x = \frac{\pi}{2}$)				or $y = \text{Cot}^{-1}x = \text{Tan}^{-1}\left(\frac{1}{x}\right) = \frac{\pi}{2} - \text{Tan}^{-1}x;$		$0 < y < \pi$
$y = \sec x, 0 \leq x \leq \pi, x \neq \frac{\pi}{2}, \|y\| \geq 1$			$x = \sec y,$	$y = \text{Arcsec } x$ or $\text{Sec}^{-1}x;$	$\|x\| \geq 1, 0 \leq y \leq \pi, y \neq \frac{\pi}{2}$ Also,	
				$y = \text{Sec}^{-1}x = \text{Cos}^{-1}\left(\frac{1}{x}\right); \|x\| \geq 1$	$0 \leq y \leq \pi, y \neq \frac{\pi}{2}$	
$y = \csc x, -\frac{\pi}{2} \leq x \leq \frac{\pi}{2}, x \neq 0, \|y\| \geq 1$			$x = \csc y,$	$y = \text{Arccsc } x$ or $\text{Csc}^{-1}; \|x\| \geq 1,$	$-\frac{\pi}{2} \leq y \leq \frac{\pi}{2}, y \neq 0$	
(Also: $\text{Sec}^{-1}x + \text{Csc}^{-1}x = \frac{\pi}{2}$)			Also, $y = \text{Csc}^{-1}x = \text{Sin}^{-1}\left(\frac{1}{x}\right); \|x\| \geq 1,$		$-\frac{\pi}{2} \leq y \leq \frac{\pi}{2}, y \neq 0$	

Inverse Operations

Finding the **sine** of an angle or a number and finding the inverse sine or **arcsine** are inverse processes of each other in much the same way that **multiplication and division** are inverse operations of each other. Two operations which are inverses of each other undo (reverse the action of) each other. Thus the function Arcsin x reverses the action of $\sin x$, and vice versa.

Example 2 Find Arcsin (0.866).   (Note that Arcsin (.866) is an angle or a number.)

Step 1: Let $y = \text{Arcsin } (0.866)$ $-\frac{\pi}{2} \leq y \leq \frac{\pi}{2}$.   (We will find y, the angle whose sine is .866)

Two conditions must be satisfied: 1. $\sin y = 0.866$); **2.** $-\frac{\pi}{2} \leq y \leq \frac{\pi}{2}$.

Step 2: Find the reference angle for the angle whose sine 0.866

From memory or calculator (or tables), the reference angle is $\frac{\pi}{3}$.   (Note that $\frac{\pi}{3} = 60°$)

Step 3: Find the quadrant in which the sine function is positive.
The sine function is positive in the **first** and second quadrants.

Since by definition, Arcsin x must be in either the **first** or the fourth quadrant, $(-\frac{\pi}{2} \leq y \leq \frac{\pi}{2})$,

Arcsin x is in the **first** quadrant where it is a positive angle.

Therefore, Arcsin $(0.866) = \frac{\pi}{3}$.   (Note that $-\frac{\pi}{2} \leq \frac{\pi}{3} \leq \frac{\pi}{2}$)

Example 3 Find Arcsin (-0.5).   (Note that Arcsin (- 0.5) is an angle or a number.)

Step 1:   Let $y = \text{Arcsin } (-0.5)$ $-\frac{\pi}{2} \leq y \leq \frac{\pi}{2}$.   (We will find y, the angle whose sine is -0.5)

Step 2:   Find the reference angle for the angle whose sine is - 0.5.

From tables or memory (or calculator), the reference angle is $\frac{\pi}{6}$.

Step 3: Find the quadrant in which the sine function is negative. The sine function is negative in the third and **fourth** quadrants. Since Arcsin x must be in either the first or the **fourth** quadrant,

$(-\frac{\pi}{2} \leq y \leq \frac{\pi}{2})$, Arcsin x is in the fourth quadrant where it is a negative angle.

Therefore, Arcsin $(-0.5) = -\frac{\pi}{6}$. Summarily, $\text{Sin}^{-1}(-x) = -\text{Sin}^{-1}x$

Example 4 Find $\text{Cot}^{-1}(-2.7)$ using a calculator.

Method 1: Step 1: Find $\frac{1}{2.7} = 0.37037037$ (Summarily, $\text{Cot}^{-1}(-2.7) = \pi - \text{Cot}^{-1}(2.7) = \pi - \text{Tan}^{-1}(\frac{1}{2.7})$)

Step 2: Find $\text{Tan}^{-1}\left(\frac{1}{2.7}\right) = \text{Tan}^{-1}(0.37037037) = 0.35$ <--- Reference angle in radians

The reference angle for $\text{Cot}^{-1}(-2.7) = 0.35$. By definition, when x is negative,

$\text{Cot}^{-1}x$ is in the second quadrant and the terminal side of the angle is in the second quadrant.

Therefore, $\text{Cot}^{-1}(-2.7) = \pi - 0.35 \boxed{= 2.79}$.

Method 2: Applying $\text{Cot}^{-1}x + \text{Tan}^{-1}x = \frac{\pi}{2}$ and $\text{Tan}^{-1}(-x) = -\text{Tan}^{-1}x$, $-\infty < x < \infty$, see also p.629.

$\text{Cot}^{-1}(-2.7) = \frac{\pi}{2} - \text{Tan}^{-1}(-2.7) = \frac{\pi}{2} - [-\text{Tan}^{-1}(2.7)]\boxed{= \frac{\pi}{2} + \text{Tan}^{-1}(2.7)} = 1.57 + 1.216 \boxed{= 2.79}$

General Composition of Trigonometric and Inverse Trigonometric Functions

Consider the following question at the beginning level of trigonometry.

" If $\sin x = \frac{1}{2}$, and the terminal side of x is in the first quadrant, find $\cot x$".

However, at the present level, we might pose the same question as:

" find $\cot\left(\text{Sin}^{-1}\frac{1}{2}\right)$" <--Find the cotangent of the angle between $-\frac{\pi}{2}$ and $\frac{\pi}{2}$ whose sine is $\frac{1}{2}$.

Example 5 Find the value of $\cot\left(\text{Sin}^{-1}\left(\frac{1}{2}\right)\right)$.

Solution: Same as if $\sin\theta = \frac{1}{2}$, and θ is in the first quadrant, find $\cot\theta$.

(or If $\sin\theta = \frac{1}{2}$ and $-\frac{\pi}{2} \le \theta \le \frac{\pi}{2}$, find $\cot\theta$.)   (Note $\sin\theta = \sin\frac{\pi}{6} = \frac{1}{2}$, $\theta = \frac{\pi}{6}$)

Step 1:   Draw a right triangle, and let $\text{Sin}^{-1}\left(\frac{1}{2}\right) = \theta$

Then $\sin\theta = \frac{\text{opposite side}}{\text{hypotenuse}} = \frac{1}{2}$, so that the side opposite θ is 1, and the hypotenuse is 2.

Step 2 Using the Pythagorean theorem, the side adjacent to θ is found to be $\sqrt{3}$.

Step 3: Since $\sin\theta = +\frac{1}{2}$, $\text{Sin}^{-1}\left(\frac{1}{2}\right)$ is in the first quadrant

Step 4: $\cot\theta = \frac{\text{adjacent side}}{\text{opposite side}} = \frac{\sqrt{3}}{1} = \sqrt{3}$

.Step 5: Determine the sign of $\cot\theta$. Since $\text{Sin}^{-1}\left(\frac{1}{2}\right)$ is in the first quadrant, where

$\cot\theta$ is also positive, $\cot\left(\text{Sin}^{-1}\left(\frac{1}{2}\right)\right) = \sqrt{3}$.

Example 6 Find the value of $\cot\left(\text{Sin}^{-1}\left(-\frac{1}{2}\right)\right)$. Same as if $\sin\theta = -\frac{1}{2}$, and θ is in the fourth

quadrant, find $\cot\theta$.(or If $\sin\theta = -\frac{1}{2}$ and $-\frac{\pi}{2} \le \theta \le \frac{\pi}{2}$, find $\cot\theta$)   See also p.629

Step 1: Draw a right triangle, and let $\text{Sin}^{-1}\left(\frac{1}{2}\right) = \theta$. Then $\sin\theta = \frac{\text{opposite side}}{\text{hypotenuse}} = \frac{1}{2}$,

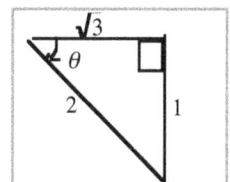

so that the side opposite θ is 1, and the hypotenuse is 2.

Step 2 Using the Pythagorean theorem, the side adjacent to θ is found to be $\sqrt{3}$

Step 3: $\sin\theta = -\frac{1}{2}$ (negative), $\text{Sin}^{-1}\left(-\frac{1}{2}\right)$ is in the fourth quadrant, See also p.629

Step 4: $\cot\theta = \dfrac{\text{adjacent side}}{\text{opposite side}} = \dfrac{\sqrt{3}}{1} = \sqrt{3}$.

Step 5: Determine the sign of $\cot\theta$. Since $\text{Sin}^{-1}\left(-\frac{1}{2}\right)$ is in the fourth quadrant, where the

cotangent is negative, $\cot\left(\text{Sin}^{-1}\left(-\frac{1}{2}\right)\right) = -\sqrt{3}$.

Special Cases of Composition of Trigonometric Functions

Example The function $y = \text{Sin}^{-1}x$ and the restricted sine function $y = \sin x$ are inverses of each other, and as such, each reverses the action of the other,
Definitions

		Note
$\text{Sin}^{-1}(\sin x) = x;$ $-\frac{\pi}{2} \le x \le \frac{\pi}{2}$ <-domain of the inside function $\sin x$		$\pi \approx 3.14$
$\sin(\text{Sin}^{-1}x) = x;$ $-1 \le x \le 1$ <-domain of the inside function $\sin^{-1}x$		$\frac{\pi}{2} \approx \frac{3.142}{2} = 1.57$
$\cos(\text{Cos}^{-1}x) = x;$ $-1 \le x \le 1$ <-domain of the inside function $\cos^{-1}x$		
$\text{Cos}^{-1}(\cos x) = x;$ $0 \le x \le \pi$ <-domain of the inside function $\cos x$		
$\tan(\text{Tan}^{-1}x) = x;$ $-\infty < x < \infty$ <-domain of the inside function $\tan^{-1}x$		
$\text{Tan}^{-1}(\tan x) = x;$ $-\frac{\pi}{2} < x < \frac{\pi}{2}$ <-domain of the inside function $\tan x$		

Examples:
Case 1

Example 1 Evaluate: $\text{Sin}^{-1}(\sin\frac{\pi}{6})$ (A)

Step 1: Evaluate the inside function. From memory or tables $\sin\frac{\pi}{6} = \frac{1}{2}$

Step 2: Evaluate $\text{Sin}^{-1}\left(\frac{1}{2}\right)$

Step 3: Find the angle in radians or number between $-\frac{\pi}{2}$ and $\frac{\pi}{2}$ whose sine is $\frac{1}{2}$

From memory or tables the angle is $\frac{\pi}{6}$.

Therefore $\text{Sin}^{-1}(\sin\frac{\pi}{6}) = \frac{\pi}{6}$ (B)

Observe that we could have obtained the angle $\frac{\pi}{6}$ by inspection and applying the definition,

$\sin^{-1}(\sin x) = x;$ if $-\frac{\pi}{2} \le x \le \frac{\pi}{2}$ or $-1.57 \le x \le 1.57$ (C)

Here, $x = \frac{\pi}{6} \approx 0.52$ and $-\frac{\pi}{2} \le \frac{\pi}{6} \le \frac{\pi}{2}$ or $-1.57 \le 0.52 \le 1.57$

Example 2: Evaluate $\text{Sin}^{-1}(\sin\frac{\pi}{4})$

The argument $\frac{\pi}{4}$ of the inside function is between $-\frac{\pi}{2}$ and $\frac{\pi}{2}$, and satisfies the domain

$-\frac{\pi}{2} \le x \le \frac{\pi}{2}$ of the restricted sine function.

Therefore, the domain of the function is satisfied, and definition (C) above is applicable, and

$\text{Sin}^{-1}(\sin\frac{\pi}{4}) = \frac{\pi}{4}$ (The inverse sine function undoes what the sine function does)

Case 2 Example 3 Evaluate $\sin\left(\operatorname{Sin}^{-1}\frac{1}{2}\right)$

The argument $\frac{1}{2}$ of the inside function is between -1 and 1, and satisfies the domain $-1 \le x \le 1$. Therefore, $\sin\left(\operatorname{Sin}^{-1}\frac{1}{2}\right) = \frac{1}{2}$. (The sine function undoes what the inverse sine function does.)

Example 4 Evaluate $\sin\left(\operatorname{Sin}^{-1}4\right)$

The argument 4 of the inside function does not satisfy the domain $-1 \le x \le 1$ of $\operatorname{Sin}^{-1}x$. Therefore, $\sin\left(\operatorname{Sin}^{-1}4\right)$ is not defined. There is no angle or number between $-\frac{\pi}{2}$ and $\frac{\pi}{2}$ whose sine is 4.

Case 3 Example 5 Evaluate $\operatorname{Sin}^{-1}(\sin\frac{3\pi}{4})$

The argument $\frac{3\pi}{4}$ of the inside function is **not** between $-\frac{\pi}{2}$ and $\frac{\pi}{2}$ does not satisfy the domain $-\frac{\pi}{2} \le x \le \frac{\pi}{2}$ and therefore $\operatorname{Sin}^{-1}(\sin x) = x$ is not applicable. However, $\sin\frac{3\pi}{4} = 0.7071$

$\operatorname{Sin}^{-1}(0.7071) = 0.7854$ (Note: $\frac{3\pi}{4} \approx 2.3562$, and therefore, $\sin^{-1}(\sin\frac{3\pi}{4})$ is not equal to $= \frac{3\pi}{4}$)

Therefore, $\operatorname{Sin}^{-1}(\sin\frac{3\pi}{4}) = 0.7854$. **Clarification of the ranges of inverse trig functions follows:**

Range of Inverse Trig Functions when $x \ge 0$		**Range of Inverse Trig Functions when $x < 0$**	
$0 \le \operatorname{Sin}^{-1}x \le \frac{\pi}{2}$ (Ist qd)	$0 < \operatorname{Cot}^{-1}x \le \frac{\pi}{2}$ (Ist qd)	$-\frac{\pi}{2} \le \operatorname{Sin}^{-1}x < 0$ (4th qd)	$\frac{\pi}{2} < \operatorname{Cot}^{-1}x < \pi$ (2nd qd)
$0 \le \operatorname{Cos}^{-1}x \le \frac{\pi}{2}$ (Ist qd)	$0 \le \operatorname{Sec}^{-1}x < \frac{\pi}{2}$ (Ist qd)	$\frac{\pi}{2} < \operatorname{Cos}^{-1}x \le \pi$ (2nd qd)	$\frac{\pi}{2} < \operatorname{Sec}^{-1}x \le \pi$ (2nd qd))
$0 \le \operatorname{Tan}^{-1}x < \frac{\pi}{2}$ (Ist qd)	$0 < \operatorname{Csc}^{-1}x \le \frac{\pi}{2}$ (Ist qd)	$-\frac{\pi}{2} < \operatorname{Tan}^{-1}x < 0$ (4th qd)	$-\frac{\pi}{2} \le \operatorname{Csc}^{-1}x < 0$ (4th qd)

qd = quadrant

Inverse Cofunction Identities

1. $\operatorname{Sin}^{-1}x + \operatorname{Cos}^{-1}x = \frac{\pi}{2}$, $[-1,1]$; **2.** $\operatorname{Tan}^{-1}x + \operatorname{Cot}^{-1}x = \frac{\pi}{2}$, $(-\infty,\infty)$; **3** $\operatorname{Sec}^{-1}x + \operatorname{Csc}^{-1}x = \frac{\pi}{2}$, $|x| \ge 1$

Each identity is subject to the domain of definition, Application: $\operatorname{Cot}^{-1}x = \frac{\pi}{2} - \operatorname{Tan}^{-1}x$

For proofs of these identities, see Appendix B, p.660.

Negative angle relationships

1. $\operatorname{Sin}^{-1}(-x) = -\operatorname{Sin}^{-1}x$; 3. $\operatorname{Tan}^{-1}(-x) = -\operatorname{Tan}^{-1}x$; 5. $\operatorname{Sec}^{-1}(-x) = \pi - \operatorname{Sec}^{-1}x$;
2. $\operatorname{Cos}^{-1}(-x) = \pi - \operatorname{Cos}^{-1}x$; 4. $\operatorname{Cot}^{-1}(-x) = \pi - \operatorname{Cot}^{-1}x$; 6. $\operatorname{Csc}^{-1}(-x) = -\operatorname{Csc}^{-1}x$.

Lesson 96 Exercises

Find the following

1. $\operatorname{Arcsin}\frac{1}{2}$; **2.** $\operatorname{Arctan}\sqrt{3}$; **3.** $\operatorname{Arccos}\frac{1}{2}$

4. $\operatorname{Arcsin}\left(-\frac{\sqrt{3}}{2}\right)$; **5.** $\operatorname{Sin}^{-1}1$; **6.** $\operatorname{Arccot}(-0.33)$

Without using tables or a calculator find :

7. $\cot\left(\operatorname{Arcsin}\left(-\frac{1}{4}\right)\right)$; **8.** $\operatorname{Arccos}\left(\cos\left(\frac{15\pi}{11}\right)\right)$

11. Express $\tan\left(\operatorname{Arcsin}x\right)$ in terms of x.

Is each of the following true for all x:? Why?

9. $\operatorname{Arcsin}\left(\sin x\right) = x$; **10.** $\sin\left(\operatorname{Arcsin}x\right) = x$

12. Express $\cos\left(\operatorname{Arctan}x\right)$ in terms of x.

13. True or false $\tan(\operatorname{Tan}^{-1}x) = x$ for all x.

Answers : 1. $\frac{\pi}{6}$; 2. $\frac{\pi}{3}$; 3. $\frac{\pi}{3}$; 4. $-\frac{\pi}{3}$; 5. $\frac{\pi}{2}$; 6. 1.88; 7. $-\sqrt{15}$; 8. $\frac{7\pi}{11} \approx 1.999$; 9. No : $-\frac{\pi}{2} \le x \le \frac{\pi}{2}$;

10. No : $-1 \le x \le 1$; 11. $\frac{x}{\sqrt{1-x^2}}$; 12. $\frac{1}{\sqrt{x^2+1}}$; 13. True. $-\infty < x < \infty$

Note: In Problems **11** and **12,** the composition of a trigonometric function with an inverse trigonometric function is an algebraic expression.

Lesson 97

How to Sketch the Graph of $y = $ Arcsin x

The graph of $y = \sin x$ is not a one-to-one function by "the horizontal line" rule since a horizontal line drawn crosses the graph of $y = \sin x$ for $0 \leq x \leq 2\pi$ in more than one point and therefore, the inverse relation of $x = \sin y$ is generally not a function. In sketching the graphs of the inverse trigonometric functions, we will be guided by the domains and ranges agreed to previously in the definitions of the inverse trigonometric functions. See page 625.

To sketch the graph of $y = $ Arcsin x we will sketch the graph of the equivalent equation

$x = \sin y$ for $-\frac{\pi}{2} \leq y \leq \frac{\pi}{2}$, by obtaining important points and connecting them.

2. The important points include 2 end points (a starting point and a stopping point), 1 mid-point and two quarter-points are needed. The points include the x- and the y--intercepts

Example 1 Sketch the graph of $y = $ Arcsin x

Solution We will sketch the graph of the equivalent equation

$$x = \sin y \text{ for } -\frac{\pi}{2} \leq y \leq \frac{\pi}{2}$$

Step 1: Choose convenient y-values from $-\frac{\pi}{2}$ to $\frac{\pi}{2}$, and calculate the corresponding x-values

using the equation $x = \sin y$ to obtain ordered pairs. It is more convenient to choose y and calculate x. The first time we chose y and calculated x was when we graphed logarithmic functions such as $y = \log_2 x$). We must be careful of how we list x- and y-values obtained from the calculations, since we are choosing y and calculating x. A table containing these values are shown below.

$x = \sin y$	-1	-.71	0	.71	1
y	$-\frac{\pi}{2}$	$-\frac{\pi}{4}$	0	$\frac{\pi}{4}$	$\frac{\pi}{2}$

Step 2: Plot the points $(-1, -\frac{\pi}{2})$, $(-.71, -\frac{\pi}{4})$, $(0, 0)$, $(.71, \frac{\pi}{4})$, $(1, \frac{\pi}{2})$. (Points on $x = \sin y$)

Step 3: Connect the points from Step 2 by a smooth curve to obtain the characteristic shape.

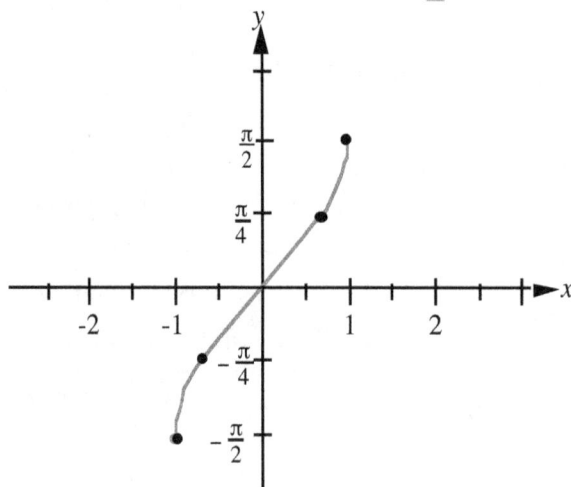

Figure: The graph of $y = $ Arcsin x or Sin^{-1}x

Practice sketching this graph and include it among your "trigonometric vocabulary".

OR   Alternatively,

Step 2: Draw one sine curve between $-\pi \le y \le \pi$ **along the y-axis** using dotted (broken lines) lines, noting that the curve is in the 1st and 3rd quadrants.

Step 3: Darken the part of the curve for which $-\frac{\pi}{2} \le y \le \frac{\pi}{2}$

Example 2   Sketch the graph of $y = \text{Arcsin } 3x$
Method 1
Solution

Step 1: Solve for x by applying the definition, if $\text{Arcsin } x = y$ then $x = \sin y$
   If $y = \text{Arcsin } 3x$ then
   $\sin y = 3x$ ("Taking the sine of both sides of the equation")
   $\frac{1}{3}\sin y = x$ where $-\frac{\pi}{2} \le y \le \frac{\pi}{2}$ Note that $\sin(\arcsin 3x) = 3x$.

Step 2: Choose convenient y-values from $-\frac{\pi}{2}$ to $\frac{\pi}{2}$, and calculate the corresponding values of x.

   using the equation $x = \frac{1}{3}\sin y$ to obtain ordered pairs. A table of the results is shown below.

$x = \frac{1}{3}\sin y$	-.33	-.24	0	.24	.33
y	$-\frac{\pi}{2}$	$-\frac{\pi}{4}$	0	$\frac{\pi}{4}$	$\frac{\pi}{2}$

Step 3: Plot the points $(-.33, -\frac{\pi}{2})$, $(-.24, -\frac{\pi}{4})$, $(0,0)$, $(.24, \frac{\pi}{4})$, $(.33, \frac{\pi}{2})$. (Points on $x = \frac{1}{3}\sin y$)

Step 4: Connect the points from Step 3 by a smooth curve to obtain the characteristic shape.

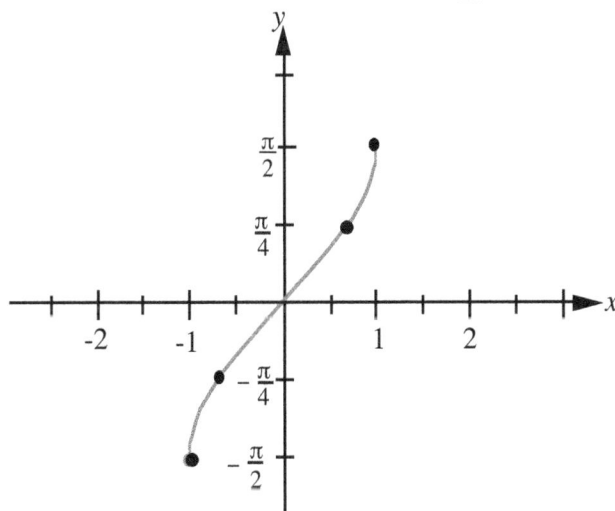

Graph of $y = \text{Arcsin } 3x$

Method 2 (Transformation Method)

   Study the next example, Example 3, and imitate.

Example 3   Sketch the graph of $y = 4\,\text{Arcsin } 2x$

Method 1

Step 1   Solve $y = 4\,\text{Arcsin } 2x$ for $\text{Arcsin } 2x$, followed by solving for x by applying $x = \sin y$

$$\frac{y}{4} = \text{Arcsin } 2x$$

. $$\sin\frac{y}{4} = 2x \qquad \text{("Taking the sine of both sides of the equation")}$$

$$x = \frac{1}{2}\sin\frac{y}{4} \qquad \text{where} \qquad -\frac{\pi}{2} \le \frac{y}{4} \le \frac{\pi}{2} < \text{note the restricted range}$$

$$x = \frac{1}{2}\sin\frac{y}{4} \quad \text{where} - 2\pi \le y \le 2\pi \; <---\; -\frac{\pi}{2}\cdot\frac{4}{1}\cdot\frac{2}{1} \le \frac{y}{4}\cdot\frac{4}{1} \le \frac{\pi}{2}\cdot\frac{4}{1}\cdot\frac{2}{1} \; \text{(solving for } y)$$

Step 2: Sketch the principal branch for $x = \frac{1}{2}\sin\frac{y}{4}$ for $-2\pi \le y \le 2\pi$.

Step 3: Determine the y-values to use to calculate x.

The distance from -2π to 2π is $2\pi - (-2\pi) = 4\pi$

Divide 4π by 4 to obtain the interval increment , π. Add π successfully to the y-values, beginning from -2π.

Step 4: Calculate the corresponding x-values, using $x = \frac{1}{2}\sin\frac{y}{4}$. (Results in the table below)

$x = \frac{1}{2}\sin\frac{y}{4}$	$-\frac{1}{2}$	-.35	0	.35	$\frac{1}{2}$
y	-2π	$-\pi$	0	π	2π

Step 5: Using this table plot and connect the points to obtain the characteristic shape.

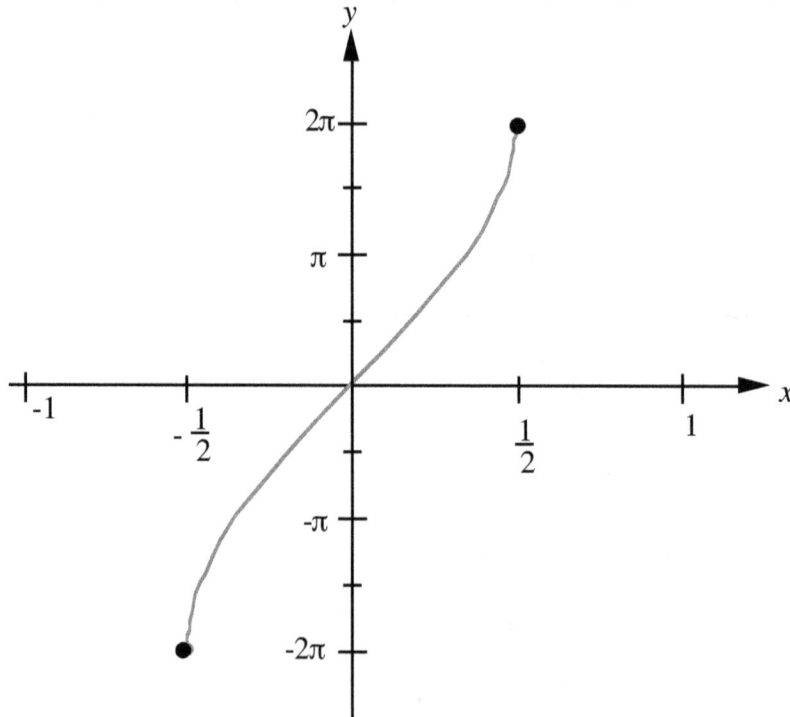

Figure 3: The Graph of $y = 4\,\text{Arcsin } 2x$

Method 2 (Transformation method)

Given $y = \text{Arcsin } x$, draw the graph of $y = 4\,\text{Arcsin } 2x$

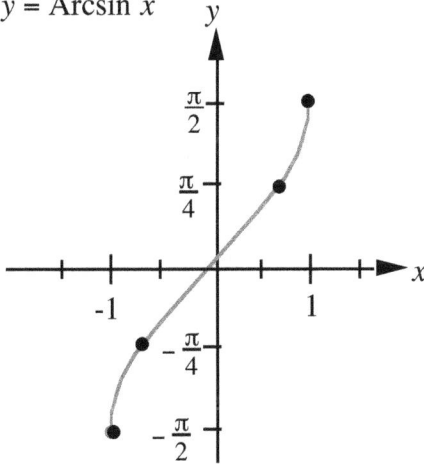

Rewrite as $y = 4\,\text{Arcsin } 2(x - 0)$. Here, $k = 0$, $h = 0$, $a = 4$, $b = 2$,

$x = \sin y$		$y = 4\,\text{Arcsin } 2(x - 0)$			
x_0	y_0	$x_n = \frac{x_0}{b} + h$	$y_n = y_0(a) + k$	(x_n, y_n)	**Step 1:** Quickly prepare a table of values for x and y,
-1	$-\frac{\pi}{2}$	$x_n = \frac{-1}{2} + 0 = -\frac{1}{2}$	$-\frac{\pi}{2}(4) + 0 = -2\pi$	$(-\frac{1}{2}, -2\pi)$	using $x = \sin y$ and convenient y-values.
-0.71	$-\frac{\pi}{4}$	$\frac{-0.71}{2} + 0 = -0.36$	$-\frac{\pi}{4}(4) + 0 = -\pi$	$(-0.36, -\pi)$	**Step 2:** Using the table of values from Step 1, determine the
0	0	$\frac{0}{2} + 0 = 0$	$0(4) + 0 = 0$	$(0, 0)$	new coordinates x_n and y_n.
0.71	$\frac{\pi}{4}$	$\frac{0.71}{2} + 0 = 0.36$	$\frac{\pi}{4}(4) + 0 = \pi$	$(0.36, \pi)$	**Step 3:** Plot the new points and connect them by a solid curve
1	$\frac{\pi}{2}$	$\frac{1}{2} + 0 = \frac{1}{2}$	$\frac{\pi}{2}(4) + 0 = 2\pi$	$(\frac{1}{2}, 2\pi)$	(Figure 3 above)

Lesson 97 Exercises

Sketch the graphs of the following: **1.** $y = \text{Arcsin } x$;   **2.** $y = 2\,\text{Arcsin } 4x$;   .

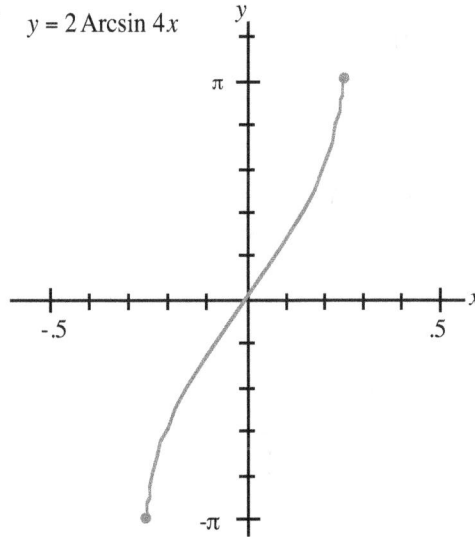

Solution:

Lesson 98 634

How to Sketch the Graph of $y = \text{Arccos } x$

To sketch the graph of $y = \text{Arccos } x$ we will sketch the graph of the equivalent equation
$$x = \cos y \text{ for } 0 \le y \le \pi, \text{ by obtaining important points and connecting them.}$$
The important points include 2 end points (a starting point and a stopping point), 1 mid-point and two quarter-points are needed. The points include the x- and y-intercepts.

Example 1   Sketch the graph of $y = \text{Arccos } x$

Solution   We will sketch the graph of the equivalent equation
$$x = \cos y \text{ for } 0 \le y \le \pi$$

Step 1: Choose convenient y -values from 0 to π, and calculate the corresponding values of x. using the equation $x = \cos y$ to obtain ordered pairs. It is more convenient to choose y and calculate x. We must be careful of how we list the ordered pairs obtained from the calculations, since we are choosing y and calculating x. A table containing these important points are shown below.

$x = \cos y$	1	.71	0	-.71	-1
y	0	$\frac{\pi}{4}$	$\frac{\pi}{2}$	$\frac{3\pi}{4}$	π

Step 2: Plot the points $(1, 0), (.71, \frac{\pi}{4}), (0, \frac{\pi}{2}), (-.71, \frac{3\pi}{4}). (-1, \pi)$ (Points on $x = \cos y$)

Step 3: Connect the points from Step 2 by a smooth curve to obtain the characteristic shape.

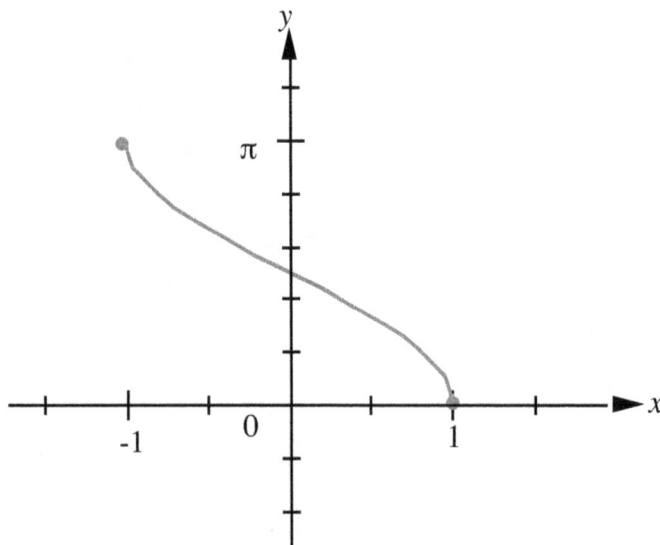

Graph of $x = \cos y$

Example 2 Sketch the graph of $y = \text{Arccos } 2x + \frac{\pi}{2}$

We cover two methods
Method 1

Step 1: Solve $y = \text{Arccos } 2x + \frac{\pi}{2}$ for Arccos $2x$,

    Then $y - \frac{\pi}{2} = \text{Arccos } 2x$ (1)

Step 2: Solve (1) for $2x$ by applying if $y = \text{Arccos } x$ then $x = \cos y$, $0 \le y \le \pi$
    Then, we obtain

    $2x = \cos(y - \frac{\pi}{2})$, $0 \le y - \frac{\pi}{2} \le \pi$

Step 3: Solve for x: $x = \frac{1}{2}\cos(y - \frac{\pi}{2})$ $0 \le y - \frac{\pi}{2} \le \pi$ <-- Note the form of the range

                    $x = \frac{1}{2}\cos(y - \frac{\pi}{2})$ $\frac{\pi}{2} \le y \le \frac{3\pi}{2}$ <---- (solving the inequality for y)

    (The amplitude $= \frac{1}{2}$)

Step 4: Sketch the graph of the principal branch: $x = \frac{1}{2}\cos(y - \frac{\pi}{2})$ where $\frac{\pi}{2} \le y \le \frac{3\pi}{2}$

    Choose convenient y-values from $\frac{\pi}{2}$ to $\frac{3\pi}{2}$, and calculate the corresponding x-values

    of x. using the equation $x = \frac{1}{2}\cos(y - \frac{\pi}{2})$ to obtain ordered pairs.

    Starting y-value $= \frac{\pi}{2}$; end-point y-value $= \frac{3\pi}{2}$; incremental y-value $= \frac{1}{4}(\frac{3\pi}{2} - \frac{\pi}{2}) = \frac{\pi}{4}$.

    Add $\frac{\pi}{4}$ successfully to the y-values, beginning from $y = \frac{\pi}{2}$.

Step 5: Plot and connect the points obtained in Step 4 by a smooth curve.

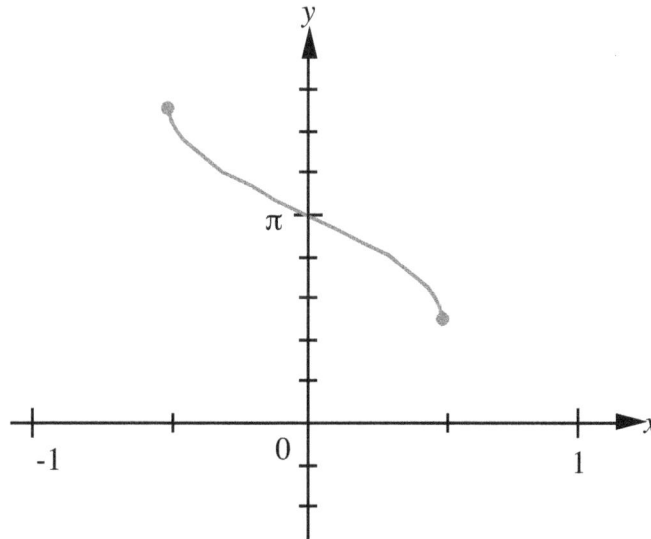

Figure 2: Graph of $y = \text{Arccos } 2x + \frac{\pi}{2}$

Method 2 (Transformation Method)

Given $y = \text{Arccos } x$, draw the graph of $y = \text{Arccos } 2x + \frac{\pi}{2}$ (see Example 1, for tables for $y = \text{Arccos } x$).

Rewrite as $y - \frac{\pi}{2} = \text{Arccos } 2(x - 0)$. Here, $k = \frac{\pi}{2}$, $h = 0$, $a = 1$, $b = 2$.

$x = \cos y$		$y - \frac{\pi}{2} = 4\,\text{Arccos } 2x$		
x_0	y_0	$x_n = \frac{x_0}{b} + h$	$y_n = y_0(a) + k$	(x_n, y_n)
1	0	$x_n = \frac{1}{2} + 0 = \frac{1}{2}$	$0(4) + \frac{\pi}{2} = \frac{\pi}{2}$	$(\frac{1}{2}, \frac{\pi}{2})$
0.71	$\frac{\pi}{4}$	$\frac{0.71}{2} + 0 = 0.36$	$\frac{\pi}{4}(1) + \frac{\pi}{2} = \frac{\pi}{4} + \frac{\pi}{2} = \frac{3\pi}{4}$	$(0.36, \frac{3\pi}{4})$
0	$\frac{\pi}{2}$	$\frac{0}{2} + 0 = 0$	$\frac{\pi}{2}(1) + \frac{\pi}{2} = \frac{\pi}{2} + \frac{\pi}{2} = \pi$	$(0, \pi)$
−0.71	$\frac{3\pi}{4}$	$\frac{-0.71}{2} + 0 = -0.36$	$\frac{3\pi}{4}(1) + \frac{\pi}{2} = \frac{3\pi}{4} + \frac{\pi}{2} = \frac{5\pi}{4}$	$(-0.36, \frac{5\pi}{4})$
-1	π	$\frac{-1}{2} + 0 = -\frac{1}{2}$	$\pi(1) + \frac{\pi}{2} = \pi + \frac{\pi}{2} = \frac{3\pi}{2}$	$(-\frac{1}{2}, \frac{3\pi}{2})$

Step 1: Quickly prepare a table of values for x and y, using $x = \cos y$ and convenient y-values.

Step 2: Using the table of values from Step 1, determine the new coordinates x_n and y_n.

Step 3: Plot the new points and connect them by a solid curve (Figure 2 above)

Lesson 98 Exercises

Sketch the graphs of (a) $y = \text{Arccos } x$; (b) $y = \text{Arccos } 2x + \frac{\pi}{2}$ (try the transformation method)

Solution:

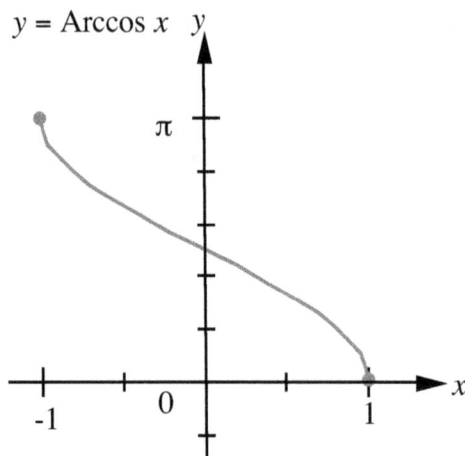

$y = \text{Arccos } x$

Lesson 99
How to Sketch the Graph of $y = \text{Arctan } x$

To sketch the graph of $y = \text{Arctan } x$ we will sketch the graph of the equivalent **equation**

$$x = \tan y \text{ for } -\frac{\pi}{2} < y < \frac{\pi}{2}$$

The curve is not defined where $\cos x$ is zero: That is it is not defined at $y = -\frac{\pi}{2}$ and $y = \frac{\pi}{2}$. The curve is discontinuous at these points and the curve is asymptotic to the horizontal lines $y = -\frac{\pi}{2}$, and $\frac{\pi}{2}$. The tangent curve has points of inflection at the y-intercept.

Example 1 Sketch the graph of $y = \text{Arctan } x$

Solution We will sketch the graph of the equivalent equation

$$x = \tan y \text{ for } -\frac{\pi}{2} < y < \frac{\pi}{2}.$$

Step 1: Choose convenient y-values from $-\frac{\pi}{2}$ to $\frac{\pi}{2}$, and calculate the corresponding x-values using the equation $x = \tan y$ to obtain a table of x- and y-values. We must be careful of how we list x- and y-values obtained from the calculations, since we are choosing y and calculating x. A table containing these values are shown below.

$x = \tan y$	undefined	-1	0	1	undefined
y	$-\frac{\pi}{2}$	$-\frac{\pi}{4}$	0	$\frac{\pi}{4}$	$\frac{\pi}{2}$

Step 2: Plot the points $(-1, -\frac{\pi}{4})$, $(0, 0)$, $(1, \frac{\pi}{4})$, and draw the horizontal asymptotes

$y = -\frac{\pi}{2}$ and $y = \frac{\pi}{2}$.

Step 4: Connect the points from Step 2 by a smooth curve, noting the asymptotic behavior near

$y = -\frac{\pi}{2}$ and $y = \frac{\pi}{2}$.

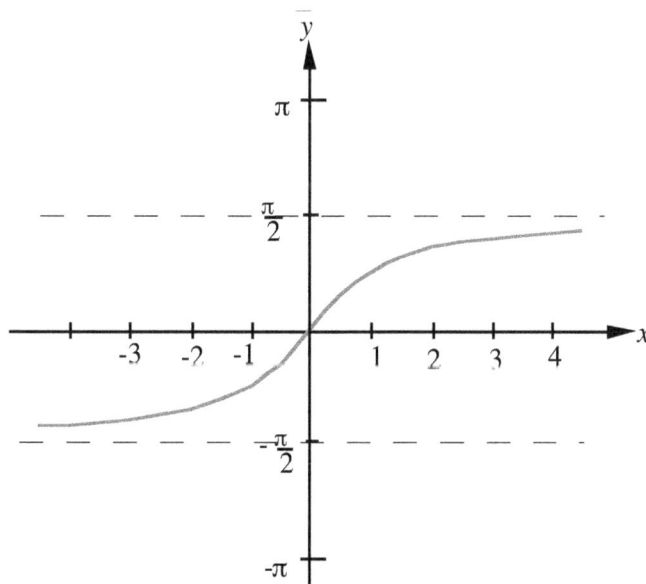

Figure $y = \text{Arctan } x$

Lesson 99 Exercises

Sketch the graphs of **1.** $y = 3\,\text{Arctan}\,x$.; . **2.** $y = 2\,\text{Arcsin}\,4x$; (Try the transformation method)

Solution:

$y = 3\,\text{Arctan}\,x$

$y = 2\,\text{Arcsin}\,4x$

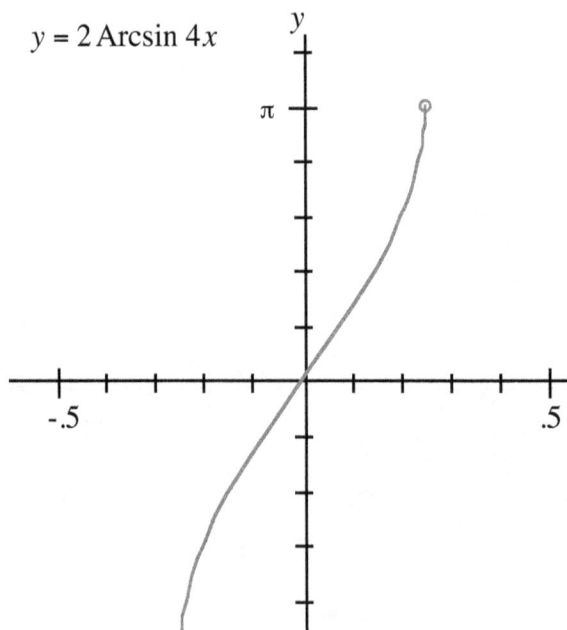

CHAPTER 38

Lesson 100: **Trigonometric Identities**
Lesson 101: **Proving Trigonometric Identities;**
Applications of Sum and Difference Identities

Lesson 100
Trigonometric Identities

Introduction

1. Suppose we are asked to sketch the graph of $f(x) = \csc x \sec x$ (1).

   At first glance, we might say that we do not know how to sketch its graph.

   However, if we express this function in terms of a single basic function, we can readily sketch it.

$$f(x) = \csc x \sec x$$

$$= \frac{1}{\sin x \cos x}$$

$$= \frac{1}{\frac{1}{2}\sin 2x} \qquad (\sin x \cos x = \tfrac{1}{2}\sin 2x \text{, from the \textbf{identity,} } 2\sin x \cos x = \sin 2x).$$

$$= \frac{2}{\sin 2x}$$

$$= 2\csc 2x \qquad (\text{Note: } \frac{1}{\sin 2x} = \csc 2x)$$

$$f(x) = 2\csc 2x$$

Since we have shown that given function is identical with this familiar cosecant function, to sketch $f(x) = \sec x \csc x$, we shall sketch $f(x) = 2\csc 2x$.

2. We may also use trigonometric identities in solving trigonometric equations. Here, we use the trigonometric identities to express a given trigonometric equation of more than one function in terms of a single function or in terms of the product of factors, each factor containing only a single function: Example $\cos x (\sin x - 1) = 0$.

Trigonometric Identities

An **identity** is an equation that is true for all permissible replacements of the independent variable.

Trigonometric identities are based on eight fundamental identities which are grouped as reciprocal identities, ratio identities and the Pythagorean identities.

Reciprocal Identities

1. $\sec\theta = \dfrac{1}{\cos\theta}$ $\qquad\qquad (\cos\theta \ne 0)$

2. $\csc\theta = \dfrac{1}{\sin\theta}$ $\qquad\qquad (\sin\theta \ne 0)$

3. $\cot\theta = \dfrac{1}{\tan\theta}$ $\qquad\qquad (\tan\theta \ne 0)$

Ratio Identities

1. $\tan\theta = \dfrac{\sin\theta}{\cos\theta}$ $\qquad\qquad (\cos\theta \ne 0)$

2. $\cot\theta = \dfrac{\cos\theta}{\sin\theta}$ $\qquad\qquad (\sin\theta \ne 0)$

Pythagorean Identities

1. $\sin^2\theta + \cos^2\theta = 1$

2. $1 + \tan^2\theta = \sec^2\theta$

3. $1 + \cot^2\theta = \csc^2\theta$

For the Pythagorean identities, identities **2** and **3** can be derived from identity **1** by dividing identity **1** by $\cos^2\theta$ and by $\sin^2\theta$ respectively. Keep this in mind when trying to memorize these identities.

Sum and Difference Identities (Addition and Subtraction Formulas)

1. $\cos(A - B) = \cos A\cos B + \sin A\sin B$

2. $\cos(A + B) = \cos A\cos B - \sin A\sin B$

3. $\sin(A + B) = \sin A\cos B + \sin B\cos A$

4. $\sin(A - B) = \sin A\cos B - \sin B\cos A$

5. $\tan(A - B) = \dfrac{\tan A - \tan B}{1 + \tan A\tan B}$

6. $\tan(A + B) = \dfrac{\tan A + \tan B}{1 - \tan A\tan B}$

Double Angle Identities

1. $\sin 2\theta = 2\sin\theta\cos\theta$

2. $\cos 2\theta = \cos^2\theta - \sin^2\theta$

3. $\cos 2\theta = 2\cos^2\theta - 1$

4. $\cos 2\theta = 1 - 2\sin^2\theta$

5. $\tan 2\theta = \dfrac{2\tan\theta}{1 - \tan^2\theta}$

Half Angle Identities

1. $\cos\frac{1}{2}\theta = \pm\sqrt{\dfrac{1+\cos\theta}{2}}$ <---	$+$ sign for $\frac{\theta}{2}$ in quad I or IV but $-$ sign for $\frac{\theta}{2}$ in quad II or III
2. $\sin\frac{1}{2}\theta = \pm\sqrt{\dfrac{1-\cos\theta}{2}}$ <---	$+$ sign for $\frac{\theta}{2}$ in quad I or II but $-$ sign for $\frac{\theta}{2}$ in quad III or IV
3. $\tan\frac{1}{2}\theta = \dfrac{1-\cos\theta}{\sin\theta} = \dfrac{\sin\theta}{1+\cos\theta}$	$+$ sign for $\frac{\theta}{2}$ in quad I or III but $-$ sign for $\frac{\theta}{2}$ in quad II or IV

The $\pm$ sign indicates the sign to use. The sign depends upon the quadrant in which $\frac{\theta}{2}$ is located.

For example, if $\frac{\theta}{2}$ is in the 1st or 4th quadrant, where the cosine function is positive, then

$$\cos\frac{1}{2}\theta = +\sqrt{\frac{1+\cos\theta}{2}}$$

However, if $\frac{\theta}{2}$ is in the 2nd or 3rd quadrant, where the cosine function is negative, then

$$\cos\frac{1}{2}\theta = -\sqrt{\frac{1+\cos\theta}{2}}$$

We will use either the plus sign or the minus sign but **never both** in any particular identity. Note also that the usage of the plus or minus sign is different from the $\pm$ signs involved in solving quadratic equations in algebra.

The following formulas will be useful in integral calculus.

Product or Product as a Sum Formulas

$A: 2\sin A\sin B = \cos(A-B) - \cos(A+B)$ $\Big\}$ $<$ -- cosine sums
$B: 2\cos A\cos B = \cos(A-B) + \cos(A+B)$

    (Derived from sum and difference identities)

$C: 2\sin A\cos B = \sin(A+B) + \sin(A-B)$ $\Big\}$ $<$ -- sine sums$\Big\}$
$D: 2\cos A\sin B = \sin(A+B) - \sin(A-B)$

    (Derived from sum and difference identities)

A **memory** device to help recall the above identities follows.

1. For **A** and **B** (sin sin or cos cos, the right hand side of each identity is in terms of **cos** function only.
   Note in **B** that in some books, the order of the right side may be reversed
2. For **C**, and **D** (sin cos,) the right hand side is in terms of **sin** function only,
   Note above also that when the left side is in terms of the same function , the right side is in terms of **cos** only, while if the left side is in terms of cos and sin, the right side is in terms of **sin** only.

Sum Identities

1. $\cos A + \cos B = 2\cos\dfrac{A+B}{2}\cos\dfrac{A-B}{2}$ $\Big\}$ For the cosine

2. $\cos A - \cos B = -2\sin\dfrac{A+B}{2}\sin\dfrac{A-B}{2}$

3. $\sin A + \sin B = 2\sin\dfrac{A+B}{2}\cos\dfrac{A-B}{2}$ $\Big\}$ For the sine

4. $\sin A - \sin B = 2\sin\dfrac{A-B}{2}\cos\dfrac{A+B}{2}$

Cofunction Relationships

$$\left.\begin{array}{l} \sin A = \cos(90^\circ - A) \\ \cos A = \sin(90^\circ - A) \end{array}\right\} \leftarrow \text{The sine and the cosine are cofunctions.}$$

$$\left.\begin{array}{l} \tan A = \cot(90^\circ - A) \\ \cot A = \tan(90^\circ - A) \end{array}\right\} \leftarrow \text{The tangent and the cotangent are cofuntions.}$$

$$\left.\begin{array}{l} \sec A = \csc(90^\circ - A) \\ \csc A = \sec(90^\circ - A) \end{array}\right\} \leftarrow \text{The secant and the cosecant are cofuntions.}$$

Note above that: $(90^\circ - A) = (\frac{\pi}{2} - A)$

Other identities (Useful when finding functional values of negative angles)

For any θ .

1. $\cos(-\theta) = \cos\theta$ **4.** $\sec(-\theta) = \sec\theta$

2. $\sin(-\theta) = -\sin\theta$ **5.** $\csc(-\theta) = -\csc\theta$

3. $\tan(-\theta) = -\tan\theta$ **6.** $\cot(-\theta) = -\cot\theta$

In the last six identities, note that in finding the trigonometric functional value of a negative angle using these identities:

Step 1: Ignore the minus sign and find the trigonometric functional value as usual (this result may be positive or negative).

Step 2: If the trigonometric function is that of the cosine, or the secant, then the result in Step 1 is the answer; but if the trigonometric function is that of the sine, the tangent, the cosecant or the cotangent, then find the negative of the result from Step 1.

Examples   **1.** $\cos(-60^\circ) = \cos 60^\circ = \frac{1}{2}$ (applying $\cos(-\theta) = \cos\theta$)

         2. $\sin(-30^\circ) = -\sin 30^\circ = -\left(\frac{1}{2}\right) = -\frac{1}{2}$ (applying $\sin(-\theta) = -\sin\theta$)

         3. $\sin(-135^\circ) = -\sin 135^\circ = -(.7071) = -.7071$ (applying $\sin(-\theta) = -\sin\theta$)

         4. $\sin(-210^\circ) = -\sin 210^\circ = -(-\frac{1}{2}) = +\frac{1}{2}$ Note: $\sin 210 = -\sin 30 = -\frac{1}{2}$

Notwithstanding the methods of the last examples, we must note that in finding functional values, we may still use the previous methods of drawing the negative angle, finding the reference angle and then finding the functional value according to the quadrant of the negative angle..

Lesson 100 Exercises

From memory, after practice, write the following down on paper:
Reciprocal identities; Pythagorean identities; Sum and difference identities; Double angle identities; half angle identities; and cofunction relationships.

Lesson 101
Proving Trigonometric Identities and
Applications of the sum and difference identities

In proving trigonometric identities, the following guidelines will be helpful:

1. Express all functions in terms of sines and cosines and simplify.
   (However, sometimes, this may not be necessary.)

2. Work **separately** on the left-hand-side and **separately** on the right-hand side.

   Do **not** work on the equation as you would operate on both sides of a conditional equation
   This is not an "if-then statement" proof.

3. Apply the appropriate trigonometric identities or formulas (from above). For example, if there are double angles, use the double angle identities to change to single angles.
   (However, sometimes, this may not be necessary.)

4 . If there are indicated operations such as addition or subtraction of fractions, combine these fractions and simplify. (You may have to go back and review how to add rational fractions)
   You may sometimes do the opposite of **4** as in **5** below.

5. If the numerator of a fraction consists of two or more terms and the denominator consists of only one term (especially, one function only), divide every term in the numerator by the denominator.

6. If there is a binomial in either the numerator or the denominator, multiply the binomial by its conjugate

7. Only practice will make the above guidelines meaningful. Do not be afraid to try and err.
   You will learn from your mistakes. After some mastery, you may even begin to enjoy proving trigonometric identities.

Example 1   Prove the following identity: $\dfrac{\sin x}{\tan x} + \dfrac{\cos x}{\cot x} = \cos x + \sin x$.

(**Note** that the right-hand side (RHS) is in terms sines and cosines. Therefore, we do not have to do any work on the RHS. We therefore work on the LHS and express the LHS in terms of sines and cosines and simplify.)

$$\frac{\sin x}{\tan x} + \frac{\cos x}{\cot x} \stackrel{?}{=} \cos x + \sin x \quad \text{(We will remove the question mark" ? " when we have}$$

$$\text{exactly the same expression on both sides of the equation)}$$

$$\frac{\sin x}{\frac{\sin x}{\cos x}} + \frac{\cos x}{\frac{\cos x}{\sin x}} \stackrel{?}{=} \cos x + \sin x \qquad \left(\tan x = \frac{\sin x}{\cos x}\, ; \cot x = \frac{\cos x}{\sin x}\right)$$

$$\frac{\sin x}{1} \cdot \frac{\cos x}{\sin x} + \frac{\cos x}{1} \cdot \frac{\sin x}{\cos x} \stackrel{?}{=} \cos x + \sin x \qquad \text{(Inverting the divisor and multiplying)}$$

$$\frac{\cancel{\sin x}}{1} \cdot \frac{\cos x}{\cancel{\sin x}} + \frac{\cancel{\cos x}}{1} \cdot \frac{\sin x}{\cancel{\cos x}} \stackrel{?}{=} \cos x + \sin x$$

$$\cos x + \sin x = \cos x + \sin x.$$

QED

Lesson 101: Proving Trigonometric Identities; Sum and Difference Identities

Example 2 Prove that $\dfrac{\sin 2x}{\sin x} = \sec x + \dfrac{\cos 2x}{\cos x}$

(In this problem, we will work on both sides.)

$$\dfrac{\sin 2x}{\sin x} \overset{?}{=} \sec x + \dfrac{\cos 2x}{\cos x}$$

$$\dfrac{2\sin x \cos x}{\sin x} \overset{?}{=} \dfrac{1}{\cos x} + \dfrac{\cos^2\theta - \sin^2\theta}{\cos x} \qquad (\sin 2x = 2\sin x \cos x \text{ for the LHS})$$

$$\dfrac{2\cancel{\sin x}\cos x}{\cancel{\sin x}} \overset{?}{=} \dfrac{1}{\cos x} + \dfrac{\cos^2 x - \sin^2 x}{\cos x} \qquad (\cos 2\theta = \cos^2\theta - \sin^2\theta \text{ for the RHS})$$

$$2\cos x \overset{?}{=} \dfrac{1 + \cos^2 x - \sin^2 x}{\cos x} \qquad (\text{Adding the fractions})$$

$$2\cos x \overset{?}{=} \dfrac{1 - \sin^2 x + \cos^2 x}{\cos x} \qquad (\text{Rewriting})$$

$$2\cos x \overset{?}{=} \dfrac{\cos^2 x + \cos^2 x}{\cos x} \qquad (1 - \sin^2\theta = \cos^2\theta)$$

$$2\cos x \overset{?}{=} \dfrac{2\cos^2 x}{\cos x}$$

$$2\cos x \overset{?}{=} \dfrac{2\cos x \cancel{\cos x}}{\cancel{\cos x}}$$

$$2\cos x = 2\cos x \qquad (\text{Note the LHS = the RHS})$$

QED

Equivalents in Trigonometry

Quantities = 1	Quantities = -1	Quantities = 0	Quantities = 2
$\sin\frac{\pi}{2} = 1;$	$\sin\frac{3\pi}{2} = -1$	$\sin 0 = 0$	$\csc\frac{\pi}{6} = 2$
$\cos 0 = 1$	$\cos\pi = -1$	$\sin\pi = 0$	$\sec\frac{\pi}{3} = 2$
$\csc\frac{\pi}{2} = 1;$	$\sec\pi = -1$	$\cos\frac{\pi}{2} = \cos\frac{3\pi}{2} = 0$	
$\tan\frac{\pi}{4} = 1$	$\csc\frac{3\pi}{2} = -1$	$\cot\frac{\pi}{2} = \cot\frac{3\pi}{2} = 0$	Quantities = $\sqrt{3}$
$\cot\frac{\pi}{4} = 1$	Quantities = $\frac{1}{2}$	$\tan\pi = 0$	$\tan\frac{\pi}{3} = \sqrt{3}$
$\sin^2\theta + \cos^2\theta = 1$	$\sin\frac{\pi}{6} = \frac{1}{2};\ \cos\frac{\pi}{3} = \frac{1}{2}$	$\tan 0 = 0$	Also, If $A = B$, then
$\sec^2\theta - \tan^2\theta = 1$			$A - B = 0$
$\csc^2\theta - \cot^2\theta = 1$			
$\cos 2\theta + 2\sin^2\theta = 1$			
$2\cos^2\theta - \cos 2\theta = 1$			

Proofs of some fundamental identities

The proofs of the fundamental identities can be constructed using the circle, below, and expressing the trigonometric functions in terms of x, y, and r. The techniques to be illustrated in the following example can also be applied to prove similar identities.

Example 1 Prove the reciprocal identity, $\sec\theta = \dfrac{1}{\cos\theta}$.

Step 1: Consider a point $P(x,y)$ at a distance r from the origin and on the terminal side of an angle θ, where θ is in standard position and also $r \neq 0$.

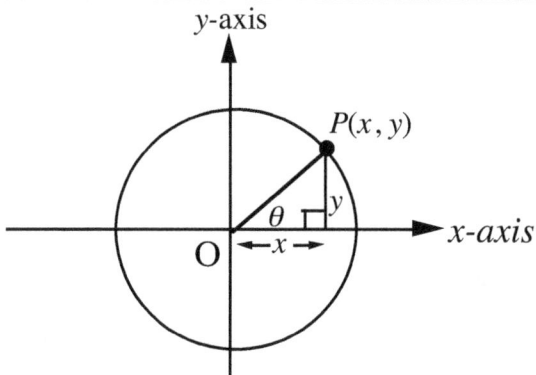

Proof: By definition, $\cos\theta = \dfrac{x}{r}$ (1)

$$\sec\theta = \frac{r}{x}\quad (2)\ \text{(the reciprocal of }\cos\theta)$$

$$\frac{1}{\cos\theta} = \frac{1}{\frac{x}{r}}\quad \text{(from (1))}$$

$$\frac{1}{\cos\theta} = \frac{r}{x}\quad (3)$$

The right-hand sides of (2) and (3) are identical.

Therefore $\sec\theta = \dfrac{1}{\cos\theta}$

(Quantities equal to the same quantity are equal to each other)

Example 2: Show that $\cos(A - B) = \cos A\cos B + \sin A\sin B$

Step 1: We first derive a formula for the length of a chord in a unit circle (Figure below)

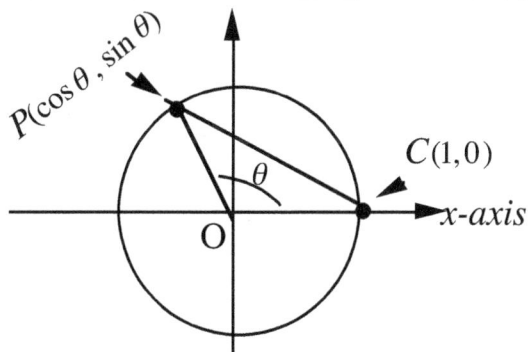

Consider a unit circle on the left. We find the length of chord PC whose corresponding arc subtends an angle θ at the center of the circle. Using the distance formula, the length of the PC is given by

$$PC = \sqrt{(\cos\theta - 1)^2 + (\sin\theta - 0)^2}$$
$$= \sqrt{\cos^2\theta - 2\cos\theta + 1 + \sin^2\theta}$$
$$= \sqrt{\cos^2\theta + \sin^2\theta - 2\cos\theta + 1}$$
$$= \sqrt{1 - 2\cos\theta + 1} = \boxed{PC = \sqrt{2 - 2\cos\theta}}\quad (1)$$

Step 2: Apply the formula derived in **Step 1** and also apply the distance formula in the figure below.

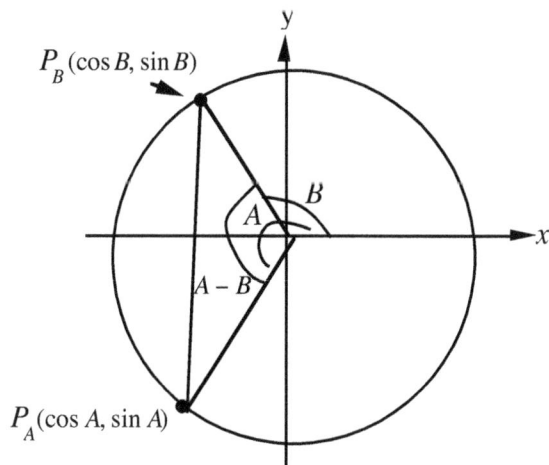

By applying (1) to chord $P_A P_B$,

$$P_A P_B = \sqrt{2 - 2\cos(A - B)}, \text{ where } \theta = A - B\quad (2)$$

Also, applying the distance formula to the chord $P_A P_B$
The length of the chord is given by

$$P_A P_B = \sqrt{(\cos B - \cos A)^2 + (\sin B - \sin A)^2}$$
$$= \sqrt{1 + 1 - 2\sin A\sin B - 2\cos A\cos B}\ \text{(simplifying)}$$
$$= \sqrt{2 - 2\sin A\sin B - 2\cos A\cos B}\quad (3)$$

Now, we equate the right hand-sides of (2) and (3)

$$\sqrt{2 - 2\cos(A - B)} = \sqrt{2 - 2\sin A\sin B - 2\cos A\cos B}$$
$$\boxed{\cos(A - B) = \cos A\cos B + \sin A\sin B}\ \text{(simplifying)}$$

The other fundamental identities of the reciprocal identities, the ratio identities, and the Pythagorean identities can be similarly proved by imitating the above example. Try to prove some of them.

Derivation of other identities

From the difference identity, $\cos(A - B) = \cos A \cos B + \sin A \sin B$ we can derive the other addition formulas as well as all the remaining identities (double-angle and half-angle identities; product as a sum identities, as well as the identities in the next section).

Applications of the sum and difference identities

From the sum and difference identities, we can prove the following identities

1. $\sin(-\theta) = -\sin\theta$

2. $\cos(-\theta) = \cos\theta$

3. $\tan(-\theta) = -\tan\theta$

We can also prove the following cofunctions which were derived previously (see. 510-511.)

$\left.\begin{array}{l} \sin A = \cos(90° - A) \\ \cos A = \sin(90° - A) \end{array}\right\}$ ← The sine and the cosine are cofunctions.

$\left.\begin{array}{l} \tan A = \cot(90° - A) \\ \cot A = \tan(90° - A) \end{array}\right\}$ ← The tangent and the cotangent are cofuntions.

$\left.\begin{array}{l} \sec A = \csc(90° - A) \\ \csc A = \sec(90° - A) \end{array}\right\}$ ← The secant and the cosecant are cofuntions.

Note above that: $(90° - A) = (\frac{\pi}{2} - A)$

Application of $\cos(A - B) = \cos A \cos B + \sin A \sin B$] to derive
$$\cos(A + B) = \cos A \cos B - \sin A \sin B$$
$\cos(A + B) = \cos[A - (-B)]$

$\qquad\qquad = \cos A \cos(-B) + \sin A \sin(-B)$ (applying $\cos(A - B) = \cos A \cos B + \sin A \sin B$)

$\boxed{\cos(A + B) = \cos A \cos B - \sin A \sin B}$ (Noting that $\cos(-B) = \cos B$; $\sin(-B) = -\sin B$)

Derivation of $\sin(A + B) = \sin A \cos B + \cos A \sin B$

Recall that $\sin\theta = \cos(\frac{\pi}{2} - \theta)$

let $\theta = (A + B)$

Then $\sin(A + B) = \cos[(\frac{\pi}{2} - (A + B)]$

$\sin(A + B) = \cos(\frac{\pi}{2} - A - B)$

$\sin(A + B) = \cos[(\frac{\pi}{2} - A) - B]$

$\sin(A + B) = \cos(\frac{\pi}{2} - A)\cos B + \sin(\frac{\pi}{2} - A)\sin B$ (applying $\cos(A - B) = \cos A \cos B + \sin A \sin B$)

$\boxed{\sin(A + B) = \sin A \cos B + \cos A \sin B}$ (Note : $\cos(\frac{\pi}{2} - A) = \sin A$; $\sin(\frac{\pi}{2} - A) = \cos A$)

Extra Example 1: Using $\tan\frac{\pi}{4} = 1$ and $\tan(A + B) = \frac{\tan A + \tan B}{1 - \tan A \tan B}$, show that

$$\ln\left|\frac{1+\tan\frac{x}{2}}{1-\tan\frac{x}{2}}\right| = \ln\left|\tan\left(\frac{\pi}{4} + \frac{x}{2}\right)\right|$$

Solution By applying $1 = \tan\frac{\pi}{4}$ to

$$\ln\left|\frac{1+\tan\frac{x}{2}}{1-\tan\frac{x}{2}}\right|,$$

$$\ln\left|\frac{1+\tan\frac{x}{2}}{1-\tan\frac{x}{2}}\right| = \ln\left|\frac{\tan\frac{\pi}{4}+\tan\frac{x}{2}}{1-\tan\frac{x}{2}\tan\frac{\pi}{4}}\right| \quad (A)$$

Similarly,, by applying

$$\tan(A + B) = \frac{\tan A + \tan B}{1 - \tan A \tan B} \quad \text{to}$$

$$\ln\left|\tan\left(\frac{\pi}{4} + \frac{x}{2}\right)\right|$$

$$\ln\left|\tan\left(\frac{\pi}{4} + \frac{x}{2}\right)\right| = \ln\left|\frac{\tan\frac{\pi}{4}+\tan\frac{x}{2}}{1-\tan\frac{x}{2}\tan\frac{\pi}{4}}\right| \quad (B)$$

Since the right sides of (A) and (B) are identical, the left sides are equal, and

$$\ln\left|\frac{1+\tan\frac{x}{2}}{1-\tan\frac{x}{2}}\right| = \ln\left|\tan\left(\frac{\pi}{4} + \frac{x}{2}\right)\right|$$

$$\therefore \ \ln\left|\frac{1+\tan\frac{x}{2}}{1-\tan\frac{x}{2}}\right| = \ln\left|\tan\left(\frac{\pi}{4} + \frac{x}{2}\right)\right|$$

Example 2

To exaggerate the substitution axiom:

If the cost of 2 books = $100,

then, the cost of $\sin\frac{\pi}{2}$ book = $ 50

(same as the cost of one book = $50; since $\sin\frac{\pi}{2} = 1$)

We could also say that

the cost of $\tan\frac{\pi}{4}$ book = $50, since

$\tan\frac{\pi}{4} = 1.$

Similarly, the cost of x^0 book = $50.

Also $x^0 = \sin\frac{\pi}{2} = \tan\frac{\pi}{4} = 7 - 6 = \frac{5}{5} = 1.$

$(x \neq 0)$

Extra: Similarly,

$(\sin^2\theta + \cos^2\theta)$ book costs $50.

since $\sin^2\theta + \cos^2\theta = 1.$

Lesson 101 Exercises

A Prove the following identities:

1. $\cot x + \tan x = \sec x \csc x$

2. $\tan x + \csc x = \dfrac{\sin x + \cot x}{\cos x}$

3. $\dfrac{\cos 2x}{\cos x} + \sec x - \dfrac{\sin 2x}{\sin x}$;

4. $\dfrac{\cos x}{\sec x} - \dfrac{\sin x}{\cot x} = \dfrac{\cos x \cot x - \tan x}{\csc x}$

5. $\csc^2 x \sec^2 x - \csc^2 x = \sec^2 x$;

6. $\sec x - \csc^2 x \sec x = -\cot x \csc x.$

7. $\dfrac{\tan x}{\sec x - 1} = \csc x + \cot x.$

B Given that $\sin A = \frac{1}{6}$ with the terminal side of $\angle A$ being in the second quadrant and $\cos B = \frac{1}{2}$ with the terminal side of $\angle B$ being in the fourth quadrant, Find
(a) $\cos(A - B)$; (b) $\sin(A + B)$; and (c) $\sin 2A.$

Answers: (a) $-\dfrac{\sqrt{35} + \sqrt{3}}{12}$; (b) $\dfrac{1 + \sqrt{105}}{12}$; (c) $-\dfrac{\sqrt{35}}{18}$

CHAPTER 39

Lesson 102

Solutions of Trigonometric Equations

A trigonometric equation is an equation containing a trigonometric function of an unknown angle. A trigonometric equation is a conditional equation. By conditional we mean the equation is true for some but not all permissible replacements of the variable. We can solve a trigonometric equation either graphically or algebraically. We will cover only algebraic methods although on occasion we may complement the algebraic solutions by graphical methods. In solving trigonometric equations, we will make use of trigonometric identities as well as inverse trigonometric relations.
We will cover two types of trigonometric equations, namely,

1. The trigonometric equation which is linear in one trigonometric function of one variable.

2. The trigonometric equation which is quadratic in one variable of the function.
   We will restrict our solutions for an unknown angle x such that $0 \le x < 360°$ ($0 \le x < 2\pi$).

General Procedure for solving trigonometric equations:

Step 1: Collect all terms on one side of the equation, preferably on the left-hand side of the equation.

Step 2: Solve the equation for the trig function as exemplified below:
   Examples in which the functions have been solved for: **1**. $\sin x = .5$
   2. $\tan x = 1$
   Examples in which the functions have **not** been solved for: 1. $4 \sin^2 x = 1$
   2. $\cos^2 x = .5$

In trying to solve for the functions we may have to apply trigonometric identities if the equation contains multiple angles and / or more than one type of a trig function. If there are multiple angles such as double angles, we may apply the double angle identities or we may substitute a new variable for the multiple angle, solve for the function in terms of the new variable, and reconvert to the original variable after having found the measure of the angle.

Step 3: Solve for the angle: Use inverse trigonometric functions (from tables, memory or calculator) to find the measures of the angles involved.

Step 4: Check the solutions if the equation was squared or raised to a power, since we might have introduced extraneous solutions.

Example 1   Solve for x: $\sin x = \frac{1}{2}$ or $.5$ $0 \le x < 360°$
Solution

Step 1: Find the angle whose sine is $\frac{1}{2}$ (from tables or calculator). The reference angle is $30°$

Step 2: Determine the quadrants in which the sine is positive.
     The sine is positive in quadrants I and II.   ($\frac{1}{2}$ is positive)
Step 3 : Specify the measures of the angles (in standard position).
     $x = 30°$ or $150°$ (The angles are measured from the positive x-axis counterclockwise)
The solution set is $\{30°, 150°\}$ or $\left\{\frac{\pi}{6}, \frac{5\pi}{6}\right\}$ in radians.

Example 2   Solve for x: $2\sin x - 1 = 0$ $0 \leq x < 360^\circ$

Step 1: Solve for $\sin x$
$$2\sin x = 1$$
$$\sin x = \tfrac{1}{2} \qquad\qquad\qquad (1)$$

Step 2: Equation (1) of Step 1 is the same as Example 1 above. We therefore repeat the solution:

The solution set is $\{30^\circ, 150^\circ\}$ or $\left\{\dfrac{\pi}{6},\ \dfrac{5\pi}{6}\right\}$ in radians

Example 3   Solve for x: $\cos 2x = \tfrac{1}{2}$.

Method 1

Step 1: Let $2x = \theta$

Then, $\cos 2x = \tfrac{1}{2}$ becomes
$$\cos \theta = \tfrac{1}{2}$$

Step 2: Determine the reference angle for θ.

The reference angle is 60° (From tables, memory or calculator)

Step 3: Determine the quadrant in which $\cos \theta$ is positive.

The cosine is positive in quadrant I and IV.   ($\tfrac{1}{2}$ is positive)

Step 4: Specify the measures of the angles   (in standard position).
$$\theta = 60^\circ, \theta = 300^\circ$$

Step 5:   The general solutions are $\theta = 60^\circ + 360n$ or $\theta = 300^\circ + 360n$ ($n = 0, 1$ for this problem)

Step 6: Change back to x. That is, replace θ by $2x$

$$2x = 60^\circ + n\cdot 360^\circ \qquad \text{or} \qquad 2x = 300^\circ + n\cdot 360^\circ$$

$$x = 30^\circ + n\cdot 180^\circ \qquad \text{or} \qquad x = 150^\circ + n\cdot 180^\circ \qquad \text{(Dividing by 2 and solving for } x\text{)}$$

For $n = 0, x = \mathbf{30^o}$ or $x = \mathbf{150^o}$ (When $n = 0, n\cdot 180 = 0$)

For $n = 1, x = 30^\circ + 180^\circ$ or $x = 150^\circ + 180^\circ$

$$x = \mathbf{210^o} \qquad \text{or} \qquad x = \mathbf{330^o}$$

The solution set is $\{30^\circ, 150^\circ , 210^\circ, 330^\circ\}$

Note that in Step 6, we considered two values of n ($n = 0, 1$) because of the "2" in $\cos 2x$. If we were given $\cos 3x$, we would have considered three values of n ($n = 0, 1, 2$) because of the "3" in $\cos 3x$.

Method 2: Using the double angle identity, $\cos 2x = 2\cos^2 x - 1$

The given equation $\cos 2x = \tfrac{1}{2}$ becomes

$$2\cos^2 x - 1 = \tfrac{1}{2}$$
$$2\cos^2 x = \tfrac{1}{2} + 1$$
$$2\cos^2 x = \tfrac{3}{2}$$
$$\cos^2 x = \tfrac{3}{4}$$
$$\cos x = \pm \sqrt{\tfrac{3}{4}}$$

$$\cos x = \pm \frac{\sqrt{3}}{2}$$

$$\cos x = + \frac{\sqrt{3}}{2} \quad \text{or} \quad \cos x = - \frac{\sqrt{3}}{2}$$

For $\cos x = + \frac{\sqrt{3}}{2}$, the reference angle is 30°.

$\cos x$ is positive in quadrants I and IV.

Therefore, $x = \mathbf{30^\circ}$ **or** $\mathbf{330^\circ}$ (angles in standard position).

Similarly, for $\cos x = - \frac{\sqrt{3}}{2}$, the reference angle is 30°.

$\cos x$ is negative in quadrants II and III.

Therefore, $x = \mathbf{150^\circ}$ **or** $\mathbf{210^\circ}$ (angles in standard position).

The solution set is $\{30^\circ, 150^\circ, 210^\circ, 330^\circ\}$.

Again, we obtain the same solution set as by Method 1. You decide which method you prefer.

Example 4 Solve for x: $\quad 2 \cos^2 x + 3 \sin x = 3, \qquad 0 \le x < 360^\circ$

Solution $\quad 2 \cos^2 x + 3 \sin x - 3 = 0$

Step 1: $\quad 2(1 - \sin^2 x) + 3 \sin x - 3 = 0 \qquad\qquad\qquad (\cos^2 x = 1 - \sin^2 x)$

$\qquad\qquad 2 - 2 \sin^2 x + 3 \sin x - 3 = 0$

$\qquad\qquad -2 \sin^2 x + 3 \sin x - 1 = 0$

$\qquad\qquad 2 \sin^2 x - 3 \sin x + 1 = 0 \qquad\qquad\qquad\qquad (1)$

*Step 2: Let $\sin x = t$, Then equation (1) becomes

$\qquad 2t^2 - 3t + 1 = 0$ <--------quadratic equation.

We solve by factoring since the factors are easily recognizable.

$\qquad\qquad (2t - 1)(t - 1) = 0$

$\qquad\qquad 2t - 1 = 0$ or $t - 1 = 0$

$\qquad\qquad t = \frac{1}{2}$ or $t = 1$

Now, replace t by $\sin x$.

Then $\sin x = \frac{1}{2}$ or $\sin x = 1$

Step 3: Solve for the angles:

$\qquad$ When $\sin x = \frac{1}{2}$,the reference angle is 30°,

$\qquad$ and $x = \mathbf{30^\circ}$ or $\mathbf{150^\circ}$

$\qquad$ When $\sin x = 1, x = \mathbf{90^\circ}$

The solution set is $\{30^\circ, 90^\circ, 150^\circ\}$.

* Note that in Step 2, we could avoid substitution by doing the following: $(2 \sin x - 1)(\sin x - 1) = 0$.

Then either $2 \sin x - 1 = 0$ or $\sin x - 1 = 0$.

Solving, $\sin x = \frac{1}{2}$ or $\sin x = 1$

Example 5: Solve for x: $\tan x = 1$ $\qquad\qquad\qquad$ $0° \le x < 360°$ $\qquad\qquad$

Solution

Step 1: Find the angle (or number) whose tangent is 1.

From tables or memory , the reference angle is $45°$ $\frac{\pi}{4}$.

Step 2: Determine the quadrant in which the tangent function is positive.

The tangent function is positive in the first and third quadrants

Step 3: Specify the measures of the angles (in standard position).

The solution set is $\left\{45°, 225°\right\}$ or $\left\{\frac{\pi}{4}, \frac{5\pi}{4}\right\}$

Extra

Solve the following system of equation simultaneously by graphing:

$$\left.\begin{array}{l} y = \tan x \\ y = 1 \end{array}\right\}$$

Hint: Sketch the graphs of $y = \tan x$ and $y = 1$ on the same set of rectangular axes and read the coordinates of the point of intersection of the two curves.

A note about **Solutions of Trigonometric Equations and Inverse Functions**

Generally, when we find the inverse of a trigonometric function we shall restrict the domain of the function so that there will be only one value of x. Note that the value we read from tables is a reference angle and x must be expressed in standard position.

However, in solving the trigonometric equations, we usually restrict our solutions to $0 \le x < 2\pi$ or $0° \le x < 360°$, and here we usually have multiple solutions.

Comparison of equivalent functions, inverse functions and reciprocal functions

Equivalent functions

Inverses

1. $y = 2x$ is equivalent to $\frac{y}{2} = x$

1. $y = 2x$ and $y = \frac{x}{2}$ are inverses.

2.   Arcsin $x = y$ is equivalent to $x = \sin y$

2. $y = \sin x$ and $y = \arcsin x$ are inverses.

3. $\log_a x = y$ is equivalent to $a^y = x$

3. $y = a^x$ and $y = \log_a x$ are inverses.$(y = a^x$ and $x = a^y)$

4. $y = x^2$, $x \ge 0$ and $x = \sqrt{y}$ are equivalent

4. $y = x^2$, $x \ge 0$ and $y = \sqrt{x}$ are inverses.

Reciprocal functions

1. $y = x$ and $y = \frac{1}{x}$ are reciprocal functions.

2. $y - a^x$ and $y - \frac{1}{a^x} = a^{-x}$ are reciprocal functions.

3. $y = \sin x$ and $y = \frac{1}{\sin x} = \csc x$ are reciprocal functions.

4. $y = 2x$ and $y = \frac{1}{2x}$ are reciprocal functions.

5. $y = x^2$ and $y = \frac{1}{x^2}$ or x^{-2} are reciprocal functions

A Solve for x: 1. $4\cos^2 x - 8\cos x + 3 = 0$; **2.** $2\cos^2 x - \cos x = 1$

Answers: **1.** $\{60^\circ, 300^\circ\}$ or $\left\{\dfrac{\pi}{3}, \dfrac{5\pi}{3}\right\}$ **2.** $\{0^\circ, 120^\circ, 240^\circ\}$ or $\left\{0, \dfrac{2\pi}{3}, \dfrac{4\pi}{3}\right\}$

B 1. Solve for x: $\cos^2 x - 1 = -\cos x - \cos^2 x$ for $0 \le x < 2\pi$

2. Solve for x: $2\cos^2 x = \sin x + 1$ for $0 \le x < 2\pi$.

Answers: **1.** $\left\{\dfrac{\pi}{3}, \dfrac{5\pi}{3}, \pi\right\}$; **2.** $\left\{\dfrac{\pi}{6}, \dfrac{5\pi}{6}, \dfrac{3\pi}{2}\right\}$

Extra

Example 6 Solve for x: $2\tan^{-1}(x) = \tan^{-1}\left(\dfrac{1}{4x}\right)$ (A)

Let $\tan^{-1} x = \theta$

Then (A) becomes

$2\theta = \tan^{-1}\left(\dfrac{1}{4x}\right)$

taking ("tan") of both sides to undo "$\tan^{-1}$"

Left side   Right side

$\Downarrow$ $\Downarrow$ Trig Identity

$\tan 2\theta = \tan\left(\tan^{-1}\dfrac{1}{4x}\right)$

$\dfrac{2\tan\theta}{1 - \tan^2\theta} = \tan\left(\tan^{-1}\dfrac{1}{4x}\right)$ $\tan 2\theta = \dfrac{2\tan\theta}{1 - \tan^2\theta}$

$\dfrac{2\tan[\tan^{-1}x]}{1 - [\tan(\tan^{-1}x)]^2} = \tan\left(\tan^{-1}\dfrac{1}{4x}\right)$

$\dfrac{2x}{1 - x^2} = \dfrac{1}{4x}$ $2\tan^{-1}(x) = \tan^{-1}\left(\dfrac{1}{4x}\right)$

$8x^2 = 1 - x^2$ $2\tan^{-1}\left(\dfrac{1}{3}\right) \overset{?}{=} \tan^{-1}\left(\dfrac{1}{\frac{4}{3}}\right)$

$9x^2 = 1$;

$x^2 = \dfrac{1}{9}$; $2\tan^{-1}\left(\dfrac{1}{3}\right) \overset{?}{=} \tan^{-1}\left(\dfrac{3}{4}\right)$

$x = \pm\sqrt{\dfrac{1}{9}}$;

$x = \pm\dfrac{1}{3}$

Example 7 Solve for x: $2\tan^{-1}\sqrt{x - x^2} = \tan^{-1}x + \tan^{-1}(1 - x)$

Solution

Step 1: $2\tan^{-1}\sqrt{x - x^2} = \tan^{-1}x + \tan^{-1}(1 - x)$ (A)

$\tan(2\tan^{-1}\sqrt{x - x^2}) = \tan[\tan^{-1}x + \tan^{-1}(1 - x)]$ (B)

(taking ("tan") of both sides to undo "$\tan^{-1}$"

Step 2: Let $\tan^{-1}\sqrt{x - x^2} = \theta_1$

Let $\tan^{-1}x = \theta_2$; Let $\tan^{-1}(1 - x) = \theta_3$

Then (B) becomes

Left side   Right side

$\Downarrow$ $\Downarrow$

$\tan 2\theta_1 = \tan(\theta_2 + \theta_3)$

$\dfrac{2\tan\theta_1}{1 - \tan^2\theta_1} = \dfrac{\tan\theta_2 + \tan\theta_3}{1 - \tan\theta_2\tan\theta_3}$

Trig. Identities

$\tan 2A = \dfrac{2\tan A}{1 - \tan^2 A}$

$\tan(A + B) = \dfrac{\tan A + \tan B}{1 - \tan A\tan B}$

Step 3:

$\dfrac{2\tan[\tan^{-1}\sqrt{x - x^2}]}{1 - [\tan(\tan^{-1}\sqrt{x - x^2})]^2} = \dfrac{\tan(\tan^{-1}x) + \tan[\tan^{-1}(1 - x)]}{1 - \tan(\tan^{-1}x) \bullet \tan[\tan^{-1}(1 - x)]}$

$\dfrac{2\sqrt{x - x^2}}{1 - [x - x^2]} = \dfrac{x + 1 - x}{1 - [x(1 - x)]}$

$\dfrac{2\sqrt{x - x^2}}{1 - x + x^2} = \dfrac{1}{1 - [x - x^2]}$

$\dfrac{2\sqrt{x - x^2}}{1 - x + x^2} = \dfrac{1}{1 - x + x^2}$

$2\sqrt{x - x^2} = 1$

(Equating the numerators since the denominators are equal)

$4(x - x^2) = 1$

(squaring both sides of the equation

Step 4: Solve the resulting quadratic equation by factoring, since the factors are easily recognizable.

$4x - 4x^2 - 1 = 0$

$4x^2 - 4x + 1 = 0$

$(2x - 1)(2x - 1) = 0$

$2x - 1 = 0$

$2x = 1$

$x = \dfrac{1}{2}$

Solution check

Substitute $x = \frac{1}{2}$ in (A)

Left side   Right side

$\Downarrow$ $\Downarrow$

$2\tan^{-1}\sqrt{\frac{1}{2} - (\frac{1}{2})^2} \overset{?}{=} \tan^{-1}\frac{1}{2} + \tan^{-1}(1 - \frac{1}{2})$

$2\tan^{-1}\sqrt{\frac{1}{2} - \frac{1}{4}} \overset{?}{=} \tan^{-1}\frac{1}{2} + \tan^{-1}(\frac{1}{2})$

$2\tan^{-1}\sqrt{\frac{1}{4}} \overset{?}{=} 2\tan^{-1}\frac{1}{2}$

$2\tan^{-1}\frac{1}{2} \overset{?}{=} 2\tan^{-1}\frac{1}{2}$ Yes.

A person standing on top of a building 80 feet tall, throws a base ball vertically upwards with an initial velocity of 24 feet per second. The equation of the vertical distance from the ground is given by $S(t) = -8t^2 + V_0t + S_0$. where t is the time in seconds, V_0 is the initial velocity, and S_0 is the initial height.

(a) Find the time the ball takes to reach its maximum height.
(b) What is the maximum height the ball reaches?
(c) Find the time the ball takes to return to the ground.

Solution

(a) Substituting 24 for V_0, and 80 for S_0, in $S(t) = -8t^2 + V_0t + S_0$, we obtain the equation

$S(t) = -8t^2 + 24t + 80$ (compare with

$f(x) = -8x^2 + 24x + 80$)
(a) since the given equation is quadratic in t. The maximum height is at the vertex of the inverted parabola. The t–coordinate of the vertex is given by $t = -\dfrac{b}{2a}$; where $a = -8$,

$b = 24$, $t = -\dfrac{24}{2(-8)} = \dfrac{3}{2}$,

The maximum height is attained after $\frac{3}{2}$ seconds

Extra: When you learn calculus and physics, you will be able to find the time taken as follows:

The velocity V(t) is given by the derivative of $S(t) = -8t^2 + 24t + 80$

$V(t) = \dfrac{dS(t)}{dt} = -16t + 24$

Since at the maximum height, the velocity is temporarily zero.. $-16t + 24 = 0$
$16t = 24$;
$t = \dfrac{24}{16}$
$t = \dfrac{3}{2}$

(b) The maximum height is found by substituting $\frac{3}{2}$ in $S(t) = -8t^2 + 24t + 80$.

$S(\frac{3}{2}) = -8(\frac{3}{2})^2 + 24(\frac{3}{2}) + 80$

$= -8(\frac{9}{4}) + 36 + 80$
$= -2(9) + 116$
$= -18 + 116$
$= 98$
The maximum height is 98 feet.

(c) when the ball returns to the ground ,
$S(t) = 0$, and $0 = -8t^2 + 24t + 80$
$-8t^2 + 24t + 80 = 0$
$8t^2 - 24t - 80 = 0$
$t^2 - 3t - 10 = 0$
$(t + 2)(t - 5) = 0$
$t + 2 = 0$ or $t - 5 = 0$
$t = -2$ or $t = 5$
We reject -2 and accept 5 since the time cannot be negative in this problem.
The ball takes 5 seconds to return to the ground.

Appendix A

How to change a terminating decimal to a rational number

Procedure: Write each decimal as a decimal fraction and reduce to its lowest terms.

Example 1:

(a) $.5 = \dfrac{5}{10} = \dfrac{1}{2}$

(b) $.25 = \dfrac{25}{100} = \dfrac{1}{4}$

(c) $.16 = \dfrac{16}{100} = \dfrac{4}{25}$

(d) $.125 = \dfrac{125}{1000} = \dfrac{1}{8}$

How to Change a repeating decimal to a rational number
(On p.476, we converted a repeating decimal to a rational fraction, using geometric series)

Example 2   Change the following to fractions;

(a) .333...,   (b) .666..., (b) ..232323......, (c) .166...

Solution

(a)   Step 1: Let $x = .333$ (1)

Step 2: Multiply equation (1) by 10 (Generally, the exponent on the 10 equals the number of digits in the repeating block.)

$10x = 3.333$ (2)

Step 3: Subtract equation (1) from equation (2) and solve for x.
$10x - x = 3.333... - .333...$

$9x = 3$

$x = \dfrac{3}{9}$

$x = \dfrac{1}{3}$

Therefore $.333... = \dfrac{1}{3}$

(b)   Step 1: Let $x = .666...$ (1)

  Step 2: Multiply equation (1) by 10
    $10x = 6.666...$ (2)

  Step 3: Subtract equation (1) from equation (2) and solve for x:
    $10x - x = 6.666... - .666...$

        $9x = 6$

        $x = \dfrac{6}{9}$

        $x = \dfrac{2}{3}$

Therefore $.666... = \dfrac{2}{3}$

(c) Step 1: Let $x = .232323...$ (1)

  Step 2: Multiply equation (1) by 10^2 or 100 (The exponent on the 10 is 2 since there are two digits in the repeating block.)

        $100x = 23.232323...$ (2)

  Step 3: Subtract equation (1) from equation (2) and solve for x:

    $100x - x = 23.232323... - .232323...$

        $99x = 23$

        $x = \dfrac{23}{99}$

d) **Method 1**
      Let $x = .166...$ (1)

Step 2: Multiply equation (1) by 10.
      $10x = 1.666$ (2)   (Note: $1.66... = 1.666...$))

Step 3: Subtract equation (1) from equation (2) and solve for x:
      $10x - x = 1.666 - .166$

        $9x = 1.5$

        $90x = 15$ (Multiplying by 10 to eliminate the decimal point)

        $x = \dfrac{15}{90}$

        $x = \dfrac{1}{6}$

Note: $.166... = .1666...$

Method 2   Let $x = .166...$ (1)

Step 2: Multiply equation (1) by 10.
$$10x = 1.66$$ (2)
Also multiplying (1) by 100,
$$100x = 16.66...$$ (3)

Step 3: Subtract (2) from (3) and solve for *x:*

$$100x - 10x = 16.666... - 1.6666...$$
$$90x = 15$$
$$x = \frac{15}{90}$$
$$x = \frac{1}{6}$$

Example 3b: (From p.136) Solve for x: $|x - 2| = |x + 6|$. We consider four cases.

Case 1: Both expressions within the absolute value bars are positive.

Then $x - 2 = x + 6$ (A)

Case 2: Both expressions within the absolute value bars are negative.

Then $-(x - 2) = -(x + 6)$ (B)

Case 3: The expression within the absolute value bars on the left is positive and that on the right is negative.

Then $x - 2 = -(x + 6)$ (C)

Case 4: The expression within the absolute bars on the left is negative and that on the right is positive.

Then $-(x - 2) = x + 6$ (D)

On simplifying, (B) reduces (A).

Similarly, (C) and (D) are equivalent, ignoring the domains of the definitions.

We therefore solve only Case 1, $x - 2 = x + 6$

and Case 3, $x - 2 = -(x + 6)$

Solving for Case 1, $x - 2 = x + 6$

$-2 = 6$, which is a contradiction.

There is no solution for Cases 1 and 2.

For Cases 3 and 4, $x - 2 = -(x + 6)$

$$x - 2 = -x - 6$$
$$x = -2.$$

We check for $x = -2$.

Then $|-2 - 2| \overset{?}{=} |-2 + 6|$; $|-4| \overset{?}{=} |4|$; $4 \overset{?}{=} 4$. Yes.

$\boxed{\text{The solution is } -2.}$

Appendix B

Definitions of Inverse Trigonometric Functions

We summarize the definitions of the **inverse trigonometric functions** below
Note that the restrictions made in obtaining the above inverse functions are arbitrary, and that other branches may be used. However, the branches presented here are the ones that are usually used.

Restricted Function	Domain	Range	Inverse Notation	Domain	Range
$y = \sin x,$	$-\frac{\pi}{2} \le x \le \frac{\pi}{2},$	$-1 \le y \le 1$	$x = \sin y,\ \ y = \text{Arc}\sin x \text{ or } \text{Sin}^{-1}x;$	$-1 \le x \le 1;$	$-\frac{\pi}{2} \le y \le \frac{\pi}{2}$
$y = \cos x,$	$0 \le x \le \pi,$	$-1 \le y \le 1$	$x = \cos y,\ \ y = \text{Arc}\cos x \text{ or } \text{Cos}^{-1}x;$	$-1 \le x \le 1;$	$0 \le y \le \pi$
$y = \tan x,$	$-\frac{\pi}{2} < x < \frac{\pi}{2},$	$-\infty < y < \infty$	$x = \tan y,\ \ y = \text{Arc}\tan x \text{ or } \text{Tan}^{-1}x;$	$-\infty < x < \infty;$	$-\frac{\pi}{2} < y < \frac{\pi}{2}$
$y = \cot x,$	$0 < x < \pi,$	$-\infty < y < \infty$	$x = \cot y,\ \ y = \text{Arc}\cot x \text{ or } \text{Cot}^{-1}x;$	$-\infty < x < \infty;$	$0 < y < \pi$
(Also: $\text{Tan}^{-1}x + \text{Cot}^{-1}x = \frac{\pi}{2}$)			or $y = \text{Cot}^{-1}x = \text{Tan}^{-1}\left(\frac{1}{x}\right) = \frac{\pi}{2} - \text{Tan}^{-1}x;$		$0 < y < \pi$
$y = \sec x, 0 \le x \le \pi, x \ne \frac{\pi}{2},\ \|y\| \ge 1$			$x = \sec y,\ y = \text{Arc}\sec x \text{ or } \text{Sec}^{-1}x;\ \|x\| \ge 1, 0 \le y \le \pi, y \ne \frac{\pi}{2}$		
			Also, $y = \text{Sec}^{-1}x = \text{Cos}^{-1}\left(\frac{1}{x}\right);\ \|x\| \ge 1 \quad 0 \le y \le \pi,\ y \ne \frac{\pi}{2}$		
$y = \csc x, -\frac{\pi}{2} \le x \le \frac{\pi}{2}, x \ne 0, \|y\| \ge 1$			$x = \csc y, y = \text{Arc}\csc x \text{ or } \text{Csc}^{-1};\ \|x\| \ge 1, -\frac{\pi}{2} \le y \le \frac{\pi}{2}, y \ne 0$		
(Also: $\text{Sec}^{-1}x + \text{Csc}^{-1}x = \frac{\pi}{2}$)			Also, $y = \text{Csc}^{-1}x = \text{Sin}^{-1}\left(\frac{1}{x}\right);\ \|x\| \ge 1,\ -\frac{\pi}{2} \le y \le \frac{\pi}{2}, y \ne 0$		

More on quadrants for the definitions of the inverse functions

$\text{Sin}^{-1}x$, $\text{Csc}^{-1}x$ and $\text{Tan}^{-1}x$ are defined in the **first** and **fourth** quadrants or appropriately quadrantal..
If $\text{Sin}^{-1}x$, $\text{Csc}^{-1}x$, or $\text{Tan}^{-1}x$ is in the fourth quadrant, then it is a **negative** angle or appropriately quadrantal. $\text{Cos}^{-1}x$, $\text{Sec}^{-1}x$, and $\text{Cot}^{-1}x$ are defined in the **first** and **second** quadrants or appropriately quadrantal. $\text{Cos}^{-1}x$, $\text{Sec}^{-1}x$, and $\text{Cot}^{-1}x$ are **never** negative.

Note: Saying that $-\frac{\pi}{2} \le y \le \frac{\pi}{2}$ implies that y is either in the first quadrant or in the fourth quadrant or appropriately quadrantal. Also, saying that $0 \le y \le \pi$ implies that y is either in the first quadrant or in the second quadrant or appropriately quadrantal.

Note the restricted domains and ranges of these inverse functions. In all operations involving inverse functions as well as in sketching their graphs, we will be guided by these domains and ranges.

For example, $\text{Sin}^{-1}x$ (read: inverse sine x) means the angle (in radians) or the number between $-\pi/2$ and $\pi/2$, whose sine is x. Similarly, $\text{Cos}^{-1}x$ (read: inverse cosine x) means the angle (in radians) or the number between 0 and π, whose cosine is x.. In each definition, two conditions must be satisfied simultaneously. For example, the conditions for $\text{Sin}^{-1}x$ are: **1**. The sine of the angle (or number) $= x$, **2**. The angle must be between $-\frac{\pi}{2}$ and $\frac{\pi}{2}$.

Inverse Cofunction Identities

1. $\text{Sin}^{-1}x + \text{Cos}^{-1}x = \frac{\pi}{2}$, $[-1,1]$; **2.** $\text{Tan}^{-1}x + \text{Cot}^{-1}x = \frac{\pi}{2}$, $(-\infty,\infty)$; **3** $\text{Sec}^{-1}x + \text{Csc}^{-1}x = \frac{\pi}{2}$, $|x| \geq 1$

Each identity is subject to the domain of definition

We prove these identities assuming $0 \leq x \leq 1$.

Proof:

Let θ and β be two angles of a right triangle.

$\sin\theta = \frac{x}{1} = x$

and $\sin^{-1}x = \theta$.   (A)

Similarly, $\cos\beta = \frac{x}{1} = x$ and

$\cos^{-1}x = \beta$ (B)

$\theta + \beta = \frac{\pi}{2}$ (C) (Sum of the other two angles of a right triangle)

$\sin^{-1}x + \cos^{-1}x = \theta + \beta$ (A) + (B)

$\sin^{-1}x + \cos^{-1}x = \frac{\pi}{2}$ $\left(\text{From (C)} \quad \theta + \beta = \frac{\pi}{2}\right)$

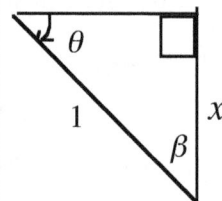

Proof:

Let θ and β be two angles of a right triangle.

$\tan\theta = \frac{x}{1} = x$

and $\tan^{-1}x = \theta$.   (A)

Similarly, $\cot\beta = \frac{x}{1} = x$ and

$\cot^{-1}x = \beta$ (B)

$\theta + \beta = \frac{\pi}{2}$ (C) (Sum of the other two angles of a right triangle)

$\tan^{-1}x + \cot^{-1}x = \theta + \beta$ (A) + (B)

$\text{Tan}^{-1}x + \text{Cot}^{-1}x = \frac{\pi}{2}$ $\left(\text{From (C)} \quad \theta + \beta = \frac{\pi}{2}\right)$

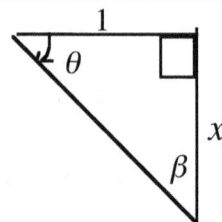

Proof:

Let θ and β be two angles of a right triangle.

$\sec\theta = \frac{x}{1} = x$

and $\sec^{-1}x = \theta$.   (A)

Similarly, $\csc\beta = \frac{x}{1} = x$ and

$\csc^{-1}x = \beta$ (B)

$\theta + \beta = \frac{\pi}{2}$ (C) (Sum of the other two angles of a right triangle)

$\sec^{-1}x + \csc^{-1}x = \theta + \beta$ (A) + (B)

$\text{Sec}^{-1}x + \text{Csc}^{-1}x = \frac{\pi}{2}$ $\left(\text{From (C)} \quad \theta + \beta = \frac{\pi}{2}\right)$

Appendix C
Area and Perimeter of a Circle; Arc Length and Sector of a Circle; Area and Perimeter of Composite Figures

Area and Perimeter of a Circle

Area of a circle
The area, A, of a circle of radius, r, is given by $A = \pi r^2$.
π is approximately equal to 3.1416; in calculations, we will leave our answers in terms of π, unless instructed otherwise.

Note also that the radius of a circle $= \frac{1}{2}$ of the diameter of the circle.

Area of a semicircle (half-circle) $= \frac{1}{2}$ of the area of the given circle.

Area of a quarter-circle $= \frac{1}{4}$ of the area of the given circle.

Perimeter (or Circumference) of a Circle

The perimeter of a circle is the distance around the circle (i.e., the length of the circle).
The perimeter, P, of a circle of radius, r, is given by $P = 2\pi r$.

The perimeter of a semicircle $= \frac{1}{2}$ of the perimeter of the given circle. However, if there is a diameter

joining the end points of the curved part, then add the diameter. Therefore, the shape of the figure determines whether or not to add the diameter. The circle formula is for the curved part only.

The perimeter of a quarter circle is $\frac{1}{4}$ of the perimeter of the given circle plus $2r$ if two radii connect the

end points of the curved part.
Example Find the area and the perimeter of a circle of radius 6 units.

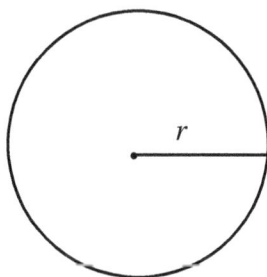

Finding the area
$r = 6$, Area of a circle $= \pi r^2$.
Area of the given circle $= \pi(6)^2 = \pi(36) = 36\pi$ sq. units

Finding the perimeter
$r - 6$ units
Perimeter of a circle $= 2\pi r$
Perimeter of the given circle $= 2\pi(6) = 12\pi$ units

Note: In the above problem, the area of a **semicircle** of radius, $r = 6$ is $\frac{1}{2}(36\pi) = 18\pi$ sq. units; and the

perimeter of the semicircle (curved part only) of radius, $r = 6$, is $\frac{1}{2}(12\pi) = 6\pi$ units. If the diameter

connects the end points of the curved part, then add the diameter, $2(6)$ to obtain $(6\pi + 12)$ units.

Arc Length and Sector of a Circle

Arc length: Arc length is the length of a part of a circle between any two points on the circle. In a circle of radius r, the arc length s cut off by a central angle θ (θ in degrees) is given by

$$s = \frac{\theta}{360} \text{ of the circumference of the circle.}$$

$$\boxed{s = \frac{\theta}{360}(2\pi r) \text{ or } \frac{\theta\pi r}{180}}.$$ (From the proportion, s is to θ as $2\pi r$ is to 360)

However, if the central angle is in radians, replace 360 by 2π and simplify to obtain
$$(360^\circ = 2\pi \text{ radians, or } 180^\circ = \pi \text{ radians})$$

$$\boxed{s = r\theta}$$

Sector of a circle: A sector of a circle is the part of the interior of the circle bounded by two radii and the intercepted arc.
In a circle of radius r, the area A of a sector with central angle θ (θ in degrees) is given by

$$A = \frac{\theta}{360} \text{ of the area of the circle.}$$

$$\boxed{A = \frac{\theta}{360}(\pi r^2)}$$ (From the proportion, A is to θ as πr^2 is to 360)

However, if θ is in radians, replace 360 by 2π and simplify to obtain

$$\boxed{A = \frac{1}{2}r^2\theta}$$

Example In the circle shown below, if the diameter is 6 and the central angle is 30°,
  (a) Find the circumference of the circle.
  (b) Find the arc length of CB
  (c) Find the area of sector COB.

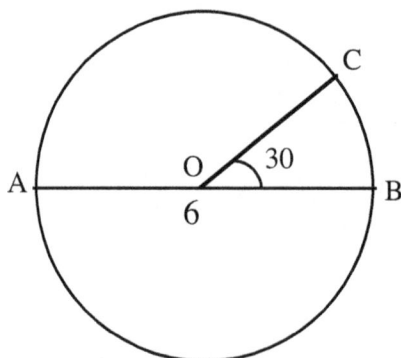

Figure 1

Solution
(a) Circumference of a circle of radius r is given by $2\pi r$.

  Radius of a circle $= \frac{1}{2}$ of the diameter of the circle.

  Therefore, $r = \frac{1}{2}(6) = 3$
  Circumference of the given circle is $2\pi r. = 2\pi(3) = 6\pi$ units.

(b) arc length, s, of CB $= \dfrac{\theta}{360}$ of the circumference of the circle

$$s = \dfrac{30}{360} \cdot 2\pi(3) = \dfrac{\pi}{2} \text{ units}$$

OR if θ is in radians, $s = r\theta = 3 \cdot \dfrac{\pi}{6}$ (**Note**: $\theta = 30° = \dfrac{\pi}{6}$; $r = 3$.)

$$s = \dfrac{\pi}{2} \text{ units.}$$

(c) Area, A , of the shaded sector $= \dfrac{30°}{360°}$ of the area of the circle.

$$A = \dfrac{30°}{360°}(\pi r^2)$$
$$= \dfrac{30°}{360°}(\pi \cdot 3^2)$$
$$= \dfrac{30}{360}(9\pi)$$
$$= \dfrac{3\pi}{4} \text{ sq. units}$$

OR if θ is in radians, $A = \dfrac{1}{2}r^2\theta$ (**Note**: $\theta = 30° = \dfrac{\pi}{6}$; $r = 3$.)

$$= \dfrac{1}{2} \cdot 3^2 \cdot \dfrac{\pi}{6}$$
$$= \dfrac{1}{2} \cdot \dfrac{9}{1} \cdot \dfrac{\pi}{6}$$
$$= \dfrac{3\pi}{4} \text{ sq. units.}$$

Areas and Perimeters of Composite Figures

Example 1   In the figure below:
  (a) Find the area of the figure.
  (b) Find the perimeter of figure.

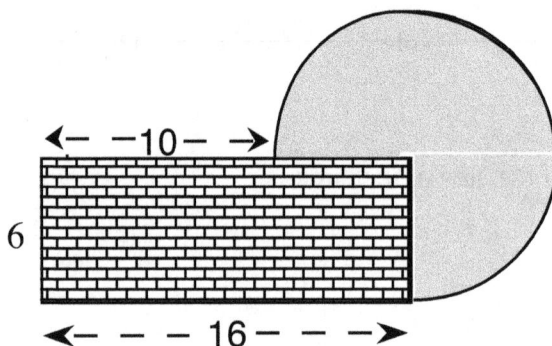

Solution

(a) **Finding the area**

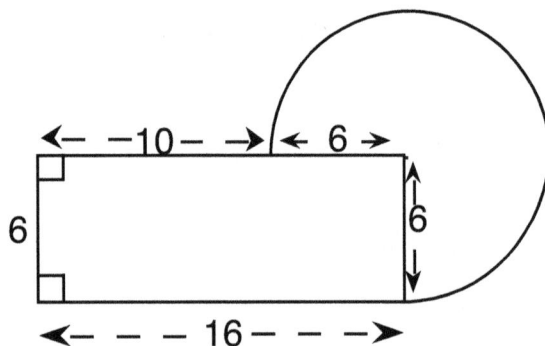

Radius, r, of the three-quarter circle $= 6$

Area of the three-quarter circle $= \dfrac{3}{4}\pi r^2$ ($\dfrac{3}{4}$ of the area of a circle of radius 6 units)

$$= \dfrac{3}{4}\pi(6)^2$$

$$= \dfrac{3}{4}\pi(36)^{9}$$

$$= 27\pi \text{ sq. units}$$

Area of the 16 by 6 rectangle $= 16 \times 6$ sq. units
$$= 96 \text{ sq. units}$$

Area of the **whole figure** = area of rectangle + area of three-quarter circle
$$= (96 + 27\pi) \text{ sq. units}$$

(b) **Finding the perimeter**

Perimeter of the three-quarter circle $= \frac{3}{4} 2\pi r$

$$= \frac{3}{4}(2\pi)(6)$$

$$= 9\pi \text{ units}$$

Perimeter of the whole figure $= 6 + 16 + 9\pi + 10$

$$= (32 + 9\pi) \text{ units}$$

Begin here and go around counterclockwise

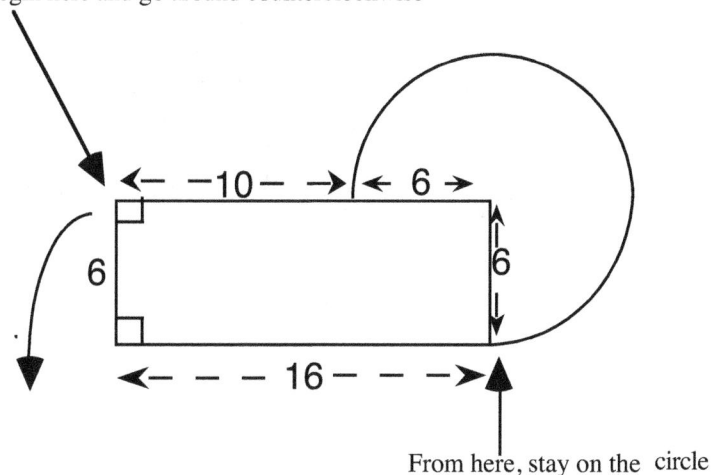

From here, stay on the circle

Note that in finding the perimeter we did **not** add the parts (6 and 6) of the rectangle which are also radii of the three-quarter circle, because the perimeter is the distance around the figure.

Appendix C Exercises

A

In the circle below, if the diameter of the circle is 18 units and the central angle has a measure $60°$,

(a) Find the circumference of the circle.

(b) Find the area of sector COB.

(c) Find the arc length of CB.

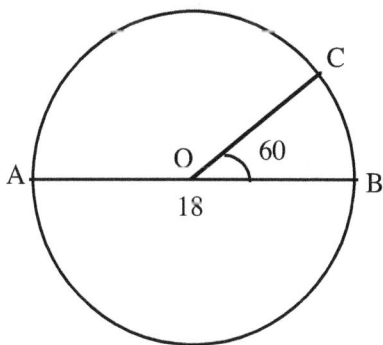

Answers (a) 18π units; (b) $\frac{27\pi}{2}$ sq. units; (c) 3π units

B

(a) Find the area of the figure.

(b) Find the perimeter of figure.

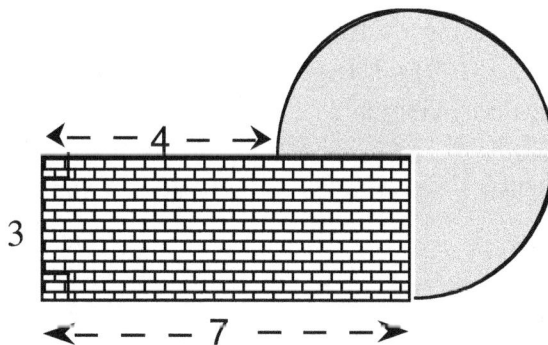

Answers: **(a)** $(21 + \frac{27\pi}{4})$ sq. units; **(b)**. $(14 + \frac{9\pi}{2})$ units

Applications of Exponential Functions
Exponential Decay

The exponential decay model is given by the equation $P(t) = P_0 e^{kt}$, where P_0 is the initial value, $P(t)$ is the value at time t. Here, $k < 0$, P is said to decay exponentially, and k is then called the decay constant.

Example 1: At the beginning of an experiment, the population was 4,500. Three hours later, the population was 3,000. Assume exponential decay.
(a) Determine the decay constant.   (b) Determine the population after 6 hours.
(c) Determine when the population will be 1,500.

(a) Step 1:

$P(t) = P_0 e^{kt}$ ($P(3) = 3000$)

$3000 = 4500 e^{k(3)}$

$\dfrac{3000}{4500} = e^{3k}$ ($P_0 = 4500$)

$\dfrac{2}{3} = e^{3k}$

$\ln \dfrac{2}{3} = \ln e^{3k}$

$\qquad = 3k \ln e$

Step 2: $\ln \dfrac{2}{3} = 3k$ $(\ln e = \log e_e = 1)$

$\left(\ln \dfrac{2}{3} \right) \div 3 = k$ and $k = -0.1352$

The decay constant $= -0.1352$

(Note the minus sign for this constant)

(b) $P(6) = 4500 e^{[(\ln 2/3)/3](6)}$

$P(6) = 4500 e^{2(\ln 2/3)}$

$\qquad = 4500(.444)$

$\qquad (k = \ln(2/3) \div 3 = -0.1352)$

$\qquad = 2000$

After 6 hours, the population is 2000.

(c) $1,500 = 4500 e^{-0.1352t}$

$\dfrac{1500}{4500} = e^{-0.1352t}$

$\dfrac{1}{3} = e^{-0.1352t}$ ($k = -0.1352$)

$\ln \dfrac{1}{3} = \ln e^{-0.1352t}$

$\ln 3^{-1} = -0.1352t \ln e$

$-\ln 3 = -0.1352t$

$\dfrac{-\ln 3}{-0.1352} = t$

$8.1 = t$

The population will reach 1500 in 8.1 hours.

Example 2: The half-life of a radioactive substance is 100 years. If 300 mg were initially present, how much of the substance remains after 500 years.

(a) Step 1:
A half-life of 100 years means the time for 300 mg to become 300/2 or 150 mg is 100 years..
Then substituting $t = 100$,
$P_0 = 300$, $P(100) = 150$,
to find the decay constant,

$150 = 300 e^{k(100)}$

$\dfrac{150}{300} = e^{k(100)}$

$\dfrac{1}{2} = e^{k(100)}$

$\ln \dfrac{1}{2} = \ln e^{k(100)}$

Step 2: $\ln 2^{-1} = \ln e^{k(100)}$

$-\ln 2 = 100k \ln e$

$-\ln 2 = 100k$

$\dfrac{-\ln 2}{100} = k$

$-0.0069 = k$

Step 3: Now, $P_0 = 300$;

$\qquad k = -0.0069$

$P(500) = 300 e^{-0.0069(500)}$

$\qquad = 300 e^{-3.45}$

$\qquad = 300(.03175)$

$\qquad = 9.53$

After 500 years, 9.53 mg remain.

Example 3: A substance decays at the rate of 0.25% per year. If initially, there were 200 mg, what is the half-life of the substance?

Solution: $k = -0.0025$ ($k < 0$)

$100 = 200 e^{-0.0025t}$ ($P(t) = P_0 e^{kt}$)

$\dfrac{100}{200} = e^{-0.0025t}$ (Note the sign of k)

$\dfrac{1}{2} = e^{-0.0025t}$

$2^{-1} = e^{-0.0025t}$

$\ln 2^{-1} = \ln e^{-0.0025t}$

$-\ln 2 = -0.0025t \ln e$

$-\ln 2 = -0.0025t$

$\dfrac{-\ln 2}{-0.0025} = t$

$277.3 = t$

The half-life is 277.3 years.

Appendix D
About Measurements

Standard Unit, Error, Rounding-off Numbers, Significant Digits, Scientific Notation

To determine the size of a physical quantity, we compare its size with a standard quantity called a unit. Example: To determine the length of the cover of a book in inches or in meters, we can use a ruler with its scale in inches or in meters.

A measurement is the ratio of the magnitude of a physical quantity to that of a standard unit

Standard unit

A standard unit is a measure with which other quantities are compared. A standard unit of measure is defined by a legal authority (such as the US Bureau of Weights and Measures) or by a conference of scientists.

Some universally accepted standards:
1. For mass, the standard (primary standard) is the kilogram (kg).
2. For length, the standard is the meter (m).
3. For time, the standard is the second (s)

Rounding-Off Numbers

The rules for rounding-off a number may differ slightly depending upon the field. For instance, in accounting the rule may be slightly different from the rule in chemistry.

1. If the digit or group of digits to be dropped is more than 500...,hen drop that digit or group and add 1 to the last digit retained.

2. If the digit or group of digits to be dropped is less than 500...,then drop that digit or group and leave the last digit retained unchanged.

3. If the digit or group of digits to be dropped is exactly 500..., then drop that portion and add 1 to the last digit retained if this digit is odd but if this digit is even, then this digit remains unchanged.

Rounding off Whole numbers
Procedure:

Step 1: Locate the digit in the round-off place.
  (The round-off place is the place to which we want to round-off the number)

Step 2: Drop all digits to the right of the round-off place, and if the digit immediately to the right of the round-off place is more than 5 or is 5 followed by non-zero digits, , add 1 to the round-off place digit (i.e. we round-up); but if the digit immediately to the right of the round-off place is less than 5, the round-off place digit remains unchanged. However, if the digit immediately to the right of the round-off place digit is 5 or 5 followed by zeros, we add 1 to the round-off place digit if it is odd, but if it is even, it remains unchanged.(i.e. we round-down). Also, replace each digit dropped by a zero.

Rounding-off Decimals

The procedure is the same as that for rounding-off whole numbers, except that after the decimal point, we do not replace any digits dropped by zeros.

Procedure:

Step 1: Locate the digit in the round-off place.
  (The round-off place is the place to which we want to round-off the number)

Step 2: Drop all digits to the right of the round-off place, and if the digit immediately to the right $\quad$ 6 6 8
of the round-off place is more than 5 or is 5 followed by non-zero digits, add 1 to the
round-off place digit (i.e., we round-up); but if the digit immediately to the right of the
round-off place is less than 5, the round-off place digit remains unchanged. However, if the
digit immediately to he right of the round-off place digit is 5 or 5 followed by zeros, we add 1
to the round-off place digit if it is odd, but if it is even, it remains unchanged.(i.e., we round-down).

Rounding-off (Alternatively)

When we round-off a number, we drop some of the digits explicitly or implicitly specified. We must
distinguish between rounding-off to a specified number of decimal places (or significant digits) and the
implicit rounding-off which we must determine from the numbers involved in the calculation.

The rules for rounding-off a number may differ slightly depending upon the field. For instance, in
accounting the rule may be slightly different from the rule in chemistry.

1. If the digit or group of digits to be dropped is more than 500...then drop that
digit or group and add 1 to the last digit retained.

2. If the digit or group of digits to be dropped is less than 500...then drop that digit or
group and leave the last digit retained unchanged.

3. If the digit or group of digits to be dropped is exactly 500..., then drop that portion and
add 1 to the last digit retained if this digit is odd but if this digit is even, then this digit remains
unchanged.

Example: The following have been rounded-off to three decimal places.

(1) .4398. ≈ **.440**

(2) .43652 ≈ **.437**

(3) .43637 ≈ **.436**

(4) .43750 ≈ **.438**

(5) .43650 ≈ **.436**

(6) .43946 ≈ **.439**

(7) .43650001 ≈ **.437**

(8) .4365001 ≈ **.437**

Example Round-off **85376.7463** to the following places, using the simple "5 or greater or less than 5
rule"

1. 85376.7463 to the nearest **thousandth** becomes **85376.746** (We do **not** replace the 3 dropped by a zero)

2. 85376.7463 to nearest **hundredth** becomes **85376.75** (We added 1 to the digit in the round-off place)

3. 85376.7463 to the nearest **tenth** becomes **85376.7** (The 7 is unchanged since the 4 dropped is less than 5)

4. 85376.7463 the nearest **unit** becomes **85377.** (Adding 1 to the 6)

5. 85376.7463 to the nearest **ten** becomes **85380.** (Replacing the 6 dropped by a zero)

6. 85376.7463 to the nearest **hundred** becomes **85400.** (Replacing the digits (6 and 7) dropped by zeros)

7. 85376.7463 to the nearest **whole number** becomes **85377.** (same as to the nearest unit)

Estimation

In estimation, we round-off the numbers before carrying out the operations of addition, subtraction,
multiplication , division etc. For convenience, we will round-off each number to the first non-zero digit,
unless specified.

Approximate Numbers, Significant Digits, Scientific Notation,   669

A measurement consists of a numerical value and a unit of that measurement. Example: 4 kilograms, where the 4 is the numerical value and the kilograms is the unit of the measurement.

Numbers obtained from a measurement are never exact (i.e., are approximate) due to the limitations of the measuring instrument as well as the skill of the person making the measurement. As such, when one records a measurement, one should indicate the reliability of the measurement. All measurements may be assumed to have an uncertainty in at least one unit in the last digit of the measurement, since in making a measurement, we usually estimate the last digit.

Results obtained from calculations using measurements are also as uncertain as the measurements themselves.

In summary, the numbers that we deal with in calculations are obtained from observations. Some of the numbers are exact and some are approximate, The approximate numbers are those numbers obtained from making measurements.

Exact Numbers: An exact number is a number that contains no uncertainties. It is assumed to be infinitely accurate. We can obtain exact numbers from definitions and from direct count. For example, the number of students in a math class by count is 25. In this case, there is no uncertainty, since we know that there are exactly 25 students. Similarly, when one counts 200 dollars, one knows that one has exactly 200 dollars, and there is therefore, no uncertainty. Also by definition, 60 minutes = 1 hour; 2.54 centimeters = 1 inch. Since these numbers are defined, this 60 and 2,54 are exact and contain no uncertainties. We can also add that this 60 has an infinite number of significant digits, (we can write 60 as 60.000...) and therefore the zero in the 60 is significant. However if you make your own measurement and by coincidence obtain 2.54 centimeters, then this 2.54 would not be exact. We can generalize that all the conversion factors (from tables) are exact.

Significant Digits (Significant Figures), Digits obtained in a measurement

A significant digit (or figure) is one which is known to be reasonably reliable (or correct).
When we make measurements, the digits we read and estimate on a scale are also called significant digits (or significant figures). These digits include digits that we are certain of, and one additional digit that we are uncertain of. This uncertain digit is obtained by the estimation of the fractional part of the smallest subdivision on the scale being used. As such, the rightmost digit is assumed to be uncertain.

Significant figure notation is an approximate method of indicating the uncertainty of a measurement. when recording a measurement.
We agree to the following:

1. The digits 1, 2, 3, 4, 5, 6, 7, 8, 9 are always significant.

2. The digit zero, 0. may or may not be significant according to its position in the number as follows:

  (a) Zeros before the first non-zero digits are **not** significant.
For example: (i) .0450 has **three** significant digits: The first zero is not significant; but the last zero is significant since if it were not we would not write it.

   (ii) .0012 has **two** significant digits. The first two zeros are not significant

  .   (b) Zeros between non-zero digits **are** significant.

Example: 3.045 has **four** significant digits. The zero between 3 and 4 is significant.
         40.240 has **five** significant digits. The last zero is significant because if it were not we would not write it.

More examples: The numbers referred to below are assumed to have been obtained from measurements.   670
23.00 has four significant digits: the zeros in this case are significant since we do not have to write the zeros if they were not significant. If the zeros were not significant we would have written 23.
2300. has two significant digits, and 600 has one significant digit; however, in each of these two examples, the number of significant digits is sometimes ambiguous.
It is suggested that when the number of significant digits is in doubt , the maximum number of significant digits is to be assumed. Also, in recording data, if we know the number of significant digits, we will use the scientific notation.
Note above that if 600 and 2300 had been obtained by counting, or by definition, all the zeros would have been significant. As a reminder, significant notation generally pertains to numbers obtained from measurements.

Using **scientific notation** avoids all ambiguities with respect to the number of significant digits..
 For instance, if we know that the above number, 600 were measured to two significant digits, then we would write 6.0×10^2. If the measurement were to three significant digits, we would write 6.00×10^2
When we deal with very large or very small numbers we prefer to write the numbers in scientific notation form. In this form, the significant digits (digits) including the zeros can be unambiguously indicated. In scientific notation:

(1)   125000 would be written 1.25×10^5

(2)   1467 would be written 1.467×10^3

(3)   .032500 would be written 3.2500×10^{-2}

(4)   .0325 would be written 3.25×10^{-2}

(5)   If 125000 were known to four significant digits it would be written 1.250×10^5.

Note: Some authors indicate which zeros are significant by underlining the last significant zero.
For example, in 23000 the first two zeros are significant but the last zero is not significant.
Note also that in some books, a decimal point placed after the last zero makes all the zeros significant. For example, 23000. has five significant digits, but 23000 has two significant digits. However, but there may still be ambiguity if the number is at the end of a sentence.

Accuracy and Precision in Measurements
Two contributions to uncertainty in measurement are limitations of precision, and limitations of accuracy.
Accuracy indicates how close a measured value is to the true value but **precision** indicates how close two measurements of the same quantity are close to each other. Generally, more precision implies more accuracy. However, there are instances in which numbers may be more precise, but may not be more accurate.. For example, if a measuring device is incorrectly calibrated (having incorrect scale).

Accuracy and Precision in Calculations
With respect to significant digits, **accuracy** refers to the number of significant digits but **precision** refers to the number of decimal places. The larger the number of significant digits in a number, the more accurate the number. The larger the number of decimal places, the more precise the number.

Note: In calculations, the approximate numbers determine how the rounding-off is done

Rounding-off to Significant Digits or Figures in Arithmetic Operations
(Implicit Specification)

1. In **multiplication and division** involving significant digits, the product or quotient (answer) should be rounded-off so that the number of significant digits in the answer is equal to the number of significant digits of the number with the least number of significant digits. (In other words, the answer should not be more accurate than the number with the least accuracy)

Example 1: Multiply 2.34 cm by 5.6 cm

Step 1: $2.34 \times 5.6 = 13.104$

Step 2: The number with the fewest number of significant digits is 5.6 and it has two significant digits. Therefore, the product (answer) should contain only two significant digits.

Thus, 13.104 becomes 13. Answer: 13 cm^2
In the above case the "4" in 2.34 is considered to be reasonably reliable. The " 6" in 5.6 is considered to be reasonably reliable.

Example 2: Maria determined the length of a piece of wood to be 6.47 yards. What is the length of this wood in feet?
Solution
By definition, 3 feet = 1 yard
6.47 yards is used to determine the number of significant digits in the answer,
(The 3 feet is exact and has an infinite number of significant digits)

$$\frac{6.47 \text{ yards}}{1} \times \frac{3 \text{ feet}}{1 \text{ yard}}$$

$= 19.41$ feet

$= 19.4$ feet (6.47 has three significant figures)

2. In **addition and subtraction**, we shall round-off so that the number of decimal place in the answer equals the number of decimal places in the number with the least number of decimal places.
(In other words, the answer should not be more precise than the number with the least precision)

Example 1: Add: 143.54, 172.3, and 64.62
Solution
143.54 < - - - - has two decimal places
172.3 < - - - - -has one decimal place (determines the number of decimal places in answer)
64.62 < - - - - - has two decimal places
380.46
Answer: 380.5 (has one decimal place)

Example 2 Subtract 12.4 from 143.63
Solution
143.63 < - - - - has two decimal places

12.4 < - - - - -has one decimal place (determines the number of decimal places in answer)

131.23
Answer: 131.2 (has one decimal place)

3. In finding powers and roots, the root or power should be rounded-off so that the number of significant digits in the answer is equal to the number of significant digits in the number.

Example : Find $\sqrt{26.9}$

Radicand has three significant digits , and therefore the root should have three significant digits.

From a calculator, $\sqrt{26.9} = 5.1865$

$\qquad \sqrt{26.9} = 5.19$ (has three significant digits. Same as in the radicand)

Note: When two or more different operations are involved, the final operation determines how the final result is rounded-off.

Example Simplify: $38.3 + 12.9(3.58)$

Solution According to order of operations, we multiply 12.9 by 3.58 first and then add 38.3

$\qquad 38.3 + 12.9(3.58)$

$= 38.3 + 46.182$

$= 84.485$

$= 84.5$ (has one decimal place as in 38.3)

Since addition was the last step, we use precision (the number of decimal places) to round-off.

Addition and Subtraction Involving Scientific Notation

Before adding or subtracting the numbers must have the **same powers** of 10. We will rewrite the expression so that the power of 10 is that of the highest power in the expression

Example: **1.** $4 \times 10^2 + 3 \times 10^2 = (4 + 3) \times 10^2$

$\qquad\qquad\qquad\qquad\qquad = 7 \times 10^2$

Example: **2.** $\quad 4 \times 10^2 + 2 \times 10^3 = 0.4 \times 10^3 + 2 \times 10^3$

$\qquad\qquad\qquad\qquad\qquad = (0.4 + 2) \times 10^3$

$\qquad\qquad\qquad\qquad\qquad = 2.4 \times 10^3$

$\qquad$ or

$\qquad 4 \times 10^2 + 2 \times 10^3 = 4 \times 10^2 + 20 \times 10^2$

$\qquad\qquad\qquad\qquad\qquad = (4 + 20) \times 10^2$

$\qquad\qquad\qquad\qquad\qquad = 24 \times 10^2$

$\qquad\qquad\qquad\qquad\qquad = 2.4 \times 10^3$ (Again, we obtain the same result)

INDEX

A

absolute minimum 249
Absolute Value 3, 135, 196
Absolute value equations 135
absolute value function 208
Absolute Value Functions 195
Absolute Value Inequalities 142, 204
Absolute value quadratics 203
Acute triangle 500
Addition
    of complex numbers 345
    of Radicals 8
    of vectors 560
Amplitude 585
Angle 526
Angle of depression .
    angle of depression 520
Angle of elevation 520
angle of rotation 376
Angles
    angle of elevation 520
    complementary angles 510
    in standard position 527, 528
Arc length 571, 662
Arithmetic Mean 468
Arithmetic Progression 468
Arithmetic Sequence 468, 469, 471
Asymptotes 273
    horizontal asymptote 273
    oblique 276
    vertical asymptote 273
Asymptotes, drawing 278
Asymptotic Formula 277
Augmented matrix 329
Axiom 492
Axis of symmetry
    of a parabola 249

B

Base formula, logarithms 309
Bearing 521
Binomial coefficient 481, 482
Binomial expansion 481
Binomial Theorem 481
    applications 482, 484
Bisector
    perpendicular bisector 116

C

Calculators
    using calculators 507
Case 4
    Graphs of Inequalities Involving the Absolute Values of two 204
Center
    of a circle 394
    of ellipse 415
Central angle 571
Change of base formula 309
Circle 391
    arc length 662
    as a special ellipse 419
    sector 662
    equation of 391
Circular Functions 572
Closed interval 126
Coefficient 483
Coefficient matrix 328
Cofactor
    of an element 332
Cofunction Relationships 641
Cofunctions 511
Column Matrix 327, 329 Column of constants 329
Combination 486, 487
Common difference 468
Common Logarithms 302
Complementary angles 510, 511
Complementary functions 511
Complex Conjugates 348
Complex Numbers 343
    addition 345
    division 365
    exponential form 364
    graphical addition and subtraction 356
    graphical representation 355
    multiplication 346
    polar representation 359
    powers 367
    subtraction 345
    trigonometric representation 359
Composite Functions 71
composite numbers 1
Composition of functions
    applications 72
Composition of trigonometric 627
Compound Inequalities 130
Compound Inequality 132
Compound Interest 316, 477
Concavities of Curves 214
Condensed method
    for solving compound inequalities 131

Condensed Method 131
Conic Sections 434
Conics
    definitions 435
    identifying conic equations 431
conjugate 13, 348
consistency 339
consistent 322, 330
constant function 212
constraints 166
Continuous Functions 270
Contraction of a Curve
    in the x-direction 233
    in the y-direction 234
Convex polygon 168
Coordinate Axes 582
Coterminal Angles 540
Counting
    fundamental principle 486
Counting numbers 2
Cramer's Rule 336
Circular cone 434
Critical Points 216
Crossing or Touching
    of Lines by Curves 262
Cube Roots 6
Cubic equation 263
Curves, expansion 232
Curves, rose 382
Curves, contraction 233

D

Decimals, Terminating & Repeating 655
De Moivre's Theorem 367
    applications 369
Decreasing Function 212
Deductive reasoning 492
Degenerate conics 434
Denominators
    rationalization 11
Dependency 339
Dependent 330
Descartes' Rule of Signs 443
Determinants 331
    expansion 333
Directed distance 376
Direction
    of resultant 569
Discontinuous Functions 270
Discriminant 33
Distance
    between two points on a line 110
    from a point to a line 113

Distance Formula 112
Division
      of a Radical 11
      of Complex Numbers 348
      of Complex Numbers in Polar Form 365
Domain 54, 250
      explicit specification 192
      implicit specification 192
      restricted domain 192
Double Angle Identities 640

E

Eccentricity, of conics 435
Edge 390
Elements
      of a Matrix 328
Ellipse 412, 434
      graph 419
      graphs 414
      properties 412
End behavior of polynomial functions 264, 269
Entries 328
Equality
      of two complex numbers 351
Equation
      center-radius form 392
      general form 392
      of a horizontal line 100
      of a vertical line 102
      of ellipse 419
Equations
      exponential 22
      exponential equations 305
      general conic equation 432
      identifying conic equations 431
      involving complex numbers 354
      literal 22
      logarithmic 22, 307
      nonlinear equations 41
      of a circle 391
      of parabolas 399
      of vertical asymptote 273
      principles for solving equations 23
      quadratic equations 26
      radical equations 22, 44
      roots of polynomial equations 448
      second degree equations 21
      solving factorable polynomial equations 40
      trigonometric equations 22, 648
      types of equations 20
Equations of lines
      finding 103
Equilateral triangle 501

Equivalent functions 651
Essential discontinuity 270
Even function 239
even roots 7
Excluded Values 54, 57, 271
Expansion of a determinant 334
Expansion of a Curve
    in the x-direction 232
    in the y-direction 234
Exponential Equations 305
Exponential expressions
    converting to radicals 14
exponential function 476
Exponential functions
    graphs 311
Exponential growth and dcay 316,478
Exponents 4
Extraneous Roots 24
extraneous solutions 44

F

Factor Theorem 441
Factorial Notation n! 479
    operations 479
Factors
    finding all factors of an integer 445
Finite Discontinuity 270
Finite Intervals 125, 127
finite sequence. 459
Fractional Exponents 14
Function 48, 49
Functional notation 52
Functional Values 440
Functions
    algebra of functions 68, 652
    comparison with relations 49
    composite functions 71
    constant functions 212
    containing square root of polynomials 63
    continuous functions 270
    discontinuous functions 270
    even functions 239
    function composition applications 71
    graphs of reciprocals 283
    greatest integer function 207
    increasing functions 211
    inverse functions 73
    inverse relations 73
    negative functions 210
    odd functions 239
    one-one functions 68
    polynomial functions 57
    positive functions 210

rational functions 58
sign function 208
specification by multipart rules   .
    specification by piecewise rules 194
step function 208
unit step function 208
Fundamental principle of counting 486

G

Gaussian Elimination 323
Geometric Mean 474
Geometric Progression 473
Geometric sequence 473, 474, 476
Geometric Series 474
Graph
    of rational functions 283
Graphing
    in the Rectangular Coordinate System of Axes 84
    intercepts method 88
    Ordered Pairs 85
    slope-intercept method 89
Graphing linear inequalities   146
Graphs
    of hyperbola 427
    of absolute value functions 197
    of absolute value quadratics 202
    of circles 391
    of ellipse 419
    of exponential functions 311
    of factorable polynomials 266
    of inverse trigonometric functions 624, 630
    of Inverse Trigonometric functions 630
    of logarithmic functions 317
    of Polar Equations 381
    of Polynomial Functions 241, 243
    of straight lines 86
    of y = Arcsin x 630
    parabolas 251, 403
    polar coordinates 376
    polynomial functions 259
    rational function 289
    rational functions 271
    shifting or translating 228
    sketching inverses 79
    vector quantities 558
    y = cos x 590
    y = cot x 600
    y = csc x 597
    y = sec x 599
    y = sin x 584, 586
    y = tan x 594, 595
Greatest integer function 207

H

Half Angle Identities 641
Half-closed interval 126
half-open Interval 126
half-plane 146, 147
    determining solution 150
Holes, functions 270
Horizontal Asymptote 273
horizontal asymptotes 274
Horizontal line
    slope 98, 100
Horizontal line test 68, 75
Horizontal shrinking 234
Horizontal stretching 234
Hyperbola 423, 434
    equation 426
    graphs 427
    properties 423
    transverse axis 425
Hypotenuse 502

I

Inconsistency, 339
Inconsistent 322
Increasing Functions 211
Independence 339
Index
    reduction 17
    same index 10
Inductive reasoning 492
Inequalities
    compound inequalities 130
    factored cubic inequalities 183
    graphing 146
    involving absolute of rational expressions 189
    involving rational expressions 185
    solving linear equations 129
    special cases, distinction 157
    word problems 132
Inequalities, quadratic 175
Inequality 123
    sense 123
Infinite discontinuity 270
Infinite Intervals 126, 127
infinite number of solutions 330
Infinite Sequence 459
Integers 2
Intercepts
    x- and y-intercepts 88
Intersection 167
Intersection of two sets 124
Intervals
    algebraic notation 127
    closed intervals 126

 graphical illustration 127
 open intervals 125
Inverse Functions 73, 651
Inverse functions, finding 76
Inverse of a function
 tests for 74
Inverse operations 23
 trigonometric functions 626
Inverse relation 257, 413, 425, 427
Inverse Relations 73
Inverse trigonometric functions 623,-629
Inverse Variation 120
Irrational number 2
Irrational numbers 1, 5
Isosceles triangle 501

L

Laplace expansion
 of a determinant 333
Law of Cosines 546, 555
Law of Sines 549, 555
Leading coefficient test 264
Lesson 1 1
Lesson 10 52
Lesson 100 639
Lesson 101 642
Lesson 11 68
Lesson 12 73
Lesson 13 83
Lesson 14 92
Lesson 15 96
Lesson 16 106
Lesson 17 110
Lesson 18 118
Lesson 19 123
Lesson 2 4
Lesson 20 129
Lesson 22 142
Lesson 23 146
Lesson 24 164
Lesson 25 166
Lesson 26 175
Lesson 27 183
Lesson 28 Domain Specification 192
Lesson 29 195
Lesson 3 20
Lesson 30 207
Lesson 31 210
Lesson 32 214
Lesson 33 217
Lesson 34 227
Lesson 35 232
Lesson 36 236
Lesson 37 241

Lesson 38 244
Lesson 39 259
Lesson 4 23
Lesson 40 266
Lesson 41 270
Lesson 42 273
Lesson 43 283
Lesson 44 298
Lesson 45 300
Lesson 46 305
Lesson 47 307
Lesson 48 311
Lesson 49 317
Lesson 5 26
Lesson 50 322
Lesson 51 327
Lesson 52 331
Lesson 53 343
Lesson 54 351
Lesson 55 355
Lesson 56 358
Lesson 57 367
Lesson 58 375
Lesson 59 380
Lesson 6 38
Lesson 60 390
Lesson 61 397
Lesson 62 412
Lesson 63 423
Lesson 64 438
Lesson 65 443
Lesson 66 452
Lesson 67 458
Lesson 68 463
Lesson 69 468
Lesson 7 40
Lesson 70 473
Lesson 71 479
Lesson 72 481
Lesson 73 486
Lesson 74 489
Lesson 75 491
Lesson 77 499
Lesson 78 503
Lesson 8 44
Lesson 80 526
Lesson 81 532
Lesson 82 535
Lesson 83 542
Lesson 84 544
Lesson 85 546
Lesson 86 552
Lesson 87 557
Lesson 88 560

Lesson 89 570
Lesson 9 48
Lesson 90 579
Lesson 91 581
Lesson 92 586
Lesson 93 590
Lesson 94 594
Lesson 95 602
Lesson 96 623
Lesson 97 630
Lesson 98 633
Lesson 99 637
Lesson 21 135
Like radicals 8
linear equation 21
Linear equations 83
Linear inequalities
    graphing 146
    systems 164
Linear Interpolation 521
Linear permutation 489
Linear Programming 166
    applications 169
Linear System of Equations
    solving by determinants 336
Lines
    perpendicular lines 107
Literal equations 22
Locus 390
Logarithmic Equations 307
Logarithmic functions
    graphs 317
Logarithms 298
    basic properties 298
    change of base formula 309
    common logarithms 302
    evaluation 301
    writing as a single log term 304
Lost Roots 24
Lower triangular 322

M

Mathematical Induction 492
    axiom of 492
Matrix 327
Matrix Methods 327
Maxima Problems 257
Maximum Point 216
Mean, Geometric 474
Mid-point
    of a line 110
Minima Problems 257
Minimum Point 216
    of a Parabola 249

Minor
    of an element of a determinant 332
Multiplication
    of Complex Numbers 346
    of Complex Numbers in Polar Form 364
    of Radicals 10
    of the square roots of negative numbers 347

N

Natural Logarithms 304
Natural numbers 2
Negative Functions 210
Negative integers 1, 2
Negative real roots 443
Non-negative integers 3
Non-negative real numbers 3
Non-positive real numbers 3
Non-real numbers 1
Number Flow Chart 1
Number line
    labeling 583

O

Objective function 166
Oblique (slant) asymptotes 276
Obtuse triangle 500
Odd functions 239
Odd roots 7
One-to-One Functions 68
Open interval 125
Ordered pair
    definition 48

P

Parabola 244, 434
    as a Conic Section 397
    axis of symmetry 249
    domain 250
    maximum point 249
    minimum point 249
    properties 244, 397
    range 250
    vertex 249
Parallel Lines
    slopes and equations 106
Partial Fractions 452
Perfect cube 12
Periodicity
    of trigonometric functions 584
Permutation 486
    cyclical permutation 490
    linear permutation 489

Permutations 489
Perpendicular Bisector 116
Perpendicular lines
    Slopes and equations 107
Piecewise Rules 194
Plotting Points 85
Point of Inflection 216
Points on a line 90
Polar Coordinate System 375
Polar Coordinates 375
    graphing 376
Polar form
    of complex numbers 360
Polygon, convex 168
Polynomial functions 57, 241
    end behavior 264, 269
    sketching graphs 243
    zeros 443
Positive functions 210
Positive integers 2
Positive real roots 443
Powers
    of Complex Numbers 367
Primary representations
    of polar coordinates 376
Prime numbers 1
Principal nth root 6
Principal square root of a number (nonzero number) is one of the two equal positive factors of that number. 6
Principle of powers 44
Principle of zero products 26
Principles, for Solving Equations 23
Product or Product as a Sum Formulas 641
Projection
    of a point on a line 390
Proving Trigonometric Identities 643
Pythagorean Identities 640
Pythagorean Theorem 502, 628
Pythagorean theorem, 555

Q

Quadrants for signs of trig functions 530
Quadratic equation 262
Quadratic equations 21, 26
    .solving by factoring 26
    discriminant 33
    quadratic-like equations 38
    relationship between a, b, c, and the roots of the equation 33
    solving by completing the square 29
    solving by square root method 28
    solving by the quadratic formula 31
Quadratic Formula 31
    as an aid in factoring 34
Quadratic Functions, Applications: Maxima and Minima Problems. 257

Quadratic inequalities
    solving by completing the square 181
    solving by graphing 182
    solving by point-testing of intervals 175
    solving by rules of signs 176
    solving by sign diagrams 177
Quadratic trinomials
    factorability 28
Quartic equation 263

R

Radian measure 507, 570
Radical 6
Radical equations 22, 44
Radicals 4, 5
    addition 8
    converting to exponential expressions 14
    division 11
    like radicals 8
    simplifying 8
    unlike radicals 8
Range 54, 250
Ratio 2
Rational 5
Rational equations 21
Rational Expression, improper 271
Rational Expression, proper 271
Rational functions 58
    excluded values 271
    graphs 283
    irreducible 289, 290
    reducible 289
Rational number 2
rational numbers 1
Rational Root 446
Rational root rule 446
Rational root theorem 446
    applications 446
Rational Zcro 446
Rationalization
    of Denominators 11
Real number line 3, 125
Real numbers 1, 125, 572
Real numbers. 2
Real-valued function 54
Reciprocal functions 651
Reciprocal Identities 640
Rectangular components 560
Rectangular Coordinate System of Axes .i 84
Rectangular form
    of complex numbers 360
Recursive formula
    of a sequence 459

Reduction of Indices 17
Reference Angle 528
Reference number 575
Reflection of a curve
    equation of 224
    or a line .about the line y = x 223
Reflection of a given curve
    about the origin 222
    about the x-axis 222
    about the y-axis 222
Reflection of a point
    about a vertical line 219
    about the horizontal line 221
    about the origin 219
    about the vertical line
    about the x-axis 218
    about the y-axis 217
Reflection of Points 217
Relation 48
Remainder Theorem 440
Resultant 560
    direction of 568
right circular cone. 434
Right triangle 500
Roots 6
    extraneous root 24
    lost roots 25
    of Complex Numbers 351, 368
    of complex numbers in polar form 369
    of polynomial equations 448
    plotting in the complex plane 373
Rose Curves 382
Roster method 123
Row Matrix 327
row operations 329
Rule of Sarrus
    for expanding a determinant 335
    Rule of signs, Descartes' rule 443

S

Same index 10
Scalar Quantity 557
Scalene triangle 501
Sector of a circle 662
Seesaw-View method 150
Sense of an inequality 123
Sequence 458
    finite sequence 459
    general term 458, 460
    infinite sequence 459
    range 458
    recursive formula 459
    specification 458
    term 458

Series 469
Set 123
Set notation 123
Set Operations
    involving inequalities 123
Set-builder notation 123
Sets: intersection 124
        union 124
Sigma notation 464
Sign diagram 333
Sign Function 208
Signed minor 332
Signed Numbers 3
Similar Radicals 8
Simple interest 477
 Slope of a Horizontal Line 94
Slope of a line 93
Slope-Intercept method 89
Slopes of lines
    signs 98
Solution half-plane 146
Solution plane 167
Square matrix 327
Square roots
    of negative numbers 344
Standard forms
    equations of conics 433
    of a system of linear equations 339
Standard notation
    for labeling a triangle 499
Standard position 527
    of an angle 528
    of an angle in radians 580
Standard-position angle 529
Step Function 207
Stretching, Vertical 232
Subtraction
    of Complex Numbers 345
Sum
    of vectors 569
Sum and Difference Identities 640
    applications 645
Sum Identities 641
Summation Index 465
Summation Notation 464
    properties 465
Symmetry
    about the origin 238
    about the x-axis 237
    about the y-axis 236
Synthetic Division 440
Synthetic Division, 438
System of Linear Equations 322
System of simultaneous linear equations 21

T

Terminating & Repeating Decimals 653
Terminology 1
Transformation Method 225,-228, 231, 610
Translation of Axes 229
Translation of Axes 229
    applications 230
Translation of Points 227
Triangles 499
    30-60-90 triangle 512
    45-45-90 triangle 513
    acute triangle 500
    classification 500, 501
    equilateral triangle 501
    isosceles triangle 501
    obtuse triangle 500
    right triangle 500, 502
    scalene triangle 501
    solutions of right triangles 514
    standard notation for labeling 499
Triangular form
    of a system of equations 322
Trigonometric Equations
    solutions 648
Trigonometric function
    of a real number 574
Trigonometric Functional Value
    given one functional value and no specification of the quadrant of the terminal side.) 544
    given one functional value and the quadrant of the terminal side of the angle 543
    given the measure of an angle 530
    given the coordinates of a point on the terminal side of an angle 532
    of any angle 526
Trigonometric functional values
    of a real number S 580
    of an angle in radians 579
Trigonometric Functions 503
    of Quadrantal Angles 536
Trigonometric graphs 581
Trigonometric Identities 639
    proving 643
Trigonometric ratios 503, 518
Triple of numbers 322
Turning Points 216
Types of equations 20
Types of Numbers 1

U

Union of two sets 124
Unit circle 572
Unit Step Function 208
Unlike radicals 8
Upper triangular 322

V

Variation 118
    combined variation 120
    direct variation 118
    inverse variation 118
Vector quantities
    graphs 558
Vector Quantity 557
Vectors 557
    addition 564
    resolution into components 560
Vertex 249, 390
    of a parabola 249, 390
Vertical Asymptote 273
Vertical line
    slope 98, 101
Vertical line test 49, 50, 77
Vertical Stretching 232
Vertical Shrinking 233

W

Whole numbers 2

X

X-and Y-intercepts method 88
X-intercept 92, 241

Y

Y-intercept 92, 94, 241

Z

Zero 443
Zeros 241, 443
    of polynomial functions 443

Mathematical Modeling

Some Reciprocal Relationships

1. Arithmetic If A working alone can do a piece of work in time t_A; B working alone can do the same work in time t_B; C working alone can do the same work in time t_C, and if A, B, and C working together, can do the same work in time t_{ABC}, then

$$\frac{1}{t_{ABC}} = \frac{1}{t_A} + \frac{1}{t_B} + \frac{1}{t_C}$$

That is, the reciprocal of the working-together time equals the sum of the reciprocals of working-alone times (individual times).

2. Geometry: For any triangle, the reciprocal of the inradius (R) equals the sum of the reciprocals of the exradii (r_1, r_2, and r_3).

Thus
$$\frac{1}{R} = \frac{1}{r_1} + \frac{1}{r_2} + \frac{1}{r_3}$$

3. Physics (Electricity) For electrical resistances in parallel (in an electric circuit), the reciprocal of the combined resistance, R, equals the sum of the reciprocals of the separate resistances, r_1, r_2, and r_3.

Thus
$$\frac{1}{R} = \frac{1}{r_1} + \frac{1}{r_2} + \frac{1}{r_3}$$

4. Physics (Optics)

For two thin lenses in contact, the reciprocal of the combined focal length, F, equals the sum of the reciprocals of the separate focal lengths, f_1 and f_2, .

Thus
$$\frac{1}{F} = \frac{1}{f_1} + \frac{1}{f_2}$$

5. Physics (Optics) For spherical mirrors and thin lenses, the reciprocal of the focal length F equals the sum of the reciprocals of the object distance, d_o and the image distance d_i.

Thus
$$\frac{1}{F} = \frac{1}{d_o} + \frac{1}{d_i}$$

6. Physics (Mechanics). If two bubbles of radii r_1, r_2, coalesce into a double bubble, the radius, R, of the

partition is given by
$$\frac{1}{R} = \frac{1}{r_1} - \frac{1}{r_2}$$

Change of base formulas

For Exponents	For logarithms
1. $b^x = a^{x\log_a b}$ 2. $b^x = e^{x\log_e b} = e^{x\ln b}$ 3. $b^x = e^{x\log_e b} = e^{x\ln b}$	$\log_b M = \dfrac{\log_c M}{\log_c b}$

Inverse Cofunction Identities

1. $\mathrm{Sin}^{-1}x + \mathrm{Cos}^{-1}x = \frac{\pi}{2}, \ [-1, 1];$ **2.** $\mathrm{Tan}^{-1}x + \mathrm{Cot}^{-1}(x) = \frac{\pi}{2}, (-\infty, \infty);$

3 $\mathrm{Sec}^{-1}x + \mathrm{Cos}^{-1}x = \frac{\pi}{2}, \ |x| \geq 1$

Negative angle relationships for inverse trig functions

1. $\mathrm{Sin}^{-1}(-x) = -\mathrm{Sin}^{-1}x;$ 3. $\mathrm{Tan}^{-1}(-x) = -\mathrm{Tan}^{-1}x;$ 5. $\mathrm{Sec}^{-1}(-x) = \pi - \mathrm{Sec}^{-1}x;$

2. $\mathrm{Cos}^{-1}(-x) = \pi - \mathrm{Cos}^{-1}x;$ 4. $\mathrm{Cot}^{-1}(-x) = \pi - \mathrm{Cot}^{-1}x;$ 6. $\mathrm{Csc}^{-1}(-x) = -\mathrm{Csc}^{-1}x$

Range of Inverse Trig Functions when $x \geq 0$

$0 \leq \mathrm{Sin}^{-1}x \leq \frac{\pi}{2}$--Ist quadrant

$0 \leq \mathrm{Cos}^{-1}x \leq \frac{\pi}{2}$--Ist quadrant

$0 \leq \mathrm{Tan}^{-1}x < \frac{\pi}{2}$--Ist quadrant

$0 < \mathrm{Cot}^{-1}x \leq \frac{\pi}{2}$--Ist quadrant

$0 \leq \mathrm{Sec}^{-1}x < \frac{\pi}{2}$--Ist quadrant

$0 < \mathrm{Csc}^{-1}x \leq \frac{\pi}{2}$--Ist quadrant

Range of Inverse Trig Functions when $x < 0$

$-\frac{\pi}{2} \leq \mathrm{Sin}^{-1}x < 0$ <--4th quadrant

$\frac{\pi}{2} < \mathrm{Cos}^{-1}x \leq \pi$ <-- 2nd quadrant

$-\frac{\pi}{2} < \mathrm{Tan}^{-1}x < 0$ <--4th quadrant

$\frac{\pi}{2} < \mathrm{Cot}^{-1}x < \pi$ <--2nd quadrant

$\frac{\pi}{2} < \mathrm{Sec}^{-1}x \leq \pi$ <--2nd quadrant

$-\frac{\pi}{2} \leq \mathrm{Csc}^{-1}x < 0$ <--4th quadrant

Memory Reminder

Compare the **change of base formulas** for logarithms and exponents, and observe how for the logarithms, the conversion factor, $\ln b$, divides; but for the exponents, $\ln b$ multiplies the exponent.

For **logarithms:** $\log_b x = \dfrac{\ln x}{\ln b} = \left(\dfrac{\log_e x}{\log_e b}\right)$

For **exponents:** $b^x = e^{x \ln b}$

Range of Inverse Trig Functions when $x \geq 0$		Range of Inverse Trig Functions when $x < 0$	
$0 \leq \mathrm{Sin}^{-1}x \leq \frac{\pi}{2}$ (Ist qd)	$0 < \mathrm{Cot}^{-1}x \leq \frac{\pi}{2}$ (Ist qd)	$-\frac{\pi}{2} \leq \mathrm{Sin}^{-1}x < 0$ (4th qd)	$\frac{\pi}{2} < \mathrm{Cot}^{-1}x < \pi$ (2nd qd)
$0 \leq \mathrm{Cos}^{-1}x \leq \frac{\pi}{2}$ (Ist qd)	$0 \leq \mathrm{Sec}^{-1}x < \frac{\pi}{2}$ (Ist qd)	$\frac{\pi}{2} < \mathrm{Cos}^{-1}x \leq \pi$ (2nd qd)	$\frac{\pi}{2} < \mathrm{Sec}^{-1}x \leq \pi$ (2nd qd)
$0 \leq \mathrm{Tan}^{-1}x < \frac{\pi}{2}$ (Ist qd)	$0 < \mathrm{Csc}^{-1}x \leq \frac{\pi}{2}$ (Ist qd)	$-\frac{\pi}{2} < \mathrm{Tan}^{-1}x < 0$ (4th qd)	$-\frac{\pi}{2} \leq \mathrm{Csc}^{-1}x < 0$ (4th qd)

qd = quadrant